# Origins of Life Volume II

In *Origins of Life Volume II*, life and its origin are inspected from traditional and unexpected points of view. The book takes an interdisciplinary approach, discussing astrobiology; chemical evolution; and how the Universe accommodates life, molecular biology, and philosophy. It is an open-minded approach, fully referenced throughout, and each chapter includes a further reading section for anyone wishing to learn more about that perspective on the origins of life.

First, everything started, with the Big Bang, from nothing. It appears now that everything was aiming toward our existence, some $13.75 \times 10^9$ years later, being capable of understanding it all. We did it using powerful tools: science, philosophy, and religion. Although we appreciate the contributions made by philosophy and religion, our contemplations and doubts are based on the plentiful scientific evidence provided.

The reader will be guided from make-up of the "life stage" (Universe), tools and materials needed for the living matter to be formed in the small part of the Universe, which one could call Human Neighborhood, or the Local Universe. It contains galaxies, galaxy clusters, and voids, and the Milky Way and its satellites influencing each other during this time span.

The book is easy to read, accompanied by numerous references; it could be of use to the expert in the field as well as for curious minds with a scientific, philosophical, or religious background.

**Dr. Vladivoj Valkovic (Vlado Valkovic)**, born 1939, a retired professor of physics, is a fellow of the American Physical Society and the Institute of Physics (London). He has authored 24 books, from *Trace Elements* published by Taylor and Francis, Ltd., International Scientific Publishers, in 1975 to *Low Energy Particle Accelerator-Based Technologies and Their Applications* published by CRC Press, Taylor and Francis Group, in 2022. He is also the author or coauthor of 410 scientific and technical papers in international scientific journals in the research areas of nuclear physics, instrumentation application of nuclear techniques to problems in biology, medicine, environmental research, and trace element analysis. Dr. Vlado Valkovic has also worked as a professor of physics, the head of laboratories, and a scientific advisor for respected institutions around the world.

# Origins of Life Volume II

## Contemplations and Doubts

Vlado Valkovic

CRC Press
Taylor & Francis Group
Boca Raton London New York

CRC Press is an imprint of the
Taylor & Francis Group, an **informa** business

Designed cover image: NASA

First edition published 2025
by CRC Press
2385 NW Executive Center Drive, Suite 320, Boca Raton FL 33431

and by CRC Press
4 Park Square, Milton Park, Abingdon, Oxon, OX14 4RN

*CRC Press is an imprint of Taylor & Francis Group, LLC*



*Library of Congress Cataloging-in-Publication Data*
Names: Valkovic, Vlado, author.
Title: Origins of life: musings from nuclear physics, astrophysics and astrobiology / Vladivoj Valkovic.
Description: First edition. | Boca Raton : CRC Press, 2022. | Includes bibliographical references and index.
Identifiers: LCCN 2021024074 (print) | LCCN 2021024075 (ebook) | ISBN 9781032010571 (hardback) | ISBN 9781032019741 (paperback) | ISBN 9781003181330 (ebook)
Subjects: LCSH: Life—Origin.
Classification: LCC QH325 .V35 2022 (print) | LCC QH325 (ebook) | DDC 576.8—dc23
LC record available at https://lccn.loc.gov/2021024074
LC ebook record available at https://lccn.loc.gov/2021024075

ISBN: 9781032432731 (hbk)
ISBN: 9781032432816 (pbk)
ISBN: 9781003366607 (ebk)

DOI: 10.1201/9781003366607

Typeset in Minion Pro
by Apex CoVantage, LLC

# Contents

# Introduction

The aim of this book is to look at the subject of origin of life in the Universe as we humans observe it. Therefore, I shall first discuss the stage arrangement for the life to emerge; it appears that life itself is universal. Next, I shall to try to illustrate how life found a sustainable place for its evolution. In order to help answer some of the difficult questions, an impressive amount of knowledge had to be accumulated by some very smart researchers. Nowadays they are helped by some very sophisticated instruments, both on the ground or up in space.

The National Academies' decadal surveys are widely recognized among policymakers and program managers as a key resource in determining where a field of research is and where it is headed. A recent decadal survey from the National Academies of Sciences, Engineering, and Medicine (2022) provided a clear exposition of the following:

1. An overview of planetary science, astrobiology, and planetary defense;
2. A broad survey of the current state of knowledge of the solar system;
3. The most compelling science questions, goals, and challenges, which should motivate future strategy in planetary science, astrobiology, and planetary defense;
4. A coherent and consistent traceability of recommended research and missions to objectives and goals;
5. A comprehensive research strategy to advance the frontiers of planetary science, astrobiology, and planetary defense during the period 2023–2032;
6. Recommendations for decision rules for the comprehensive research strategy that can accommodate significant deviations in the projected budget by new discoveries or technological developments;
7. An awareness of the science and space mission plans and priorities of NASA human space exploration programs;
8. Opportunities for collaborative research that are relevant to science priorities between Science Mission Directorate's (SMD's) four science divisions; and
9. The state of the profession, including issues of diversity, inclusion, equity, and accessibility; the creation of safe workspaces; and recommended policies and practices to improve the state of the profession.

Subjects elaborated upon in great detail include an introduction to planetary science, astrobiology, and planetary defense and a tour of the solar system—a transformative decade of exploration.

New discoveries are coming so fast that it is very difficult to follow the new developments, to assess their importance and value, and to evaluate the interpretations and conclusions made. Earlier centuries and decades needed to make a step in the right direction; nowadays we are bombarded with "important" discoveries on almost daily basis. For example, here is a list of record-breaking discoveries from the year 2023.

Albert et al. (2023) reported the first detection of a TeV γ-ray flux from the solar disk (6.3σ), based on 6.1 years of data from the High Altitude Water Cherenkov (HAWC) observatory. The 0.5–2.6 TeV spectrum is well fitted by a power law, $dN/dE = A(E/1\ \text{TeV})^{-\gamma}$, with $A = (1.6 \pm 0.3) \times 10^{-12}\ \text{TeV}^{-1}\,\text{cm}^{-2}\,\text{s}^{-1}$ and $\gamma = 3.62 \pm 0.14$. The flux shows a strong indication of anticorrelation with solar activity. These results extend the bright, hard GeV emission from the disk observed with Fermi-LAT, seemingly due to hadronic galactic cosmic rays showering on nuclei in the solar atmosphere. However, current theoretical models are unable to explain the details of how solar magnetic fields shape these interactions. HAWC's TeV detection thus deepens the mysteries of the solar-disk emission.

Several record-breaking gamma rays were detected in 2023, with 20 TeV photons detected coming from the pulsar within the Vela supernova remnant. It looks like TeV pulsar astronomy started, according to the H.E.S.S. Collaboration (2023). The High Energy Stereoscopic System (HESS) observatory in Namibia indicates that Vela unleashed the highest-energy radiation ever seen coming from a pulsar. Specifically, this object seems to have emitted gamma ray photons with at least 20 teraelectronvolts (TeVs), or $20 \times 10^{12}$ electronvolts.

Next is the most intense, long-lasting, and powerful explosion ever seen, ten times brighter than any known supernova and still erupting even now, discovered in a galaxy whose light has been traveling to us for $8 \times 10^9$ years, according to Wiseman et al. (2023). They presented observations from X-ray to mid-infrared wavelengths of the most energetic non-quasar transient ever observed, AT2021lwx. Data show a single optical brightening by a factor >100 to a luminosity of $7 \times 10^{45}$ erg $s^{-1}$ and a total radiated energy of $1.5 \times 10^{53}$ erg, both greater than any known optical transient. The decline is smooth and exponential, and the ultraviolet–optical spectral energy distribution resembles a blackbody with a temperature of $1.2 \times 10^4$ K. Tentative X-ray detections indicate a secondary mode of emission, while a delayed mid-infrared flare points to the presence of dust surrounding the transient. The spectra are similar to recently discovered optical flares in known active galactic nuclei but lack some characteristic features. The lack of emission for the previous 7 years is inconsistent with the short-term, stochastic variability observed in quasars, while the extreme luminosity and long time-scale of the transient disfavor the disruption of a single solar-mass star. The luminosity could be generated by the disruption of a much more massive star, but the likelihood of such an event occurring is small. A plausible scenario is the accretion of a giant molecular cloud by a dormant black hole of $10^8$–$10^9$ solar masses. AT2021lwx thus represents an extreme extension of the known scenarios of black hole accretion.

Next, the most distant fast radio burst (FRB) ever detected was also detected in 2023 by Ryder et al. The $8 \times 10^9$–year-old blast is the most ancient and distant rapid burst of radio waves ever seen originating from colliding galaxies. The authors observed the burst FRB 20220610A and localized it to a morphologically complex host galaxy system at redshift $1.016 \pm 0.002$. The burst redshift and dispersion measure are consistent with passage through a substantial column of plasma in the intergalactic medium and extend the relationship between those quantities measured at lower redshift. The burst shows evidence for passage through additional turbulent magnetized plasma, potentially associated with the host galaxy. They use the burst energy of $2 \times 10^{42}$ erg to revise the empirical maximum energy of an FRB.

The most distant detection of radio emissions associated with neutral hydrogen gas was also achieved in 2023, with the discovery of radio waves from a galaxy that we see as it existed in the Universe $8.8 \times 10^9$ years ago; see Chakraborty and Roy (2023). The 21-cm line is the most fundamental wavelength in the whole of radio astronomy. It is used to study the distribution of hydrogen gas in galaxies throughout the cosmos.

Also, the fastest runaway stars ever seen speeding through our galaxy were revealed in July 2023 by astronomers poring over data of stellar motions collected by the European Space Agency's Gaia satellite; see El-Badry et al. (2023). They reported a spectroscopic search for hypervelocity white dwarfs (WDs), which are runaways from Type Ia supernovae (SNe Ia) and related thermonuclear explosions. In their study, candidates are selected from Gaia data with high tangential velocities and blue colors. They found six new runaways, including four stars with radial velocities (RVs) > 1000 km $s^{-1}$ and total space velocities $\gtrsim$ 1300 km $s^{-1}$. These are most likely the surviving donors from double-degenerate binaries in which the other WDs exploded. The other two objects have lower minimum velocities, $\gtrsim$ 600 km $s^{-1}$, and may have formed through a different mechanism, such as pure deflagration of a WD in a Type Iax supernova. The four fastest stars are hotter and smaller than the previously known "D6 stars", with effective temperatures ranging from ~20,000 to ~130,000 K and radii of ~0.02–0.10 $R_\odot$. Three of these have carbon-dominated atmospheres, and one has a helium-dominated atmosphere. The high observed velocities suggest that a dominant fraction of the observed hypervelocity WD population comes from double-degenerate binaries whose total mass significantly exceeds the Chandrasekhar limit. However, the two nearest and faintest D6 stars have the lowest velocities and masses, suggesting that observational selection effects favor rarer, higher-mass stars.

Also, it was reported that the James Webb Space Telescope (JWST), spotted a black hole in a galaxy called CEERS 1019, which we see as it existed about 13.3 billion years ago (just 570 million years after the Big Bang). The mass of the black hole is about $9 \times 10^6$ times the mass of the sun, or about twice as massive as the supermassive black hole at the center of our Milky Way galaxy (Kocevski et al. 2023). They reported discovery of two low-luminosity, broad-line active galactic nuclei (AGN) at $z > 5$ identified using JWST NIRSpec spectroscopy from the Cosmic Evolution Early Release Science (CEERS) Survey. They detected broad Hα emission in the spectra of both sources, with FWHM of $2038 \pm 286$ and $1807 \pm 207$ km $s^{-1}$, resulting in virial black hole (BH) masses that are 1–2 dex below that of existing samples of luminous quasars at $z > 5$. The first source, reported as CEERS 1670 at $z = 5.242$, is 2–3 dex fainter than known quasars at similar redshifts and was previously identified as a candidate low-luminosity AGN based on its morphology and rest-frame optical spectral energy distribution (SED). They measured a BH mass of $M_{BH} = (1.3 \pm 0.4) \times 10^7\ M_{\odot}$, confirming that this AGN is powered by the least-massive BH known in the Universe at the end of cosmic reionization. The second source, CEERS 3210 at $z = 5.624$, is implied to be a heavily obscured, broad-line AGN caught in a transition phase between a dust-obscured starburst and an unobscured quasar. They estimated its BH mass to be in the range of $M_{BH} \simeq (0.9 - 4.7) \times 10^7\ M_{\odot}$, depending on the level of dust obscuration assumed.

Luhman et al. (2024), using the JWST, found a brown dwarf, along with other brown dwarfs holding less than eight times the mass of Jupiter, in the star cluster IC 348. IC 348 is found a thousand light years away in the Perseus Molecular Cloud. Based on the low extinctions of those objects and the strengths of the overtone and combination bands, they concluded that hydrocarbon resides in their atmospheres rather than in foreground material. Their detections of the 3.4-μm feature are the first in atmospheres outside of the solar system. The presence of this hydrocarbon is not predicted by any atmospheric models of young brown dwarfs. Based on its luminosity and evolutionary models, the faintest new member of IC 348 has an estimated mass of 3–4 $M_{Jup}$, making it a strong contender for the least massive free-floating brown dwarf that has been directly imaged to date.

Next, we shall describe some of the facilities that have contributed to recent advances of the knowledge described in first two chapters. They are not sorted chronologically nor according to their merits. Throughout the history of science, revolutionary instruments have propelled our understanding with their landmark discoveries.

**The Hubble Space Telescope** is a testament to that concept. Its design, technology, and serviceability have made it one of NASA's most transformative observatories. From determining the atmospheric composition of planets around other stars to discovering dark energy, Hubble has changed humanity's understanding of the Universe since its 1990 launch. Hubble's unique characteristics, its broad wavelength coverage, its position beyond Earth's atmosphere, and its assortment of science instruments have made it one of astronomy's most important tools. Hubble was designed as a general-purpose observatory, meant to explore the Universe in visible, ultraviolet, and infrared wavelengths. To date, the telescope has studied a vast array of cosmic objects, providing views that astronomers were unable to capture from the ground.

The Hubble Space Telescope realized is space on May 19, 2009 by the space shuttle Atlantis following the repairs and upgrades of Servicing Mission 4. Above the Earth's atmosphere, Hubble is more powerful than any ground-based telescope. Hubble has been in low-Earth orbit, approximately 326 miles (525 km) above the surface. Floating at this position, Hubble does not have to contend with any form of light pollution or weather conditions. This gives it a permanent, clear dark sky, which, combined with its sharp vision, allows it to see objects 10 times fainter than those that can be observed from Earth even by the largest telescopes. This vantage point above the atmosphere gives Hubble the ability to take breathtaking images not available from the ground. Hubble has extremely high angular resolution, the ability to distinguish between two objects that are very close together. For Hubble, this means it can see fine details in star-forming nebulae, galaxies, and other cosmic objects.

Hubble has affected every area of astronomy. Its most notable scientific discoveries reflect the broad range of research and the breakthroughs it has achieved. Hubble has made over 15 million observations since its launch. So far, there are more than 20,000 published scientific papers using Hubble data coauthored by over 23,000 scientists. Hubble still transmits about 150 gigabits of raw scientific data to Earth every week.

**The James Webb Space Telescope** is the successor to the Hubble Space Telescope. The James Webb Space Telescope keeps finding galaxies that should not exist. Six of the earliest and most massive galaxies that NASA's breakthrough telescope has seen so far appear to be bigger and more mature than they should be given where they are in the Universe, researchers have warned. The new findings build on previous research where scientists reported that despite coming from the very beginnings of the Universe, the galaxies were as mature as our own Milky Way.

Boylan-Kolchin (2023), in his paper on stress-testing ΛCDM with high-redshift galaxy candidates, suggests that the information from the JWST poses a profound dilemma for scientists. The data indicate there might be something wrong with the dark energy and cold dark matter paradigm, or ΛCDM (the model that has been guiding cosmology for decades).

On the more practical side, exploring the galactic chemical composition across the Milky Way, and specifically across the solar neighborhood, provides insights into the chemical evolution of the Universe. Since the formation of the first stars some hundred million years after the Big Bang (BB), heavier elements are synthesized in different stellar production processes at the expense of lighter elements. When the relative abundances of life-forming elements evaluated for the last universal common ancestor (LUCA) are compared with the solar neighborhood stellar abundances, a striking similarity occurs. Valković and Obhođaš (2022) tested the hypothesis that in some particular regions and at some particular time, the abundance curve of the first living matter and the Universe coincided. Indeed, the best agreement between the two curves was obtained for $(4 \pm 1) \times 10^9$ years after the BB, indicating the time of the origin of life. All organisms evolved on the Earth independently of place and time lead to the LUCA and involve chiral molecules such as L amino acids and D sugars in fundamental life processes. The growing evidence from carbonaceous meteorite analysis shows an excess of L-type amino acids and D-type sugars, suggesting that the increase in L-type or D-type molecular chirality is the process that takes place in planetary and stellar forming systems; thus the life emerging from interstellar molecular clouds (IMCs) had to be chiral.

The **radio telescope array Atacama Large Millimeter/Submillimeter Array** (ALMA) project began when Europe, North America, and Eastern Asia each worked in isolation to propose a telescope that would receive light in millimeter and submillimeter wavelengths in large magnitudes. After its design, one of the most important success factors for this type of radioastronomy project was in its location. Low-frequency waves are better received at high altitudes and in dry climates, as this reduces the amount of noise. In addition, the surface area should be large and flat enough for distribution of the antennas. Several areas in New Mexico, Hawaii (Mauna-Kea), France (Plateau de Bure), and northern Africa were explored. However, none of these sites measured up to the conditions in northern Chile, specifically the Chajnantor Plateau, located in the heart of the Atacama Desert.

In 1995, the three partners, represented by NRAO, ESO, and NAOJ, ran tests in the Chilean plains, with positive results. Consequently, in 1999, Europe and North America signed a memorandum of understanding, and two years later in Tokyo, a resolution was signed to support the joint intent to construct ALMA between Europe, North America, and Japan. At the end of 2003, after a series of tests, the first stone was placed in what was to become the most ambitious radio observatory on Earth.

ALMA will operate for many decades, and it will be used by several generations, including those who are currently elementary and junior high school students. Ten percent of observation time is allocated to the host country, Chile, and 90% for the partners, according to their financial contribution to ALMA. All partners contribute observing time toward "Open Skies" such that any astronomer, regardless of their affiliation, can apply for ALMA time. Another important aspect is that ALMA is designed to give calibrated data products to its users. Therefore, even non-expert researchers can use ALMA!

The data produced during an ALMA observing run, as well as their calibrations, are stored in the ALMA Archive, located at the headquarters in Santiago, and copied fully by the ARCs. Once the observatory reaches its full capability, storage will increase to a range of 80 gigabytes per day. Stored data will eventually become public, but not immediately. Scientists who are allocated observing time with ALMA are granted exclusive access to the gathered data for 1 year. This period is known as a "proprietary period" and starts when the data are distributed to the principal investigator. After this period has elapsed, the data become public at the ALMA Archive, and any investigator can retrieve them through the ALMA science website.

ALMA has 66 antennas: fifty-four 12-meter-diameter antennas and twelve 7-meter-diameter antennas. The radio telescope combines signals from each, working as an interferometer, in other words, a single giant telescope equal in size to the total array. With perfect parabolic shapes and a precision equivalent to within a fraction of the thickness of a human hair, this is one of the most precise antennas in the world. They do not need a dome or cupola to protect them. Although their reflectors are exposed to extreme conditions at an altitude of 5000 meters above sea level in the Chajnantor Plateau, facing gusty winds and temperatures that vary from 20 to –20 degrees Celsius, and even snow, the antennas are designed to withstand these conditions.

The most visible part of an antenna is the reflector. Each reflector performs the same function as a mirror in an optical telescope: capturing radiation from distant astronomical objects and directing it towards a front end that measures the levels of that radiation. The distinguishing feature of all telescopes is the wavelength of the radiation absorbed. Visible light captured by optical telescopes, with wavelengths between 380 and 750 nanometers, makes up only a small fraction of the spectrum of electromagnetic radiation. When it has all its receiver bands enabled, ALMA will capture radiation from space in longer wavelengths, from a few hundred micrometers to approximately 1 millimeter (close to 1000 times longer than visible light waves).

**The Chandra X-ray Observatory** is NASA's premier X-ray astronomy mission. It was named the Chandra X-ray Observatory in honor of the late Indian-American Nobel laureate Subrahmanyan Chandrasekhar. Chandra was designed to provide order-of-magnitude advances over previous X-ray astronomy missions with regard to spatial and spectral resolution. Chandra's High Resolution Mirror Assembly produces images of the point spread function (PSF) of <0.5 arcsec. Two grating systems—the Low Energy Transmission Grating (LETG) and the High Energy Transmission Grating (HETG)—offer resolving powers well in excess of 500 over much of their bandwidth and together cover the range from 0.1 to 10 keV. The principal elements of the observatory are:

1. The High Resolution Mirror Assembly (HRMA)—The HRMA consists of a nested set of four paraboloid-hyperboloid (Wolter-1) grazing-incidence X-ray mirror pairs, with the largest having a diameter of 1.2 m (twice that of the Einstein Observatory). The focal length is 10 m.

2. The Advanced CCD Imaging Spectrometer (ACIS)—The Advanced CCD Imaging Spectrometer offers the ability to simultaneously acquire high-resolution images and moderate resolution spectra. The instrument can also be used in conjunction with the High Energy Transmission Grating to obtain very high resolution spectra. ACIS contains 10 planar 1024 × 1024 pixel CCDs (Figure 2), four arranged in a 2 × 2 array (ACIS-I) used for imaging and six arranged in a 1 × 6 array (ACIS-S) used either for imaging or as a grating readout.

3. The High Resolution Camera (HRC)—The HRC is composed of two microchannel plate (MCP) imaging detectors: HRC-I designed for wide-field imaging and HRC-S designed to serve as a readout for the LETG. HRC-I is placed at right angles to the optical axis, tangent to the focal surface. HRC-S is made of three flat elements, the outer two of which are tilted to approximate the LETG Rowland circle. The HRC detectors have the highest spatial resolution on Chandra, matching the HRMA point spread function most closely.

4. The High Energy Transmission Grating Spectrometer (HETGS)—In operation with the High Resolution Mirror Assembly and a focal plane imager, the complete instrument is referred to as the HETGS.

5. Low Energy Transmission Grating Spectrometer (LETGS)—The LETGS consists of a low energy transmission grating, a focal plane imaging detector (HRC-S or ACIS-S), and the High Resolution Mirror Assembly. The LETGS provides high-resolution spectroscopy (>1000) between 80 and 175 angstroms (0.07–0.15 keV) and moderate resolving power at shorter wavelengths. The nominal LETGS wavelength range accessible with the HRC-S is 1.2–175 angstroms (0.07–10 keV); ACIS-S coverage is 1.2–65 Å (0.20–10 keV).

**Herschel** is an ESA space observatory with science instruments provided by European-led principal investigator consortia and with important participation from NASA. Herschel was launched on May 14, 2009, on an Ariane 5 from ESA's Spaceport in Kourou, French Guiana. The Herschel Space Observatory was approximately 7.5 m high and 4 m wide, with a launch mass of 3400 kg. The primary mirror was 3.5 m in diameter and accompanied by three science instruments: two cameras and imaging spectrometers—the Photoconductor Array Camera and Spectrometer (PACS) and Spectral and Photometric Imaging REceiver (SPIRE)—and a very high-resolution spectrometer, the Heterodyne Instrument for the Far Infrared (HIFI). Together, these instruments cover approximately 55–671 microns. The detectors in these instruments were cooled to temperatures close to absolute zero by a sophisticated cryogenic system. Herschel operated autonomously, receiving commands and returning acquired data to Earth in a 3-hour communications period every day.

Herschel reached its operational orbit about 100 days after launch. It operated from a largo halo orbit around the second Lagrangian point of the Sun–Earth system (L2), a virtual point located 1.5 million km from Earth in the direction opposite to the Sun. The satellite's average distance from L2 was 800,000 km. Hershel's primary ground station was ESA's deep space antenna in New Norcia, Australia. The mission operated until the cryostat ran out of helium, about 4 years after launch.

The **Laser Interferometer Space Antenna** (LISA) is a space-based gravitational wave observatory building on the success of the LISA Pathfinder and LIGO. Led by ESA, the LISA mission is a collaboration of ESA, NASA, and an international consortium of scientists.

LISA consists of three spacecraft that are separated by millions of miles and trailing tens of millions of miles, more than 100 times the distance to the Moon, behind the Earth as we orbit the Sun. These three spacecraft relay laser beams back and forth between the different spacecraft, and the signals are combined to search for gravitational wave signatures that come from distortions of space-time. We need a giant detector bigger than the size of Earth to catch gravitational waves from orbiting black holes millions of times more massive than our sun. NASA is a major collaborator in the European Space Agency–led mission, which is scheduled to launch in the early 2030s, and we are getting ready for it now! For more, see https://lisa.nasa.gov/.

**Planck**, a European Space Agency satellite, launched on May 14, 2009, and measured the cosmic microwave background (CMB), the residual radiation left over from the Big Bang, at a much greater sensitivity and resolution than was provided by the US Wilkinson Microwave Anisotropy Probe (WMAP). It was named in honor of German physicist Max Planck, a pioneer in quantum physics and in the theory of blackbody radiation. It was launched on an Ariane 5 rocket that also carried Herschel, an infrared space telescope.

Canadians are part of an international scientific team that has been poring over a new survey of the entire sky taken in microwave wavelengths by Planck, a mission launched in 2009 by the European Space Agency and supported by the Canadian Space Agency. They have assembled a unique catalogue of thousands of new objects ranging from some of the largest structures in the Universe—clusters of galaxies—to the very smallest: cold dust, the actual building blocks that help form new generations of stars.

Planck's supersensitivity to microwaves has allowed it to detect 189 gigantic clusters of galaxies, including 20 that had never been seen before. When this ancient shockwave of microwave background travels through a cluster of galaxies, it shifts the energy levels of the radiation, leaving behind a distinctive spectral signature that Planck can see. Hopes are now high that this new hunting technique will lead to the discovery of thousands of clusters, which could shed light on the overall structure and evolution of the Universe.

By mapping the sky in these longest of wavelengths (beyond what the human eye can see), Planck is also providing a detailed and first-of-its-kind inventory of a menagerie of foreground structures—cosmic stuff in the foreground of this relic radiation emitted some 13 billion years ago.

Among the most peculiar finds is the presence of extremely rapidly spinning dust grains scattered across the star-forming regions of the Milky Way and neighboring satellite galaxies. They not only emit microwaves like most other dust encountered so far, but they also send out radio waves as they spin. Astronomers are now struggling to understand how the smallest dust grains in the Universe, made up of only 10 to 50 atoms, can rotate 10 billion times a second.

Another Planck discovery puzzling astronomers is something called "dark gas", a new type of molecular gas found clinging to the edges of giant molecular clouds within the Milky Way. Dark gas may have an impact on galaxy formation and evolution. In the most detailed survey of cold dust in and around the Milky Way ever conducted, Planck has also catalogued nearly a thousand clumps of gas and dust, dubbed "cold cores", that are scattered across our local stellar neighborhood. The coldest known objects in the Universe, cold cores are star factories in the making and as such will play a key role in our understanding of the very earliest stages of star birth.

**The South Pole Telescope (SPT)** is a 10-meter-diameter microwave/millimeter/sub-millimeter telescope located at the NSF Amundsen-Scott South Pole Station, which is the best currently operational site on Earth for mm-wave survey observations due to its stable, dry atmosphere. With its unique location, the SPT plays a critical role in achieving the goal of the Event Horizon Telescope (EHT) to image the event horizon around the black hole at the center of our galaxy.

The initial goal of the SPT is to explore the nature of dark energy, an unexplained phenomenon responsible for the observed acceleration in the expansion of the Universe. The SPT will search for massive clusters of galaxies by looking for spectral distortions in the cosmic microwave background. Dark energy inhibits the growth of galaxy clusters, so studying the population of clusters through cosmic time will constrain models of dark energy. One can virtually visit the South Pole Telescope and the Amundsen-Scott South Pole Station, which is operated by the National Science Foundation, via panoramic Google Street View format.

This research is a collaboration among nine US and Canadian institutions.

Chapter 1, "Universe", covers our knowledge of and provides a number of discussions about the beginning of our Universe, the Big Bang. In particular, the following subjects are covered in some detail: the early Universe, the Universe in time, the Universe now and the cosmological (Hubble) constant. The chapter follows with discussions of Dark matter dogma and Dark energy, ending with a report on many observations and property determinations of Black holes and "dark stars". The rest of the chapter is devoted to subjects within astronomy: galaxies, dwarf galaxies, globular clusters, galaxy formation, dynamics, mergers, collisions, and galaxy groups. Cosmic rays are of interest because of their role in the origin of life process, in particular cosmic ray–induced chemistry on dust particles and cosmic ray–induced physics processes.

Stars are the subject of extensive discussion because of a variety of problems that can be considered, including star formation, pulsars, neutron stars, binary and quadruple star systems, free-floating stars, exploding stars, supernovae, red giants, and white dwarfs.

Chapter 2, "Local Universe", concentrates attention on a small part of the Universe, which one could call the Human Neighborhood, or the Local Universe. The Local Universe contains galaxies, galaxy clusters, and voids; it also consists of the Milky Way and its satellites: the Local Group, the Virgo Supercluster, and beyond. We shall try to present the properties and their changes in the time for majority of our neighbors. One has to be very careful since the influence of one part of our neighborhood on another has changed during the time span of $(13.787 \pm 0.020) \times 10^9$ years, the calculated life of the Universe by Planck Collaboration (2020). Not only the size but also all other characteristics have been changing as a function of time, in particular elemental composition. The formation of stars and their aging has played an important role in chemical element formation. For life, essential elements were formed in the required concentration ratios during this period. The galactic interstellar medium, housing a variety of dust particle clouds, was harboring a variety of physics and chemistry processes on the surfaces of dust grains, resulting in what we call molecular clouds. This is the perfect environment for the generation of molecules required for the process called the origin of life!

Most of the attention is paid to the Milky Way galaxy, its galactic disk, bulge, center, halo, and elemental composition. Our interest is in galactic interstellar medium, dust particles, agglomeration, physics and chemistry on the surface of dust grains, and molecular clouds. Our sun neighborhood is called the solar neighborhood. We discuss the history of the solar neighborhood as well as two important subjects: magnetic fields in the solar neighborhood and time variation of elemental composition.

Next we discuss places where life can develop to its full extent: planets, planet formation processes, protoplanetary disks, exoplanets, and free-floating planets. Afterwards, we discuss the solar system, its history, habitable Earth formation, Present: sun, planets and their moons, asteroids, Earth and chemical element abundances.

Chapter 3, "Life", summarizes different thoughts and answers to the question *What is life?* Different viewpoints, many with deep philosophical and historical roots, have shaped the scientific study of the origin of life. Some of these argue that primeval life was based on simple anaerobic microorganisms able to use a wide inventory of abiotic organic materials (i.e. a heterotrophic origin), whereas others invoke a more sophisticated organization, one that thrived on simple inorganic molecules (i.e. an autotrophic origin). While many scientists assume that life started as a self-replicative molecule, the first gene, a primitive self-catalytic metabolic network has also been proposed as a starting point. Even the emergence of the cell itself is a contentious issue: did boundaries and compartments appear early or late during life's origin? Starting with a recent definition of life, based on concepts of autonomy and open-ended evolution, it is proposed here that, first, organic molecules self-organized in a primordial metabolism located inside protocells. The flow of matter and energy across those early molecular systems allowed the generation of more ordered states, forming the cradle of the first genetic records. Thus, the origin of life was a process initiated within ecologically interconnected autonomous compartments that evolved into cells with hereditary and true Darwinian evolutionary capabilities. In other words, the individual existence of life preceded its historical-collective dimension (Peretó 2005).

We only scratch the surface of the subject of thermodynamic considerations of life because of conceptual difficulties and the complexity of the subject; detailed discussions would make the book thicker and more difficult to read. Instead, we discuss in more detail cells: the fundamental units of life as we know it. All living things show behavior—they respond to stimuli from the environment. The environment is dynamic, and the behavior is adaptive. Because of this, the behavior has a temporal component—the organism anticipates its future in relation to the present. The success of this behavioral anticipation is essential to survival. Therefore, we discuss to some extent life, time, and memory. Of great interest to us are the biological effects of magnetic fields, in particular, concentration factors for elements essential for life as functions of magnetic field intensity.

Element concentration factors (ECF) describe life forms' interaction with their environment. ECF is a function of many environmental parameters, including element concentrations, magnetic field intensity, and temperature, to mention only some. We discuss the ECFs for primitive eukaryotic, bacterial, and archaeal taxa, starting with an overview of knowledge on the elemental composition of Earth's LUCA. Sporulation is an interesting phenomenon in some bacteria as a response to changes in environment. We introduce a concept of memory as required for the survival of species. These memory effects may be more than incidental to the regulatory mechanisms controlling acclimation or the status of metabolic stores. They may be regulated by the cell and confer fitness to the organism in the evolutionary game it participates in.

Chirality of molecules is deeply rooted in the functioning of life processes. We discuss the origins of chirality, assuming an important role of preferential destruction, namely, preferential destruction of enantiomers due to p-p scattering. We propose experimental verification of preferential destruction of one isomer due to asymmetry in [p (in cosmic rays) – p (in amino acid)] scattering. Preferential destruction seems to be a fundamental bio-process. The role of learning chemical networks should be studied in more detail.

The importance of metals and heavy elements for life processes has been studied for some time; attention is paid to the study of their role in the functioning of molecules essential for life. Evolution is a process that requires more attention, especially given that more and more bio-molecules are being detected in space.

Chapter 4, "How the Universe Accommodated Life—Origin of Life", considers these two related problems. In order to shed some light on the question of how the Universe accommodated life, we discuss chemical abundances in the Universe as well as related properties of all bodies, such as metallicity. Of interest is mass–metallicity time dependence since it can help answer the question on origin of life time. It is of importance to understand the processes leading to the formation of elements required by life, in particular processes influencing C/O ratios and processes responsible for the presence of heavy life-required elements (Mo, I, W).

For this reason, our attention is also directed towards heavy metal stars, which can help in answering the question of when and where life originated. It could be that the Big Bang itself and the origin of life are somehow related! In this respect, the possibility of life on rogue planets could be of special interest.

The best estimate for $T_{origin}$ in the solar neighborhood is discussed in detail in the paper by Valković and Obhođaš (2022), who propose that the origins of chiral life in interstellar molecular clouds is at $(4 \pm 1) \times 10^9$ years after the Big Bang. An evaluation of possible errors in determination of this value is discussed in some detail.

We introduce a first form of life in the molecular cloud and call it the first universal stardust ancestor (FUSDA). An obvious transformation is FUSDA into Earth's LUCA, as discussed in detail. The rest of the chapter is devoted to the properties of LUCA as we know them, as well as the effects of the environment on them, in particular the effects of Earth's magnetic field and some other environmental parameters.

Chapter 5, "Present Life in the Universe", begins by discussing the anthropic principle, in particular cosmic fine-tuning, in some detail. It tries to handle a difficult question: is life universal? The answer is in accepting an idea about the existence of life in molecular clouds—rather a provocative proposal but based on some experimental evidence that the environment and physical and chemical processes could allow this to happen.

Next we discuss life in the solar system, the origin of life on planet Earth, evolution, and other issues, which leads to yet another question: are we alone? To answer this, we have to cover a large list of issues, including discussions of habitable zones, in particular galactic habitable zones. Some details are given for issues like interstellar panspermia and role old M stars could play. Today a search for life on exoplanets is an accepted activity, and lots of effort is going towards the possible detection of extraterrestrial life and discussions of what should be appropriate bio-signatures. Of course, there is always a possibility of finding life in other galaxies!

Chapter 6, "Physical; Philosophical, and Religious Considerations: End of Life, Death of the Universe", covers some very difficult and complex issues. Many questions that were once considered to fall into the realm of theology, where we lacked definitive knowledge, have now become scientific questions that have definitive answers. Scientifically, we now know how planet Earth emerged during the formation of our solar system some 4.5 billion years ago. We also know how life evolved and various plants and animals emerged through the ages on planet Earth and how recent and ancient events shaped our planet's geological, atmospheric, and hydrological history. We also know how the stars, galaxies, and larger structures in our Universe formed and grew from a more uniform, smaller, denser, hotter past. Yet between the interface of these two fields, science and theology, beyond our definite knowledge but without an appeal to an authoritative source, lies philosophy.

Among physical considerations, homochirality has received special attention because of its critical role in life processes. The philosophical consequences of experimental confirmation of proposed hypotheses are always difficult to forecast; they always result in surprises. One such issue is the proposed universality of life.

Religious considerations include a discussion of astrotheology as a contemporary theology. We discuss in some detail a response to the possible discovery of extraterrestrial life. Among other issues of interest is also the end of life on Earth, because it is obvious that Earth cannot support life forever. A related question is knowledge on how the Sun will age.

Somehow, related issue is also death of Universe, what happens after. There are two possibilities: the cold death of the Universe and the heat death of the Universe. Both are discussed briefly, as well as the issue of the end of time.

Since the author is not going in depth on all mentioned discoveries, each chapter contains a comprehensive list of references and a suggested list of additional readings.

I would like to express my gratitude to my wife, who let me write this book when she needed me most in fighting for the return to normal life after being born again. I am also thankful to my young colleague, Mr. Andrija Vinkovic, who found enough time to make all the drawings in this book in spite of the many obligations he had both at work and at home.

## REFERENCES

Albert, A., Alfaro, R., Alvarez, C., et al. (HAWC Collaboration). 2023. Discovery of gamma rays from the quiescent sun with HAWC. Physical Review Letters 131(5): 051201. https://doi.org/10.1103/PhysRevLett.131.051201.

Chakraborty, A., and Roy, N. 2023. Detection of H I 21 cm emission from a strongly lensed galaxy at $z \sim 1.3$. Monthly Notices of the Royal Astronomical Society 519(3): 4074–4081. https://doi.org/10.1093/mnras/stac3696.

El-Badry, K., Shen, K. J., Chandra, V., et al. 2023. The fastest stars in the galaxy. (30 pages). arXiv:2306.03914v2 [astro-ph.SR] 25 Jul.

Kocevski, D., Onoue, M., Inayshi, K., et al. 2023. Hidden Little Monsters: Spectroscopic Identification of Low-Mass, Broad-Line AGN at $z > 5$ with CEERS. (16 pages). arXiv:2302.00012v1 [astro-ph.GA].

Luhman, K. L., Alves de Oliveira, C., Baraffe, I., et al. 2024. A JWST survey for planetary mass brown dwarfs in IC 348*. The Astronomical Journal 167: 19 (15 pages). https://doi.org/10.3847/1538-3881/ad00b7.

National Academies of Sciences, Engineering, and Medicine. 2022. Origins, Worlds, and Life: A Decadal Strategy for Planetary Science and Astrobiology 2023–2032. Washington, DC: The National Academies Press. https://doi.org/10.17226/26522.

Peretó, J. 2005. Controversies on the origin of life. International Microbiology 8: 23–31. www.im.microbios.org.

Ryder, D., Bannister, K. W., Bhandari, S., et al. 2023. A luminous fast radio burst that probes the universe at redshift 1. Science 382(6668): 294–299. https://doi.org/10.1126/science.adf2678.

The H.E.S.S. Collaboration, et al. 2023. Discovery of a radiation component from the vela pulsar reaching 20 teraelectronvolts. Nature Astronomy 7: 1341–1350. https://doi.org/10.1038/s41550-023–02052–3.

Valkovic, V. 2021. Origins of Life: Musings from Nuclear Physics. In Astrophysics and Astrobiology. Boca Raton, FL: CRC Press, Taylor and Francis Group.

Valković, V., and Obhođaš, J. 2022. Origins of chiral life in interstellar molecular clouds. The Astronomical Journal 163: 270 (14 pages). https://doi.org/10.3847/1538-3881/ac63d0.

Wiseman, P., Wang, Y., Hönig, S., et al. 2023. Multiwavelength observations of the extraordinary accretion event AT2021lwx. Monthly Notices of the Royal Astronomical Society 522(3): 3992–4002. https://doi.org/10.1093/mnras/stad1000.

CHAPTER 1

# Universe

## 1.1 BIG BANG

Modern cosmology is rooted in the early observations of an expanding Universe (Hubble 1929) and the nearly uniform 3 K cosmic microwave background (CMB) radiation assumed to be the result of the Big Bang (Penzias and Wilson 1965). These two discoveries are the foundation of the "standard" six-parameter cosmological constant (Lambda–Λ) cold dark matter (CDM) model, which has been successfully applied to the interpretation of a range of cosmologically oriented observations.

It is a widespread belief that before the Big Bang, there was no time or space. The Big Bang marked the origin of the Universe: it all started from the singularity, a single point that was infinitely small, infinitely hot, and infinitely dense. According to the ΛCDM model, this happened about $13.75 \times 10^9$ years ago. However, the James Webb Space Telescope (JWST) revealed large galaxies existing at $\sim 0.3 \times 10^9$ after the Big Bang, requiring a more rapid star formation in the earliest galaxies than permitted by ΛCDM model simulation, that is, contradicting the accepted age-redshift relation; see Melia (2023) and Gupta (2023).

An interesting idea is that the Universe could be spontaneously created from nothing, but no rigorous proof has been given. He et al. 2014 presented such a proof based on the analytic solutions of the Wheeler-DeWitt equation (WDWE). Explicit solutions of the WDWE for the special operator ordering factor $p = -2$ (or 4) show that, once a small true vacuum bubble is created by quantum fluctuations of the metastable false vacuum, it can expand exponentially no matter whether the bubble is closed, flat, or open. The exponential expansion will end when the bubble becomes large, and thus the early Universe appears. With the de Broglie-Bohm quantum trajectory theory, they showed explicitly that it is the quantum potential that plays the role of the cosmological constant and provides the power for the exponential expansion of the true vacuum bubble. So it is clear that the birth of the early Universe completely depends on the quantum nature of the theory.

From $t = 0$ to $t = 10^{-43}$ s, the four forces (electromagnetic, weak nuclear force, strong nuclear force, and gravitational force) were combined into a single unified force. This period is called the Planck era, from time zero to approximately $10^{-43}$ seconds. During this era, the temperature was approximately $T = 10^{32}$ K.

Actually, the US National Institute for Standards and Technology gives the value of the Planck time as $5.391247 \times 10^{-44}$ s. In other sources, including Planck's original paper, one may find a slightly bigger value, around $1.35 \times 10^{-43}$ s. This is due to the use of two different versions of Planck's constant: Planck's original quantity, h, and h divided by $2\pi$.

The Planck scale is the physical scale beyond which current physical theories do not apply and cannot be used to calculate what happened. The problem of the Universe emerging out of the Planck epoch is discussed by Sivaram (1986). He pointed out that an earlier exponential expansion phase well before the onset of the Grand Unified Theory (GUT) phase transition is essential. Such an expansion can occur owing to the breaking of scale invariance at Planck energies in a unified theory of gravity with other interactions. At this time scale, every physical quantity, like temperature, energies, and so on, was in the range of Planck unit values. During the Planck epoch, the temperature and average energies within the Universe were so high that even subatomic particles could not form and even the present four fundamental forces were combined and formed one unified fundamental force (see Figure 1.1).

DOI: 10.1201/9781003366607-1

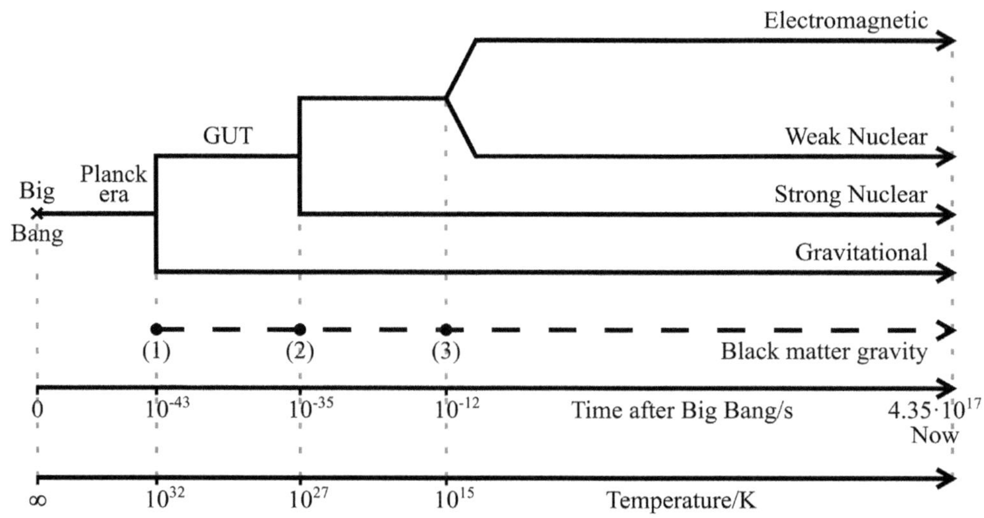


FIGURE 1.1 Today, four fundamental forces exist: the strong nuclear force, the weak nuclear force, the electromagnetic force, and the gravitational force. According to unified theories, early in the Universe's existence, these forces began as a single force, with each separating out at a specific temperature as the cosmos cooled down.

Traditional Big Bang cosmology predicts a gravitational singularity before this time, but this theory relies on the theory of general relativity, which is thought to break down for this epoch due to quantum effects. The immeasurably hot and dense state of the Planck epoch is succeeded by the grand unification epoch, where gravitation force is separated from the unified fundamental force.

The period of cosmic history ranging from $10^{-43}$ to $10^{-35}$ s after the Big Bang is almost unimaginable. During this time, gravity had become its own distinct force, but those forces now experienced individually as the separate strong, weak, and electromagnetic forces were all essentially indistinguishable. This is called the GUT era, following Grand Unified Theory attempts to describe all forces except gravity in a single framework (for example, string theories). This era began with gravity's separation from other three forces and ended with the separation of the strong nuclear force from the electroweak force; see Figure 1.1.

In this era, structures larger than Planck scale began to form stringlike objects consisting of nodes organized along one dimension of space. With the information (provided by their links) about the seven compact dimensions, they took on the properties of the individual particles recognized in the Standard Model. Huge ensembles of these strings began to behave as organized quantum fields. The way in which one string interacted with another is described by the way in which one enormous collection of nodes changed into another collection as part of a spin foam pattern.

During all these phase transitions, the amount of information coded into the network of nodes steadily increased. This had a profound effect upon how precisely the mathematical relationships between nodes along the emerging time axis could be specified. These relationships are now called the physical laws of nature and are described in the details of the Standard Model. So, during this later stage of the Planck era, not only did time emerge but also the laws of nature governing present space-time.

We assume that at that time, black matter gravity separated as an independent force, marked (1) in Figure 1.1, which continued as such, allowing dark matter particles generated in (1) to interact only by gravitational force. At times marked (2) and (3), more complex dark matter particles appear, keeping the same properties, interacting only via gravity.

The GUT Era is when three of the four fundamental forces are combined, but gravity has become distinct. There are a class of theories called Grand Unified Theories that attempt to describe all forces except gravity in a single framework. The leading type are so-called string theories, and some are only partially successful. Theorists would say that in the GUT Era, the gravity force "froze out" of the Universe. Near the end of this era, Grand Unified Theories predict that the Universe cooled to the point that the nuclear strong force began to freeze out, leaving three fundamental forces: gravity, the strong force, and the still-combined electroweak force. This "phase transition" released a huge amount of energy, causing space to undergo rapid inflation. At only $10^{-36}$ s, pieces of the Universe the size of an atomic nucleus might have grown to the size of our solar system. Note that this inflation is extremely large compared to the speed of light; however, since space itself is what is expanding, it does not have to obey the speed limit of the speed of light. During this inflationary phase, the Universe cools to approximately $T = 10^{27}$ K. Expansion of the Universe from the Big Bang until today is presented in Figure 1.2.

What followed is known as the Electroweak Era or the Age of Leptons. During this era, only electromagnetic and nuclear weak forces were still combined. There were no ordinary particles yet, just photons. By the time of $10^{-10}$ s, the temperature cools below $10^{15}$ K, allowing electromagnetic and nuclear weak forces to separate. The first protons formed at about $10^{-6}$ s, starting the era of nucleons. Just milliseconds after the Universe's Big Bang, chaos reigned. Atomic nuclei fused and broke apart in hot, frenzied motion. Incredibly strong pressure waves built up and squeezed matter so tightly together that black holes formed, so-called primordial black holes (PBHs). Here is an interesting question: did primordial black holes help or hinder formation of the Universe's first stars, eventually born about 100 million years later? Effects of stellar-mass primordial black holes on first star formation were discussed by Liu et al. (2022) using cosmological hydrodynamic zoom-in simulations and semi-analytical models.

Models developed by Liu et al. (2022) self-consistently combine two competing effects: initial (isocurvature) perturbations induced by PBHs and BH accretion feedback. Focusing on PBHs with masses $\sim 30M_{\odot}$, they found that the standard picture of first star formation in molecular-cooling minihaloes is not changed by PBHs, as the simulated star-forming gas clouds in the central parsec are very similar to those in the ΛCDM case when PBHs make up $f_{PBH} \sim 10^{-4}$–0.1 of dark matter.

The NASA/LAMBDA Archive Team (2023) has illustrated the evolution of the Universe with five theorized stages, providing a context for discussing ΛCDM parameters. In this scheme, the infant Universe is an extremely hot, dense, nearly homogeneous mixture of photons and matter, tightly coupled as a plasma. An approximate timeline of its theoretical evolution is presented by the following explanatory text:

1. The initial conditions of this early plasma are currently thought to have been established during a period of rapid expansion known as inflation. Density fluctuations in the primordial plasma are seeded by quantum fluctuations in the field driving inflation. The amplitude of the primordial gravitational potential fluctuations is nearly the same on all spatial scales.
2. Small perturbations propagate through the plasma collisionally as a sound wave, producing under- and overdensities in the plasma with simultaneous changes in density of matter and radiation. CDM does not share in these pressure-induced oscilla-

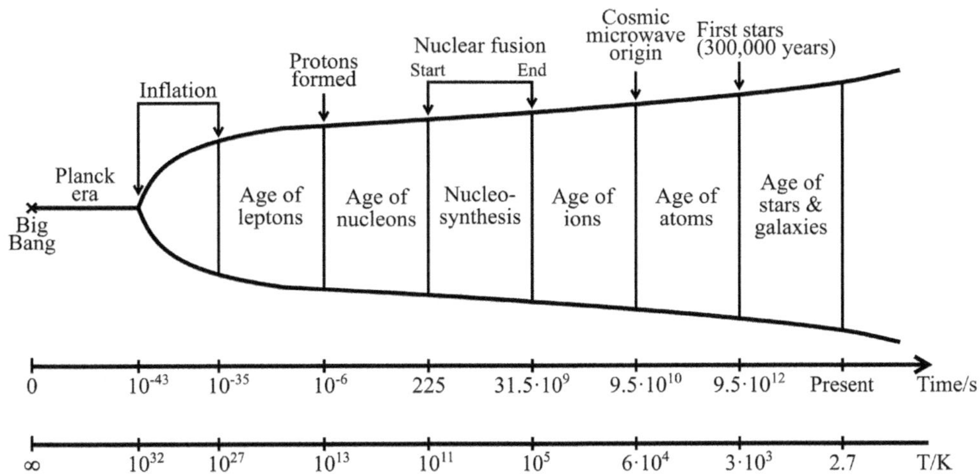


FIGURE 1.2 Expansion of the Universe from the Big Bang until today.

tions but does act gravitationally, either enhancing or negating the acoustic pattern for the photons and baryons.

3. Eventually physical conditions in the expanding, cooling plasma reach the point where electrons and baryons are able to stably recombine, forming atoms, mostly in the form of neutral hydrogen. The photons decouple from the baryons as the plasma becomes neutral, and perturbations no longer propagate as acoustic waves: the existing density pattern becomes "frozen". This snapshot of the density fluctuations is preserved in the CMB anisotropies and the imprint of baryon acoustic oscillations (BAOs) observable today in large-scale structures.

4. Recombination produces a largely neutral Universe that is unobservable throughout most of the electromagnetic spectrum, an era sometimes referred to as the "Dark Ages". During this era, CDM begins gravitational collapse in overdense regions. Baryonic matter gravitationally collapses into these CDM halos, and "Cosmic Dawn" begins with the formation of the first radiation sources such as stars. Radiation from these objects reionizes the intergalactic medium.

5. The structure continues to grow and merge under the influence of gravity, forming a vast cosmic web of dark matter density. The abundance of luminous galaxies traces the statistics of the underlying matter density. Clusters of galaxies are the largest bound objects. Despite this reorganization, galaxies retain the BAO correlation length that was established in the era of the CMB.

6. As the Universe continues to expand over time, the negative pressure associated with the cosmological constant (the form of dark energy in ΛCDM) increasingly dominates over opposing gravitational forces, and the expansion of the Universe accelerates (NASA/LAMBDA Archive Team 2023).

An article by Zyla et al. (2020) lists, evaluates, and presents the average measured properties of gauge bosons and the recently discovered Higgs boson, leptons, quarks, mesons, and baryons. It also summarizes searches for hypothetical particles such as supersymmetric particles, heavy bosons, axions, and dark photons. All known particle properties and search limits are also listed.

The visible world is founded on the proton, the only composite building block of matter that is stable in nature. Consequently, understanding the formation of matter relies on explaining the dynamics and the properties of the proton's bound state. According to Li et al. (2022a), a fundamental property of the proton involves the response of the system to an external electromagnetic field. It is characterized by the electromagnetic polarizabilities (Zyla et al. 2020) that describe how easily the charge and magnetization distributions inside the system are distorted by the electromagnetic field. Moreover, the generalized polarizabilities (Fonvieille et al. 2020) map out the resulting deformation of the densities in a proton subject to an electromagnetic field. They disclose essential information about the underlying system dynamics and provide a key for decoding the proton structure in terms of the theory of the strong interaction that binds its elementary quark and gluon constituents.

The decay $n \rightarrow p + e^- + \bar{\nu}_e$ is the simplest example of nuclear β-decay. The mean neutron lifetime $\tau_n$ provides a key input for predicting primordial light element abundances. The lifetime, combined with decay correlations, tests the V-A structure of the weak interaction in the Standard Model (SM) at low energy while avoiding nuclear structure corrections. Gonzalez et al. (2021) reported an improved measurement of the free neutron lifetime $\tau_n$ using the UCNτ apparatus at the Los Alamos Neutron Science Center. They counted a total of approximately $38 \times 10^6$ surviving ultracold neutrons (UCNs) after storing in UCNτ's magnetogravitational trap over two data acquisition campaigns in 2017 and 2018. They extracted $\tau_n$ from three blinded, independent analyses by both pairing long and short storage time runs to find a set of replicate $\tau_n$ measurements and by performing a global likelihood fit to all data while self-consistently incorporating the β-decay lifetime. Both techniques achieve consistent results and find a value $\tau_n = 877.75 \pm 0.28$s. With 0.039% total uncertainty, this is the first experiment to determine the neutron lifetime with uncertainty smaller than the average 0.09(2)% shift predicted from recent theoretical work on radiative corrections. This result, combined with forthcoming experimental and lattice QCD determinations of the axial coupling constant, will independently probe the disagreement in first-row Cabibbo-Kobayashi-Maskawa

(CKM) unitarity while avoiding nuclear structure corrections required for superallowed nuclear decays.

The neutron mean lifetime $\tau_n$ is an important parameter for the weak interaction, and more precise knowledge of its value is needed in particle physics, nuclear physics, and cosmology. Specifically, $\tau_n$ is a key input to calculations of primordial helium abundance, and the uncertainties in these predictions are presently dominated by those on $\tau_n$. Primordial nucleosynthesis is one of the major lines of evidence for the Big Bang, along with the observed expansion of the Universe and cosmic microwave background. Additionally, in combination with β-decay correlations, $\tau_n$ is used to test the unitarity of the Cabibbo-Kobayashi-Maskawa matrix, which represents the most important low-energy tests of the standard model.

Currently, there are two competing values for $\tau_n$ based on the results of two different classes of laboratory experiments. The "bottle" experiments involve counting the number of neutrons that survive within a material, magnetic, and/or gravitational trap as a function of time. The "beam" experiments involve measuring the rate of production of β-decay products in a neutron beam passing through a trapping region. The average beam measurement $\tau_{\text{beam n}} = 888 \pm 2$ s differs by about 4σ from the more precise ultra-cold trapped neutron average $\tau_{\text{bottle n}} = 879.4 \pm 0.6$ s. This discrepancy, which has persisted for 15 years, has become known as the "neutron lifetime puzzle". The most likely explanation for the discrepancy is the presence of an unaccounted for systematic error in one, or both, classes of experiment. However, given the direction of the disagreement, a physical explanation is possible where the neutron decays to unobserved particles outside of the standard model with a branching fraction of approximately 1%.

Wilson et al. (2021) used data from the Lunar Prospector Neutron Spectrometer to make the second space-based measurement of the free neutron lifetime, finding $\tau_n = 887 \pm 14_{\text{stat}}{}^{+7}{}_{-3\,\text{syst}}$ s, which is within 1σ of the accepted value. This measurement expands the range of planetary bodies where the neutron lifetime has been quantified from space and by extending the modeling to account for nonuniform elemental composition. Their modeling moves space-based neutron lifetime measurement towards the ultimate goal of reducing the magnitude of the systematics on a future space measurement to the level of those seen in laboratory-based experiments.

The Hot Big Bang is often considered the origin of all matter and radiation in the Universe. Primordial nucleosynthesis (BBN) provides strong evidence that the early Universe contained a hot plasma of photons and baryons with a temperature T > MeV. However, the earliest probes of dark matter originate from much later times, around the epoch of structure formation. Freese and Winkler (2023) described a scenario in which dark matter (and possibly dark radiation) can be formed around or even after BBN in a second Big Bang, which they dub the "Dark Big Bang". The latter occurs through a phase transition in the dark sector, which transforms dark vacuum energy into a hot dark plasma of particles; in their paper, they focus on a first-order phase transition for the Dark Big Bang. The correct dark matter abundance can be set by dark matter cannibalism or by pair annihilation within the dark sector followed by a thermal freeze-out. Alternatively ultra-heavy "darkzilla" dark matter can originate directly from bubble collisions during the Dark Big Bang. They show that the Dark Big Bang is consistent with constraints from structure formation and the cosmic microwave background if it occurred when the Universe was less than 1 month old, corresponding to a temperature in the visible sector above O (keV). While dark matter evades direct and indirect detection, the Dark Big Bang gives rise to striking gravity wave signatures to be tested in pulsar timing array experiments. Furthermore, the Dark Big Bang allows for realizations of self-interacting and/or warm dark matter, which suggest an exciting discovery potential in future small-scale structure observations.

It is of interest to mention Penrose's (2018) considerations of the Big Bang and its dark matter content. This is not explained by current conventional inflationary cosmology. Therefore, he introduces conformal cyclic cosmology, providing another perspective on these issues, one consequence being the necessary initial presence of a dominant scalar material that interacts only gravitationally but which must ultimately slowly decay away in a novel but perhaps detectable way. According to this theory, our current Universe picture provides but one eon of an unending succession of expanding eons, each having an initial Big Bang, which is the conformal continuation of the remote exponential expansion of its previous eon.

### 1.1.1 Early Universe

Although most primordial black holes have just tens of solar masses, a very small fraction will be far heavier, ranging from hundreds to tens of thousands of solar masses. Born less than a second after the Big Bang, these monstrous objects would then act as giant seeds for the formation of the first galaxies and quasars, which would rapidly develop suppermassive black holes (SMBHs) at their centers. Such seeds could also account for the existence of intermediate-mass black holes possessing 1000 to a million solar masses observed orbiting SMBHs and at the centers of globular clusters of stars. In short, primordial black holes may be the missing link between conventional stellar-mass black holes and SMBHs. The observational case for this scenario is building rapidly: recent detections of unexpectedly abundant X-ray sources in the early Universe are most easily explained by large numbers of primordial black holes producing X-rays as they gorge on gas less than 1 billion years after the Big Bang. The theorists need to push their simulations to ever-greater realism, but they also need to formulate robust, first- or zero-order, empirical tests, providing observers with some clearly defined targets (Bromm 2013).

From the beginning, the Universe has gone through several characteristic phases. Immediately after the Big Bang, the Universe was a hot, dense ionized plasma; it is assumed that it required some 380,000 years to cool down enough that the protons and electrons that filled the Universe could combine into neutral hydrogen atoms. This period is called the "Dark Ages" since there was no source of light anywhere in the Universe.

Roughly $10^8$ years after Big Bang, the first stars and galaxies were formed, and gas became gradually ionized. This process separated electrons from protons, leaving them as free particles—this period is known as "Cosmic dawn". Today, intergalactic hydrogen is fully ionized and forms gas clouds. According to Bosman et al. (2022), the end of the reionization phase is put $1.1 \times 10^9$ years after the Big Bang. Their published results indicate that reionization-related fluctuations persist in the intergalactic medium until at least $z = 5.3$ ($t = 1.1 \times 10^9$ years after the Big Bang.

Pressure waves generated in the hot plasma of the early Universe become imprinted in baryon fluctuations approximately 390,000 years after the hot Big Bang. The remnants of these waves create a ruler that, observed across time in the evolving Universe, provides constraints on the physics governing cosmic evolution. The theory of the physics of the early hot Universe leads to a prediction of baryon acoustic oscillations that has received confirmation from the pairwise separations of galaxies in samples of hundreds of thousands of objects. Tully et al. (2023) presented evidence for the discovery of a remarkably strong individual contribution to the BAO signal at $z = 0.068$, an entity that is given the name Ho'oleilana. The radius of the 3D structure is 155 $h_{75}^{-1}$ Mpc. At its core is the Boötes supercluster. The Sloan Great Wall, Center for Astrophysics Great Wall, and Hercules complex all lie within the BAO shell. The interpretation of Ho'oleilana as a BAO structure with their preferred analysis implies a value of the Hubble constant of $76.9^{+8.2}{}_{-4.8}$ km $s^{-1}$ $Mpc^{-1}$.

The Early Release Observations (EROs) of JWST beautifully demonstrate the promise of JWST in characterizing the Universe at Cosmic Dawn. Rhoads et al. (2023) analyzed the Near Infrared Spectrograph ERO spectra of three $z \sim 8$ galaxies to determine their metallicities, gas temperatures, and ionization. These galaxies offer the first opportunity to understand the physical properties of Epoch-of-Reionization galaxies through detailed rest-optical emission-line spectroscopy. We show that these objects have metal abundances $12 + \log[O/H] \approx 6.9\text{–}8.3$, based on both the $T_e$ method and on a recent calibration of the $R_{23}$ metallicity indicator. Since the spectra are some of the earliest science data from JWST, Rhoads et al. (2023) compared several line ratios with values expected from robust physics to validate our measurement procedures. They compared the abundances and emission-line ratios to a nearby sample of Green Pea galaxies—a population of nearby emission-line galaxies whose UV properties resemble Epoch-of-Reionization galaxies and which often have large Lyman continuum escape fractions. The JWST data show striking similarities between these high-redshift galaxies and nearby Green Peas. The $z \sim 8$ galaxies span the metallicity range covered by Green Peas. They also show the compact morphology that is typical of emission-line–dominated galaxies at all redshifts. Based on these similarities with Green Peas, it is likely that these are the first rest-optical spectra of galaxies that are actively driving cosmological reionization.

Note that Green Pea galaxies were named because of their color and small size. They are just 5% of the size of the Milky Way and 1% of the mass but produce stars at an enormous rate—roughly 100 times as fast as expected given their mass. They also seem to contain relatively few heavy elements. Their greenish hue comes from the glow of ionized oxygen heated by newborn stars. Green Peas have fascinated astronomers because, even though they have been observed only in the nearby Universe, their properties resemble those expected for early galaxies.

Galaxies with stellar masses as high as ~$10^{11}$ solar masses have been identified at redshifts z ~ 6, approximately $1 \times 10^9$ years after the Big Bang. Labbé et al. (2023) have recently published a paper on a population of red candidate massive galaxies ~$600 \times 10^6$ years after the Big Bang. They make use of the 1–5 μm coverage of the JWST early release observations to search for intrinsically red galaxies in the first ≈ 750 million years of cosmic history. In the survey area, they found six candidate massive galaxies (stellar mass > $10^{10}$ solar masses) at $7.4 \leq z \leq 9.1$, $(500–700) \times 10^6$ years after the Big Bang, including one galaxy with a possible stellar mass of ~$10^{11}$ solar masses. If verified with spectroscopy, the stellar mass density in massive galaxies would be much higher than anticipated from previous studies based on rest-frame ultraviolet-selected samples.

The radio telescope array Atacama Large Millimeter/Submillimeter Array (ALMA) has pinpointed the exact cosmic age of a distant JWST-identified galaxy, GHZ2/GLASS-z12, at $367 \times 10^6$ years after the Big Bang. ALMA's deep spectroscopic observations revealed a spectral emission line associated with ionized oxygen near the galaxy, which has been shifted in its observed frequency due to the expansion of the Universe since the line was emitted. A new study led by a joint team at Nagoya University and the National Astronomical Observatory of Japan has measured the cosmic age of a very distant galaxy, Bakx et al. (2022). The team used the ALMA radio telescope array to detect a radio signal that has been traveling for approximately 97% of the age of the Universe. This discovery confirms the existence of galaxies in the very early Universe found by the James Webb Space Telescope.

These ALMA observations provide robust evidence of the existence of galaxies within the first few hundred million years after the Big Bang and confirm the surprising results from the Webb observations. The combined power of Webb and the radio telescope array ALMA gives us the confidence to push our cosmic horizons ever closer to the dawn of the Universe.

The detection of z ~ 13 galaxy candidates has opened a new window on galaxy formation at an era only $330 \times 10^6$ years after the Big Bang. The physical nature of these sources was investigated to resolve the dilemma of witnessing star-forming galaxies or quasars at such early epochs. If powered by star formation, the observed ultraviolet (UV) luminosities and number densities can be jointly explained if:

1. these galaxies are extreme star-formers with star formation rates 5–24× higher than those expected from extrapolations of average lower-redshift relations;
2. the star formation efficiency increases with halo mass and is countered by increasing dust attenuation from z ~ 10–5; and
3. they form stars with an extremely top-heavy initial mass function. The quasar hypothesis is also plausible, with the UV luminosity produced by black holes of ~ 108 Msun accreting at or slightly above the Eddington rate (fEdd ~ 1.0).

This black hole mass at z ~ 13 would require very challenging, but not implausible, growth parameters. If spectroscopically confirmed, these two sources will represent a remarkable laboratory to study the Universe at previously inaccessible redshifts.

Observations show that supermassive black holes with a mass of ~$10^9$ $M_\odot$ exist when the Universe is just 6% of its current age. Feng et al. (2021) proposed a scenario where a self-interacting dark matter halo experiences gravothermal instability and its central region collapses into a seed black hole. The presence of baryons in protogalaxies could significantly accelerate the gravothermal evolution of the halo and shorten collapse timescales. The central halo could dissipate its angular momentum remnant via viscosity induced by the self-interactions. The host halo must be on high tails of density fluctuations, implying that high-z SMBHs are expected to be rare in this scenario. They further derive conditions for triggering the general relativistic instability of the collapsed region. Their results indicate that self-interacting dark matter can provide a unified explanation for diverse dark matter distributions in galaxies today and the origin of SMBHs at redshifts z ~ 6–7.

Along with $H_2$, HD has been found to play an important role in the cooling of the primordial gas for the formation of the first stars and galaxies. It has also been observed in a variety of cool molecular astrophysical environments. The rate of cooling by HD molecules requires knowledge of collisional rate coefficients with the primary impactors, H, He, and $H_2$. To improve knowledge of the collisional properties of HD, Nolte et al. (2012) presented rate coefficients for the He–HD collision system over a range of collision energies from $10^{-5}$ to $5 \times 10^3$ $cm^{-1}$. Fully quantum mechanical scattering calculations were performed for initial HD rovibrational states of j = 0 and 1 for v = 0–17, which utilized accurate diatom rovibrational wave functions. Rate coefficients of all $\Delta v = 0, -1$, and $-2$ transitions are reported. Significant discrepancies with previous calculations, which adopted a small basis and harmonic HD wave functions for excited vibrational levels, were found for the highest previously considered vibrational state of v = 3.

The emergence of the first black holes during the first $10^9$ years of cosmic history marks a key event in cosmology. Their formation is part of the overall process of ending the cosmic dark ages, when the first stars appeared in low-mass dark matter haloes about a few $100 \times 10^6$ years after the Big Bang. The first stars, galaxies, and black holes transformed the Universe from its simple initial state into one of ever-increasing complexity. Smith and Bromm (2019) reviewed recent progress on our emerging theoretical picture of how the first black holes appeared on the cosmic scene and how they affected the subsequent history of the Universe. Their focus was on supermassive black holes, in particular assessing possible pathways to the formation of the billion-solar-mass black holes implied to power the luminous quasars at high redshifts. They concluded with a discussion of upcoming empirical probes, such as the James Webb Space Telescope and the Laser Interferometer Space Antenna (LISA).

The visibility of high-redshift Lyman-alpha emitting galaxies (LAEs) provides important constraints on galaxy formation processes and the Epoch of Reionization (EoR). However, predicting realistic and representative statistics for comparison with observations represents a significant challenge in the context of large-volume cosmological simulations. The THESAN project offers a unique framework for addressing such limitations by combining state-of-the-art galaxy formation (IllustrisTNG) and dust models with the AREPO-RT radiation-magnetohydrodynamics solver. In their initial study, Smith et al. (2022) presented Lyman-alpha centric analysis for the flagship simulation that resolves atomic cooling haloes throughout a $(95.5cMpc)^3$ region of the Universe. To avoid numerical artifacts, they devised a novel method for accurate frequency-dependent line radiative transfer in the presence of continuous Hubble flow, transferable to broader astrophysical applications as well. Their scalable approach highlights the utility of LAEs and red damping-wing transmission as probes of reionization, which reveal nontrivial trends across different galaxies, sightlines, and frequency bands that can be modeled in the framework of covering fractions. In fact, after accounting for environmental factors influencing large-scale ionized bubble formation such as redshift and UV magnitude, the variation across galaxies and sightlines mainly depends on random processes, including peculiar velocities and self-shielded systems that strongly affect unfortunate rays more than others do. Throughout the EoR, local and cosmological optical depths are often greater than or less than unity, such that the $\exp^{-\tau}$ behavior leads to anisotropic and bimodal transmissivity.

Over the past decade, the search for high-redshift (high-z) galaxies using deep Hubble Space Telescope (HST) images in fields on both "blank sky" and lensing clusters seems to suggest that the number density of galaxies at $z > 6$ sharply decreases toward higher and higher redshifts and even a dearth of galaxies at $z > 10$. This, if true, would create a severe problem for our understanding of the cosmic hydrogen reionization. There has been a consensus that the reionization ended at $z \approx 6.2$. There is also a piece of evidence from the detection of 1.420 GHz (21 cm) H I absorption of the cosmic microwave background that the first stars came into being at $z \approx 17.2$ (spanning $z = 13.7$–$22.9$), which is to say that the reionization should begin at the same redshift. While this result is still under debate, an independent line of evidence also points to a similar redshift as the epoch of the first stars.

One example is the gravitationally lensed galaxy MACS 1149-JD1, which is at $z = 9.1096$ based on the Atacama Large Millimeter/Submillimeter Array detection of its [O III] 88-μm line. The very existence of oxygen means that the galaxy has already been polluted by the previous generation of stars. To reproduce its spectral energy distribution, especially its Balmer break, these authors found that the onset of the galaxy must be $\sim 290 \times 10^6$ years earlier, that is, at the formation redshift $z_f \approx 15$. It is therefore conceivable that first stars should be formed at an even earlier time (by a few tens of $10^6$ years), and so should the beginning of the reionization. All this, however, is inconsistent with the suggestion that the number of galaxies diminishes at $z > 10$ because this would imply that the reionization could not have happened; see Yan et al. (2023).

On July 13, 2022, NASA released to the whole world the data obtained by the James Webb Space Telescope Early Release Observations. These are the first set of science-grade data from this long-awaited facility, marking the beginning of a new era in astronomy. In the study of the early Universe, JWST will allow us to push far beyond $z \approx 11$, the redshift boundary previously imposed by the 1.7-μm red cutoff of the Hubble Space Telescope. In contrast, JWST's NIRCam reaches ~5 μm. Among the JWST ERO targets, there is a nearby galaxy cluster, SMACS 0723–73, which is a massive cluster and has been long recognized as a potential "cosmic telescope" in amplifying background galaxies. The ERO six-band NIRCam observations on this target have covered an additional flanking field not boosted by gravitational lensing, which also sees far beyond HST. Yan et al. (2023) reported the result from the search of candidate objects at $z > 11$ using these ERO data. In total, there are 87 such objects identified by using the standard "dropout" technique. These objects are all detected in multiple bands and therefore cannot be spurious. For most of them, their multiband colors are inconsistent with known types of contaminants. If the detected dropout signature is interpreted as the expected Lyman break, it implies that these objects are at $z \approx 11$–20. The large number of such candidate objects at such high redshifts was not expected from the previously favored predictions.

The first stars were born from chemically pristine gas. They were likely massive, and thus they rapidly exploded as supernovae, enriching the surrounding gas with the first heavy elements. In the Local Group, the chemical signatures of the first stellar population were identified among low-mass, long-lived, very metal-poor ($[Fe/H] < -2$) stars, characterized by high abundances of carbon over iron ($[C/Fe] > +0.7$): the so-called carbon-enhanced metal-poor stars. Conversely, a similar carbon excess caused by first-star pollution was not found in dense neutral gas traced by absorption systems at different cosmic times. Saccardi et al. (2023) presented the detection of 14 very metal-poor, optically thick absorbers at redshift $z \sim 3$–4. Among these, three are carbon enhanced and reveal an overabundance with respect to Fe of all the analyzed chemical elements (O, Mg, Al, and Si). Their relative abundances show a distribution with respect to [Fe/H] that is in very good agreement with those observed in nearby very metal-poor stars. All the performed tests support the idea that these C-rich absorbers preserve the chemical yields of the first stars. Their findings suggest that the first-star signatures can survive in optically thick but relatively diffuse absorbers, which are not sufficiently dense to sustain star formation and hence are not dominated by the chemical products of normal stars.

On June 5, NASA released a new composite image from the JWST. It shows the Universe when it was less than $600 \times 10^6$ years old. In this image, one can explore more than 45,000 baby galaxies, especially by zooming into the large interactive image. Since the Universe is $13.75 \times 10^9$ years old, at $600 \times 10^6$ years, the cosmos was in its infancy. The galaxies had just formed from the expanding cloud of mainly hydrogen gas, moving outward from the Big Bang. The statement accompanying this Webb image release said: "The sheer number of these galaxies was far beyond predictions from observations made before Webb's launch". This image is part of the JWST Advanced Deep Extragalactic Survey (JADES) program. It focuses on an area in Fornax called GOODS-South, part of the Great Observatories Origins Deep Survey (GOODS). Because of its dim stars, Fornax the Furnace is a hard-to-find constellation in the Southern Hemisphere. None of the stars in Fornax is particularly bright, Its brightest star, Alpha Fornacis, is magnitude 3.9. This star lies about 46 lightyears from Earth. The second brightest star, Beta Fornacis, at magnitude 4.4, lies 168 lightyears away; the constellation hold in its darker depths at least six star systems with exoplanets, the Fornax Dwarf galaxy, which is a satellite galaxy of the Milky Way, as well as the Fornax Cluster of galaxies.

### 1.1.2 The Universe in Time

An observer moving with respect to the cosmic rest frame should observe a concentration and brightening of galaxies in the direction of motion and a spreading and dimming in the opposite direction. The velocity inferred from this dipole should match that of the cosmic microwave background temperature dipole if galaxies are on average at rest with respect to the CMB rest frame. However, recent studies have claimed a many-fold enhancement of galaxy counts and flux in the direction of the solar motion compared to the CMB expectation, calling into question the standard cosmology. Darling (2022) showed that the sky distribution and brightness of extragalactic radio sources are consistent with the CMB dipole in direction and velocity.

Protoclusters are defined as overdense regions in the high-z Universe and are expected to evolve into massive, virialized clusters of galaxies at the present epoch; see Overzier (2016) and references therein. Finding and characterizing protoclusters is key to studying the large-scale structure of the Universe and the transformational processes that affect star formation and nuclear activity in the member galaxies, such as powerful mergers, gas cooling, and feedback effects. In practice, although a protocluster is observationally identified as a high-z structure that is overdense compared to its surroundings, the dynamical state of the structure, and therefore whether it will evolve into a low-z cluster, is usually highly uncertain (Muldrew et al. 2015). At present, there are no standard methods to systematically search for bona fide protoclusters, and this is why biased-tracer techniques are often used to identify interesting targets. One of these consists in searching around high-redshift, powerful radio galaxies, which are often found to be a beacon for overdense regions (Miley et al. 2006). Independently of the selection method, in order to confirm their nature and to trace the activity of several processes occurring during their rapid evolution, protocluster candidates must be followed up intensively with multiwavelength campaigns.

Here we present a couple of randomly chosen reports as an indication of efforts done in the field. Tozzi et al. (2022) presented an analysis of the 700 ks Chandra ACIS-S observation of the field around the Spiderweb Galaxy at $z = 2.156$, focusing on the nuclear activity in the associated large-scale environment. They identified unresolved X-ray sources down to flux limits of $1.3 \times 10^{-16}$ and $3.9 \times 10^{-16}$ erg/s/cm$^2$ in the soft and hard parts of the spectra, respectively. They searched for counterparts in the optical, NIR, and submm bands to identify X-ray sources belonging to the protocluster. They detected $10^7$ X-ray unresolved sources within 5 arcmin (corresponding to 2.5 Mpc) of J1140–2629, among which 13 have optical counterparts with spectroscopic redshift $2.11 < z < 2.20$, and one source with photometric redshift consistent with this range. Their 0.5–10 keV rest frame luminosities are larger than $2 \times 10^{43}$ erg/s, significantly greater than X-ray luminosities expected from star formation activity. The X-ray luminosity function of AGN in the volume associated with the Spiderweb protocluster in the range $10^{43} < L_X < 10^{44.5}$ erg/s is at least 10 times higher than that in the field at the same redshift and significantly flatter. The X-ray AGN fraction is measured to be $(25.5 \pm 4.5)\%$ in the stellar mass range $\log(M_*/M_{sun}) > 10.5$, corresponding to an enhancement of $6.0^{+9.0}_{-3.0}$ with respect to the COSMOS field at comparable redshifts and stellar mass range. The conclusion is that the galaxy population in the Spiderweb Protocluster is characterized by enhanced X-ray nuclear activity triggered by environmental effects on Mpc scales.

Recently, Peng et al. (2023) reported the discovery of two companion sources to a strongly lensed galaxy SPT0418–47 at redshift 4.225, targeted by the JWST Early Release Science program. We confirm that these sources are at a similar redshift to the ring based on Hα detected in the NIRSpec spectrum and [CII] λ158-μm line from the ALMA. Using multiple spectral lines detected in JWST/NIRSpec, the rest-frame optical to infrared images from NIRCam, and MIRI and far-infrared dust continuum detected by ALMA, the authors argue that the newly discovered sources are actually lensed images of the same companion galaxy, SPT0418-SE, hereafter referred to "SE", located within 5 kpc in the source plane of the ring. The star formation rate derived using [CII] and the dust continuum puts a lower limit of 17 $M_\odot$ yr$^{-1}$, while the SFRHα is estimated to be >2 times lower, thereby confirming that SE is a dust-obscured star-forming galaxy. Analysis using optical strong line diagnostics suggests that SE has near-solar elemental abundance, while the ring appears to have supersolar metallicity O/H and N/O. The authors also attempted to reconcile the high metallicity in this system by invoking early onset of star formation with continuous high star-forming efficiency or by suggesting that optical strong line diagnostics need revision at high redshift. They suggest that SPT0418–47 resides in a massive dark-matter halo with yet-to-be discovered neighbors. This work highlights the importance of joint analysis of JWST and ALMA data for a deep and complete picture of the early Universe.

### 1.1.3 The Universe Now

Our current understanding of physics is based on four fundamental forces: the strong and weak nuclear forces, electromagnetism, and gravity. These must ultimately be different aspects of the same unified fundamental interaction. However, the long-sought unification with gravity has been hampered by theoretical difficulties and its relative weakness, which makes it extremely hard to build accurate gravitational experiments. Gravity is so weak that two electrons 1000 km apart repel each other by $2.5 \times 10^{-10}$ m/s$^2$, slightly more than the Galactocentric orbital acceleration of the solar system due to the combined gravitational pull of the entire galaxy (Gaia Collaboration 2021). The gravitational fields of individual particles are thus immeasurably small; see Banik and Zhao (2022).

The relative weakness of gravity means that one must by necessity consider astronomical observations in order to understand it empirically, which is probably the first step to a deeper theoretical understanding. A field theoretical basis was put forth in terms of the Poisson equation, which clarified that the inverse square decline of both gravity and the electrostatic force can be understood in terms of field lines spreading in 3D. Newtonian gravity is an excellent description of non-relativistic motions in the solar system, so much so that discrepancies it faced with the motion of Uranus were attributed to a previously unknown planet that was subsequently discovered, Neptune.

The small but statistically significant 43 arcsec per century discrepancy between observations and the Newtonian expectation for its rate of perihelion precession was for a long time also attributed to an undiscovered planet. The anomalous perihelion precession of Mercury was actually an important clue towards a fully relativistic theory of gravity, namely General Relativity (GR); see Einstein (1915). This was designed to reduce to Newtonian dynamics in the appropriate limit of low velocities and shallow potentials with escape velocities much smaller than c, the speed of light. The predictions of GR have been verified one after another over the past. Observations of the star S2 allow particularly stringent constraints because it approaches within 120 AU of the galactic center black hole. Even in this strong field, GR correctly predicts the gravitational redshift of S2; see GRAVITY Collaboration (2018) and its pericenter precession (GRAVITY Collaboration 2020), which is much more extreme than for Mercury. Especially striking have been the recent observations of a black hole shadow; see The Event Horizon Telescope, EHT, Collaboration (2019) and the measurement that gravitational waves (GWs) travel at c to a precision of $\sim 10^{-15}$; see LIGO and Virgo Collaborations (2017).

Despite these impressive successes, it is less clear whether GR can be extrapolated to the non-relativistic motions in the outskirts of galaxies and galaxy clusters. The unexpectedly high velocity dispersion of galaxies relative to each other in the Coma Cluster could be the first evidence for a breakdown of Newtonian gravity, or it could indicate missing mass—although even a combination of the two should not be lightly discarded.

Although structures in the Universe form on a wide variety of scales, from small dwarf galaxies to large super clusters, the generation of angular momentum across these scales is poorly understood. Wang et al. (2021a) investigated the possibility that filaments of galaxies—cylindrical tendrils of matter hundreds of millions of light years across—are themselves spinning. By stacking thousands of filaments together and examining the velocity of galaxies perpendicular to the filament's axis (via their redshift and blueshift), they find that these objects display vortical motion consistent with rotation, making them the largest objects known to have angular momentum. The strength of the rotation signal is directly dependent on the viewing angle and the dynamical state of the filament. Filament rotation is more clearly detected when viewed edge-on. In addition, the more massive the haloes that sit at either end of the filaments, the more rotation is detected. These results signify that angular momentum can be generated on unexpectedly large scales.

The current paradigm of structure formation predicts that most of the gas in the intergalactic medium (IGM) is organized in a "cosmic web" composed of nonuniform gaseous filaments connecting galaxies on scales of megaparsecs. These filaments feed the circumgalactic medium (CGM), the gaseous component that is responsible for regulating the gas exchange between galaxies and the surrounding IGM. This complex interplay between the infall of gas from filaments through the CGM and the ejection of matter from the feedback processes is expected to play a key role in the regulation of galaxy growth through cosmic time.

The IGM has mainly been explored over the past decades using absorption line spectroscopy, which provides a powerful way to trace the neutral hydrogen observed in Lyman-alpha (Ly$\alpha$) absorption against bright background quasars. However, it has not been possible to obtain a detailed picture of these filaments, as information is limited to one dimension along the line-of-sight to the background source. The low sky density of sufficiently bright background sources prevents the study of the cosmic web on scales smaller than a few megaparsecs. Only very recently have sparse two-dimensional constraints on the IGM structure at a transverse spatial sampling of a few megaparsecs started to become available.

Bacon et al. (2021) reported the discovery of diffuse extended Ly$\alpha$ emission from redshift 3.1 to 4.5, tracing cosmic web filaments on scales of 2.5–4 cMpc. These structures have been observed in overdensities of Ly$\alpha$ emitters in the MUSE Extremely Deep Field, a 140-h-deep MUSE observation located in the Hubble Ultra-Deep Field. Among the 22 overdense regions identified, five are likely to harbor very extended Ly$\alpha$ emission at high significance with an average surface brightness of $5 \times 10^{-20}$ erg s$^{-1}$ cm$^{-2}$ arcsec$^{-2}$. Remarkably, 70% of the total Ly$\alpha$ luminosity from these filaments comes from beyond the circumgalactic medium of any identified Ly$\alpha$ emitter. Fluorescent Ly$\alpha$ emission powered by the cosmic UV background can only account for less than 34% of this emission at $z \approx 3$ and for not more than 10% at higher redshift. The authors found that the bulk of this diffuse emission could be reproduced by the unresolved Ly$\alpha$ emission of a large population of ultra low-luminosity Ly$\alpha$ emitters ($<10^{40}$ erg s$^{-1}$), provided that the faint end of the Ly$\alpha$ luminosity function is steep ($\alpha \lessapprox -1.8$), it extends down to luminosities lower than $10^{38}$–$10^{37}$ erg s$^{-1}$, and the clustering of these Ly$\alpha$ emitters is significant (filling factor < 1/6). If these Ly$\alpha$ emitters are powered by star formation, then this implies their luminosity function needs to extend down to star formation rates $< 10^{-4} M_{\odot}$ yr$^{-1}$. These observations provide the first detection of the cosmic web in Ly$\alpha$ emission in typical filamentary environments and the first observational clue indicating the existence of a large population of ultra low-luminosity Ly$\alpha$ emitters at high redshift.

The direct detection of gravitational waves, which are the ripples of spacetime predicted by general relativity, opens a new era of the exploration of the early Universe. The monumental discovery was the detection of the merger of two black holes by LIGO, which provides a new way to observe the Universe, complementing traditional electromagnetic observations. SGWBs are formed from the superpositions of many unresolved GW sources of both astrophysical and cosmological origins, which can be detected by searching for correlated signals between multiple detectors. Searching for such stochastic gravitational wave backgrounds (SGWBs) has been the main scientific goal of multiband GW observers, including LIGOVirgo-KAGRA, LISA, Taiji, TianQin, and pulsar timing array (PTA) experiments.

According to Mu et al. (2023), ultra-slow-roll (USR) inflation predicts an exponential amplification of scalar perturbations at small scales, which leads to a stochastic gravitational wave background through the coupling of the scalar and tensor modes at the second-order expansion of the Einstein equation. The authors searched for such a scalar-induced SGWB from the NANOGrav 15-year (NG15) dataset and found that the SGWB from USR inflation could explain the observed data. They placed constraints on the amplitude of the scalar power spectrum to $P_{Rp} > 10^{-1.80}$ at a 95% confidence level (C.L.) at the scale of $k \sim 20$ pc$^{-1}$. They find that $\log_{10} P_{Rp}$ degenerates with the peak scale $\log_{10} k_p$. They also obtain the parameter space allowed by the data in the USR inflationary scenario, where the e-folding numbers of the duration of the USR phase has a lower limit $\Delta N > 2.80$ (95% C.L.) when the USR phase ends at $N \approx 20$. Since the priors for the model parameters are uncertain, they do not calculate the Bayes factors. Instead, to quantify the goodness of fit, they calculate the maximum values of the log-likelihood for USR inflation, bubble collision of the cosmological phase transition, and inspiraling supermassive black hole binaries (SMBHBs), respectively. Their results imply that the SGWB from USR inflation can fit the data better than the one from SMBHBs.

### 1.1.4 Cosmological (Hubble) Constant

It was the discovery of the red shift of light from distant objects and consequently the expansion of the Universe by astronomer Edwin Hubble in the 1930s (Hubble 1929; Hubble and Humason 1931) that forced Einstein to drop a factor from his equations called the cosmological constant, represented by the Greek letter lambda ($\Lambda$).

When Einstein formulated general relativity in 1915, he had been surprised that it indicated the Universe should be expanding or contracting. Since he favored the idea of a steady-state Universe, as did many at the time, so this was a problem.

To combat this, Einstein introduced $\Lambda$—a self-admitted "fudge factor" he is later said to have described as his "greatest blunder"—as a form of "anti-gravity" to balance gravity and ensure the Universe he modeled was a steady-state one and was not expanding or contracting.

The cosmological constant was thus confined to the cosmic dustbin, but it would not stay there for long. The revelation that the expansion of the Universe is accelerating was even more astounding than Hubble's discovery, and ironically, it forced cosmologists to rescue the cosmological constant $\Lambda$. Today $\Lambda$ is used to represent the effect of dark energy, a new form of "anti-gravity" driving the Universe apart rather than keeping it stable (Lea 2022).

Unfortunately, the cosmological constant Λ is as much of a headache for cosmologists now as it was for Einstein—maybe more. The prime suspect accounting for Λ currently is the vacuum energy of space itself actually exerting negative pressure on cosmic objects. This would imply that dark energy is the same everywhere, but there is a major problem with this explanation. There exists a huge disparity between the large value of vacuum energy suggested by quantum theory and the value of Λ given by observation. The theoretical estimate for this energy of empty space from quantum field theory is somewhere in the order of $1 \times 10^{120}$ larger than the value of Λ astronomers observe out in the cosmos by looking at the redshift of supernovas. It's for good reason, then, that estimations of Λ from quantum field theory have been called "the worst theoretical prediction in the history of physics" by some scientists.

Lambda cold dark matter and modified Newtonian dynamics (MOND) are the two major hypotheses for the discrepancies between the observed properties of galaxies and the current laws of physics. The former predicts the existence of cosmological large-scale structures, while the latter is more advantaged in providing an accurate account for many galactic observations. MOND is challenged for its empirical model that lacks a complete theory.

The so-called Hubble constant ($H_0$) is the measure of the rate of expansion of the Universe. Cosmology's standard model prediction for this rate is significantly slower than the rate found by recent precise local measurements. This problem is subject to many discussions, and many cosmologists are trying to solve it by proposing changes in accepted cosmological model. Cyr-Racine et al. (2022) found that a uniform scaling of the gravitational free-fall rates and photon-electron scattering rate left most dimensionless cosmological observables nearly invariant. In addition, this led them to an interesting conclusion about the existence of a mirror Universe similar to ours but invisible to us except through gravitational impact on our world. A "mirror world" dark sector allows for effective scaling of the gravitational free-fall rates while respecting the measured mean photon density today.

The present expansion rate of the Universe, $H_0$, sets its size and age scale, relating redshift to distance and time. The value of $H_0$ could be determined locally with measurements of distances and redshifts, and it can be predicted from a cosmological model calibrated in the early Universe with measurements of the cosmic microwave background. The comparison of measured and predicted values of $H_0$ provides a test of the validity of cosmological models, from the early times, when the Universe was dense and dominated by dark matter and radiation, to the present, when it is dilute and dominated by dark energy.

Riess et al. (2022) reported calibration of $H_0$ by observation of Cepheid variables in the host galaxies of 42 Type Ia supernovae from Hubble Space Telescope. Their baseline result from the Cepheid SN Ia sample is $H_0 = 73.04 \pm 1.04$ km $s^{-1}$ $Mpc^{-1}$. They reported a 5σ difference with prediction of $H_0$ Planck Cosmic Microwave Background observations under ΛCDM, with no indication that the discrepancy arouse from measurement uncertainties or possible analysis variations. The source of this discrepancy between direct and cosmological routes of determining the Hubble constant remains unknown.

Joint analysis of Dark Energy Survey (DES) Year 3 data and CMB lensing from the South Pole Telescope (SPT) and Planck has been presented in three papers by DES and SPT Collaborations. In the first paper, DES and SPT Collaborations' (2023a) construction of CMB lensing maps and modeling choices are discussed. In the second paper, DES and SPT Collaborations' (2023b) cross-correlation measurements and cosmological constraints are discussed. In the third paper, DES and SPT Collaborations' (2023c) combined cosmological constraints are presented. Joint analyses of cross-correlations between measurements of galaxy positions, galaxy lensing, and lensing of the cosmic microwave background offer powerful constraints on the large-scale structure of the Universe. Special attention was paid to the cosmological constraints from the analysis of such cross-correlations measured using Year 3 data from the Dark Energy Survey and CMB data from the South Pole Telescope and Planck. Two key ingredients of analysis are presented: (1) an improved CMB lensing map in the SPT-SZ survey footprint and (2) the analysis methodology that will be used to extract cosmological information from the cross-correlation measurements. Relative to previous lensing maps made from the same CMB observations, authors have implemented techniques to remove contamination from the thermal Sunyaev Zel'dovich effect, enabling the extraction of cosmological information from smaller angular scales of the cross-correlation measurements than in previous analyses with DES Year 1 data. The expected cosmological constraints from the galaxy survey-CMB lensing auto and cross-correlations were forecasted. It was found that the galaxy-CMB lensing and galaxy shear-CMB lensing correlations will on their own provide a constraint on $S_8 = \sigma_8(\Omega_m/0.3)^{½}$ at the few percent level, providing a powerful consistency check for the DES-only constraints. It was found that the joint analysis of CMB lensing cross-correlations can provide constraints on the shear calibration amplitude at the 5% to 10% level, after DES and SPT Collaborations (2023a).

The Hubble constant ($H_0$) tension between Type Ia supernovae (SNe Ia) and Planck measurements ranges from 4 to 6$\sigma$. To investigate this tension, Dainotti et al. (2021) estimated $H_0$ in the ΛCDM and $w_0 w_a$ CDM (cold dark matter) models by dividing the Pantheon sample, the largest compilation of SNe Ia, into 3, 4, 20, and 40 bins. They fit the extracted $H_0$ values with a function mimicking the redshift evolution: $g(z) = H_0(z) = \breve{H}_0/(1+z)^{\alpha}$, where $\alpha$ indicates an evolutionary parameter and $\breve{H}_0 = H_0$ at $z = 0$. The absolute magnitude of SNe Ia is set so that $H_0 = 73.5$ km s$^{-1}$ Mpc$^{-1}$. They found that $H_0$ evolves with redshift, showing a slowly decreasing trend, with $\alpha$ coefficients consistent with zero only from 1.2 to 2.0 $\sigma$. Although the $\alpha$ coefficients are compatible with zero in 3$\sigma$, this may affect cosmological results. They measured locally a variation of $H_0(z=0) - H_0(z=1) = 0.4$km s$^{-1}$ Mpc$^{-1}$ in three and four bins. Extrapolating $H_0(z)$ to $z = 1100$, the redshift of the last scattering surface, values of $H_0$ compatible in 1$\sigma$ with Planck measurements independent of the cosmological models and number of investigated bins are obtained. They have reduced the $H_0$ tension in the range from 54% to 72% for both cosmological models.

A range of cosmological observations demonstrate an accelerated expansion of the Universe, and the most likely explanation of this phenomenon is a cosmological constant. Given the importance of understanding the underlying physics, it is relevant to investigate alternative models. The article by Loeve et al. (2021) uses numerical simulations to test the consistency of one such alternative model. Specifically, this model has no cosmological constant; instead, the dark matter particles have an extra force proportional to the velocity squared, somewhat reminiscent of the magnetic force in electrodynamics. The constant strength of the force is the only free parameter. Because bottom-up structure formation creates cosmological structures whose internal velocity dispersions increase in time, this model may mimic the temporal evolution of the effect from a cosmological constant. It is shown that models with force linearly proportional to internal velocities, or models proportional to velocity to power 3 or more, cannot mimic the accelerated expansion induced by a cosmological constant. However, models proportional to velocity squared are still consistent with the temporal evolution of a Universe with a cosmological model.

Brandenberger et al. (2023) studied the parametric resonance excitation of the electromagnetic field by a gravitational wave. They showed that there is narrow band resonance. For an electromagnetic field in the vacuum, the resonance occurs only in the second band, and its strength is thus suppressed by two powers of amplitude of the gravitational wave. On the other hand, in the case of an electromagnetic field in a medium with the speed of light smaller than 1 (in natural units), there is a band of Fourier modes which undergo resonance in the first band, allowing graviton to photon conversion via parametric resonance.

In a recent paper, Boylan-Kolchin (2023) reported on stress testing ΛCDM with high-redshift galaxy candidates. Early data from the JWST have revealed a lot of high-redshift galaxy candidates with unexpectedly high stellar masses. An immediate concern is the consistency of these candidates with galaxy formation in the standard ΛCDM cosmological model, wherein the stellar mass ($M_\star$) of a galaxy is limited by the available baryonic reservoir of its host dark matter halo. The mass function of dark matter haloes therefore imposes an absolute upper limit on the number density n ($>M_\star$, z) and stellar mass density $\rho_\star$ ($>M_\star$, z) of galaxies more massive than $M_\star$ at any epoch z. The most massive galaxy candidates in JWST observations at $z \approx 7$–10 lie at the very edge of these limits, indicating an important unresolved issue with the properties of galaxies derived from the observations, how galaxies form at early times in ΛCDM, or this standard cosmology itself.

The Hubble constant is the current rate of expansion of the Universe. Predictions for this rate—from cosmology's standard model—are significantly slower than the rate found by most precise local measurements. This discrepancy is one that many cosmologists have been attempting to solve by changing the current cosmological model. The challenge is to do so without ruining the agreement between standard model predictions and many other cosmological phenomena, such as the cosmic microwave background.

The absolute calibration of the classical Cepheid luminosity scale is fundamental for a distance estimation in the nearby Universe and the accurate measurement of Hubble's constant, $H_0$. The third data release (DR3) of the ESA mission Gaia has provided astrometry of unprecedented quantity and quality (The Gaia Collaboration 2018, 2021) for approximately 1.5 billion stars in the magnitude range $3 < G < 21$, including 14,992 classical Cepheid stars with an average parallax uncertainty of 70 μas. Because the parallax is generally considered the gold standard of geometric distance measurements, the Gaia parallaxes are of crucial importance for the absolute calibration of Leavitt's law (LL); see Leavitt and Pickering (1912), also known as the period-luminosity relation, and they are of great interest for all further applications of Cepheids as distance tracers.

Reyes and Anderson (2023) have conducted a search for open clusters near classical galactic Cepheids based on high-quality astrometry from the third data release (DR3) of the ESA mission Gaia to improve the calibration of Leavitt's law. Their approach required no prior knowledge of existing clusters, allowing the authors to both detect new host clusters and cross-check previously reported associations. Their gold sample consists of 34 Cepheids residing in 28 open clusters, including 27 fundamental mode and seven overtone Cepheids. Three new bona fide cluster Cepheids are reported (V0378 Cen, ST Tau, and GH Lup), and the host cluster identifications for three others (VW Cru, IQ Nor, and SX Vel) are corrected. Nonvariable cluster members allowed determination of cluster parallaxes to ~ 7 μas in the range $12.5 < G < 17$ mag.

Redshifts used in current cosmological supernova samples are measured using two primary techniques, one based on well-measured host galaxy spectral lines and the other based on supernova-dominated spectra. Steinhardt et al. (2020) constructed an updated Pantheon catalog with revised redshifts, redshift sources, and estimated uncertainties for the entire sample to investigate whether these two techniques yield consistent results. The best-fit cosmological parameters using these two measurement techniques disagree, with a supernova-only sample producing $\Omega_m$ 3.2σ higher and $H_0$ 2.5σ lower than a host-only sample, and explored several possible sources of bias that could result from using the lower-precision supernova-dominated redshifts. In a pilot study, we show that using a host redshift-only subsample will generically produce lower $\Omega_m$ and matter density $\Omega_m h^2$ and slightly higher $H_0$ than previous analysis, which, for the Pantheon data set, could result in supernova and cosmic microwave background measurements agreeing on $\Omega_m h^2$ despite tension in $H_0$. To obtain rigorous results, though, the Pantheon catalog should be improved by obtaining host spectra for supernovae that have faded, and future surveys should be designed to use host galaxy redshifts rather than lower-precision methods.

The accuracy of the Hubble constant tested with high-resolution observations of Cepheids by JWST has been recently reported by Riess et al. (2023). Over the last decade, an intriguing tension has emerged between the value of the Hubble constant ($H_0$) measured directly from redshifts and distances, following its local definition and independent of models, and the same parameter inferred from the cosmological model, calibrated in the early Universe. This "Hubble tension" is seen with high significance (>5σ confidence) across a wide range of local, independent distance indicators and among experiments measuring the cosmic microwave background and thus requires a solution. The James Webb Space Telescope provides new capabilities to scrutinize and refine some of the strongest observational evidence for this tension. The most significant differences are seen between measurements of local Type Ia supernovae (SNe Ia) calibrated by Cepheid variables, which yield $H_0 = 73.0 \pm 1.0$ km s$^{-1}$ Mpc$^{-1}$ (Riess et al. 2022), and the analysis of Planck observations of the cosmic microwave background (Planck Collaboration 2020), which predict $H_0 = 67.4 \pm 0.5$ km s$^{-1}$ Mpc$^{-1}$ in conjunction with ΛCDM. Cepheids have been the preferred primary distance indicator for these studies because they are extraordinarily luminous (MH ~ −7 mag at a period of 30 days), intrinsically precise (~3% in distance per star), reliably identifiable due to their periodicity since Leavitt and Pickering (1912), and well understood since Eddington (1917).

High-resolution JWST observations can test confusion-limited HST observations for a photometric bias that could affect extragalactic Cepheids and the determination of the Hubble constant. Riess et al. (2023) presented JWST NIRCAM observations in two epochs and three filters of >320 Cepheids in NGC 4258 (which has a 1.5% maser-based geometric distance) and in NGC

5584 (host of SN Ia 2007af), near the median distance of the SH0ES HST SN Ia host sample and with the best leverage among them to detect such a bias. JWST provides far superior source separation from line-of-sight companions than HST in the NIR to largely negate confusion or crowding noise at these wavelengths, where extinction is minimal. The result is a remarkable >2.5× reduction in the dispersion of the Cepheid P–L relations, from 0.45 to 0.17 mag, improving individual Cepheid precision from 20% to 7%. Two-epoch photometry confirmed identifications, tested JWST photometric stability, and constrained Cepheid phases. The P–L relation intercepts are in very good agreement, with differences (JWST–HST) of 0.00 ± 0.03 and 0.02 ± 0.03 mag for NGC 4258 and NGC 5584, respectively. The difference in the determination of $H_0$ between HST and JWST from these intercepts is 0.02 ± 0.04 mag, insensitive to JWST zero points or count-rate non-linearity thanks to error cancellation between rungs. The authors explored a broad range of analysis variants (including passband combinations, phase corrections, measured detector offsets, and crowding levels), indicating robust baseline results. These observations provide the strongest evidence yet that systematic errors in HST Cepheid photometry do not play a significant role in the present Hubble tension. Upcoming JWST observations of >12 SN Ia hosts should further refine the local measurement of the Hubble constant.

Current cosmological data exhibit discordance between indirect and some direct inferences of the present-day expansion rate $H_0$. Early dark energy (EDE), which briefly increases the cosmic expansion rate prior to recombination, is a leading scenario for resolving this "Hubble tension" while preserving a good fit to cosmic microwave background data. However, this comes at the cost of changes in parameters that affect structure formation in the late-time Universe, including the spectral index of scalar perturbations $n_S$. Goldstein et al. (2023) present the first constraints on axionlike EDE using data from the Lyman-α forest, that is, absorption lines imprinted in background quasar spectra by neutral hydrogen gas along the line of sight. We consider two independent measurements of the one-dimensional Ly-α forest flux power spectrum from the Sloan Digital Sky Survey (SDSS eBOSS) and from the MIKE/HIRES and X-Shooter spectrographs. They combine these with a baseline dataset composed of Planck CMB data and baryon acoustic oscillation measurements. Combining the eBOSS Ly-α data with the CMB and BAO dataset reduces the 95% confidence level upper bound on the maximum fractional contribution of EDE to the cosmic energy budget $f_{EDE}$ from 0.07 to 0.03 and constrains $H_0 = 67.9_{-0.4}^{+0.4}$ km/s/Mpc (68% C.L.), with maximum *a posteriori* value $H_0 = 67.9$ km/s/Mpc. Similar results are obtained for the MIKE/HIRES and X-Shooter Ly-α data. Their Ly-α-based EDE constraints yield $H_0$ values that are in > 4σ tension with the SH0ES distance-ladder measurement and are driven by the preference of the Ly-α forest data for $n_S$ values lower than those required by EDE cosmologies that fit Planck CMB data. Taken at face value, the Ly-α forest severely constrains canonical EDE models that could resolve the Hubble tension.

Keeley and Shafieloo (2023) are ruling out new physics at low redshift as a solution to the $H_0$ tension. To robustly answer this question, they use a very flexible parametrization for the dark energy equation of state such that every cosmological distance still allowed by data exists within this prior volume. To then answer whether there exists a satisfactory solution to the $H_0$ tension within this comprehensive parametrization, they constrained the parametric form using different partitions of the Planck cosmic microwave background, SDSS-IV/eBOSS DR16 baryon acoustic oscillation, and Pantheon supernova datasets. When constrained by just the cosmic microwave background dataset, there exists a set of equations of state which yields high $H_0$ values, but these equations of state are ruled out by the combination of the supernova and baryon acoustic oscillation datasets. In other words, the constraint from the cosmic microwave background, baryon acoustic oscillation, and supernova datasets together does not allow for high $H_0$ values and converges around an equation of state consistent with a cosmological constant. Thus, since this very flexible parametrization does not offer a solution to the $H_0$ tension, there can be no solution to the $H_0$ tension that adds physics at only low redshifts. This is directly related to the expansion history of the Universe and its geometrical properties.

The Λ cold dark matter standard cosmological model is in severe tension with several cosmological observations. Foremost is the Hubble tension, which exceeds 5σ confidence. Galaxy number counts show the Keenan–Barger–Cowie (KBC) supervoid, a significant underdensity out to 300 Mpc that cannot be reconciled with ΛCDM cosmology. Haslbauer et al. (2020) previously showed that a high local Hubble constant arises naturally due to gravitationally driven outflows from the observed KBC supervoid. The main prediction of model by Mazurenko et al. (2023) is that peculiar velocities are typically much larger than expected in the ΛCDM framework. This agrees with the recent discovery by Watkins et al. (2023) that galaxies in the CosmicFlows-4 catalogue have significantly faster bulk flows than expected in the ΛCDM model on scales of Mpc. The rising bulk flow curve is unexpected in standard cosmology, causing 4.8σ tension at Mpc. In this work, the authors determined what the semi-analytic void model of Haslbauer et al. predicts for the bulk flows on these scales. They found qualitative agreement with the observations, especially if their vantage point is chosen to match the observed bulk flow on a scale of Mpc. This represents a highly non-trivial success of a previously published model that was not constrained by bulk flow measurements but was shown to solve the Hubble tension and explain the KBC void consistently with the peculiar velocity of the Local Group. Their results suggest that several cosmological tensions can be simultaneously resolved if structure grows more efficiently than in the ΛCDM paradigm on scales of tens to hundreds of Mpc.

## 1.2 DARK MATTER DOGMA

The standard cold dark matter plus cosmological constant model predicts that galaxies form within dark-matter haloes and that low-mass galaxies are more dark-matter dominated than massive ones. The unexpected discovery of two low-mass galaxies lacking dark matter immediately provoked concerns about the standard cosmology and ignited explorations of alternatives, including self-interacting dark matter and modified gravity. Apprehension grew after several cosmological simulations using the conventional model failed to form adequate numerical analogs with comparable internal characteristics (stellar masses, sizes, velocity dispersions, and morphologies). The standard paradigm naturally produces galaxies lacking dark matter with internal characteristics in agreement with observations. Using a state-of-the-art cosmological simulation and a meticulous galaxy-identification technique, Moreno et al. (2022) found that extreme close encounters with massive neighbors can be responsible for this. They predict that ~30% of massive central galaxies (with at least $10^{11}$ solar masses in stars) harbor at least one dark-matter–deficient satellite (with $10^8$–$10^9$ solar masses in stars). This distinctive class of galaxies provides an additional layer in understanding of the role of interactions in shaping galactic properties.

The Coma Cluster of galaxies, as studied by a composite of modern space and ground-based telescopes. The infrared data come from the Spitzer Space telescope, while ground-based data come from the Sloan Digital Sky Survey. The Coma Cluster is dominated by two giant elliptical galaxies, with over 1000 other spirals and ellipticals inside. The speed of the individual galaxies within the Coma Cluster is too great for the cluster to remain a bound entity based on its normal matter content alone. Only unless a substantial amount of additional matter, that is, a source of dark matter, exists throughout this cluster can it remain a bound object under Einstein's laws of General Relativity. Possible solutions include either unseen matter or modifying Einstein's gravity.

The extended rotation curve of M33, also known as the Triangulum galaxy is discussed next. These rotation curves of spiral galaxies ushered in the modern astrophysics concept of dark matter to the general field. The galaxies without dark matter represent fewer than 1% of galaxies. Dark matter is not the only possible explanation for this observation; modified gravity can account for this, and other observations of similar objects on galaxy scales, just as successfully. Corbelli and Salucci (2000) presented rotation curve of the nearby galaxy M33 out to a galactocentric distance of 16 kpc (13 disk scale lengths). The rotation curve keeps rising out to the last measured point and implies a dark halo mass >~5 × 1010Msolar. The stellar and gaseous disks provide virtually equal contributions to the galaxy gravitational potential at large galactocentric radii, but no obvious correlation is found between the radial distribution of dark matter and the distribution of stars or gas. Results of the best fit to the mass distribution in M33 show a dark halo which controls the gravitational potential from 3 kpc outward, with a matter density which decreases radially as R-1.3. The density profile is consistent with the theoretical predictions for structure formation in hierarchical clustering cold dark matter models and favors lower mass concentrations than those expected in the standard cosmogony.

Since its discovery (Zwicky 1933), dark matter has been shown by numerous observations to be the Universe's dominant source of gravity and composed of undiscovered, nonbaryonic material. Its nature and composition, however, remain unknown, since the

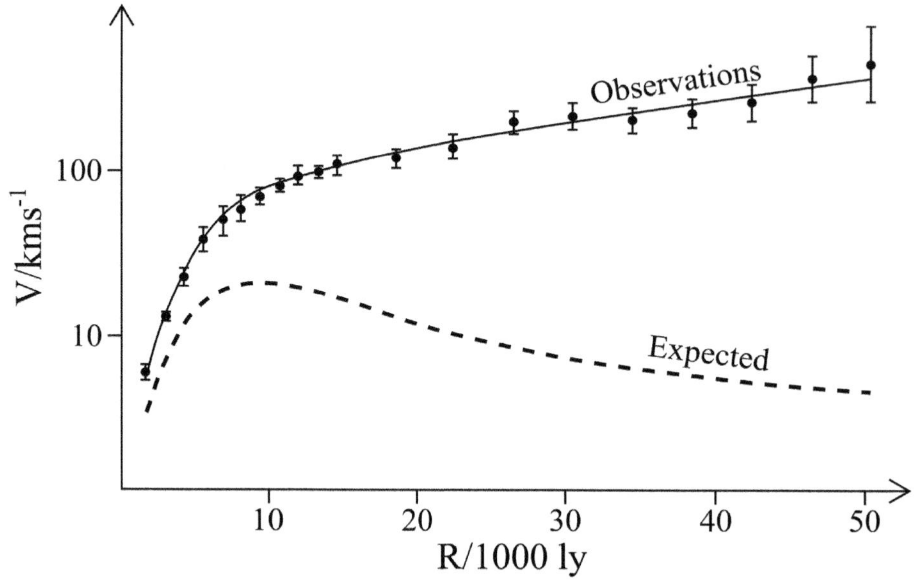


FIGURE 1.3 The extended rotation curve of M33, the Triangulum galaxy, as obtained by observation compared to the expectations from the visible disk.

Standard Model does not offer any viable dark matter candidate.

The solution to the dark matter problem could lie in neutrino cosmology. The three flavors of Standard Model neutrinos are massless and display only left-handed chirality. It is now known that neutrinos oscillate between flavors and therefore are not massless, in contrast to the Standard Model, and the probability of oscillation between flavors can be described by the mixing angle, $\theta$. There is currently no explanation for the collective phenomenon of neutrino mass and oscillation, but a possible solution is the existence of hypothetical right-handed neutrinos. This could give neutrinos more direct correspondence to other known fermions—all of which can exhibit both right- and left-handed chirality—and resolve major problems in modern physics in addition to the neutrino mass problem. Since the weak interaction only couples to left-handed neutrinos, right-handed neutrinos are referred to as sterile neutrinos due to their resulting lack of interaction via any forces besides gravity. In accordance with this naming convention, left-handed neutrinos are known as active neutrinos. Incidentally, the sterile neutrino could also solve the matter–antimatter asymmetry problem in the early Universe by giving rise to baryons through the process of leptogenesis. The most relevant feature of the sterile neutrino to this problem is its status as a dark matter candidate. Active neutrino masses are well known for being too small to constitute dark matter, but sterile neutrinos can have much larger masses. In particular, sterile neutrino dark matter is thought to have mass in the keV range. Also, in a process separate from neutrino oscillation, any neutrino can decay into another neutrino with lower mass. The probability of this decay depends on the difference in masses of the neutrinos in question, so it rarely happens to active neutrinos.

Sicilian et al. (2020) presented a comprehensive search for the 3.5 keV line using ~51 Ms of archival Chandra observations peering through the Milky Way's dark matter halo from across the entirety of the sky, gathered via the Chandra Source Catalog Release 2.0. They considered the data's radial distribution, organizing observations into four data subsets based on angular distance from the galactic center. All data were modeled using both background-subtracted and background-modeled approaches to account for the particle instrument background, demonstrating the statistical limitations of the currently available ~1 Ms of particle background data. A nondetection is reported in the total data set, allowing us to set an upper limit on 3.5 keV line flux and constrain the sterile neutrino dark matter mixing angle. The upper limit on $\sin^2(2\theta)$ is $2.58 \times 10^{-11}$ (though systematic uncertainty may increase this by a factor of ~2), corresponding to an upper limit on 3.5 keV line flux of $2.34 \times 10^{-7}$ photons $s^{-1}$ $cm^{-2}$. These limits show consistency with recent constraints and several prior detections. Nondetections are reported in all radial data subsets, allowing constraint of the spatial profile of 3.5 keV line intensity, which does not conclusively differ from Navarro–Frenk–White predictions. Thus, while offering heavy constraints, the authors do not entirely rule out the sterile neutrino dark matter scenario or the more general decaying dark matter hypothesis for the 3.5 keV line. They have also used the nondetection of any unidentified emission lines across the continuum to further constrain the sterile neutrino parameter space.

The missing gravity in galaxies requires dark matter or alternatively a modification of gravity or inertia. These theoretical possibilities of fundamental importance may be distinguished by the statistical relation between the observed centripetal acceleration of particles in orbital motion and the expected Newtonian acceleration from the observed distribution of baryons in galaxies. In a paper by Chae (2022), predictions of cold dark matter halos, modified gravity, and modified inertia are compared and tested by a statistical sample of galaxy rotation curves from the Spitzer Photometry and Accurate Rotation Curves database. Modified gravity under an estimated mean external field correctly predicts the observed statistical relation of accelerations from both the inner and outer parts of rotation curves. Taken at face value, there is a 6.9,s difference between the inner and outer parts on an acceleration plane, which would be inconsistent with current proposals of modified inertia. Removing galaxies with possible systematic concerns such as central bulges or special inclinations does not change this trend. Cold dark matter halos predict a systematically deviating relation from the observed one. All aspects of rotation curves appear to be most naturally explained by modified gravity.

Closer to home, the observation that the Andromeda galaxy (M31) is approaching the Milky Way (MW) despite their presumed recession shortly after the Big Bang also argues for significant missing gravity on Mpc scales, a line of evidence known as the Local Group (LG) timing argument. Moreover, the rotation curves (RCs) of galaxies do not follow the expected Keplerian decline beyond the extent of their luminous matter distributions. This led to the widespread acceptance that astronomical systems require significantly more gravity than can be explained using the conventional gravity of their detectable mass. Since one possible explanation is the existence of large amounts of invisible mass, this problem is usually called the dark matter (DM) problem. However, DM is by no means the only possible explanation. Therefore, we recommend the terminology "missing gravity problem", with the missing gravity arising from DM, a modification to GR in a regime where it has not been tested, or both; see Banik and Zhao (2022).

Some authors argue that dark matter may not exist; they favor a new theory of gravity called Milgromian dynamics or MOND, requiring no invisible matter (Milgrom 1983). MOND's main postulate is that when gravity becomes very week (i.e. near the edge of galaxies), it starts behaving differently from Newtonian physics. The basis of the modification is the assumption that in the limit of small acceleration, a very low $a_0$, the acceleration of a particle at distance r from a mass M satisfies approximately $a^2/a_0 \approx MGr^{-2}$, where $a_0$ is a constant of the dimensions of an acceleration. In this way, it is possible to explain why stars, planets, and gas in the outskirts of over 150 galaxies rotate faster than expected based on just their visible mass. MOND predictions are more accurate than the prediction of the models with the dark matter. The available observations strongly favor MOND over the standard cosmological model.

A very interesting paper by Farnes (2018) describes a unifying theory of dark energy and dark matter: negative masses and matter creation within a modified ΛCDM framework. He has constructed a toy model which suggests that both dark phenomena can be unified into a single negative mass fluid. The model is a modified ΛCDM cosmology and indicates that continuously created negative masses can resemble the cosmological constant and flatten the rotation curves of galaxies. A schematic cartoon of gravitational interactions between positive and negative mass particles is shown in Figure 1.4. The model leads to a cyclic Universe with a time-variable Hubble parameter, potentially providing compatibility with the current tension that is emerging in cosmological measurements. In the first three-dimensional N-body simulations of negative mass matter in the scientific literature, this exotic material naturally forms haloes around galaxies that extend to several galactic radii. These haloes are not cuspy. The proposed cosmological model is therefore able to predict the observed distribution of dark matter in galaxies from first principles. The model makes several testable predictions and seems to have the potential to be consistent with observational evidence from distant supernovae, the cosmic microwave background, and galaxy clusters. These findings may imply that negative masses

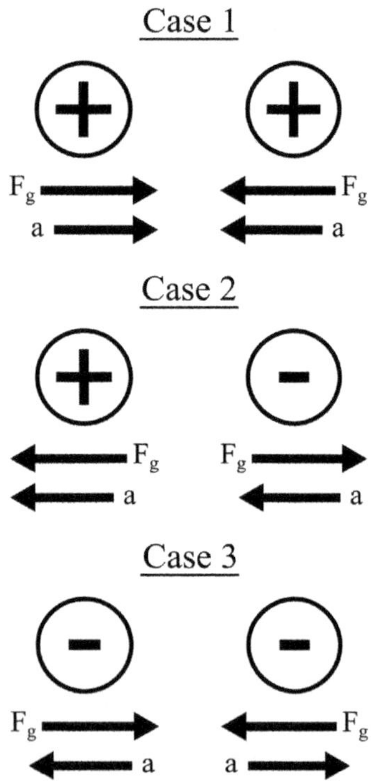


FIGURE 1.4 Schematic cartoon of gravitational interactions between positive and negative mass particles. Vectors indicate the direction of the gravitational force, $F_g = -GM_1M_2/r^2$, that is experienced by a given particle, while lower vectors indicate the direction of the acceleration, $a = F_g/M$, that is experienced by a given particle when the weak equivalence principle holds. There are three possible cases:

Case 1: the familiar positive–positive mass interaction, in which both particles accelerate towards one another via gravitational attraction,

Case 2: the positive–negative mass interaction, in which both particles accelerate in the same direction—pointing from the negative mass towards the positive mass. This case is sometimes referred to as runaway motion, and

Case 3: the negative–negative mass interaction, in which both particles accelerate away from one another via gravitational repulsion.

are a real and physical aspect of our Universe, or alternatively may imply the existence of a superseding theory that in some limit can be modeled by effective negative masses. Both cases lead to the surprising conclusion that the compelling puzzle of the dark Universe may have been due to a simple sign error (Farnes 2018).

Chung (2020) has modified Farnes' unifying theory of dark energy and dark matter discussed previously; that is, dark energy and dark matter can be unified into a single negative-mass dark fluid which is created continuously from the negative-mass Universe to our positive-mass Universe. The model is a modified ΛCDM cosmology and indicates that continuously created negative masses can resemble the cosmological constant and flatten the rotation curves of galaxies. In the first three-dimensional N-body simulations of negative mass dark matter in the scientific literature, dark matter naturally forms halos around galaxies that extend to several galactic radii. Chung's paper modifies and reinterprets Farnes' unifying theory of dark energy and dark matter in the positive-negative mass Universe pair model where positive-mass and negative-mass Universes are symmetrical. The first modification is about dark energy. With the symmetry between the positive-mass Universe and negative-mass Universe, the portion of dark energy to be created continuously from the negative-mass Universe to our positive-mass Universe cannot start from zero to more than 50%. Since the observed portion of dark energy is 68.6% (Planck Collaboration 2020) for a symmetrical positive-negative mass Universe pair, dark energy is more than negative mass from the symmetrical negative-mass Universe. As a result, the source of dark energy needs modification. Another modification is about dark matter. Evidence for the early existence of dark matter comes from measurements on cosmological scales of anisotropies in the cosmic microwave background (Planck Collaboration 2016). The CMB provides irrefutable evidence for the early existence of dark matter. Farnes' unifying theory does not allow the early existence of positive-mass dark matter, so dark matter needs modification.

According to the theoretical calculation, the calculated percentages of dark energy, dark matter, and baryonic matter are 68.6 (input from the observation), 26, and 5.2, respectively, in agreement with observed 68.6, 26.5, and 4.9, respectively, and dark energy started in $4.33 \times 10^9$ years ago in agreement with the observed $(4.71 \pm 0.98) \times 10^9$ years ago. In conclusion, the modified Farnes' unifying theory reinterprets Farnes' equations and is a unifying theory of dark energy, dark matter, and baryonic matter in the positive-negative mass Universe pair. The unifying theory explains protogalaxy and galaxy evolutions in agreement with the observations (Chung 2020).

Dark matter resides in "halos" surrounding galaxies, and information on the evolution of this invisible substance can be obtained by examining galaxies over a wide range of cosmic time. Fortunately, there is another way to probe the dark matter around such galaxies: via the imprint it leaves on the pattern of cosmic microwave background temperature fluctuations through gravitational lensing (Marques and Huffenberger 2022). Figure 1.5 shows how Miyatake et al. (2022) used measurements of this lensing by the Planck satellite and measurements of optical-wavelength light (orange arrow) from such distant galaxies by the Subaru telescope on Earth to determine the distribution of dark matter around the galaxies.

The Standard Model of cosmology explains many observed properties of the Universe and says that a Big Bang created all matter, space, and time. According to the Standard Model of cosmology, most galaxies should be surrounded by a halo of dark matter which interacts with other objects only via gravitational force. However, some studies (see Asencio et al. (2022)) concluded that the observed deformations of dwarf galaxies in the Fornax Cluster and the lack of low surface brightness dwarfs towards its center are incompatible with Lambda cold dark matter standard model of cosmology expectations but well consistent with Milgromian dynamics (MOND). The study is based on analysis of dwarf galaxies in the Formax Cluster, the second-closest large galaxy cluster to Milky Way. According to the Standard Model, these little galaxies should have dark matter halos, but the observations do not find them.

According to Asencio et al. (2023), the El Gordo Galaxy Cluster challenges ΛCDM for any plausible collision velocity. El Gordo (ACT-CL J0102–4915) is an extraordinarily large and bright galaxy cluster collision. Previously, they found that El Gordo is in $6.2\sigma$ tension with the Λ cold dark matter standard model when assuming the nominal mass and infall velocity values from the hydrodynamical simulations of Zhang et al. (2015). The recent weak-lensing study of Kim et al. (2021) showed that the mass of El Gordo is actually $(2.13 \pm 0.24) \times 10^{15}\ M_{sun}$. The authors of this article explored the level of tension between El Gordo and ΛCDM for the new mass estimate, assuming several Vinfall values. They found that in order to reduce the tension below the $5\sigma$ level, the El Gordo subclusters should have $V_{infall} < 2300$ km s–1 ($V_{infall} < 1800$ km $s^{-1}$ when considering the combined tension with the Bullet Cluster). To the best of our knowledge, the El Gordo hydrodynamical simulations conducted so far require $V_{infall} \geq 2500$ km $s^{-1}$ to simultaneously reproduce its morphology and its high X-ray luminosity and temperature. The authors concluded that El Gordo still poses a significant challenge to ΛCDM cosmology. Whether the properties of El Gordo can be reconciled with a lower $V_{infall}$ should be tested with new hydrodynamical simulations that explore different configurations of the interaction

For more than two decades, the DAMA/LIBRA experiment at the Gran Sasso National Laboratory near L'Aquila, Italy, has been reporting yearly fluctuations in flashes registered by the sodium iodide crystals of its detector. These events peak in June and bottom out in December, as physicists would expect if dark-matter particles were spread through the Milky Way. Adhikari et al. (2022) have shown that an underground experiment in South Korea can "see" dark matter streaming through Earth—or not, depending on how its data are sliced. The results cast further doubt on a decades-old claim that another experiment has been detecting the mysterious substance.

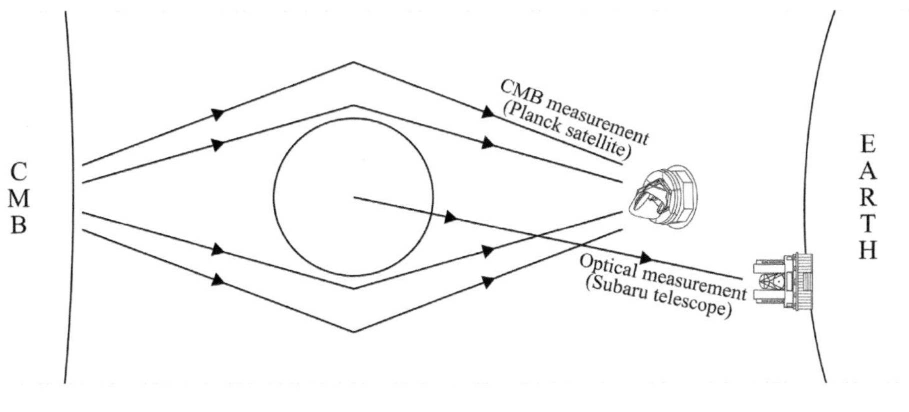


FIGURE 1.5 A distant galaxy and its surrounding dark-matter halo (blue) can bend light rays (white arrows) from the cosmic microwave background through gravitational lensing.

But none of the many other dark-matter experiments has so far seen signals compatible with DAMA/LIBRA's. Now one of these—the COSINE-100 detector—has shown that, by using a certain type of analysis on its own data, it can produce similar seasonal fluctuations. That the data analysis created a spurious result could be likely explanation for the DAMA/LIBRA claims, reported dark-matter signal could be due to analysis error. Observations that physicists have so far failed to replicate could be the result of misinterpreted data, Castelvecchi (2022a).

Impulsive supernova feedback and nonstandard dark matter models, such as self-interacting dark matter (SIDM), are the two main contenders for the role of the dominant core formation mechanism at the dwarf galaxy scale. Burger et al. (2022) show that the impulsive supernova cycles that follow episodes of bursty star formation leave distinct features in the distribution function of stars: groups of stars with similar ages and metallicities develop overdense shells in phase space. If cores are formed through supernova feedback, they predicted the presence of such features in star-forming dwarf galaxies with cored host halos. Their systematic absence would favor alternative dark matter models, such as SIDM, as the dominant core formation mechanism.

According to Calmet and Kuipers (2021), the quantum gravity leads to lower and upper bounds on the masses of dark matter candidates. These bounds depend on the spins of the dark matter candidates and the nature of interactions in the dark matter sector. For example, for singlet scalar dark matter, they found a mass range $10^{-3}$ eV $\lesssim m_{\phi} \lesssim 10^{7}$ eV. The lower bound comes from limits on fifth force–type interactions and the upper bound from the lifetime of the dark matter candidate.

The nature of dark matter remains unknown to date, although several candidate particles are being considered in a dynamically changing research landscape. Scalar field dark matter is a prominent option that is being explored with precision instruments, such as atomic clocks and optical cavities. Vermeulen et al. (2021) described a direct search for scalar field dark matter using a gravitational-wave detector, which operates beyond the quantum shot-noise limit. They set new upper limits on the coupling constants of scalar field dark matter as a function of its mass by excluding the presence of signals that would be produced through the direct coupling of this dark matter to the beam splitter of the GEO600 interferometer. These constraints improve on bounds from previous direct searches by more than six orders of magnitude and are, in some cases, more stringent than limits obtained in tests of the equivalence principle by up to four orders of magnitude. Their work demonstrates that scalar field dark matter can be investigated or constrained with direct searches using gravitational-wave detectors and highlights the potential of quantum-enhanced interferometry for dark matter detection.

In conclusion, astronomical observations have firmly established that a significant fraction of the mass energy in the Universe is in the form of dark matter, which interacts only through gravity. Fundamental particles that are beyond the Standard Model of particle physics are the most popular candidate for dark matter. However, such particles have so far evaded confident detection through direct or indirect methods. Massive astrophysical compact halo objects (MACHOs), in particular primordial black holes, are also potential candidates of dark matter. PBHs could be formed via the collapse of large overdensities in the early Universe. Their abundance is heavily constrained by the non-observation of their signatures in a variety of astronomical probes. These include the effects of Hawking evaporation, microlensing of stars and supernovae, gravitational wave observations, accretion effects from X-ray binaries, distortions of cosmic microwave background, dynamic effects such as the stability of certain wide binaries and stellar clusters, and formation of large-scale structures. Nevertheless, the possibility of them contributing to the dark matter cannot be ruled out in several mass windows. Recent observations of GWs from massive black hole binaries have resulted in renewed interest in PBHs.

Basak et al. (2022) presented constraints on the abundance of MACHOs through the non-observation of gravitational microlensing signatures in the GW signals detected by LIGO and Virgo. If a significant fraction of dark matter is in the form of compact objects, they will cause microlensing effects in the gravitational wave signals observable by LIGO and Virgo. From the non-observation of microlensing signatures in the binary black hole events from the first two observing runs and the first half of the third observing run, we constrain the fraction of compact dark matter in the mass range $10^{2}$–$10^{5}$ $M_{sun}$ to be less than $\approx 50-80\%$ (details depend on the assumed source population properties and the Bayesian priors). These modest constraints will be significantly improved in the next few years with the expected detection of thousands of binary black hole events, providing a new avenue to probe the nature of dark matter.

On other hand, the aim of a paper by Sipols and Pavlovich (2020) is to test the need for non-baryonic dark matter in the context of galactic rotation and the apparent difference between distributions of galactic mass and luminosity. They present a set of rotation curves and 3.6 μm surface brightness profiles for a diverse sample of 214 galaxies. Using rotation curves as the sole input into their Newtonian disk model, they compute non-parametric radial profiles of surface mass density. All profiles exhibit lower density than parametric models with dark halos and provide a superior fit with observed rotation curves. Assuming all dynamical mass is in main-sequence stars, they estimated radial distributions of characteristic star mass implied by the corresponding pairs of density and brightness profiles. They found that for 132 galaxies, or 62% of the sample, the relation between density and brightness can be fully explained by a radially declining stellar mass gradient. Such an idealized stellar population fitting can also largely address density and brightness distributions of the remaining 82 galaxies, but their periphery shows, on average, 14 $M_{\odot}/pc^2$ difference between total density and light-constrained stellar density. The authors discussed how this density gap can be interpreted by considering a low-luminosity baryonic matter, observational uncertainties, and visibility cutoffs for red dwarf populations. Last, they reported tight correlation between radial density and brightness trends and the discovered flattening of surface brightness profiles—both being evidence against dark matter. Their findings make non-baryonic dark matter unnecessary in the context of galactic rotation.

In their paper, Mancera Piña et al. (2021) argue that there is no need for dark matter: resolved kinematics of the ultra-diffuse galaxy AGC 114905. They present new HI interferometric observations of the gas-rich ultra-diffuse galaxy AGC 114905, which previous work, based on low-resolution data, identified as an outlier of the baryonic Tully-Fisher relation. The new observations, at a spatial resolution ~2.5 times higher than before, reveal a regular HI disk rotating at about 23 km $s^{-1}$. Their kinematic parameters, recovered with a robust 3D kinematic modeling fitting technique, show that the flat part of the rotation curve is reached. Intriguingly, the rotation curve can be explained almost entirely by the baryonic mass distribution alone. They showed that a standard cold dark matter halo that follows the concentration-halo mass relation fails to reproduce the amplitude of the rotation curve by a large margin. Only a halo with an extremely low concentration reaches agreement with the data. They also found that the rotation curve of AGC 114905 deviates strongly from the predictions of modified Newtonian dynamics. The inclination of the galaxy, which is measured independently from our modeling, remains the largest uncertainty in their analysis, but the associated errors are not large enough to reconcile the galaxy with the expectations of cold dark matter or modified Newtonian dynamics.

The axion has emerged in recent years as a leading particle candidate to provide the mysterious dark matter in the cosmos. Chadha-Day et al. (2022) described the historical roots of the axion in the Standard Model of particle physics and the problem of charge-parity invariance of the strong nuclear force. The authors also discussed how the axion emerges as a dark matter candidate and how it is produced in the early Universe. The symmetry properties of the axion dictate the form of its interactions with ordinary matter. Astrophysical considerations restrict the particle mass and interaction strengths to a limited range, which facilitates the planning of experiments to detect the axion. A companion review discusses the exciting prospect that the axion could be detected in the near term in the laboratory.

Semertzidis and Youn (2022) review the experimental aspects of axion physics and discuss the past, present, and future of the direct search programs. The axion is a highly motivated elementary particle that could address two fundamental questions in physics—the strong charge-parity (CP) problem and the dark matter mystery. Experimental searches for this hypothetical particle started reaching theoretically interesting sensitivity levels, particularly in the micro-electron volt (gigahertz) region. They rely on microwave resonators in strong magnetic fields with signals read out by quantum noise limited amplifiers. Concurrently, there have been intensive experimental efforts to widen the search range by devising various techniques and to enhance sensitivities by implementing advanced technologies.

Modifications of General Relativity generically contain additional degrees of freedom that can mediate forces between matter particles. One of the common manifestations of a fifth force in alternative gravity theories is a difference between the gravitational potentials felt by relativistic and non-relativistic particles, also known as "the gravitational slip". In contrast, a fifth force between dark matter particles, due to dark sector interaction, does not cause a gravitational slip, making the latter a possible smoking gun of modified gravity. Bonvin and Pogosian (2023) point out that a force acting on dark matter particles, as in models of coupled quintessence, would also manifest itself as a measurement of an effective gravitational slip by cosmological surveys of large-scale structure. This is linked to the fact that redshift-space distortions due to peculiar motion of galaxies do not provide a measurement of the true gravitational potential if dark matter is affected by a fifth force. Hence, it is extremely challenging to distinguish a dark sector interaction from a modification of gravity with cosmological data alone. Future observations of gravitational redshift from galaxy surveys can help to break the degeneracy between these possibilities, by providing a direct measurement of the distortion of time.

Benetti et al. (2023) explored the possibility that the dark matter component in galaxies may originate fractional gravity. In such a framework, the standard law of inertia continues to hold, but the gravitational potential associated with a given DM density distribution is determined by a modified Poisson equation including fractional derivatives (i.e., derivatives of noninteger type) that are meant to describe nonlocal effects. They analytically derived the expression of the potential that in fractional gravity corresponds to various spherically symmetric density profiles, including the Navarro–Frenk–White (NFW) distribution that is usually exploited to describe virialized halos of collisionless DM as extracted from N-body cosmological simulations. They show that in fractional gravity, the dynamics of a test particle moving in a cuspy NFW density distribution are substantially altered with respect to the Newtonian case, mirroring what in Newtonian gravity would instead be sourced by a density profile with an inner core. They tested the fractional gravity framework on galactic scales, showing that:

(i) it can provide accurate fits to the stacked rotation curves of spiral galaxies with different properties, including dwarfs; (ii) it can reproduce to reasonable accuracy the observed shape and scatter of the radial acceleration relation over an extended range of galaxy accelerations; and (iii) it can properly account for the universal surface density and the core radius versus disk scale length scaling relations.

Acevedo et al. (2023) showed that Milky Way white dwarfs are excellent targets for dark matter detection. Using Fermi and H.E.S.S. galactic center gamma-ray data, they investigated sensitivity to DM annihilating within white dwarfs into long-lived or boosted mediators and producing detectable gamma rays. Depending on the galactic DM distribution, they set new constraints on the spin-independent scattering cross section down to 10–45–10–41 $cm^2$ in the sub-GeV DM mass range, which is multiple orders of magnitude stronger than existing limits. For a generalized NFW DM profile, they find that their white dwarf constraints exceed spin-independent direct detection limits across most of the sub-GeV to multi-TeV DM mass range, achieving sensitivities as low as about 10–46 $cm^2$. In addition, they improve earlier versions of the DM capture calculation in white dwarfs by including the low-temperature distribution of nuclei when the white dwarf approaches crystallization. This yields smaller capture rates than previously calculated by a factor of a few up to two orders of magnitude, depending on white dwarf size and the astrophysical system.

Currently, the Λ cold dark matter model, which relies on the existence of cold dark matter and a cosmological constant Λ, best describes the Universe. However, we lack information in the high-redshift (z) region between Type Ia supernovae (SNe Ia; up to $z = 2.26$) and the cosmic microwave background ($z = 1100$), an interval crucial to test cosmological models and their possible evolution. Dainotti et al. (2023) defined a sample of 983 quasars up to $z = 7.54$ with a reduced intrinsic dispersion $\delta = 0.007$, which determines the matter density parameter ΩM with the same precision of SNe Ia. Although previous analyses have used quasars as cosmological tools, this is the first time that high redshift sources, in this case quasars, as standalone cosmological probes has yielded such tight constraints on ΩM. Their results show the importance of correcting cosmological relationships for selection biases and redshift evolution and how the choice of a golden sample reduces considerably the intrinsic scatter. This proves the reliability of quasars as standard cosmological candles.

Gehrman et al. (2023) outlined a new production mechanism for dark matter that they dub "recycling": dark sector particles are kinematically trapped in the false vacuum during a dark phase transition; the false pockets collapse into primordial black holes, which ultimately evaporate before Big Bang nucleosynthesis to reproduce the dark sector particles. The requirement that all PBHs evaporate prior to BBN necessitates high-scale phase transitions and hence high-scale masses for the dark sector particles in the true vacuum. Their mechanism is therefore particularly suited for the production of ultra heavy dark matter (UHDM) with masses above $\sim 10^{12}$ GeV. The correct relic density of UHDM is obtained because of the exponential suppression of the false pocket number density. Recycled UHDM has several novel features: the dark sector today consists of multiple decoupled species that were once in thermal equilibrium, and the PBH formation stage has extended mass functions whose shape can be controlled by IR operators coupling the dark and visible sectors.

Inoue et al. (2023) showed that the lensing power spectra for gravitational potential, astrometric shift, and convergence perturbations are powerful probes to investigate dark matter structures on small scales. They reported the first lower and upper bounds of these lensing power spectra on angular scale $\sim 1''$ toward the anomalous quadruply lensed quasar MG J0414+0534 at a redshift $z = 2.639$. To obtain the spectra, they conducted observations of MG J0414+0534 using the Atacama Large Millimeter/Submillimeter Array with high angular resolution (0."02–0."05). They developed a new partially nonparametric method in which Fourier coefficients of potential perturbation are adjusted to minimize the difference between linear combinations of weighted mean de-lensed images. Using positions of radio-jet components, extended dust emission on scales >1 kpc, and mid-infrared flux ratios, the range of measured convergence, astrometric shift, and potential powers at an angular scale of ~1."1 (corresponding to an angular wavenumber of $l = 1.2 \times 10^6$ or ~9 kpc in the primary lens plane) within $1\sigma$ are $\Delta_\kappa = 0.02–0.028$, $\Delta_\alpha = 7–9$ mas, and $\Delta_\psi = 1.2–1.6$ mas$^2$, respectively. Their partially nonparametric lens models suggest the presence of a clump in the vicinity of object Y, a possible dusty dwarf galaxy, and some small clumps in the vicinity of other lensed quadruple images. Although much fainter than the previous report, they detected weak continuum emission possibly from object Y with a peak flux of ~100 μJy beam$^{-1}$ at the ~$4\sigma$ level.

One particularly important question on dark matter is how it interacts with Standard Model particles beyond the gravitational effect. From the perspective of particle physics, one popular construction involves a dark force carrier that mediates the interactions between DM and SM particles. Huang et al. (2023) reported results of a search for dark-matter–nucleon interactions via a dark mediator using optimized low-energy data from the PandaX-4T liquid xenon experiment. With the ionization-signal-only data and utilizing the Migdal effect, they set the most stringent limits on the cross section for dark matter masses ranging from 30 MeV/$c^2$ to 2 GeV/$c^2$. Under the assumption that the dark mediator is a dark photon that decays into scalar dark matter pairs in the early Universe, we rule out the significant parameter space of such a thermal relic dark-matter model.

Collider searches for dark matter have often focused on scenarios in which DM particles are produced in association with heavy Standard Model particles or photons or jets. However, no experimental evidence confirming the existence of DM has been found so far. Several proposed models include a strongly coupled dark sector, giving rise to unexplored collider-event topologies and thus motivating the search reported by the ATLAS Collaboration (2024). Semi-visible jets, with a significant contribution to the event's missing transverse momentum, can arise in strongly interacting dark sectors. This results in an event topology where one of the jets can be aligned with the direction of the missing transverse momentum. The Collaboration presented the first search for semi-visible jets produced via a *t*-channel mediator exchange. The analysis uses proton-proton collisions with an integrated luminosity of 139 fb$^{-1}$ and a center-of-mass energy of 13 TeV, collected with the ATLAS detector during the Run 2 of the LHC. No excess over Standard Model predictions was observed. Assuming a coupling strength of unity between the mediator, a Standard Model quark, and a dark quark, mediator masses up to 2.7 TeV were excluded at the 95% confidence level. Upper limits on the coupling strength were also derived.

### 1.2.1 Dark Energy

Observations of Type Ia supernovae (SNe Ia) at redshift $z < 1$ provide evidence that the expansion of the Universe at the present time is accelerating, behavior attributed to "dark energy" with negative pressure. Dark energy is introduced to explain the phenomenon of the Universe acceleration, which has been experimentally shown from the observation of Ia supernovae at redshift $z < 1$, Riess et al. (1998). The magnitude of the observed acceleration was not anticipated by theory and continues to defy post facto explanation.

Several authors (see Petit and D'Agostini (2021) and references therein) suggested that the observed phenomenon could be the effect of a negative pressure associated with a negative mass content. In this theory all particles may have negative mass. Matter-antimatter duality holds in the negative mass world. The negative matter, which acts as "dark energy", is composed by negative energy particles plus their antiparticles.

The dominating form of energy in the cosmos, dark energy is driving the accelerating expansion of the Universe, but its nature remains a complete mystery; see Lea (2022) and references therein. Dark energy is a hypothetical form of energy that is proposed to explain why the Universe is expanding at an accelerating rate. Dark energy is an "anti-gravity" force providing a negative pressure that fills the Universe and stretches the very fabric of spacetime. As it does so, dark energy drives cosmic objects apart at an increasingly rapid rate rather than drawing them together as gravity does. Dark energy is estimated to account for between around 68% to 72% of the Universe's total energy and matter—its matter/energy budget—meaning it heavily dominates both dark matter and everyday matter.

There are some leading candidates to account for dark energy. These include the vacuum energy of space—particles literally popping in and out of existence in empty space—and a "fifth force" responsible for the negative pressure that might cause the accelerated expansion of the Universe. Other possibilities a range of different "flavors" of fields that could account for dark energy, such as a low-energy field dubbed "quintessence" and fields of tachyons—hypothetical particles that travel faster than light and thus back in time. All of these remain purely hypothetical, meaning that the only way we can really "know" dark energy currently is via its effect on the Universe.

Gravity still overruns dark energy at small scales. Dark energy only seems to act on the largest scales of the Universe, with the expansion of the Universe a phenomenon that can only be measured by observing galaxies and other cosmic objects that are separated by massive gulfs of space on the order of millions, billions, and even tens of billions of lightyears apart and away from us. And the greater the distance that separates these cosmic objects, the more rapidly they race away from each other.

The main way we know dark matter exists is from its gravitational effect of holding galaxies together. Without the gravitational influence of dark matter, galaxies are swirling so rapidly that the gravitational influence of their visible matter—stars, planets, gas, and dust—would be insufficient to prevent them from flying apart.

That means as dark energy is pushing things apart on a large scale, dark matter is holding galaxies together on a smaller scale. In that respect, dark energy and dark matter could almost be considered to have opposite effects in the Universe.

Observations have found black holes spanning 10 orders of magnitude in mass across most of cosmic history. The Kerr black hole solution is, however, provisional, as its behavior at infinity is incompatible with an expanding Universe. Black hole models with realistic behavior at infinity predict that the gravitating mass of a black hole can increase with the expansion of the Universe independently of accretion or mergers, in a manner that depends on the black hole's interior solution. Farrah et al. (2023) tested this prediction by considering the growth of supermassive black holes in elliptical galaxies over $0 < z \leq 2.5$. They found evidence for cosmologically coupled mass growth among these black holes, with zero cosmological coupling excluded at 99.98% confidence. The redshift dependence of the mass growth implies that, at $z \leq 7$, black holes contribute an effectively constant cosmological energy density to Friedmann's equations. The continuity equation then requires that black holes contribute cosmologically as vacuum energy. They further showed that black hole production from the cosmic star formation history gives the value of $\Omega_\Lambda$ measured by Planck while being consistent with constraints from massive compact halo objects. They proposed that stellar remnant black holes are the astrophysical origin of dark energy, explaining the onset of accelerating expansion at $z \sim 0.7$.

Andreia et al. (2022) claim that rapidly descending dark energy will result in the end of cosmic expansion. If dark energy is a form of quintessence driven by a scalar field $\varphi$ evolving down a monotonically decreasing potential $V(\varphi)$ that passes sufficiently below zero, the Universe is destined to undergo a series of smooth transitions. The currently observed accelerated expansion will cease; soon thereafter, expansion will come to an end altogether, and the Universe will pass into a phase of slow contraction. In their paper, Andreia et al. (2022) consider how short the remaining period of expansion can be given current observational constraints on dark energy. They also discuss how this scenario fits naturally with cyclic cosmologies and recent conjectures about quantum gravity.

### 1.2.2 Dark Stars

We shall touch this subject only briefly, although it may be of great importance in the early period of the Universe. The James Webb Space Telescope is poised to revolutionize our understanding of the formation and properties of first luminous objects in the Universe. Since beginning to take data, JWST has discovered a surprising number of extremely bright high redshift galaxy candidates (see the list of early reports: Naidu et al. (2022a) and (2022b), Finkelstein et al. (2022), Donnan et al. (2022), and Labbé et al. (2023), which are difficult to reconcile with expectations from numerical simulations of the Universe in the canonical ΛCDM scenario). Ilie et al. (2023) showed that some of the JWST high redshift galaxy candidates could instead be Dark Stars (DSs), early stars made almost entirely of hydrogen and helium but powered by dark matter heating rather than by fusion. Dark Stars provide a good match to JWST data, both in terms of their spectra (good fits to JWST photometry) and in that JWST (given its angular resolution) cannot rule out a point source interpretation of many of these candidates.

Prior to JWST there were only very limited data on the period when the first stars and galaxies formed. Therefore, numerical simulations were the primary tool to describe the properties of the first stars and galaxies in the Universe. In the standard picture of the first stars formed roughly (100–400) × $10^6$ years after the Big Bang ($z \sim 20$–10) as a consequence of the gravitational collapse of pristine, zero metallicity, molecular hydrogen clouds at the center of $10^6$–$10^8 M_{sun}$ minihaloes. This population of stars grew via accretion, reaching masses of (at most) $10^3 M_{sun}$, and populated the first galaxies. A different type of first star, so-called Dark Stars, early stars powered by dark matter heating rather than by fusion, were proposed previously by Spolyar et al. (2008) and (2009). They proposed that the first phase of stellar evolution in the history of the Universe may be Dark Stars, powered by dark matter heating rather than by nuclear fusion, and in their papers examined the history of these Dark Stars. The power source is annihilation of weakly interacting massive particles (WIMPs), which are their own antiparticles.

Dark Stars should be observationally distinct from standard Pop III stars. In addition, Dark Stars avoid the (unobserved) element enrichment produced by the standard first stars. Once the dark matter fuel is exhausted, the Dark Star becomes a heavy main sequence star; these stars eventually collapse to form massive black holes that may provide seeds for the supermassive black holes and intermediate black holes.

According to Sutter (2023), dark matter may have its own "invisible" periodic table of elements. The Universe may have produced dark matter in the first few minutes of the Big Bang, according to new research. Those particles then got trapped into ultradense pockets. Some of those pockets splintered off to become black holes, which then dissolved into a shower of multiple dark matter particle "species", creating a "dark matter periodic table" of invisible elements, the study authors suggest.

Previous work assumed that dark matter was simple: a single kind of extremely lightweight particle, flooding the Universe, that hardly ever interacted with normal matter. But intense searches for new such particles have come up empty. That led cosmologists to wonder if dark matter particles might be much rarer but also much heavier. However, theoretical physicists have struggled to find models that would seed the early Universe with the right number of such heavy particles. Gehrman et al. (2023) outlined a new production mechanism for dark matter that they dub "recycling": dark sector particles are kinematically trapped in the false vacuum during a dark phase transition; the false pockets collapse into primordial black holes, which ultimately evaporate before Big Bang nucleosynthesis to reproduce dark sector particles. The requirement that all PBHs evaporate prior to BBN necessitates high-scale phase transitions and hence high-scale masses for the dark sector particles in the true vacuum. Their mechanism is therefore particularly suited for the production of ultraheavy dark matter with masses above $\sim 10^{12}$ GeV. The correct relic density of UHDM is obtained because of the exponential suppression of the false pocket number density. Recycled UHDM has several novel features: the dark sector today consists of multiple decoupled species that were once in thermal equilibrium, and the PBH formation stage has extended mass functions whose shape can be controlled by IR operators coupling the dark and visible sectors.

## 1.3 BLACK HOLES

Stellar-mass black holes are the final remnants of stars born with more than 15 solar masses Some $10^9$ of them are expected to reside in the Local Group, but only a few are known by detection of X-rays emitted as they accrete material from a companion star. Shenar et al. (2022) reported an object called VFTS 243, which is a massive X-ray faint binary in the Large Magellanic Cloud. It consists of an O-type star ($M = 25 \times M_{solar}$) with an orbital period of 10.4 days and an unseen companion of at least $M = 9 \times M_{solar}$ assumed to be a black hole This is the first X-ray-quiet black hole unambiguously detected outside our galaxy. From the kinematics of VFTS 243, the authors concluded that the collapse of the progenitor into a black hole was associated with little or no ejected material or black hole kick.

Supermassive black holes occupy the center of galaxies, with masses ranging from 1 million to 10 billion solar masses. Some SMBHs are in a bright phase called active galactic nuclei. AGNs will eventually burn out since there is a maximum mass limit for SMBHs; scientists have long since pondered when that will be.

A question of how light in the form of a wave would behave under the influence of a gravitational field, with Albert Einstein's General Relativity Theory in 1915, was answered by Schwarzschild's solution to this problem a year later. Schwarzschild (1916) also predicted the existence of a critical circumference of a body beyond which light would be unable to cross: the Schwarzschild radius (see Figure 1.6). This idea was similar to that of Michell (1783/1784), but now this critical circumference was understood as an impenetrable barrier. It was Lemaître (1933) who showed that this impenetrability was only an illusion that a distant observer would have.

Black holes destroy thousands of stars to fuel growth. Massive black hole formation in dense stellar formations studied by Baldassare et al. (2022) by analyzing Chandra X-ray Observatory imaging of 108 galaxies hosting nuclear star clusters (NSCs) to search for signatures of massive black holes. NSCs are extremely dense stellar environments with conditions that can theoretically facilitate massive BH formation. Stone et al. (2017) found that sufficiently dense NSCs should be unstable to the runaway growth of a stellar-mass BH into a massive BH via tidal captures. Furthermore, there is a velocity dispersion threshold (40 km $s^{-1}$) above which NSCs should inevitably form a massive BH. To provide an observational test of these theories, Baldassare at al. 2022 measured X-ray emission from NSCs and compared it to the measured velocity dispersion and tidal capture runaway timescale. They found that NSCs above the 40 km $s^{-1}$ threshold are X-ray detected at roughly twice the rate of those below (after accounting for contamination from X-ray binaries). These results are consistent with a scenario in which dense, high-velocity NSCs can form massive BHs, providing a formation pathway that does not rely on conditions found only at high redshift.

Hawking (1971) proposed that the Sun may harbor a primordial black hole whose accretion supplies some of the solar luminosity. Such an object would have formed within the first 1 s after the Big Bang with the mass of a moon or an asteroid. These light black holes are a candidate solution to the dark matter problem, and could grow to become stellar-mass black holes if captured by stars. Bellinger et al. (2023) computed the evolution of stars having such a BHs at their center. They found that such objects can be surprisingly long lived, with the lightest black holes having no influence over stellar evolution, while more massive ones consume the star over time to produce a range of observable consequences. Models of the Sun born about a BH whose mass has since grown to approximately $10-6\ M_\odot$ are compatible with current observations. In this scenario, the Sun would first dim

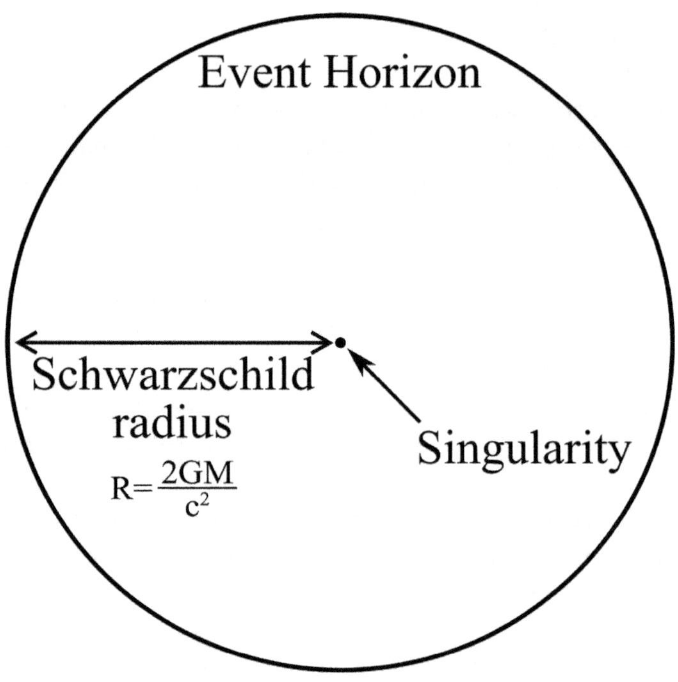


FIGURE 1.6 A critical circumference of a body, beyond which light would be unable to cross: the Schwarzschild radius.

to half its current luminosity over a span of 100 Myr as the accretion starts to generate enough energy to quench nuclear reactions. The Sun would then expand into a fully convective star, where it would shine luminously for potentially several Gyr with an enriched surface helium abundance, first as a sub-subgiant star and later as a red straggler, before becoming a sub-solar-mass BH. They also presented results for a range of stellar masses and metallicities. The unique internal structures of stars harboring BHs may make it possible for asteroseismology to discover them, should they exist.

Sweeney et al. (2022) chartered the galactic distribution of neutron stars and black holes and found it to exhibit a fundamentally different distribution and structure to the visible galaxy. This is due mainly to the natal kick received by the remnant at the event of its supernova birth. Due to this kick, they found that 30% of remnants have sufficient kinetic energy to entirely escape the galactic potential, leading to a galactic mass loss.

Lam et al. (2022) presented results from a search for stellar mass black holes in the Milky Way using gravitational microlensing. They analyzed five Milky Way black hole candidates that represent a theoretical expectation of the number of black holes in the Milky Way. Some of the studied objects turned out to be white dwarfs or neutron stars. Combining HST astrometry and densely sampled ground-based photometry, they derived masses for these five lenses as well as their probabilities of being BHs. Of the five targets, they made one definitive > 1 mas detection of astrometric microlensing (OB110462). The mass of the lens in OB110462 is in the range 1.6–4.4 $M_{\odot}$, making it the first detection of a compact object through astrometric microlensing.

Active galactic nucleus feedback during galaxy mergers has been the most favored model to explain black hole–galaxy co-evolution. However, how the AGN-driven jet-wind-radiation is coupled with the gas of the merging galaxies, which leads to positive feedback, momentarily enhanced star formation, and subsequently negative feedback, a decline in star formation, is poorly understood. Only a few cases are known where the jet and companion galaxy interaction leads to minor off-axis distortions in the jets and enhanced star formation in the gas-rich minor companions. Hota et al. (2022) reported one extraordinary case, RAD12, discovered by the RAD@home citizen science collaboratory, where for the first time a radio jet-driven bubble (~ 137 kpc) showed a symmetric reflection after hitting an incoming galaxy that was not a gas-rich minor but a gas-poor early-type galaxy in a major merger. Surprisingly, neither positive feedback nor any radio lobe on the counter jet side, if any, was detected. It is puzzling if RAD12 is a genuine one-sided jet or a case of radio lobe trapped, compressed, and re-accelerated by shocks during the merger.

All big galaxies have supermassive black holes in their centers. These enormous supermassive black holes have grown from smaller "seed" black holes, intermediate mass black holes (IMBHs), still having a mass of $10^3$ to $10^5$ solar masses. Some of the smaller galaxies, like dwarf galaxies, have IMBHs in their center. However, it is difficult to detect them since dwarf galaxies make stars at a high rate, which emits a lot of light and can mimic the appearance of having a bright, feeding black hole. Polimera et al. (2022) recently improved an older method so that these two processes can be separated.

Miki et al. (2021) discuss the destruction of the central black hole gas reservoir through head-on galaxy collision. A massive black hole exists in almost every galaxy. They occasionally radiate a vast amount of light by releasing gravitational energy of accreting gas, with a cumulative active period of only a few $10^8$ years, so-called the duty cycle of the active galactic nuclei. That is, many galaxies today host a starving massive black hole. Although galaxy collisions have been thought to enhance nucleus activity, the origin of the duty cycle, especially the shutdown process, is a still critical issue. The authors showed that galaxy collisions are also capable of suppressing black hole fueling by using an analytic model and three-dimensional hydrodynamic simulations, applying the well-determined parameter sets for the galactic collision in the Andromeda galaxy. A model presented by Miki et al. (2021) demonstrates that a central collision of galaxies can strip the torus-shaped gas surrounding the massive black hole, the putative fueling source. The derived condition for switching off black hole fueling indicates that a significant fraction of currently bright nuclei can become inactive, reminiscent of fading/dying active nucleus phenomena associated with galaxy merging events. Galaxy collisions may therefore be responsible for both switching off and turning on nucleus activity, depending on the collision orbit (head-on or far off center).

Understanding how super-massive black holes form and grow in the early Universe has become a major challenge since it was discovered that luminous quasars existed only 700 million years after the Big Bang (Fujimoto et al. 2022). Simulations indicate an evolutionary sequence of dust-reddened quasars emerging from heavy dust-obscured starburst that then transition to unobscured luminous quasars by expelling gas and dust. Although the last phase has been identified out to a redshift of 7.6, a transitioning quasar has not been found at similar redshifts owing to their faintness at optical and near-infrared wavelengths. Fujimoto et al. (2022) reported observations of an ultraviolet compact object, GNz7q, associated with a dust-enshrouded starburst at a redshift of $7.1899 \pm 0.0005$. The host galaxy is more luminous in dust emission than any other known object at this epoch, forming 1,600 solar masses of stars per year within a central radius of 480 parsecs. A red point source in the far ultraviolet is identified in deep, high-resolution imaging and slitless spectroscopy. GNz7q is extremely faint in X-rays, which indicates the emergence of a uniquely ultraviolet compact star-forming region or a Compton-thick super-Eddington black hole accretion disk at the dusty starburst core. In the latter case, the observed properties are consistent with predictions from cosmological simulations and suggest that GNz7q is an antecedent to unobscured luminous quasars at later epochs.

Tidal disruption events (TDEs) are bursts of electromagnetic energy that are released when supermassive black holes at the centers of galaxies violently disrupt a star that passes too close. TDEs provide a window through which to study accretion onto supermassive black holes; in some rare cases, this accretion leads to launching of a relativistic jet, but the necessary conditions are not fully understood. The best-studied jetted TDE so far is Swift J1644 + 57, which was discovered in γ-rays but was too obscured by dust to be seen at optical wavelengths. Andreoni et al. (2022) reported the optical detection of AT2022cmc, a rapidly fading source at cosmological distance (redshift $z = 1.19325$), the unique light curve of which transitioned into a luminous plateau within days. Observations of a bright counterpart at other wavelengths, including X-ray, submillimeter, and radio, support the interpretation of AT2022cmc as a jetted TDE containing a synchrotron "afterglow", probably launched by a supermassive black hole with spin greater than approximately 0.3. Using four years of Zwicky Transient Facility survey data, the authors calculated a rate of $0.02^{+0.04}_{-0.01}$ Gpc$^{-3}$ yr$^{-1}$ for on-axis jetted TDEs based on the luminous, fast-fading red component, thus providing a measurement complementary to the rates derived from X-ray and radio observations. Correcting for beaming angle effects, this rate confirms that approximately 1% of TDEs have relativistic jets. Optical surveys can use AT2022cmc as a prototype to unveil a population of jetted TDEs.

Black holes that eventually merge in AGN disks can be brought into the disk through gas capture from the surrounding nuclear star cluster8 or can be produced through in-situ star formation. Once a black hole is in the disk, it will undergo radial migration and can as a result pair up with another black hole to form a binary. Recent studies show that interactions between such migrating binary black holes and other single black holes in the AGN disk, referred to as binary-single interactions, likely provide the main pathway for bringing binaries to mergers.

There is some weak evidence that the black hole merger named GW190521 had a non-zero eccentricity. In addition, the component black holes' masses exceeded the limit predicted by stellar evolution. The large masses can be explained by successive mergers, which may be efficient in gas disks surrounding active galactic nuclei, but it is difficult to maintain an eccentric orbit all the way to the merger, as basic physics would argue for circularization. Samsing et al. (2022) show that AGN-disk environments can lead to an excess of eccentric mergers if the interactions between single and binary black holes are frequent and occur with mutual inclinations of less than a few degrees. They further illustrated that this eccentric population has a different distribution of the inclination between the spin vectors of the black holes and their orbital angular momentum at merger, referred to as the spin-orbit tilt, compared to the remaining circular mergers.

Occasionally a star gets too close to a supermassive black hole, at which point tidal forces on the star exceed its self-gravity, tearing apart the ill-fated interloper. Tidal disruption events—as these transient events are called—occur when a star wanders within the tidal radius, $r_t \approx R_\star(M_{BH}/M_\star)^{1/3}$, of the SMBH. Owing to the large spread in specific orbital energy acquired by the post-disruption stellar debris, roughly half of the debris is bound in a range of highly eccentric orbits around the SMBH, whereas the rest goes unbound. The bound material is expected to fall toward the SMBH, forming an accretion disk. The gravitational potential energy lost by the infalling matter is converted into electromagnetic radiation, resulting in a luminous flare that enables the detection of TDEs.

Patra et al. (2022) presented optical spectropolarimetry of the tidal disruption event AT 2019qiz on days +0 and +29 relative to maximum brightness. Continuum polarization, which informs the shape of the electron-scattering surface, was found to be consistent with 0% at peak brightness. On day +29, the continuum polarization rose to ~1%, making this the first reported spectropolarimetric evolution of a TDE. These findings are incompatible with a naked eccentric disk that lacks significant mass outflow. Instead, the spectropolarimetry paints a picture wherein, at maximum brightness, high-frequency emission from the accretion disk is reprocessed into the optical band by a nearly spherical, optically thick, electron-scattering photosphere located far away from the black hole. The authors estimated the radius of the scattering photosphere to be ~100 au at maximum brightness—significantly larger than the tidal radius (~1 au) and the thermalization radius (~30 au) where the optical continuum is formed. A month later, as the fallback rate drops and the scattering photosphere recedes, the continuum polarization increases, revealing a moderately aspherical interior. They also saw evidence for smaller-scale density variations in the scattering photosphere, inferred from the scatter of the data in the Stokes $q - u$ plane. On day +29, the H$\alpha$ emission-line peak is depolarized to ~0.3% (compared to ~1% continuum polarization) and displays a gradual rise toward the line's redder wavelengths. This observation indicates the H$\alpha$ line formed near the electron-scattering radius.

The self-lensing of a massive black hole binary, which occurs when the two BHs are aligned close to the line of sight, is expected to produce periodic, short-duration flares. Davelaar and Haiman (2022a) and (2022b) studied the shapes of self-lensing flares (SLFs) via general-relativistic ray tracing in a superimposed binary BH metric, in which the emission is generated by geometrically thin accretion flows around each component. The suite of models covers eccentric binary orbits, black hole spins, unequal mass binaries, and different emission model geometries. They explored this parameter space and reported how the light curves change as a function of, for example, binary separation, inclination, and eccentricity. They also compared their light curves to those in the microlensing approximation and showed how strong deflections, as well as time-delay effects, changed the size and shape of the SLF.

If gravitational waves from the inspiraling MBHB are observed by LISA, SLFs can help securely identify the source, localize it on the sky, and constrain the graviton mass by comparing the phasing of the SLFs and the GWs. Additionally, when these systems are viewed edge-on, the SLF shows a distinct dip that can be directly correlated with the BH shadow size. This opens a new way to measure BH shadow sizes in systems that are unresolvable by current VLBI facilities.

Sahu et al. (2022) reported the first unambiguous detection and mass measurement of an isolated stellar-mass black hole. They used the Hubble Space Telescope to carry out precise astrometry of the source star of the long-duration ($t_E$ ~ 270 days), high-magnification microlensing event MOA-2011-BLG191/OGLE-2011-BLG-0462 (hereafter designated as MOA-11-191/OGLE-11-462) in the direction of the galactic bulge. HST imaging, conducted at eight epochs over an interval of 6 years, reveals a clear relativistic astrometric deflection of the background star's apparent position. Ground-based photometry of MOA-11-191/OGLE-11-462 shows a parallactic signature of the effect of the Earth's motion on the microlensing light curve. Combining the HST astrometry with the ground-based light curve and the derived parallax, they obtain a lens mass of $7.1 \pm 1.3\ M_{sun}$ and a distance of $1.58 \pm 0.18$ kpc. They showed that the lens emits no detectable light, which, along with having a mass higher than is possible for a white dwarf or neutron star, confirms its BH nature. Their analysis also provided an absolute proper motion for the BH. The proper motion is offset from the mean motion of galactic-disk stars at similar distances by an amount corresponding to a transverse space velocity of ~45 km $s^{-1}$, suggesting that the BH received a "natal kick" from its supernova explosion. Previous mass determinations for stellar-mass BHs have come from radial-velocity measurements of galactic X-ray binaries and from gravitational radiation emitted by merging BHs in binary systems in external galaxies. Their mass measurement is the first for an isolated stellar-mass BH using any technique.

Saracino et al. (2021) reported detection of a black hole in the ~100-Myr-old massive cluster NGC 1850 in the Large Magellanic Cloud. It is in a binary system with a main-sequence turn-off star ($4.9 \pm 0.4\ M_{\odot}$), which is starting to fill its Roche lobe and is becoming distorted. Using 17 epochs of Very Large Telescope/Multi-Unit

Spectroscopic Explorer observations, they detected radial velocity variations exceeding 300 $km\,s^{-1}$ associated with the target star, linked to the ellipsoidal variations measured by the fourth phase of the Optical Gravitational Lensing Experiment in the optical bands. Under the assumption of a semidetached system, the simultaneous modelling of radial velocity and light curves constrains the orbital inclination of the binary to 38° ± 2°, resulting in a true mass of the unseen companion of $11.1^{+2.1}_{-2.4} M_{\odot}$. This represents the first direct dynamical detection of a black hole in a young massive cluster, opening up the possibility of studying the initial mass function and the early dynamical evolution of such compact objects in high-density environments.

Recently, Farrah et al. (2023) provided observational evidence for cosmological coupling of black holes and its implications for an astrophysical source of dark energy. Observations have found black holes spanning 10 orders of magnitude in mass across most of cosmic history. The Kerr black hole solution is, however, provisional, as its behavior at infinity is incompatible with an expanding Universe. Black hole models with realistic behavior at infinity predict that the gravitating mass of a black hole can increase with the expansion of the Universe independently of accretion or mergers in a manner that depends on the black hole's interior solution. Farrah et al. (2023) tested this prediction by considering the growth of supermassive black holes in elliptical galaxies over $0 < z < 2.5$. We find evidence for cosmologically coupled mass growth among these black holes, with zero cosmological coupling excluded at 99.98% confidence. The redshift dependence of the mass growth implies that, at $z < 7$, black holes contribute an effectively constant cosmological energy density to Friedmann's equations. The continuity equation then requires that black holes contribute cosmologically as vacuum energy. We further show that black hole production from the cosmic star formation history gives the value of $\Omega_{\Lambda}$ measured by Planck while being consistent with constraints from massive compact halo objects. We thus propose that stellar remnant black holes are the astrophysical origin of dark energy, explaining the onset of accelerating expansion at $z \sim 0.7$.

Moore and Vecchio (2021) have discussed ultra-low-frequency from cosmological and astrophysical processes. Gravitational waves at ultra-low frequencies ($\lesssim$100 nHz) are key to understanding the assembly and evolution of astrophysical black hole binaries with masses $\sim 10^6$–$10^9\, M_{\odot}$ at low redshifts. These gravitational waves also offer a unique window into a wide variety of cosmological processes. Pulsar timing arrays are beginning to measure this stochastic signal at ~1–100 nHz, and the combination of data from several arrays is expected to confirm a detection in the next few years. The dominant physical processes generating gravitational radiation at nHz frequencies are still uncertain. Pulsar timing array observations alone are currently unable to distinguish a binary black hole astrophysical foreground from a cosmological background due to, say, a first-order phase transition at a temperature ~1–100 MeV in a weakly interacting dark sector. The authors explore the extent to which incorporating integrated bounds on the ultra-low-frequency gravitational wave spectrum from any combination of cosmic microwave background, Big Bang nucleosynthesis or astrometric observations can help to break this degeneracy.

Bright flaring from accreting black holes is seen at all wavelengths, but the mechanism powering high-energy flares is still a topic of major debate. Rapid γ-ray flares have been observed from active galactic nuclei, in the form of very-high-energy (>100 GeV) emission. The variability timescale of the flares can be shorter than the light crossing time of the event horizon, constraining the emitting region to be on the order of a Schwarzschild radius. Bright TeV flares are also periodically observed from the supermassive black hole M87, in the center of the Messier 87 galaxy. The flares show a flux rise and decay timescale of 1–3 days, emitting $>10^{41}$ erg $s^{-1}$, which is nonnegligible compared to the total jet power of $10^{42}$–$10^{44}$ erg $s^{-1}$.

Magnetic reconnection can power bright, rapid flares originating from the inner magnetosphere of accreting black holes. Ripperda et al. (2022) conducted extremely high-resolution (5376 × 2304 × 2304 cells) general-relativistic magnetohydrodynamics simulations, capturing plasmoid-mediated reconnection in a 3D magnetically arrested disk for the first time. They showed that an equatorial, plasmoid-unstable current sheet forms in a transient, nonaxisymmetric, low-density magnetosphere within the inner few Schwarzschild radii. Magnetic flux bundles escape from the event horizon through reconnection at the universal plasmoid-mediated rate in this current sheet. The reconnection feeds on the highly magnetized plasma in the jets and heats the plasma that ends up trapped in flux bundles to temperatures proportional to the jet's magnetization. The escaped flux bundles can complete a full orbit as low-density hot spots, consistent with Sgr A* observations by the GRAVITY interferometer. Reconnection near the horizon produces sufficiently energetic plasma to explain flares from accreting black holes, such as the TeV emission observed from M87. The drop in the mass accretion rate during the flare and the resulting low-density magnetosphere make it easier for very high-energy photons produced by reconnection-accelerated particles to escape. The extreme-resolution results in a converged plasmoid-mediated reconnection rate that directly determines the timescales and properties of the flare.

Massive black holes at the centers of galaxies can launch powerful wide-angle winds that, if sustained over time, can unbind the gas from the stellar bulges of galaxies. These winds may be responsible for the observed scaling relation between the masses of the central black holes and the velocity dispersion of stars in galactic bulges. Propagating through the galaxy, the wind should interact with the interstellar medium, creating a strong shock, similar to those observed in supernovae explosions, which is able to accelerate charged particles to high energies. Ajello et al. (2021) used data from the Fermi Large Area Telescope to search for γ-ray emission from galaxies with an ultrafast outflow (UFO): a fast ($v \sim 0.1$ c), highly ionized outflow, detected in absorption at hard X-rays in several nearby active galactic nuclei. Adopting a sensitive stacking analysis, they were able to detect the average γ-ray emission from these galaxies and rule out that it is due to processes other than UFOs. Moreover, their analysis shows that the γ-ray luminosity scales with the AGN bolometric luminosity and that these outflows transfer ~0.04% of their mechanical power to γ-rays. Interpreting the observed γ-ray emission as produced by cosmic rays (CRs) accelerated at the shock front. They found that the γ-ray emission may attest to the onset of the wind–host interaction and that these outflows can energize charged particles up to the transition region between galactic and extragalactic CRs.

Black hole accretion disks formed in compact-object mergers or collapsars may be major sites of the rapid-neutron-capture (r-)process, but the conditions determining the electron fraction ($Y_e$) remain uncertain given the complexity of neutrino transfer and angular-momentum transport. After discussing relevant weak-interaction regimes, Just et al. (2022) studied the role of neutrino absorption for shaping $Y_e$ using an extensive set of simulations performed with two-moment neutrino transport and again without neutrino absorption. They varied the torus mass, BH mass, and spin and examined the impact of rest-mass and weak-magnetism corrections in the neutrino rates. They also tested the dependence on the angular-momentum transport treatment by comparing axisymmetric models using the standard α-viscosity with viscous models assuming constant viscous length-scales ($l_t$) and 3D magnetohydrodynamic (MHD) simulations. Finally, they discussed the nucleosynthesis yields and basic kilonova properties. They found that absorption pushes $Y_e$ towards ~0.5 outside the torus while inside increasing the equilibrium value $Y_e^{eq}$ by ~0.05–0.2. Correspondingly, a substantial ejecta fraction is pushed above $Y_e = 0.25$, leading to a reduced lanthanide fraction and a brighter, earlier, and bluer kilonova than without absorption. More compact tori with higher neutrino optical depth, τ, tend to have lower $Y_e^{eq}$ up to $\tau \sim 1$–10, above which absorption becomes strong enough to reverse this trend. Disk ejecta are less (more) neutron rich when employing an lt = constant viscosity (MHD treatment). The solar-like abundance pattern found for our MHD model marginally supports collapsar disks as major r-process sites, although a strong r-process may be limited to phases of high mass-infall rates, $M^{\cdot} >\sim 2 \times 10-2\ M_{sun}\ s^{-1}$.

A candidate runaway supermassive black hole identified by shocks and star formation in its wake is discussed by Van Dokkum et al. (2023). The interaction of a runaway supermassive black hole with the circumgalactic medium can lead to the formation of a wake of shocked gas and young stars behind it. The authors reported the serendipitous discovery of an extremely narrow linear feature in HST/ACS images that may be an example of such a wake. The feature extends 62 kpc from the nucleus of a compact star-forming galaxy at $z = 0.964$. Keck LRIS spectra show that the $[O_{III}]/H_\beta$ ratio varies from ~ 1 to ~ 10 along the feature, indicating a mixture of star formation and fast shocks. The feature terminates in a bright $[O_{III}]$ knot with a luminosity of $\approx 1.9 \times 10^{41}$ ergs $s^{-1}$. The stellar continuum colors vary along the feature and are well fitted by a simple model that has a monotonically increasing age with distance from the tip. The line ratios, colors, and overall morphology are consistent with an ejected SMBH moving through the CGM at high speed while triggering star formation. The best-fit time since ejection is ~39 Myr, and the implied velocity is $v_{BH} \sim 1600$ km $s^{-1}$. The feature is not perfectly straight in the HST images, and we show that the amplitude of the observed spatial variations is consistent with the runaway SMBH interpretation. Opposite the primary wake is a fainter and shorter feature, marginally detected in [$O_{III}$] and the rest-frame far-ultraviolet. This feature may be shocked gas behind a binary SMBH that was ejected at the same time as the SMBH that produced the primary wake.

Every galaxy with stellar mass above $\sim 10^{10}$ $M_{sun}$ seems to host a massive black hole ($M_{BH} > 10^5$ $M_{sun}$) at its center. It remains unclear when and how these massive BHs form, although some sort of exotic high-mass seed formation mechanism may be required by the existence of the brightest high-redshift quasars. Formation channels for massive BH seeds can be divided into two categories: those that operate only at high redshift and channels which continue to produce massive BH seeds throughout cosmic time.

Black holes are the extreme laboratories to study accretion and ejection physics and to eventually test theories of gravity. In addition to the supermassive BHs at the centers of galaxies, there are also stellar-mass BHs, which are mostly discovered when they are X-ray bright in systems called X-ray binaries (XRBs). These are binary systems composed of a compact object accreting material from a companion star; see Remillard and McClintock (2006) for a review and references therein. If the mass of the companion star is lower than roughly one solar mass, it is called a low-mass XRB (LMXB). Depending on the nature of the central compact object, the source is categorized as either a black hole XRB (BHXB) or a neutron star XRB (NSXB).

Wang et al. (2022) performed the first systematic search of all NICER archival observations of black hole (and candidate) low-mass X-ray binaries for signatures of reverberation. Reverberation lags result from the light travel time difference between the direct coronal emission and the reflected disk component, and therefore their properties are a useful probe of the disk-corona geometry. They detected new signatures of reverberation lags in eight sources, increasing the total sample from 3 to 11, and studied the evolution of reverberation lag properties as the sources evolve in outbursts. They found that in all nine sources with more than one reverberation lag detection, the reverberation lags become longer and dominate at lower Fourier frequencies during the hard-to-soft state transition. This result shows that the evolution in reverberation lags is a global property of the state transitions of black hole low-mass X-ray binaries, which is valuable in constraining models of such state transitions. The reverberation lag evolution suggests that the corona is the base of a jet that vertically expands and/or gets ejected during state transition. They also discovered that in the hard state, the reverberation lags get shorter, just as the quasiperiodic oscillations (QPOs) move to higher frequencies, but then in the state transition, while the QPOs continue to higher frequencies, the lags get longer. They also discuss the implications of the coronal geometry and physical models of QPOs in light of this new finding.

Abbott et al. (2021) reported the observation of gravitational waves from two compact binary coalescences in LIGO's and Virgo's third observing run with properties consistent with neutron star–black hole (NSBH) binaries. The two events are named GW200105_162426 and GW200115_042309, abbreviated as GW200105 and GW200115; the first was observed by LIGO Livingston and Virgo and the second by all three LIGO–Virgo detectors. The source of GW200105 has component masses $8.9^{+1.2}_{-1.5}$ and $1.9^{+0.3}_{-0.2}$ $M_{sun}$, whereas the source of GW200115 has component masses $5.7^{+1.8}_{-2.1}$ and $1.5^{+0.7}_{-0.3}$ $M_{sun}$ (all measurements quoted at the 90% credible level). The probability that the secondary's mass is below the maximal mass of a neutron star is 89%–96% and 87%–98%, respectively, for GW200105 and GW200115, with the ranges arising from different astrophysical assumptions. The source luminosity distances are $280 \pm 110$ and $300^{+150}_{-100}$ Mpc, respectively. The magnitude of the primary spin of GW200105 is less than 0.23 at the 90% credible level, and its orientation is unconstrained. For GW200115, the primary spin has a negative spin projection onto the orbital angular momentum at 88% probability. We are unable to constrain the spin or tidal deformation of the secondary component for either event. We infer an NSBH merger rate density of $45^{+75}_{-33}$ $Gpc^{-3}$ $yr^{-1}$ when assuming that GW200105 and GW200115 are representative of the NSBH population or $130^{+112}_{-69}$ $Gpc^{-3}$ $yr^{-1}$ under the assumption of a broader distribution of component masses.

A supra-massive population of stellar-mass black holes in the globular cluster Palomar 5 is discussed by Gieles et al. (2021). Palomar 5 is one of the sparsest star clusters in the galactic halo and is best known for its spectacular tidal tails, spanning over 20° across the sky. With N-body simulations, Gieles et al. (2021) showed that both distinguishing features can result from a stellar-mass black hole population, constituting

~20% of the present-day cluster mass. In this scenario, Palomar 5 formed with a "normal" black hole mass fraction of a few percent, but stars were lost at a higher rate than black holes, such that the black hole fraction gradually increased. This inflated the cluster, enhancing tidal stripping and tail formation. A billion years from now, the cluster will dissolve as a 100% black hole cluster. Initially denser clusters end up with lower black hole fractions, smaller sizes, and no observable tails. Black hole–dominated, extended star clusters are therefore the likely progenitors of the recently discovered thin stellar streams in the galactic halo.

The innermost regions of accretion disks around black holes are strongly irradiated by X-rays that are emitted from a highly variable, compact corona, in the immediate vicinity of the black hole. The X-rays that are seen reflected from the disk, and the time delays, as variations in the X-ray emission echo or "reverberate" of the disk, provide a view of the environment just outside the event horizon. I Zwicky 1 (I Zw 1) is a nearby narrow-line Seyfert 1 galaxy. Previous studies of the reverberation of X-rays from its accretion disk revealed that the corona is composed of two components: an extended, slowly varying component extending over the surface of the inner accretion disk, and a collimated core, with luminosity fluctuations propagating upwards from its base, which dominates the more rapid variability. Wilkins et al. (2021) reported observations of X-ray fares emitted from around the supermassive black hole in I Zw 1. X-ray refection from the accretion disk is detected through a relativistically broadened iron K line and Compton hump in the X-ray emission spectrum. Analysis of the X-ray fares reveals short flashes of photons consistent with the re-emergence of emission from behind the black hole. The energy shifts of these photons identify their origins from different parts of the disk. These are photons that reverberate of the far side of the disk and are bent around the black hole and magnified by the strong gravitational field. Observing photons bent around the black hole confirms a key prediction of general relativity.

Fernández et al. (2019) investigated the long-term evolution of black hole accretion disks formed in neutron star mergers. These disks expel matter that contributes to an r-process kilonova and can produce relativistic jets powering short gamma-ray bursts. They reported the results of a three-dimensional, general-relativistic magnetohydrodynamic (GRMHD) simulation of such a disk, which is evolved for long enough (~9 s, or $\sim 6 \times 10^5$ $r_g/c$) to achieve completion of mass ejection far from the disk. Their model starts with a poloidal field and fully resolves the most unstable mode of the magnetorotational instability. They parametrize the dominant microphysics and neutrino cooling effects and compare with axisymmetric hydrodynamic models with shear viscosity. The GRMHD model ejects mass in two ways: a prompt MHD-mediated outflow and a late-time, thermally driven wind once the disk becomes advective. The total amount of unbound mass ejected (0.013 $M_{sun}$, or ≈ 40% of the initial torus mass) is twice as much as in hydrodynamic models, with higher average velocity (0.1c) and a broad electron fraction distribution with a lower average value (0.16). Scaling the ejected fractions to a disk mass of ~0.1 $M_{sun}$ can account for the red kilonova from GW170817 but underpredicts the blue component. About $\sim 10^{-3}$ $M_{sun}$ of material should undergo neutron freezeout and could produce a bright kilonova precursor in the first few hours after the merger. With their idealized initial magnetic field configuration, we obtain a robust jet and sufficient ejecta with Lorentz factor ~1 – 10 to (over) produce the non-thermal emission from GW1708107.

Black holes may gain mass from the expansion of the Universe itself. There is overwhelming evidence for the existence of ultrarelativistic compact objects, both as the end stages of stellar collapse and as the supermassive compact objects in the centers of galaxies. These compact objects are interpreted within the framework of General Relativity, which has withstood tests on scales ranging from cosmological to the binary pulsar. A contemporary triumph in this field, following the development of numerical methods for modeling gravitational radiation from compact object mergers, was the detection of such mergers by gravitational-wave observatories. The LIGO Scientific Collaboration and the Virgo Collaboration (LVC) have currently published over 50 confident detections of compact binary mergers, the majority being consistent with Kerr black holes.

Croker et al. (2021) demonstrated a single-parameter route for reproducing higher-mass objects, as observed in the LIGO–Virgo mass distribution, using only the isolated-binary stellar evolution channel. This single parameter encodes the cosmological mass growth of compact stellar remnants that exceed the Tolman–Oppenheimer–Volkoff limit. Cosmological mass growth appears in known solutions to General Relativity with cosmological boundary conditions. They considered the possibility of solutions with cosmological boundary conditions, which reduce to Kerr on timescales short compared to the Hubble time. They also discussed complementary observational signatures of these solutions that can confirm or invalidate their astrophysical relevance.

The first direct dynamical detection of a dual supermassive black hole system at sub-kiloparsec separation has been reported by Voggel et al. (2022). They investigated whether the two recently discovered nuclei in NGC 7727 both host a super-massive black hole. They used the high spatial resolution mode of the integral-field spectrograph MUSE on the VLT in adaptive optics mode to resolve the stellar kinematics within the sphere of influence of both putative black holes. They combined the kinematic data with an HST-based mass model and used Jeans models to measure their SMBH mass. They report the discovery of a dual SMBH system in NGC 7727. They detect a SMBH in the photometric center of the galaxy in Nucleus 1, with a mass of $M_{SMBH} = 1.54^{+0.18}{}_{-0.15} \times 10^8\ M_{sun}$. In the second nucleus, which is 500 pc offset from the main nucleus, they also find a clear signal for a SMBH with a mass of $M_{BH} = 6.33^{+3.32}{}_{-1.40} \times 10^6\ M_{sun}$. Both SMBHs are detected at high significance. The off-axis nature of Nucleus 2 makes modeling the system challenging; however, a number of robustness tests suggest that a black hole is required to explain the observed kinematics. The SMBH in the offset Nucleus 2 makes up 3.0% of its total mass, which means its SMBH is over-massive compared to the MBH–MBulge scaling relation. This confirms it as the surviving nuclear star cluster of a galaxy that has merged with NGC 7727. This discovery is the first dynamically confirmed dual SMBH system with a projected separation of less than a kiloparsec and the nearest dynamically confirmed dual SMBH at a distance of 27.4 Mpc. The second nucleus is in an advanced state of inspiral, and it will eventually result in a 1:24 mass ratio SMBH merger. Optical emission lines suggest Nucleus 2 is a Seyfert galaxy, making it a low-luminosity active galactic nuclei. There are likely many more quiescent SMBHs as well as dual SMBH pairs in the local Universe that have been missed by surveys that focus on bright accretion signatures.

Larson et al. (2023) reported the discovery of an accreting supermassive black hole at $Z = 8.679$. This galaxy, denoted here as CEERS_1019, was previously discovered as a Ly$\alpha$-break galaxy by Hubble with a Ly$\alpha$ redshift from Keck. As part of the Cosmic Evolution Early Release Science (CEERS) survey, they have observed this source with JWST/NIRSpec spectroscopy, MIRI imaging, NIRCam imaging, and NIRCam/WFSS slitless spectroscopy. They used the NIRSpec R ~ 1000 medium-resolution grating spectra covering 1–5 $\mu$m to uncover a plethora of significantly detected emission lines. The extremely strong [O III] emission line, observed at 4.8473 $\mu$m, confirms the ground-based Ly$\alpha$ redshift. They detected a significant broad (FWHM ~ 1200 km s–1) component in the H$\beta$ emission line, which most likely originates in the broad-line region of an active galactic nucleus as the lack of a similar broad component in other, stronger, forbidden lines rejects an outflow origin. This hypothesis is supported by the presence of weak high-ionization lines (N V, N IV], C III]), some of which have broad components, as well as a spatial point-source component embedded within a smoother surface brightness profile. The implied mass of the black hole is $\log (M_{BH}/M_{sun}) = 6.95 \pm 0.37$ and estimated to be accreting at 1.2 (± 0.5) × the Eddington limit. The 1–8 $\mu$m photometric spectral energy distribution (SED) from NIRCam and MIRI shows a continuum dominated by starlight and constrains the host galaxy to be massive ($\log M/M_{sun} \sim 9.5$) and highly star-forming (SFR ~ 30 $M_{sun}\ yr^{-1}$; log sSFR ~ $-7.9\ yr^{-1}$). The ratios of detected strong emission lines show that the gas in this galaxy is metal-poor ($Z/Z_{sun} \sim 0.1$), dense ($n_e \sim 10^3\ cm^{-3}$), and highly ionized (log U ~ −2.1), consistent with the general galaxy population observed with JWST at such high redshifts that do not display significant AGN signatures. They used this presently highest-redshift AGN discovery to place constraints on black hole seeding models and find that a combination of either super-Eddington accretion from stellar seeds or Eddington accretion from very massive black hole seeds is required to form this object by the observed epoch.

Powerful relativistic jets are one of the ubiquitous features of accreting black holes at all scales. GRS 1915 + 105 is a well-known fast-spinning black-hole X-ray binary with a relativistic jet, termed a “microquasar”, as indicated by its superluminal motion of radio emission. It has exhibited persistent X-ray activity over the last 30 years, with quasiperiodic oscillations of approximately 1–10 Hz and 34 and 67 Hz in the X-ray band. These oscillations probably originate in the inner accretion disk, but other origins have been considered. Radio observations found variable light curves with quasiperiodic flares or oscillations with periods of approximately 20–50 min. Tian et al. (2023) reported two instances of approximately 5-Hz transient periodic oscillation features from the source detected in the 1.05- to 1.45-GHz radio band that occurred in January 2021 and June 2022. Circular polarization was also observed during the oscillation phase.

The detection of starlight from the host galaxies of quasars during the reionization epoch ($z > 6$) has been elusive, even with deep Hubble Space Telescope observations. The current highest redshift quasar host detected, at $z = 4.5$, required the magnifying effect of a foreground lensing galaxy. Low-luminosity quasars from the Hyper Suprime-Cam Subaru Strategic Program (HSC-SSP) mitigate the challenge of detecting their underlying, previously undetected host galaxies. Ding et al. (2023) reported rest-frame optical images and spectroscopy of two HSC-SSP quasars at $z > 6$ with the JWST. Using near-infrared camera imaging at 3.6 and 1.5 μm and subtracting the light from the unresolved quasars, we find that the host galaxies are massive (stellar masses of $13 \times 10^{10}$ and $3.4 \times 10^{10}$ $M_{\odot}$, respectively), compact and disk-like. Near-infrared spectroscopy at medium resolution shows stellar absorption lines in the more massive quasar, confirming the detection of the host. Velocity-broadened gas in the vicinity of these quasars enables measurements of their black hole masses ($1.4 \times 10^9$ and $2.0 \times 10^8$ $M_{\odot}$, respectively). Their location in the black hole mass–stellar mass plane is consistent with the distribution at low redshift, suggesting that the relation between black holes and their host galaxies was already in place less than a $10^9$ years after the Big Bang.

Astrophysical models of binary-black hole mergers in the Universe require a significant fraction of stellar-mass black holes to receive negligible natal kicks to explain the gravitational wave detections. This implies that BHs should be retained even in open clusters with low escape velocities (1 km $s^{-1}$). Torniamenti et al. (2023) searched for signatures of the presence of BHs in the nearest open cluster to the Sun—the Hyades—by comparing density profiles of direct N-body models to data from Gaia. The observations were best reproduced by models with 2–3 BHs at present. Models that never possessed BHs have an half-mass radius ~30% smaller than the observed value, while those where the last BHs were ejected recently (150 Myr ago) can still reproduce the density profile. In 50% of the models hosting BHs, the authors found BHs with stellar companion(s). Their period distribution peaks at ~$10^3$ yr, making them unlikely to be found through velocity variations. They look for potential BH companions through large Gaia astrometric and spectroscopic errors, identifying 56 binary candidates—none of which is consistent with a massive compact companion. Models with 2–3 BHs have an elevated central velocity dispersion, but observations cannot yet discriminate. The authors conclude that the present-day structure of the Hyades requires a significant fraction of BHs to receive natal kicks smaller than the escape velocity of ~3 km $s^{-1}$ at the time of BH formation and that the nearest BHs to the Sun are in, or near, the Hyades.

The nearby radio galaxy M87 offers a unique opportunity to explore the connections between the central supermassive black hole and relativistic jets. Previous studies of the inner region of M87 revealed a wide opening angle for the jet originating near the black hole. The Event Horizon Telescope resolved the central radio source and found an asymmetric ring structure consistent with expectations from general relativity. With a baseline of 17 years of observations, there was a shift in the jet's transverse position, possibly arising from an 8- to 10-year quasi-periodicity. However, the origin of this sideways shift remains unclear. Cui et al. (2023) reported an analysis of radio observations over 22 years that suggests a period of about 11 years for the variation in the position angle of the jet. They infer seeing a spinning black hole that induces the Lense–Thirring precession of a misaligned accretion disk. Similar jet precession may commonly occur in other active galactic nuclei but has been challenging to detect owing to the small magnitude and long period of the variation.

Schneider et al. (2023) studied bimodal black hole mass distribution and chirp masses of binary black hole mergers. In binary black hole mergers from isolated binary-star evolution, both black holes are from progenitor stars that have lost their hydrogen-rich envelopes by binary mass transfer. Envelope stripping is known to affect the presupernova core structures of such binary-stripped stars and thereby their final fates and compact remnant masses. In their paper, they show that binary-stripped stars give rise to a bimodal black hole mass spectrum with characteristic black hole masses of about 9 and 16 $M_{sun}$ across a large range of metallicities. The bimodality is linked to carbon and neon burning becoming neutrino dominated, which results in interior structures that are difficult to explode and likely lead to black hole formation. The characteristic black hole masses from binary-stripped stars have corresponding features in the chirp-mass distribution of binary black hole mergers: peaks at about 8 and 14 $M_{sun}$ and a dearth in between these masses. Current gravitational-wave observations of binary black hole mergers show evidence for a gap at 10–12 $M_{sun}$ and peaks at 8 and 14 $M_{sun}$ in the chirp-mass distribution. These features are in agreement with our models of binary-stripped stars. In the future, they may be used to constrain the physics of late stellar evolution and supernova explosions and may even help measure the cosmological expansion of the Universe.

Quasars are the most luminous subset of the active galactic nucleus population and represent the most rapid phase of supermassive black hole growth. In the standard quasar model, the accretion disk obscuration is due to the canonical dusty torus. Andonie et al. (2024) argue that a substantial part of the quasar obscuration can come from the interstellar medium (ISM) when the quasars are embedded in compact starbursts. They use an obscuration-unbiased sample of 578 infrared (IR) quasars at $z \approx 1–3$ and archival Atacama Large Millimetre/Submillimetre Array submillimeter host galaxy sizes to investigate the ISM contribution to the quasar obscuration. They calculate star formation rates (SFRs) and ISM column densities for the IR quasars and a control sample of submillimeter galaxies (SMGs) not hosting quasar activity and show that: (1) the quasar obscured fraction is constant up to SFR $\approx 300\ M_{sun}\ yr^{-1}$, and then increases towards higher SFR, suggesting that the ISM obscuration plays a significant role in starburst host galaxies, and (2) at SFR $\geq 300\ M_{sun}\ yr^{-1}$, SMGs and IR quasars have similarly compact submillimeter sizes ($R_e \approx 0.5–3$ kpc) and consequently, the ISM can heavily obscure the quasar, even reaching Compton-thick ($N_H > 10^{24}\ cm^{-2}$) levels in extreme cases. Based on our results, we infer that ≈10–30% of the IR quasars with SFR $\geq 300\ M_{sun}\ yr^{-1}$ are obscured solely by the ISM.

The growth of the first massive black holes from JWST/NIRSpec—spectroscopic redshift confirmation of an X-ray luminous AGN at $z = 10.1$ is described in a recent paper by Goulding et al. (2023). The James Webb Space Telescope is now detecting early black holes as they transition from "seeds" to supermassive BHs. Recently, Bogdan et al. (2023) reported the detection of an X-ray luminous supermassive BH, UHZ-1, with a photometric redshift at $z > 10$. Such an extreme source at this very high redshift provides new insights on seeding and growth models for BHs given the short time available for formation and growth. Harnessing the exquisite sensitivity of JWST/NIRSpec, here the authors report the spectroscopic confirmation of UHZ-1 at $z = 10.073 \pm 0.002$. They find that the NIRSpec/Prism spectrum is typical of recently discovered $z \approx 10$ galaxies, characterized primarily by star formation features. They see no clear evidence of the powerful X-ray source in the rest-frame UV/optical spectrum, which may suggest heavy obscuration of the central BH, in line with the Compton-thick column density measured in the X-rays. They perform a stellar population fit simultaneously to the new NIRSpec spectroscopy and previously available photometry. The fit yields a stellar-mass estimate for the host galaxy that is significantly better constrained than prior photometric estimates ($M_* \sim 1.4{-}^{0.3-}{}_{0.4} \times 10^8\ M_\odot$). Given the predicted BH mass ($M_{BH} \sim 10^7$–$10^8\ M_\odot$), the resulting ratio of $M_{BH}/M_*$ remains 2 to 3 orders of magnitude higher than local values, thus lending support to the heavy seeding channel for the formation of supermassive BHs within the first billion years of cosmic evolution.

El-Badry et al. (2023) report discovery of a bright, nearby ($G = 13.8$; $d = 480$ pc) Sun-like star orbiting a dark object. They identified the system as a black hole candidate via its astrometric orbital solution from the Gaia mission. Radial velocities validated and refined the *Gaia* solution, and spectroscopy ruled out significant light contributions from another star. Joint modelling of radial velocities and astrometry constrains the companion mass of $M_2 = 9.62 \pm 0.18 M_\odot$. The spectroscopic orbit alone sets a minimum companion mass of $M_2 > 5M_\odot$; if the companion were a $5M_\odot$ star, it would be 500 times more luminous than the entire system. These constraints are insensitive to the mass of the luminous star, which appears as a slowly rotating G dwarf ($T_{eff} = 5850$ K, $\log g = 4.5$, $M = 0.93\ M_\odot$), with near-solar metallicity ([Fe/H] = −0.2) and an unremarkable abundance pattern. They found no plausible astrophysical scenario that can explain the orbit and does not involve a black hole. The orbital period, $P_{orb} = 185.6$ d, is longer than that of any known stellar-mass black hole binary. The system's modest eccentricity ($e = 0.45$), high metallicity, and thin-disk galactic orbit suggest that it was born in the Milky Way disk with at most a weak natal kick. How the system formed is uncertain. Common envelope evolution can only produce the system's wide orbit under extreme and likely unphysical assumptions. Formation models involving triples or dynamical assembly in an open cluster may be more promising. This is the nearest known black hole by a factor of 3, and its discovery suggests the existence of a sizable population of dormant black holes in binaries.

## 1.4 GALAXIES

Some classes of active galactic nuclei have jets of relativistic plasma/particles emanating from the central region, which produce radio synchrotron emission. In some cases, these jets propagate to extremely large distances beyond the host galaxy and into the intergalactic medium. When the projected linear size of the jets and lobes exceeds 0.7 Mpc, such systems are referred to as giant radio galaxies (GRGs).

Delhaize et al. (2021) reported the discovery of two new giant radio galaxies using the MeerKAT International GHz Tiered Extragalactic Exploration (MIGHTEE) survey. Both GRGs were found within a ~1~1 deg$^2$ region inside the COSMOS field. They have redshifts of $z = 0.1656$ and $z = 0.3363$ and physical sizes of 2.4 and 2.0 Mpc, respectively. Only the cores of these GRGs were clearly visible in previous high-resolution Very Large Array observations, since the diffuse emission of the lobes was resolved out. However, the excellent sensitivity and *uv* coverage of the new MeerKAT telescope allowed this diffuse emission to be detected. The GRGs occupy an unpopulated region of radio power–size parameter space. Based on a recent estimate of the GRG number density, the probability of finding two or more GRGs with such large sizes at $z < 0.4$ in a ~1~1 deg$^2$ field is only $2.7 \times 10^{-6}$, assuming Poisson statistics. This supports the hypothesis that the prevalence of GRGs has been significantly underestimated in the past due to limited sensitivity to low surface brightness emission. The two GRGs presented here may be the first of a new population to be revealed through surveys like MIGHTEE that provide exquisite sensitivity to diffuse, extended emission.

The James Webb Space Telescope has discovered some massive, deep-red, disk-shaped galaxies not visible by Hubble Space Telescope. Astronomers call them HST-dark galaxies; in their paper, Nelson et al. (2022) also called these galaxies ultra-red flattened objects, or UFOs. These disk-shaped galaxies have a redshift (z) between 2 and 6, that is, as they were in the Universe (10.3 to 12.7) $\times 10^9$ years ago. JWST can see these HST-dark galaxies because it observes in infrared light, the part of the spectrum in which these galaxies shine.

The detection of two z ~ 13 galaxy candidates by Pacucci et al. (2022) has opened a new window on galaxy formation at an era only $330 \times 10^6$ years after the Big Bang. They investigated the physical nature of these sources wondering if they were witnessing star-forming galaxies or quasars at such early epochs. If powered by star formation, the observed ultraviolet luminosities and number densities can be jointly explained if:

1. these galaxies are extreme star-formers with star formation rates 5 – 24× higher than those expected from extrapolations of average lower-redshift relations;
2. the star formation efficiency increases with halo mass and is countered by increasing dust attenuation from z ~ 10 – 5;
3. they form stars with an extremely top-heavy initial mass function.

The quasar hypothesis is also plausible, with the UV luminosity produced by black holes of ~108$M_\odot$ accreting at or slightly above the Eddington rate ($f_{Edd}$~1.0). This black hole mass at z~13 would require very challenging, but not implausible, growth parameters. If spectroscopically confirmed, these two sources will represent a remarkable laboratory to study the Universe at previously inaccessible redshifts

The Virgo Cluster is a large cluster of galaxies whose center is 16.5 ± 0.1 Mpc away in the constellation Virgo (Mei et al. 2007), comprising possibly up to 2,000 galaxies. The cluster forms the heart of the larger Virgo Supercluster, of which the Local Group, containing our Milky Way galaxy, is a member. It is estimated that the Virgo Cluster's mass is $1.2 \times 10^{15}$ $M_\odot$ out to 8 degrees of the cluster's center or a radius of about 2.2 Mpc (Fouqué et al. 2001).

Jones et al. (2022) reported five blue stellar systems in the direction of the Virgo Cluster, indicating a distant group of galaxies. With the exception of one object, they are all extremely low stellar mass, blue stellar population, and high metallicity. They also reported observations of a sample of isolated blue stellar systems in the Virgo cluster: AGC 226067 and AGC 226 226178. Those are two young, blue, extremely low mass, gas-rich, metal-rich, actively star-forming stellar systems. The properties of both systems imply that they formed via in situ star formation in gaseous debris stripped from a much larger object.

The exact geometry and functioning of the central engines of active galactic nuclei are still highly debated. The long duty cycle of AGN ($10^7$-$10^9$ years) compared to human timescales is expected to prevent direct observation of the ignition or quenching of an AGN. However, recent discoveries of so-called changing-look AGN (CLAGN), have given us rare glimpses of extreme changes in the AGN state in a few months to years. Not all of these changes happen in the same way, which intimates the complexity of the physical mechanisms at work in the central engine (Laha et al. 2022).

The nearby type-II AGN 1ES1927+654 went through a violent changing-look (CL) event beginning December 2017 during which the optical and UV fluxes increased by four magnitudes over a few months, and broad emission lines newly appeared in the optical/UV. By July 2018, the X-ray coronal emission had completely vanished, only to reappear a few months later. Laha et al. (2022) reported the evolution of the radio, optical, UV, and X-rays from the pre-flare state through mid-2021 with new and archival data from the Very Long Baseline Array, the European VLBI Network, the Very Large Array, the Telescopio Nazionale Galileo, Gran Telescopio Canarias, the Neil Gehrels Swift observatory, and XMM-Newton. The main results from their work are:

1. The source has returned to its pre-CL state in optical, UV, and X-ray; the disk–corona relation has been re-established, as it has been in the pre-CL state. The optical spectra are dominated by narrow emission lines.

2. The UV light curve follows a shallower slope of $\propto$ t−0.91 ± 0.04 compared to that predicted by a tidal disruption event. It is assumed that a magnetic flux inversion event is the possible cause for this enigmatic event.

3. The compact radio emission which we tracked in the pre-CL (2014), during CL (2018), and post-CL (2021) at spatial scales < 1 pc was at its lowest level during the changing look event in 2018, nearly contemporaneous with a low 2 – 10 keV emission. The radio to X-ray ratio of the compact source LRadio/LX-ray ~ 10−5.5, follows the relation, typically found in coronally active stars, and several AGN.

4. They did not detect any presence of nascent jets at the spatial scales of ~ 5 – 10 pc.

The current paradigm of structure formation predicts that most of the gas in the intergalactic medium is organized in a "cosmic web" composed of nonuniform gaseous filaments connecting galaxies on scales of megaparsecs. These filaments feed the circumgalactic medium, the gaseous component that is responsible for regulating the gas exchange between galaxies and the surrounding IGM. This complex interplay between the infall of gas from filaments through the CGM and the ejection of matter from the feedback processes is expected to play a key role in the regulation of galaxy growth through cosmic time.

Spiral galaxies have distinct internal structures, including a stellar bulge, a disk, and spiral arms. It is unknown when in cosmic history these structures formed. Tsukui and Iguchi (2021) analyzed observations of BRI 1335–0417, an intensely star-forming galaxy in the distant Universe, at a redshift of 4.41. The $[C_{II}]$ gas kinematics shows a steep velocity rise near the galaxy center and has a two-armed spiral morphology, which extends from about 2 to 5 kiloparsecs in radius. We interpret these features as due to a central compact structure such as a bulge, a rotating gas disk, and either spiral arms or tidal tails. These features had formed within $1.4 \times 10^9$ years after the Big Bang, long before the peak of cosmic star formation.

The Milky Way disk consists of two prominent components—a thick, alpha-rich, low-metallicity component and a thin, metal-rich, low-alpha component. External galaxies have been shown to contain thin- and thick-disk components, but whether distinct components in the [α/Fe]–[Z/H] plane exist in other Milky Way–like galaxies is not yet known. Scott et al. (2021) presented Very Large Telescope (VLT)–Multi Unit Spectroscopic Explorer (MUSE) observations of UGC 10738, a nearby, edge-on Milky Way–like galaxy. They demonstrated through stellar population synthesis model fitting that UGC 10738 contains alpha-rich and alpha-poor stellar populations with similar spatial distributions to the same components in the Milky Way. They discussed how the finding that external galaxies also contain chemically distinct disk components might act as a significant constraint on the formation of the Milky Way's own thin and thick disk.

Scott et al. (2021) explore whether chemically distinct thick and thin disks are present in nearby disk galaxies that are structurally similar to our own galaxy–Milky Way analogs (MWAs). Recent improvements in integral field spectroscopy enable integrated light measurements that can sample physical scales, allowing direct comparison to Milky Way observations. They present Very Large Telescope–Multi Unit Spectroscopic Explorer observations of UGC 10738, a nearby MWA. They derive spatially resolved star formation and chemical enrichment histories and compare these to similar measurements for the Milky Way.

Observations of high-redshift galaxies provide a critical direct test to the theories of early galaxy formation, yet to date, only three have been spectroscopically confirmed at $z > 12$. Due to strong gravitational lensing over a wide area, the galaxy cluster field A2744 is ideal for searching for the earliest galaxies. Wang et al. (2023) present JWST/NIRSpec observations of two galaxies: a robust detection at $z_{spec} = 12.393^{+0.004}{}_{-001}$ and a plausible candidate at $z_{spec} = 13.079^{+0.013-}{}_{0.001}$. The galaxies are discovered in JWST/NIRCam imaging, and their distances are inferred with JWST/NIRSpec spectroscopy, all from the JWST Cycle 1 UNCOVER Treasury survey. Detailed stellar population modeling using JWST NIRCam and NIRSpec data corroborates the primeval characteristics of these galaxies: low mass ($\sim 10^8$ $M_{sun}$), young, rapidly assembling, metal-poor, and star-forming. Interestingly, both galaxies are spatially resolved, having lensing-corrected rest-UV effective radii on the order of 300–400 pc, which are notably larger than other spectroscopically confirmed systems at similar redshifts. The observed dynamic range of $z \geq 10$ sizes spans over one order of magnitude, implying a significant scatter in the size–mass relation at early times. Deep into the epoch of reionization, these discoveries elucidate the emergence of the first galaxies.

### 1.4.1 Dwarf Galaxies

During last two decades, many faint dwarf galaxies (DGs) have been discovered within the Local Group, some of them orbiting Milky Way and Andromeda galaxies, many of them being in the category of ultra-faint dwarf galaxy; see Collins et al. 2022).

Star formation histories (SFHs) reveal physical processes that influence how galaxies form their stellar mass. Olsen et al. (2021) compared the SFHs of a sample of 36 nearby (D <≈ 4 Mpc) dwarf galaxies from the ACS Nearby Galaxy Survey Treasury (ANGST), inferred from the color–magnitude diagrams (CMDs) of individually resolved stars in these galaxies, with those reconstructed by broadband spectral energy distribution fitting using the dense basis SED-fitting code. When comparing individual SFHs, Olsen et al. (2021) introduced metrics for evaluating SFH reconstruction techniques. For both the SED and CMD methods, the median normalized SFH of galaxies in the sample shows a period of quiescence at lookback times of $(3–6) \times 10^9$ years followed by rejuvenated star formation over the past $3 \times 10^9$ years that remains active until the present day. To determine if these represent special epochs of star formation in the D < 4 Mpc portion of the Local Volume, the authors break this ANGST dwarf galaxy sample into subsets based on specific star formation rate and spatial location. Modulo offsets between the methods of about $1 \times 10^9$ years, all subsets show significant decreases and increases in their median normalized SFHs at the same epochs, and the majority of the individual galaxy SFHs are consistent with these trends. These results motivate further study of potential synchronized star formation quiescence and rejuvenation in the Local Volume as well as development of a hybrid method of SFH reconstruction that combines CMDs and SEDs, which have complementary systematics.

Mutlu-Pakdil et al. (2022) presented deep Hubble Space Telescope imaging of five faint dwarf galaxies associated with the nearby spiral NGC 253 (D ≈ 3.5 Mpc). Three of these are newly discovered dwarf galaxies, while all five were found in the Panoramic Imaging Survey of Centaurus and Sculptor, a Magellan+Megacam survey to identify faint dwarfs and other substructures in resolved stellar light around massive galaxies outside of the Local Group. Their HST data reach ≈>3 magnitudes below the tip of the red giant branch for each dwarf, allowing us to derive their distances, structural parameters, and luminosities. All five systems contain mostly old, metal-poor stellar populations (age ~ $12 \times 10^9$years, [M/H] < −1.5) and have sizes ($r_h \sim$ 110–3000 pc) and luminosities ($M_V \sim$ −7 to −12 mag) largely consistent with Local Group dwarfs. The three new NGC 253 satellites are among the faintest systems discovered beyond the Local Group. They also use archival $H_I$ data to place limits on the gas content of our discoveries. Deep imaging surveys such as our program around NGC 253 promise to elucidate the faint end of the satellite luminosity function and its scatter across a range of galaxy masses, morphologies, and environments in the decade to come.

Dual AGNs are important for understanding galaxy-merger-triggered fueling of black holes and hierarchical growth of structures. The least-explored type of dual AGNs are those associated with mergers of two dwarf galaxies. According to observations and cosmological simulations, dwarf galaxies are the most abundant type of galaxies in the early Universe, and the galaxy merger rate is dominated by dwarfs. However, these mergers are generally too distant to be directly observed, and low-redshift dwarf-dwarf merger-related dual AGNs are notoriously hard to find. Mićić et al. (2023) presented the first results of our large-scale search for this elusive type of object and the first two candidates for dual AGN in dwarf-dwarf mergers. Both objects exhibit tidal features (tails and bridges) characteristic of galaxy mergers/interactions. One object is apparently in a late-stage merger with an AGN separation of $< 5$ kpc, while the second is in an early-stage merger, with interacting galaxies having established a tidal bridge. Both objects have dual, luminous X-ray sources that are most likely due to actively accreting massive black holes. Also, both objects have infrared counterparts, with colors consistent with being AGN. Follow-up observations will provide us a glimpse into key processes that govern the earliest phases of growth of galaxies, their central black holes, and merger-induced star formation.

### 1.4.2 Globular Clusters

Globular clusters are typically much larger than open clusters and are tightly gravitationally bound. They are far more densely populated, with populations ranging from tens of thousands to millions of stars. The intense gravitational attraction between the closely packed stars gives globular clusters their regular, spherical shape. They are populated by older, redder stars than open clusters (which might disperse before their stars can become really old). Their considerable gravitational attraction makes them very stable, and therefore they can be extremely long lived, surviving to be billions of years old. They are associated with all types of galaxies, and within the Milky Way, they are found in the halo and bulge. All star clusters are of great interest to astronomers, because their constituent stars all formed at approximately the same time and location and had similar initial composition. Therefore, stellar clusters offer unique insights into how stars form and evolve. It is very challenging to observe individual stars within globular clusters, however, as they are so densely packed.

Fast radio bursts (FRBs) are flashes of unknown physical origin. The majority of FRBs have been seen only once, although some are known to generate multiple flashes. Many models invoke magnetically powered neutron stars (magnetars) as the source of the emission. Recently, the discovery of another repeater (FRB 20200120E) was announced, in the direction of the nearby galaxy M81, with four potential counterparts at other wavelengths. Kirsten et al. (2022) reported observations that localized the FRB to a globular cluster associated with M81, where it is 2 parsecs away from the optical center of the cluster. Globular clusters host old stellar populations, challenging FRB models that invoke young magnetars formed in a core-collapse supernova. The authors proposed instead that FRB 20200120E originates from a highly magnetized neutron star formed either through the accretion-induced collapse of a white dwarf or the merger of compact stars in a binary system. Compact binaries are efficiently formed inside globular clusters, so a model invoking them could also be responsible for the observed bursts.

As stated by Sbordone et al. (2015), it is a fundamental prediction of models of galaxy formation based on the cold dark matter scenario that dark matter haloes of the size of that of the Milky Way grow through the accretion of smaller subsystems. The merging of minor systems is supposed to be a common event in the early stages of the galactic history, playing a role even in the formation of the stellar disk. Despite this general agreement, the processes governing galaxy formation still present many obscure aspects, and understanding them is one of the greatest challenges of modern astrophysics. For example, it has been noticed that the chemical abundance patterns of present-day dwarf spheroidal (dSph) galaxies in the Local Group are very different from those observed among stars in the galactic halo.

The tidal disruption of the Sagittarius dwarf spheroidal galaxy (Sgr dSph) is producing the most prominent substructure in the Milky Way halo, the Sagittarius Stream. Aside from field stars, it is suspected that the Sgr dSph has lost a number of globular clusters. Many galactic GCs are thought to have originated in the Sgr dSph. While for some candidates, an origin in the Sgr dSph has been confirmed owing to chemical similarities, others exist whose chemical composition has never been investigated. NGC 5053 and NGC 5634 are two of these scarcely studied Sgr dSph candidate-member clusters. To characterize their composition, Sbordone et al. (2015) analyzed one giant star in NGC 5053 and two in NGC 5634 by determining atmospheric parameters and abundances for up to 21 elements between O and Eu. The abundances are compared with those of MW halo field stars, of unassociated Milky Way halo globulars, and of the metal-poor Sgr dSph main body population. The authors derive a metallicity of [Fe $_{II}$/H] = −2.26 ± 0.10 for NGC 5053, and of [Fe $_{I}$/H] = −1.99 ± 0.075 and −1.97 ± 0.076 for the two stars in NGC 5634. This makes NGC 5053 one of the most metal-poor globular clusters in the MW. Both clusters display an α enhancement similar to the one of the halo at comparable metallicity. The two stars in NGC 5634 clearly display the Na-O anticorrelation widespread among MW globulars. Most other abundances are in good agreement with standard MW halo trends. In conclusion, the chemistry of the Sgr dSph main body populations is similar to that of the halo at low metallicity. It is thus difficult to discriminate between an origin of NGC 5053 and NGC 5634 in the Sgr dSph and one in the MW. However, the abundances of these clusters do appear closer to that of Sgr dSph than of the halo, favoring an origin in the Sgr dSph system (Sbordone et al. 2015).

Globular clusters are ancient stellar populations with no star formation or core-collapse supernovae. Several lines of evidence suggest that globular clusters are rich in planets. If so, and if advanced civilizations can develop there, then the distances between these civilizations and other stars would be far smaller than typical distances between stars in the galactic disk. The relative proximity would facilitate interstellar communication and travel. However, the very proximity that promotes interstellar travel also brings danger, since stellar interactions can destroy planetary systems. However, by modeling globular clusters and their stellar populations, Di Stefano and Ray (2016) found that large regions of many globular clusters can be thought of as "sweet spots" where habitable-zone planetary orbits can be stable for long times. The authors also compute the ambient densities and fluxes in the regions within which habitable-zone planets can survive. Globular clusters are among the best targets for searches for extraterrestrial intelligence (SETI). They used the Drake equation to compare globular clusters to the galactic disk in terms of the likelihood of housing advanced communicating civilizations. We also consider free-floating planets, since wide-orbit planets can be ejected and travel freely through the cluster. A civilization spawned in a globular cluster may have opportunities to establish self-sustaining outposts, thereby reducing the probability that a single catastrophic event will destroy the civilization or its descendants. Although individual civilizations within a cluster may follow different evolutionary paths or even be destroyed, the cluster may always host some advanced civilization once a small number of them have managed to jump across interstellar space.

GCs continue to be a topic of active research, with many fundamental questions remaining. These include: How and when did GCs form? Where did they form and how did they assemble into today's GC systems? What was their initial mass function? Answers to these questions will come from both observations and simulations. Most GCs of the Milky Way, and those around external galaxies, are very old. Although indications are that the metal-rich GCs are approximately (1–1.5) × $10^9$ years younger than the metal-poor ones, it is still difficult to rule out conclusively coeval ages for the two subpopulations. Absolute ages place the oldest GCs at around 12.5 × $10^9$ years, but uncertainties extend their ages well into the epoch of reionization ($z > 6$). Near-term prospects for making progress with absolute ages for resolved GCs include obtaining near-IR color magnitude diagrams that exploit the age sensitivity of the main sequence turnoff and a feature associated with $H_2$ opacity. For unresolved but nearby GCs, measuring relative ages with the next generation of wide-field, multiplexing spectrographs will be very challenging with 8–10-m telescopes. Progress will probably have to wait for large numbers of GCs to be observed with the 20–40-m class telescopes and will need to be accompanied by a better understanding of theoretical integrated-light age diagnostics (Forbes et al. 2018).

An exciting new avenue, which has been opened up by HST, and will no doubt be followed up by JWST, is the direct observation of proto-GC candidates at high redshift. Some fraction of these objects may have survived destruction processes to make up the present-day population of GCs. Multiple populations (within an individual GC) appear to be a common feature in old GCs—indeed, they may be a defining feature of a GC. Recently, all the hallmarks of multiple populations have been found in compact, massive star clusters of age around $2 \times 10^9$ years. Thus, the conditions necessary for multiple populations exist in the local Universe. These young GCs, forming at $z \sim 0.17$, suggest a common single channel for GC formation. Owing to internally and externally driven destruction processes, GC systems formed with much more mass than is observed today. The commonly discussed formation models for multiple populations require that all individual GCs be at least ten times more massive than they are currently in order to solve the mass budget problem (i.e. to have enough material processed through a first generation of stars to form a second generation of stars that matches the present enriched fraction ratio). However, this universal amount of mass loss is not expected from models of GC evolution; instead, it comes from fine-tuning the initial conditions in order to match the observed population ratios. Relatively extreme initial conditions must be assumed, and the present-day GC population properties are not consistent with these assumptions (Forbes et al. 2018).

Magnetic fields are fundamental to the evolution of galaxies, playing a key role in the astrophysics of the interstellar medium and star formation. Large-scale ordered magnetic fields have been mapped in the Milky Way and nearby galaxies, but it is not known how early in the Universe such structures formed. Geach et al. (2023) report the detection of linearly polarized thermal emission from dust grains in a strongly lensed, intrinsically luminous galaxy that is forming stars at a rate more than 1,000 times that of the Milky Way at redshift 2.6, within $2.5 \times 10^9$ years of the Big Bang. The polarized emission arises from the alignment of dust grains with the local magnetic field. The median polarization fraction is of the order of 1%, similar to nearby spiral galaxies. These observations support the presence of a 5-kiloparsec-scale ordered magnetic field with a strength of around 500 μG or lower, oriented parallel to the molecular gas disk. This confirms that such structures can be rapidly formed in galaxies, early in cosmic history.

Román et al. (2023) performed a study of dynamically cold stellar streams that reveals information about the gravitational potential where they reside and provides important constraints on the properties of dark matter. However, the intrinsic faintness of these streams makes their detection beyond local environments highly challenging. Here, we report the detection of an extremely faint stellar stream ($\mu_{g,max} = 29.5$ mag arcsec$^{-2}$) with an extraordinarily coherent and thin morphology in the Coma Galaxy Cluster. This Giant Coma Stream spans ~510 kpc in length and appears as a free-floating structure located at a projected distance of 0.8 Mpc from the center of Coma. They did not identify any potential galaxy remnant or core, and the stream structure appears featureless in our data. They interpreted the Giant Coma Stream as being a recently accreted, tidally disrupting passive dwarf. Using the Illustris-TNG50 simulation, they identify a case with similar characteristics, showing that, although rare, these types of streams are predicted to exist in Λ-CDM. Their work unveils the presence of free-floating, extremely faint and thin stellar streams in galaxy clusters, widening the environmental context in which these objects are found ahead of their promising future application in the study of the properties of dark matter.

### 1.4.3 Galaxy Formation, Dynamics, Mergers, Collisions

Obtaining an accurate picture of galaxy formation and evolution is amongst the most outstanding questions in astrophysics today. There have been many studies examining this process in some detail over the past 30 years or so, including extensive Hubble Space Telescope campaigns and ground-based imaging studies. One of the most critical aspects for studying galaxy formation and evolution is to determine the redshifts of objects such that their evolutionary connections can be made, either through mass or number density selections to connect systems at different points in their evolution.

Two major ways in which galaxy studies progress are through the investigation of galaxies at different redshifts and by using the emission of light (direct or indirect) from these systems to measure the amount of current star formation and past assembly of stellar mass. The process of star formation is fundamental for understanding the assembly of galaxies and how the Universe was reionized. However, our resulting understanding of the star formation history of galaxies is largely based on observations of single integrated light measures of star formation rate within entire galaxies over a variety of redshifts. While much work in this area has been done, we are just starting to understand how and why star formation is distributed within galaxies. We also know that the sources of reionization are likely galaxies at $z > 6$ (see Duncan and Conselice 2015 and Robertson et al. 2015), yet we do not know from which types of galaxies, or modes of galaxy formation, ionizing radiation is emitted from. Both of these problems can be addressed in unique ways by searching for and characterizing emission line systems in the distant Universe, and perhaps by examining analog galaxies at lower redshifts, such as the "Green Peas" and Lyman α (Lyα) emitters at z ~ 1.5, Griffiths et al. (2021).

Emission line galaxies provide a crucial tool for the study of galaxy formation and evolution, providing a means to trace a galaxy's star formation history or metal enrichment and to identify galaxies at a range of stellar masses. Griffiths et al. (2021) presented a study of emission line galaxies in the Survey for High-z Absorption Red and Dead Sources (SHARDS) Frontier Fields (FFs) medium-band survey. Through detailed flux calibrations, they combine the first results of the SHARDS-FF survey with existing Hubble Frontier Field data to select 1098 candidate emission line galaxies from the Hubble Frontier Filed clusters Abell 370 and MACS J1149.5 + 2223. Furthermore, the authors implemented this deep medium-band imaging to update photometric redshift estimates and stellar population parameters and discover 38 predominantly low-mass Hα emitters at redshifts $0.24 < z < 0.46$. Overall, 27 of these sources have corresponding ultraviolet data from the Hubble Space Telescope that allowed distinguishing these sources and investigation of the burstiness of their star formation histories. They found that more than 50% of the sample shows an enhancement in Hα over UV, suggesting recent bursts in star formation on time-scales of a few to tens of Myr. They investigated these sources and found that they are typically low-mass discy galaxies with normal sizes. Their structures and star formation suggest that they are not undergoing mergers but are bursting due to alternative causes, such as gas accretion.

After large galaxies merge, their central supermassive black holes are expected to form binary systems. Their orbital motion generates a gravitational wave background at nanohertz frequencies. Searches for this background utilize pulsar timing arrays, which perform long-term monitoring of millisecond pulsars at radio wavelengths. Fermi-LAT Collaboration used 12.5 years of Fermi Large Area Telescope data to form a gamma-ray pulsar timing array. Results from 35 bright gamma-ray pulsars place a 95% credible limit on the GWB characteristic strain of $1.0 \times 10^{-14}$ at a frequency of 1 $yr^{-1}$. The sensitivity is expected to scale with the observing time span as $t^{13/6}$. This direct measurement provides an independent probe of the GWB while offering a check on radio noise models.

There is evidence for a strong connection between major galaxy mergers (<3:1 stellar mass ratio) and supermassive black hole growth from theoretical models and computational simulations; see, for example, Blumenthal and Barnes (2018) and references therein, since they provide a very efficient mechanism to remove angular momentum and drive gas to the nuclear regions as the two black holes are dragged toward each other during the dynamical friction phase. The SMBH pairs can sometimes be seen as dual active galactic nuclei, which provide a unique signature of merger-driven black hole growth.

Koss et al. (2023) presented multi-wavelength high-spatial resolution observations of UGC 4211 at $z = 0.03474$, a late-stage major galaxy merger at the closest nuclear separation yet found in near-IR imaging. Using Hubble Space Telescope/Space Telescope Imaging Spectrograph, Very Large Telescope/MUSE+AO, Keck/OSIRIS+AO spectroscopy, and the Atacama Large Millimeter/Submillimeter Array observations, they showed that the spatial distribution, optical and near-infrared emission lines, and millimeter continuum emission are all consistent with both nuclei being powered by accreting supermassive black holes. Their data, combined with common black hole mass prescriptions, suggest that both SMBHs have similar masses. The projected separation of 230 pc (~6× the black hole sphere of influence) represents the closest-separation dual active galactic nuclei studied to date with multi-wavelength resolved spectroscopy and shows the potential of nuclear (<50 pc) continuum observations with ALMA to discover hidden growing SMBH pairs. While the exact occurrence rate of close-separation dual AGN is not yet known, it may be surprisingly high, given that UGC 4211 was found within a small, volume-limited sample of nearby hard X-ray detected AGN. Observations of dual SMBH binaries in the subkiloparsec regime at the final stages of dynamical friction provide important constraints for future gravitational wave observatories.

Jin et al. (2023) reported the discovery of a compact group of galaxies, CGG-z5, at z ~ 5.2 in the EGS field covered by the JWST/CEERS survey. CGG-z5 was selected as the highest overdensity of galaxies at $z > 2$ in recent JWST public surveys, and it consists of six candidate members lying within a projected area of 1.5″×3″ (10 × 20 $kpc^2$). All group members are HST/F435W and HST/F606W dropouts while securely detected in the JWST/NIRCam bands, yielding a narrow range of robust photometric redshifts $5.0 < z < 5.3$. The most massive galaxy in the group has a stellar mass $\log (M_*/M_\odot) \approx 9.8$, while the rest are low-mass satellites ($\log (M_*/M_\odot) \approx 8.4$–$9.2$). While several group members were already detected in the HST and IRAC bands, the low stellar masses and the compactness of the structure required the sensitivity and resolution of JWST for its identification. To assess the nature and evolutionary path of CGGz5, we searched for similar compact structures in the Eagle simulations and followed their evolution with time. We find that all the identified structures merge into a single galaxy by $z = 3$ and form a massive galaxy ($\log (M_*/M_\odot) > 11$) at $z \sim 1$. This implies that CGG-z5 could be a "proto-massive galaxy" captured during a short-lived phase of massive galaxy formation.

Collisions and mergers are the primary way galaxy clusters form into the massive structures observed today and also function as the Universe's largest particle accelerators. Of interest is the developing galaxy cluster, Abell 2256, located 780 million lightyears from Earth. Astronomers studying Abell 2256 are trying to understand the cause of its unusual structure. Each telescope provides a different aspect of the story, as galaxy clusters are some of the largest objects in the Universe, made up of hundreds or even thousands of individual galaxies (Rajpurohit et al. 2022; Rajpurohit et al. 2023).

Franco et al. (2021) discuss the ramp-up of interstellar medium enrichment at $z > 4$. Fluorine is one of the most interesting elements for nuclear and stellar astrophysics. Fluorine abundance was first measured for stars other than the Sun in 1992, then for a handful of metal-poor stars, which are likely to have formed in the early Universe. The main production sites of fluorine are under debate and include asymptotic giant branch stars, the ν-process in core-collapse supernovae, and Wolf–Rayet stars. Due to the difference in the mass and lifetime of progenitor stars, high-redshift observations of fluorine can help constrain the mechanism of fluorine production in massive galaxies. The authors reported the detection of HF (signal-to-noise ratio of 8) in absorption in a gravitationally lensed dusty star-forming galaxy at redshift $z = 4.4$ with NHF/NH2 as high as $\sim 2 \times 10^{-9}$, indicating a very quick ramp-up of the chemical enrichment in this high-z galaxy. At $z = 4.4$, asymptotic giant branch stars of a few solar masses are very unlikely to dominate the enrichment. They showed that Wolf–Rayet stars are required to produce the observed fluorine abundance at this time, with other production mechanisms becoming important at later times. These observations therefore provide an insight into the underlying processes driving the ramp-up phase of chemical enrichment alongside rapid stellar mass assembly in a young massive galaxy.

ALMA observations of $z > 6$ galaxies have revealed abnormally high $[\mathrm{OIII}]_{88\mu m}/[\mathrm{CII}]_{158\mu m}$ ratios and $[\mathrm{CII}]_{158\mu m}$ deficits compared to local galaxies. The origin of this behavior is unknown. Numerous solutions have been proposed, including differences in C and O abundance ratios, observational bias, and differences in ISM properties, including ionization parameter, gas density, or photodissociation region (PDR) covering fraction. In order to elucidate the underlying physics that drives this high redshift phenomenon, Katz et al. (2021) employed SPHINX20, a state-of-the-art, cosmological radiation-hydrodynamics simulation that resolves detailed ISM properties of thousands of galaxies in the epoch of reionization which has been post-processed with CLOUDY to predict emission lines. They found that the observed $z > 6$ $[\mathrm{OIII}]_{88\mu m}$-SFR and $[\mathrm{CII}]_{158\mu m}$-SFR relations can only be reproduced when the C/O abundance ratio is ~8× lower than solar and the total metal production is ~4× higher than that of a Kroupa IMF. This implies that high-redshift galaxies are potentially primarily enriched by low-metallicity core-collapse supernovae with a more top-heavy IMF. As AGB stars and type-Ia supernova begin to contribute to galaxy metallicity, both the $[\mathrm{CII}]_{158\mu m}$-SFR and $[\mathrm{CII}]_{158\mu m}$ luminosity functions are predicted to converge to observed values at $Z \sim 4.5$. While they demonstrated that ionization parameter, LyC escape fraction, ISM gas density, and CMB attenuation all drive galaxies towards higher $[\mathrm{OIII}]_{88\mu m}/[\mathrm{CII}]_{158\mu m}$, observed values at $z > 6$ can only be reproduced with substantially lower C/O abundances compared to solar. The combination of $[\mathrm{CII}]_{158\mu m}$ and $[\mathrm{OIII}]_{88\mu m}$ can be used to predict the values of ionization parameter, ISM gas density, and LyC escape fraction and they provided estimates of these quantities for nine observed $z > 6$ galaxies. Finally, they demonstrated that $[\mathrm{OI}]_{63\mu m}$ can be used as a replacement for $[\mathrm{CII}]_{158\mu m}$ in high-redshift galaxies where $[\mathrm{CII}]_{158\mu m}$ is unobserved and argue that more observation time should be used to target $[\mathrm{OI}]_{63\mu m}$ at $z > 6$.

Radiative and mechanical feedback of massive stars regulates star formation and galaxy evolution. Positive feedback triggers the creation of new stars by collecting dense shells of gas, while negative feedback disrupts star formation by shredding molecular clouds. Although key to understanding star formation, their relative importance is unknown. Luisi et al. (2021) reported velocity-resolved observations from the Stratospheric Observatory for Infrared Astronomy (SOFIA) legacy program FEEDBACK of the massive star-forming region RCW 120 in the [CII] 1.9-THz fine-structure line, revealing a gas shell expanding at 15 km/s. Complementary Atacama Pathfinder Experiment (APEX) CO J = 3–2 345-GHz observations exhibit a ring structure of molecular gas, fragmented into clumps that are actively forming stars. Their observations demonstrate that triggered star formation can occur on much shorter time scales than hitherto thought (<0.15 million years), suggesting that positive feedback operates on short time periods.

Galaxy mergers occur frequently in the early Universe and bring multiple supermassive black holes into the nucleus, where they may eventually coalesce. Identifying post-merger-scale (that is, less than around a few kpc) dual SMBHs is a critical pathway to understanding their dynamic evolution and successive mergers. Whereas serendipitous discovery of ~kpc-scale dual SMBHs at $z < 1$ is possible, such systems are elusive at $z > 2$ but critical in constraining the progenitors of SMBH mergers. The redshift $z \approx 2$ also marks the epoch of peak activity of luminous quasars, and therefore the probing of this spatial regime at high redshift is of particular importance in understanding the evolution of quasars. However, given stringent resolution requirements, there is currently no confirmed <10 kpc physical SMBH pair at $z > 2$. Shen et al. (2021) reported two sub-arcsec double quasars at $z > 2$ that were discovered from a targeted search with a novel astrometric technique, demonstrating a high success rate ($\gtrsim$50%) in this systematic approach. These high-redshift double quasars could be the long-sought kpc-scale dual SMBHs, or sub-arcsec gravitationally lensed quasar images. One of these double quasars (at $z = 2.95$) was spatially resolved with optical spectroscopy and slightly favors the scenario of a physical quasar pair with a projected separation of 3.5 kpc (0.46″). Follow-up observations of double quasars discovered by this targeted approach will be able to provide observational constraints on kpc-scale dual SMBHs at $z > 2$. Two sub-arcsec double quasars at $z > 2$ are discovered from a targeted search with a novel astrometric technique. They could be the long-sought kpc-scale dual supermassive black holes or sub-arcsec gravitationally lensed quasar images.

In the local Universe, SMBHs have been identified at the centers of massive galaxies, and there is a tight correlation between MBHs and the properties of the host galaxy, such as bulge mass and stellar velocity dispersion. These relations strongly suggest that the formation and growth of SMBHs and host galaxies are intimately linked, a coevolution of these two components of galaxies. Although the detailed mechanisms by which the correlation arises remain unclear, some theoretical models suggest that strong negative AGN feedback on star formation, which is connected to the merger histories of galaxies, plays a key role in driving the coevolution.

Endsley et al. (2023) presented band 6 ALMA observations of a heavily obscured radio-loud ($L_{1.4\ \mathrm{GHz}} = 10^{25.4}$ W Hz$^{-1}$) active galactic nucleus candidate at $z_{phot} = 6.83 \pm 0.06$ found in the 1.5 deg$^2$ COSMOS field. The ALMA data reveal detections of exceptionally strong [$C_{II}$]158 μm ($z_{[CII]} = 6.8532$) and underlying dust continuum emission from this object (COS-87259), where the [$C_{II}$] line luminosity, line width, and 158-μm continuum luminosity are comparable to those seen from $z \sim 7$ sub-mm galaxies and quasar hosts. The 158 m continuum detection suggests a total infrared luminosity of $9 \times 10^{12}$ $L_{sun}$ with corresponding very large obscured star formation rate (1300 $M_\odot$ yr$^{-1}$) and dust mass ($2 \times 10^9$ $M_{sun}$). The strong break seen between the VIRCam and IRAC photometry perhaps suggests that COS-87259 is an extremely massive reionization-era galaxy with $M \approx 1.7 \times 10^{11}$ $M_{sun}$. Moreover, the MIPS, PACS, and SPIRE detections imply that this object harbors an AGN that is heavily obscured with a bolometric luminosity. Future very wide area surveys with, for example, Roman and Euclid have the potential to identify many more extremely red yet UV-bright $z \gtrsim 7$ objects similar to COS-87259, providing richer insight into the occurrence of intense obscured star formation and supermassive black hole growth among this population.

The evolution of satellite galaxies is shaped by their constant interaction with the circumgalactic medium surrounding central galaxies, which in turn may be affected by gas and energy ejected from the central supermassive black hole. The nature of such a coupling between black holes and galaxies is, however, much debated, and observational evidence remains scarce. Martin-Navarro et al. (2021) reported an analysis of archival data on 124,163 satellite galaxies in the potential wells of 29,631 dark matter halos with masses between $10^{12}$ and $10^{14}$ solar masses. They found that quenched satellite galaxies are relatively less frequent along the minor axis of their central galaxies. This observation might appear counterintuitive given that black hole activity is expected to eject mass and energy preferentially in the direction of the minor axis of the host galaxy. They showed that the observed anisotropic signal results precisely from the ejective nature of black hole feedback in massive halos, as outflows powered by active galactic nuclei clear out the circumgalactic medium, reducing the ram pressure and thus preserving star formation in satellite galaxies. This interpretation is supported by the IllustrisTNG suite of cosmological numerical simulations, even though the model's sub-grid implementation of black hole feedback is effectively isotropic.

After galaxies merge, the supermassive black holes at their centers are expected to form binaries that emit gravitational waves at nanohertz frequencies. Numerous SMBH binaries throughout the Universe should combine to produce a gravitational wave background. Existing searches for this signal use radio observations of pulsars as sensitive clocks and look for small shifts in the pulse timings. The Fermi-LAT Collaboration (2022) (see also Ajello et al. 2021) implemented a pulsar timing array using gamma rays and achieved a sensitivity close to that of the radio approaches. The results set an independent upper limit on the gravitational wave background, which is subject to different noise sources.

The ultra-diffuse galaxies DF2 and DF4 in the NGC 1052 group share several unusual properties: they both have large sizes, rich populations of overluminous and large globular clusters, and very low velocity dispersions that indicate little or no dark matter. It has been suggested that these galaxies were formed in the aftermath of high-velocity collisions of gas-rich galaxies, events that resemble the collision that created the bullet cluster but on much smaller scales. The gas separates from the dark matter in the collision and subsequent star formation leads to the formation of one or more dark-matter-free galaxies. Van Dokkum et al. (2022) showed that the present-day line-of-sight distances and radial velocities of DF2 and DF4 are consistent with their joint formation in the aftermath of a single bullet-dwarf collision, around 8 billion years ago. Moreover, they found that DF2 and DF4 are part of an apparent linear substructure of 7 to 11 large, low-luminosity objects. They proposed that these all originated in the same event, forming a trail of dark-matter-free galaxies that is roughly more than two megaparsecs long and angled $7° \pm 2°$ from the line of sight. They also tentatively identified the highly dark-matter–dominated remnants of the two progenitor galaxies that are expected at the leading edges of the trail.

Galaxies can grow through their mutual gravitational attraction and subsequent union. While orbiting a regular high-surface-brightness galaxy, the body of a low-mass galaxy can be stripped away. However, the stellar heart of the infalling galaxy, if represented by a tightly bound nuclear star cluster, is more resilient. From archival Hubble Space Telescope images, Graham et al. (2021) have discovered a red, tidally stretched star cluster positioned ~5″ (~400 pc in projection) from, and pointing toward the center of, the post-merger spiral galaxy NGC 4424. The star cluster, which we refer to as Nikhuli, has a near-infrared luminosity of $(6.88 \pm 1.85) \times 10^6\ Le_{sun,F160W}$ and likely represents the nucleus of a captured/wedded galaxy. Moreover, from our Chandra X-ray Observatory image, Nikhuli is seen to contain a high-energy X-ray point source, with $L_{0.5-8\ keV} = 6.31^{+7.50}{}_{-3.77} \times 10^{38}$ erg $s^{-1}$ (90% confidence). They argue that this is more likely to be an active massive black hole than an X-ray binary. Lacking an outward-pointing comet-like appearance, the stellar structure of Nikhuli favors infall rather than ejection from a gravitational wave recoil event. A minor merger with a low-mass early-type galaxy may have sown a massive black hole, aided an X-shaped pseudobulge, and be sewing a small bulge. The stellar mass and the velocity dispersion of NGC 4424 predict a central black hole of $(0.6–1.0) \times 10^5$ Me, similar to the expected intermediate-mass black hole in Nikhuli and suggestive of a black hole supply mechanism for bulgeless late-type galaxies. We may potentially be witnessing black hole seeding by capture and sinking, with a nuclear star cluster the delivery vehicle.

Over the past decade, large, deep surveys of high-redshift galaxies have been dedicated to gaining insight into the physical processes in the formation and evolution of galaxies across the Universe. The major process in transforming baryonic matter is star formation. It converts the local gas reservoir into heavy elements that are ejected from star-forming regions by way of feedback, thereby seeding the interstellar medium and the intergalactic medium with metals. One of the main constraints on cosmological models is the so-called cosmic star formation rate density, which has been intensely studied ever since the seminal work of Madau et al. (1996), both observationally and theoretically.

It is still not understood why star-forming galaxies deviate from the ultraviolet color-attenuation relation of starburst galaxies. Previous work and models hint that the role of the shape of the attenuation curve and the age of stellar populations play an important role. In their paper, Boquien et al. (2012) aimed at understanding the fundamental reasons for this deviation. They used the CIGALE spectral energy distribution fitting code to model the far-ultraviolet to far-infrared emission of a set of seven reasonably face-on spiral galaxies from the Herschel Reference Survey on a pixel-by-pixel basis. They explored the influence of a wide range of physical parameters to quantify their influence and impact any accurate determination of the attenuation from the ultraviolet color and to discover why normal galaxies do not follow the same relation as starburst galaxies. They found that the deviation from the starburst relation can be explained best by intrinsic ultraviolet color differences between different regions in galaxies. Variations in the shape of the attenuation curve can also play a secondary role. Standard age estimators of the stellar populations, such as the D4000 index or the birthrate parameter, prove to be poor predictors of the intrinsic ultraviolet color. These results are also retrieved on a sample of 58 spiral galaxies drawn from the Herschel Reference Survey sample when considering their integrated fluxes. In conclusions, when correcting the emission of normal star-forming galaxies for attenuation, it is crucial to consider possible variations in both the intrinsic ultraviolet color of the stellar populations and the shape of the attenuation curve.

Metallicity is a key parameter for understanding galaxy evolution, as it plays an important role in many fundamental galactic physical processes, such as star formation, gas cooling and collapse, stellar evolution, and dust formation. To measure the metal abundance of various astrophysical objects, different techniques are used. A widely used metallicity proxy is oxygen abundance measured from emission lines in star-forming regions of galaxies. Since oxygen is the most abundant heavy element, the oxygen abundance is a good approximation to the total metal abundance. For non-star-forming regions without emission lines in the spectra, absorption lines and stellar population models are employed. Generally, the metallicity obtained from emission lines refers to the gas component, while the metallicity determined from stellar continuum and absorption line fitting refers to the stellar component.

Lian et al. (2018) investigated the mass–metallicity relations for the gaseous ($MZR_{gas}$) and stellar components ($MZR_{star}$) of local star-forming galaxies based on a representative sample from Sloan Digital Sky Survey Data Release 12. The mass-weighted average stellar metallicities are systematically lower than the gas metallicities. This difference in metallicity increases towards galaxies with lower masses and reaches 0.4–0.8 dex at $10^9$ $M_{sun}$ (depending on the gas metallicity calibration). As a result, the $MZR_{star}$ is much steeper than the $MZR_{gas}$. The much lower metallicities in stars compared to the gas in low-mass galaxies imply dramatic metallicity evolution with suppressed metal enrichment at early times. A paper by Lian et al. (2018) explained the observed large difference in gas and stellar metallicity and to infer the origin of the mass–metallicity relations. To this end, they developed a galactic chemical evolution model accounting for star formation, gas inflow and outflow. By combining the observed mass–metallicity relation for both gas and stellar components to constrain the models, they found that only two scenarios are able to reproduce the observations. Either strong metal outflow or a steep initial mass function (IMF) slope at early epochs of galaxy evolution is needed. Based on these two scenarios, for the first time, they successfully reproduced the observed $MZR_{gas}$ and $MZR_{star}$ simultaneously, together with other independent observational constraints in the local Universe. Their model also naturally reproduces the flattening of the $MZR_{gas}$ at the high-mass end, leaving the $MZR_{star}$ intact, as seen in observational data.

GW170817 was the first—and thus far the only—cosmic event where both gravitational waves and electromagnetic radiation, or light, were detected. This combination provides scientists with critical information about the physics of neutron star mergers and related phenomena, using observations at many different parts of the electromagnetic spectrum. Chandra is the only observatory still able to detect light from this extraordinary cosmic collision more than four years after the original event.

Galaxy mergers produce pairs of supermassive black holes, which may be witnessed as dual quasars if both SMBHs are rapidly accreting. The kiloparsec (kpc)-scale separation represents a physical regime sufficiently close for merger-induced effects to be important yet wide enough to be directly resolvable with the facilities currently available. Whereas many kpc-scale dual active galactic nuclei—the low-luminosity counterparts of quasars—have been observed in low-redshift mergers, no unambiguous dual quasar is known at cosmic noon ($z \approx 2$), the peak of global star formation and quasar activity. Chen et al. (2023) reported multiwavelength observations of SDSS J0749 + 2255 as a kpc-scale, dual-quasar system hosted by a galaxy merger at cosmic noon, $z = 2.17$. They discovered extended host galaxies associated with the much brighter compact quasar nuclei (separated by 0.46" or 3.8 kpc) and low-surface-brightness tidal features as evidence for galactic interactions. Unlike its low-redshift and low-luminosity counterparts, SDSS J0749 + 2255 is hosted by massive compact disk-dominated galaxies. The apparent lack of stellar bulges and the fact that SDSS J0749 + 2255 already follows the local SMBH mass–host stellar mass relation suggest that at least some SMBHs may have formed before their host stellar bulges. While still at kpc-scale separations where the host-galaxy gravitational potential dominates, the two SMBHs may evolve into a gravitationally bound binary system in around $0.22 \times 10^9$ years.

Komugi et al. (2022) estimated the amount of negative feedback energy injected into the interstellar medium of the host galaxy of 3C 273, a prototypical radio-loud quasar. They obtained 93, 233, and 343 GHz continuum images with the Atacama Large Millimeter/Submillimeter Array. After self-calibration and point-source subtraction, we reach an image dynamic range of ~85,000 at 93 GHz, ~39,000 at 233 GHz, and ~2500 at 343 GHz. These are currently the highest image dynamic ranges obtained using ALMA. They detected spatially extended millimeter emission associated with the host galaxy, cospatial with the extended emission line region (EELR) observed in the optical. The millimeter spectral energy distribution and comparison with centimeter data show that the extended emission cannot be explained by dust thermal emission or synchrotron or thermal bremsstrahlung arising from massive star formation. The authors interpreted the extended millimeter emission as thermal bremsstrahlung from gas directly ionized by the central source. The extended flux indicates that at least ~7% of the bolometric flux of the nuclear source was used to ionize atomic hydrogen in the host galaxy. The ionized gas is estimated to be as massive as $10^{10}$–$10^{11}$ $M_{sun}$, but the molecular gas fraction with respect to the stellar mass is consistent with other ellipticals, suggesting that direct ionization ISM by the QSO may not be sufficient to suppress star formation, or there is a short time interval before negative feedback becomes observable. The discovery of a radio counterpart to EELRs provides a new pathway to studying the QSO–host ISM interaction.

McKinney et al. (2023) reported a near-infrared-faint, far-infrared-luminous dusty galaxy at $z \sim 5$ in COSMOS-Web. A growing number of far-infrared (FIR) bright sources completely invisible in deep extragalactic optical surveys hint at an elusive population of $z > 4$ dusty, star-forming galaxies. Cycle 1 JWST surveys are now detecting their rest-frame optical light, which provides key insight into their stellar properties and statistical constraints on the population as a whole. The work by McKinney et al. (2023) presents the JWST Near Infrared Camera (NIRCam) counterpart from the COSMOS-Web survey to an FIR SCUBA-2 and Atacama Large Millimeter/Submillimeter Array source, AzTECC71, which was previously undetected at wavelengths shorter than 850 μm. AzTECC71, among the reddest galaxies in COSMOS-Web with F277W–F444W ~ 0.9, is undetected in NIRCam/F150W and F115W and fainter in F444W than other submillimeter galaxies identified in COSMOS-Web by 2–4 magnitudes. This is consistent with the system having both a lower stellar mass and higher redshift than the median dusty, star-forming galaxy. With deep ground- and space-based upper limits combined with detections in F277W, F444W, and the FIR including ALMA Band 6, the authors find a high probability (99%) that AzTECC71 is at $z > 4$ with $z_{phot} = 5.7^{+0.8-}{}_{0.7}$. This galaxy is massive ($\log M_{star}/M_{sun} \sim 10.7$) and infrared-luminous ($\log L_{IR}/L_{sun} \sim 12.7$), comparable to other optically undetected but FIR-bright dusty, star-forming galaxies at $z > 4$. This population of luminous, infrared galaxies at $z > 4$ is largely unconstrained but constitutes an important bridge between the most extreme dust-obscured galaxies and more typical high-redshift star-forming galaxies. If further FIR-selected galaxies that drop out of the F150W filter in COSMOS-Web have redshifts $z > 4$ like AzTECC71, then the volume density of such sources may be ~3–10× greater than previously estimated.

According to Guo et al. (2023), galactic outflows are believed to play a critical role in the evolution of galaxies by regulating their mass build-up and star formation. Theoretical models assume bipolar shapes for the outflows that extend well into the circumgalactic medium, up to tens of kiloparsecs perpendicular to the galaxies. They have been directly observed in the local Universe in several individual galaxies, for example, around the Milky Way and M82. At higher redshifts, cosmological simulations of galaxy formation predict an increase in the frequency and efficiency of galactic outflows owing to the increasing star-formation activity. Galactic outflows are usually of low gas density and low surface brightness and therefore difficult to observe in emission towards high redshifts. Guo et al. (2023) presented an ultra-deep Multi-Unit Spectroscopic Explorer image of the mean Mg II emission surrounding a sample of galaxies at $z \approx 1$ that strongly suggests the presence of outflowing gas on physical scales of more than 10 kpc. We find a strong dependence of the detected signal on the inclination of the central galaxy, with edge-on galaxies clearly showing enhanced Mg II emission along the minor axis, whereas face-on galaxies show much weaker and more isotropic emission. They interpreted these findings as supporting the idea that outflows typically have a bipolar cone geometry perpendicular to the galactic disk. They demonstrated that this CGM-scale outflow was prevalent among galaxies with stellar mass $M_* \gtrsim 10^{9.5} M_\odot$.

## 1.5 GALAXY GROUPS

Galaxy groups and clusters are the largest known gravitationally bound objects to have arisen thus far in the process of cosmic structure formation. The nearly 10,000 galaxies captured in the Hubble Ultra Deep Field may look like they are randomly scattered across the sky. But galaxies, including the Milky Way, are often part of larger structures and superstructures in space. Galaxy groups and clusters are collections of galaxies bound together by gravity. They are building blocks for the larger-scale structures of the Universe (NASA 2023).

Galaxy groups generally hold 100 or fewer galaxies. The Milky Way and its closest neighbors, including the Andromeda galaxy, live alongside over 50 other galaxies within the Local Group. Galaxy clusters contain many more galaxies, from hundreds to thousands. The Coma cluster, for example, houses over 1,000 individual galaxies and stretches over 20 million lightyears across. In addition to their galaxies, these clusters also contain a lot of hot gas—tens of millions of degrees—and abundant amounts of dark matter, an invisible substance that scientists know must exist but have yet to identify.

The central regions of galaxy clusters are permeated by magnetic fields and filled with relativistic electrons. When clusters merge, the magnetic fields are amplified, and relativistic electrons are re-accelerated by turbulence in the intracluster medium. These electrons reach energies of 1–10 GeV and, in the presence of magnetic fields, produce diffuse radio halos that typically cover an area of around 1 $Mpc^2$. Cuciti et al. (2022) reported observations of four clusters whose radio halos are embedded in much more extended, diffuse radio emission, filling a volume 30 times larger than that of radio halos. The emissivity in these larger features is about 20 times lower than the emissivity in radio halos. The authors concluded that relativistic electrons and magnetic fields extend far beyond radio halos and that the physical conditions in the outer regions of the clusters are quite different from those in the radio halos.

Superclusters are large collections of galaxy clusters, groups, and individual galaxies that are typically not gravitationally bound to each other. The Local Group, containing the Milky Way, is located on the outer edges of the Laniakea supercluster. Groupings of superclusters can form walls or sheets of galaxies. These walls are likely the largest-known superstructures within the observable Universe. They can stretch hundreds of millions of lightyears across but are relatively thin—only about 20 million lightyears deep. The first galactic wall astronomers discovered, called the Coma Wall, is about 500 million lightyears long, 16 million lightyears deep, and 300 million lightyears away. Scientists have since identified additional walls, most notably the Sloan Great Wall, which spans about 1.4 billion lightyears.

Galaxies, galaxy groups and clusters, superclusters, and galactic walls are arranged in twisting, threadlike structures called the cosmic web. The web forms as the gravitational attraction of the Universe's matter draws larger and larger objects together. This leads to concentrations of galaxies with voids of space in between, as if the galaxies were resting on empty bubbles. Where those concentrations intersect, they form larger structures like clusters and walls. The cosmic web forms a sort of scaffolding that cradles everything that exists within the Universe (NASA 2023).

There are many studies of the large-scale properties of the Universe. Here we shall mention only some. For example, Chowdhury et al. (2020) quantify the properties of the gas and dark matter around active galactic nuclei in simulated galaxy groups and clusters and analyze the effect of AGN feedback on the surrounding intracluster (group) medium. Their results suggest downsizing of AGN luminosity with host halo mass, supporting the results obtained from clustering studies of AGNs. By examining the temperature and density distribution of the gas in the vicinity of AGNs, they show that due to feedback from the central engine, the gas gets displaced from the center of the group/cluster, resulting in a reduction of the density but an enhancement of temperature.

McAlpine at el. (2022) presented SIBELIUS-DARK, a constrained realization simulation of the local volume to a distance of 200 Mpc from the Milky Way. SIBELIUS-DARK is the first study of the Simulations Beyond the Local Universe (SIBELIUS) project, which has the goal of embedding a model Local Group–like system within the correct cosmic environment. The simulation is dark matter only, with the galaxy population calculated using the semi-analytic model of galaxy formation, GALFORM. The authors demonstrated that the large-scale structure that emerges from the SIBELIUS constrained initial conditions matches the observational data well. The inferred galaxy population of SIBELIUS-DARK also matches the observational data well, both statistically for the whole volume and on an object-by-object basis for the most massive clusters. For example, the K-band number counts across the whole sky and when divided between the northern and southern galactic hemispheres are well reproduced by SIBELIUS-DARK. They found that the local volume is somewhat unusual in the wider context of CDM: it contains an abnormally high number of supermassive clusters, as well as an overall large-scale underdensity at the level of ≈5% relative to the cosmic mean. However, whilst rare, the extent of these peculiarities does not significantly challenge the CDM model. SIBELIUS-DARK is the most comprehensive constrained realization simulation of the local volume to date, and with this paper, they publicly released the halo and galaxy catalogues at $z = 0$, which might be useful to the wider astronomy community.

Galaxy clusters are known to harbor magnetic fields, the nature of which remains unresolved. Intracluster magnetic fields can be observed at the density contact discontinuity formed by cool and dense plasma running into hot ambient plasma, and the discontinuity exists near the second-brightest galaxy, MRC 0600–399, in the merging galaxy cluster Abell 3376 (redshift 0.0461). Elongated X-ray emission in the east–west direction shows a comet-like structure that reaches the megaparsec scale. Previous radio observations detected bent jets from MRC 0600–399, moving in same direction as the sub-cluster, against ram pressure. Chibueze et al. (2021) reported radio observations of MRC 0600–399 that have 3.4 and 11 times higher resolution and sensitivity, respectively, than the previous results. In contrast to typical jets, MRC 0600–399 shows a 90-degree bend at the contact discontinuity, and the collimated jets extend over 100 kiloparsecs from the point of the bend. They saw diffuse, elongated emission that they named "double-scythe" structures. The spectral index flattens downstream of the bend point, indicating cosmic-ray reacceleration. High-resolution numerical simulations reveal that the ordered magnetic field along the discontinuity has an important role in the change of jet direction. The morphology of the double-scythe jets is consistent with the simulations. Their results provide insights into the effect of magnetic fields on the evolution of the member galaxies and intra-cluster medium of galaxy clusters.

### 1.5.1 Metal Content

Two major astrophysical discoveries have provided key answers to the fundamental question of the origin of the chemical elements in the past century: the discovery that stellar nucleosynthesis is responsible for the production of all the heavy elements from lithium to uranium and the detection of line emission due to highly ionized iron in the X-ray spectra of the intra-cluster medium (ICM). The impact of these two discoveries was extraordinary. The first one demonstrated that all the elements (with the exception of hydrogen, helium, and traces of lithium and beryllium produced by the Big Bang nucleosynthesis) are made in the cores of stars and in supernovae (SNe) and that when a star explodes as a supernova, it enriches the surrounding interstellar medium with freshly created elements. The second one showed that galaxies lost part of their synthesized elements and that there has been a considerable exchange of chemical elements between stars, galaxies, and the hot plasma surrounding them. This also means that the chemical elements trace the formation and evolution of structure which is shaped by the physical processes occurring on a very wide range of spatial scales, from the size of single supernova remnants to cosmological volumes.

The evolution of the content of heavy elements in galaxies, the relative chemical abundances, their spatial distribution, and how these scale with various galactic properties provide unique information on the galactic evolutionary processes across the cosmic epochs. In recent years major progress has been made in constraining the chemical evolution of galaxies and inferring key information relevant to our understanding of the main mechanisms involved in galaxy evolution. Maiolino and Mannucci (2019) provide an overview of these various areas. After an overview of the methods used to constrain the chemical enrichment in galaxies and their environment, they discuss the observed scaling relations between metallicity and galaxy properties, the observed relative chemical abundances, how the chemical elements are distributed within galaxies, and how these properties evolve across the cosmic epochs. They discuss how the various observational findings compare with the predictions from theoretical models and numerical cosmological simulations. They also briefly discuss the open problems and the prospects for major progress in this field in the nearby future.

Calura and Matteucci (2004) discussed the cosmic metal production and the mean metallicity of the Universe. By means of detailed chemo-photometric models for elliptical, spiral and irregular galaxies, they evaluated the cosmic history of the production of chemical elements as well as the metal mass density of the present-day Universe. In particular, they studied the production rates of some of the most common elements (O, Mg, N, Si, Fe, and Zn), detected in local stars, galactic and extragalactic H II regions, and high-redshift objects. Such a study allows us to compute in detail the fraction of each element locked up in stars, interstellar gas, and intergalactic medium. The authors calculated the mean metal abundances for galaxies of different morphological types, along with the average metallicity of galactic matter in the Universe (stars, gas, and intergalactic medium). For the average metallicity of galaxies in the local Universe, we find $Z_{gal} = 0.0175$, that is, close to the solar value. They found the main metal production in spheroids (ellipticals and bulges) to occur at very early times, implying an early peak in the metal production and a subsequent decrease. On the other hand, the metal production in spirals and irregulars is always increasing with time. They found that the average $[O/Fe]_{star,E}$ ratio in stars in spheroids should be +0.4 dex, whereas the average $[O/Fe]_{gas,E}$ ratio in the gas should be −0.33 dex, as a result of the large amount of Fe produced in these systems by Type Ia supernovae after star formation has stopped. The same quantities for spirals are ~ +0.1 dex for the stellar component and almost solar (~ +0.01 dex) for the gas component. They suggested that a Salpeter-like initial mass function is the best candidate for the universal IMF because it allows us to reproduce the majority of observational constraints. They performed a self-consistent census of the baryons and metals in the local Universe finding that, while the vast majority of the baryons lie outside galaxies in the intergalactic medium, 52% of the metals (with the exception of the Fe-peak elements) are locked up in stars and the interstellar medium. They estimated indirectly the amount of baryons that resides in the IGM and derive its mean Fe abundance, finding a value of $X_{Fe,IGM} = 0.05\ X_{Fe,sun}$. The authors believe that this estimate is uncertain by a factor of ~2, owing to the normalization of the local luminosity function. This means that the Fe abundance of 0.3 solar inferred from X-ray observations of the hot intracluster medium is higher than the average Fe abundance of the intergalactic gas in the field.

Galaxy groups host the majority of matter and more than half of all the galaxies in the Universe. Their hot ($10^7$ K), X-ray emitting intra-group medium (IGrM) reveals emission lines typical of many elements synthesized by stars and supernovae. Because their gravitational potentials are shallower than those of rich galaxy clusters, groups are ideal targets for studying, through X-ray observations, feedback effects, which leave important marks on their gas and metal contents. Gastaldello et al. (2021) reviewed the history and present status of the chemical abundances in the IGrM probed by X-ray spectroscopy. They discussed the limitations of our current knowledge, in particular due to uncertainties in the modeling of the Fe-L shell by plasma codes, and coverage of the volume beyond the central region. They further summarize the constraints on the abundance pattern at the group mass scale and the insight it provides to the history of chemical enrichment. Parallel to the observational efforts, they reviewed the progress made by both cosmological hydrodynamical simulations and controlled high-resolution 3D simulations to reproduce the radial distribution of metals in the IGrM, the dependence on system mass from group to cluster scales, and the role of AGN and SN feedback in producing the observed phenomenology. Finally, the authors highlighted future prospects in this field, where progress will be driven both by a much richer sample of X-ray emitting groups identified with eROSITA, and by a revolution in the study of X-ray spectra expected from micro-calorimeters onboard XRISM and ATHENA.

Péroux and Howk (2020) discussed the cosmic baryon and metal cycles. Characterizing the relationship between stars, gas, and metals in galaxies is a critical component of understanding the cosmic baryon cycle. They compiled contemporary censuses of the baryons in collapsed structures and their chemical makeup and dust content. They showed the following:

> The Hi (hydrogen) mass density of the Universe is well determined to redshifts $z \approx 5$ and shows minor evolution with time. New observations of molecular hydrogen reveal its evolution mirrors that of the global star formation rate density, implying a universal cosmic molecular gas depletion timescale. The low-redshift decline of the star-formation history is thus driven by the lack of molecular gas supply due to a drop in net accretion rate related to the decreased growth of dark matter halos.

The metal mass density in cold gas ($T \le 104$ K) contains virtually all the metals produced by stars for $z \ge 2.5$. At lower redshifts, the contributors to the total amount of metals are more diverse; at $z < 1$, most of the observed metals are bound in stars. Overall, there is little evidence for a "missing metals problem" in modern censuses.

They characterize the dust content of neutral gas over cosmic time, finding the dust-to-gas and dust-to-metals ratios fall with decreasing metallicity. We calculate the cosmological dust mass density in the neutral gas up to $z \approx 5$. There is good agreement between multiple tracers of the dust content of the Universe.

Low and intermediate mass stars with super solar metallicities comprise a known portion of the Universe. Yet yields for asymptotic giant branch (AGB) stars with metallicities greater than $Z = 0.04$ do not exist in the literature. This contributes a significant uncertainty to galactic chemical evolution simulations. We present stellar yields of AGB stars for $M = 1 - 8_{sun}$ and $Z = 0.04 - 0.10$. Cinquegrana and Karakas (2021) also weighted these yields to represent the chemical contribution of a metal-rich stellar population. They found that as metallicity increases, the efficiency of the mixing episodes (known as the third dredge up) on the thermally pulsing AGB (TP-AGB) decrease significantly. Consequently, much of the nucleosynthesis that occurs on the TP-AGB is not represented on the surface of very metal-rich stars. It instead remains locked inside the white dwarf remnant. The temperatures at the base of the convective envelope also decrease with increasing metallicity. For the intermediate mass models, this results in the occurrence of only partial hydrogen burning at this location, if any burning at all. They also investigated heavy element production via the slow neutron capture process (s-process) for three $6M_{sun}$ models: $Z = 0.04$, 0.05, and 0.06. There is minor production at the first s-process peak at strontium, which decreases sharply with increasing metallicity. They found the chemical contributions of our models are dominated by proton capture nucleosynthesis, mixed to the surface during first and second dredge up events. This conclusion is mirrored in our stellar population yields, weighted towards the lower mass regime to reflect the mass distribution within a respective galaxy.

Heintz et al. (2023) presented the gas and stellar content of a metal-poor galaxy at $z = 8.496$ as revealed by JWST and ALMA. They presented a joint analysis of the galaxy S04590 at $z = 8.496$ based on NIRSpec, NIRCam, and NIRISS observations obtained as part of the Early Release Observations program of the James Webb Space Telescope and the far-infrared [C II] 158 μm emission line detected by dedicated Atacama Large Millimeter/Submillimeter Array observations. They determined the physical properties of S04590 from modeling of the spectral energy distribution and through the redshifted optical nebular emission lines detected with JWST/NIRSpec. The best-fit SED model reveals a low-mass ($M_{star} = 10^{7.2}$–$10^8$ $M_{sun}$) galaxy with a low oxygen abundance of 12+log O/H = $7.16{-_{0.12}}^{+0.10}$ derived from the strong nebular and auroral emission lines. Assuming that [C II] effectively traces the interstellar medium, we estimate the total gas mass of the galaxy to be Mgas = $(8.0 \pm 4.0) \times 10^8$ $M_{sun}$ based on the luminosity and spatial extent of [C II]. This yields an exceptionally high gas fraction, $f_{gas} = M_{gas}/(M_{gas} + M_{star}) \ge 90\%$, though one still consistent with the range expected for low metallicity. They further derived the metal mass of the galaxy based on the gas mass and gas-phase metallicity, which was found to be consistent with the expected metal production from Type II supernovae. Finally, they made the first constraints on the dust-to-gas (DTG) and dust-to-metal (DTM) ratios of galaxies in the epoch of reionization at $z \geq 6$, showing overall low mass ratios of logDTG < −3.8 and logDTM < −0.5, though they are consistent with established scaling relations and in particular with those of the local metal-poor galaxy I Zwicky 18. Their analysis highlights the synergy between ALMA and JWST in characterizing the gas, metal, and stellar content of the first generation of galaxies.

## 1.6 COSMIC RAYS

Highly energetic cosmic rays, despite representing only a microscopic fraction of particles by number density ($n_{CR}/n_{gas} \approx 10^{-9}$), are roughly in energy equipartition with thermal and magnetic energy in the Milky Way interstellar medium; see Boulares and Cox (1990). In their work, three gravitational potentials differing in the content of dark matter in the galactic plane are used to study the structure of the z-distribution of mass and pressure in the solar neighborhood. A P(0) of roughly $(3.9 \pm 0.6) \times 10^{-12}$ dyn/cm$^2$ is obtained, with roughly equal contributions from magnetic field, cosmic ray, and kinetic terms. This boundary condition restricts both the magnitude of gravity and the high z-pressure. It favors lower gravity and higher values for the cosmic ray, magnetic field, and probably the kinetic pressures than have been popular in the past. Inclusion of the warm $H^+$ distribution carries a significant mass component into the z about 1 kpc regime.

Because hadronic cosmic rays do not suffer from radiative losses (though hadronic collisions, Coulomb collisions, and "collisionless" scattering off magnetic perturbations can be large energy sinks), and they do not lose as much energy to adiabatic expansion compared to thermal gas ($P \propto \rho^{4/3}$ for a relativistic gas instead of $P \propto \rho^{5/3}$ for a nonrelativistic gas), they fundamentally alter ISM dynamics and are believed to be an important component of feedback, which regulates star formation in galaxies (Bustard and Zweibel 2021). For GeV cosmic rays, which make up the peak of the cosmic-ray energy spectrum and therefore contain most of the energy and momentum of the cosmic-ray population, the dominant transport mode is self-confinement via streaming instability. In this picture, cosmic rays with a bulk drift speed greater than the Alfvén speed can excite Alfvén waves through gyroresonance and pitch angle scatter off the waves until the cosmic rays attain isotropy in the wave frame. If a conducting liquid is placed in a constant magnetic field, every motion of the liquid gives rise to an EMF, which produces electric currents. Owing to the magnetic field, these currents give mechanical forces, which change the state of motion of the liquid. Thus a kind of combined electromagnetic-hydro-dynamic wave is produced. Scattering transfers energy from the cosmic-ray population to the waves, and subsequent wave damping heats the background gas.

Most galaxy evolution simulations that include a relativistic cosmic-ray fluid assume some combination of diffusive and streaming transport in addition to cosmic-ray advection with the nonrelativistic thermal gas. Given the canonical picture that 10% of supernova energy is converted to cosmic-ray energy through first-order Fermi acceleration, simulations with diffusive or streaming transport included generally find colder, smoother, more extended outflows than thermally driven winds, and in many cases, the nonthermal support is necessary to launch an outflow.

The bulk propagation speed of GeV-energy cosmic rays is limited by frequent scattering off hydromagnetic waves. Most galaxy evolution simulations that account for this confinement assume the gas is fully ionized and cosmic rays are well coupled to Alfvén waves; however, multiphase density inhomogeneities, frequently underresolved in galaxy evolution simulations, induce cosmic-ray collisions and ionization-dependent transport driven by cosmic-ray decoupling and elevated streaming speeds in partially neutral gas. How do cosmic rays navigate and influence such a medium, and can we constrain this transport with observations? In this chapter, we simulate cosmic-ray fronts impinging upon idealized, partially neutral clouds and lognormally distributed clumps, with and without ionization-dependent transport. With these high-resolution simulations, we identify cloud interfaces as crucial regions where cosmic-ray fronts can develop a stairstep pressure gradient sufficient to collisionlessly generate waves, overcome ion–neutral damping, and exert a force on the cloud.

Bustard and Zweibel (2021) found that the acceleration of cold clouds is hindered by only a factor of a few when ionization-dependent transport is included, with additional dependencies on magnetic field strength and cloud dimensionality. They also probe how cosmic rays sample the background gas and quantify collisional losses. Hadronic gamma-ray emission maps are qualitatively different when ionization-dependent transport is included, but the overall luminosity varies by only a small factor, as the short cosmic-ray residence times in cold clouds are offset by the higher densities that cosmic rays sample.

Castellina and Donato (2013) presented a review of the main properties of the charged component of galactic cosmic rays, particles detected at Earth with an energy spanning tens of MeV up to about $10^{19}$ eV. In their work, the properties of cosmic rays are discussed with respect to different energy ranges. The mechanisms leading to the acceleration of nuclei by supernova remnants and to their subsequent diffusion through the inhomogeneities of the galactic magnetic field are discussed, and some clue is given on the predictions and observations of fluxes of antimatter, both from astrophysical sources and from dark matter annihilation in the galactic halo. Direct methods are viable up to $\simeq 10^{14}$ eV by means of experiments flown on balloons or satellites, while above that energy cosmic rays can be studied only indirectly by exploiting the particle cascades they produce in the atmosphere. The possible physical interpretation of the peculiar features observed in the energy spectrum of galactic cosmic rays, and in particular the so-called "knee" at about $4 \times 10^{15}$ eV, is discussed. A section is devoted to the region between about $10^{18}$ and $10^{19}$ eV, which is believed to host the transition between galactic and extragalactic cosmic rays. The conclusion gives some perspectives on the cosmic ray astrophysics field. Thanks to a wealth of different experiments, this research area is living a very flourishing era. The activity is exciting from both the theoretical and the instrumental sides, and its interconnection with astronomy, astrophysics, and particle physics experiences nonstop growth.

Amenomori et al. (2021) reported, for the first time, the long-awaited detection of diffuse gamma rays with energies between 100 TeV and 1 PeV in the galactic disk. Particularly, all gamma rays above 398 TeV are observed apart from known TeV gamma-ray sources and compatible with expectations from the hadronic emission scenario in which gamma rays originate from the decay of $\pi^0$s produced through the interaction of protons with the interstellar medium in the galaxy. This is strong evidence that cosmic rays are accelerated beyond PeV energies in our galaxy and spread over the galactic disk.

George et al. (2009) reported improved measurements of elemental abundances and spectra for galactic cosmic-ray (GCR) nuclei obtained by the Cosmic Ray Isotope Spectrometer on board NASA's Advanced Composition Explorer (ACE) spacecraft during the minimum and maximum phases of solar cycle 23. They discuss results for particles with nuclear charge $5 \leq Z \leq 28$ and typical energies between 50 and 500 MeV nucleon$^{-1}$. They demonstrated that a detailed "leaky box" galactic propagation model combined with a spherically symmetric solar modulation model gives a good (but not perfect) fit to the observed spectra by using a solar modulation parameter of $\varphi = 325$ MV at solar minimum and $\varphi = 900$ MV at solar maximum. Although our results are generally consistent with previous measurements from space-based and balloon-based missions, there are significant differences. The large geometrical acceptance and excellent charge resolution of the instrument result in the most detailed and statistically significant record of GCR composition to date in this energy range. The measurements reported here serve as a high-precision baseline for continued studies of GCR composition, solar modulation over the solar cycle, space radiation hazards, and other applications.

The cosmic ray element abundances (He-Ni) measured at Earth compared to the solar system abundances, all relative to silicon, are presented for low energies, 70–280 MeV n$^{-1}$ Garcia-Munoz and Simpson (1979) and for high energies—2000 MeV n$^{-1}$. Solar system concentration values are taken from Cameron (1982). Of interest are measurements of the less abundant nuclei occurring in the mass range beyond nickel. Comparison of cosmic ray element abundances (Co-Ru) measured at Earth and calculated at the cosmic ray source with solar system material and the results of rapid neutron capture nucleosynthesis; see Binns et al. (1981) and Israel (1981).

The charge spectrum measured at Earth [or cosmic rays with $Z \geq 32$ is reported by Binns et al. (1982). For cosmic ray sources and solar system abundances, see Cameron (1982), with estimated errors by Meyer (1979). When a ratio of the galactic cosmic ray source abundances to the solar system abundances in the atomic number range 1 to 40 is considered, the ratio normalized to unity at silicon ($Z$ ~– 14), there are considerable uncertainties, especially in some of the solar system abundances and the odd-Z nuclei in the galactic cosmic rays.

The cosmic ray source abundance ratio for C/O (0.85) is a factor of ~1.5 larger than the ratio of 0.58 derived from solar spectroscopy or solar flare accelerated particles. Since both nuclei are A/Z = 2 particles and there are only small corrections to the ratio to take account of their first ionization potentials, this high ratio may indicate that the source matter is closely connected with matter from supernovae. Recent theoretical advances have not changed this conclusion.

It is now established that there are important differences in elemental and isotopic abundance between the source matter of galactic cosmic rays and the solar system matter. These large differences (more than a factor of 2 in some cases, in contrast to <1 (for isotopic anomalies in solar system matter) may become essential clues to the identification of astrophysical sites for cosmic ray matter and to the nucleosynthesis processes that preceded the acceleration of the cosmic rays or galactic chemical evolution. There are also significant similarities demonstrating that the cosmic ray source material is not predominantly freshly made, r-processed material. With our present limited knowledge of abundances—especially isotopic—for different astrophysical sites, it appears that cosmic ray source matter may be a mixture of aged r-process and interstellar medium material. Clearly, the abundances in the cosmic ray sources, the interstellar medium, supernova remnants, molecular clouds, and so forth are all essential inputs for unraveling their respective nucleosynthesis histories. For example, the recent measurements by Jura and Smith (1981) of rubidium in the interstellar medium indicate for the first time that r-process elements can be measured in the interstellar gas, Simpson (1983).

**Table 1.1** Elemental abundances in cosmic rays, SuperTIGER TOA abundances for the entire analyzed charge range ($16 \leq Z \leq 56$) are shown. Results from past analyses are included for comparison, including SuperTIGER from Murphy et al. (2016), TIGER from Rauch et al. (2009), HEAO-3 from Byrnak et al. (1983) and Engelmann et al. (1990), and solar system abundances from Lodders (2003). All abundances are normalized to $^{26}Fe = 10^6$.

On 8 December, the SuperTIGER instrument was launched from Williams Field, Antarctica, on a long-duration balloon flight that lasted 55 days and maintained a mean altitude of 125,000 feet. SuperTIGER measured the relative abundances of galactic cosmic-ray

TABLE 1.1 Elemental Abundances in Cosmic Rays

| Element/ Isotope | Element/H | Element/Fe | Reference |
|---|---|---|---|
| **H** | 1.000 | | - |
| **$^2$H** | $2.1 \times 10^{-2}$ | | Adriani et al. (2016) |
| He | $8.14 \times 10^{-2}$ | | Cummings et al. (2016) |
| $^3$He | $6.626 \times 10^{-3}$ | | Webber (2017) |
| **Li** | $9.753 \times 10^{-5}$ | | Cummings et al. (2016) |
| Be | $4.383 \times 10^{-5}$ | | Cummings et al. (2016) |
| **B** | $2.157 \times 10^{-4}$ | | Cummings et al. (2016) |
| **C** | $1.671 \times 10^{-3}$ | | Cummings et al. (2016) |
| **N** | $2.444 \times 10^{-4}$ | | Cummings et al. (2016) |
| **O** | $1.570 \times 10^{-3}$ | | Cummings et al. (2016) |
| **F** | $5.123 \times 10^{-6}$ | | Cummings et al. (2016) |
| Ne | $1.507 \times 10^{-4}$ | | Cummings et al. (2016) |
| **Na** | $1.784 \times 10^{-5}$ | | Cummings et al. (2016) |
| **Mg** | $2.264 \times 10^{-4}$ | | Cummings et al. (2016) |
| Al | $3.302 \times 10^{-5}$ | | Cummings et al. (2016) |
| **Si** | $1.898 \times 10^{-4}$ | | Cummings et al. (2016) |
| **P** | $3.036 \times 10^{-6}$ | | Cummings et al. (2016) |
| **S** | $2.087 \times 10^{-5}$ | | Cummings et al. (2016) |
| **Cl** | $1.898 \times 10^{-6}$ | | Cummings et al. (2016) |
| Ar | $4.554 \times 10^{-6}$ | | Cummings et al. (2016) |
| **K** | $2.657 \times 10^{-6}$ | | Cummings et al. (2016) |
| **Ca** | $1.195 \times 10^{-5}$ | | Cummings et al. (2016) |
| Sc | $1.898 \times 10^{-6}$ | | Cummings et al. (2016) |
| Ti | $1.101 \times 10^{-5}$ | | Cummings et al. (2016) |
| ,, | $5.123 \times 10^{-6}$ | | Cummings et al. (2016) |
| **Cr** | $1.347 \times 10^{-5}$ | | Cummings et al. (2016) |
| **Mn** | $5.693 \times 10^{-6}$ | | Cummings et al. (2016) |
| **Fe** | $1.152 \times 10^{-4}$ | $10^6$ | Cummings et al. (2016), Murphy et al. (2016) |
| **Co** | $6.519 \times 10^{-7}$ | | George et al. (2009) |
| **Ni** | $6.452 \times 10^{-6}$ | 237391 | Cummings et al. (2016), Murphy et al. (2016) |
| **Cu** | $6.091 \times 10^{-8}$ | | Murphy et al. (2016) |
| **Zn** | $7.130 \times 10^{-8}$ | 2623 | Murphy et al. (2016) |
| Ga | | 239 | Murphy et al. (2016) |
| Ge | | 354 | Murphy et al. (2016) |
| **As** | | 65 | Murphy et al. (2016) |
| **Se** | | 160 | Murphy et al. (2016) |
| Br | | 49 | Murphy et al. (2016) |
| Kr | | 91 | Murphy et al. (2016) |
| Rb | | 31 | Murphy et al. (2016) |
| Sr | | 105 | Murphy et al. (2016) |
| Y | | 30 | Murphy et al. (2016) |
| Zr | | 35 | Murphy et al. (2016) |
| **Mo** | | ~11.7 | Simpson (1983) |
| **I** | | | |
| **W** | | | |

*Source:* Murphy et al. (2016) presented abundance data of elements from Fe to Zr in the cosmic radiation measured by the Super TIGER (Trans-Iron Galactic Element Recorder) instrument during 55 days of exposure on a long-duration balloon flight over Antarctica.

nuclei with high statistical precision and well-resolved individual element peaks from $^{10}$Ne to $^{40}$Zr. SuperTIGER also made exploratory measurements of the relative abundances up to $^{56}$Ba. The SuperTIGER instrument determines the charge (Z) and energy of GCR using two independent detector stacks composed of a scintillating fiber hodoscope, two Cherenkov light detectors, and three scintillation light detectors. The SuperTIGER data are analyzed to account for instrument effects and then corrected for nuclear interactions and energy losses that occur within the instrument, atmosphere, and interstellar medium to obtain relative abundances at the GCR source.

In a recent publication, H.E.S.S. Collaboration has described time-resolved hadronic particle acceleration in the recurrent nova RS Ophiuchi (H.E.S.S. Collaboration et al. 2022). RS Ophiuchi (RS Oph) is a recurrent nova system comprising a white dwarf and a companion red giant star. Novae are a source of high-energy particles, with non-thermal gamma-ray emission in the range ~100 MeV to ~10 GeV. The RS Oph system is located approximately 1.4 kpc from Earth, but analysis of Gaia data suggests larger distances of 2.3 (Gaia Collaboration 2018) or 2.7 kpc (Lindegren et al. 2021), although the reliability of these estimates is questionable due to the orbital motion in RS Oph. The binary components have a separation of 1.48 astronomical units (au), close enough for the white dwarf to continually accrete material from its companion. At irregular intervals, enough material accumulates on the surface of the white dwarf to trigger a thermonuclear explosion, driving a quasi-spherical shock into the red giant's wind.

Recurrent novae are repeating thermonuclear explosions in the outer layers of white dwarfs due to the accretion of fresh material from a binary companion. The shock generated by ejected material slamming into the companion star's wind accelerates particles to very high energies. The H.E.S.S. Collaboration et al. (2022) reported very-high energy (VHE, ~100 GeV) gamma rays from the recurrent nova RS Ophiuchi up to a month after its 2021 outburst, using the High Energy Stereoscopic System. The VHE emission has a similar temporal profile to lower-energy GeV emission, indicating a common origin, with a 2-day delay in peak flux. These observations constrain models of time-dependent particle energization, favoring a hadronic emission scenario over the leptonic alternative. This confirms that shocks in dense winds provide favorable environments for efficient cosmic-ray acceleration to very high energies.

Cosmic rays with energies up to a few PeV are known to be accelerated within the Milky Way. Traditionally, it has been presumed that supernova remnants were the main source of these very-high-energy cosmic rays, but theoretically it is difficult to accelerate protons to PeV energies, and observationally there simply is no evidence of the remnants being sources of hadrons with energies above a few tens of TeV. One possible source of protons with those energies is the galactic center region.

Abeysekara et al. (2021) reported observations of 1–100 TeV $\gamma$ rays coming from the Cygnus Cocoon, which is a superbubble that surrounds a region of massive star formation. These $\gamma$ rays are likely produced by 10–1000 TeV freshly accelerated cosmic rays that originate from the enclosed star-forming region Cyg OB2. Until now, it was not known that such regions could accelerate particles to these energies. The measured flux likely originates from hadronic interactions. The spectral shape and the emission profile of the Cocoon changes from GeV to TeV energies, which reveals the transport of cosmic particles and historical activity in the superbubble.

The isotopes $^{2}$H and $^{3}$He in cosmic rays are generally believed to be of secondary origin, resulting mainly from the nuclear interactions of primary cosmic-ray protons and $^{4}$He with the interstellar medium. The spectral shape and composition of the secondary isotopes are therefore completely determined by the source spectrum of the parent elements and by the propagation process. Measurements of the secondary isotope spectra are then a powerful tool to constrain the parameters of the galactic propagation models. Adriani et al. (2016) reported the cosmic-ray hydrogen and helium ($^{1}$H, $^{2}$H, $^{3}$He, $^{4}$He) isotopic composition has been measured with the satellite-borne experiment PAMELA, which was launched into low-Earth orbit on board the Resurs-DK1 satellite on 15 June 2006. The isotopic composition was measured between 100 and 1100 MeV/n for hydrogen and between 100 and 1400 MeV/n for helium isotopes using two different detector systems over the 23rd solar minimum from July 2006 to December 2007.

Since August 2012, Voyager 1 has been observing the local interstellar energy spectra of galactic cosmic-ray nuclei down to 3 MeV $nuc^{-1}$ and electrons down to 2.7 MeV. According to Cummings et al. (2016), the H and He spectra have the same energy dependence between 3 and 346 MeV $nuc^{-1}$, with a broad maximum in the 10–50 MeV $nuc^{-1}$ range and a H/He ratio of 12.2 ± 0.9. The peak H intensity is ~15 times that observed at 1 AU, and the observed local interstellar gradient of 3–346 MeV H is −0.009 ± 0.055% $AU^{-1}$, consistent with models with no local interstellar gradient. The energy spectrum of electrons ($e^- + e^+$) with 2.7–74 MeV is consistent with $E^{-1.30 \pm 0.05}$ and exceeds the H intensity at energies below ~50 MeV. Propagation model fits to the observed spectra indicate that the energy density of cosmic-ray nuclei with >3 MeV $nuc^{-1}$ and electrons with >3 MeV is 0.83–1.02 eV $cm^{-3}$, and the ionization rate of atomic H is in the range of $1.51–1.64 \times 10^{-17}$ $s^{-1}$. This rate is a factor >10 lower than the ionization rate in diffuse interstellar clouds, suggesting significant spatial inhomogeneity in low-energy cosmic rays or the presence of a suprathermal tail on the energy spectrum at much lower energies. The propagation model fits also provide improved estimates of the elemental abundances in the source of galactic cosmic rays (Cummings et al. 2016).

The least interactive particles ever observed in the Universe are neutrinos. Their neutral charge makes them unaffected by electromagnetism, and their lack of color charge means that they don't participate in the strong force. They are affected by the weak force, but it has an exceedingly short range, roughly the diameter of a proton. With those properties, neutrinos can travel through normal matter more or less undetected. Most astrophysical neutrinos seen so far have come from somewhere beyond our galaxy. Now the IceCube Collaboration has identified a high-energy neutrino flux in the Milky Way that originates predominantly in the galactic plane; see IceCube Collaboration (2023).

The Amundsen–Scott South Pole Station in Antarctica hosts the IceCube Neutrino Observatory. The detector is a cubic kilometer of ice with more than 5000 spherical optical sensors, which hang on cables drilled as deep as 2500 m. Most neutrinos pass through the detector area without interacting, but when one slams into a water molecule, it produces muons and other charged particles. If any of them are energetic enough, they travel faster than light does in the ice, which results in the emission of blue-tinged Cherenkov radiation. By measuring the Cherenkov radiation picked up by the optical sensors and the blue light's spatial pattern, researchers can infer the particle's energy and from what direction it arrived.

Telescope Array Collaboration (2023) reported an extremely energetic cosmic ray observed by a surface detector array. Cosmic rays are energetic charged particles from extraterrestrial sources, with the highest-energy events thought to come from extragalactic sources. Their arrival is infrequent, so detection requires instruments with large collecting areas. In their work, they report the detection of an extremely energetic particle recorded by the surface detector array of the Telescope Array experiment. They calculate the particle's energy as ~40 joules, or exceeding ~240 exa-electron volts (EeV). Its arrival direction points back to a void in the large-scale structure of the Universe. Possible explanations include a large deflection by the foreground magnetic field, an unidentified source in the local extragalactic neighborhood, or an incomplete knowledge of particle physics.

### 1.6.1 Cosmic Ray–Induced Chemistry on Dust Particles

The transformation of the energy dependence of the cosmic ray proton flux in the keV to GeV range is investigated theoretically when penetrating inside molecular clouds by Yeghikyan (2011). He found that the energy losses of the cosmic ray particles by interaction with the matter of the molecular cloud are principally caused by inelastic interaction with electrons. The transformed energy distribution of the incoming protons is determined mainly by the column density of the absorbing medium. In other words, the author has shown that the distribution of the cosmic ray particle flux inside the molecular cloud can be calculated if the energy loss of cosmic ray protons in molecular hydrogen is known.

Tungsten isotopic compositions in stardust SiC grains from the Murchison meteorite: constraints on the s-process in the Hf-Ta-W-Re-Os region is discussed by Ávila et al. (2012). They reported the first tungsten isotopic measurements in stardust silicon carbide (SiC) grains recovered from the Murchison carbonaceous chondrite. The isotopes $^{182,183,184,186}W$ and $^{179,180}Hf$ were measured on both an aggregate (KJB fraction) and single stardust SiC grains (LS+LU fraction) believed to have condensed in the outflows of low-mass carbon-rich asymptotic giant branch (AGB) stars with close-to-solar metallicity. The SiC aggregate shows small deviations from terrestrial (= solar) composition in the $^{182}W/^{184}W$ and $^{183}W/^{184}W$ ratios, with deficits in $^{182}W$ and $^{183}W$ with respect to $^{184}W$. The $^{186}W/^{184}W$ ratio, however, shows no apparent deviation from the solar value. Tungsten isotopic measurements in single mainstream stardust SiC grains revealed lower than solar $^{182}W/^{184}W$, $^{183}W/^{184}W$, and $^{186}W/^{184}W$ ratios. The authors compared the SiC data with theoretical predictions of the evolution of W isotopic ratios in the envelopes of AGB stars. These ratios are affected by the slow neutron-capture process and match the SiC data regarding their $^{182}W/^{184}W$, $^{183}W/^{184}W$, and $^{179}Hf/^{180}Hf$ isotopic compositions, although a small adjustment in the s-process production of $^{183}W$ is needed in order to have a better agreement between the SiC data and model predictions. The models cannot explain the $^{186}W/^{184}W$ ratios observed in the SiC grains, even when the current $^{185}W$ neutron-capture cross section is increased by a factor of two. Further study is required to better assess how model uncertainties (e.g., the formation of the $^{13}C$ neutron source, the mass-loss law, the modeling of the third dredge-up, and the efficiency of the $^{22}Ne$ neutron source) may affect current s-process predictions.

Previously, Avila et al. (2008) reported the first tungsten isotopic measurements in stardust silicon carbide (SiC) grains recovered from the Murchison carbonaceous chondrite. The s-process nucleosynthesis in the Hf-Ta-W-Re-Os path has received considerable attention lately. New neutron capture cross sections for $^{174,176,177,178,179,180,182}Hf$, $^{185}W$ and $^{186,187,188}Os$ have been reported, and small anomalies in W and Os isotopes have been observed in primitive meteorites. However, as suggested by, model calculations for s-process nucleosynthesis appear to underestimate $^{182}W$ and overestimate $^{186}Os$, and this may have implications for $^{182}Hf$–$^{182}W$ and $^{187}Re$–$^{187}Os$ chronometers. Tungsten isotopes are particularly important because they are affected by several branching points ($^{182}Ta$, $^{181,182}Hf$, and $^{185}W$), which also affect Re and Os isotopes. Here we report, for the first time, W isotopic measurements in presolar SiC grains in order to provide additional constraints on s-process nucleosynthesis. $^{182,183,184,186}W$ and $^{180}Hf$ were measured with SHRIMP RG at ANU in an aggregate of presolar SiC grains (KJB fraction) extracted from the Murchison meteorite. Tungsten isotopes were measured as WO+ ions, which have a higher yield than the atomic species ($WO^+/W^+ \sim 3$). An Oprimary ion beam of 5 nA was focused to sputter an area of 20 μm in diameter. SHRIMP RG was operated at a mass resolving power of $m/\Delta m = 5000$ (at 1% peak). At this level, isobaric interferences were well resolved from the WO+ species. NISTsilicate glasses and synthetic SiC were used to monitor instrumental mass fractionation and isobaric interferences.

The W isotopic compositions are anomalous in comparison to those observed in normal solar system materials. The SiC grains appear to be enriched in $^{182}W$ and $^{184}W$ relative to $^{183}W$, as expected for s-process nucleosynthesis in AGB stars. However, an unexpected enrichment in $^{186}W$ is observed. The low $^{180}Hf/^{183}W$ ratios determined here imply a low contribution from radiogenic $^{182}W$ after SiC condensation; otherwise, the $^{182}W$ excesses would be even higher. The observed enrichment in $^{186}W$ requires the activation of the $^{185}W$ branching point during AGB thermal pulses, when marginal activation of the $^{22}Ne(\alpha,n)^{25}Mg$ source produces neutron densities as high as $N_n = 5 \times 10^9$ neutrons $cm^{-3}$ [9], bypassing $^{186}Os$. This result is in disagreement with $^{96}Zr$ depletions in SiC grains that indicate that the $^{22}Ne(\alpha,n)^{25}Mg$ source was weak in their parent stars. Production and destruction of W isotopes by cosmic rays still need to be investigated, especially for samples, such as presolar grains, exposed for a long time to galactic cosmic radiation.

### 1.6.2 Cosmic Ray–Induced Physics Processes

Cosmic rays play a decisive role within our own galaxy. They provide partial pressure support against gravity, they trace past energetic events such as supernovae, and they reveal the underlying structure of the baryonic matter distribution through their interactions. To study the impact of CRs on galaxy and cosmic structure formation and evolution, Enßlin et al. (2007) developed an approximative framework for treating dynamical and radiative effects of CRs in cosmological simulations. Their guiding principle is to try to find a balance between capturing as many physical properties of CR populations as possible while at the same time requiring the fewest extra computational resources possible. They approximated the CR spectrum of each fluid element by a single power-law, with spatially and temporally varying normalization, low-energy cut-off, and spectral index. Principles of conservation of particle number, energy, and pressure are then used to derive evolution equations for the basic variables describing the CR spectrum, both due to adiabatic and non-adiabatic processes. The processes considered include compression and rarefaction, CR injection via shocks in supernova remnants, injection in structure formation shock waves, in-situ re-acceleration of CRs, CR spatial diffusion, CR energy losses due to Coulomb interactions, ionization losses, Bremsstrahlung losses, and, finally, hadronic interactions with the background gas, including the associated γ-ray and radio emission due to subsequent pion decay. They showed that the formalism reproduces CR energy densities, pressure, and cooling rates with an accuracy of ~10% in steady state conditions where CR injection balances cooling. It is therefore a promising formulation to allow simulations where CR physics is included. Their framework seems to be well suited to be included in numerical simulation schemes of galaxy and structure formation.

Energetic cosmic rays are a major component of astrophysical plasmas next to magnetic fields, radiation, and thermal gas. Cosmic rays are usually sub-dominant in density but carry as much pressure as the thermal plasma background. In some cases, cosmic rays drift at faster speeds with respect to the normal modes' phase speeds of the background plasma. Because of this, cosmic rays are a strong source of free energy that causes new classes of kinetic or convective instabilities. Recent years have seen the development of intense analytical and numerical efforts to analyze the onset of an instability produced by the motion of these particles at fast bulk speeds: this is the streaming instability. The streaming instability has been applied to different space plasmas and astrophysical contexts like strong shocks, jets, or interstellar and intergalactic medium studies. Streaming instabilities participate in the production of magnetic turbulence at scales corresponding to the gyroradius of the particles. By scattering off their self-generated waves, cosmic rays are coupled to the background thermal plasma. This mechanism is able to self-confine cosmic rays around sources and launch winds out of the disk of the galaxy, hence impacting galactic matter dynamics and ultimately the galactic star formation rate (Marcowith et al. 2021).

The streaming instability has a resonant branch and a non-resonant branch, which have been both subjects of strong research activities during recent years in various scientific contexts: CR acceleration at space plasma and astrophysical shocks and CR propagation from their sources and in the interstellar/intergalactic gas. As GeV or multi-GeV CRs can trigger these instabilities, they should also participate in a feedback process induced by CRs over multi-scale dynamical processes in our galaxy, from molecular cloud collapse to the production of galactic winds or in galaxy clusters. The inclusion of CR streaming and the damping of self-generated turbulence as a strong heating process in fluid models seems important to account for the feedback processes necessary to properly account for the galactic star formation rate. This should pass by theoretical and numerical studies of the impact of CR streaming over thermal instability and Parker instability. In another context, the study of the dynamics of relativistic electron–positron beams in jets issued from compact objects is also an opportunity to investigate the possible role of the SI in mediating turbulence production and particle stochastic acceleration. Microscopic kinetic processes associated with non-thermal components are now entering into the game of the multi-scale physics of astrophysical complex systems (Marcowith et al. 2021).

Cosmic rays are able to heat interstellar dust grains. This may enhance molecule mobility in icy mantles that have accumulated on the grains in dark cloud cores. A three-phase astrochemical model was used by Kalvāns (2014) to investigate molecule mobility in interstellar ices. Specifically, diffusion through pores in ice between the subsurface mantle and outer surface, assisted by whole-grain heating, was considered. It was found that the pores could serve as an efficient transport route for light species. The diffusion of chemical radicals from the mantle to the outer surface are most effective. These species accumulate in the mantle because of photodissociation by the cosmic-ray induced photons. The faster diffusion of hydrogen within the warm ice enhances the hydrogenation of radicals on pore surfaces. The overall result of the whole grain heating-induced radial diffusion in ice are higher abundances of ice species, whose synthesis involves light radicals. Examples of stable species synthesized this way include complex organic molecules, OCS, $H_2O_2$, and cyanoplyynes (Kalvāns 2014).

Atomic and molecular species in the gas phase found in dense interstellar environments such as molecular cloud cores can actively condense on carbonaceous or silicate grains. If grain sizes are large enough (≥0.03 $\mu$m) to prevent stochastic impulse heating induced by energetic agents, for instance, ultraviolet photons or cosmic ray ions, the condensed species can turn into ice mantles on bare grain surfaces. The ice mantles would normally stay on grain surfaces in cold environments since efficient thermal desorption processes are possible only for sufficiently high ambient medium temperatures (≥20 K). The gas-phase chemistry cannot simply explain molecular species that have been observed in the gas phase in dense clump structures and molecular cloud cores. Therefore, non-thermal desorption mechanisms are needed to explain the observed gas-phase abundances of some molecular species in these environments with typical conditions of $T \leq 20$ K, $n_H \geq 10^4$ cm$^{-3}$, $AV \geq 10$ mag.

In molecular cloud cores, the cosmic ray–induced sputtering via CR ion-icy grain collision is one of the desorption processes for ice molecules from mantles around dust grains. The efficiency of this process depends on the incident CR ion properties as well as the physicochemical character of the ice mantle. Arslan et al.'s (2022) main objective is the examination of the sputtering efficiency for $H_2O$ and CO ices found in molecular cloud cores. In the calculation, the authors considered a multi-dimensional parameter space that consists of 30 CR ion types, five different CR ion energy flux distributions, two separate ice mantle components (pure H2O and CO), three ice formation states, and two sputtering regimes (linear and quadratic). They found that the sputtering behavior of $H_2O$ and CO ices is dominated by the quadratic regime rather than the linear regime, especially for CO sputtering. The sputtering rate coefficients for $H_2O$ and CO ices show distinct variations with respect to the adopted CR ion energy flux as well as the grain size–dependent mantle depth. The maximum radius of the cylindrical latent region is quite sensitive to the effective electronic stopping power. The track radii for CO ice are much bigger than $H_2O$ ice values. In contrast to the $H_2O$ mantle, even relatively light CR ions ($Z \geq 4$) may lead to a track formation within the CO mantle.

Cosmic rays are believed to play an important role in the dynamics of the interstellar medium. Using data from the Voyager and Pioneer spacecraft at distances up to 60 AU from the Sun to help define the effects of solar modulation, Webber (1998) made an estimate of the local interstellar cosmic-ray spectra and the associated energy density and ionization rate. The energy density, including the low-energy electron contribution, is ~1.80 eV cm$^{-3}$, equivalent energetically to that in a 7 kG magnetic field. The ionization rate from this same cosmic-ray population, including a significant contribution from low-energy electrons, is (3–4) × $10^{-17}$ s$^{-1}$ (H atom). This is most likely a minimum value and is not too different from estimates of the typical ionization rate required to heat the interstellar medium and maintain the diffuse ionized gas range from (3–30) × $10^{-17}$ s$^{-1}$ (H atom) as a function of locale in the galaxy. This suggests that cosmic rays could be a significant contributor to interstellar heating. The cosmic-ray ionization rate will be more or less uniform throughout the galactic disk and also perpendicular to the disk according to the scale height of cosmic rays, which may be ~1 kpc, thus providing a source of ionization at high Z values. The recent observation of low-energy particle acceleration at the heliospheric termination shock, which produces local ionization rates comparable to those due to galactic cosmic rays, raises the interesting possibility that "hot spots" where the ionization rate due to locally accelerated particles may exceed $10^{-16}$ s$^{-1}$ (H atom) may occur near massive stars with more powerful stellar winds.

### 1.6.3 Cosmic Ray–Induced Biological Processes

Space travel has advanced significantly over the last six decades, with astronauts spending up to 6 months at the International Space Station. Nonetheless, the living environment while in outer space is extremely challenging to astronauts. In particular, exposure to space radiation represents a serious potential long-term threat to the health of astronauts because the amount of radiation exposure accumulates during their time in space. Therefore, health risks associated with exposure to space radiation are an important topic in space travel, and characterizing space radiation in detail is essential for improving the safety of space missions. Furukawa et al. (2020) provided an overview of the space radiation environment and briefly present current and future endeavors that monitor different space radiation environments. Afterwards, they present research evaluating adverse biological effects caused by exposure to various space radiation environments and how these can be reduced. They especially consider the deleterious effects on cellular DNA and how cells activate DNA repair mechanisms. The latest technologies being developed, for example, a fluorescent ubiquitination-based cell cycle indicator to measure real-time cell cycle progression and DNA damage caused by exposure to ultraviolet radiation, are presented. Progress in examining the combined effects of microgravity and radiation to animals and plants are summarized, and our current understanding of the relationship between psychological stress and radiation is presented. They also provide details about protective agents and the study of organisms that are highly resistant to radiation and how their biological mechanisms may aid developing novel technologies that alleviate biological damage caused by radiation. Future research that furthers our understanding of the effects of space radiation on human health will facilitate risk-mitigating strategies to enable long-term space and planetary exploration.

In the hypothesis of exploration and settlement on the other planets, it is necessary to make long-duration space flights safer for astronauts. A detrimental effect often observed is indeed a state of frailty such as cardiovascular and cognitive disorders. The presence of these typical features of aging is very intriguing, and the cellular pathways involved in both cases are investigated not only to minimize the health risk of space missions but also to bring insight on the age-related mechanisms. While being aware that astronauts are subjected to numerous stress factors, one should also be concerned by the effects induced by the heavy ion component of cosmic rays, which, thanks to their high linear energy transfer (LET), are able to induce biological effects that persist over time and are transmitted through the tissue.

The increasing duration of space missions involves a progressively higher exposure of astronauts to cosmic rays, whose most hazardous component is made up of high-atomic-number and high-energy (HZE) ions. HZE ions interact along their tracks with biological molecules, inducing changes on living material qualitatively different from that observed after irradiation for therapeutic purposes or following nuclear accidents. HZE ions trigger in cells different responses initialized by DNA damage and mitochondria dysregulation, which cause a prolonged state of sterile inflammation in the tissues. These cellular phenomena may explain why spending time in space was found to cause the onset of a series of diseases normally related to aging. These changes that mimic aging but take place more quickly also make space flights an opportunity to study the mechanisms underlying aging. In a short review, Giovanetti et al. (2020) describe the biological mechanisms underlying cell senescence and aging; the peculiar characteristics of HZE ions, their interaction with living matter and the effects on the organism; and the key role of mitochondria in HZE ion-induced health effects and aging-related phenomena.

Galactic cosmic rays pose a serious threat to astronauts' health during deep space missions. The possible functional alterations of the central nervous system (CNS) under GCR exposure can be critical for mission success. According to Kokhan and Dobynde (2023), despite the obvious negative effects of ionizing radiation, a number of neutral or even positive effects of GCR irradiation on CNS functions were revealed in ground-based experiments with rodents and primates. Their article is focused on the GCR exposure effects on emotional state and cognition, emphasizing positive effects and their potential mechanisms. They integrate these data with GCR effects on adult neurogenesis and pathological protein aggregation, forming a complete picture, and conclude that GCR exposure causes multidirectional effects on cognition, which may be associated with emotional state alterations. However, the irradiation in space-related doses either has no effect or has performance-enhancing effects in solving high-level cognition tasks and tasks with a high level of motivation.

Atri (2016) discussed the possibility of galactic cosmic ray-induced radiolysis-powered life in subsurface environments in the Universe. Photosynthesis is a mechanism developed by terrestrial life to utilize the energy from photons of solar origin for biological use. Subsurface regions are isolated from the photosphere and consequently are incapable of utilizing this energy. This opens up the opportunity for life to evolve alternative mechanisms for harvesting available energy. Bacterium *Candidatus Desulforudis audaxviator*, found 2.8 km deep in a South African mine, harvests energy from radiolysis, induced by particles emitted from radioactive U, Th, and K present in surrounding rock. Another radiation source in the subsurface environments is secondary particles generated by galactic cosmic rays. Using Monte Carlo simulations, it is shown that it is a steady source of energy comparable to that produced by radioactive substances, and the possibility of a slow metabolizing life flourishing on it cannot be ruled out. Two mechanisms are proposed through which GCR-induced secondary particles can be utilized for biological use in subsurface environments: (1) GCRs injecting energy in the environment through particle-induced radiolysis and (2) organic synthesis from GCR secondaries interacting with the medium. Laboratory experiments to test these hypotheses are also proposed. Implications of these mechanisms on finding life in the solar system and elsewhere in the Universe are discussed.

## 1.7 STARS

As asymptotic giant branch stars age over $10^4$–$10^5$ year, they eject over half or more of their mass in slow winds, and then, in a short 100–1000-year period, are transformed into planetary nebulae (PN) with a dazzling variety of morphologies and widespread presence of point symmetry. Recent morphological studies with the Hubble Space Telescope support the idea that directed, high-speed outflows initiated during the very late AGB phase play a crucial role in the transformation to PN. However, evidence for such jets and disks in AGB stars is scant; this phase is presumably so short that few nearby stars are likely to be caught in the act. The carbon star V Hya (distance ~400 pc; GAIA (eDR3) parallax = 2.31 ± 0.108 mas $yr^{-1}$) is the nearest and best example of this phase and a key object for understanding the formation of bipolar and multipolar PN.

Sahai et al. (2022) have observed the mass-losing carbon star V Hya that is apparently transitioning from an asymptotic giant branch star to a bipolar planetary nebula, at an unprecedented angular resolution of ~0 4–0 6 with the Atacama Large Millimeter/Submillimeter Array. Their $^{13}CO$ and $^{12}CO$ (J = 3–2 and J = 2–1) images have led to the discovery of a remarkable set of six expanding rings within a flared, warped disk structure undergoing dynamical expansion (DUDE) that lies in the system's equatorial plane. They also found, for the first time, several bipolar, high-velocity outflows, some of which have parabolic morphologies, implying wide opening angles, while one (found previously) is clumpy and highly collimated. The latter is likely associated with the high-velocity bullet-like ejections of ionized gas from V Hya; a possible molecular counterpart to the oldest of the four bullets can be seen in the $^{12}CO$ images. They found a bright, unresolved central source of continuum emission (FWHM size ≤165 au); about 40% of this emission can be produced in a standard radio photosphere, while the remaining 60% is likely due to thermal emission from very large (millimeter-sized) grains, with mass $\geq 10^{-5}$ $M_{sun}$. They used a radiative transfer model to fit the salient characteristics of the DUDE's $^{13}CO$ and $^{12}CO$ emission out to a radius of 8″ (3200 au) with a flared disk of mass $1.7 \times 10^{-3}$ $M_{sun}$, whose expansion velocity increases very rapidly with the radius inside a central region of size ~200 au and then more slowly outside it, from 9.5 to 11.5 km $s^{-1}$. The DUDE's underlying density decreases radially, interspersed with local increases that represent the observationally well-characterized innermost three rings.

Welch et al. (2022) report a finding of a highly magnified star at redshift 6.2. Galaxy clusters magnify background objects through strong gravitational lensing. Typical magnifications for lensed galaxies are factors of a few but can also be as high as tens or hundreds, stretching galaxies into giant arcs. Individual stars can attain even higher magnifications given fortuitous alignment with the lensing cluster. Recently, several individual stars at redshift z ~ 1–1.5 have been discovered, magnified by factors of thousands, temporarily boosted by microlensing. The authors reported observations of a more distant and persistent magnified star at redshift $z_{phot} = 6.2 \pm 0.1$, $900 \times 10^6$ years after the Big Bang. This star is magnified by a factor of thousands by the foreground galaxy cluster lens WHL0137–08 (z = 0.566), as estimated by four independent lens models. Unlike previous lensed stars, the magnification and observed brightness (AB mag 27.2) have remained roughly constant over 3.5 years of imaging and follow-up. The delensed absolute UV magnitude MUV = $-10 \pm 2$ is consistent with a star of mass M > $50M_{\odot}$. Confirmation and spectral classification are forthcoming from approved observations with the James Webb Space Telescope.

Solar flares are often accompanied by filament/prominence eruptions (~$10^4$ K and ~$10^{10-11}$ $cm^{-3}$), sometimes leading to coronal mass ejections that directly affect the Earth's environment. "Superflares" are found on some active solar-type (G-type main-sequence) stars, but the filament eruption–coronal mass ejection association has not been established. Namekata et al. (2021) showed that optical spectroscopic observation of the young solar-type star EK Draconis reveals evidence for a stellar filament eruption associated with a superflare. This superflare emitted a radiated energy of $2.0 \times 10^{33}$ erg, and a blue-shifted hydrogen absorption component with a high velocity of −510 km $s^{-1}$ was observed shortly afterwards. The temporal changes in the spectra strongly resemble those of solar filament eruptions. Comparing this eruption with solar filament eruptions in terms of the length scale and velocity strongly suggests that a stellar coronal mass ejection occurred. The erupted filament mass of $1.1 \times 10^{18}$ g is ten times larger than those of the largest solar coronal mass ejections. The massive filament eruption and an associated coronal mass ejection provide the opportunity to evaluate how they affect the environment of young exoplanets/the young Earth and stellar mass/angular momentum evolution.

Stars form in galaxies from gas that has been accreted from the intergalactic medium. Simulations have shown that recycling of gas—the reaccretion of gas that was previously ejected from a galaxy—could sustain star formation in the early Universe. Zhang et al. (2023) observed the gas surrounding a massive galaxy at redshift 2.3 and detected emission lines from neutral hydrogen, helium, and ionized carbon that extend 100 kiloparsecs from the galaxy. The kinematics of this circumgalactic gas is consistent with an inspiraling stream. The carbon abundance indicates that the gas had already been enriched with elements heavier than helium, previously ejected from a galaxy. They interpreted the results as evidence of gas recycling during high-redshift galaxy assembly.

In unveiling the nature of the first stars, the main astronomical clue is the elemental compositions of the second generation of stars, observed as extremely metal-poor (EMP) stars, in the Milky Way. However, no observational constraint was available on their multiplicity, which is crucial for understanding early phases of galaxy formation. Hartwig et al. (2023) developed a new data-driven method to classify observed EMP stars into mono- or multi-enriched stars with support vector machines. They also use their own nucleosynthesis yields of core-collapse supernovae with mixing fallback that can explain many of the observed EMP stars. Their method predicts, for the first time, that 31.8% ± 2.3% of 462 analyzed EMP stars are classified as mono-enriched. This means that the majority of EMP stars are likely multi-enriched, suggesting that the first stars were born in small clusters. Lower-metallicity stars are more likely to be enriched by a single supernova, most of which have high carbon enhancement. They also found that Fe, Mg, Ca, and C were the most informative elements for this classification. In addition, oxygen was found to be very informative despite its low observability. Their data-driven method sheds new light on solving the mystery of the first stars from the complex data set of galactic archeology surveys.

### 1.7.1 Star Formation

Understanding the star formation rate dependence on time is of the utmost importance for the understanding of when the life in Universe could have originated (Livio 2007). Plots of the star formation rate as a function of the redshift have been produced by many authors. They all agree that the SFR rises from the present to about $z \approx 1$. However, Madau and Dickinson (2014) reviewed the range of experimental techniques and theoretical tools that allow astronomers to map the cosmic history of star formation, heavy element production, and reionization of the Universe from the cosmic "dark ages" to the present epoch. A consistent picture emerged, whereby the SFR density peaked approximately $3.5 \times 10^9$ years after the Big Bang, at $z \approx 1.9$, and declined exponentially at later times, with an e-folding timescale of $3.9 \times 10^9$ years. Half of the stellar mass observed today was formed before a redshift $z = 1.3$. About 25% formed before the peak of the cosmic SFR density, and another 25% formed after $z = 0.7$. Less than ~1% of today's stars formed during the epoch of reionization. Under the assumption of a universal initial mass function, the global stellar mass density inferred at any epoch matches reasonably well the time integral of all the preceding star-formation activity. The comoving rates of star formation and central black hole accretion follow a similar rise and fall, offering evidence for coevolution of black holes and their host galaxies. The rise of the mean metallicity of the Universe to about 0.001 solar by $z = 6$, 1 Gyr after the Big Bang, appears to have been accompanied by the production of fewer than ten hydrogen Lyman-continuum photons per baryon, a rather tight budget for cosmological reionization.

The cosmic history of star formation is one of the most fundamental observables in astrophysical cosmology. Under the simple assumption of a universal IMF, there is reasonable agreement between the global SMD inferred at any particular time and the time integral of all the preceding instantaneous star-formation activity, although modest offsets may still point toward systematic uncertainties. A consistent picture is emerging, whereby the SFRD peaked ~3.5 Gyr after the Big Bang and dropped exponentially at $z < 1$ with an e-folding timescale of 3.9 Gyr. The Universe was a much more active place in the past: Stars formed at a peak rate approximately nine times higher than is seen today. Approximately 25% of the present-day SMD formed at $z > 2$, before the peak of the SFRD, and another 25% formed since $z = 0.7$, that is, roughly over the last half of the Universe's age. From the peak of the SFRD at $z \approx 2$ to the present day, and perhaps earlier as well, most stars formed in galaxies that obey a relatively tight SFR-M∗ correlation, and only a small fraction formed in starbursts with significantly elevated specific SFRs. The smooth evolution of this dominant main-sequence galaxy population suggests that the evolution of the cosmic SFH is primarily determined by a balance between gas accretion and feedback processes, both closely related to galaxy mass, and that stochastic events such as merger-driven starbursts play a relatively minor role. The growth histories of the stellar component of galaxies and their central black holes are similar in shape, suggesting broad coevolution of black holes and their host galaxies. The rise of the mean metallicity of the Universe to 0.001 solar by redshift six, 1 Gyr after the Big Bang, appears to have been accompanied by the production of fewer than ten hydrogen LyC photons per baryon, indicating a rather tight budget for cosmological reionization. The SFRD at $z \approx 7$ was approximately the same as that of today, at $z \approx 0$, but only 1% of today's SMD was formed during the epoch of reionization, after Madau and Dickinson (2014).

A knowledge of the SFR as a function of redshift allows for meaningful constraints to be placed on the global emergence of carbon-based life for the first time. The main contributors of carbon to the interstellar medium are intermediate-mass (1–8 $M_{sun}$) stars through the asymptotic giant branch and planetary nebulae phases. A knowledge of the cosmic SFR history, together with a knowledge of the initial stellar mass function, therefore allows for an approximate calculation of the rate of carbon production as a function of redshift. The peak in the carbon production rate is somewhat delayed (by $\leq 1 \times 10^9$ years) with respect to the SFR peak. The decline in the carbon production rate is also shallower than the decline in the SFR for $z \leq 1$ because of the buildup of a stellar reservoir in the earlier epochs.

One would expect the emergence of most carbon-based life in the Universe to be perhaps not too far from the peak in the carbon production rate, around $z \approx 1$ (for a peak in the SFR at $z \approx 1$–2). Other considerations, related to the production of radioactive elements, lead to similar conclusions (Hogan 2000). The time scale required to develop intelligent civilizations may be within a factor 2 of the lifetime of F and K stars (one possessing continuously habitable zones), see Kasting et al. (1993)

Interstellar clouds provide the raw material, gas and dust, for the formation of stars and planetary systems (Boss 2004). A low-mass star is formed after a protostar with an accretion disk, and strong mass outflow is formed because of the gravitational collapse of a molecular cloud on a time scale of $10^4$–$10^5$ years. When a star is still in the earliest stages of formation, it does not have enough temperature in its core to ignite fusion of hydrogen and helium. Instead, the star shines with just the gravitational energy of its continue collapse. Astronomers call this pre-star a T Tauri star. Only after $10^6$–$10^7$ years does nuclear fusion start and the new star reach the main sequence, becoming a true star.

Most T Tauri stars have circumstellar disks (masses $10^{-3}$–$10^{-1}$ solar masses) of sizes in the region of 100–400 au. These disks are reservoirs of gas and dust for the formation of planetary systems.

The first generation of stellar objects, called Population III (Pop III), are massive stars born from a gas of pristine composition consisting almost exclusively of hydrogen and helium. If the initial mass of the Pop III stars extended to masses as low as ~1 $M_{sun}$, their lifetime would be as long as the age of the galaxy, and they would survive to be observed in the present galaxy (Yoshii et al. 2022). However, no single star without detectable metals has been found anywhere in the galaxy yet.

On the other hand, massive Pop II stars (mass ≥ 10 $M_{sun}$), which cause explosive nucleosynthesis, are short lived and should have already become SNe long before the Milky Way galaxy formed. Yoshii et al. 2022 studied the distant quasar J1342+0928 at $z = 7.54$, deduced the Fe and Mg abundances, and found it extremely enriched by a factor of 20 relative to the solar Fe abundances. They report a very low Mg/Fe: $[\mathrm{Mg/Fe}] = -1.11 \pm 0.12$ together with $[\mathrm{Fe/H}] = +1.36 \pm 0.19$, only 700 million years after the Big Bang.

Yoshii et al. (2022) suggested that such an unusual abundance feature could not be explained by the standard view of chemical evolution that considers only the contributions from canonical supernovae. They proposed that the large amount of iron was supplied by pair-instability supernova (PISN) caused by the explosion of a massive (100–300 $M_{sun}$) Pop III star. The upper limit of the mass range for the first stars is not known; however, the possibility of 500–1000 $M_{sun}$ has been considered by several authors. For example, Ohkubo et al. (2006) reported explosive nucleosynthesis of Pop III

SNe with 500 and 1000 $M_{sun}$ which could be a possible candidate for the iron supply to J1342 + 0928.

The properties of star formation are well defined and require environments characterized by a sufficiently low temperature (<20 K) and a high gas density (Hsieh et al. 2021). In contrast, the temperature of gaseous-dusty filaments in the galactic center exceeds a gas temperature of ~6000 K and a dust temperature of ~250 K (Moser et al. 2017). Furthermore, the velocity dispersion of all known objects close to Sgr A* outpaces typical numerical values in star formation regions by several magnitudes (Genzel et al. 2000). Due to the presence of the 4 × 106–Me supermassive black hole Sgr A*, tidal forces hinder gas clumping, which is necessary for the formation of stars. However, Yusef-Zadeh et al. (2013) found, at a distance of about 0.6 pc from Sgr A*, SiO clumps that imply high-mass star formation. The authors suggest that the observed clumps are an indication of in situ star formation that took place during the last $10^4$–$10^5$ yr. In a subsequent analysis, ALMA observations showed high-velocity clumps in the inner parsec but also within the circumnuclear disk (CND) directed toward Sgr A* (Hsieh et al. 2021). Some of these dense clumps meet the conditions necessary for star formation. In addition, Jalali et al. (2014) propose that the interplay between Sgr A* and inspiraling clumps of several 10 solar masses could trigger star formation.

The proposed observation of young stellar objects (YSOs) in the galactic center still raises the question of where and how these objects could have formed due to the violent vicinity of Sgr A*. Peißker et al. (2023) reported the multiwavelength detection of a highly dynamic YSO close to Sgr A* that might be a member of the IRS 13 cluster. They observed the previously known coreless bow-shock source X3 in the near- and mid-infrared (NIR/MIR) with SINFONI (VLT), NACO (VLT), ISAAC (VLT), VISIR (VLT), SHARP (NTT), and NIRCAM2 (KECK). In the radio domain, they used CO continuum and H30α ALMA observations to identify system components at different temperatures and locations concerning the central stellar source. It is suggested that these radio/submillimeter observations in combination with the NIR Brγ line can be associated with a protoplanetary disk of the YSO, which is consistent with manifold VISIR observations that reveal complex molecules and elements such as PAH, S IV, Ne II, and Ar III in a dense and compact region. Based on the photometric multiwavelength analysis, the authors inferred the mass of $15{+}10^{-5}$ $M_{sun}$ for the YSO with a related age of a few $10^4$ yr. Due to this age estimate and the required relaxation timescales for high-mass stars, this finding is an indication of ongoing star formation in the inner parsec. The proper motion and 3D distance imply a relation between X3 and IRS 13. It is argued that IRS 13 may serve as a birthplace for young stars that are ejected due to the evaporation of the cluster.

There is broad consensus that star formation and its quenching, as well as gas outflows from galaxies, are regulated by energy and momentum injection from young massive stars and feedback from supermassive black holes. Details and the relative role of different feedback processes, however, are still debated. In particular, cosmic rays accelerated at strong shocks formed by stellar winds and supernova explosions that accompany star formation have been the focus of recent research. CRs play a key role in regulating thermal balance in dense molecular clouds and setting conditions for star formation. Due to their long cooling times, CRs can significantly prolong the Sedov–Taylor stages of SN remnants, leading to larger momentum injected into the ISM. On larger scales, CRs can be one of the important drivers of galactic winds as suggested by a number of analytical models. Observations of the γ-ray emission around star clusters, isolated supernova remnants, and pulsar wind nebulae indicate that the cosmic-ray diffusion coefficient near acceleration sites can be suppressed by a large factor compared to the galaxy average. Semenov et al. (2021) explored the effects of such local suppression of CR diffusion on galaxy evolution using simulations of isolated disk galaxies with regular and high gas fractions. The results obtained show that while CR propagation with constant diffusivity can make gaseous disks more stable by increasing the midplane pressure, large-scale CR pressure gradients cannot prevent local fragmentation when the disk is unstable. In contrast, when CR diffusivity is suppressed in star-forming regions, the accumulation of CRs in these regions results in strong local pressure gradients that prevent the formation of massive gaseous clumps. As a result, the distribution of dense gas and star formation changes qualitatively: a globally unstable gaseous disk does not violently fragment into massive star-forming clumps but maintains a regular grand-design spiral structure. This effect regulates star formation and disk structure and is qualitatively different from and complementary to the global role of CRs in vertical hydrostatic support of the gaseous disk and in driving galactic winds.

Black-hole–driven outflows have been observed in some dwarf galaxies with active galactic nuclei and probably play a role in heating and expelling gas (thereby suppressing star formation), as they do in larger galaxies. The extent to which black-hole outflows can trigger star formation in dwarf galaxies is unclear, because work in this area has previously focused on massive galaxies, and the observational evidence is scarce. Henize 2–10 is a dwarf starburst galaxy previously reported to have a central massive black hole, although that interpretation has been disputed because some aspects of the observational evidence are also consistent with a supernova remnant. At a distance of approximately 9 Mpc, it presents an opportunity to resolve the central region and to determine if there is evidence for a black-hole outflow influencing star formation. Schutte and Reine (2022) reported optical observations of Henize 2–10 with a linear resolution of a few parsecs. We find an approximately 150-pc-long ionized filament connecting the region of the black hole with a site of recent star formation. Spectroscopy reveals a sinusoid-like position–velocity structure that is well described by a simple precessing bipolar outflow. They concluded that this black-hole outflow triggered the star formation.

Katsianis et al. (2017) investigated the evolution of the galaxy star formation rate function (SFRF) and cosmic star formation rate density (CSFRD) of $z \sim 0–8$ galaxies in the Evolution and Assembly of GaLaxies and their Environments (EAGLE) simulations. In addition, they presented a compilation of ultraviolet, infrared, and H α SFRFs and compared these with the predictions from the EAGLE suite of cosmological hydrodynamic simulations. They found that the constraints implied by different indicators are inconsistent with each other for the highest star-forming objects at $z < 2$, a problem that is possibly related to selection biases and the uncertainties of dust attenuation effects. EAGLE's feedback parameters were calibrated to reproduce realistic galaxy sizes and stellar masses at $z = 0.1$. In their work, they tested if and why those choices yield realistic star formation rates for $z \sim 0–8$ as well. The demonstrated that supernovae feedback plays a major role at setting the abundance of galaxies at all star-forming regimes, especially at high redshifts. On the contrary, active galactic nuclei feedback becomes more prominent at lower redshifts and is a major mechanism that affects only the highest star-forming systems. Furthermore, they found that galaxies with SFR ~ (1–10) $M_{sun}\ yr^{-1}$ dominate the CSFRD at redshifts $z \leq 5$, while rare high star-forming galaxies (SFR ~ (10–100) $M_{sun}\ yr^{-1}$ contribute significantly only briefly around the peak era ($z \sim 2$) and then are quenched by AGN feedback. In the absence of this prescription objects with SFR ~ (10–100) $M_{sun}\ yr^{-1}$ would dominate the CSFRD, while the cosmic budget of star formation would be extremely high. Finally, we demonstrate that the majority of the cosmic star formation occurs in relatively rare high-mass haloes (MHalo ~ $10^{11-13}\ M_{sun}$) even at the earliest epochs.

Yan et al. (2022) presented a new method for probing the star formation history of the Universe, tomographic cross-correlation between the cosmic infrared background (CIB) and galaxy samples. The galaxy samples are from the Kilo-Degree Survey (KiDS), while the CIB maps are made from Planck sky maps at 353, 545, and 857 GHz. They measured the cross-correlation in harmonic space within $100 < \ell < 2000$ with a significance of 43σ. They modeled the cross-correlation with a halo model, which links CIB anisotropies to star formation rates and galaxy abundance. They assumed that the SFR has a lognormal dependence on halo mass and that the galaxy abundance follows the halo occupation distribution (HOD) model. The cross-correlations give a best-fit maximum star formation efficiency of $\eta_{max} = 0.41^{+0.09}_{-0.14}$ at a halo mass $\log_{10} (M_{peak}/M_{sun}) = 12.14 \pm 0.36$. The derived star formation rate density is well constrained up to $z \sim 1.5$. The constraining power at high redshift is mainly limited by the KiDS survey depth. They also showed that the constraint is robust to uncertainties in the estimated redshift distributions of the galaxy sample. A combination with external SFRD measurements from previous studies gives $\log_{10} (M_{peak}/M_{sun}) = 12.42^{+0.35}_{-0.19}$. This tightens the SFRD constraint up to $z = 4$, yielding a peak SFRD of $0.09^{+0.003}_{-0.004}\ M_{sun}\ year^{-1} M_{sun}\ pc^{-3}$ at $z = 1.74^{+0.06}_{-0.02}$, corresponding to a lookback time of $10.05^{+0.12}_{-0.03} \times 10^9$ years. Both constraints are consistent, and the derived SFRD agrees with previous studies and simulations. This validates the use of CIB tomography as an independent probe of the star formation history of the Universe. Additionally, we estimate the galaxy bias, b, of KiDS galaxies from the constrained HOD parameters and obtain an increasing bias from $b = 1.1^{+0.17}_{-0.31}$ at $z = 0$ to $b = 1.96^{+0.18}_{-0.64}$ at $z = 1.5$, which highlights the potential of this method as a probe of galaxy abundance. They also provided a forecast for future galaxy surveys and conclude that, due to their considerable depth, future surveys will yield a much tighter constraint on the evolution of the SFRD.

McLeod et al. (2023) reported a probable Keplerian disk feeding an optically revealed massive young star. The canonical picture of star formation involves disk-mediated accretion, with Keplerian accretion disks and associated bipolar jets primarily observed in nearby, low-mass young stellar objects. Recently, rotating gaseous structures and Keplerian disks have been detected around several massive ($M > 8\ M_{\odot}$) YSOs (MYSOs), including several disk-jet systems. All the known MYSO systems are in the Milky Way, and all are embedded in their natal material. The authors reported the detection of a rotating gaseous structure around an extragalactic MYSO in the Large Magellanic Cloud. The gas motion indicates that there is a radial flow of material falling from larger scales onto a central disk-like structure. The latter exhibits signs of Keplerian rotation so that there is a rotating toroid feeding an accretion disk and thus the growth of the central star. The system is in almost all aspects comparable to Milky Way high-mass YSOs accreting gas from a Keplerian disk. The key difference between this source and its galactic counterparts is that it is optically revealed rather than being deeply embedded in its natal material, as is expected of such a massive young star. The authors suggest that this is the consequence of the star having formed in a low-metallicity and low-dust content environment. Thus, these results provide important constraints for models of the formation and evolution of massive stars and their circumstellar disks.

The tremendous tidal force that is linked to the supermassive black hole at the center of our galaxy is expected to strongly subdue star formation in its vicinity. Stars within 1" from the SMBH thus likely formed further from the SMBH and migrated to their current positions. In a study by Nishiyama et al. (2024), spectroscopic observations of the star S0–6/S10, one of the closest (projected distance from the SMBH of ~0".3) late-type stars were conducted. Using metal absorption lines in the spectra of S0–6, the radial velocity of S0–6 from 2014 to 2021 was measured, and a marginal acceleration was detected, which indicated that S0–6 is close to the SMBH. The S0–6 spectra were employed to determine its stellar parameters including temperature, chemical abundances ([M/H], [Fe/H], [He/Fe], [Ca/Fe], [Mg/Fe], [Ti/Fe]), and age. As suggested by the results of this study, S0–6 is very old ($\geq 10 \times 10^9$ years) and has an origin different from that of stars born in the central pc region.

There is no proof that black holes contain singularities when real physical bodies generate them. Penrose (1965) claimed 60 years ago that trapped surfaces inevitably lead to light rays of finite affine length (FALLs). Hawking (1972) then asserted that these must end in actual singularities. When they could not prove this, they decreed it to be self evident. It is shown that there are counterexamples through every point in the Kerr metric. These are asymptotic to at least one event horizon and do not end in singularities.

### 1.7.2 Pulsars

Although neutron stars were long predicted in astrophysical theory, it wasn't until 1967 that the first was discovered, as a pulsar, by Dame Jocelyn Bell Burnell. Since then, we know of hundreds more. Professor Dame Jocelyn Bell Burnell discovered pulsars while she was a postgraduate student at New Hall (now Murray Edwards College) carrying out research at Cambridge's Cavendish Laboratory with Antony Hewish. The Nobel Prize for Physics was awarded for the discovery of pulsars to Antony Hewish and Martin Ryle in 1974.

Fast radio bursts are mysterious millisecond-duration transients prevalent in the radio sky. Rapid accumulation of data in recent years has facilitated an understanding of the underlying physical mechanisms of these events. Knowledge gained from the neighboring fields of gamma-ray bursts and radio pulsars also offered insight. Zhang (2020) recently reviewed developments in this fast-moving field. Two generic categories of radiation model invoking either magnetospheres of compact objects (neutron stars or black holes) or relativistic shocks launched from such objects have been much debated. The recent detection of a galactic fast radio burst in association with a soft gamma-ray repeater suggests that magnetar engines can produce at least some, and probably all, fast radio bursts. Other engines that could produce fast radio bursts are not required but are also not impossible. Long-duration gamma-ray bursts are produced during the collapse of massive stars. It is assumed that they carry information about the properties of progenitor stars. Gottlieb et al. (2022) were the first to present 3D general-relativity-magneto hydrodynamic simulations spanning six orders of magnitude in the space-time continuum. They argue that numerical tools are essential for modeling the jet evolution before and after it breaks out from the star.

For example, the CHIME/FRB Collaboration detected a persistent radio signal from a faraway galaxy (named FRB 20191221A) that appears to flash with surprising regularity. The signal could come from either a radio pulsar or a magnetar, both of which are types of extremely dense, rapidly spinning collapsed cores of giant stars—neutron stars.

In the work reported by Niu et al. (2022) the dispersive sweep of fast radio bursts (FRBs) has been used to probe the ionized baryon content of the intergalactic medium, which is assumed to dominate the total extragalactic dispersion. Although the host-galaxy contributions to the dispersion measure appeared to be small for most FRBs, in at least one case there was evidence for an extreme magneto-ionic local environment and a compact persistent radio source. Niu et al. (2022) reported the detection and localization of the repeating FRB 20190520B, which is co-located with a compact, persistent radio source and associated with a dwarf host galaxy of high specific-star-formation rate at a redshift of $0.241 \pm 0.001$. The estimated host-galaxy dispersion measure of proximately $903^{+72}_{-111}$ parsecs/cm$^3$, which is nearly an order of magnitude higher than the average of FRB host galaxies, far exceeds the dispersion-measure contribution of the intergalactic medium.

Radio pulsars are rotation-powered neutron stars, which emit coherent beams of radio emission generated by highly relativistic particles in regions above their magnetic poles. Their known spin periods (P) range from 1.4 ms to 23.5 s, and they are divided into various sub-classes (e.g. rotating radio transients, millisecond pulsars, magnetars)—depending on their observational properties. Particle acceleration and abundant electron–positron pair production is postulated to be an essential condition for the coherent radio emission from pulsars, with the particle acceleration potential expected to be lower for larger spin periods. As seen in most neutron stars, the radio emission is also expected to be strongly inhibited or cease if the magnetic field configuration and strength, exceed the quantum critical field ($B_{cr} = 4.413 \times 10^{13}$ G). Caleb et al. (2022) presented the discovery of a highly magnetized 75.88 s period radio-emitting neutron star, PSR J0901–4046, which challenges these conditions for, and the nature of, the radio emission and raises questions about the spin evolution of neutron stars in general.

Usually, the radio-emitting neutron star population encompasses objects with spin periods ranging from milliseconds to tens of seconds. As they age and spin more slowly, their radio emission is expected to cease. Caleb et al. (2022) discovered an ultra-long period radio-emitting neutron star, PSR J0901–4046, with spin properties distinct from the known spin and magnetic-decay powered neutron stars. With a spin period of 75.88 s, a characteristic age of 5.3 Myr, and a narrow pulse duty-cycle, it is uncertain how radio emission is generated and challenges our current understanding of how these systems evolve. The radio emission has unique spectro-temporal properties such as quasi-periodicity and partial nulling that provide important clues to the emission mechanism. Detecting similar sources is observationally challenging, which implies a larger undetected population. Their discovery establishes the existence of ultra-long period neutron stars, suggesting a possible connection to the evolution of highly magnetized neutron stars, ultra-long period magnetars, and fast radio bursts.

Chirenti et al. (2023) reported kilohertz quasiperiodic oscillations in short gamma-ray bursts. Short gamma-ray bursts (GRBs) are associated with binary neutron star mergers, which are multimessenger astronomical events that have been observed both in gravitational waves and in the multiband electromagnetic spectrum. Depending on the masses of the stars in the binary and on details of their largely unknown equation of state, a dynamically evolving and short-lived neutron star may be formed after the merger, existing for approximately 10–300 ms before collapsing to a black hole. Numerical relativity simulations across different groups consistently show broad power spectral features in the 1–5-kHz range in the post-merger gravitational-wave signal, which is inaccessible by current gravitational-wave detectors but could be seen by future third-generation ground-based detectors in the next decade. This implies the possibility of quasiperiodic modulation of the emitted gamma rays in a subset of events in which a neutron star is formed shortly before the final collapse to a black hole. In their paper, Chirenti et al. (2023) presented two such signals identified in the short bursts GRB 910711 and GRB 931101B from archival Burst and Transient Source Experiment (BATSE) data, which are compatible with the predictions from numerical relativity.

Gamma-ray bursts are flashes of high-energy radiation arising from energetic cosmic explosions. Bursts of long (greater than 2 s) duration are produced by the core collapse of massive stars and those of short (less than 2 s) duration by the merger of compact objects, such as two neutron stars. A third class of events with hybrid high-energy properties was identified but never conclusively linked to a stellar progenitor. The lack of bright supernovae rules out typical core-collapse explosions, but their distance scales prevent sensitive searches for direct signatures of a progenitor system. Troja et al. (2022) reported observations of the exceptionally bright GRB 211211A that classify it as a hybrid event and constrain its distance scale to only 346 megaparsecs. Our measurements indicate that its lower-energy (from ultraviolet to near infrared) counterpart is powered by a luminous (approximately $10^{42}$ erg per second) kilonova possibly formed in the ejecta of a compact object merger.

Gamma-ray bursts are divided into two populations: long GRBs that derive from the core collapse of massive stars and short GRBs that form in the merger of two compact objects. Although it is common to divide the two populations at a gamma-ray duration of 2 s, classification based on duration does not always map to the progenitor. Notably, GRBs with short ($\lesssim$2 s) spikes of prompt gamma-ray emission followed by prolonged, spectrally softer extended emission (EE-SGRBs) have been suggested to arise from compact object mergers. Compact object mergers are of great astrophysical importance as the only confirmed site of rapid neutron capture (*r*-process) nucleosynthesis, observed in the form of so-called kilonovae. Rastinejad et al. (2022) reported the discovery of a possible kilonova associated with the nearby (350 Mpc) minute-duration GRB 211211A. The kilonova implies that the progenitor is a compact object merger, suggesting that GRBs with long, complex light curves can be spawned from merger events. The kilonova of GRB 211211A has a similar luminosity, duration, and color to that which accompanied the gravitational wave–detected binary neutron star (BNS) merger GW170817.

Fast radio bursts are extragalactic radio flashes of unknown physical origin. Their high luminosities and short durations require extreme energy densities, such as those found in the vicinity of neutron stars and black holes. Studying the burst intensities and polarimetric properties on a wide range of timescales, from milliseconds down to nanoseconds, is key to understanding the emission mechanism. However, high-time-resolution studies of FRBs are limited by their unpredictable activity levels, available instrumentation, and temporal broadening in the intervening ionized medium. Nimmo et al. (2022) showed that the repeating FRB 20200120E can produce isolated shots of emission as short as about 60 nanoseconds in duration, with brightness temperatures as high as $3 \times 10^{41}$ K (excluding relativistic effects), comparable with "nano-shots" from the Crab pulsar. Comparing both the range of timescales and luminosities, they found that FRB 20200120E observationally bridges the gap between known galactic young pulsars and magnetars and the much more distant extragalactic FRBs. This suggests a common magnetically powered emission mechanism spanning many orders of magnitude in timescale and luminosity. In their article, they probe a relatively unexplored region of the short-duration transient phase space; they highlight that there probably exists a population of ultrafast radio transients at nanosecond to microsecond timescales, to which current FRB searches are insensitive.

The LOw Frequency ARray (LOFAR) (see Van Haarlem et al. 2013) Two-Meter Sky Survey (LoTSS) (see Shimwell et al. 2017) is one of several ongoing very wide area deep radio wavelength sky surveys. Second data release has been recently published (Shimwell et al. 2022), This data release from the ongoing LOFAR Two-Meter Sky Survey presents 120–168 MHz images covering 27% of the northern sky. Coverage is split into two regions centered at approximately 12h45m +44°30′ and 1h00m +28°00′ and spanning 4178 and 1457 square degrees, respectively. The images were derived from 3451 h (7.6 PB) of LOFAR high band antenna data which were corrected for direction-independent instrumental properties as well as direction-dependent ionospheric distortions during extensive, but fully automated, data processing. A catalogue of 4,396,228 radio sources is derived from total intensity maps, where the majority of these have never been detected at radio wavelengths before. At 6″ resolution, the full bandwidth continuum maps with a central frequency of 144 MHz have a median rms sensitivity of 83 μJy beam$^{-1}$, a flux density scale accuracy of approximately 10%, and an astrometric accuracy of 0.2″, and we estimate the point-source completeness to be 90% at a peak brightness of 0.8 mJy beam$^{-1}$. By creating three 16-MHz bandwidth images across the band, they were able to measure the in-band spectral index of many sources, albeit with an error on the derived spectral index of >±0.2, which is a consequence of the flux-density scale accuracy and small fractional bandwidth.

Recent observations of the radio sky by Sweijen et al. (2022) show that the vast majority of sources detected at 144 MHz are unresolved at the typical resolution of a few arcseconds, demonstrating the need for sub-arcsecond-resolution surveys to make detailed studies. At low radio frequencies, high spatial resolution is challenged by the ionosphere and by the propagation delay of radio waves that it induces. If not adequately corrected for, this blurs the images to arcsecond or even arcminute scales. In addition, the required image size to map the degree-scale field of view of low-frequency radio telescopes at sub-arcsecond resolution is far greater than what typical software and hardware are currently capable of handling. The authors presented deep degree-scale sub-arcsecond imaging at low radio frequencies. They derived ionospheric corrections in 44 directions on individual sources with compact sub-arcsecond structures. This has yielded a sensitive 6.6 deg$^2$ 144 MHz map with a resolution of $0.38'' \times 0.30''$ and a sensitivity of 25 μJy per beam near the phase center. This allowed mapping of the entire northern low-frequency sky at sub-arcsecond resolution.

Wang et al. (2021b) reported the discovery of a highly polarized, highly variable, steep-spectrum radio source, ASKAP J173608.2–321635, located ~4° from the galactic center in the galactic plane. The source was detected six times between January and September 2020 as part of the Australian Square Kilometre Array Pathfinder Variables and Slow Transients (ASKAP VAST) survey at 888 MHz. It exhibited a high degree (~25%) of circular polarization when it was visible. They monitored the source with the MeerKAT telescope from November 2020 to February 2021 on a 2–4 week cadence. The source was not detected with MeerKAT before 7 February 2021 when it appeared and reached a peak flux density of 5.6 mJy. The source was still highly circularly polarized, but also showed up to 80% linear polarization and then faded rapidly with a timescale of 1 day. The rotation measure of the source varied significantly, from $-11.8 \pm 0.8$ to $-64.0 \pm 1.5$ rad m–2 over 3 days. No X-ray counterpart was found in follow-up Swift or Chandra observations about a week after the first MeerKAT detection, with upper limits of ~5.0 × 1031 erg s–1 (0.3–8 keV, assuming a distance ~10 kpc). No counterpart is seen in new or archival near-infrared observations down to J = 20.8 mag. The authors discuss possible identifications for ASKAP J173608.2–321635, including a low-mass star/substellar object with extremely low infrared luminosity, a pulsar with scatter-broadened pulses, a transient magnetar, or a galactic center radio transient: none of these fully explains the observations, which suggests that ASKAP J173608.2–321635 may represent part of a new class of objects being discovered through radio imaging surveys.

Observations of binary pulsars and pulsars in globular clusters suggest that at least some pulsars must receive weak natal kicks at birth. If all pulsars received strong natal kicks above 50 km s$^{-1}$, those born in globular clusters would predominantly escape, while wide binaries would be disrupted. On the other hand, observations of transverse velocities of isolated radio pulsars indicate that only $5 \pm 2\%$ have velocities below 50 km s$^{-1}$. Willcox et al. (2021) explored this apparent tension with rapid binary population synthesis modeling. They proposed a model in which supernovae with characteristically low natal kicks (e.g., electron-capture supernovae) only occur if the progenitor star has been stripped via binary interaction with a companion. They showed that this model naturally reproduces the observed pulsar speed distribution without reducing the predicted merging double neutron star yield. The authors estimated that the zero-age main-sequence mass range for noninteracting progenitors of electron-capture supernovae should be no wider than $\approx 0.2\ M_\odot$.

The high-frequency radio sky is bursting with synchrotron transients from massive stellar explosions and accretion events, but the low-frequency radio sky has, so far, been quiet beyond the galactic pulsar population and the long-term scintillation of active galactic nuclei. The low-frequency band, however, is sensitive to exotic coherent and polarized radio-emission processes, such as electron-cyclotron maser emission from faring M dwarfs, stellar magnetospheric plasma interactions with exoplanets, and a population of steep-spectrum pulsars, making galactic-plane searches a prospect for blind-transient discovery. Here we report an analysis of archival low-frequency radio data that reveals a periodic, low-frequency radio transient. Hurley-Walker et al. (2022) found that the source pulses every 18.18 min, an unusual periodicity that has not been observed previously. The emission is highly linearly polarized, bright, persists for 30–60 s on each occurrence, and is visible across a broad frequency range. At times, the pulses comprise short-duration (<0.5 s) bursts; at others, a smoother profile is observed. These profiles evolve on timescales of hours. By measuring the dispersion of the radio pulses with respect to frequency, they have localized the source to within our own galaxy and suggest that it could be an ultra-long-period magnetar.

Martijn et al. (2022) discovered Alcyoneus, a low-excitation radio galaxy with a projected proper length $l_p = 4.99 \pm 0.04$ Mpc. Its jets and lobes are all four detected at very high significance, and the SDSS-based identification of the host, at spectroscopic redshift $z_{spec} = 0.24674 \pm 6{\cdot}10^{-5}$, is unambiguous. The total luminosity density at ν = 144 MHz is $L_\nu = 8 \pm 1{\cdot}10^{25}$ W $Hz^{-1}$, which is below average, though near median (percentile 45 ± 3%), for GRGs. The host is an elliptical galaxy with a stellar mass $M_{star} = 2.4 \pm 0.4 \cdot 10^{11}$ $M_{sun}$ and a supermassive black hole mass $M_{star} = 4 \pm 2{\cdot}10^8 M_{sun}$, both of which tend towards the lower end of their respective GRG distributions (percentiles 25 ± 9% and 23 ± 11%). The host resides in a filament of the Cosmic Web. Through a new Bayesian model for radio galaxy lobes in three dimensions, the authors estimated the pressures in the $Mpc^3$-scale northern and southern lobe to be $P_{min,1} = 4.8 \pm 0.3{\cdot}10^{-16}$ Pa and $P_{min,2} = 4.9 \pm 0.6{\cdot}10^{-16}$ Pa, respectively. The corresponding magnetic field strengths are $B_{min,1} = 46 \pm 1$ pT and $B_{min,2} = 46 \pm 3$ pT.

Over a dozen millisecond pulsars are ablating low-mass companions in close binary systems. In the original "black widow", the 8-hour orbital period eclipsing pulsar PSR J1959 + 2048 (PSR B1957 + 20), high-energy emission originating from the pulsar is irradiating and may eventually destroy a low-mass companion. These systems are not only physical laboratories that reveal the interesting results of exposing a close companion star to the relativistic energy output of a pulsar but are also believed to harbor some of the most massive neutron stars, allowing for robust tests of the neutron star equation of state. Burdge et al. (2022) reported observations of ZTF J1406 + 1222, a wide hierarchical triple hosting a 62-minute orbital period black widow candidate, the optical flux of which varies by a factor of more than ten. ZTF J1406 + 1222 pushes the boundaries of evolutionary models, falling below the 80-minute minimum orbital period of hydrogen-rich systems. The wide tertiary companion is a rare low-metallicity cool subdwarf star, and the system has a galactic halo orbit consistent with passing near the galactic center, making it a probe of formation channels, neutron star kick physics, and binary evolution.

Odd radio circles (ORCs) are recently discovered faint diffuse circles of radio emission of unknown cause, surrounding galaxies at moderate redshift ($z \sim 0.2$—$0.6$). Norris et al. (2022) presented detailed new MeerKAT radio images at 1284 MHz of the first ORC, originally discovered with the Australian Square Kilometre Array Pathfinder, with higher resolution (6 arcsec) and sensitivity (~ 2.4 μJy/beam). In addition to the new images, which reveal a complex internal structure consisting of multiple arcs, they also presented polarization and spectral index maps. Based on these new data, we consider potential mechanisms that may generate the ORCs.

### 1.7.3 Neutron Stars

Because neutron stars are so dense, they have intense gravitational and magnetic fields. The gravity of a neutron star is about a thousand billion times stronger than that of the Earth. Thus, the surface of a neutron star is exceedingly smooth; gravity does not permit anything tall to exist. Neutron stars may have "mountains", but they are only inches tall. Neutron stars with abnormally strong magnetic fields are known as magnetars. The origin of these abnormal stars with ultra-powerful magnetic fields is unknown.

Neutron stars are the collapsed cores of formerly massive stars that have been crushed to an extreme density by supernova explosions. A neutron star is not as dense as a black hole, but it's denser than any other known type of star.

Neutron stars rotate extremely rapidly, and we can use the radio beams of a pulsar to measure just how fast. The fastest rotating neutron star discovered rotates 716 times per second, Hessels et al. (2006), about a quarter of the speed of light.

Magnetic fields on the neutron star focus radio waves into two beams fired into space from its magnetic poles. If the neutron star is oriented precisely so that these beams become visible from our earthly viewpoint, we see flashes of radio light at regular and extremely exact intervals. Although neutron stars were long predicted in astrophysical theory, it wasn't until 1967 that the first was discovered, as a pulsar; see Bell Burnell (1969). Since then, at least hundreds more have been discovered A neutron star does not generate any light or heat of its own after its formation. Over millions of years its latent heat will gradually cool from an initial 600,000 K, eventually ending its life as the cold, dead remnant of a star.

Magnetars are young neutron stars with high magnetic fields that are usually observed at X-ray wavelengths. Isolated neutron stars (NSs) with extremely strong magnetic fields are referred to as magnetars.

There are about 30 confirmed magnetars known, many of which are detectable only during periods of enhanced activity. Magnetar emission is powered by the magnetic field, producing bursts of hard (≈10 to 100 keV) X-rays, with luminosity $L \approx 10^{38}$ to $10^{47}$ erg $s^{-1}$ and duration ≈ 0.1 to 100 s. Magnetars also exhibit persistent X-ray emission at $L \approx 10^{33}$ to $10^{35}$ erg $s^{-1}$, which is pulsed at spin frequencies $f \approx 0.1$ to 10 Hz with spin-down rates $f \approx$ ($10^{16}$ to $10^{8}$) Hz. These properties indicate high magnetic fields $B \lesssim 10^{15}$ G.

The emission mechanism and geometry of the emitting region have been unclear. Taverna et al. (2022) measured the X-ray polarization of the magnetar 4U 0142 + 61. The polarization degree and angle change as a function of X-ray energy, indicating two different emission regions. The authors preferred a model in which most of the X-rays are emitted by an equatorial band on the surface of the neutron star, with some of the photons then being scattered to higher energies by collisions with electrons in the surrounding magnetic field.

Neutron stars are extreme astrophysical objects whose interiors may contain exotic new forms of matter. It appears that the structure and size of neutron stars are linked to the thickness of the neutron skin in atomic nuclei via the neutron-matter equation of state, Horowitz and Piekarewicz (2001). It turns out that the nucleus $^{208}Pb$ is an attractive target for studying this link both experimentally and theoretically owing to the large excess of neutrons and its simple structure.

Magnetars are strongly magnetized, isolated neutron stars with magnetic fields up to around $10^{15}$ gauss, luminosities of approximately $10^{31}$–$10^{36}$ erg $s^{-1}$, and rotation periods of about 0.3–12.0 s. Very energetic giant flares from galactic magnetars (peak luminosities of $10^{44}$–$10^{47}$ erg $s^{-1}$, lasting approximately 0.1 s) have been detected in hard X-rays and soft γ-rays, and only one has been detected from outside our galaxy. During such giant flares, quasi-periodic oscillations with low (less than 150 Hz) and high (greater than 500 hertz) frequencies have been observed, but their statistical significance has been questioned. High-frequency QPOs have been seen only during the tail phase of the flare. Castro-Tirado et al. (2021) reported the observation of two broad QPOs at approximately 2,132 and 4,250 Hz in the main peak of a giant γ-ray flare in the direction of the NGC 253 galaxy, disappearing after 3.5 ms. The flare was detected on 15 April 2020 by the Atmosphere–Space Interactions Monitor instrument aboard the International Space Station, which was the only instrument that recorded the main burst phase (0.8–3.2 ms) in the full energy range ($50 \times 10^{3}$ to $40 \times 10^{6}$ eV) without suffering from saturation effects such as deadtime and pile-up. Along with sudden spectral variations, these extremely high-frequency oscillations in the burst peak are a crucial component that will aid our understanding of magnetar giant flares.

Hu et al. (2022) developed a unified ab initio framework, using nuclear forces from chiral effective field theory (EFT) to link the physics of nucleon-nucleon scattering and few-nucleon systems to properties of heavy-mass nuclei up to $^{208}Pb$ and ultimately to the nuclear-matter equation of state near saturation density. According to the authors, information contained in nucleon-nucleon scattering significantly constrains the properties of neutron matter. This same information constrains neutron skins, which provide a non-trivial empirical check on the reliability of ab initio predictions for the neutron-matter equation of state.

Collision of two neutron stars provides a burst of gravitational waves. This event is relatively rare. In addition, analysis of the debris showed elements as strontium and zirconium as well as the very heavy lanthanide group of elements such as cerium and neodymium.

Optical superluminal motion measurement in neutron star merger GW170817 is reported by Mooley et al. (2022). They reported report on superluminal motion measurement, at seven times the speed of light, leveraging Hubble Space Telescope precision astrometry and previous radio VLBI data for GW170817. They thereby obtained a measurement of the Lorentz factor of the wing of the structured jet, as well as substantially improved constraints on the viewing angle (19–25 degrees) and the initial Lorentz factor of the jet core (more than 40).

The only way to directly measure the magnetic field near the surface of a neutron star is by detection of cyclotron resonance scattering features (CRSF) (Kong et al. 2022). The authors measured energy of 146 keV during the 2017. outburst of pulsar Swift J0243.6 + 6124, close to its peak luminosity of $\sim 2 \times 10^{39}$ erg $s^{-1}$. They also estimated a surface magnetic field of $\sim 1.6 \times 10^{13}$ G due to the presence of multipole field components close to the surface of the neutron star.

Jordana-Mitjans et al. (2022) observed a short gamma-ray burst from a protomagnetar remnant by the contemporaneous detection of gravitational waves and gamma rays from GW170817/GRB 170817A, followed by kilonova emission a day after. Their observations confirmed compact binary neutron star mergers as progenitors of short duration gamma-ray bursts and cosmic sources of heavy r-process nuclei. However, the nature (and life span) of the merger remnant and the energy reservoir powering these bright gamma-ray flashes remains debated, while the first minutes after the merger are unexplored at optical wavelengths. In the article, the authors reported the earliest discovery of bright thermal optical emission associated with short GRB 180618A with extended gamma-ray emission—with ultraviolet and optical multicolor observations starting as soon as 1.4 minutes post-burst. The spectrum is consistent with a fast-fading afterglow and emerging thermal optical emission 15 minutes post-burst, which fades abruptly and chromatically (flux density $F_\nu \propto t^{-\alpha}$, $\alpha = 4.6 \pm 0.3$) just 35 minutes after the GRB. Their observations from gamma rays to optical wavelengths are consistent with a hot nebula expanding at relativistic speeds, powered by the plasma winds from a newborn, rapidly spinning, and highly magnetized neutron star (i.e., a millisecond magnetar), whose rotational energy is released at a rate $L_{th} \propto t^{-(2.22 \pm 0.14)}$ to reheat the unbound merger-remnant material. These results suggest that such neutron stars can survive the collapse to a black hole on timescales much larger than a few hundred milliseconds after the merger and power the GRB itself through accretion.

Neutron stars are "cosmic zombies"—corpses of massive stars that collapsed in violent explosions after running out of fuel. By studying these ultradense objects, researchers hope to understand the behavior of matter under extreme conditions, which might help in cracking some of the biggest mysteries in physics. As announced at a press conference during the 2021 APS April meeting, NASA's Neutron star Interior Composition Explorer, an X-ray telescope on the International Space Station, has measured the size of the heaviest known neutron star. The surprisingly large radius measured for this star implies a stiffer-than-expected state of matter in the core, disfavoring models that predict a "squishy" center.

Neutron stars are the densest observable objects in the cosmos—packing twice the mass of the Sun in a sphere as wide as a large city. In the outer core of the star, the large pressure breaks up nuclei into nucleons and crushes protons and electrons together, leaving behind a sea of mostly neutrons. Researchers are unsure, however, about what happens in the inner core of the star. Do neutrons persist or decompose into their quark constituents? Do these quarks interact to form exotic particles?

Since no laboratory experiment can reproduce neutron star conditions, the only option for studying this exceptional state of matter is to observe neutron stars themselves—inferring what is going on inside the star from basic properties such as mass and size. Such measurements, however, are not easily done. Up until today, about 2000 neutron stars have been discovered, but only a handful of them have been sized up, typically by monitoring the X-ray emission of gas surrounding the star.

NICER, an International Space Station payload, has developed a unique sizing method applicable to rapidly rotating neutron stars known as pulsars. As pulsars rotate, hot spots on their surface emit X-rays that scan the cosmos like lighthouse beams. The experiment monitors the pulsar's oscillatory X-ray brightness, "time stamping" the arrival of each X-ray photon with a precision of about 100 ns. The path that these photons take is distorted by the gravitational warping of spacetime around the star, allowing some hot spots to remain visible even as they rotate to the far side of the star.

From the X-ray time-stamped data, the researchers reconstruct the gravitational potential and, in turn, infer the star size. The combination of X-ray spectroscopy capabilities with timing is a unique feature of NICER, which allows researchers to fully exploit information on the star's spin to constrain its properties. The NICER Collaboration first used this method in 2019 to measure PSR J0030, a pulsar 1000 light years from Earth. Weighing 1.4 solar masses, J0030 was found to have a diameter of about 26 km. In the new measurement, the collaboration turned to the most massive known neutron star, PSR J0740. Nearly four times more distant than J0030, J0740 is 20 times fainter and was thus a more difficult experiment. The collaboration tasked the data analysis to two independent teams, who used different assumptions on, for instance, the X-ray background in the sky and the instrument calibration. The teams, led by Anna Watts of the University of Amsterdam and Cole Miller of the University of Maryland in College Park, came up with similar values for the most likely diameter of the star: 25 and 27 km, respectively, both close to that of the previously measured lighter pulsar.

The size measurement will allow researchers to evaluate different options for the star interior. In some models, neutrons break apart into free-roaming quarks, which leads to a squishy, compressible core. These models deliver the counterintuitive prediction that neutron stars should become smaller as their mass increases. Other models, in which some neutrons persist, or the quarks interact strongly, predict a harder-to-compress form of matter.

Pulsars are usually depicted as perfectly symmetric magnetic dipoles, with their hot spots at each of the pulsar's poles. For both stars investigated by NICER, however, the hot spots appear to lie on the same hemisphere, suggesting a much more complex and asymmetric field configuration. Understanding the nature of matter inside these stars will have implications on many fundamental questions, as neutron stars play a role in the synthesis of the heaviest elements, and they are promising targets for even spotting dark matter (Rini 2021).

Nova explosions are caused by global thermonuclear runaways triggered in the surface layers of accreting white dwarfs. It has been predicted that localized thermonuclear bursts on white dwarfs can also take place, similar to type-I X-ray bursts observed in accreting neutron stars. Unexplained rapid bursts from the binary system TV Columbae, in which mass is accreted onto a moderately strong magnetized white dwarf from a low-mass companion, have been observed on several occasions in the past 40 years. During these bursts, the optical/ultraviolet luminosity increases by a factor of more than three in less than an hour and fades in around 10 hours. Fast outflows have been observed in ultraviolet spectral lines, with velocities of more than 3500 kilometers per second, comparable to the escape velocity from the white dwarf surface.

### 1.7.4 Binary Quadruple Star Systems

Binary stars are foundational to modern astrophysics. They underpin precision measurements of stellar structure, age, and composition; they provide stringent tests of general relativity; they make possible the study of faint and rare objects such as black holes and neutron stars; they are the progenitors of gravitational wave events. The components of binaries often interact, dramatically changing their evolution and giving rise to a bewildering zoo of astrophysical phenomenology. Binaries are not rare: about half of all Sun-like stars are binaries, and the binary fraction is even higher for massive stars. To understand stars—especially massive stars—it is thus critical to understand binaries. In the last decade, industrial-scale stellar surveys have provided exquisite spectroscopic, photometric, and astrometric data for hundreds of millions of stars in the Milky Way. The goal of this thesis has been to leverage this deluge of data toward new insight into the formation, evolution, and cosmic importance of binaries. On the one hand, binary population demographics represent the end state of the star formation process and thus inform theoretical star formation models. On the other, binary population demographics are an important input for stellar population models: stars in interacting binaries follow qualitatively different evolutionary histories from single stars of similar mass and composition. An improved census of the binary population thus has broad utility across astrophysics. Among other results, this thesis expands the total number of binaries known by more than a factor of 10. The methods developed here are expected to continue to bear fruit in the coming decades as panchromatic surveys look ever deeper into the galaxy (El-Badry 2021).

A key aspect of the formation of stars and their planets regards the pristine chemical composition of their parental cloud of gas and dust. Various connections between the chemical composition of planet–host stars and planet frequency are observed and proposed, such as primordial abundance anomalies, planet engulfment events, retention of refractory elements by rocky planets and/or cores of giant planets, and due to the formation of Jupiter analogs, which could trap significant quantity of dust exterior to their orbits preventing large amounts of refractory elements from being accreted to the forming star.

Abundances of light elements in dwarf stars of different ages are important constraints for stellar yields, galactic chemical evolution and exoplanet chemical composition studies. Botelho et al. (2020) have measured C and N abundances and $^{12}C/^{13}C$ ratios for a sample of 63 solar twins spanning a wide range in age, based on spectral synthesis of a comprehensive list of CH A-X and CN B-X features using HARPS spectra. The analysis of 55 thin-disk solar twins confirms the dependences of [C/Fe] and [N/Fe] on [Fe/H]. [N/Fe] is investigated as a function of [Fe/H] and age for the first time for these stars. Their derived correlation [C/Fe]–age agrees with works for solar-type stars and solar twins, but the [N/Fe]–age correlation does not. The relations [C, N/Fe]–[Fe/H] and [C, N/Fe]–age for the solar twins lay under solar. $^{12}C/^{13}C$ is found to be correlated with [Fe/H] and seems to have decreased along the evolution of the local thin disk. Predictions from chemical evolution models for the solar vicinity corroborate the relations [C, N/Fe]–[Fe/H], $^{12}C/^{13}C$–age, and [N/O]–[O/H] but do not for the $^{12}C/^{13}C$–[Fe/H] and [C/O]–[O/H] relations. The N/O ratio in the Sun is placed at the high end of the homogeneous distribution of solar twins, which suggests uniformity in the N–O budget for the formation of icy planetesimals, watery superearths, and giant planets. C and N had different nucleosynthetic origins along the thin-disk evolution, as shown by the relations of [C/N], [C/O], and [N/O] against [O/H] and age. [C/N] and [C/O] are particularly observed increasing in time for solar twins younger than the Sun (Botelho et al. 2020).

NASA's Transiting Exoplanet Survey Satellite (TESS) has found over 5000 candidate exoplanet candidates and 197 confirmed exoplanets since its mission began in late 2018. TESS is good at finding exoplanets, but the spacecraft is a powerful scientific platform, and it's made other discoveries, too. Scientists working with TESS recently announced 97 quadruple star candidates, nearly doubling the number of known quadruple systems. Kostov et al. (2022) presented a catalog of 97 uniformly vetted candidates for quadruple star systems. The candidates were identified in TESS full-frame image data from sectors 1–42 through a combination of machine-learning techniques and visual examination, with major contributions from a dedicated group of citizen scientists. All targets exhibit two sets of eclipses with two different periods, both of which pass photocenter tests confirming that the eclipses are on target. This catalog outlines the statistical properties of the sample, nearly doubles the number of known multiply eclipsing quadruple systems, and provides the basis for detailed future studies of individual systems. Several important discoveries have already resulted from this effort, including the first sextuply eclipsing sextuple stellar system and the first transiting circumbinary planet detected from one sector of TESS data.

Hansen (2021) presented a catalog of unbound stellar pairs within 100 pc of the Sun that are undergoing close, hyperbolic encounters. The data are drawn from the GAIA EDR3 catalogue, and the limiting factors are errors in the radial distance and unknown velocities along the line of sight. Such stellar pairs have been suggested to be possible events associated with the migration of technological civilizations between stars. As such, this sample may represent a finite set of targets for a SETI search based on this hypothesis. This catalog contains a total of 132 close passage events, featuring stars from across the entire main sequence, with 16 pairs featuring at least one main sequence star of spectral type between K1 and F3. Many of these stars are also in binaries, so that we isolate eight single stars as the most likely candidates to search for an ongoing migration event—HD 87978, HD 92577, HD 50669, HD 44006, HD 80790, SPM J2126 + 5338, LSPM J0646 + 1829, and HD 192486. Amongst host stars of known planets, the stars GJ 433 and HR 858 are the best candidates.

### 1.7.5 Exploding Stars and Supernovae

Kobayashi et al. (2000) made a prediction of the cosmic supernova rate history as a composite of the supernova rates in spiral and elliptical galaxies. They included the metallicity effect on the evolution of Type Ia supernova (SN Ia) progenitors and construct detailed models for the evolutions of spiral and elliptical galaxies in clusters and the field to meet the latest observational constraints. In the cluster environment, the synthesized cosmic star formation rate has an excess at corresponding to the early starburst in ellipticals $z > 3$ and a shallower slope from the present to the peak at the redshift of $z \sim 1.4$ compared with Madau's plot. In the field environment, we assume that ellipticals form over a wide range of redshifts as $1 < z < 4$. The synthesized cosmic SFR has a broad peak around $z \sim 3$, which is in good agreement with the observed one. The resultant cosmic SFRs lead to the following predictions for the cosmic SN Ia rate: (1) the SN Ia rate in spirals has a break at $z \sim 2$ because of the low-metallicity inhibition of SNe Ia, regardless of whether the galaxies are in clusters or in the field; (2) at high redshifts, the SN Ia rate has a strong peak around $z \sim 3$ in clusters, whereas in the field, a much lower rate is expected, reflecting the difference in the formation epochs of ellipticals.

Chen et al. (2022) reported shock cooling of a red-supergiant supernova at redshift 3 in lensed images. The core-collapse supernova of a massive star rapidly brightens when a shock, produced following the collapse of its core, reaches the stellar surface. As the shock-heated star subsequently expands and cools, its early-time light curve should have a simple dependence on the size of the progenitor and therefore final evolutionary state. Measurements of the radius of the progenitor from early light curves exist for only a small sample of nearby supernovae, and almost all lack constraining ultraviolet observations within a day of explosion. The several-day time delays and magnifying ability of galaxy-scale gravitational lenses, however, should provide a powerful tool for measuring the early light curves of distant supernovae and thereby studying massive stellar populations at high redshift.

Chen et al. (2022) analyzed individual rest-frame exposures in the ultraviolet to the optical taken with the Hubble Space Telescope, which simultaneously capture, in three separate gravitationally lensed images, the early phases of a supernova at redshift $z \approx 3$ beginning within $5.8 \pm 3.1$ hours of explosion. The supernova, seen at a lookback time of approximately $11.5 \times 10^9$ years, is strongly lensed by an early-type galaxy in the Abell 370 cluster. The authors constrained the pre-explosion radius to be $533^{+154}{}_{-119}$ solar radii, consistent with a red supergiant. Highly confined and massive circumstellar material at the same radius can also reproduce the light curve, but because no similar low-redshift examples are known, this was assumed to be unlikely.

From the early radiation of type II-P supernovae (SNe), it has been claimed that the majority of their red supergiant (RSG) progenitors are enshrouded by large amounts of circumstellar material at the point of explosion. The inferred density of this CSM is orders of magnitude above that seen around RSGs in the field and is therefore indicative of a short phase of elevated mass-loss prior to explosion. It is not known over what time-scale this material gets there: is it formed over several decades by a "superwind" with mass-loss rate $\dot{M} \sim 10^{-3} M_{\odot} yr^{-1}$; or is it formed in less than a year by a brief "outburst" with $\dot{M} \sim 10^{-1} M_{\odot} yr^{-1}$? Davies et al. (2022) simulated spectra for RSGs undergoing such mass-loss events and demonstrated that in either scenario, the CSM suppressed the optical flux by over a factor of 100 and that of the near-IR by a factor of 10. They argue that the "superwind" model can be excluded, as it causes the progenitor to be heavily obscured for decades before explosion and is strongly at odds with observations of II-P progenitors taken within 10 yr of core-collapse. Instead, our results favor abrupt outbursts < 1 yr before explosion as the explanation for the early optical radiation of II-P SNe. We therefore predict that RSGs will undergo dramatic photometric variability in the optical and infrared in the weeks to months before core collapse.

Nova explosions are caused by global thermonuclear runaways triggered in the surface layers of accreting white dwarfs. It has been predicted that localized thermonuclear bursts on white dwarfs can also take place, similar to type-I X-ray bursts observed in accreting neutron stars. Unexplained rapid bursts from the binary system TV Columbae, in which mass is accreted onto a moderately strong magnetized white dwarf from a low-mass companion, have been observed on several occasions in the past 40 years, as reported by Scaringi et al. (2022). During these bursts, the optical/ultraviolet luminosity increases by a factor of more than three in less than an hour and fades in around 10 hours. Fast outflows have been observed in ultraviolet spectral lines, with velocities of more than 3,500 kilometers per second, comparable to the escape velocity from the white dwarf surface. Here we report on optical bursts observed in TV Columbae and in two additional accreting systems, EI Ursae Majoris and ASASSN-19bh. The bursts have a total energy of approximately $10^{-6}$ times than those of classical nova explosions (micronovae) and bear a strong resemblance to type-I X-ray bursts. We exclude accretion or stellar magnetic reconnection events as their origin and suggest thermonuclear runaway events in magnetically confined accretion columns as a viable explanation.

Type Ia supernovae (SNe Ia) are thought to result from thermonuclear explosions of white dwarfs in a binary system. Since the peak luminosity in the optical band is almost uniform among the objects, SNe Ia can be utilized as distance indicators in cosmology. SNe Ia also play an important role as major suppliers of iron, contributing to the chemical enrichment of the Milky Way and galaxy clusters. Despite this importance, however, many fundamental aspects of these explosions remain elusive. Despite their importance, a detailed understanding of Type Ia supernovae (SNe Ia) remains elusive. X-ray measurements of the element distributions in supernova remnants (SNRs) offer important clues for understanding the explosion and nucleosynthesis mechanisms for SNe Ia. However, it is challenging to observe the entire ejecta mass in X-rays for young SNRs, because the central ejecta may not have been heated by the reverse shock yet. Fukushima et al. (2020) presented over 200 kilosecond Chandra observations of the Type Ia SNR G344.7–0.1, whose age is old enough for the reverse shock to have reached the SNR center, providing an opportunity to investigate the distribution of the entire ejecta mass. They revealed a clear stratification of heavy elements with a centrally peaked distribution of the Fe ejecta surrounded by intermediate-mass elements (IMEs: Si, S, Ar, Ca) with an arc-like structure. The centroid energy of the Fe K emission is marginally lower in the central Fe-rich region than in the outer IME-rich regions, suggesting that the Fe ejecta were shock-heated more recently. These results are consistent with the prediction for standard SN Ia models, where the heavier elements are synthesized in the interior of an exploding white dwarf. They find, however, that the peak location of the Fe K emission is slightly offset to the west with respect to the geometric center of the SNR. This apparent asymmetry is likely due to the inhomogeneous density distribution of the ambient medium, consistent with our radio observations of the ambient molecular and neutral gas.

In the transitional mass range (~8–10 solar masses) between white dwarf formation and iron core-collapse supernovae, stars are expected to produce an electron-capture supernova. Theoretically, these progenitors are thought to be super-asymptotic giant branch stars with a degenerate O + Ne + Mg core, and electron capture onto Ne and Mg nuclei should initiate core collapse. However, no supernovae have unequivocally been identified from an electron-capture origin, partly because of uncertainty in theoretical predictions. Hiramatsu et al. (2021) presented six indicators of electron-capture supernovae and show that supernova 2018zd is the only known supernova with strong evidence for or consistent with all six: progenitor identification, circumstellar material, chemical composition, explosion energy, light curve, and nucleosynthesis. For supernova 2018zd, they inferred a super-asymptotic giant branch progenitor based on the faint candidate in the pre-explosion images and the chemically enriched circumstellar material revealed by the early ultraviolet colors and flash spectroscopy. The light-curve morphology and nebular emission lines can be explained by the low explosion energy and neutron-rich nucleosynthesis produced in an electron-capture supernova. This identification provides insights into the complex stellar evolution, supernova physics, cosmic nucleosynthesis, and remnant populations in the transitional mass range.

Novae are caused by runaway thermonuclear burning in the hydrogen-rich envelopes of accreting white dwarfs, which leads to a rapid expansion of the envelope and the ejection of most of its mass. Theory has predicted the existence of a "fireball" phase following directly on from the runaway fusion, which should be observable as a short, bright, and soft X-ray flash before the nova becomes visible in the optical. König et al. (2022) reported observations of a bright and soft X-ray flash associated with the classical galactic nova YZ Reticuli 11 h before its 9 mag optical brightening. No X-ray source was detected 4 h before and after the event, constraining the duration of the flash to shorter than 8 h. In agreement with theoretical predictions, the source's spectral shape is consistent with a black-body of $3.27^{+0.11}_{-0.33} \times 10^5$ K ($28.2^{+0.9}_{-2.8}$ eV) or a white dwarf atmosphere, radiating at Eddington luminosity, with a photosphere that is only slightly larger than a typical white dwarf.

Luminous red variables are most likely eruptions that are the outcome of stellar mergers. V838 Mon is one of the best-studied members of this class, representing an archetype for stellar mergers resulting from B-type stars. As result of the merger event, "nova-like" eruptions occur, driving mass loss from the system. As the gas cools, considerable circumstellar dust is formed. V838 Mon erupted in 2002 and is undergoing very dynamic changes in its dust composition, geometry, and infrared luminosity, providing a real-time laboratory to validate mineralogical condensation sequences in stellar mergers and evolutionary scenarios. Woodward et al. (2021) discussed recent NASA Stratospheric Observatory for Infrared Astronomy 5–38 μm observations combined with archival NASA Spitzer spectra that document the temporal evolution of the freshly formed (within the last ≤20 yr) circumstellar material in the environs of V838 Mon. Changes in the 10-μm spectral region are strong evidence that we are witnessing a classical dust condensation sequence expected to occur in oxygen-rich environments, where alumina formation is followed by that of silicates at the temperature cools.

The behavior of massive stars in their final years of evolution is almost entirely unconstrained. However, we can probe these terminal phases of stellar evolution prior to the core-collapse of massive stars >8 $M_{sun}$ by understanding the composition and origin of the high-density, circumstellar material surrounding these stars at the time of explosion. This CSM can be composed of primordial stellar material or elements synthesized during different stages of nuclear burning and is enriched as the progenitor star loses mass via wind and violent outbursts. Early-time optical observations of young ($t <$ 10 days since shock breakout; SBO) Type II supernovae (SNe II) are one such probe of the final stages of stellar evolution. In the era of all-sky transient surveys, rapid ("flash") spectroscopic observations have become a powerful tool for understanding the very nearby circumstellar environment of pre-SN progenitor systems in the final days to months before explosion. Obtaining spectra of young SNe II in the hours to days following SBO allows us to identify prominent emission lines in very early-time SN spectra that result from the recombination of unshocked, photoionized CSM (Jacobson-Galán et al. 2022).

Supernovae are among the most energetic cataclysms in the cosmos, shining for a period of weeks or months, and in some rare cases emitting more light than an entire galaxy. Supernova explosions comprise several types, but the most common occurs at the end of the life of a very large star—one somewhere between 8 and 140 times the mass of the Sun. The star runs out of fuel for the nuclear fusion that had been powering it, leaving behind an inert core of iron and nickel in a state of plasma. The outer layers of the star begin to fall inwards, and the core starts to collapse. In a span of milliseconds, most of the matter in the core gets so compressed that protons and electrons combine to form neutrons. The core's density suddenly rises by several orders of magnitude, because neutrons take up much less space than plasma. The neutrons pack into a denser ball—as dense as the laws of physics permit, forming what Habig calls a proto-neutron star inside the core (Castelvecchi 2022b).

The formation of each neutron releases a neutrino, and so the core's collapse releases a brief initial burst of neutrinos. After falling for thousands of kilometers in an intense gravitational field, the material hits the hard surface of the neutron core, bouncing back with a shock wave that propagates outwards. The shock wave is so violent that the rest of the star disintegrates, leaving only the neutron star as a remnant. During the collapse itself, the energy released by the falling matter smashes elementary particles together, as happens in a high-energy collider, continuously turning energy into particles of all kinds. When a collision produces a neutrino, that particle will have a high chance of escaping into outer space without hitting anything else. As a result, many neutrinos are produced over a period of 10 seconds or more. Researchers estimate that SN 1987A ejected $10^{58}$ of these particles. On these time scales, neutrinos are by far the dominant way in which the supernova dissipates energy. Although the shock wave can take many hours to make it through the outer layers of the star and to become visible, neutrinos come out right away, practically at the speed of light. More than 99% of the energy from a core-collapse supernova escapes not as light but as neutrinos. Eventually, most of the star's original mass disperses into interstellar space. On average, one or two Milky Way stars per century undergo core collapse, according to the most recent estimates. Yet throughout history, only five supernovae have been recorded as being visible with the naked eye, with two thought to be of the core-collapse type. There are various reasons for this discrepancy. If enough mass concentrates in the collapsing core, it forms a black hole without producing much of a light show. In perhaps the majority of cases, an explosion does happen but remains hidden from view by thick interstellar dust in the plane of the Milky Way, where massive stars reside.

Neutrino alert systems are not new: one has existed for nearly two decades. The Supernova Early Warning System (SNEWS) is a network involving Super-K and several other neutrino observatories. It includes IceCube, an array of light sensors embedded in a cubic kilometer of Antarctica's ice, and KM3NeT, a similar array submerged in the Mediterranean Sea. Large neutrino facilities now under construction in the United States and China are expected to join in the next few years, and Japan is building Hyper-Kamiokande, which will be five times larger than Super-K. The main idea of SNEWS is to combine signals to improve the confidence in a detection, even if the individual ones look marginal at best. Each detector runs software that notifies a central SNEWS server of any unusual activity. SNEWS sends an alert to astronomers only if neutrino detectors in two separate geographical areas see a spike in activity within 10 seconds of each other.

Jacobson-Galán et al. (2022) presented panchromatic observations and modeling of supernova 2020tlf, the first normal Type II-P/L SN with confirmed precursor emission, as detected by the Young Supernova Experiment transient survey. Pre-SN activity was detected in riz-bands at −130 days and persisted at relatively constant flux until first light. Soon after discovery, "flash" spectroscopy of SN 2020tlf revealed narrow, symmetric emission lines that resulted from the photoionization of circumstellar material shed in progenitor mass-loss episodes before explosion. Surprisingly, this novel display of pre-SN emission and associated mass loss occurred in a red supergiant progenitor with zero-age main-sequence mass of only 10–12 $M_{sun}$, as inferred from nebular spectra. Modeling of the light curve and multi-epoch spectra with the non-LTE radiative-transfer code CMFGEN and radiation hydrodynamical code HERACLES suggests a dense CSM limited to $r \approx 10^{15}$

cm and mass-loss rate of $10^{-2}$ $M_{sun}$ $yr^{-1}$. The luminous light-curve plateau and persistent blue excess indicate an extended progenitor, compatible with an RSG model with $R_{star} = 1100\ R_{sun}$. Limits on the shock-powered X-ray and radio luminosity are consistent with model conclusions and suggest a CSM density of $\rho < 2 \times 10^{-16}$ g $cm^{-3}$ for distances from the progenitor star of $r \approx 5 \times 10^{15}$ cm, as well as a mass-loss rate of $< 1.3 \times 10^{-5}\ M_{\odot}yr^{-1}$ at larger distances. A promising power source for the observed precursor emission is the ejection of stellar material following energy disposition into the stellar envelope as a result of gravity waves emitted during either neon/oxygen burning or a nuclear flash from silicon combustion.

Dupree et al. (2023) reported on the great dimming of Betelgeuse: A surface mass ejection and its consequences. The bright supergiant Betelgeuse (Alpha Orionis, HD 39801) underwent a historic optical dimming during 27 January–13 February 2020. Many imaging and spectroscopic observations across the electromagnetic spectrum were obtained prior to, during, and subsequent to this dimming event. These observations of Betelgeuse reveal that a substantial surface mass ejection (SME) occurred and moved out through the extended atmosphere of the supergiant. A photospheric shock occurred in January–March 2019, progressed through the extended atmosphere of the star during the following 11 months, and led to dust production in the atmosphere. Resulting from the substantial mass outflow, the stellar photosphere was left with lower temperatures and the chromosphere with a lower density. The mass ejected could represent a significant fraction of the total annual mass-loss rate from the star, suggesting that episodic mass-loss events can contribute an amount comparable to that of the stellar wind. Following the SME, Betelgeuse was left with a cooler average photosphere, an unusual short photometric oscillation, reduced velocity excursions, and the disappearance of the ~400-day pulsation in the optical and radial velocity for more than 2 years following the Great Dimming.

### 1.7.6 Red Giants

A star expands to become a red giant when it has fused all the hydrogen in its core into helium. If the star is in a binary system, its envelope can overflow onto its companion or be ejected into space, leaving a hot core and potentially forming a subdwarf-B star. However, most red giants that have partially transferred envelopes in this way remain cool on the surface and are almost indistinguishable from those that have not. Among ~7000 helium burning red giants observed by NASA's Kepler mission, we use asteroseismology to identify two classes of stars that must have undergone dramatic mass loss, presumably due to stripping in binary interactions. The first class comprises about seven underluminous stars with smaller helium-burning cores than their single-star counterparts. Theoretical models show that these small cores imply the stars had much larger masses when ascending the red giant branch. The second class consists of 32 red giants with masses down to 0.5 $M_{sun}$, whose implied ages would exceed the age of the Universe had no mass loss occurred. The numbers are consistent with binary statistics, and our results open up new possibilities to study the evolution of post-mass-transfer binary systems (Li et al. 2022b).

Red giants are stars in the late stages of stellar evolution. Because they have exhausted the supply of hydrogen in their core, they burn the hydrogen in the surrounding shell. Once the helium in the core starts fusing, the star enters the clump phase, which is identified as a striking feature in the color-magnitude diagram. Since clump stars share similar observational properties, they are heavily used in astrophysical studies as probes of distance, extinction through the galaxy, galaxy density, and stellar chemical evolution. Vrard et al. (2022) performed a detailed observational characterization of the deepest layers of clump stars using asteroseismic data from Kepler. They found evidence for large core structural discontinuities in about 6.7% of the stars in our sample, implying that the region of mixing beyond the convective core boundary has a radiative thermal stratification. These stars are otherwise similar to the remaining stars in our sample, which may indicate that the building of the discontinuities is an intermittent phenomenon.

Betelgeuse ($\alpha$ Orionis; HD 39801) is a well-known bright red supergiant that shows semi-regular variations with four approximate periods of 2200, 420, 230, and 185 days. While the longest period was customarily regarded as a long secondary period (LSP) of unknown origin, Saio et al. (2023) identified the ~2200-day period as the radial fundamental mode and the three shorter periods as the radial first, second, and third overtones. From a nonadiabatic pulsation analysis including the pulsation/convection coupling, they found that these radial pulsation modes are all excited in the envelope of a model in a late stage of the core-carbon burning. Models with similar pulsation property have masses around 11 $M_\odot$ (19 $M_\odot$ at ZAMS) with luminosities (log $L/L_\odot$ = 5.27~5.28) and effective temperatures (log$T_{eff} \approx$ 3.53) that are consistent with the range of the observational determinations. They also found that a synthetic light curve obtained by adding the fundamental and the first-overtone mode qualitatively agrees with the light curve of Betelgeuse up to the Great Dimming. Their conclusion is that Betelgeuse is in the late stage of core carbon burning and a good candidate for the next galactic supernova.

Planetary nebulae, the ejected envelopes of red giant stars, provide us with a history of the last, mass-losing phases of 90% of stars initially more massive than the Sun. In their article, De Marco et al. (2022) analyze images of the planetary nebula NGC 3132 from the James Webb Space Telescope Early Release Observations. A structured, extended hydrogen halo surrounding an ionized central bubble is imprinted with spiral structures, probably shaped by a low-mass companion orbiting the central star at about 40–60 au. The images also reveal a mid-infrared excess at the central star, interpreted as a dusty disk, which is indicative of an interaction with another closer companion. Including the previously known A-type visual companion, the progenitor of the NGC 3132 planetary nebula must have been at least a stellar quartet. The JWST images allow us to generate a model of the illumination, ionization, and hydrodynamics of the molecular halo, demonstrating the power of JWST to investigate complex stellar outflows. Furthermore, new measurements of the A-type visual companion allow authors to derive the value for the mass of De Marco et al. (2022) the progenitor of a central star with excellent precision: 2.86 ± 0.06 $M_\odot$. These results serve as pathfinders for future JWST observations of planetary nebulae, providing unique insight into fundamental astrophysical processes, including colliding winds and binary star interactions, with implications for supernovae and gravitational-wave systems. JWST observations of the planetary nebula NGC 3132 reveal a rich molecular environment containing evidence of multiple stellar interactions on a variety of scales.

### 1.7.7 White Dwarfs

White dwarf stars are the most common stellar fossils. When in binaries, they make up the dominant form of compact object binary within the galaxy and can offer insight into different aspects of binary formation and evolution.

When they reach the end of their long evolutions, smaller stars—those up to eight times as massive as our sun—typically become white dwarfs. These ancient stars are incredibly dense. White dwarfs typically have a radius just 1% that of our sun; their mass is about the same. Eventually—over tens or even hundreds of billions of years—a white dwarf cools until it becomes a black dwarf, which emits no energy. Because the Universe's oldest stars are only 10 to 20 billion years old, there are no known black dwarfs—yet.

There are six known white dwarfs within 18 light-years or so of the Earth. Their basic parameters are given in Table 1.2. Apparent magnitude refers to brightness as seen from Earth on a logarithmic scale using the fifth root of 100 as the base.

Sirius B is the companion to Sirius, which is the brightest star in Earth's sky. As can be seen from the chart, Sirius B is unusually hot, and that means it is unusually young, probably around only 250 × $10^6$ years. It is also unusually massive, being the remnant of a giant B-class star that probably weighed in at over 5 solar masses when it was born.

Procyon B is the companion to Procyon, a bright first-magnitude star not far at all in the sky from Sirius. Sirius and Procyon are the luminaries of the adjacent constellations Canis Major and Canis Minor, respectively, otherwise known as the Big Dog and the Little Dog. Procyon B orbits very closely to Procyon and is a real challenge to see except with large telescopes. Van Maanen's Star is a lonely solo star and is quite old, as can be seen from its unusually cool temperature. It is so cool, it is only about 400K hotter than the Sun, which means that it is more of a yellow-white dwarf than it is a white dwarf. It takes a long time to cool off this much, probably 9 billion years or more.

TABLE 1.2 The Six Known White Dwarfs within 18 Lightyears of the Earth

| Name | Distance /ly | T(K) | Mass /Solar | App Mag |
|---|---|---|---|---|
| Sirius B | 8.58 | 25,193 | 1.01 | 8.44 |
| Procyon B | 11.43 | 7,740 | 0.60 | 10.81 |
| Van Maanen's Star | 14.04 | 6,220 | 0.64 | 12.40 |
| GJ 440 | 15.09 | 7,910 | 0.62 | 11.75 |
| 40 Eridani B | 16.25 | 16,176 | 0.57 | 9.52 |
| Stein 2051 B | 18.06 | 7,120 | 0.68 | 12.43 |

The white dwarf known mostly by the rather anonymous catalog number GJ 440 is also a solo star. Its parent star was about double the mass of the Sun when it was young, and the dwarf is neither especially new nor especially old, probably coming in at about 3 billion years. 40 Eridani B is a member of a triple star system and was once the brightest and most massive of the three, since the other two are relatively cool K-class and M-class stars, respectively. Precisely because it was the most massive, it was the first to make it to the white dwarf stage, probably less than a billion years ago, judging from its high temperature. It is also the lightest white dwarf in this group, indicating that its parent star was only somewhat more massive than the Sun.

Stein 2051 B is one-half of a red dwarf/white dwarf double star, with the red dwarf being so small and so dim that it isn't much brighter than the white dwarf.

Three types of X-ray sources are known to be associated with white dwarfs:

1. Photospheric X-ray emission from a WD itself;
2. Accretion of material from a close binary companion, as in cataclysmic variables; and
3. Coronal X-ray emission from a late-type binary companion, such as dMe stars.

The latter two types of sources require the WDs to be in binary systems, and their observed X-ray spectra commonly peak near 1 keV or higher, consistent with thermal emission from plasma with temperatures of a few ×106 K. In contrast, the photospheric X-ray emission from a WD is soft and detectable only at photon energies <0.5 keV, Chu et al. (2021).

With an effective temperature of ≈200,000 K, KPD 0005 + 5106 is one of the hottest white dwarfs. ROSAT unexpectedly detected "hard" (~1 keV) X-rays from this apparently single WD. Chu et al. (2021) obtained Chandra observations that confirm the spatial coincidence of this hard X-ray source with KPD 0005 + 5106. They also obtained XMM-Newton observations of KPD 0005 + 5106, as well as PG 1159–035 and WD 0121–756, which are also apparently single and whose hard X-rays were detected by ROSAT at 3σ–4σ levels. The XMM-Newton spectra of the three WDs show remarkably similar shapes that can be fitted by models including a blackbody component for the stellar photospheric emission, a thermal plasma emission component, and a power-law component. Their X-ray luminosities in the 0.6–3.0 keV band range from $4 \times 10^{29}$ to $4 \times 10^{30}$ erg $s^{-1}$. The XMM-Newton EPIC-pn soft-band (0.3–0.5 keV) light curve of KPD 0005 + 5106 is essentially constant, but the hard-band (0.6–3.0 keV) light curve shows periodic variations. An analysis of the generalized Lomb–Scargle periodograms for the XMM-Newton and Chandra hard-band light curves finds a convincing modulation (false-alarm probability of 0.41%) with a period of 4.7 ± 0.3 hours. Assuming that this period corresponds to a binary orbital period, the Roche radii of three viable types of companion have been calculated: M9V star, T brown dwarf, and Jupiter-like planet. Only the planet has a size larger than its Roche radius, although the M9V star and T brown dwarf may be heated by the WD and inflate past the Roche radius. Thus, all three types of companion may be donors to fuel accretion–powered hard X-ray emission.

Pelisoli et al. (2022) presented optical photometry of the cataclysmic variable LAMOST J024048.51 + 195226.9 taken with the high-speed, five-band CCD camera HiPERCAM on the 10.4m Gran Telescopio Canarias. They detected pulsations originating from the spin of its white dwarf, finding a spin period of 24.9328(38) s. The pulse amplitude is of the order of 0.2% in the *g* band, below the detection limits of previous searches. This detection establishes LAMOST J024048.51 + 195226.9 as only the second white dwarf magnetic propeller system, a twin of its long-known predecessor, AE Aquarii. At 24.93 s, the white dwarf in LAMOST J024048.51 + 195226.9 has the shortest known spin period of any cataclysmic variable star. The white dwarf must have a mass of at least $0.7 M_{\odot}$ to sustain so short a period. The observed faintest *u*-band magnitude sets an upper limit on the white dwarf's temperature of ~25000~25000 K. The pulsation amplitudes measured in the five HiPERCAM filters are consistent with an accretion spot of ~30000~30000K covering ~2% of the white dwarf's visible area, although spots that are hot and smaller or cooler and larger cannot be ruled out.

Pelisoli et al. (2023) also reported the discovery of a sibling of AR Sco, J191213.72–441045.1, which harbors a white dwarf in a 4.03 h orbit with an M dwarf and exhibits pulsed emission with a period of 5.30 min. This discovery establishes binary white dwarf pulsars as a class and provides support for proposed formation models for white dwarf pulsars.

Elms et al. (2022) identified two ultra-cool ($T_{eff} < 4000$ K) metal-polluted (DZ) white dwarfs WD J2147–4035 and WD J1922+0233 as the coolest and second coolest DZ stars known to date with $T_{eff} \approx 3050$ K and $T_{eff} \approx 3340$ K, respectively. Strong atmospheric collision-induced absorption (CIA) causes the suppression of red optical and infra-red flux in WD J1922+0233, resulting in an unusually blue color given its low temperature. WD J2147−4035 has moderate infra-red CIA yet has the reddest optical colors known for a DZ white dwarf. Microphysics improvements to the non-ideal effects and CIA opacities in our model atmosphere code yield reasonable solutions to observations of these ultra-cool stars. WD J2147−4035 has a cooling age of over $10 \times 10^9$ years, which is the largest known for a DZ white dwarf, whereas WD J1922+0233 is slightly younger, with a cooling age of $9 \times 10^9$ years. Galactic kinematics calculations from precise Gaia EDR3 astrometry reveal these ultra-cool DZ stars as likely members of the galactic disk; thus they could be pivotal objects in future studies constraining an upper age limit for the disk of the Milky Way. They presented intermediate-resolution spectroscopy for both objects, which provides the first spectroscopic observations of WD J2147−4035. Detections of sodium and potassium are made in both white dwarfs, in addition to calcium in WD J1922+0233 and lithium in WD J2147−4035. They also identified the magnetic nature of WD J2147−4035 from Zeeman splitting in the lithium line and also made a tentative detection of carbon, so we classify this star as DZQH. WD J1922+0233 likely accreted planetary crust debris, while the debris composition that polluted WD J2147−4035 remains unconstrained.

The initial–final mass relation (IFMR) links the birth mass of a star to the mass of the compact remnant left at its death. While the relevance of the IFMR across astrophysics is universally acknowledged, not all of its fine details have yet been resolved. A new analysis of a few carbon–oxygen white dwarfs in old open clusters of the Milky Way led Marigo et al. (2020) to identify a kink in the IFMR located over a range of initial masses, $1.65 \lesssim Mi/M_{\odot} \lesssim 2.10$. The kink's peak in white dwarf mass of about $0.70-0.75\,M_{\odot}$ is produced by stars with Mi $\approx 1.8-1.9\,M_{\odot}$, corresponding to ages of about $1.8-1.7$ Gyr. Interestingly, this peak coincides with the initial mass limit between low-mass stars that develop a degenerate helium core after central hydrogen exhaustion and intermediate-mass stars that avoid electron degeneracy. Marigo et al. (2020) interpret the IFMR kink as the signature of carbon star formation in the Milky Way. This finding is critical to constraining the evolution and chemical enrichment of low-mass stars and their impact on the spectrophotometric properties of galaxies.

## REFERENCES


Abbott, R., Abbott, T. D., Abraham, S., et al. 2021. Observation of gravitational waves from two neutron star-black hole coalescences. The Astrophysical Journal Letters 915: L5 (24 pages). https://doi.org/10.3847/2041-8213/ac082e.

Abeysekara, A. U., Albert, A., Alfaro, R., et al. 2021. HAWC observations of the accelerator of very-high-energy cosmic rays in the Cygnus Cocoon. Nature Astronomy Letters. https://doi.org/10.1038/s41550-021-01318-y.

Acevedo, J. F., Leane, R. K., and Santos-Olmsted, L. 2023. Milky Way white dwarfs as sub-GeV to multi-TeV dark matter detectors. arXiv:2309.10843v1 [hep-ph]. https://doi.org/10.48550/arXiv.2309.10843.

Adhikari, G., Carlin, N., Choi, J. J., et al. 2022. An induced annual modulation signature in COSINE-100 data by DAMA/LIBRA's analysis method. (9 pages). https://doi.org/10.48550/arXiv.2208.05158.

Adriani, O., Barbarino, G. C., Bazilevskaya, G. A., et al. 2016, Measurements of cosmic-ray hydrogen and helium isotopes with the PAMELA experiment. The Astrophysical Journal 818: 68 (14 pages). https://doi.org/10.3847/0004-637X/818/1/68.

Ajello, M., Baldini, L., Baller, J., et al. 2021. Gamma rays from fast black-hole winds. The Astrophysical Journal 921: 144 (14 pages). https://doi.org/10.3847/1538-4357/ac1bb2.

Alfvén, H. 1942. Existence of electromagnetic-hydrodynamic waves. Nature 150(3805): 405–406.

Amenomori, M., Bao, Y. W., Bi, X. J., et al. 2021. First detection of sub-PeV diffuse gamma rays from the galactic disk: Evidence for ubiquitous galactic cosmic rays beyond PeV energies. Physical Review Letters 126: 141101 (7 pages).

Andonie, C., Alexander, D. M., Claire Greenwell, C., et al. 2024. Obscuration beyond the nucleus: Infrared quasars can be buried in extreme compact starbursts. Monthly Notices of the Royal Astronomical Society L144–L150. https://doi.org/10.1093/mnrasl/slad144.

Andreia, C., Ijjasb, A., and Steinhardt, P. J. 2022. Rapidly descending dark energy and the end of cosmic expansion. PNAS 119(15): e2200539119 (4 pages). https://doi.org/10.1073/pnas.2200539119.

Andreoni, I., Coughlin, M. W., Perley, D. A., et al. 2022. A very luminous jet from the disruption of a star by a massive black hole. Nature 612: 430–434. https://doi.org/10.1038/s41586-022-05465-8.

Areal, M. B., Paron, S., Ortega, M. E., and Duvidovich, L. 2019. Study of the 13CO/C18O abundance ratio towards filamentary infrared dark cloud IRDC 34.43 + 0.24. Publications of Astronomical Society of Australia 36: e049 (8 pages). https://doi.org/10.1017/pasa.2019.44.

Arslan, Ö., Hocuk, S., Paola Caselli, P., and Küçük, I. 2022. The cosmic-ray induced sputtering process on icy grains. arXiv:2211.01511v2 [astro-ph.EP], 15 Nov. https://doi.org/10.1093/mnras/stac3196.

Asencio, E., Banik, I., and Kroupa, P. 2023. The El Gordo galaxy cluster challenges ΛCDM for any plausible collision velocity. The Astrophysical Journal 954: 162 (6 pages). https://doi.org/10.3847/1538-4357/ace62a.

Asencio, E., Banik, I., Mieske, S., et al. 2022. The distribution and morphologies of Fornax Cluster dwarf galaxies suggest they lack dark matter. Monthly Notices of the Royal Astronomical Society 515(2): 2981–3013.

The ATLAS Collaboration. 2024. Search for dark matter produced in association with a dark Higgs boson in the $bb^{-}$ final state using $\sqrt{s}$ = 13 TeV collisions recorded with the ATLAS detector. ATLAS CONF Note ATLAS-CONF-2024-004. CERN 27. March 2024.

Atri, D. 2016. On the possibility of galactic cosmic ray-induced radiolysis-powered life in subsurface environments in the universe. Journal of the Royal Society, Interface 13: 20160459 (8 pages). http://dx.doi.org/10.1098/rsif.2016.0459.

Avila, J. N., Ireland, T. R., Holden, P., et al. 2009. Tungsten isotopic compositions in presolar silicon carbide grains. 71st Annual Meteoritical Society Meeting, July 28–August 1, Matsue: Abstract 5120.

Ávila, J. N., Lugaro, M., Ireland, T. R., et al. 2012. Tungsten isotopic compositions in stardust SiC grains from the Murchison meteorite: Constraints on the s-process in the Hf–Ta–W–Re–Os region. The Astrophysical Journal 744(1): 49 (13 pages). https://doi.org/10.1088/0004-637X/744/1/49.

Bacon, R., Mary, D., Garel, T., et al. 2021. The MUSE extremely deep field: The cosmic web in emission at high redshift. Astronomy & Astrophysics 647: A107 (26 pages). https://doi.org/10.1051/0004-6361/202039887.

Bakx, T. J. L. C., Zavala, J. A., Mitsuhashi, I., et al. 2022. Deep ALMA redshift search of a z ~ 12 GLASS-JWST galaxy candidate. Monthly Notices of the Royal Astronomical Society 519(4): 5076–5085. https://doi.org/10.1093/mnras/stac3723.

Baldassare, V. F., Stone, N. C., Foord, A., et al. 2022. Massive black hole formation in dense stellar environments: Enhanced X-ray detection rates in high-velocity dispersion nuclear star clusters. The Astrophysical Journal 929: 84 (11 pages). https://doi.org/10.3847/1538-4357/ac5f51.

Banik, I., and Zhao, H. 2022. From galactic bars to the Hubble tension: Weighing up the astrophysical evidence for Milgromian gravity. Symmetry 14: 1331 (147 pages). https://doi.org/10.3390/sym14071331.

Basak, S., Ganguly, A., Haris, K., et al. 2022. Constraints on compact dark matter from gravitational wave microlensing. The Astrophysical Journal Letters 926: L28 (9 pages). https://doi.org/10.3847/2041-8213/ac4dfa.

Bell Burnell, J. 1969. The measurement of radio source diameters using a diffraction method. Repository.cam.ac.uk (PhD thesis), University of Cambridge. https://doi.org/10.17863/CAM.4926.

Bellinger, E. P., Caplan, M. E., Ryu, T., et al. 2023. Solar evolution models with a central black hole. The Astrophysical Journal 959: 113 (14 pages). https://doi.org/10.3847/1538-4357/ad04de.

Benetti, F., Lapi, A., Gandolfi, G., et al. 2023. Dark matter in fractional gravity. I. Astrophysical tests on galactic scales. The Astrophysical Journal 949: 65 (20 pages). https://doi.org/10.3847/1538-4357/acc8ca.

Binns, W. R., Bose, R. G., Braun, D. L., et al. 2014. The SuperTIGER instrument: Measurement of elemental abundances of ultra-heavy galactic cosmic rays. The Astrophysical Journal 788: 18 (11 pages). https://doi.org/10.1088/0004-637X/788/1/18.

Binns, W. R., Fickle, R. K., Garrard, T. L., et al. 1981. Cosmic-ray abundances of elements with atomic number 26 < z < 40 measured on HEAO 3. Astrophysical Journal Letters 247: Ll15–L118.

Blum, J., Levasseur-Regourd, A.-C., Muñoz, O., et al. Dust in space. Europhysics News 39(3): 27–29. http://dx.doi.org/10.1051/epn:2008303.

Blumenthal, K. A., and Barnes, J. E. 2018. Go with the flow: Understanding inflow mechanisms in galaxy collisions. Monthly Notices of the Royal Astronomical Society 479(3): 3952–3965. https://doi.org/10.1093/mnras/sty1605.

Bogdan, A., Goulding, A., Natarajan, P., et al. 2023. Evidence for heavy seed origin of early supermassive black holes from a z~10 X-ray quasar. (27 pages). arXiv:2305.15458v2 [astro-ph.GA], 25 Sep.

Bonvin, C., and Pogosian, L. 2023. Modified Einstein versus modified Euler for dark matter. arXiv:2209.03614v2 [astro-ph.CO], 7 Jun.

Boquien, M., Buat, V., Boselli, A., et al. 2012. The IRX-β relation on subgalactic scales in star-forming galaxies of the *Herschel* reference survey. Astronomy & Astrophysics 539: A145 (17 pages). https://doi.org/10.1051/0004-6361/201118624.

Bosman, S. E. I., Davies, F. B., Becker, G. D., et al. 2022. Hydrogen reionization ends by z = 5.3: Lyman-α optical depth measured by the XQR-30 sample. Monthly Notices of the Royal Astronomical Society 514: 55–76.

Boss, A. P. 2004. From Molecular Clouds to Circumstellar Disks. In M. Festou, H. U. Keller, and H. A. Weaver (eds.), Comets II (pp. 67–80). Tucson, AZ: University of Arizona Press.

Botelho, R. B., Milone, A. de C., Meléndez, J., et al. 2020. Carbon, isotopic ratio $^{12}C/^{13}C$, and nitrogen in solar twins: Constraints for the chemical evolution of the local disc. Monthly Notices of the Royal Astronomy Society 499: 2196–2213.

Boulares, A., and Cox, D. P. 1990. Galactic hydrostatic equilibrium with magnetic tension and cosmic-ray diffusion. Astrophysical Journal 365: 544–558.

Boylan-Kolchin, M. 2023. Stress testing ΛCDM with high-redshift galaxy candidates. Nature Astronomy. https://doi.org/10.1038/s41550-023-01937-7.

Brandenberger, R., Delgado, P. C. M., Ganz, A., and Lin, C. 2023. Graviton to photon conversion via parametric resonance. Physics of the Dark Universe 40: 101202 (5 pages). https://doi.org/10.1016/j.dark.2023.101202.

Bromm, V. 2013. Formation of the first stars. arXiv:1305.5178v2 [astro-ph.CO]. https://doi.org/10.48550/arXiv.1305.5178.

Burdge, K. B., Marsh, T. R., Fuller, J., et al. 2022. A 62-minute orbital period black widow binary in a wide hierarchical triple. Nature 605: 41–45.

Burger, J. D., Zavala, J., Sales, L. V., et al. 2022. Kinematic signatures of impulsive supernova feedback in dwarf galaxies. Physical Review Letters 129(19): 191103 (7 pages). https://link.aps.org/doi/10.1103/PhysRevLett.129.191103.

Bustard, M. C., and Zweibel, E. G. 2021. Cosmic-ray transport, energy loss, and influence in the multiphase interstellar medium. The Astrophysical Journal 913: 106 (26 pages). https://doi.org/10.3847/1538-4357/abf64c.

Byrnak, B., Lund, N., Rasmussen, I. L., et al. 1983. The Abundances of the Elements with Z>26 in the Cosmic Radiation. Proceedings of the 18th International Cosmic Ray Conference, held in Bangalore, India, 22 August—3 September. Volume 2 (OG Sessions), p. 29.

Caleb, M., Heywood, I., Rajwade, K., et al. 2022. Discovery of a radio-emitting neutron star with an ultra-long spin period of 76 s. Nature Astronomy. https://doi.org/10.1038/s41550-022-01688-x.

Calmet, X., and Kuipers, F. 2021. Theoretical bounds on dark matter masses. Physics Letters B 814: 136068 (3 pages).

Calura, F., and Matteucci, F. 2004. Cosmic metal production and the mean metallicity of the Universe. Monthly Notices of the Royal Astronomical Society 350(1): 351–364. https://doi.org/10.1111/j.1365-2966.2004.07648.x.

Cameron, A. G. W. 1982. Elemental and Nuclidic Abundances in the solar system. In C. Barnes, R. N. Clayton, and D. N. Schramm (eds.), Essays in Nuclear Astrophysics (p. 23). Cambridge: Cambridge University Press.

Castellina, A., and Donato, F. 2013. Astrophysics of Galactic Charged Cosmic Rays. In T. D. Oswalt and G. Gilmore (eds.), Planets, Stars and Stellar Systems (pp 725–788). Dordrecht: Springer. https://doi.org/10.1007/978-94-007-5612-0_14.

Castelvecchi, D. 2022a. Notorious dark-matter signal could be due to analysis error. https://doi.org/10.1038/d41586-022-02222-9.

Castelvecchi, D. 2022b. A supernova could light up the Milky Way at any time. Astronomers will be watching. Nature 602(7898): 562–565. https://doi.org/10.1038/d41586-022-00425-8.

Castro-Tirado, A. J., Ostgaard, N., Göğüş, E., et al. 2021. Very-high-frequency oscillations in the main peak of magnatar giant flare. Nature 600: 621–624.

Chadha-Day, F., Ellis, J., and Marsh, D. J. E. 2022. Axion dark matter: What is it and why now? Science Advances 8: eabj3618 (19 pages) https://doi.org/10.1126/sciadv.abj3618.

Chae, K.-H. 2022. Distinguishing dark matter, modified gravity, and modified inertia with the inner and outer parts of galactic rotation curves. The Astrophysical Journal 941: 55 (18pages). https://doi.org/10.3847/1538-4357/ac93f.

Chen, W., Kelly, P. L., Oguri, M., et al. 2022. Shock cooling of a red-supergiant supernova at redshift 3 in lensed images. Nature 611: 256–259. https://doi.org/10.1038/s41586-022-05252-5.

Chen, Y.-C., Liu, X., Foord, A., et al. 2023. A close quasar pair in a disk-disk galaxy merger at z = 2.17. Nature 616(7955): 45–49. https://doi.org/10.1038/s41586-023-05766-6.

Chibueze, J. O., Sakemi, H., Ohmura, T., et al. 2021. Jets from MRC 0600–399 bent by magnetic fields in the cluster Abell3376. Nature 593: 47–50.

Chirenti, C., Dichiara, S., Lien, A., et al. 2023. Kilohertz quasiperiodic oscillations in short gamma-ray bursts. Nature 613: 253–256. https://doi.org/10.1038/s41586-022-05497-0.

Chowdhury, R. K., Chatterjee, S., Lonappan, A. I., et al. 2020. Cosmological simulation of galaxy groups and clusters. I. Global effect of feedback from active galactic nuclei. The Astrophysical Journal 889: 60 (12 pages). https://doi.org/10.3847/1538-4357/ab5b96.

Chu, Y.-H., Toalá, J. A., Guerrero, M. A., et al. 2021. Hard X-Ray emission associated with white dwarfs. IV. Signs of accretion from substellar companions. The Astrophysical Journal 910: 119 (10 pages). https://doi.org/10.3847/1538-4357/abe5a5.

Chung, D. 2020. A unifying theory of dark energy, dark matter, and baryonic matter in the positive-negative mass Universe pair: Protogalaxy and galaxy evolutions. Journal of Modern Physics 11: 1091–1122. https://doi.org/10.4236/jmp.2020.117069.

Cinquegrana, G. C., and Karakas, A. I. 2021. The most metal-rich stars in the universe: Chemical contributions of low and intermediate mass asymptotic giant branch stars with metallicities between $0.04 \leq Z \leq 0.10$. (20 pages). arXiv:2111.09527v1 [astro-ph.SR], 18 Nov.

Collins, M. L. M., Charles, E. J. E., Martinez-Delgado, D., et al. 2022. Pegasus V—a newly discovered ultra-faint dwarf galaxy on the outskirts of andromeda. arXiv:2204.09068v1 [astro-ph.GA], 19 Apr.

Corbelli, E., and Salucci, P. 2000. The extended rotation curve and the dark matter halo of M33. Monthly Notices of the Royal Astronomical Society 311(2): 441–447. https://doi.org/10.1046/j.1365-8711.2000.03075.x.

Croker, K. S., Zevin, M., Farrah, D., et al. 2021. Cosmologically coupled compact objects: A single-parameter model for LIGO–Virgo mass and redshift distributions. The Astrophysical Journal Letters 921: L22 (6 pages). https://doi.org/10.3847/2041-8213/ac2fad.

Cuciti, V., de Gasperin, F., Brüggen, M., et al. 2022. Galaxy clusters enveloped by vast volumes of relativistic electrons. Nature 609: 911–914. https://doi.org/10.1038/s41586-022-05149-3.

Cui, Y., Hada, K., Kawashima, T., et al. 2023. Precessing jet nozzle connecting to a spinning black hole in M87. Nature 621: 711–715. https://doi.org/10.1038/s41586-023-06479-6.

Cummings, A. C., Stone, E. C., Heikkila, B. C., et al. 2016. Galactic cosmic rays in the local interstellar medium: VOYAGER 1 observations and model results. The Astrophysical Journal 831: 18 (21 pages). https://doi.org/10.3847/0004-637X/831/1/18.

Cyr-Racine, F.-Y., Ge, F., and Knox, L. 2022. Symmetry of cosmological observables, a mirror world dark sector, and the Hubble constant. Physical Review Letters 128(20): 201301. https://doi.org/10.1103/PhysRevLett.128.201301.

Dainotti, M. G., Bargiacchi, G., Lenart, A. L., et al. 2023. Quasars: Standard candles up to z=7.5 with the precision of Supernovae Ia. The Astrophysical Journal 950: 45 (8 pages). https://doi.org/10.3847/1538-4357/accea0.

Dainotti, M. G., De Simone, B., Schiavone, T., et al. 2021. On the Hubble constant tension in the SNe Ia Pantheon sample. The Astrophysical Journal 912: 150 (15 pages). https://doi.org/10.3847/1538-4357/abeb73.

Darling, J. 2022. The Universe is Brighter in the direction of our motion: Galaxy counts and fluxes are consistent with the CMB dipole. The Astrophysical Journal Letters 931: L14 (8 pages). https://doi.org/10.3847/2041-8213/.

Davelaar, J., and Haiman, Y. 2022a. Self-lensing flares from black hole binaries: Observing black hole shadows via light curve tomography. Physical Review Letters 128: 191101. https://doi.org/10.1103/PhysRevLett.128.191101.

Davelaar, J., and Haiman, Y. 2022b. Self-lensing flares from black hole binaries: General-relativistic ray tracing of black hole binaries. Physical Review D 105(10): 103010 (18 pages). https://link.aps.org/doi/10.1103/PhysRevD.105.103010.

Davies, B., Plez, B., and Petrault, M. 2022. Explosion imminent: The appearance of red supergiants at the point of core-collapse. Monthly Notices of the Royal Astronomical Society 517(1): 1483–1490. https://doi.org/10.1093/mnras/stac2427.

Delhaize, J. Heywood, I., Prescott, M., et al. 2021. MIGHTEE: Are giant radio galaxies more common than we thought? Monthly Notices of the Royal Astronomical Society 501: 3833–3845.

De Marco, O., Akashi, M., Akras, S., et al. 2022. The messy death of a multiple star system and the resulting planetary nebula as observed by JWST. Nature Astronomy 6: 1421–1432.

DES and SPT Collaborations, Abbott, T. M. C., Aguena, M., Alarcon, A., et al. 2023c. Joint analysis of dark energy survey year 3 data and CMB lensing from SPT and Planck. III. Combined cosmological constraints. Physical Review D 107: 023531. https://doi.org/10.1103/PhysRevD.107.023531.

DES and SPT Collaborations, Chang, C., et al. 2023b. Joint analysis of Dark Energy Survey Year 3 data and CMB lensing from SPT and Planck. II. Cross-correlation measurements and cosmological constraints. Physical Review D 107: 023530. https://doi.org/10.1103/PhysRevD.107.023530.

DES and SPT Collaborations, Omori, Y., et al. 2023a. Joint analysis of dark energy survey year 3 data and CMB lensing from SPT and Planck. I. Construction of CMB lensing maps and modeling choices. Physical Review D 107: 023529. https://doi.org/10.1103/PhysRevD.107.023529.

Ding, X., Onoue, M., Silverman, J. D., et al. 2023. Detection of stellar light from quasar host galaxies at redshifts above 6. Nature. article (5 pages). https://doi.org/10.1038/s41586-023-06345-5.

Di Stefano, R., and Ray, A. 2016. Globular clusters as cradles of life and advanced civilizations. The Astrophysical Journal 827: 54 (12 pages). https://doi.org/10.3847/0004-637X/827/1/54.

Donnan, C. T., McLeod, D. J., Dunlop, J. S., et al. 2022. The evolution of the galaxy UV luminosity function at redshifts z ~8–15 from deep JWST and ground-based near-infrared imaging. arXiv:2207.12356 [astro-ph.GA].

Duncan, K., and Conselice, C. J. 2015. Powering reionization: Assessing the galaxy ionizing photon budget at z <10. Monthly Notices of the Royal Astronomical Society 451(2): 2030–2049. https://doi.org/10.1093/mnras/stv1049.

Dupree, A. K., Strassmeier, K. G., Calderwood, T., et al. 2023. The great dimming of Betelgeuse: A surface mass ejection and its consequences. The Astrophysical Journal 936: 18 (9 pages). https://doi.org/10.3847/1538-4357/ac7853.

Eddington, A. S. 1917. The pulsation theory of Cepheid variables. The Observatory 40: 290–293.

Einstein, A. 1915. The field equations of gravitation. Sitzungsber. Preuss.Akad.Wiss.Berlin (Math.Phys.) 1915: 844–847.

El-Badry, K., Rix, H.-W., and Eliot Quataert, E. 2023. A Sun-like star orbiting a black hole. Monthly Notices of the Royal Astronomical Society 518(1): 1057–1085. https://doi.org/10.1093/mnras/stac3140.

El-Badry, K. J. 2021. Thesis: Binary stars across the Milky Way: Probes of star formation and evolution. https://escholarship.org/uc/item/1bn3w3tc.

Elms, A. K., Tremblay, P.-E., Gänsicke, B. T., et al. 2022. Spectral analysis of ultra-cool white dwarfs polluted by planetary debris. (18 pages). arXiv:2206.05258v2 [astro-ph.SR], 7 Oct.

Enßlin, T. A., Pfrommer, C., Springel, V., and Jubelgas, M. 2007. Cosmic ray physics in calculations of cosmological structure formation. Astronomy and Astrophysics 473: 41–57. https://doi.org/10.1051/0004-6361:20065294.

Endsley, R., Stark, D. P., Lyu, J., et al. 2023. ALMA confirmation of an obscured hyperluminous radio-loud AGN at z = 6.853 associated with a dusty starburst in the 1.5 deg$^2$ COSMOS field. Monthly Notices of the Royal Astronomical Society 520(3): 4609–4620. https://doi.org/10.1093/mnras/stad266.

Engelmann, J. J., Ferrando, P., Soutoul, A., et al. 1990. Charge composition and energy spectra of cosmic-ray nuclei for elements from Be to Ni. Results from HEAO-3-C2. Astronomy and Astrophysics 233: 96–111.

Farnes, J. S. 2018. A unifying theory of dark energy and dark matter: Negative masses and matter creation within a modified ΛCDM framework. Astronomy and Astrophysics (20 pages).

Farrah, D., Croker, K. S., Zevin, M., et al. 2023. Observational evidence for cosmological coupling of black holes and its implications for an astrophysical source of dark energy. The Astrophysical Journal Letters 944: L31 (9pages). https://doi.org/10.3847/2041-8213/acb704.

Feng, W.-X., Yu, H.-B., and Zhong, Y.-M. 2021. Seeding supermassive black holes with self-interacting dark matter: A unified scenario with baryons. The Astrophysical Journal Letters 914: L26 (9 pages). https://doi.org/10.3847/2041–8213/ac04b0.

Fernández, R., Tchekhovskoy, A., Quataert, E., et al. 2019. Log-term GRMHD simulations of neutron star merger accretion discs: Implications for electromagnetic counterparts. MNRAS 482: 3373–3393.

Finkelstein, S. L., Bagley, M. B., Arrabal Haro, P., et al. 2022. A long time ago in a galaxy far, far away: A Candidate z ~14 galaxy in early JWST CEERS imaging. arXiv:2207.12474 [astro-ph.GA].

Fonvieille, H., Pasquini, B., and Sparveris, N. 2020. Virtual Compton scattering and nucleon generalized polarizabilities. Progress in Particle and Nuclear Physics 113: 103754. https://doi.org/10.1016/j.ppnp.2020.103754.

Forbes, D. A., Bastian, N., Gieles, M., et al. 2018. Globular cluster formation and evolution in the context of cosmological galaxy assembly: Open questions. Proceedings of the Royal Society A: Mathematical, Physical and Engineering Sciences 474(2210): 20170616 (27 pages). https://doi.org/10.1098/rspa.2017.0616.

Fouqué, P., Solanes, J. M., Sanchis, T., and Balkowski, C. 2001. Structure, mass and distance of the Virgo cluster from a Tolman-Bondi model. A&A 375(3): 770–780. https://doi.org/10.1051/0004-6361:20010833.

Franco, M., Coppin, K. E. K., Geach, J. E., et al. 2021. The ramp-up of interstellar medium enrichment at z > 4. Nature Astronomy Letters. https://doi.org/10.1038/s41550-021-01515-9.

Freese, K., and Winkler, M. W. 2023. Dark matter and gravity waves from a dark big bang. (46 pages). arXiv:2302.11579v1 [astro-ph.CO], 22 Feb.

Fujimoto, S., Brammer, G. B., Waison, D., et al. 2022. A dusty compact object bridging galaxies and quasars at cosmic dawn. Nature 604: 261–265.

Fukushima, K., Yamaguchi, H., Slane, P. O., et al. 2020. Element stratification in the middle-aged SNIa remnant G344.7–0.1. The Astrophysical Journal 897: 62 (13 pages). https://doi.org/10.3847/1538-4357/ab94a6.

Furukawa, S., Nagamatsu, A., Nenoi, M., et al. 2020. Space radiation biology for "living in space". BioMed Research International 2020: 4703286 (25 pages). https://doi.org/10.1155/2020/4703286.

Gaia Collaboration. 2018. Gaia data release 2. Summary of the contents and survey properties. Astronomy & Astrophysics 616: A1 (22 pages). https://doi.org/10.1051/0004-6361/201833051.

Gaia Collaboration. 2021. Gaia early data release 3: Acceleration of the solar system from Gaia astrometry. Astronomy & Astrophysics 649: A9 (19 pages). https://doi.org/10.1051/0004-6361/202039734.

Garcia-Munoz, M., Simpson, J. A. 1979. Proceeding 16th International Cosmic Ray Conference (Kyoto) 1: 270.

Gastaldello, F., Simionescu, A., Mernier, F., et al. 2021. The metal content of the hot atmospheres of galaxy groups. Universe 7: 208 (48 pages). https://doi.org/10.3390/universe7070208.

Geach, J. E., Lopez-Rodriguez, E., M. J. Doherty, M. J., et al. 2023. Polarized thermal emission from dust in a galaxy at redshift 2.6. Nature 621: 483–486. https://doi.org/10.1038/s41586-023-06346-4.

Gehrman, T. C., Es Haghi, B. S., Sinha, K., and Xu, T. 2023. Recycled dark matter. (23 pages). arXiv:2310.08526v2 [hep-ph]. https://doi.org/10.48550/arXiv.2310.08526.

Genzel, R., Pichon, C., Eckart, A., et al. 2000. Stellar dynamics in the galactic centre: Proper motions and anisotropy. Monthly Notices of the Royal Astronomical Society 317(2): 348-374. https://doi.org/10.1046/j.1365–8711.2000.03582.x.

George, J. S., Lave, K. A., Wiedenbeck, M. E., et al. 2009. Elemental composition and energy spectra of galactic cosmic rays during solar cycle 23. The Astrophysical Journal 698: 1666–1681. https://doi.org/10.1088/0004-637X/698/2/1666.

Gieles, M., Erkal, D., Antonini, F., et al. 2021. A supra-massive population of stellar-mass black holes in the globular cluster Palomar 5. Nature Astronomy. https://doi.org/10.1038/s41550-021-01392-2.

Giovanetti, A., Tortolici, F., and Rufini, S. 2020. Why do the cosmic rays induce aging? Frontiers in Physiology 11: 955 (6 pages). https://doi.org/10.3389/fphys.2020.00955.

Goldstein, S., Hill, J. C., Iršič, V., and Sherwin, B. D. 2023. Canonical Hubble-tension-resolving early dark energy cosmologies are inconsistent with the Lyman-α forest. Physical Review Letters 131(20): 201001. https://doi.org/10.1103/PhysRevLett.131.201001.

Gonzalez, F. M., Fries, E. M., Cude-Woods, C., et al. 2021. Improved neutron lifetime measurements with UCNτ. Physical Review Letters 127: 162501 (6 pages).

Gottlieb, O., Liska, M., Tchekhovskoy, A., et al. 2022. Black hole to photosphere: 3DS GRMHD simulations of collapsars reveal wobbling and hybrid composition jets. The Astrophysical Journal Letters 933: L9 (12 pages). https://doi.org/10.3847/2041-8213/ac7530.

Goulding, A. D., Greene, J. E., Setton, D. J., et al. 2023. UNCOVER: The growth of the first massive black holes from JWST/NIRSpec—spectroscopic redshift confirmation of an X-ray luminous AGN at z=10.1. The Astrophysical Journal Letters 955: L24 (8 pages). https://doi.org/10.3847/2041-8213/acf7c5.

GRAVITY Collaboration. 2017. First light for GRAVITY: Phase referencing optical interferometry for the very large telescope interferometer. Astronomy & Astrophysics 602: A94 (23 pages). https://doi.org/10.1051/0004-6361/201730838.

GRAVITY Collaboration. 2018. Detection of the gravitational redshift in the orbit of the star S2 near the galactic centre massive black hole. Astronomy & Astrophysics 615: L15 (10 pages). https://doi.org/10.1051/0004-6361/201833718.

GRAVITY Collaboration. 2020. Detection of the Schwarzschild precession in the orbit of the star S2 near the galactic centre massive black hole. Astronomy & Astrophysics 636: L5 (14 pages). https://doi.org/10.1051/0004-6361/202037813.

Griffiths, A., Conselice, C. J., Ferreira, L., et al. 2021. Emission line galaxies in the SHARDS Frontier Fields—I. Candidate selection and the discovery of bursty Hα emitters. Monthly Notices of the Royal Astronomical Society 508: 3860–3876.

Guo, Y., Bacon, R., Bouché, N. F., et al. 2023. Bipolar outflows out to 10 kpc for massive galaxies at redshift $z \approx 1$. Nature 624: 53–56. https://doi.org/10.1038/s41586-023-06718-w.

Gupta, R. 2023. JWST early Universe observations and ΛCDM cosmology. Monthly Notices of the Royal Astronomical Society 524(3): 3385–3395, Article stad2032. https://doi.org/10.1093/mnras/stad2032.

Hansen, B. M. S. 2021. Unbound close stellar encounters in the solar neighborhood. arXiv:2112.00852v1 [astro-ph. SR], 1 Dec.

Haslbauer, M., Banik, I., and Kroupa, P. 2020. The KBC void and Hubble tension contradict ΛCDM on a Gpc scale – Milgromian dynamics as a possible solution. MNRAS 499: 2845–2883. https://doi.org/10.1093/mnras/staa2348.

Hawking, S. 1971. Gravitationally collapsed objects of very low mass. Monthly Notices of the Royal Astronomical Society 152(1): 75–78. https://doi.org/10.1093/mnras/152.1.75.

He, D., Gao, D., and Cai, Q-y. 2014. Spontaneous creation of the universe from nothing. arXiv:1404.1207v1 [gr-qc], 4 Apr.

Heintz, K. E., Giménez-Arteaga, C., Fujimoto, S., et al. 2023. The gas and stellar content of a metal-poor galaxy at z=8.496 as revealed by JWST and ALMA. The Astrophysical Journal Letters 944: L30 (13 pages). https://doi.org/10.3847/2041-8213/acb2cf.

H.E.S.S. Collaboration et al. 2022. Time-resolved hadronic particle acceleration in the recurrent nova RS Ophiuchi. Science. https://doi.org/10.1126/science.abn0567.

Hessels, J. W. T., Ransom, S. M., Stairs, I. H., et al. 2006. A radio pulsar spinning at 716 Hz. Science 311(5769): 1901–1904.

Hiramatsu, D., Howell, D. A., Van Dyk, S. D., et al. 2021. The electron-capture origin of supernova 2018zd. Nature Astronomy Letters. https://doi.org/10.1038/s41550-021-01384-2.

Hogan, C. J. 2000. Why the universe is just so. Review of Modern Physics 72(4): 1149–1161. https://doi.org/10.1103/RevModPhys.72.1149.

Horowitz, C. J., and Piekarewicz, J. 2001. Neutron star structure and the neutron radius of $^{208}$Pb. Physical Review Letters 86: 3647–3650.

Hota, H., Dabhade, P., Sravani Vaddi, S., et al. 2022. RAD@home citizen science discovery of an active galactic nucleus spewing a large unipolar radio bubble on to its merging companion galaxy. Monthly Notices of the Royal Astronomical Society: Letters 517(1): L86–L91. https://doi.org/10.1093/mnrasl/slac116.

Hsieh, P.-Y., Koch, P. M., Kim, W.-T., et al. 2021. The circumnuclear disk revealed by ALMA. I. Dense clouds and tides in the galactic center. The Astrophysical Journal 913: 94 (22 pages). https://doi.org/10.3847/1538-4357/abf4cd.

Hu, B., Jiang, W., Miyagi, T., et al. 2022. Ab initio predictions link the neutron skin of $^{208}$Pb to nuclear forces. Nature Physics 18: 1196–1200.

Huang, D., Abdukerim, A., Bo, Z., et al. 2023. Search for dark-matter–nucleon interactions with a dark mediator in PandaX-4T. Physical Review Letters 131: 191002 (6 pages). https://doi.org/10.1103/PhysRevLett.131.191002.

Hubble, E. 1929. A relation between distance and radial velocity among extra-galactic nebulae. Proceedings of the National Academy of Sciences of the United States of America 15(3): 168–173. https://doi.org/10.1073/pnas.15.3.168.

Hubble, E., and Humason, M. L. 1931. The velocity-distance relation among extra-galactic nebulae. Astrophysical Journal 74: 43–80. https://doi.org/10.1086/143323.

Hurley-Walker, N., Zhang, X., Bahramian, A., et al. 2022. A radio transient with unusually slow periodic emission. Nature 601: 526–530. https://doi.org/10.1038/s41586-021-04272-x.

IceCube Collaboration. 2023. Observation of high-energy neutrinos from the galactic plane. Science 380(6652): 1338–1343. https://doi.org/10.1126/science.adc9818.

Ilie, C., Paulin, J., and Freese, K. 2023. Supermassive Dark Star candidates seen by JWST. PNAS 120(30): e2305762120 (6 pages). https://doi.org/10.1073/pnas.2305762120.

Inoue, K. T., Minezaki, T., Matsushita, T., and Nakanishi, K. 2023. ALMA measurement of 10 kpc scale lensing-power spectra toward the lensed quasar MG J0414+0534. The Astrophysical Journal 954: 197 (33 pages). https://doi.org/10.3847/1538-4357/aceb5f.

Israel, M. H. 1981. Proceeding 17th International Cosmic Ray Conference (Paris) 12: 53.

Jacobson-Galán, W. V., Dessart, L., Jones, D. O., et al. 2022. Final moments I: Precursor emission, envelope inflation, and enhanced mass loss preceding the luminous type II supernova 2020df. The Astrophysical Journal 924: 15 (25 pages). https://doi.org/10.3847/1538-4357/ac3f3a.

Jalali, B., Pelupessy, F. I., Eckart, A., et al. 2014. Star formation in the vicinity of nuclear black holes: Young stellar objects close to Sgr A*. Monthly Notices of the Royal Astronomical Society 444(2): 1205–1220. https://doi.org/10.1093/mnras/stu1483.

Jin, S., Sillassen, N. B., Magdis, G. E., et al. 2023. Massive galaxy formation caught in action at z ~5 with JWST. Astronomy and Astrophysics 670: L11 (8 pages) https://doi.org/10.1051/0004-6361/202245724.

Jones, M. G., Sand, D. J., Bellazzini, M., et al. 2022. Young, blue, and isolated stellar systems in the Virgo cluster. II. A new class of stellar system. arXiv:2205.01695v1 [astro-ph.GA], 3 May.

Jordana-Mitjans, N., Mundell, C. G., Guidorzi, C., et al. 2022. A short gamma-ray burst from a protomagnetar remnant. The Astrophysical Journal 939: 106 (17 pages). https://doi.org/10.3847/1538–4357/ac972b.

Jura, M., and Smith, W. H. 1981. Discovery of interstellar rubidium. The Astrophysical Journal 251: L43–L44.

Just, O., Goriely, S., Janka, H.-Th., et al. 2022. Neutrino absorption and other physics dependences in neutrino-cooled black hole accretion discs. Monthly Notices of the Royal Astronomical Society 509: 1377–1412.

Kalvāns, J. 2014. Cosmic-ray induced diffusion in interstellar ices. Baltic Astronomy 23: 137– 149. https://doi.org/10.1515/astro-2017-0178.

Kasting, J. F., Whitmore, D. P., and Reynoldes, R. T. 1993. Habitable zones around main sequence stars. Icarus 101(1): 108–128. https://doi.org/10.1006/icar.1993.1010.

Katz, H., Rosdahl, J., Kimm, T., et al. 2021. The nature of high [OIII]88μm/[CII]158μm galaxies in the epoch of reionization: Low carbon abundance and top-heavy IMF? arXiv:2108.01074 [astro-ph.GA], 02 Aug.

Keeley, R. E., and Shafieloo, A. 2023. Ruling out new physics at low redshift as a solution to the $H_0$ tension. Physical Review Letters 131(11): 111002 (8 pages). https://doi.org/10.1103/PhysRevLett.131.111002.

Kerr, R. P. 2023. Do black holes have singularities? (19 pages). arXiv:2312.00841v1 [gr-qc], 1 Dec.

Kim, J., Jee, M. J., Hughes, J. P., et al. 2021. Head-to-toe measurement of El Gordo: Improved analysis of the galaxy cluster ACTCL J0102–4915 with new wide-field Hubble space telescope imaging data. The Astrophysical Journal 923: 101 (20 pages). https://doi.org/10.3847/1538-4357/ac294f.

Kirsten, F., Marcote, B., Nimmo, K., et al. 2022. A repeating fast radio burst source in a globular cluster. Nature 602: 585–589.

Kobayashi, C., Karakas, A. I., and Lugaro, M. 2020b. The origin of elements from carbon to uranium. The Astrophysical Journal 900(2): 179 (33 pages).

Kobayashi, C., Tsujimoto, T., and Nomoto, K. 2000. The history of the cosmic supernova rate derived from the evolution of the host galaxies. The Astrophysical Journal 539: 26–38.

Kokhan, V. S., and Dobynde, M. I. 2023. The effects of galactic cosmic rays on the central nervous system: From negative to unexpectedly positive effects that astronauts may encounter. Biology 12(3): 400 (28 pages). https://doi.org/10.3390/biology12030400.

Komugi, S., Toba, Y., Matsuoka, Y., et al. 2022. Detection of extended millimeter emission in the host galaxy of 3C 273 and its implication for QSO feedback via high dynamic range ALMA imaging. The Astrophysical Journal 930: 3 (7 pages). https://doi.org/10.3847/1538-4357/ac616e.0.

Kong, L.-D., Zhang, S., Zhang, S.-N., et al. 2022. Insight-HXMT discovery of the highest-energy CRSF from the first galactic ultraluminous X-ray pulsar Swift J0243.6 + 6124. The Astrophysical Journal Letters 933: L3 (9 pages). https://doi.org/10.3847/2041-8213/ac7711.

König, O., Wilms, J., Arcodia, R., et al. 2022. X-ray detection of a nova in the fireball phase. Nature 605: 248–250.

Koss, M. J., Treister, E., Kakkad, D., et al. 2023. UGC 4211: A confirmed dual active galactic nucleus in the local Universe at 230 pc nuclear separation. The Astrophysical Journal Letters 942: L24 (15pages). https://doi.org/10.3847/2041-8213/aca8f0.

Kostov, V. B., Powell, B. P., and Rappaport, S. A. 2022. Ninety-seven eclipsing quadruple star candidates discovered in TESS full-frame images. The Astrophysical Journal Supplement Series 259: 66 (27 pages). https://doi.org/10.3847/1538-4365/ac5458.

Labbé, I., van Dokkum, P., Nelson, E., et al. 2023. A population of red candidate massive galaxies ~600 Myr after the Big Bang. Nature (Accelerated Article Preview). https://doi.org/10.1038/s41586-023-05786-2.

Laha, S., Meyer, E., Roychowdhury, A., et al. 2022. A radio, optical, UV and X-ray view of the enigmatic changing look active galactic nucleus 1ES 1927 + 654 from its pre- to post-flare states. arXiv:2203.07446v1 [astro-ph.HE], 14 Mar.

Lam, C. Y., Lu, J. R., Udalski, A., et al. 2022. An isolated mass gap black hole or neutron star detected with astrometric microlensing. arXiv:2202.01903v2 [astro-ph.GA], 31 May.

Larson, R. L., Finkelstein, S. L., Kocevski, D. D., et al. 2023. A CEERS discovery of an accreting supermassive black hole 570 Myr after the big bang: Identifying a progenitor of massive z > 6 quasars. arXiv:2303.08918v1 [astro-ph.GA]. https://doi.org/10.48550/arXiv.2303.08918.

Lea, R. 2022. What is dark energy? www.space.com/dark-energy-what-is-it. Published November 24. Accessed 01.06.2023.

Leavitt, H. S., and Pickering, E. C. 1912. Periods of 25 variable stars in the small magellanic cloud. Harvard College Observatory Circular 173: 1–3.

Lemaître, G. 1933. Discussion sur l'évolution de l'univers; The extension in universe. Annals of the Brussels Science Society 53A: 51–83.

Li, R., Sparveris, N., Atac, H., et al. 2022a. Measured proton electromagnetic structure deviates from theoretical predictions. Nature. https://doi.org/10.1038/s41586-022-05248-1.

Li, Y., Bedding, T. R., Murhy, S. J., et al. 2022b. Discovery of post-mass-transfer helium-burning red giants using astroseismology. Nature Astronomy Letters. https://doi.org/10.1038/s41550-022-01648-5.

Lian, J., Thomas, D., and Maraston, C. 2018. The mass–metallicity relations for gas and stares in star-forming galaxies: Strong outflow versus variable IMF. Monthly Notices of the Royal Astronomical Society 474: 1143–1164.

LIGO and Virgo Collaborations. 2017. GW170817: Observation of gravitational waves from a binary neutron star inspiral. Physical Review Letters 119: 161101 (18 pages). https://doi.org/10.1103/PhysRevLett.119.161101.

Lindegren, L., Bastian, U., Biermann, M., et al. 2021. Gaia early data release 3. Astronomy & Astrophysics 649: A4 (31 pages). https://doi.org/10.1051/0004-6361/202039653.

Liu, B., Zhang, S., and Bromm, V. 2022. Effects of stellar-mass primordial black holes on first star formation. Monthly

Notices of the Royal Astronomical Society 514(2): 2376–2396. https://doi.org/10.1093/mnras/stac1472.

Livio, M. 2007. The Interconnections Between Cosmology and Life. In J. D. Barrow, et al. (eds.), Fitness of the Cosmos for Life: Biochemistry and Fine-Tuning. Cambridge: Cambridge University Press.

Lodders, K. 2003. Solar system abundances and condensation temperatures of the elements. The Astrophysical Journal 591(2): 1220–1247. https://doi.org/10.1086/375492.

Loeve, K., Nielsen, K. S., and Hansen, S. H. 2021. Consistency analysis of a dark matter velocity dependent force as an alternative to the cosmological constant. The Astrophysical Journal 910: 98 (4pages). arXiv: 2102.07792v1 [astro-ph.CO], 15 Feb; April 1. https://doi.org/10.3847/1538-4357/abe5a2.

Lopatka, A. 2023. High-energy neutrinos originate in our own galaxy. Physics Today 76 (9): 13–15. https://doi.org/10.1063/PT.3.5303.

Luisi, M., Anderson, L. D., Schneider, N., et al. 2021. Stellar feedback and triggered star formation in the prototypical bubble RCW 120. Science Advances 7: eabe9511 (8 pages). http://advances.sciencemag.org/.

Madau, P., and Dickinson, M. 2014. Cosmic star-formation rate. Annual Review of Astronomy and Astrophysics 52: 415–486. https://doi.org/10.1146/annurev-astro-081811-125615.

Madau, P., Ferguson, H. C., Dickinson, M. E., et al. 1996. High-redshift galaxies in the Hubble Deep Field: Colour selection and star formation history to z ~ 4. Monthly Notices of the Royal Astronomical Society 283: 1388–1404.

Maiolino, R., and Mannucci, F. 2019. De re metallica: The cosmic chemical evolution of galaxies. The Astronomy and Astrophysics Review 27: 3 (187 pages). https://doi.org/10.1007/s00159-018-0112-2.

Mancera Piña, P. E., Fraternali, F., Oosterloo, T., et al. 2021. No need for dark matter: Resolved kinematics of the ultra-diffuse galaxy AGC 114905. arXiv:2112.00017v1 [astro-ph.GA] 30 Nov. (Original work published 2015).

Marcowith, A., van Marle, A. J., and Plotnikov, I. 2021. The cosmic ray-driven streaming instability in astrophysical and space plasmas. Physics of Plasmas 28: 080601 (16 pages). https://doi.org/10.1063/5.0013662.

Marigo, P., Cummings, J. D., Curtis, J. L., et al. 2020. Carbon star formation as seen through the non-monotonic initial–final mass relation. Nature Astronomy 4: 1102–1110. https://doi.org/10.1038/s41550-020-1132-1.

Marques, G., and Huffenberger, K. 2022. Dark matter mapped around distant galaxies. Physics 15: 117 (3 pages). https://doi.org/10.1103/Physics.15.117.

Martijn, S. S. L., van Weeren, R. J., Hardcastle, M. J., et al. 2022. The discovery of a radio galaxy of at least 5 Mpc. arXiv:2202.05427v1 [astro-ph.GA], 11 Feb.

Martin-Navarro, I., Pillepich, A., Nelson, D., et al. 2021. Anisotropic satellite galaxy quenching modulated by black hole activity. Nature 594: 187–190.

Mazurenko, S., Banik, I., Kroupa, P., and Haslbauer, M. 2023. A simultaneous solution to the Hubble tension and observed bulk flow within 250 $h^{-1}$ Mpc. Monthly Notices of the Royal Astronomical Society 527(3): 4388–4396. https://doi.org/10.1093/mnras/stad3357.

McAlpine, S., Helly, J. C., Schaller, M., et al. 2022. SIBELIUS-DARK: A galaxy catalogue of the local volume from a constrained realization simulation. arXiv:2202.04099v1 [astro-ph.GA], 08 Feb.

McKinney, J., Manning, S. M., Cooper, O. R., et al. 2023. A near-infrared-faint, far-infrared-luminous dusty galaxy at z~5 in COSMOS-Web. The Astrophysical Journal 956: 72 (12 pages). https://doi.org/10.3847/1538-4357/acf614.

McLeod, A. F., Klaassen, P. D., Reiter, M., et al. 2023. A probable Keplerian disk feeding an optically revealed massive young star. Nature (11 pages). https://doi.org/10.1038/s41586-023-06790-2.

Mei, S., Blakeslee, J. P., Cötè, P., et al. 2007. The ACS Virgo Cluster Survey. XIII. SBF distance catalog and the three-dimensional structure of the Virgo Cluster. The Astrophysical Journal 655: 144–162.

Melia, F. 2023. The cosmic timeline implied by the JWST high-redshift galaxies. Monthly Notices of the Royal Astronomical Society Letters 521(1): L85–L89. https://doi.org/10.1093/mnrasl/slad025.

Meyer, J. P. 1979. Proc. 16th Int. Conf. Cosmic Rays (Kyoto) 2:115.

Michell, J. 1784. On the Means of Discovering the Distance, Magnitude, Etc., of the Fixed Stars. In Consequence of the Diminution of Their Light, in Case Such a Diminution Should Be Found to Take Place in Any of Them, and Such Other Data Should Be Procured from Observations, as Would Be Further Necessary for That Purpose (vol. 74, p. 35). London: Philosophical Transactions of the Royal Society of London. Presented to the Royal Society. (Original work published November 27, 1783).

Mićić, M., Holmes, O. I., Wells, B. N., et al. 2023. Two candidates for dual AGN in dwarf-dwarf galaxy merger. The Astrophysical Journal 944: 160 (10 pages). https://doi.org/10.3847/1538-4357/acalbb.

Miki, Y., Mori, M., and Kawaguchi, T. 2021. Destruction of the central black hole gas reservoir through head-on galaxy collisions. Nature Astronomy Letters https://doi.org/10.1038/s41550-050-01286-9.

Miley, G. K., Overzier, R. A., Zirm, A. W., et al. 2006. The Spiderweb galaxy: A forming massive cluster galaxy at z ~ 2. The Astrophysical Journal 650(1): L29–L32. https://doi.org/10.1086/508534.

Milgrom, M. 1983. A modification of the Newtonian dynamics as a possible alternative to the hidden mass hypothesis. Astrophysical Journal 270: 365–370. https://doi.org/10.1086/161130.

Miyatake, H., Harikane, Y., Ouchi, M., et al. 2022. First identification of a CMB lensing signal produced by 1.5 million galaxies at z ~ 4: Constraints on matter density fluctuations at high redshift. Physical Review Letters 129: 061301. https://doi.org/10.1103/PhysRevLett.129.061301.

Mooley, K. P., Anderson, J., and Lu, W. 2022. Optical superluminal motion measurement in the neutron-star merger GW170817. Nature 610: 273–276. https://doi.org/10.1038/s41586-022-05145-7.

Moore, C. J., and Vecchio, A. 2021. Ultra-low-frequency gravitational waves from cosmological and astrophysical processes. Nature Astronomy Letters. https://doi.org/10.1038/s41550-021-01489-8.

Moreno, J., Danieli, S., Bullock, J. S., et al. 2022. Galaxies lacking dark matter produced by close encounters in a cosmological simulation. Nature Astronomy 6: 496–502. https://doi.org/10.1038/s41550-021-01598-4.

Moser, L., Sánchez-Monge, Á., Eckart, A., et al. 2017. Approaching hell's kitchen: Molecular daredevil clouds in the vicinity of Sagittarius A*. Astronomy & Astrophysics 603: A68 (54 pages). https://doi.org/10.1051/0004-6361/201628385.

Mu, B., Liu, J., Cheng, G., and Guo, Z.-K. 2023. Constraints on ultra-slow-roll inflation with the NANOGrav 15-year dataset. (6 pages). arXiv:2310.20564v1 [astro-ph.CO], 31 Oct.

Muldrew, S. I., Hatch, N. A., and Cooke, E. A. 2015. What are protoclusters?—Defining high-redshift galaxy clusters and protoclusters. Monthly Notices of the Royal Astronomical Society 452(3): 2528–2539. https://doi.org/10.1093/mnras/stv1449.

Murphy, R. P., Sasaki, M., Binns, W. R., et al. 2016. Galactic cosmic ray origins and OB associations: Evidence from SuperTIGER observations of elements $_{26}$Fe through $_{40}$Zr. The Astrophysical Journal 831: 148 (7 pages). https://doi.org/10.3847/0004-637X/831/2/148.

Mutlu-Pakdil, B., Sand, D. J., Crnojević, D., et al. 2022. Hubble space telescope observations of NGC 253 dwarf satellites: Three ultra-faint dwarf galaxies. The Astrophysical Journal 926: 77 (9 pages). https://doi.org/10.3847/1538-4357/ac4418.

Naidu, R. P., Oesch, P. A., van Dokkum, P., et al. 2022a. Two remarkably luminous galaxy candidates at $z \approx 11 - 13$ revealed by JWST. arXiv:2207.09434 [astro-ph.GA].

Naidu, R. P., Oesch, P. A., D. J. Setton, D. J., et al. 2022b. Schrodinger's galaxy candidate: Puzzlingly luminous at $z \approx 17$, or dusty/quenched at $z \approx 5$? arXiv:2208.02794 [astro-ph.GA].

Namekata, K., Maehara, H., Honda, S., et al. 2021. Probable detection of an eruptive filament from a superflare on a solar-type star. Nature Astronomy Letters. https://doi.org/10.1038/s41550-021-01532-8.

NASA. 2023. Galaxies. https://universe.nasa.gov/galaxies/large-scale-structures/. Downloaded 30.08.2023.

NASA/LAMBDA Archive Team. 2023. ΛCDM Model of Cosmology. https://lambda.gsfc.nasa.gov/education/graphic_history/univ_evol.html. Downloaded 13.07.2023.

Nelson, E. J., Suess, K. A., Bezanson, R., et al. 2022. JWST reveals a population of ultra-red, flattened disk galaxies at $2 \leq z \leq 6$ previously missed by HST. (13 pages). arXiv:2208.01630v1 [astro-ph.GA], 2 Aug.

Nimmo, K., Hessels, J. W. T., Kirsten, F., et al. 2022. Burst timescales and luminosities as links between young pulsars and fast radio burst. Nature Astronomy. https://doi.org/10.1038/s41550-021-01569-9.

Nishiyama, S., Kara, T., Thorsbro, B., et al. 2024. Origin of an orbiting star around the galactic supermassive black hole. Proceedings of the Japan Academy, Series B: Physical and Biological Sciences (14 pages). https://doi.org/10.2183/pjab.100.007.

Niu, C.-H., Aggarwal, K., Li, D., et al. 2022. A repeating fast radio burst associated with a persistent radi source. Nature. article (5 pages). https://doi.org/10.1038/s41586-022-04755-5.

Nolte, J. L., Stancil, P. C., Lee, T.-G., et al. 2012. Rovibrational quenching rate coefficients of HD in collisions with He. The Astrophysical Journal 744: 62 (7 pages). https://doi.org/10.1088/0004-637X/744/1/62.

Norris, R. P., Collier, J. D., Crocker, R. M., et al. 2022. MeerKAT uncovers the physics of an odd radio circle. Monthly Notices of the Royal Astronomical Society 513(1): 1300–1316. https://doi.org/10.1093/mnras/stac701.

Ohkubo, T., Umeda, H., Maeda, K., et al. 2006. Core-collapse very massive stars: Evolution, explosion, and nucleosynthesis of population III 500–1000 $M_{\odot}$ stars The Astrophysical Journal 645: 1352–1372.

Olsen, C., Gawiser, E., Iyer, K., et al. 2021. Star formation histories from spectral energy distribution and color-magnitude diagrams agree: Evidence for synchronized star formation in local volume dwarf galaxies over the past 3 Gyr. The Astrophysical Journal 913: 45 (19 pages). https://doi.org/10.3847/1538-4357/abf3c2.

Overzier, R. A. 2016. The realm of the galaxy protoclusters. Astronomy and Astrophysics Review 24: paper no.14. https://doi.org/10.1007/s00159-016-0100-3.

Pacucci, F., Dayal, P., Harikane, Y., et al. 2022. Are the newly-discovered $z \sim 13$ drop-out sources starburst galaxies or quasars? Monthly Notices of the Royal Astronomical Society: Letters 514(1): L6–L10 (6 pages). https://doi.org/10.1093/mnrasl/slac035.

Patra, K. C., Lu, W., Brink, T. G., et al. 2022. Spectropolarimetry of the tidal disruption event AT 2019qiz: A quasispherical reprocessing layer. Monthly Notices of the Royal Astronomical Society 515(1): 138–145. https://doi.org/10.1093/mnras/stac1727.

Peißker, F., Zajaček, M., Sabha, N. B., et al. 2023. X3: A high-mass young stellar object close to the supermassive black hole Sgr A*. The Astrophysical Journal 944: 231 (28 pages). https://doi.org/10.3847/1538-4357/aca977.

Pelisoli, I., Marsh, T. R., Buckley, D. A. H., et al. 2023. A 5.3-min-period pulsing white dwarf in a binary detected from radio to X-rays. Nature Astronomy. https://doi.org/10.1038/s41550-023-01995-x.

Pelisoli, I., Marsh, T. R., Dhillon, V. S., et al. 2022. Found: A rapidly spinning white dwarf in LAMOST J024048.51+195226.9. Monthly Notices of the Royal Astronomical Society: Letters 509(1): L31–L36. https://doi.org/10.1093/mnrasl/slab116.

Peng, B., Vishwas, A., Stacey, G., et al. 2023. Discovery of a dusty, chemically mature companion to a z~4 starburst galaxy in JWST ERS data. The Astrophysical Journal Letters 944: L36 (8 pages). https://doi.org/10.3847/2041-8213/acb59c.

Penrose, R. 1965. Gravitational collapse and space-time singularities. Physical Review Letters 14(3): 57. https://doi.org/10.1103/PhysRevLett.14.57.

Penrose, R. 2018. The big bang and its dark-matter content: Whence, whither, and wherefore. Foundations of Physics 48: 1177–1190. https://doi.org/10.1007/s10701-018-0162-3.

Penzias, A. A., and Wilson, R. W. 1965. A measurement of excess antenna temperature at 4080 Mc/s. Astrophysical Journal 142: 419–421. https://doi.org/10.1086/148307.

Péroux, C., and Howk, J. C. 2020. The cosmic baryon and metal cycles. The Annual Review of Astronomy and Astrophysics 58: 363–406. https://doi.org/10.1146/annurev-astro-021820-120014.

Petit, J.-P., and d'Agostini, G. 2021. Negative mass hypothesis in cosmology and the nature of dark energy. Astrophysics and Space Science 354: 611–615.

Planck Collaboration. 2016. Planck 2015 results XIII. Cosmological parameters. Astronomy & Astrophysics 594: A13 (63 pages). https://doi.org/10.1051/0004-6361/201525830.

Planck Collaboration. 2020. Planck 2018 results VI. Cosmological parameters. Astronomy & Astrophysics 641: A6 (67 pages). https://doi.org/10.1051/0004-6361/201833910.

Polimera, M. S., Kannappan, S. J., Richardson, C. T., et al. 2022. RESOLVE and ECO> finding low-metallicity z ~ 0 dwarf AGN candidates using optimized emission-line diagnostics. The Astrophysical Journal 931: 44 (23 pages). https://doi.org/10.3847/1538-4357/ac 6595.

Prasad, S. S., and Tarafdar, S. P. 1983. UV radiation inside dense clouds—its possible existence and chemical implications. Astrophysical Journal 267: 603–609.

Rajpurohit, K., Osinga, E., Brienza, M., et al. 2023. Deep low-frequency radio observations of Abell 2256. Astronomy & Astrophysics 669: A1 (19 pages). https://doi.org/10.1051/0004-6361/202244925.

Rajpurohit, K., van Weeren, R. J., Hoeft, M., et al. 2022. Deep low-frequency radio observations of A2256. I. The filamentary radio relic. The Astrophysical Journal 927: 80 (21 pages). https://doi.org/10.3847/1538–4357/ac4708.

Randich, S., and Magrini, L. 2021. Light elements in the Universe. Frontiers in Astronomy and Space Science 8: 616201 (28 pages). https://doi.org/10.3389/fspas.2021.616201.

Rastinejad, J. C., Gompertz, B. P., Levan, A. J., et al. 2022. A kilonova following a long-duration gamma-ray burst at 350 Mpc. Nature 612: 223–227. https://doi.org/10.1038/s41586-022-05390-w.

Rauch, B. F., Link, J. T., Lodders, K., et al. 2009. Cosmic ray origin in OB associations and preferential acceleration of refractory elements: Evidence from abundances of elements $^{26}$Fe through $^{34}$Se. The Astrophysical Journal 697(2): 2083–2088.

Reich, W., and Wielebinski, R. Copernicus GmbH—Provided by the NASA Astrophysics Data System.

Remillard, R. A., and McClintock, J. E. 2006. X-Ray properties of black-hole binaries. Annual Review of Astronomy and Astrophysics 44: 49–92.

Reyes, M. C., and Anderson, R. I. 2023. A 0.9% calibration of the galactic Cepheid luminosity scale based on Gaia DR3 data of open cluster and Cepheids. Astronomy and Astrophysics 672: A85 (22 pages). https://doi.org/10.1051/0004-6361/202244775.

Rhoads, J. E., Wold, I. G. B., Harish, S., et al. 2023. Finding peas in the early Universe with JWST. The Astrophysical Journal Letters 942: L14 (10 pages). https://doi.org/10.3847/2041-8213/acaaaf.

Riess, A. G., Anand, G. S., Yuan, W., et al. 2023. Crowded no more: The accuracy of the Hubble Constant tested with high resolution observations of Cepheids by JWST. (20 pages). arXiv:2307.15806v1 [astro-ph.CO], 28 Jul.

Riess, A. G., Filippenko, A. V., Challis, P., et al. 1998. Observational evidence from supernovae for an accelerating universe and a cosmological constant. The Astronomical Journal 116(3): 1009–1038.

Riess, A. G., Yuan, W., Macri, L. M., et al. 2022. A comprehensive measurement of the local value of the Hubble constant with 1 km $s^{-1}$ $Mpc^{-1}$ uncertainty from the Hubble space telescope and the shoes team. arXiv:2112.04510v2 [astro-ph.CO], 10 Jan.

Rini, M. 2021. Sizing up the most massive neutron star. Physics 14: 64 (3 pages). https://doi.org/10.1103/Physics.14.64.

Ripperda, B., Liska, M., Chatterjee, K., et al. 2022. Black hole flares: Ejection of accreted magnetic flux through 3D plasmoid-mediated reconnection. The Astrophysical Journal 924: L32 (15 pages). https://doi.org/10.3847/2041-8213/ac46a1.

Robertson, B. E., Ellis, R. S., Furlanetto, S. R., and Dunlop, J. S. 2015. Cosmic reionization and early star-forming galaxies: A joint analysis of new constraints from Planck and the Hubble space telescope. The Astrophysical Journal Letters 802: L19 (5 pages). https://doi.org/10.1088/2041-8205/802/2/L19.

Román, J., Rich, R. M., Ahvazi, N., et al. 2023. A giant thin stellar stream in the Coma Galaxy Cluster. Astronomy & Astrophysics 679: A157 (14 pages). https://doi.org/10.1051/0004-6361/202346780.

Saccardi, A., Salvadori, S., D'Odorico, V., et al. 2023. Evidence of first stars-enriched gas in high-redshift absorbers. The Astrophysical Journal 948: 35 (16 pages). https://doi.org/10.3847/1538-4357/acc39f.

Sahai, R., Huang, P.-S., S. Scibelli, S., et al. 2022. The rapidly evolving asymptotic giant branch star, V Hya: ALMA finds a multiring circus with high-velocity outflows. The Astrophysical Journal 929: 59 (29 pages). https://doi.org/10.3847/1538-4357/ac568a.

Sahu, K. C., Anderson, J., Casertano, S., et al. 2022. An isolated stellar-mass black hole detected through astrometric microlensing. arXiv:2201.13296v1 [astro-ph. SR], 31 Jan.

Saio, H., Nandal, D., Georges Meynet, G., and Ekstöm, S. 2023. The evolutionary stage of Betelgeuse inferred from its pulsation periods. arXiv:2306.00287v1 [astro-ph.SR], 1 Jun (11 pages).

Samsing, J., Bartos, I., D'Orazio, D. J., et al. 2022. AGN as potential factories for eccentric black hole merger. Nature 603: 237–240.

Saracino, S., Kamann, S., Guarcello, M. G., et al. 2021. A black hole detected in the young massive LMC cluster NGC 1850. Monthly Notices of the Royal Astronomical Society 511(2): 2914–2924. https://doi.org/10.1093/mnras/stab3159.

Sbordone, L., Monaco, L., Moni Bidin, C., et al. 2015. Chemical abundances of giant stars in NGC 5053 and NGC 5634, two globular clusters associated with Sagittarius dwarf spheroidal galaxy. Astronomy & Astrophysics 579: A104 (12 pages).

Scaringi, S., Groot, P. J., Knigge, C., et al. 2022. Localized thermonuclear bursts from accreting magnetic white dwarfs. Nature 604: 447–450.

Schneider, F. R. N., Podsiadlowski, P., and Laplace, E. 2023. Bimodal black hole mass distribution and chirp masses of binary black hole mergers. The Astrophysical Journal Letters 950: L9 (10 pages). https://doi.org/10.3847/2041-8213/acd77a.

Schutte, Z., and Reines, A. E. 2022. Black-hole-triggered star formation in the dwarf galaxy Henize 2–10. Nature 601: 329–333. https://doi.org/10.1038/s41586-021-04215-6.

Schwarzschild, K. 1916. Über das Gravitationsfeld einer Kugel aus inkompressibler Flüssigkeit nach der Einsteinschen Theorie (pp. 424–434). Berlin: Sitzungsberichte der Königlich Preussischen Akademie der Wissenschaften zu Berlin. Physics & Maths. Klasse.

Scott, N., van de Sande, J., Sharma, S., et al. 2021. Identification of an [α/Fe]—enhanced thick disk component in an edge-on Milky Way analog. The Astrophysical Journal Letters 913: L11 (6 pages). https://doi.org/10.3847/2041-8213/abfc57.

Semenov, V. A., Kravtsov, A. V., and Caprioli, D. 2021. Cosmic-ray diffusion suppression in star-forming regions inhibits clump formation in gas-rich galaxies. The Astrophysical Journal 910: 126 (20 pages). https://doi.org/10.3847/1538-4357abe2a6.

Semertzidis, Y. K., and Youn, S. W. 2022. Axion dark matter: How to see it? Science Advances 8: eabm9928 (13 pages). https://doi.org/10.1126/sciadv.abm9928.

Shen, Y., Chen, Y.-C., Hwang, H.-C., et al. 2021. A hidden population of high-redshift double quasars unveiled by astrometry. Nature Astronomy Letters. https://doi.org./10.1038/s41550-021-01323-1.

Shenar, T., Sana, H., Mahz, L., et al. 2022. An X-ray-quiet black hole born with a negligible kick in a massive binary within the Large Magellanic Cloud. Naure Astronomy. https://doi.org/10.1038/s41550–022–01730-y.

Shimwell, T. W., Hardcastle, M. J., Tasse, C., et al. 2022. The LOFAR two-metre sky survey. V. Second dana release. Astronomy & Astrophysics 659: A1 (27 pages). https://doi.org/10.1051/0004-6361/202142484.

Shimwell, T. W., Röttgering, H. J. A., Best, P. N., et al. 2017. The LOFAR two-metre sky survey. I. Survey description and preliminary data release. Astronomy & Astrophysics 598: A104 (22 pages). https://doi.org/10.1051/0004–6361/201629313.

Sicilian, D., Cappelluti, N., Bulbul, E., et al. 2020. Probing the Milky Way's dark matter halo for the 3.5 keV line. Astrophysical Journal 905: 146 (30 pages). https://doi.org/10.3847/1538-4357/abbee9.

Simpson, J. A. 1983. Elemental and isotopic composition of the galactic cosmic rays. Annual Review of Nuclear and Particle Science 33: 323–381.

Sipols, A., and Pavlovich, A. 2020. Dark matter dogma: A study of 214 galaxies. Galaxies 8: 36 (32 pages). https://doi.org/10.3390/galaxies8020036.

Sivaram, C. 1986. Evolution of the Universe through the Planck epoch. Astrophysics and Space Science 125: 189–199. https://doi.org/10.1007/BF00643984.

Smith, A., and Bromm, V. 2019. Supermassive black holes in the early Universe. arXiv:1904.12890v1 [astro-ph.GA], 29 Apr.

Smith, A., Kannan, R., and Garaldi, E. 2022. The THESAN project: Lyman-α emission and transmission during the Epoch of reionization. Monthly Notices of the Royal Astronomical Society 512(3): 3243–3265. https://doi.org/10.1093/mnras/stac713.

Spolyar, D., Bodenheimer, P., Freese, K., and Gondolo, P. 2009. Dark stars: A new look at the first stars in the Universe. The Astrophysical Journal 705: 1031. arXiv: 0903.3070 [astro-ph.CO].

Spolyar, D., Freese, K., and Gondolo, P. 2008. Dark matter and the first stars: A new phase of stellar evolution. Physical Review Letters 100: 051101. arXiv:0705.0521 [astro-ph].

Steinhardt, C. L., Sneppen, A., and Sen, B. 2020. Effects of supernova redshift uncertainties on the determination of cosmological parameters. The Astrophysical Journal 902: 14 (11 pages). https://doi.org/10.3847/1538-4357/abb140.

Stone, N. C., Küpper, A. H. W., and Ostriker, J. P. 2017. Formation of massive black holes in galactic nuclei: Runaway tidal encounters. Monthly Notices of the Royal Astronomical Society 467(4): 4180–4199. https://doi.org/10.1093/mnras/stx097.

Sutter, P. M. 2023. Dark matter may have its own 'invisible' periodic table of elements. LiveScience. November 10. www.livescience.com/space/cosmology/dark-matter-may-have-its-own-invisible-periodic-table-of-elements.

Sweeney, D., Tuthill, P., Sharma, S., and Hirai, R. 2022. The galactic underworld: The spatial distribution of compact remnants. Monthly Notices of the Royal Astronomical Society 516(4): 4971–4979.

Sweijen, F., van Weeren, R. J., Röttgering, H. J. A., et al. 2022. Deep sub-arcsecond wide-field imaging of the Lockman Hole field at 144 MHz. Nature Astronomy Letters 6: 350–356.

Taverna, R., Turolla, R., Muleri, F., et al. 2022. Polarized X-rays from a magnetar. Science 378: 646–650.

Telescope Array Collaboration. 2023. An extremely energetic cosmic ray observed by a surface detector array. Science 382(6673): 903–907. https://doi.org/10.1126/science.abo5095.

The CHIME/FRB Collaboration, Anderson, B. C., Bandura, K., et al. 2022. Sub-second periodicity in a fast radio burst. Nature 607: 256–259.

The Event Horizon Telescope Collaboration. 2019. First M87 event horizon telescope results. IV. Imaging the central supermassive black hole. The Astrophysical Journal Letters 875(1): L4 (52 pages). https://doi.org/10.3847/2041-8213/ab0e85.

The Fermi-LAT Collaboration. 2022. A gamma-ray pulsar timing array constrains the nanohertz gravitational wave background. Science. https://doi.org/10.1126/science.abm3231 (10 pages).

Tian, P., Zhang, P., Wang, W., et al. 2023. Subsecond periodic radio oscillations in a microquasar. Nature. article (18 pages). https://doi.org/10.1038/s41586-023-06336-6.

Torniamenti, S., Gieles, M., Penoyre, Z., et al. 2023. Stellar-mass black holes in the Hyades star cluster? Monthly Notices of the Royal Astronomical Society 524: 1965–1986. https://doi.org/10.1093/mnras/stad1925.

Tozzi, P., Pentericci, L., Gilli, R., et al. 2022. The 700 ks *Chandra* spiderweb field I: Evidence for widespread nuclear activity in the protocluster. Astronomy & Astrophysics. (34 pages). arXiv:2203.02208v1 [astro-ph.GA], 4 Mar.

Troja, E., Fryer, C. L., O'Connor, B., et al. 2022. A nearby long gamma-ray burst from a merger of compact objects. Nature 612: 228–231. https://doi.org/10.1038/s41586-022-05327-3.

Tsukui, T., and Iguchi, S. 2021. Spiral morphology in an intensely star-forming disk galaxy more than 12 billion ago. Science Reports (8 pages). https://doi.org/10.1126/science.abe9680.

Tully, R. B., Howlett, C., and Pomarède, D. 2023. Ho'oleilana: An individual baryon acoustic oscillation? The Astrophysical Journal 954: 169 (14 pages). https://doi.org/10.3847/1538-4357/aceaf3.

Van Dokkum, P., Pasha, I., Buzzo, M. L., et al. 2023. A candidate runaway supermassive black hole identified by shocks and star formation in its wake. arXiv:2302.04888v1 [astro-ph.GA], 9 Feb.

Van Dokkum, P., Shen, Z., Keim, M. A., et al. 2022. A trail of dark-matter-free galaxies from a bullet-dwarf collision. Nature 605: 435–439. https://doi.org/10.1038/s41586-022-04665-6.

Van Haarlem, M. P., Wise, M. W., Gunst, A. W., et al. 2013. LOFAR: The low-frequency array. Astronomy and Astrophysics 556: A2 (53 pages). https://doi.org/10.1051/0004-6361/201220873.

Vermeulen, S. M., Relton, P., Grote, H., et al. 2021. Direct limits for scalar field dark matter from a gravitational-wave detector. Nature 600: 424–428.

Voggel, K. T., Seth, A. C., Baumgardt, H., et al. 2022. First direct dynamical detection of a dual supermassive black hole system at sub-kiloparsec separation. Astronomy and Astrophysics 658: A152 (15 pages). https://doi.org/10.1051/0004-6361/202140827.

Vrard, M., Cunha, M. S., Bossini, D., et al. 2022. Evidence of structural discontinuities in the inner core of red-giant stars. Nature Communications 13: 7553 (9 pages). https://doi.org/10.1038/s41467-022-34986-z.

Wang, B., Fujimoto, S., Labbé, I., et al. 2023. UNCOVER: Illuminating the early Universe—JWST/NIRSpec confirmation of z > 12 galaxies. The Astrophysical Journal Letters 957: L34 (13pages). https://doi.org/10.3847/2041-8213/acfe07.

Wang, J., Kara, E., Lucchini, M., et al. 2022. The NICER "reverberation machine": A systematic study of time lags in black hole X-ray binaries. The Astrophysical Journal 930: 18 (24 pages). https://doi.org/10.3847/1538-4357/ac6262.

Wang, P., Libeskind, N. I., Tempel, E., et al. 2021a. Possible observational evidence for cosmic filament spin. Nature Astronomy Articles. https://doi.org/10.1038/s41550-021-01380-6.

Wang, Q. D. 2021. Chandra large-scale mapping of the galactic center: Probing high-energy structures around the central molecular zone. Monthly Notices of the Royal Astronomical Society 504(2): 1609–1618.

Wang, Z., Kaplan, D. L., Murphy, T., et al. 2021b. Discovery of ASKAP J173608.2–321635 as a highly polarized transient point source with the Australian SKA pathfinder. The Astrophysical Journal 920: 45 (19 pages). https://doi.org/10.3847/1538-4357/ac2360.

Webber, W. R. 1998. A new estimate of the local interstellar energy density and ionization rate of galactic cosmic rays. The Astrophysical Journal 506: 329–334.

Webber, W. R. 2017. GeV/nuc and the cosmic ray He/C nuclei ratio. https://arh.org/1711.11584.

Welch, B., Coe, D., Broadhurst, T., et al. 2022. A highly magnified star at redshift 6.2. Nature 603: 815–818.

Wilkins, D. R., Gallo, L. C., Costantini, E., et al. 2021. Light bending and X-ray echoes from behind a supermassive black hole. Nature 595: 657–660.

Will, P., Busemann, H., Riebe, M. E. I., and Maden, C. 2022. Indigenous noble gases in the moon's interior. Science Advances 8: eabl4920 (8 pages). https://doi.org/10.1126/sciadv.abl4920.

Willcox, R., Mandel, I., Thrane, E., et al. 2021. Constraints on weak supernova kicks from observed pulsar velocities. The Astrophysical Letters 920: L37 (7 pages). https://doi.org/10.3847/2041-8213/ac2cc8.

Wilson, J. T., Lawrence, D. J., and Peplowski, P. N. 2021. Measurement of the free neutron lifetime using the neutron spectrometer on NASA's Lunar Prospector mission. Physical Review C 104: 045501 (10 pages). https://doi.org/10.1103/PhysRevC.104.045501.

Witze, A. 2022. Landmark WEBB telescope releases first science images—astronomers are in awe. Nature 607: 429–430.

Woodward, C. E., Evans, A., Banerjee, D. P. K., et al. 2021. The infrared evolution of dust in V838 monocerotis. The Astronomical Journal 162: 183 (9 pages). https://dio.org/10.3847/1538-3881/ac1f1e.

Yan, H., Ma, Z., Ling, C., et al. 2023. First batch of z≈11–20 candidate objects revealed by the James Webb Space—telescope early release observations on SMACS 0723–73.

The Astrophysical Journal Letters 942: L9 (20 pages). https://doi.org/10.3847/2041-8213/aca80c.

Yan, Z., van Waerbeke, L., Wright, A. H., et al. 2022. Cosmic star formation history with tomographic cosmic infrared background-galaxy cross-correlation. arXiv:2204.01649v3 [astro-ph.GA], 18 Nov.

Yeghikyan, A. G. 2011. On the cosmic ray-induced ionization rate in molecular clouds. ISRN Astronomy and Astrophysics. article ID 905015 (10 pages).

Yoshii, Y., Sameshima, H., Tsujimoto, T., et al. 2022. Potential signature of population III pair-instability supernova ejecta in the BLR gas of the most distant quasar at z=7.54. The Astrophysical Journal 937: 61 (12 pages). https://doi.org/10.3847/1538-4357/ac8163.

Yusef-Zadeh, F., Royster, M., Wardle, M., et al. 2013. ALMA observations of the galactic center: SiO outflows and high-mass star formation near Sgr A*. The Astrophysical Journal Letters 767: L32 (7 pages). https://doi.org/10.1088/2041-8205/767/2/L32.

Zhang, B. 2020. The physical mechanisms of fast radio burst. Nature 587: 45–53.

Zhang, C., Yu, Q., and Lu, Y. 2015. Simulating the galaxy cluster "El Gordo" and identifying the merger configuration. The Astrophysical Journal 813: 129 (17 pages). https://doi.org/10.1088/0004-637X/813/2/129.

Zhang, S., Cai, Z., Xu, D., et al. 2023. Inspiraling streams of enriched gas observed around a massive galaxy 11 billion years ago. Science 380(6644): 494–498. https://doi.org/10.1126/science.abj9192.

Zhang, Y., Snellen, I. A. G., Bohn, A. J., et al. 2021. The $^{13}CO$-rich atmosphere of a young accreting super-Jupiter. Nature 595: 370–372.

Zwicky, F. 1933. The redshift of extragalactic nebulae (Die Rotverschiebung von extragalaktischen Nebeln). Helvetica Physica Acta 6: 110–127. https://arxiv.org/abs/1612.00805.

Zyla, P. A., Particle Dana Group, et al. 2020. Review of particle physics. Progress of Theoretical and Experimental Physics 2020: 083C01. https://doi.org/10.1093/ptep/ptaa104.

## ADDITIONAL READING

Abbott, R., Abbott, T. D., Acernese, F., et al. 2021. GWTC-3: Compact binary coalescences observed by LIGO and Virgo during the second part of the third observing run. arXiv:2111.03606v2 [gr-qc], 17 Nov.

Afek, G., Carney, D., and Moore, D. C. 2022. Coherent scattering of low mass dark matter from optically trapped sensors. Physical Review Letters 128: 101301 (6 pages). https://doi.org/10.1103/PhysRevLett.128.101301.

Agostini, M., Altenmüller, K., Appel, S., et al. 2020. Improved measurement of 8B solar neutrinos with 1.5 kt·y of Borexino exposure. Physical Review D 101: 062001 (14 pages). https://doi.org/10.1103/PhysRevD.101.062001.

Alexander, S. 2005. A quantum gravitational relaxation of the cosmological constant. arXiv:hep-th/0503146v1, 18 Mar.

Baumann, D., Bertone, G., Stout, J., and Tomaselli, G. M. 2022. Sharp signals of boson clouds in black hole binary inspirals. arXiv:2206.01212v1 [gr-qc]. https://doi.org/10.48550/arXiv.2206.01212.

Bolton, J. S., Caputo, A., Liu, H., and Viel, M. 2022. Comparison of low-redshift Lyman-α forest observations to hydrodynamical simulations with dark photon dark matter. Physical Review Letters 129: 211102 (8 pages). https://doi.org/10.1103/PhysRevLett.129.211102.

Buonocore, S., and Sen, M. 2021. Anomalous diffusion of cosmic rays: A geometric approach. AIP Advances 11: 055221 (10 pages). https://doi.org/10.1063/5.0049401.

Calmet, X., Casadio, R., Hsu, S. D. H., et al. 2022. Quantum hair from gravity. Physical Review Letters 128: 111301 (6 pages). https://doi.org/10.1103/PhysRevLett.128.111301.

Castro Segura, N., Knigge, C., Long, K. S., et al. 2022. A persistent ultraviolet outflow from an accreting neutron star binary transient. Nature 603: 52–57. https://doi.org/10.1038/s41586-021-04324-2.

Chakraborty, A., and Roy, N. 2022. Detection of HI 21 cm emission from a strongly lensed galaxy at z ~ 1.3. Monthly Notices of the Royal Astronomical Society 519: 4074–4081.

Chen, N. M., Tucker, M. A., Hoyer, N., et al. 2023. Serendipitous nebular-phase JWST imaging of SN Ia SN 2021aefx: Testing the confinement of $^{56}Co$ decay energy. The Astrophysical Journal Letters 944: L28 (9 pages). https://doi.org/10.3847/2041-8213/acb6d8.

De Miguel, J. 2021. A dark matter telescope probing the 6 to 60 GHz band. Journal of Cosmology and Astroparticle Physics 04: 075 (24 pages). https://doi.org/10.1088/1475-7516/2021/04/075.

De Vries, M., and Romani, R. W. 2022. The long filament of PSR J2030+4415. arXiv:2202.03506v1 [astro-ph.HE], 07 Feb.

Garg, D., and Dodin, I. Y. 2022. Gravitational wave modes in matter. Journal of Cosmology and Astroparticle Physics 08: 017 (23 pages). https://doi.org/10.1088/1475-7516/2022/08/017.

Giovanetti, C., Lisanti, M., Liu, H., et al. 2022. Joint cosmic microwave background and big bang nucleosynthesis constraints on light dark sectors with dark radiation. Physical Review Letters 129: 021302 (8 pages). https://doi.org/10.1103/PhysRevLett.129.021302.

Giulietti, M., Lapi, A., Massardi, M., et al. 2023. ALMA resolves the first strongly lensed optical/near-IR-dark galaxy. The Astrophysical Journal 943: 151 (15 pages). https://doi.org/10.3847/1538-4357/aca53f.

Goldstein, S., Colin Hill, J., Iršič, V., and Sherwin, B. D. 2023. Canonical Hubble-tension-resolving early dark energy cosmologies are inconsistent with the Lyman-α forest. Physical Review Letters 131: 201001.

Graham, A. W., Soria, R., Ciambur, B. C., et al. 2021. Potential black hole seeding of the spiral galaxy NGC4424 via an infalling cluster. The Astrophysical Journal 923: 146 (14 pages). https://doi.org/10.3847/1538-4357/ac235b.

Gullieusznik, M., Giunchi, E., Poggianti, B. M., et al. 2023. UV and Hα HST observations of six GASP jellyfish galaxies.

The Astrophysical Journal 945: 54 (13 pages). https://doi.org/10.3847/1538-4357/acb59b.

Guo, Y., Jogee, S., Finkelstein, S. L., et al. 2022. First look at z > 1 bars in the rest-frame near-infrared with JWST early CEERS imaging. arXiv:2210.08658v2 [astro-ph.GA], 12 Dec.

Harikane, Y., Inoue, A. K., Mawatari, K., et al. 2022. A search for H-Dropout Lyman break galaxies at z ~ 12–16. arXiv:2112.09141v4 [astro-ph.GA], 12 Feb.

Hartwig, T., Ishigaki, M. N., Kobayashi, C. 2023. Machine learning detects multiplicity of the first stars in stellar archaeology data. The Astrophysical Journal 946: 20 (18 pages). https://doi.org/10.3847/1538-4357/acbcc6.

Hawking, S. W. 1972. Black holes in general relativity. Communications in Mathematical Physics 25: 152–166.

Izumi, T., Matsuoka, Y., Fujimoto, S., et al. 2021. Subaru high-z exploration of low-luminosity quasars (SHELLQs). XIII: Large-scale feedback and star formation in a low-luminosity quasar at z=7.07 on the local black hole to host mass relation. arXiv:2104.05738v2 [astro-ph.GA], 20 May.

Katsianis, A., Blanc, G., Lagos, C. P., et al. 2017. The evolution of the star formation rate function in the EAGLE simulations: A comparison with UV, IR and Hα observations from z ~ 8 to z ~ 0. Monthly Notices of Royal Astronomical Society 472: 919–939. https://doi.org/10.1093/mnras/stx2020.

Long, X., Patnaude, D. J., Plucinski, P. P., et al. 2022. The proper motion of the pulsar J1124-5916 in the galactic supernova remnant G292.0 + 1.8. arXiv:2205.07951v1 [astro-ph.HE], 16 May.

Malanushenko, A., Cheung, M. C. M., DeForest, C. E., et al. 2022. The coronal veil. The Astrophysical Journal 927: 1 (27 pages). https://doi.org/10.3847/1538–4357/ac3df9.

Mendelsohn, E. J., Newberg, H. J., Shelton, S., et al. 2022. Estimate of the mass and radial profile of the orphan-chenab stream's dwarf-galaxy progenitor using Milky Way@home. The Astrophysical Journal 926: 106 (22 pages). https://doi.org/10.3847/1538-4357/ac498a.

Niedermann, F., and Sloth, M. S. 2021. New early dark energy. Physical Review D 103(4): L041303 (7 pages). https://doi.org/10.1103/PhysRevD.103.L041303.

Otsuka, T., Abe, T., Yoshida, T., et al. 2022. α-Clustering in atomic nuclei from first principles with statistical learning and the Hoyle state character. Nature Communications 13: 2234 (10 pages). https://doi.org/10.1038/s41467-022-29582-0.

Pattavina, L., Iachellini, N. F., Pagnanini, L., et al. 2021. RES-NOVA sensitivity to core-collapse and failed core-collapse supernova neutrinos. arXiv:2103.08672v3 [astro-ph.IM], 20 Oct.

Payne, A. V., Shappee, B. J., Hinkle, J. T., et al. 2021. ASASSN-14ko is a periodic nuclear transient in ESO 253-G003. The Astrophysical Journal 910: 125 (20 pages). https://doi.org/10.3847/1538-4357/abe38d.

Paynter, J., Webster, R., and Thrane, E. 2021. Evidence for an intermediate-mass black hole from a gravitationally lensed gamma-ray burst. Nature Astronomy Letters. https://doi.org/10.1038/s41550-021-01307-1.

Perera, B. B. P., DeCesar, M. E., Demorest, P. B., et al. 2019. The international pulsar timing array: Second data release. Monthly Notices of the Royal Astronomical Society 490: 4666–4687.

Potapov, A., Jäger, C., and Henning, T. 2020. Ice coverage of dust grains in cold astrophysical environments. Physical Review Letters 124(22): 221103 (7 pages). https://link.aps.org/doi/10.1103/PhysRevLett.124.221103.

Poutanen, J., Veledina, A., Berdyugin, A. V., et al. 2022. Black hole spin-orbit misalignment in the X-ray binary MAXI J1820 + 070. Science 375: 874–876.

Prantzos, N. 2000. Yields of massive stars and their role in galactic chemical evolution studies. New Astronomy Reviews 44: 303–313.

Ratzinger, W., and Schwaller, P. 2021. Whispers from the dark side: Confronting light new physics with NANOGrav data. SciPost Physics 10: 047. https://doi.org/10.21468/SciPostPhys.10.2.047.

Reiter, M., Morse, J. A., Smith, N., et al. 2022. Deep diving off the 'cosmic cliffs': Previously hidden outflows in NGC 3324 revealed by JWST. Monthly Notices of the Royal Astronomical Society 517(4): 5382–5405. https://doi.org/10.1093/mnras/stac2820.

Ridden-Harper, R., Tucker, B. E., Garnavich, P., et al. 2019. Discovery of a new WZ Sagittae-type cataclysmic variable in the Kepler/K2data. Monthly Notices of the Royal Astronomical Society 490: 5551–5559.

Rosas, V. G., Isbell, J. W., Jaffe, W., et al. 2022. Thermal imaging of dust hiding the black hole in NGC1068. Nature 602: 403–407. https://doi.org/10.1038/s41586-021-04311-7.

Scalo, J., and Elmegreen, B. G. 2004. Interstellar turbulence II: Implications and effects. Annual Review of Astronomy and Astrophysics 42: 275–316.

Shen, Z., Danieli, S., van Dokkum, P., et al. 2021. A tip of the red giant branch distance of 22.1 ± 1.2 Mpc to the dark matter deficient galaxy NGC 1052-DF2 from 40 orbits of Hubble space telescope imaging. The Astrophysical Journal Letters 914: L12 (9 pages). https://doi.org/10.3847/2041-8213/ac0335.

Smith, L. C., Koposov, S. E., Lucas, P. W., et al. 2021. WWW-WIT-08: The giant that blinked. Monthly Notices of the Royal Astronomical Society 505(2): 1992–2008.

Stanke, T., Arce, H. G., Bally, J., et al. 2022. The APEX large CO heterodyne Orion legacy survey (ALCOHOLS). I. Survey overview. (37 pages). ArXiv:2201.00463v1 [astro-ph.GA], 3 Jan.

Stassun, K. G., Torres, G., Johnston, C., et al. 2021. Discovery and characterization of a rare magnetic hybrid β Cephei slowly pulsating B-type star in an eclipsing binary in the young open cluster NGC 6193. The Astrophysical Journal 910: 133 (12 pages). https://doi.org/10.3847/1538-4357/abc5a8.

Takhistov, V., Fuller, G. M., and Kusenko, A. 2021. Test for the origin of solar mass black holes. Physical Review Letters

126(7): 071101 (6 pages). https://doi.org/10.1103/PhysRevLett.126.071101.

The KATRIN Collaboration. 2022. Direct neutrino-mass measurement with sub-electronvolt sensitivity. Nature Physics 18: 160–166.

The LIGO Scientific Collaboration, The Virgo Collaboration and The KAGRA Collaboration. 2021. All-sky search for gravitational wave emission from scalar boson clouds around spinning black holes in LIGO O3 data. arXiv:2111.15507v1 [astro-ph.HE], 30 Nov.

Ubertosi, F., Gitti, M., Brighenti, F., et al. 2021. The deepest CHANDRA view of RBS 797: evidence for two pairs of equidistant X-ray cavities. arXiv:2111.03679v1 [astro-ph.GA], 5 Nov.

Vagnozzi, S., Visinelli, L., Brax, P., et al. 2021. Direct detection of dark energy: The XENONIT excess and future prospects. Physical Review D 104: 063023 (24 pages).

Vopson, M. M. 2021. Estimation of the information contained in the visible matter of the universe. AIP Advances 11: 105317 (5 pages). https://doi.org/10.1063/5.0064475.

Watkins, R., Allen, T., Bradford, C. J., et al. 2023. Analysing the large-scale bulk flow using cosmicflows4: Increasing tension with the standard cosmological model. Monthly Notices of the Royal Astronomical Society 524(2): 1885–1892. https://doi.org/10.1093/mnras/stad1984.

Yao, Y., Ho, A. Y. Q., Medvedev, P., et al. 2021. The X-ray and radio loud fast blue optical transient AT2020mrf: Implications for an emerging class of engine-driven massive star explosions. arXiv:2112.0076v1 [astro-ph.HE], 1 Dec.

Yung, L. Y. A., Somerville, R. S., Finkelstein, S. L., et al. 2023. Semi-analytic forecasts for roman—the beginning of a new era of deep-wide galaxy surveys. Monthly Notices of the Royal Astronomical Society 519(1): 1578–1600. https://doi.org/10.1093/mnras/stac3595.

Zhang, Y., Yanny, B., Palmese, A., et al. 2019. Dark energy survey year 1 results: Detection of intracluster light at redshift ~ 0.25. The Astrophysical Journal 874: 165 (19 pages). https://doi.org/10.3847/1538-4357/ab0dfd.

CHAPTER 2

# Local Universe

## 2.1 INTRODUCTION

In this chapter, we shall concentrate our attention on the small part of the Universe, which we could call the human neighborhood, or the Local Universe. The Local Universe contains galaxies, galaxy clusters, and voids. It consists of the Milky Way and its satellites: the Local Group, the Virgo Supercluster, and beyond. We shall try to present the properties and their changes over time for majority of our neighbors.

We have to be very careful, since the influence of one part of our neighborhood on another has changed during the time span of $(13.787 \pm 0.020) \times 10^9$ years, the life of the Universe calculated by the Planck Collaboration (2020). Not only the size but also all other characteristics have been changing as a function of time, in particular the elemental composition. The formation of stars and their aging has played an important role in chemical element formation. Elements essential for life were formed during this period in the required concentration ratios.

The galactic interstellar medium, housing a variety of dust particle clouds, harbored a variety of physics and chemistry processes on the surfaces of dust grains, resulting in what we call molecular clouds. This is the perfect environment for the generation of the molecules required for the process called the origin of life! The galactic interstellar medium played a very important role in the process of the origin of life.

This chapter starts with a detailed description of Milky Way galaxy. Its properties and characteristics are described, in particular the galactic disk, bulge, center, halo, and its elemental composition. The Milky Way is a spiral disk galaxy containing a few hundred billion stars. The supermassive black hole, Sgr A*, and surrounding regions of active star formation power an outflowing multiphase wind from the center of the Milky Way. Special attention is paid to the galactic interstellar medium, dust particles and their agglomeration, and physics and chemistry on the surface of dust grains and the molecular clouds they form.

The solar neighborhood is studied in some detail, in particular its history, since it is believed that it hides information about the origin of life. Two of its properties are of importance for the study of the origin of life: (1) magnetic fields in the solar neighborhood and (2) time variation of elemental composition. Once life is formed, it needs a planet environment for its development. Therefore, we discuss next the objects called planets, starting from planet formation processes in protoplanetary disks. This is of importance since the life formed in molecular clouds had to survive the planet formation process. Huge efforts in the detection and study of the properties of exoplanets are described in detail. We also mention interesting objects: free-floating planets.

Our solar system is the only home of intelligent life known to us; therefore, it is of importance to understand its history and in particular the habitable Earth formation. To accomplish this; we have to start from the present and accumulate knowledge about the Sun; planets and their moons; asteroids; and especially the planet Earth and its physical and chemical properties, including chemical element abundances. All of this is done by citing the relevant published works.

## 2.2 MILKY WAY GALAXY

Our galaxy is a complex system in which several processes operate simultaneously: metal-poor gas is accreted, is chemically enriched by dying stars, and then drifts inwards, surrendering its angular momentum to

DOI: 10.1201/9781003366607-2

stars; new stars are formed on nearly circular orbits in the equatorial plane and then diffuse through orbit space to eccentric and inclined orbits; the central stellar bar surrenders angular momentum to the surrounding disk and dark halo while acquiring angular momentum from inspiralling gas; and the outer parts of the disk are constantly disturbed by satellite objects, both luminous and dark, as they sweep through pericenter. Binney (2013) reviewed the conceptual tools required to bring these complex happenings into focus. Our first concern must be the construction of equilibrium models of the galaxy, for upon these hang our hopes of determining the galaxy's mean gravitational field, which is required for every subsequent step. Ideally, this equilibrium model should be formulated so that the secular evolution of the system can be modeled with perturbation theory. Such a theory can be used to understand how stars diffuse through orbit space from either the thin gas disk in which we presume disk stars formed or the debris of an accreted object, the presumed origin of many halo stars. Coupling this understanding with the still very uncertain predictions of the theory of stellar evolution and nucleosynthesis, one can finally extract a complete model of the chemodynamic evolution of our reasonably generic galaxy. Binney (2013) discussed the relation of such a model to cosmological simulations of galaxy formation, which provided general guidance but cannot be relied on for quantitative detail.

The prehistoric stars whose formation epochs lie beyond the redshift accessible by the Hubble Ultra Deep Field have been found en masse in little satellites around the Milky Way. Other less fortunate dwarf galaxies have been pulled apart by gravity to furnish the diffuse galactic halo. These recently uncovered relics of the ancient dwarf galaxy population may play a vital role in the pursuit of reconstructing the formation of the galaxy. Its path from the distant pre-reionization era, through the most active growth periods to the present day, can be gleaned by studying the chemical composition and the phase-space density distribution of these halo denizens. How this galactic archaeological record is collected and analyzed and what aspects of galaxy (dwarf and otherwise) formation and evolution it illuminates are the topics of the review by Binney (2013).

Battaglia et al. (2013) reviewed current understanding of the internal dynamic properties of the dwarf galaxies surrounding the Milky Way, including Carina, Draco, Fornax, Leo I, Leo II, Sculptor, Sextans, and Ursa Minor. These are the most dark matter–dominated galaxies, and as such may be considered ideal laboratories to test the current concordance cosmological model and in particular provide constraints on the nature of the dominant form of dark matter. The authors discuss the latest observations of the kinematics of stars in these systems and how these may be used to derive their mass distribution. They describe the various dynamic techniques used, with emphasis on complementarity and limitations, and discuss what the results also imply in the context of cosmological models.

The mass budget for the ISM in the Milky Way (MW) galaxy is shown in Figure 2.1.

Here are some details of Milky Way characteristics for star formation, which has been more or less steady for past approximately 8 × 109 years (Rocha-Pinto et al. 2000), for $M > M_{sun}$ stars are dying at ~same rates as being formed.

In the Milky Way, $M_{ISM} \approx 5 \times 10^9\ M_{sun}$. Sources and sinks are as follows:

MISM ≈ +1 Msun/year due to infall,

MISM ≈ −3 Msun/year due to star formation, and

MISM ≈ +1 Msun/year due to stellar outflows.

Net: ~1 Msun/year

- ISM is declining on the timescale MISM/|MISM| ≈ $5 \times 10^9$ years.
- Atom (or grain) in ISM incorporated in a star on timescale

MISM/SFR ≈ $(5 \times 10^9$ Msun)/(3Msun/year) ≈ $1.5 \times 10^9$ years.

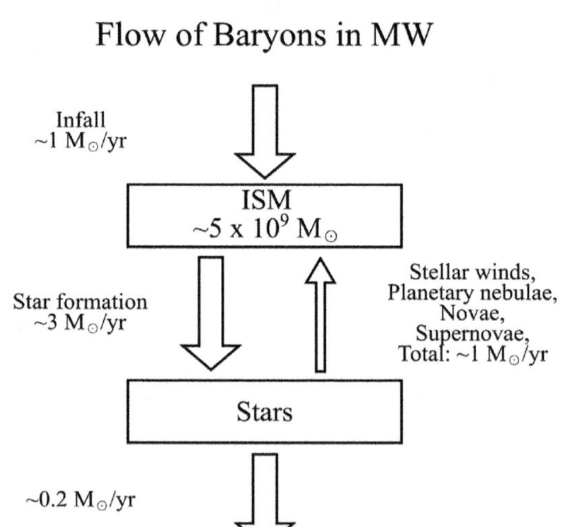


FIGURE 2.1 Flow of baryons in Milky Way galaxy.

From the archaeological point of view, the local dwarf galaxies are unique objects in which the imprint of the conditions that shaped the early structure formation can be studied today at high resolution. Over the last decade, this new window into the high redshift Universe has started to be exploited using deep wide-field imaging, high-resolution spectroscopy, and cutting edge N-body and hydro-dynamical simulations. Belokurov (2013) reviewed the recent advances in the observational studies of the Milky Way dwarf galaxies, with the aim of understanding the properties of the population as a whole and to assist in an objective comparison between the models and the data.

Bland-Hawthorn and Gerhard (2016) reviewed the structure and stellar populations of the galaxy, as shown in Figure 2.2. The Milky Way is a spiral disk galaxy containing a few hundred billion stars. It has a total dark matter mass of $(1.3 \pm 0.3) \times 10^{12}$ $M_{sun}$ and a total stellar mass of $(5 \pm 1) \times 10^{10}$ $M_{sun}$. The Milky Way and Andromeda, its slightly more massive sister galaxy, are by far the most massive galaxies in the Local Group. They are surrounded by approximately 50 known much smaller dwarf galaxies (with dark matter masses of $10^{6}$–$10^{10}$ $M_{sun}$). The galactic disk, which has a radius of 10–15 kpc, consists of several components: a thin disk with a vertical scale height of 300 ± 50 pc, a thick disk with a scale height of 900 ± 180 pc, and an even more vertically extended so-called metal-weak thick disk. The central ~3 kpc region of the galaxy, the somewhat X-shaped bulge, is a very dense and luminous region of ongoing star formation that surrounds the Milky Way's central supermassive black hole of $(4.2 \pm 0.2) \times 10^{6}$ $M_{sun}$. Both the disk and, especially, the bulge are old structures; however, they contain predominantly younger stars with near-solar abundances, although the bulge might contain a small number of the oldest stars left over from the Milky Way's earliest formation phase.

The supermassive black hole, Sgr A*, and surrounding regions of active star formation power an outflowing multiphase wind from the center of the Milky Way. Outflows play a critical role in the baryon cycle, the cycling of gas into and out of galaxies, which helps regulate the evolution of galaxies. The MW offers us a front-row view to study outflows over a range of wavelengths and phases and understand their impact on galaxy evolution

Cashman et al. (2023) characterized the chemical and physical conditions in an outflowing high-velocity cloud (HVC) in the inner galaxy. They report a super-solar metallicity of [O/H] = +0.36 ± 0.12 for the HVC at $v_{LSR} = 125.6$ km s$^{-1}$ toward the star HD 156359 (l = 328°. 7, b = −14°.5, d = 9 kpc, z = −2.3 kpc). Using archival observations from the Far Ultraviolet Spectroscopic Explorer (FUSE), the Hubble Space Telescope Imaging Spectrograph, and the European Southern Observatory Fiber-Fed Extended Range Optical Spectrograph, they measured high-velocity absorption in $H_{I}$, $O_{I}$, $C_{II}$, $N_{II}$, $Si_{II}$, $Ca_{II}$, $Si_{III}$, $Fe_{III}$, $C_{IV}$, $Si_{IV}$, $N_{V}$, and $O_{VI}$. The team measured a low $H_{I}$ column density of log N (H I) = 15.54 ± 0.05 in the HVC from multiple unsaturated H I Lyman series lines in the FUSE data. They determined a low dust depletion level in the HVC from the relative strength of silicon, iron, and calcium absorption relative to oxygen, with [Si/O] = −0.33 ± 0.14, [Fe/O] = −0.30 ± 0.20, and [Ca/O] = −0.56 ± 0.16. Analysis of the high ion absorption using collisional ionization models indicates that the hot plasma is multiphase, with the C IV and Si IV tracing $10^{4.9}$ K gas and $N_{V}$ and $O_{VI}$ tracing $10^{5.4}$ K gas. The cloud's metallicity, dust content, kinematics, and close proximity to the disk are all consistent with a galactic wind origin. As the HD 156359 line of sight probes the inner galaxy, the HVC appears to be a young cloud caught in the act of being entrained in a multiphase galactic outflow and driven out into the halo.

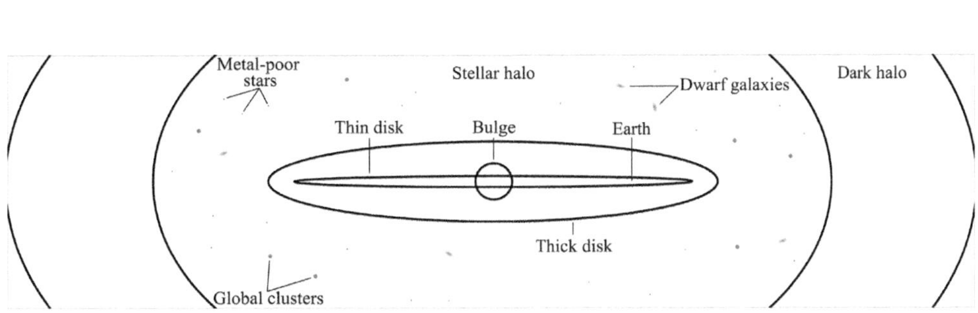


FIGURE 2.2 Structure and stellar populations of the Milky Way. Stellar archaeology is based on old, metal-poor halo stars. Dwarf galaxy archaeology utilizes stars in satellite dwarf galaxies that orbit the Milky Way.

Here are some Milky Way facts (see: https://nineplanets.org/milky-way/). Since our solar system is part of the Milky Way galaxy that contains our solar system, these are based upon the galaxy's appearance from Earth.

The galactic center of the Milky Way is about 25,000 lightyears away from our Sun. In the galactic center is Sagittarius A*, a supermassive black hole that has a mass of about 4.3 million times that of our Sun. The Milky Way has been estimated to have around 100 million black holes, but not as gigantic as the one in the center.

The mass of the Milky Way is believed to be around 100 billion times that of the Sun, which would translate to the fact that our galaxy might have around 100–400 billion stars. The radius of the Milky Way is estimated to be around 52.850 lightyears. The Milky Way is believed to be around 12 billion years old.

The Milky Way is made up of about 90% dark matter and about 10% "luminous matter", or matter. This large quantity of dark matter causes an invisible halo that has been demonstrated by simulations of how the Milky Way spins.

Our galaxy belongs to a small cluster of galaxies called the Local Group, which is one of many other groups and clusters of galaxies located near the outer edge of the Virgo supercluster. Our galaxy grew by devouring other small galaxies. Such a process is still ongoing even now. The Large and Small Magellanic Clouds are orbiting the Milky Way galaxy, while the Sagittarius dwarf galaxy is in the process of merging with the main body of the Milky Way galaxy. Estimated calculations predict that our galaxy will also collide with the Andromeda galaxy in around 8 billion years.

The Milky Way galaxy is moving at a velocity of 600 $kms^{-1}$ towards a densely packed region of space called the Shapley Attractor. In addition to the pulling effect of the Shapley Attractor, the Milky Way galaxy is also pushed by the Dipole Repeller—an empty void. About 10–15% of the Milky Way's visible matter is made of dust and gas, with the rest being stars.

How the Milky Way formed is still a matter of debate. The Big Bang theory suggests that after the explosion, the universe was composed of radiation and subatomic particles.

There are two strong viewpoints here:

- Small particles slowly teamed up and gradually formed stars, star clusters, and eventually galaxies.
- The universe first organized as immense clumps of matter that later subdivided into galaxies.

It is still up to debate. However, we know that the Milky Way galaxy grew by merging with other galaxies. Almost half of the matter in the Milky Way is believed to have come from other distant galaxies. These very old stars and clusters now constitute the galactic halo of our galaxy. It continued to grow by merging and because of the accretion of gas that came directly from the galactic halo.

When the first stars of the Milky Way were born, the mass was large enough that the galaxy was spinning relatively quickly. This led the gaseous interstellar medium to collapse from a roughly spheroidal shape to that of a disk. Later generations of stars formed in this spiral disk, with the younger stars, including our Sun, currently observed to be in the disk.

The Milky Way galaxy and the Andromeda galaxy are a binary system of giant spiral galaxies belonging to a group of 50 closely bound galaxies known as the Local Group. The Local Group is surrounded by a Local Void, itself being part of the Virgo Supercluster of galaxies. The Virgo Supercluster is also surrounded by giant voids, such as the Microscopium Void, Sculptor Void, Bootes Void, and Canes-Major Void. The voids change shape over time, creating filamentous structures of galaxies. The Virgo Supercluster, for instance, is being drawn towards the Great Attractor, which in turn forms part of a greater structure named Laniakea.

The Milky Way's galactic disk is surrounded by a spheroidal halo of old stars and globular clusters. About 90% of them lie within 100,000 lightyears of the galactic center. Almost 40% of the clusters in our galaxy are on retrograde orbits, moving in the opposite direction from the Milky Way rotation.

Deason et al. (2020) used cosmological simulations of isolated Milky Way mass galaxies, as well as Local Group (LG) analogs, to define the "edge"—a caustic manifested in a drop in density or radial velocity—of galactic-sized haloes, both in dark matter and in stars. In the dark matter, we typically identify two caustics: the outermost caustic located at $\sim 1.4r_{200m}$, corresponding to the "splashback" radius, and a second caustic located at $\sim 0.6r_{200m}$, which likely corresponds to the edge of the virialized material that has completed at least two pericentric passages. The splashback radius is ill defined in LG-type environments where the haloes of the two galaxies overlap. However, the second caustic is less affected by the presence of a companion and is a more useful definition for the boundary of the MW halo. Curiously, the stellar distribution also has a clearly defined caustic, which, in most cases, coincides with the second caustic of the dark matter. This can be identified in both radial density and radial velocity profiles and should be measurable in future observational programs. Finally, the authors showed that the second caustic can also be identified in the phase–space distribution of dwarf galaxies in the LG. Using the current dwarf galaxy population, they predict the edge of the MW halo to be 292 ± 61 kpc.

There are several pieces of evidence that the compact radio source with the designation Sagittarius A∗ (Sgr A∗) is coincident with the dynamical center of the galaxy and marks the position of a supermassive black hole with a mass of $4.3 \times 10^6$ $M_\odot$ at a distance of 8.2 kpc. Surrounding Sgr A∗ out to radii of ~100–200 pc is a twisted torus-like structure detected at far-infrared wavelengths, rich in molecular gas and dust known as the Central Molecular Zone (CMZ), thought to be driven and sustained by the innermost region of the galaxy's barred potential. The environment within the CMZ is, by numerous metrics, extreme compared to that of the disk. The density, temperature, and turbulent velocity of the gas in this region are all between 1–3 orders of magnitude higher than the average of the material in the disk, and the cosmic ray energy density is also 2–3 orders of magnitude higher than that of the galactic disk.

The region is host to several clusters of young, massive stars, including the well-known Arches and Quintuplet clusters, as well as the stars that reside in the young, massive nuclear star cluster within the central ~0.3 pc surrounding Sgr A∗. The formation of these is assumed to be caused by the episodic infall of molecular clouds from the innermost orbits of the CMZ towards the vicinity of Sgr A∗. Some of the giant molecular cloud complexes in the GC are sites of vigorous star formation (e.g. Sagittarius B2); however, this is not true for all of them, with the surprising lack of star formation in some cloud complexes being explained either due to the young ages of the clouds or the formation of higher-density cores being prevented due to turbulence and strong tidal effects.

The Milky Way galaxy is over $13 \times 10^9$ years old, and in that time it has had more than one galactic core. It is predicted that the Milky Way will collide with the Andromeda galaxy some time in $4.5 \times 10^9$ years.

The Λ cold dark matter (ΛCDM) model, where Λ is the cosmological constant, is the paradigm for galaxy and structure formation. A key ΛCDM prediction is that galaxies such as the Milky Way should be surrounded by a dispersion-supported dark matter halo and by satellite galaxies formed within its substructures. However, while several ΛCDM predictions, including the discovery of dozens of MW satellites, have since been borne out, the "plane of satellites problem" has emerged as its most persistent challenge.

The Milky Way is surrounded by 11 "classical" satellite galaxies (LMC, SMC, Sagittarius, Fornax, Leo I, Sculptor, Leo II, Sextans, Carina, Draco, Ursa Minor) in a remarkable configuration: a thin plane that is possibly rotationally supported. Such a structure is thought to be highly unlikely to arise in the standard cosmological model. While other apparent discrepancies between predictions and observations of Milky Way satellite galaxies may be explained either through baryonic effects or by invoking alternative forms of dark matter particles, there is no known mechanism for making rotating satellite planes within the dispersion-supported dark matter haloes predicted to surround galaxies such as the Milky Way. This is the so-called "plane of satellites problem", which challenges not only the ΛCDM model but also the entire concept of dark matter. In their paper, Sawala et al. (2022) showed that the reportedly exceptional anisotropy of the Milky Way satellites could be explained, in large part, by their lopsided radial distribution combined with the temporary conjunction of the two most distant satellites, Leo I and Leo II. Using Gaia proper motions, the authors have shown that the orbital pole alignment is much more common than previously reported and revealed the plane of satellites to be transient rather than rotationally supported. Compared with new simulations, where such short-lived planes are common, they found the Milky Way satellites compatible with standard model expectations.

It appears that the standard Lambda cold dark matter model is currently the best model we have for our universe, and solving the plane of satellites problem resolved yet another challenge to the model.

Zhou et al. (2022b) placed the Milky Way in the context of similar-looking galaxies in terms of their star formation and chemical evolution histories. They selected a sample of 138 Milky Way analogs (MWAs) from the SDSS-IV/MaNGA survey based on their masses, Hubble types, and bulge-to-total ratios. To compare their chemical properties to the detailed spatially resolved information available for the MW, they used a semi-analytic spectral fitting approach, which fits a self-consistent chemical evolution and star formation model directly to the MaNGA spectra. The authors modeled the galaxies' inner and outer regions, assuming that some of the material lost in stellar winds falls inwards. They also incorporated chemical enrichment from type II and Ia supernovae to follow the alpha-element abundance at different metallicities and locations. They found some MWAs where the stellar properties closely reproduce the distribution of age, metallicity, and alpha enhancement at both small and large radii in the MW. In these systems, the match is driven by the longer timescale for star formation in the outer parts and the inflow of enriched material to the central parts. However, other MWAs have very different histories. These divide into two categories: self-similar galaxies where the inner and outer parts evolve identically and centrally quenched galaxies where there is very little evidence of late-time central star formation driven by material accreted from the outer regions. They found that, although selected to be comparable, there are subtle morphological differences between galaxies in these different classes and that the centrally quenched galaxies formed their stars systematically earlier (Zhou et al. 2022b).

Sagittarius A* is the supermassive black hole located in the center of Milky Way, at a distance of 8 kpc from Earth. Its estimated mass is ~$4 \times 10^6$ $M_{solar}$. The so-called galactic central molecular zone is composed of massive clouds in the inner core of the galaxy. The zone displays strong and diffuse X-ray emission between 2 and 10 keV. There is a clear 6.4-keV iron line correlated with molecular structures. Clavel et al. (2013) observed short and strong variations in the X-ray emission from several clouds in the Sagittarius A complex related to its past activity. In addition, rapid variability across the electromagnetic spectrum was observed, in particular in the infrared, during which the observed flux density rises by 1–2 orders of magnitude.

Studying hot spots associated with Sgr A* has been done by several groups, summarized in recent work by Wielgus et al. (2022). Contrary to GRAVITY Collaboration near-IR observation reports (see GRAVITY Collaboration 2017, 2018), IR observations by Wielgus et al. (2022) suggest shorter periods and a brightness centroid loop shifted from the gravity center.

Eckart and Genzel (1996) identified a few dozen stars orbiting Sgr A* and named the related stellar members "S stars". These S stars belong to the "S cluster" and can be characterized by velocities up to several $10^3$ $kms^{-1}$, some of them moving along highly eccentric orbits (Ali et al. 2020). Peißker et al. (2023) presented a finding of a new stellar source, which they called S4716, orbiting Sgr A* in about 4.0 years with a periapse distance of about 100 AU. The authors concluded that the star they found, S4716, shares comparable orbital properties with star S4711.

Stars orbiting the compact radio source, is Sgr A* in the galactic center, also serve as precision probes of the gravitational field around the closest massive black hole. Data measured for S stars like S2, S29, S38, S55, S4711, and recent discovery S4716 allow for a high-precision determination of gravitational potential around Sgr A*; see GRAVITY Collaboration (2022b).

The Milky Way is the galaxy for which we have the most detailed observational data, with extremely precise astrometric and kinematic data from the Gaia satellite, as well as accurate chemical abundances for large numbers of stars from the Apache Point Observatory Galactic Evolution Experiment, GALactic Archaeology with HERMES, and the LAMOST Experiment for Galactic Understanding and Exploration surveys. Striking features of the Milky Way are the chemically distinct thin and thick disks. Thick-disk stars are enhanced in [α/Fe] with respect to thin-disk stars of a similar metallicity, with the overall metallicity lower than that of the thin disk. The two disks have distinct spatial distributions, with the thick disk having a larger scale height of ~900 pc, compared to the thin disk's ~300 pc (for details, see Scott et al. (2021).

Donlon et al. (2020) identified shell structures in the Milky Way for the first time. They found two shells in the Virgo Overdensity region and two shells in the Hercules Aquila Cloud region using Sloan Digital Sky Survey, Gaia, and LAMOST data. These shell stars are a subset of the substructure previously identified as the Virgo Radial Merger (VRM). Timing arguments for these shells indicate that their progenitor dwarf galaxy passed through the galactic center 2.7 ± 0.2 Gyr ago. Based on the time of collision, it is also possible that the VRM is related to the phenomenon that created phase-space spirals in the vertical motion of the disk and/or the Splash and could have caused a burst of star formation in the inner disk. The authors analyzed phase mixing in a collection of radial merger N-body simulations and find that shell structure similar to that observed in Milky Way data disappears by $5 \times 10^9$ years after collision with the galactic center. The method used to calculate the merger time of the VRM was able to reliably recover the correct merger times for these simulations. Previous work supported the idea that the VRM and the Gaia Sausage/Gaia–Enceladus Merger are the same. However, the Gaia Sausage is widely believed to be (8–11) × $10^9$ years old. The disparate ages could be reconciled if the larger age is associated with an infall time when the progenitor crossed the virial radius; they do not constrain the time at which the progenitor became bound to the Milky Way. Alternatively, the Gaia Sausage could be younger than previously thought.

The standard cosmological model predicts that galaxies are built through hierarchical assembly on cosmological timescales. The Milky Way, like other disk galaxies, underwent violent mergers and accretion of small satellite galaxies in its early history. Owing to Gaia Data Release 23 and spectroscopic surveys, the stellar remnants of such mergers have been identified. The chronological dating of such events is crucial to uncover the formation and evolution of the galaxy at high redshift, but it has so far been challenging due to difficulties in obtaining precise ages for these oldest stars. Here we combine asteroseismology—the study of stellar oscillations—with kinematics and chemical abundances to estimate precise stellar ages (~11%) for a sample of stars observed by the Kepler space mission. Crucially, this sample includes not only some of the oldest stars that were formed inside the galaxy but also stars formed externally and subsequently accreted onto the Milky Way. Leveraging this resolution in age, Montalbán et al. (2021) provided compelling evidence in favor of models in which the galaxy had already formed a substantial population of its stars (which now reside mainly in its thick disk) before the infall of the satellite galaxy Gaia-Enceladus/Sausage around 10 billion years ago.

According to various ground-based photometric surveys in the past two decades, as well as the recent Gaia data releases, the number of known MW stellar streams has increased substantially to over 60. Some of these streams clearly emanate from known DGs or GCs (e.g., the Sagittarius and Palomar 5 streams), but most streams lack an obvious progenitor. While imaging surveys can provide on-sky positions, distances, and proper motions (PMs), spectroscopic observations can reveal line-of-sight velocities and chemical properties of the streams. However, spectroscopic observations of individual stars in stellar streams are extremely challenging due to the low stellar density and large angular extents (>10o) of these systems, particularly for the more distant (and thus fainter) streams beyond a heliocentric distance of 10 kpc. Prior to 2018, only a handful of stellar streams had spectroscopic measurements: Sagittarius (for details and references, see Li et al. (2022)), Cetus, GD-1, Orphan, Ophiuchus, and the 300 km s–1 stream. Even for these spectroscopically confirmed streams, spectroscopic measurements are often quite sparse due to the challenges of efficiently targeting member stars. This has made it hard to answer some of the most basic questions, such as whether these streams originated from disrupting GCs or dissolving DGs.

Li et al. (2022) reported the kinematic, orbital, and chemical properties of 12 stellar streams with no evident progenitors using line-of-sight velocities and metallicities from the Southern Stellar Stream Spectroscopic Survey ($S^5$), proper motions from Gaia EDR3, and distances derived from distance tracers or the literature. This data set provides the largest homogeneously analyzed set of streams with full 6D kinematics and metallicities. All streams have heliocentric distances between ~10 and 50 kpc. The velocity and metallicity dispersions show that half of the stream progenitors were disrupted dwarf galaxies, while the other half originated from disrupted globular clusters, hereafter referred to as DG and GC streams. Based on the mean metallicities of the streams and the mass–metallicity relation, the luminosities of the progenitors of the DG streams range between Carina and Ursa Major I ($-9.5 \approx< M_V \approx< 5.5$). Four of the six GC streams have mean metallicities of [Fe/H] < –2, more metal-poor than typical Milky Way GCs at similar distances. Interestingly, the 300S and Jet GC streams are the only streams on retrograde orbits in our dozen stream sample.

Naidu et al. (2021) reconstructed the last major merger of the Milky Way by using the H3 Survey. Several lines of evidence suggest that the Milky Way underwent a major merger at $z \sim 2$ with the Gaia-Sausage Enceladus (GSE) galaxy. The authors used H3 Survey data to argue that GSE entered the galaxy on a retrograde orbit based on a population of highly retrograde stars with chemistry similar to the largely radial GSE debris. They presented the first tailored N-body simulations of the merger. From a grid of ≈500 simulations, they found that a GSE with $M_{star} = 5 \times 10^8\ M_{sun}$, $M_{DM} = 2 \times 10^{11}\ M_{sun}$ best matches the H3 data. This simulation shows that the retrograde stars are stripped from GSE's outer disk early in the merger. Despite being selected purely on angular momenta and radial distributions, this simulation reproduces and explains the following phenomena:

1. the triaxial shape of the inner halo, whose major axis is at ≈35° to the plane and connects GSE's apocenters;
2. the Hercules-Aquila Cloud and the Virgo Overdensity, which arise due to apocenter pileup; and
3. the 2 × 109–year lag between the quenching of GSE and the truncation of the age distribution of the in situ halo, which tracks the lag between the first and final GSE pericenters.

Naidu et al. (2021) made the following predictions:

1. the inner halo has a "double-break" density profile with breaks at both ≈15–18 kpc and 30 kpc, coincident with the GSE apocenters, and
2. the outer halo has retrograde streams awaiting discovery at >30 kpc that contain ≈10% of GSE's stars. The retrograde (radial) GSE debris originates from its outer (inner) disk—exploiting this trend, we reconstruct the stellar metallicity gradient of GSE (−0.04 ± 0.01 dex r50−1).

These simulations imply that GSE delivered ≈20% of the Milky Way's present-day dark matter and ≈50% of its stellar halo.

Meingast et al. (2021) presented a novel view on the morphology and dynamical state of ten prominent nearby (≲500 pc), young (~(30–300) × $10^9$ years) open star clusters with Gaia DR2: α Per, Blanco 1, IC 2602, IC 2391, Messier 39, NGC 2451A, NGC 2516, NGC 2547, Platais 9, and the Pleiades. They introduced a pioneering member-identification method that is informed by cluster bulk velocities and deconvolves the spatial distribution with a mixture of Gaussians. Our approach enables inferring the true spatial distribution of the clusters by effectively filtering field star contaminants while at the same time mitigating the effect of positional errors along the line of sight. This first application of the method reveals vast stellar coronae that extend for ≳100 pc and surround the cluster cores, which are comparatively tiny and compact. The coronae and cores form intertwined, coeval, and comoving extended cluster populations, each encompassing tens of thousands of cubic parsecs and stretching across tens of degrees in the sky. Their analysis shows that the coronae are gravitationally unbound but largely make up the bulk of the stellar mass of the populations. Most systems are in a highly dynamic state, showing evidence of expansion and sometimes simultaneous contraction along different spatial axes. The velocity field of the extended populations for the cluster cores appears asymmetric but is aligned along a spatial axis unique to each cluster. The overall spatial distribution and the kinematic signature of the populations are largely consistent with the differential rotation pattern of the Milky Way. This finding underlines the important role of global galactic dynamics in the fate of stellar systems. Their results highlight the complexity of the Milky Way's open cluster population and call for a new perspective on the characterization and dynamical state of open clusters.

Using numerical simulations, Shirazi et al. (2022) studied the escape of globular clusters from the satellite dwarf spheroidal galaxies (dSphs) of the Milky Way. We start by following the orbits of a large sample of GCs around dSphs in the presence of the MW potential field. They then obtain the fraction of GCs leaving their host dSphs within a Hubble time. They model dSphs by a Hernquist density profile with masses between $10^7$ and $7 \times 10^9$ $M_\odot$. All dSphs lie on the galactic disk plane, but they have different orbital eccentricities and apogalactic distances. They compute the escape fraction of GCs from 13 of the most massive dSphs of the MW, using their realistic orbits around the MW (as determined by Gaia). The escape fraction of GCs from 13 dSphs is in the range 12% to 93%. The average escape time of GCs from these dSphs was less than 8 × $10^9$ years, indicating that the escape process of GCs from dSphs was over. They then adopt a set of observationally constrained density profiles for specific case of the Fornax dSph. According to their results, the escape fraction of GCs shows a negative correlation with both the mass and the apogalactic distance of the dSphs, as well as a positive correlation with the orbital eccentricity of dSphs. In particular, they find that the escape fraction of GCs from the Fornax dSph is between 13% and 38%. Finally, they observe that when GCs leave their host dSphs, their final orbit around the MW does not differ much from their host dSphs.

Ou et al. (2023) construct the circular velocity curve of the Milky Way out to ~30 kpc, providing an updated model of the dark matter density profile. They derive precise parallaxes for 120,309 stars with a data-driven model, using APOGEE DR17 spectra combined with GaiaDR3, 2MASS, and WISE photometry. At outer galactic radii up to 30 kpc, they find a significantly faster decline in the circular velocity curve compared to the inner parts. This decline is better fit with a cored Einasto profile with a slope parameter $0.91^{+0.04}_{-0.05}$ than a generalized Navarro-Frenk-White (NFW) profile. The virial mass of the best-fit dark matter halo profile is only $1.81^{+0.06}_{-0.05} \times 10^{11}\ M_\odot$, significantly lower than what a generalized NFW profile delivers. They present a study of the potential systematics, affecting mainly large radii. Such a low mass for the galaxy is driven by the functional forms tested, given that it probes beyond our measurements. It is found to be in tension with mass measurements from globular clusters, dwarf satellites, and streams. Their best-fit profile also lowers the expected dark matter annihilation signal flux from the galactic center by more than an order of magnitude, compared to an NFW profile-fit.

Our position inside the galactic disk has previously prevented us from establishing an accurate rotation curve (RC). The advent of Gaia and its third data release (Gaia DR3) made it possible to specify the RC up to twice the optical radius. Jiao et al. (2023) aim to establish a new RC of the galaxy from the Gaia DR3 by drastically reducing systematic uncertainties. Their goal is to provide a new estimate of the mass of the galaxy. They compared different estimates, established a robust assessment of the systematic uncertainties, and addressed differences in methodologies, particularly regarding distance estimates. They find a sharply decreasing RC for the Milky Way; the decrease in velocity between 19.5 and 26.5 kpc is approximately 30 km $s^{-1}$. They identify, for the first time, a Keplerian decline of the RC, starting at ~19 kpc and ending at ~26.5 kpc from the galaxy center, while a flat RC is rejected with a significance of 3σ. The total mass is revised downwards to $2.06^{+0.24}_{-0.13} \times 10^{11}\ M_\odot$, which is in agreement with the absence of a significant mass increase at radii larger than 19 kpc. They evaluated the upper limit on the total mass by considering the upper values of velocity measurements, which leads to a strict, unsurpassable limit of $5.4 \times 10^{11}\ M_\odot$.

The total mass of Sgr A∗, also referred to as the dynamical mass, $M_{dyn}$, is known to high accuracy; see GRAVITY Collaboration (2019). Both the black hole irreducible mass, $M_{irr}$, and the rotational mass, $M_{rot}$, contribute to the dynamical black hole mass: $M^2_{dyn} = M^2_{irr} + M^2_{rot}$. The spin mass-energy available for extraction is $M_{spin} = M_{dyn} - M_{irr}$; this is the spin mass energy that is available to, and could in principle, power outflows and jets, for example. Extraction of spin mass energy from a black hole can have a significant impact on the black hole environment and can decrease the black hole dynamical mass. In addition, the gravitational impact of a spinning black hole on bodies and material in the immediate environment of the black hole is significantly different from that of a non-spinning black hole. For these reasons, it is interesting and important to empirically determine the spin properties of Sgr A*.

In the work by Daly et al. (2023), six archival Chandra observations are matched with eight sets of radio data and studied in the context of the outflow method to measure and study the spin properties of Sgr A*. Three radio and X-ray data sets obtained simultaneously, or partially simultaneously, are identified as preferred for the purpose of measuring the spin properties of Sgr A*. Similar results are obtained with other data sets. Results obtained with the preferred data sets are combined and indicate a weighted mean value of the spin function of F = 0.62 ± 0.10 and dimensionless spin angular momentum of $a_* = 0.90 \pm 0.06$. The spin function translates into measurements of the black hole rotational mass, $M_{rot}$, irreducible mass, $M_{irr}$, and spin mass-energy available for extraction, $M_{spin}$, relative to the total black hole dynamical mass, $M_{dyn}$. Weighted mean values of $(M_{ro}t/M_{dyn}) = (0.53 \pm 0.06)$, $(M_{irr}/M_{dyn}) = (0.85 \pm 0.04)$, $(M_{spin}/M_{dyn}) = (0.15 \pm 0.04)$, $M_{rot} = (2.2 \pm 0.3) \times 10^6\ M_\odot$, $M_{irr} = (3.5 \pm 0.2) \times 10^6\ M_\odot$, and $M_{spin} = (6.2 \pm 1.6) \times 10^5\ M_\odot$ are obtained; of course $(M_{rot}/M_{irr}) = (0.62 \pm 0.10)$ since $(M_{rot}/M_{irr}) = F$. Values obtained for Sgr A* are compared with those obtained for M87 based on the published spin function which indicate that M87 carries substantially more rotational energy and spin mass energy relative to the total (i.e., dynamical) black hole mass, the irreducible black hole mass, and in absolute terms.

### 2.2.1 Galactic Disk, Bulge, Center, and Halo

Recent observations have revealed interstellar features that apparently connect energetic activity in the central region of our galaxy to its halo. The nature of these features, however, remains largely uncertain. Wang (2021) presented a Chandra mapping of the central 2°×4° field of the galaxy, revealing a complex of X-ray-emitting threads plus plume-like structures emerging from the galactic center. This mapping shows that the northern plume or fountain is offset from a well-known radio lobe (or the GCL), which, however, may represent a foreground $H_{II}$ region, and that the southern plume is well wrapped by a corresponding radio lobe recently discovered by MeerKAT. In particular, we find that a distinct X-ray thread, G0.17 – 0.41, is embedded well within a non-thermal radio filament, which is locally inflated. This thread with a width of ~1.6 arcsec (FWHM) is ~2.6 arcmin or 6 pc long at the distance of the GC and has a spectrum that can be characterized by a power law or an optically thin thermal plasma with temperature 3 keV. The X-ray-emitting material is likely confined within a strand of magnetic field with strength 1 mG, not unusual in such radio filaments. These morphological and spectral properties of the radio/X-ray association suggest that magnetic field re-connection is the energy source. Such re-connection events are probably common when flux tubes of anti-parallel magnetic fields collide and/or become twisted in and around the diffuse X-ray plumes, representing blowout superbubbles driven by young massive stellar clusters in the GC. The understanding of the process, theoretically predicted in analog to solar flares, can have strong implications for the study of interstellar hot plasma heating, cosmic ray acceleration and turbulence (Wang 2021).

A halo is a spherical cloud of stars surrounding a star; more than 200 distant variable stars known as RR Lyrae stars have been discovered in the Milky Way's stellar halo. The most distant of these stars is more than a million light years from Earth, almost half the distance to our neighboring galaxy, Andromeda, which is about 2.5 million light years away. The stellar halo component of our galaxy is much bigger than the disk, which is about 100,000 light years across. Our solar system resides in one of the spiral arms of the disk. In the middle of the disk is a central bulge, and surrounding it is the halo, which contains the oldest stars in the galaxy.

The disk of the Milky Way galaxy is typically modeled with two double-exponential density functions—one for the thin disk and one for the thick disk. This density function is a good fit to the star counts regardless of whether there are two distinct components or a continuum from lower-scale height and higher metallicity to higher-scale height and lower metallicity. However, recent studies of the disk component of the Milky Way have uncovered disk warps, asymmetries in the star counts above and below the disk, and bulk motion of the disk that is inconsistent with our exponential disk profile. Xu et al. (2015) showed that in the anticenter region, between galactic longitudes of $110° < l < 229°$, there is an oscillating asymmetry in the main-sequence star counts on either side of the galactic plane using data from the Sloan Digital Sky Survey. This asymmetry oscillates from more stars in the north at distances of about 2 kpc from the Sun to more stars in the south at 4–6 kpc from the Sun to more stars in the north at distances of 8–10 kpc from the Sun. They also see evidence that there are more stars in the south at distances of 12–16 kpc from the Sun. The three more distant asymmetries form roughly concentric rings around the galactic center, opening in the direction of the Milky Way's spiral arms. The northern ring, 9 kpc from the Sun, is easily identified with the previously discovered Monoceros Ring. Parts of the southern ring at 14 kpc from the Sun have previously been identified as related to the Monoceros Ring, and others have been called the Triangulum Andromeda Overdensity. The two nearer oscillations are approximated by a toy model in which the disk plane is offset by the order of 100 pc up and then down at different radii. They also show that the disk is not azimuthally symmetric around the galactic anticenter and that there could be a correspondence between our observed oscillations and the spiral structure of the galaxy. Their observations suggest that the TriAnd and Monoceros Rings (which extend to at least 25 kpc from the galactic center) are primarily the result of disk oscillations.

Stellar orbits at the galactic center provide a very clean probe of the gravitational potential of the supermassive black hole. They can be studied with unique precision, beyond the confusion limit of a single telescope, with the near-infrared interferometer GRAVITY. Imaging is essential to search the field for faint, unknown stars on short orbits which potentially could constrain the black hole spin. Furthermore, it provides the starting point for astrometric fitting to derive highly accurate stellar positions. The GRAVITY Collaboration (2022a) presented $G^R$, a new imaging tool specifically designed for galactic center observations with GRAVITY. The algorithm is based on a Bayesian interpretation of the imaging problem, formulated in the framework of information field theory and building upon existing works in radio-interferometric imaging. Its application to GRAVITY observations from 2021 yields the deepest images to date of the galactic center on scales of a few milliarcseconds. The images reveal the complicated source structure within the central 100 mas around Sgr A*, where they detected the stars S29 and S55 and confirmed S62 on its trajectory, slowly approaching Sgr A*. Furthermore, they were able to detect S38, S42, S60, and S63 in a series of exposures for which they offset the fiber from Sgr A*. They provide an update on the orbits of all the aforementioned stars. In addition to these known sources, the images also reveal a faint star moving to the west at a high angular velocity. They could not find any coincidence with any known source and, thus, they refer to the new star as S300. From the flux ratio with S29, they estimate its K-band magnitude as $m_K$ (S300) '19.0 – 19.3. Images obtained with CLEAN confirm the detection. To assess the sensitivity of our images, we note that fiber damping reduces the apparent magnitude of S300, and the effect increases throughout the year as the star moves away from the field center. Furthermore, we performed a series of source injection tests. Under favorable circumstances, sources well below a magnitude of 20 can be recovered, while 19.7 is considered the more universal limit for a good data set.

Gravitational interactions between the Large Magellanic Cloud and the stellar and dark matter halo of the Milky Way are expected to give rise to disequilibrium phenomena in the outer Milky Way. A local wake is predicted to trail the orbit of the LMC, and a large-scale over-density is predicted to exist across a large area of the northern galactic hemisphere. Conroy et al. (2021) reported the detection of both the local wake and northern over-density (hereafter the "collective response") in a map of the galaxy based on 1,301 stars at galactocentric distances between 60 and 100 kiloparsecs. The location of the wake is in good agreement with an N-body simulation that includes the dynamic effect of the LMC on the Milky Way halo. The density contrast of the wake and collective response are stronger in the data than in the simulation. The detection of a strong local wake is independent evidence that the Magellanic clouds are on their first orbit around the Milky Way. The wake traces the path of the LMC, which will provide insight into the orbit of the LMC, which in turn is a sensitive probe of the mass of the LMC and the Milky Way. These data demonstrate that the outer halo is not in dynamical equilibrium, as is often assumed. The morphology and strength of the wake could be used to test the nature of dark matter and gravity.

Laporte et al. (2022) used the Gaia EDR3 data and legacy spectroscopic surveys to map the Milky Way disk substructure towards the galactic anticenter at heliocentric distances d ≥ 10 kpc. They report the discovery of multiple previously undetected new filaments embedded in the outer disk in highly extincted regions. Stars in these overdensities have the distance gradients expected for disk material and move on disk-like orbits with $v_\phi \sim$ 170–230 $kms^{-1}$, showing small spreads in energy. Such a morphology argues against a quiescently growing galactic thin disk. Some of these structures are interpreted as excited outer disk material, kicked up by satellite impacts and currently undergoing phase mixing (disk streams, "feathers"). Due to the long timescale in the outer disk regions, these structures can stay coherent in configuration space over several Gyrs. The authors nevertheless note that some of these structures could also be folds in the perturbed disk seen in projection from the Sun's location. A full 6D phase–space characterization and age dating of these structures should help distinguish between the two possible morphologies.

Russeil (2003) attempted to interpret the complexes' distribution in terms of spiral structure by fitting models with two, three, and four logarithmic spiral arms. The four-arm model seems more appropriate to represent the grand design of our galaxy; see Figure 2.3. In this model, the Norma arm and the external arm appear as the two extremities of a single arm called the Norma-Cygnus arm.

The X-ray satellite eROSITA has recently revealed two gigantic bubbles extending to ~80° above and below the galactic center. The morphology of these "eROSITA bubbles" bears a remarkable resemblance to the Fermi bubbles previously discovered by the Fermi Gamma-Ray Space Telescope and its counterpart, the microwave haze. The physical origin of these striking structures has been intensely debated; however, because of their symmetry about the galactic center, they probably originate from some energetic outbursts from the galactic center in the past. Here we propose a theoretical model in which the eROSITA bubbles, Fermi bubbles, and microwave haze could be simultaneously explained by a single event of jet activity from the central supermassive black hole a few million years ago. Using numerical simulations, we show that this model could successfully reproduce the morphology and multi-wavelength spectra

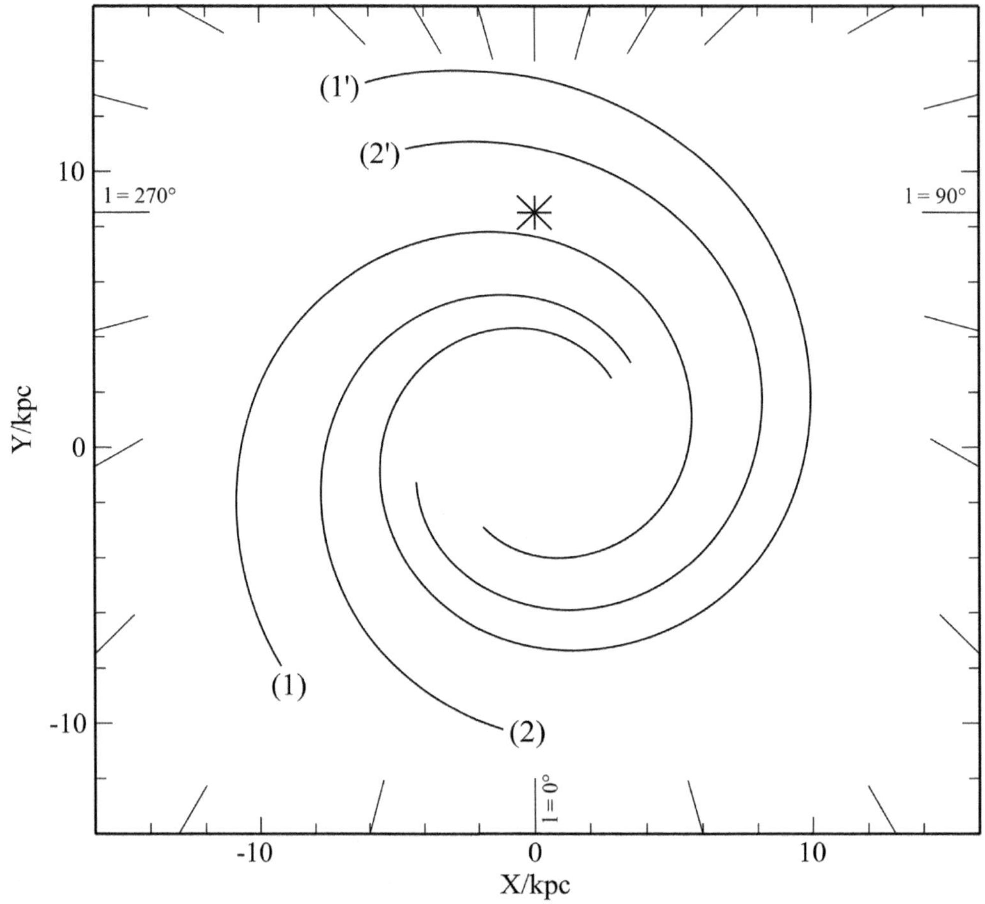


FIGURE 2.3 The adopted four-arm model. The symbol size is proportional to the excitation parameter. The Sun position is given by the large star symbol. All the complexes are plotted. 1: Sagittarius-Carina arm, 2: Scutum-Crux arm, 1': Norma-Cygnus arm and 2': Perseus arm.

of the observed bubbles and haze, which allows us to derive critical constraints on the energetics and timescales of the outburst. This study serves as an important step forward in our understanding of the past galactic center activity of our own galaxy and may bring valuable insights into the broader picture of supermassive-black-hole–galaxy co-evolution in the context of galaxy formation (Yang et al. 2022).

More than 35 years have passed since the discovery of the magnetized radio filaments associated with the galactic center Radio Arc near l~0.2°. These observations showed linear, magnetized features running mainly perpendicular to the galactic plane. Their morphology was unique and different from the shell-like or jet-like nonthermal radio continuum sources that had been observed. This was the first indication that the nucleus of our galaxy harbors energetic activity that produces relativistic particles along straight filaments with no obvious source of acceleration. Dozens of nonthermal radio filaments with similar characteristics to the prototype filaments in the Radio Arc have been discovered in the intervening years, indicating narrow synchrotron structures, and have been confirmed by polarization measurements.

Yusef-Zadeh et al. (2022) presented high-pass filtered continuum images of the inner 3.5o × 2.5o of the galactic center at 20 cm with 6.4″ resolution. These mosaic images are taken with MeerKAT and reveal a large number of narrow filaments, roughly an order of magnitude increase in their numbers compared to past measurements. For the first time, they carried out population studies of the spectral index and magnetic field of the entire region. The mean spectral indices of the filaments are steeper than supernova remnants (SNRs) (−0.62) with a value of $\alpha \sim -0.83$. The variation in α is much larger than for the SNRs, suggesting that these characteristics have a different origin. A large-scale cosmic ray–driven wind has recently been proposed to explain the origin of filaments and the large-scale 430 pc bipolar radio and X-ray structure. This favors the possibility that the large-scale bipolar radio/X-ray structure is produced by past activity of Sgr A* rather than coordinated burst of supernovae. A trend of steeper indices is also noted with increasing distance from the galactic plane. This could be explained by either synchrotron cooling or weak shocks accelerating cosmic-ray particles in the context of the cosmic ray–driven wind. The mean magnetic field strengths along the filaments ranges from ~100 to 400 μG depending on the assumed ratio of cosmic-ray protons to electrons. Given that there is a high cosmic ray pressure in the galactic center, the large equipartition magnetic field implies that the magnetic field is weak in most of the interstellar volume of the galactic center.

The inner ~200-pc region of the galaxy contains a 4-million $M_\odot$ supermassive black hole (SMBH), significant quantities of molecular gas, and star formation and cosmic ray energy densities that are roughly two orders of magnitude higher than the corresponding levels in the galactic disk. At a distance of only 8.2 kpc, the region presents astronomers with a unique opportunity to study a diverse range of energetic astrophysical phenomena, from stellar objects in extreme environments to the SMBH and star-formation driven feedback processes that are known to influence the evolution of galaxies as a whole. Heywood et al. (2022) presented a new survey of the galactic center conducted with the South African MeerKAT radio telescope. Radio imaging offers a view that is unaffected by the large quantities of dust that obscure the region at other wavelengths, and a scene of striking complexity is revealed. They produced total intensity and spectral index mosaics of the region from 20 pointings (144 hours on target in total), covering 6.5 square degrees with an angular resolution of 4", at a central frequency of 1.28 GHz. Many new features are revealed for the first time due to a combination of MeerKAT's high sensitivity, exceptional u,v-plane coverage, and geographical vantage point. They highlight some initial survey results, including new supernova remnant candidates, many new non-thermal filament complexes, and enhanced views of the Radio Arc bubble, Sgr A, and Sgr B regions.

The C-19 stellar stream is a grouping of stars in the Milky Way halo that share a common orbital motion around the galaxy and was discovered through the application of the STREAMFINDER algorithm to the astrometric data of Gaia Early Data Release 3. The stellar structure has a very low density and extends over approximately 15° on the sky. The reality of the stellar stream was confirmed via the photometric metallicities (solar-scaled iron content, [Fe/H]) of the Pristine Survey 10, which revealed the coherent and extremely low metallicity of candidate member stars. Especially for the brighter stars, a regime with little expected contamination, the Pristine photometric metallicities exhibit a very coherent metallicity at an extremely metal-poor level. Selecting stars with Gaia proper motions similar to those of the STREAMFINDER sample yields the color-magnitude diagram (CMD) of stars associated with the structure at high significance. The CMD of the stream displays a sparsely populated but well-defined red giant branch that extends into a main-sequence turn-off at the faint end of the data, along with a clear, and very blue, horizontal branch. These features are typical of old and metal-poor globular clusters. Correcting for stars that are fainter than the magnitude limit of the sample, the authors estimate that the total luminosity and stellar mass of the C-19 progenitor are at least $3.9 \times 10^3 L_\odot$ and $0.8 \times 10^4 M_\odot$, respectively, at the favored heliocentric distance of 18 kpc. This can only be a lower limit, as the C-19 stream may extend over a larger region of the sky, but with lower densities that prevent its detection at this stage. This mass is, however, quite typical for halo globular clusters (Martin et al. 2022b).

Stellar ejecta gradually enrich the gas out of which subsequent stars form, making the least chemically enriched stellar systems direct fossils of structures formed in the early Universe. Although a few hundred stars with metal content below 1/1000th of the solar iron content are known in the galaxy, none of them inhabit globular clusters, some of the oldest known stellar structures. These show metal content of at least approximately 0.2% of the solar metallicity ([Fe/H] $\gtrsim$ −2.7). This metallicity floor appears universal, and it has been proposed that protogalaxies that merged into the galaxies we observe today were simply not massive enough to form clusters that survived to the present day. Here we report observations of a stellar stream, C-19, whose metallicity is less than 0.05% of the solar metallicity ([Fe/H] = $-3.38 \pm 0.06$ (statistical) $\pm$ 0.20 (systematic)). The low-metallicity dispersion and the chemical abundances of the C-19 stars show that this stream is the tidal remnant of the most metal-poor globular cluster ever discovered and is significantly below the purported metallicity floor: clusters with significantly lower metallicities than observed today existed in the past and contributed their stars to the Milky Way halo (Martin et al. 2022a).

Globular clusters are tightly bound groupings of stars found in the outer halo of the Milky Way. They can contain millions of stars that are older and of lower metallicity than the general population. Their origins are unclear, but the Milky Way contains at least 150 globular clusters and probably many more. Omega Centauri is the brightest globular cluster in the night sky. It holds about 10 million stars and is the most massive globular cluster in the Milky Way. Their cores are densely packed with stars, and the dense environment can lead to powerful gravitational interactions. Those interactions can propel individual stars to extremely high velocities. Their high velocities can break the MW's gravitational hold on them, and they can escape into intergalactic space.

The dense environments in the cores of globular clusters facilitate many strong dynamical encounters among stellar objects. These encounters have been shown capable of ejecting stars from the host GC, whereupon they become runaway stars, or hypervelocity stars if unbound to the galactic potential. Cabrera and Rodriguez (2023) studied high-speed stellar ejecta originating from GCs by using Monte Carlo N-body models, in particular focusing on binary-single encounters involving compact objects. They paired their model-discriminated populations with observational catalogs of Milky Way GCs to compose a present day galactic population of stellar ejecta. They found that these kinds of encounters can accelerate stars to velocities in excess of 2000 km $s^{-1}$, to speeds beyond the previously predicted limits for ejecta from star-only encounters and in the same regime of galactic center ejections. However, the same ejections can only account for 1.5–20% of the total population of stellar runaways and only 0.0001–1% of hypervelocity stars, with similar relative rates found for runaway white dwarfs. We also provide credible regions for ejecta from 149 Milky Way GCs, which could be useful as supplementary evidence when pairing runaway stars.

Useful information about Milky Way globular clusters can be found on (https://spider.seds.org/spider/MWGC/mwgc.html): this list contains all 157 Milky Way globular clusters known at the time of this writing (May–June 2011). These objects belong to the halo of our own galaxy, the Milky Way. The purpose of this web/service is to provide images, basic data, and links to further information for all Milky Way globular clusters, as far as available. Similar info is provided for former globular cluster candidates. For each object, authors give positional data (right ascension [RA] and declination [Dec] for epoch J2000.0 as well as the constellation), distance from the Sun and the galactic center in thousands of light years (kilo light years, kly), apparent diameter (in arc minutes), and apparent visual brightness (in magnitudes) as well as radial velocity, linear diameter, and absolute brightness. Standard links include Messier and NGC online catalogs, an image from the STScI Digitized Sky Survey (DSS), the SIMBAD Astronomical Database, and Marco Castellani's Galactic Globular Cluster database (GGC-DB).

Modern galactic surveys have revealed an ancient merger that dominates the stellar halo of our galaxy (Gaia–Sausage–Enceladus). Using chemical abundances and kinematics from the H3 Survey, Han et al. (2022) identified 5559 halo stars from this merger in the radial range rGal = 6–60kpc. They forward model the full selection function of H3 to infer the density profile of this accreted component of the stellar halo. They consider a general ellipsoid with principal axes allowed to rotate with respect to the galactocentric axes, coupled with a multiply broken power law. The best-fit model is a triaxial ellipsoid (axes ratios 10:8:7) tilted 25° above the galactic plane toward the Sun and a doubly broken power law with breaking radii at 12 and 28 kpc. The doubly broken power law resolves a long-standing dichotomy in literature values of the halo breaking radius at either ~15 or ~30 kpc assuming a singly broken power law. N-body simulations suggest that the breaking radii are connected to apocenter pile-ups of stellar orbits, and so the observed double-break provides new insight into the initial conditions and evolution of the GSE merger. Furthermore, the tilt and triaxiality of the stellar halo could imply that a fraction of the underlying dark matter halo is also tilted and triaxial. This has important implications for dynamical mass modeling of the galaxy as well as direct dark matter detection experiments.

Astronomers find dozens of massive stars fleeing the Milky Way. Nobody knows how many runaway stars are on the way out of our galaxy, but astronomers keep finding more of them. Some estimates say there are 10 million runaway stars fleeing the Milky Way, but we don't know for sure. It may depend on the mechanism that drives them away, and that is something astrophysicists don't fully understand; see Carretero-Castrillo et al. (2023). There are two competing theories that attempt to explain runaway stars, and both involve massive stars. One is the dynamical ejection scenario (DES), and the other is the binary supernova scenario (BSS).

OB stars often form in binary pairs. In the BSS, one star explodes as a supernova, and the explosion kicks the other star. If the situation is right, the surviving star is given enough energy in the right direction that it can escape from its bond with its partner, which is now a neutron star or a black hole. It can also escape the gravitational pull of the Milky Way. If that happens, it begins its long journey into intergalactic space. In the DES, there's no dramatic supernova explosion. Instead, a star in a compact, densely packed region experiences gravitational interactions with other stars. Encounters between binary and single stars can produce runaways, and so can encounters between two binary pairs. The OB associations where O-type and B-type stars tend to form are the types of dense environments that can trigger runaway stars. Since most of these stars are massive, most of the runaway stars are, too; see Gough (2023).

Carretero-Castrillo et al. (2023) presented a two-dimensional method in the velocity space to discover runaway stars as those that deviate significantly from the velocity distribution of field stars. Field stars are considered to follow the galactic rotation curve. They found 106 O runaway stars, 42 of which were not previously identified as runaways. They found 69 Be runaway stars, 47 of which were not previously identified as runaways. The dispersion of runaway stars is a few times higher in Z and b than that of field stars. This is explained by the ejections they underwent when they became runaways. The percentage of runaways is 25.4% for O-type stars, and it is 5.2% for Be-type stars. In addition, they conducted simulations in three dimensions for their catalogs. They revealed that these percentages could increase to ~30% and ~6.7%, respectively. Their runaway stars include seven X-ray binaries and one gamma-ray binary. Moreover, they obtain velocity dispersions of ~5 km s–1 perpendicular to the galactic plane for O- and Be-type field stars. These values increase in the galactic plane to ~7 km s–1 for O-type stars due to uncertainties and to ~9 km s–1 for Be-type stars due to galactic velocity diffusion.

### 2.2.2 Elemental Composition

Significant progress has been made in the field of galactic chemical evolution (GCE) since the first postulation of stellar nucleosynthesis by Eddington (1920). The first techniques to determine element abundances in both gas (e.g. Aller 1942) and stars (e.g. Chamberlain and Aller 1951) were developed, and the theory of stellar nucleosynthesis was given a more formal footing by Burbidge et al. (1957). Later, more sophisticated studies of GCE were stimulated by the celebrated review by Tinsley (1980). Today, it is accepted that a galaxy's metallicity is related to its luminosity, age, and stellar mass and that different types of stars contribute to GCE in different ways; see Salaris and Cassisi (2005).

Yates et al. (2013) updated the treatment of chemical evolution in the Munich semi-analytic model, L-GALAXIES. Their new implementation includes delayed enrichment from stellar winds, Type II supernovae (SNe-II), and Type Ia supernovae (SNe-Ia), as well as metallicity-dependent yields and a reformulation of the associated supernova feedback. Two different sets of SN-II yields and three different SN-Ia delay-time distributions (DTDs) are considered, and 11 heavy elements (including O, Mg, and Fe) are self-consistently tracked. They compared the results of this new implementation with data on (1) local, star-forming galaxies; (2) Milky Way disk G dwarfs; and (3) local, elliptical galaxies. They find that the $z = 0$ gas-phase mass–metallicity relation is very well reproduced for all forms of DTD considered, as is the [Fe/H] distribution in the Milky Way disk. The [O/Fe] distribution in the Milky Way disk is best reproduced when using a DTD with ≤50% of SNe-Ia exploding within $\sim 400 \times 10^6$ years. Positive slopes in the mass – [α/Fe] relations of local ellipticals are also obtained when using a DTD with such a minor "prompt" component. Alternatively, metal-rich winds that drive light α elements directly out into the circumgalactic medium also produce positive slopes for all forms of DTD and SN-II yields considered. Overall, they find that the best model for matching the wide range of observational data considered here should include a power-law SN-Ia DTD, SN-II yields that take account of prior mass-loss through stellar winds, and some direct ejection of light α elements out of galaxies.

Due to their production sites, as well as to how they are processed and destroyed in stars, the light elements are excellent tools to investigate a number of crucial issues in modern astrophysics: from stellar structure and non-standard processes at work in stellar interiors to age dating of stars; from pre-main sequence evolution to the star formation histories of young clusters and associations and to multiple populations in globular clusters; from Big Bang nucleosynthesis to the formation and chemical enrichment history of the Milky Way galaxy and its populations, just to cite some relevant examples. In their paper, Randich and Magrini (2021) focus on lithium, beryllium, and boron (LiBeB) and on carbon, nitrogen, and oxygen (CNO). LiBeB are rare elements, with negligible abundances with respect to hydrogen; on the contrary, CNO are among the most abundant elements in the Universe, after H and He. Pioneering observations of light-element surface abundances in stars started almost 70 years ago, and huge progress has been achieved since then. Indeed, for different reasons, precise measurements of LiBeB and CNO are difficult, even in our Sun; however, the advent of state-of-the-art ground- and space-based instrumentation has allowed the determination of high-quality abundances in stars of different types belonging to different galactic populations, from metal-poor halo stars to young stars in the solar vicinity and from massive stars to cool dwarfs and giants. Noticeably, the recent large spectroscopic surveys performed with multi-fiber spectrographs have yielded detailed and homogeneous information on the abundances of Li and CNO for statistically significant samples of stars; this has allowed us to obtain new results and insights and, at the same time, raise new questions and challenges. A complete understanding of the light-element patterns and evolution in the Universe has not yet been achieved. Perspectives for further progress will open up soon thanks to the new-generation instrumentation that is under development and will come online in the coming years, after Randich and Magrini (2021).

Carbon, nitrogen, and oxygen are among the most abundant elements in the Universe. Their abundance ratios give important insights into both stellar physics and galactic chemical evolution. Their origin, together with other elements, has been extensively discussed in recent reviews and books. An effective illustration of the differences in the production sites and time scales of the CNO elements and Fe is in the review by Maiolino and Mannucci (2019). On the one hand, oxygen, like most of the α elements, is produced by massive stars ($M > 8\ M_\odot$), which terminate their lives as core-collapse SNe, with short time scales for its release in the interstellar medium. On the other hand, a lower fraction of carbon and nitrogen is produced by massive stars. Most of their production is due to intermediate-mass stars ($2\ M_\odot < M < 8\ M_\odot$), with longer time scales and thus contributing to ISM enrichment at later times. In this general overview, there are still several open issues concerning the nucleosynthesis of the CNO elements. As mentioned, carbon is produced by both massive and LIMS stars. However, the relative importance of the two major sites of production and their contribution to the global evolution of carbon over time is still unclear. For instance, some works are in favor of a dominant contribution from metal-rich massive stars, while others find more evidence for the production of C in LIMS during the latest stages of their evolution.

Oxygen is almost entirely produced by massive stars and ejected into the interstellar medium via the explosion of core collapse (type II) SNe. However, convection, rotation, and mass loss play a major role largely affecting the resulting yields. Recent constraints of the origin of C from chemical evolution are presented by Franchini et al. (2020), who, thanks to a sample of F-G-K stars from the GES DR5, complemented with Gaia DR2 astrometry, investigated the behaviors of [C/H], [C/Fe], and [C/Mg] vs. [Fe/H], [Mg/H], and age. The detected trends pose important constraints to the origin of C, suggesting that it is primarily produced in massive stars. In addition, the increase of [C/Mg] for young thin-disk stars might indicate a contribution from low-mass stars, polluting the ISM at later epochs, or increased production from massive stars at high metallicity, due to the enhanced mass loss.

The GALAH survey is providing detailed abundances for a large sample of stars (70,000 stars in their DR2). Griffith et al. (2019) presented C abundances from about 12,000 stars in GALAH DR2, separating the two disk components in the plane [C/Mg] vs. [Mg/H], and suggested that about 75% of solar C comes from core-collapse supernovae, while the remaining fraction is due to delayed mechanisms. Romano et al. (2020) compared the data from both GALAH and GES with the results of galactic chemical evolution models, concluding that more than 60% of the solar C abundance comes from massive stars. Their results are more generally extended, as they claim that the majority of C in the Universe comes from massive (fast) rotators, with a non-negligible contribution from intermediate-mass stars. The data from GES also contribute to our understanding of the evolution of oxygen. Franchini et al. (2021) analyzed the oxygen abundances of a stellar sample representative of the thin and thick disks, aiming at investigating possible differences in their oxygen content and at understanding the origin of the galactic oxygen enrichment. They found a systematic difference between the [O/Fe] content in the thin and thick disk populations at a given [Fe/H], with the thick disk being more enhanced in [O/Fe] and [O/H] with respect to thin disk stars, and a monotonic decrease of [O/Fe] with increasing metallicity, even at very high metallicity. Their result suggests that oxygen enrichment is mostly due to massive stars, through core-collapse supernovae, with no evidence of contributions from SNIa or AGB stars. In addition, they found that oxygen and magnesium do not follow the same evolution, since they do not evolve in lockstep, and thus they might have a different origin. Interestingly, they found that the abundance ratio [Mg/O] correlates with stellar ages, and thus it might be used as a good indicator of ages. Finally, new NLTE abundances for carbon and oxygen have been provided for the GALAH survey (Amarsi et al. 2020). For C and O, the NLTE corrections have minor effects with respect to the LTE abundances. The observed trend in the GALAH survey of [C/Fe] vs. [Fe/H] indicates a decrease that reflects the different timescales for the production of iron (from SNIa) and of carbon (from AGBs and massive stars). For O, there is a step linear decrease in [O/Fe] that reflects the two main mechanisms of production of these elements: SNIa for iron and massive stars from oxygen. The results from GALAH are in general good agreement with previous high-spectral resolution observations, but they are not completely consistent with the results of the APOGEE survey. Hayes et al. (2018) that indicates a plateau in the [O/Fe] and even a slight increase in [C/Fe] at super solar metallicities. This could be due to the determination of C and O abundances in APOGEE from molecular lines, which can be affected by important systematics.

Massive sets of stellar spectroscopic observations are rapidly becoming available, and these can be used to determine the chemical composition and evolution of the galaxy with unprecedented precision. One of the major challenges in this endeavor involves constructing realistic models of stellar spectra with which to reliably determine stellar abundances. At present, large stellar surveys commonly use simplified models that assume that the stellar atmospheres are approximately in local thermodynamic equilibrium (LTE). To test and ultimately relax this assumption, Amarsi et al. (2020) performed non-LTE calculations for 13 different elements (H, Li, C, N, O, Na, Mg, Al, Si, K, Ca, Mn, and Ba) using recent model atoms that have physically motivated descriptions for the inelastic collisions with neutral hydrogen, across a grid of 3756 1D MARCS model atmospheres that spans $3000 \leq T_{eff}/K \leq 8000$, $-0.5 \leq \log g/cm\ s^{-2} \leq 5.5$, and $-5 \leq [Fe/H] \leq 1$. They present the grids of departure coefficients that have been implemented into the GALAH DR3 analysis pipeline in order to complement the extant non-LTE grid for iron. They also present a detailed line-by-line re-analysis of 50 126 stars from GALAH DR3. They found that relaxing LTE can change the abundances by between −0.7 and +0.2 dex for different lines and stars. Taking departures from LTE into account can reduce the dispersion in the [A/Fe] versus [Fe/H] plane by up to 0.1 dex, and it can remove spurious differences between the dwarfs and giants by up to 0.2 dex. The resulting abundance slopes can thus be qualitatively different in non-LTE, possibly with important implications for the chemical evolution of our galaxy. The grids of departure coefficients are publicly available and can be implemented into LTE pipelines to make the most of observational data sets from large spectroscopic surveys.

Accurate determinations of stellar parameters and distances for large complete samples of stars are keys for conducting detailed studies of the formation and evolution of our galaxy. Huang et al. (2022) presented stellar atmospheric parameters (effective temperature, luminosity classifications, and metallicity) estimates for some 24 million stars determined from the stellar colors of SMSS DR2 and Gaia EDR3, based on training data sets with available spectroscopic measurements from previous high/medium/low-resolution spectroscopic surveys. The number of stars with photometric-metallicity estimates is 4–5 times larger than that collected by the current largest spectroscopic survey to date—LAMOST—over the course of the past decade. External checks indicate that the precision of the photometric-metallicity estimates are quite high, comparable to or slightly better than that derived from spectroscopy, with typical values around 0.05–0.15 dex for both dwarf and giant stars with $[Fe/H] > -2.010$, 0.10–0.20 dex for giant stars with $-2.0 < [Fe/H] \leq -1.0$ and 0.20–0.25 dex for giant stars with $[Fe/H] \leq -2.0$, and include estimates for stars as metal-poor as $[Fe/H] \sim -3.5$, substantially lower than previous photometric techniques. Photometric-metallicity estimates are obtained for an unprecedented number of metal-poor stars, including a total of over three million metal-poor (MP; [Fe/H] ≤ −1.0) stars, over half a million very metal-poor (VMP; [Fe/H] ≤ −2.0) stars, and over 25,000 extremely metal-poor (EMP; [Fe/H] ≤ −3.0) stars. Moreover, distances are determined for over 20 million stars in their sample. For the over 18 million sample stars with accurate Gaia parallaxes, stellar ages are estimated by comparing with theoretical isochrones. Astrometric information is provided for the stars in our catalog, along with radial velocities for ~10% of our sample stars, taken from completed/ongoing large-scale spectroscopic surveys.

On average, one or two Milky Way stars per century undergo core collapse, according to the most recent estimates; see Castelvecchi (2022). Yet throughout history, only five supernovae have been recorded as being visible with the naked eye, with two thought to be of the core-collapse type. There are various reasons for this discrepancy. If enough mass concentrates in the collapsing core, it forms a black hole without producing much of a light show. In perhaps the majority of cases, an explosion does happen but remains hidden from view by thick interstellar dust in the plane of the Milky Way, where massive stars reside.

Fortunately, the same physics that lets neutrinos escape a star's core will also let them cross the dusty galactic center unimpeded. This means that neutrino detectors on Earth will pick up a shower of neutrinos no matter what, and so will record collapsing stars that would not have been detected by any other means.

The detection of a silicon carbide feature in the infrared spectrum of the planetary nebula IC 2501 suggests that the nebular envelope is carbon rich. Goharji and Adams (1984) have observed IC 2501 with the IUE satellite and have combined these observations with optical measurements in order to analyze the physical conditions and abundances in the nebula. The derived value of the C/O abundance ratio is 1.2, a result which is insensitive to the adopted value of the electron temperature in the nebula.

Neutron-star mergers were recently confirmed as sites of rapid-neutron-capture (r-process) nucleosynthesis. However, in galactic chemical evolution models, neutron-star mergers alone cannot reproduce the observed element abundance patterns of extremely metal-poor stars, which indicates the existence of other sites of r-process nucleosynthesis. These sites may be investigated by studying the element abundance patterns of chemically primitive stars in the halo of the Milky Way, because these objects retain the nucleosynthetic signatures of the earliest generation of stars. Yong et al. (2021) reported the element abundance pattern of the extremely metal-poor star SMSS J200322.54−114203.3. They observed a large enhancement in r-process elements, with very low overall metallicity. The element abundance pattern is well matched by the yields of a single 25-solar-mass magneto-rotational hyper-nova. Such a hyper-nova could produce not only the r-process elements but also light elements during stellar evolution and iron-peak elements during explosive nuclear burning. Hyper-novae are often associated with long-duration γ-ray bursts in the nearby Universe. This connection indicates that similar explosions of fast-spinning strongly magnetized stars occurred during the earliest epochs of star formation in our galaxy.

The mergers of binary compact objects such as neutron stars and black holes are of central interest to several areas of astrophysics, including as the progenitors of gamma-ray bursts (GRBs), sources of high-frequency gravitational waves (GW), and likely production sites for heavy element nucleosynthesis via rapid neutron capture (the r-process). Levan et al. (2023) presented observations of the exceptionally bright gamma-ray burst GRB 230307A. They show that GRB 230307A belongs to the class of long-duration gamma-ray bursts associated with compact object mergers and contains a kilonova similar to AT2017gfo, associated with the gravitational-wave merger GW170817. They obtained James Webb Space Telescope mid-infrared imaging and spectroscopy 29 and 61 days after the burst. The spectroscopy shows an emission line at 2.15 microns, which we interpret as tellurium (atomic mass A = 130), and a very red source, emitting most of its light in the mid-IR due to the production of lanthanides. These observations demonstrate that nucleosynthesis in GRBs can create r-process elements across a broad atomic mass range and play a central role in heavy element nucleosynthesis across the Universe.

### 2.2.3 Galactic Interstellar Medium

The interstellar medium comprises gases at different temperatures and densities, including ionized, atomic and molecular species, and dust particles. The neutral ISM is dominated by neutral hydrogen and has ionization fractions of up to 8%. The concentration of chemical elements heavier than helium (metallicity) spans orders of magnitudes in galactic stars, because they formed at different times. However, the gas near the Sun is assumed to be well mixed and to have a solar metallicity in traditional chemical evolution models. The ISM chemical abundances can be accurately measured with ultraviolet absorption-line spectroscopy. However, the effects of dust depletion—which removes part of the metals from the observable gaseous phase and incorporates it into solid grains—have prevented, until recently, a deeper investigation of the ISM metallicity. De Cia et al. (2021) reported the dust-corrected metallicity of the neutral ISM measured towards 25 stars in our galaxy. They found large variations in metallicity over a factor of ten (with an average of 55 ± 7% solar metallicity and a standard deviation of 0.28 dex), including many regions of low metallicity, down to about 17% solar metallicity and possibly below. Pristine gas falling onto the galactic disk in the form of high-velocity clouds can cause the observed chemical inhomogeneities on scales of tens of parsecs. Their results suggest that this low-metallicity accreting gas does not efficiently mix into the ISM, which may help us understand metallicity deviations in nearby coeval stars.

Here we present some details of De Cia et al.'s (2021) work: they analyzed Hubble Space Telescope (HST) Space Telescope Imaging Spectrograph (STIS) near-ultraviolet spectra of 25 bright type-O and type-B stars in the galaxy. When available, they included archival Very Large Telescope (VLT) Ultraviolet and Visual Echelle Spectrograph (UVES) high-resolution optical spectra of these targets. They identified and analyzed the absorption lines from Mg i, Al ii, Si ii, Cr ii, Fe ii, Co ii, Ni ii, Zn ii, and Ti ii. They reported 3$\sigma$ significance levels for the quoted limits and 1$\sigma$ for the quoted errors, unless otherwise stated. They determined the neutral gas metallicity using two independent approaches and with different assumptions, which were called the relative method and the F* method. Both methods aim to characterize the strength of dust depletion along galactic lines of sight. They used both methods to cross-check their results. In brief, the F* and relative methods use either the gas-phase or relative abundances to estimate by how much the observations are affected by dust depletion. Each method uses a specific parameter to estimate the overall amount of dust depletion in a system, either [Zn/Fe] fit or F*. They measured total metallicities $[M/H]_{tot}$ ranging between −0.76 and +0.26 dex, where about two-thirds of their sample show subsolar metallicities. The average metallicity in their sample is −0.26 ± 0.06 dex (that is, 55 ± 7% solar), with a standard deviation of 0.28 dex. The most notable result is that the maximum variations between lines of sight are more than an order of magnitude, mostly subsolar. It is possible that the low metallicities that they measured represent a mix of two different gases, namely a nearly solar interstellar medium component with high depletion levels with the injection of significant amounts of pristine gas with zero depletion and metallicities even lower than what was measured.

The interstellar medium contains approximately a few percent of the mass of the galaxy. In addition to main constituents, H and He gas, it contains submicron dust particles composed mainly of silicates and carbonaceous material (~1% relative to interstellar gas). Dark clouds ($T < 50$ K, $\rho = (10^3–10^5)$ cm$^{-3}$) offer a protected environment for the formation of larger molecules. Those regions experience a limited radiation fields induced by cosmic rays. Icy surfaces of small dust grains offer a surface for catalytic reactions to form carbon bearing ice molecules (CO, $CO_2$, $CH_3OH$, and more). Observations at infrared, radio, millimeter, and sub-millimeter frequencies show that a large variety of organic molecules are present in the dense interstellar gas.

Dust controls the temperature of the interstellar medium by accounting for most of the elements that provide cooling, as well by providing heating through electrons ejected photoelectrically from grains (Mathis 1990). Dust grains play a central role in the physics and chemistry of interstellar and circumstellar environments. They influence the thermal properties of the medium by absorption and emission of stellar light;

TABLE 2.1 Injection of Gas and Dust from Stellar Sources*

| Gas MSun/year | Dust MSun/year | Stellar Source |
|---|---|---|
| 0.4 | 0.002 | Planetary nebulae (~0.3/year) |
| 0.5 | 0.0025 | Red giant, red supergiant, C star winds |
| 0.06 | <0.0001 (?) | OB, WR, other warm/hot star winds |
| 0.25 | 0.0002 (?) | SNe (1/100 year, ~$10^{-2}$ $M_{Sun}$ dust/Sn?) |
| 0.01 | 0.00001 | Novae (100/year, $10^{-7}$ $M_{Sun}$ dust/nova?) |
| ~1.2 | ~0.005 | All stellar sources |

* Information from Draine (2010).

provide surfaces for chemical reactions responsible for the synthesis of a major part of important astronomical molecules from the simplest ones such as $H_2$ and $H_2O$ to complex organic molecules, such as sugars and alcohols; and are building blocks of comets, asteroids, planetesimals, and planets.

Ginsburg et al. (2023) reported JWST NIRCam observations of G0.253 + 0.016, the molecular cloud in the Central Molecular Zone known as "The Brick", with the F182M, F187N, F212N, F410M, F405N, and F466N filters. They catalog 56,146 stars detected in all six filters using the crowdsource package. Stars within and behind The Brick exhibit prodigious absorption in the F466N filter that is produced by a combination of CO ice and gas. In support of this conclusion, and as a general resource, they present models of CO gas and ice and $CO_2$ ice in the F466N, F470N, and F410M filters. Both CO gas and ice contribute to the observed stellar colors. They showed that CO gas does not absorb the Pf$\beta$ and Hu $\in$ lines in F466N but that these lines show excess absorption, indicating that CO ice is present and contributes to observed F466N absorption. The most strongly absorbed stars in F466N are extincted by ~2 mag, corresponding to >80% flux loss. This high observed absorption requires very high column densities of CO and thus a total CO column that is in tension with standard CO abundance and/or gas-to-dust ratios. This result suggests the $CO/H_2$ ratio and dust-to-gas ratio are greater in the galactic center than in the galactic disk. Ice and/or gas absorption is observed even in the cloud outskirts, implying that additional caution is needed when interpreting stellar photometry in filters that overlap with ice bands throughout galactic centers.

### *2.2.3.1 Dust Particles, Agglomeration*

Here we discuss some of the dust properties and their behavior. All heavy elements enter ISM via stellar outflows/ejecta. Among other elements, silicon (Si) is leaving stars condensed into dust grains silicates, SiC. If nothing else happens, stardust grain will reside in the ISM ~$1.5 \times 10^9$ years before being incorporated into protostar or protoplanetary disk. During the stay of stardust grains (and the Si atoms in them) in the ISM, following could happen:

- chemical reaction with reactive species (H, O);
- UV photolysis (silicate immune?);
- CR damage (amorphization); CR flux is uncertain;
- sputtering in hot gas;
- grain–grain collisions;
- coagulation at low velocity;
- shattering at intermediate velocity;
- vaporization at high velocity;
- cratering of large grains by small grains.

Grain lifetime against destruction $t_{dest} \approx 5 \times 109M_{\odot}/(103M_{\odot}/102$ years) $\approx 5 \times 108$ years. $t_{dest}$ is short compared to $1.5 \times 109$ years residence time. Strong UV presumably controls type(s) of materials formed: amorphous and hydrocarbons. The Milky Way galaxy is not special: most dust in star-forming galaxies is presumably also grown in the MW. Large amounts of dust seen in high-z systems (like J114816 z $\approx$ 6.4) are likely to have been grown mainly in the ISM: SNe, AGB stars need to provide only a fraction of the dust—seed material, after Draine (2010).

The coagulation of cosmic dust grains is a fundamental process which takes place in astrophysical environments, such as presolar nebulae and circumstellar and protoplanetary disks. Cosmic dust grains can become charged through interaction with their plasma environment or other processes, and the resultant electrostatic force between dust grains can strongly affect their coagulation rate. Since ions and electrons are collected on the surface of the dust grain at random time intervals, the electrical charge of a dust grain experiences stochastic fluctuations. In Matthews et al.'s (2013) study, a set of stochastic differential equations was developed to model these fluctuations over the surface of an irregularly shaped aggregate. Then, by employing the data produced, the influence of the charge fluctuations on the coagulation process and the physical characteristics of the aggregates formed was examined. It was shown that dust with small charges (due to the small size of the dust grains or a tenuous plasma environment) is affected most strongly.

The coagulation of cosmic dust grains is a fundamental process which takes place in astrophysical environments, such as presolar nebulae and circumstellar and protoplanetary disks. Submillimeter observations of T Tauri stars give evidence that as young stellar objects (YSOs) and associated disk material evolve, there is a greater population of large dust grains in the disk, which presumably grow through the agglomeration of smaller grains (Andrews and Williams 2005). The processes by which these aggregates grow to kilometer-sized planetesimals, the precursors to planets, are still poorly understood, though much numerical and experimental work has been done to provide a physical description of the earliest stages of planetesimal formation.

One of the key factors in the growth of aggregate grains is their internal structure. At low velocities, colliding dust particles tend to stick at their point of contact without restructuring (Dominik et al. 2007), producing fluffy aggregates which are strongly coupled to the ambient gas. Brownian motion, vertical settling, and turbulence are mechanisms that provide a relative velocity of the grains with respect to the gas and to each other, allowing the grains to collide and grow. This growth may continue until the aggregates are up to centimeter sizes, at which point the collision energy is enough to cause compaction.

Grün et al. (2019) reviewed the development of dust science from the first ground-based astronomical observations of dust in space to compositional analysis of individual dust particles and their source objects. A multitude of observational techniques is available for the scientific study of space dust: from meteors and interplanetary dust particles collected in the upper atmosphere to dust analyzed in situ or returned to Earth. In situ dust detectors have been developed from simple dust impact detectors determining the dust hazard in Earth orbit to dust telescopes capable of providing compositional analysis and accurate trajectory determination of individual dust particles in space. The concept of dust astronomy has been developed, recognizing that dust particles, like photons, carry information from remote sites in space and time. From knowledge of the dust particles' birthplace and their bulk properties, we learn about the remote environment out of which the particles were formed. Dust observatory missions like Cassini, Stardust, and Rosetta study Saturn's satellites and rings and the dust environments of comet Wild 2 and comet Churyumov-Gerasimenko, respectively. Supplemented by simulations of dusty processes in the laboratory, we are beginning to understand the dusty environments in space.

Researchers have summarized the long history of dust research starting from the perception by many scientists that dust is dangerous to encounter, dirty and impure in its properties, and difficult to quantify individually and collectively. In our present understanding, dust is a treasure trove of information on its sources and the processes that shape its properties throughout space and time.

Millimeter and submillimeter particle aggregation in microgravity has been studied in some informal experiments on the International Space Station (Love et al. 2014). The most important observation was that submillimeter particles rapidly and spontaneously form clusters. The aggregates can be strong enough to survive turbulence in a protoplanetary nebula. Although these simple experiments did not formally separate the parameters, the authors concluded that there is no simple relationship between the speed and strength of aggregation and the variables like number density, composition, shape, and surface roughness. However, larger particles appeared to aggregate less strongly than smaller ones, as expected for a mechanism that plays surface-dominated electrostatic forces against mass-proportional inertial ones. Round, smooth particles aggregated weakly or not at all. In mixtures of particles, certain phases clumped more strongly than others, creating some selection effects that controlled the composition of the aggregates.

Many models of dust growth predict the development of fractal structures that evolve into non-fractal, porous dust pebbles representing the main component for planetesimal accretion. Mannel et al. (2016) reported the study of cometary dust particles collected by ESA's Rosetta, the first spacecraft to orbit a cometary nucleus. Rosetta monitored comet 67P/Churyumov-Gerasimenko's evolution during its closest approach to the Sun and beyond. Its Philae probe made the first successful landing on the surface a comet and began sending back images and data. In addition, the Rosetta mission offered the opportunity to investigate almost pristine cometary dust particles.

All particles analyzed by Mannel et al. (2016) have been found to be hierarchical agglomerates, most showing compact packing. One of the particles was found to be extremely porous, allowing a fractal analysis with density-density correlation function yielding a fractal dimension of $Df = 1.70 \pm 0.1$. This and similar fractal particles have been interpreted as pristine agglomerates built into the protoplanetary disk and preserved in the comet. The subunit size distribution of the fractal particle and the compact agglomerate particle are strikingly similar. Mannel et al. (2016) interpreted the fractal particles as preserved material from the early growth phases in the protoplanetary disk. The hierarchical construction of these particles is probably a result of the process dominated by a slow cluster-cluster agglomeration. A subsequent compaction and fragmentation phase evolves the majority of the fractal particles to compacted, non-fractal, centimeter-sized pebbles that represent the final stage of dust aggregation before a new process leads to integration into cometesimals and comets.

Although a fractal in the strict mathematical sense does not exist in nature, there are natural objects that can be described well by a fractal dimension, one of them being cometary dust. Most of the fractals in nature show self-similarity only in a statistical way, and their scale invariance is limited to a small range of magnification factors; however, fractal dust agglomerates show a statistical fractal structure which extends over a few levels of magnification.

Wurm and Blum (2000) experimentally studied the structure of cosmic dust aggregates and their alignment by motion relative to gas. They studied the shape of dust grains grown in a cluster-cluster type of aggregation (CCA) and derived characteristic axial ratios to describe the nonsphericity. CCAs might be described by an axial ratio $\rho_{CCA} = r_{g,max}/r_{g,min} \approx 2.0$ in the limit of large aggregates, where $r_{g,min}$ and $r_{g,max}$ describe the minimum and maximum radius of rotation, while small aggregates show a somewhat larger value in their mean axial ratio up to $\rho_{CCA} \approx 3.0$ but rapidly decrease to the limit $\rho_{CCA} \approx 2.0$. The axial ratios for large aggregates are in agreement with the general findings of different authors for axial ratios of interstellar dust grains that are generally described by rods or spheroids. Beyond this kind of agreement, our approach does not necessarily require a special shape for individual dust grains but rather offers a physical process to generate nonsphericity. Although simple shapes might be sufficient for first-order applications and are easier to handle analytically, the results offer a firm ground of special axial ratios for rods or spheroids on a more physical basis apart from any ad hoc assumptions. The authors also found an alignment of the aggregates during sedimentation in a gas along the drift axis leading to an axial ratio of with respect to the drift axis and an axis perpendicular to this drift. This result, $\rho_{CCA,align} = 1.21 \pm 0.02$, is directly applicable to dust grains in protoplanetary disks and planetary atmospheres; see also Blum et al. (2008).

Formation of kilometer-sized planetesimals is an important step during growth from dust grains to planets in protoplanetary disks. Formation mechanisms are still under debate because of some barriers against dust growth. One of the barriers is the "radial drift barrier". Dust particles are subject to significant radial drift toward a central star as a result of aerodynamical interaction with gas once they grow into intermediate-sized grains. The sizes of those grains are on the order of centimeters to meters in minimum-mass solar nebula. Fast radial drift generally prevents further dust growth, although it depends on disk models including an initial dust-to-gas mass ratio.

Tominaga et al. (2021) presented a new instability driven by a combination of coagulation and radial drift of dust particles. They refer to this instability as "coagulation instability" and regard it as a promising mechanism to concentrate dust particles and assist planetesimal formation in the very early stages of disk evolution. Because of dust-density dependence of collisional coagulation efficiency, dust particles efficiently (inefficiently) grow in a region of positive (negative) dust density perturbations, leading to a small radial variation of dust sizes and as a result radial velocity perturbations. The resultant velocity perturbations lead to dust concentration and amplify dust density perturbations. This positive feedback makes a disk unstable. The growth timescale of coagulation instability is a few tens of orbital periods, even when the dust-to-gas mass ratio is on the order of $10^{-3}$. In a protoplanetary disk, radial drift and coagulation of dust particles tend to result in dust depletion. The present instability locally concentrates dust particles even in such a dust-depleted region. The resulting concentration provides preferable sites for dust–gas instabilities to develop, which leads to further concentration. Dust diffusion and aerodynamical feedback tend to stabilize short-wavelength modes but do not completely suppress the growth of coagulation instability. Therefore, coagulation instability is expected to play an important role in setting up the next stage for other instabilities, such as streaming instability or secular gravitational instability, to further develop toward planetesimal formation.

Dust grains and polycyclic aromatic hydrocarbons (PAHs) in the interstellar medium play a central role in shaping the radiation field in galaxies, converting short-wavelength light from stars and other emitters into long-wavelength emission in the infrared. The mid-IR emission observed by the James Webb Space Telescope (JWST) is generated by small dust grains, which are heated stochastically to temperatures of ≥ 100–150 K to emit photons at mid-IR ≤ (5 $\lambda/\mu m < 60$) wavelengths. The PAHs reemit absorbed radiation in discrete spectral band features in the range $3 < \lambda/\mu m \leq 21$, which corresponds to the stretching and bending modes of bonds in the large molecules. Because the IR radiation is significantly less affected by absorption, dust and PAH emission provide a vital low-extinction view into the densest regions of galaxies. Hassani et al. (2023) used PHANGS–James Webb Space Telescope data to identify and classify 1271 compact 21 μm sources in four nearby galaxies using MIRI F2100W data. They identified sources using a dendrogram-based algorithm and measured the background-subtracted flux densities for JWST bands from 2 to 21 μm. Using the spectral energy distribution (SED) in JWST and HST bands plus ALMA and MUSE/VLT observations, they classified the sources by eye. They used this classification to define regions in color–color space and so establish a quantitative framework for classifying sources. They identified 1085 sources as belonging to the ISM of the target galaxies, with the remainder being dusty stars or background galaxies. These 21 μm sources are strongly spatially associated with H II regions (>92% of sources), while 74% of the sources are coincident with a stellar association defined in the HST data. Using SED fitting, they found that the stellar masses of the 21 μm sources span a range of $10^2$–$10^4$ $M_{sun}$, with mass-weighted ages down to 2 Myr. There is a tight correlation between attenuation-corrected Hα and 21 μm luminosity for $L_{\nu,F2100W}$ $10^{19} W Hz^{-1}$. Young embedded source candidates selected at 21 μm are found below this threshold and have $M_{star} < 10^3 M_{sun}$.

### 2.2.3.2 *Physics and Chemistry on the Surface of Dust Grains*

It has been known for some time that cosmic ray particles impacting interstellar dust grains are able to induce grain heating; see de Jong and Kamijo (1973) and Kalvāns (2018). The energy deposited by a cosmic ray particle in the grain smoothly dissipates as heat over the whole volume of the grain. This process allows molecules on the grain surface to overcome energy barriers for evaporation, diffusion, and chemical reactions. Many researchers, as summarized by Kalvāns (2018), have studied the subject of cosmic ray induced whole grain heating. As an example, let us mention that an olivine grain with radius of 0.1 μm can be heated to the temperature of approximately 100 K.

The origin and depletion of iron differ from all other abundant refractory elements that make up the composition of the interstellar dust. Iron is primarily synthesized in Type Ia supernovae (SNe Ia) and in core collapse supernovae (CCSN) and is present in the outflows from AGB stars. Only the latter two are observed to be sources of interstellar dust, since searches for dust in SN Ia have provided strong evidence for the absence of any significant mass of dust in their ejecta. Consequently, more than 65% of the iron is injected into the ISM in gaseous form. Yet ultraviolet and X-ray observations along many lines of sight in the ISM show that iron is severely depleted in the gas phase compared to expected solar abundances. The missing iron, comprising about 90% of the total, is believed to be locked up in interstellar dust. This suggests that most of the missing iron must have precipitated from the ISM gas by cold accretion onto preexisting silicate, carbon, or composite grains. Iron is thus the only element that requires most of its growth to occur outside the traditional stellar condensation sources. This is a robust statement that does not depend on our evolving understanding of the dust destruction efficiency in the ISM. Reconciling the physical, optical, and chemical properties of such composite grains with their many observational manifestations is a major challenge for understanding the nature and origin of interstellar dust (Dwek 2016).

Dust grains play a central role in the physics and chemistry of cosmic environments. They influence the optical and thermal properties of the medium due to their interaction with stellar radiation; provide surfaces for the chemical reactions that are responsible for the synthesis of a significant fraction of key astronomical molecules; and are the building blocks of pebbles, comets, asteroids, planetesimals, and planets. Matter in the Universe evolves through the so-called cosmic life cycle of matter. The word "interstellar" is attached to this matter, as it describes its affiliation to the interstellar medium—the space between the stars with surrounding gravitationally attached material. The cosmic life cycle can be thought to begin with the diffuse interstellar medium, where number densities, n, are typically 50 $cm^{-3}$ and temperatures are around 80 K, which is dominated by neutral hydrogen atoms. The diffuse ISM evolves into translucent clouds and then into cold, dense interstellar molecular clouds (which may or may not contain prestellar cores) with a typical n of $10^4$ $cm^{-3}$ and temperatures as low as 10 K, where molecular hydrogen and many other molecular species are found. Dense clouds collapse into denser cores forming central stars surrounded by protostellar envelopes that evolve into protoplanetary (or planet-forming) disks with large number density gradients ranging from $10^4$ to $10^{12}$ $cm^{-3}$ and temperature gradients from 10 to several 1000 K. Planet-forming disks evolve into planetary systems containing planets and debris disks (remnants of protoplanetary disks after planet formation). Central stars at the end of their life (so-called evolved stars) expand, absorbing the circumstellar material, and, finally, release their shells into the ISM, feeding the diffuse interstellar medium.

Dust is one of the fundamental players in the physical and chemical processes taking place in the cosmic environments mentioned previously. Beginning its lifecycle in the envelopes of evolved stars and, in their later evolutionary stages, planetary nebulae and supernovae, dust enters the ISM, where it can be destroyed and re-condensed, and travels through diffuse and dense clouds, protostellar envelopes, and planet-forming disks. Dust influences the thermodynamic properties of the medium by absorption and emission of stellar radiation; serves as a third body and a catalyst for key chemical reactions; participates in the formation of planets and their atmospheres and, possibly, in the delivery of the chemical precursors to life to those planets; and ends its lifecycle in the circumstellar media of dying stars.

Carbonaceous and siliceous materials are the major components of the interstellar dust. Other types of solids, such as metallic oxides, carbides, sulfides, and even metals themselves, are minor components. The formation pathways of interstellar dust remain a matter of significant scientific discussion. Primary dust formation occurs via gas-phase reactions in high-temperature circumstellar envelopes of evolved stars, mainly of asymptotic giant branch stars. AGB stars have extended atmospheres rich in oxygen, silicon, and carbon. These elements are manufactured in the stellar core and are dredged up to the surface by convection currents in the hot stellar plasma. In oxygen-rich AGB stars, where $n(O)/n(C) > 1$ and carbon is predominantly trapped in carbon monoxide (CO), oxygen atoms react in the stellar atmosphere with silicon and any other metal atoms to form amorphous and crystalline oxide and silicate grains. Carbon-rich stars, where $n(O)/n(C) < 1$, give rise to carbon particles in the form of graphite or amorphous carbon grains. These primary dust particles or grains have sizes from a few to tens of nanometers, after Potapov and McCoustra (2021).

After formation in circumstellar shells of evolved stars, dust grains are blown away from their parent stars by radiation pressure into the ISM, where they can be completely destroyed by supernova shocks. Estimates show that only a part of the dust produced by stars survives into the ISM. However, observations demonstrate the presence of dust grains in high concentrations in the ISM. Thus, an efficient formation mechanism is required to explain the difference between stellar formation and interstellar destruction rates. This alternative mechanism for dust grain formation and growth may be cold condensation under interstellar conditions. Laboratory experiments have demonstrated efficient formation of nanometer-sized silicate and carbon grains at low temperatures. In this process, hydrogen molecules may play a moderating role.

In the ISM, dust grain temperatures drop to as low as 10–20 K as the density increases, and they can acquire additional hydrogen, oxygen, carbon, nitrogen, and sulfur atoms, which have also escaped from stars. Here, intense dust growth takes place, mainly defined by two mechanisms: the addition of gas-phase atoms and molecules and coagulation via grain–grain collisions. Consequently, dust grains in the ISM reach micrometer sizes. In addition, molecule formation on dust grain surfaces takes place, with mobile atoms travelling across the surfaces to meet each other and react giving birth to basic astrophysical molecules such as $H_2$ and $H_2O$.

Many interstellar complex organic molecules (COMs) are believed to be produced on the surfaces of icy grains at low temperatures. Atomic carbon is considered responsible for the skeletal evolution processes, such as C–C bond formation, via insertion or addition reactions. However, before reactions, C atoms must diffuse on the surface to encounter reaction partners; therefore, information on their diffusion process is critically important for evaluating the role of C atoms in the formation of COMs. Tsuge et al. (2023) reported in situ detection of C atoms on ice achieved by a combination of photostimulated desorption and resonance-enhanced multiphoton ionization methods. They found that C atoms weakly bound to the ice surface diffused above approximately 30 K and produced $C_2$ molecules. The activation energy for C-atom surface diffusion was experimentally determined to be 88 meV (1,020 K), indicating that the diffusive reaction of C atoms is activated at approximately 22 K on interstellar ice. The facile diffusion of C atoms at temperatures above 22 K on interstellar ice opens a previously overlooked chemical regime where the increase in complexity of COMs is driven by C atoms. Carbon addition chemistry can be an alternative source of chemical complexity in translucent clouds and protoplanetary disks, with crucial implications in our current understanding on the origin and evolution of organic chemistry in the Universe.

In denser regions of the ISM, molecular clouds, small molecules, such as CO and $N_2$, adsorb on grain surfaces and, together with surface synthesized molecules, build up icy mantles. These mantles include $H_2O$ (the main constituent accounting for more than 60% of the ice in most lines-of-sight), CO, $CO_2$, $NH_3$, $CH_4$, $CH_3OH$, and other minor components. Rich chemistry takes place in and on these icy mantles in dense, cold environments (molecular clouds, protostellar envelopes, and planet-forming disks beyond the snowline), which ultimately results in the formation of complex organic molecules such as alcohols, amines, aldehydes, and ketones, which are considered precursors to prebiotic species. Astrophysically, COMs are C-bearing molecules consisting of six or more atoms. So, for example, methanol ($CH_3OH$) is a COM from the astrophysical/astrochemical point of view. This is quite different from what is understood by "complex molecules" in chemistry and biology. Prebiotic molecules have not yet been unambiguously detected in the ISM, protostellar envelopes, or planet-forming disks, but many precursors have been identified. However, amino acids and other key biological molecule precursors such as simple sugars, purines, and pyrimidines have been detected in comets and meteorites. Amino acids are the building blocks of proteins and, thus, are critical to life on Earth. Results of laboratory experiments mimicking the physicochemical processes in the ISM show that the generation of amino acids in the ISM is possible, supporting the hypothesis of their formation in the ISM and subsequent exogenous delivery to Earth via comets, asteroids, and their meteoritic remains (Potapov and McCoustra 2021).

In protostellar envelopes and planet-forming disks (circumstellar media of young stars), grains continue to grow reaching millimeter sizes. In disks, grains are believed to stick together, held by molecular forces, and form kilometer-sized planetesimals that, in turn, accrete and form planets due to their gravity. Another mechanism driving planetary growth linked to the observational detection of large amounts of pebbles in protoplanetary disks and supported by theoretical calculations is considered pebble accretion revolving around millimeter- to centimeter-sized pebbles. Chemistry in envelopes and disks continues to be very complex. The high densities and temperatures and intense UV irradiation from a central star make them, possibly, more appropriate environments for the formation of prebiotic molecules than molecular clouds. In reference to the formation of planets, in particular extrasolar planets, dust grains play an important role in the formation of not only planetary bodies but also planetary atmospheres, defining their dynamics and chemistry. The formation of planets is also often accompanied by the formation of debris disks around a central star. In debris disks, small dust grains can be formed top-down through collision cascades started by collisions of planetesimals.

Interstellar dust grains comprise only around 1% of the mass of the ISM, or about 0.1% of the galactic mass, but play a fundamental role in its evolution. Most of the dust found in space comes from AGB stars which have moved off the main sequence and entered the red giant phase of their evolution (Potapov and McCoustra 2021).

#### *2.2.3.3 Molecular Clouds*

Giant molecular clouds (GMCs) and their stellar offspring are the building blocks of galaxies. The physical characteristics of GMCs and their evolution are tightly connected to galaxy evolution. The macroscopic properties of the interstellar medium propagate into the properties of GMCs condensing out of it, with correlations between, for example, the galactic and GMC scale gas pressures, surface densities, and volume densities. That way, the galactic environment sets the initial conditions for star formation within GMCs. After the onset of massive star formation, stellar feedback from, for example, photoionization, stellar winds, and supernovae eventually contributes to dispersing the parent cloud, depositing energy, momentum, and metals into the surrounding medium, thereby changing the properties of galaxies. This cycling of matter between gas and stars, governed by star formation and feedback, is therefore a major driver of galaxy evolution. Much of the recent debate has focused on the durations of the various evolutionary phases that constitute this cycle in galaxies and what these can teach us about the physical mechanisms driving the cycle. We review results from observational, theoretical, and numerical work to build a dynamical picture of the evolutionary lifecycle of GMC evolution, star formation, and feedback in galaxies (Chevance et al. 2020).

The size of molecular clouds can be between a few light years and up to 600 light years. Their total mass can reach several million solar masses. Molecular clouds with dimensions of more than 15 light years are sometimes called giant molecular clouds. The physical processes which govern the behavior of giant molecular clouds are large-scale processes acting on galactic scale. The lack of observational constraints results in major uncertainties in simulations of galaxy formation and evolution (Chevance et al. 2020).

The processes of star formation, as well as their feedback, happen in clouds within galaxies (at scale of ~100 pc). In these regions, cold dust and gases coalesce into the seeds that will give rise to stars and possibly planets. Complex organic molecules that could serve as building blocks of life may be already present in the clouds. It should be noted that stellar feedback deposits energy, momentum, mass, and metals into the surrounding interstellar medium through photoionization, stellar winds, and supernovae preserving the continuous evolution of the ISM of galaxies.

Scibelli and Shirley (2020) conducted a large sample survey of 31 starless and pre-stellar cores in the Taurus molecular cloud, detecting methanol ($CH_3OH$) in 100% of the cores targeted and acetaldehyde ($CH_3CHO$) in 70%. At least two transition lines of each molecule were measured, allowing placing tight constraints on excitation temperature, column density, and abundance. The authors found that complex organic molecules are detectable in the gas phase and are being formed early, at least hundreds of thousands of years prior to star and planet formation.

Molecular clouds are important because they are the raw materials of stars and planets. A dark nebula, a dark cloud, is a very dense part of a bigger molecular cloud. The high density and the presence of interstellar dust cause the light extinction. These are the regions where new stars are forming.

Observations of magnetic fields in molecular clouds are essential for understanding their role in the evolution of dense clouds and in the star formation process. The predictions of two extreme paradigms for what drives star formation—magnetic and thermal support with ambipolar diffusion and compressible turbulence with negligible magnetic fields—are first discussed in order to formulate three tests of the two models that can be applied using observations of magnetic field strengths and morphologies. Techniques for measuring magnetic fields in molecular clouds are described. Three examples of the observational results are discussed in the paper by Crutcher (2003)—the low-mass starless core L 183, the young stellar system and molecular outflow NGC 1333 IRAS4A, and the high-mass star formation region DR 21 (OH). Finally, these and all other results of observations of magnetic fields in molecular clouds are used to apply the three tests. The conclusion is that the magnetic support model is consistent with the data, but both turbulence and strong magnetic fields are important in the physics of molecular clouds and in the star formation process.

Far-infrared and millimeter wavelength observations of linearly polarized dust emission may be used to map the morphology of the magnetic field projected onto the plane of the sky, $B_{pos}$. Linear polarization arises from elongated grains with their short axes aligned, generally with B (Lazarian 2003). The position angle of maximum emission will be perpendicular to $B_{pos}$. However, it is not possible to measure directly the strength of $B_{pos}$, since weak magnetic fields can align grains, so the degree of polarization is not a measure of field strength.

The study of molecular abundances towards molecular clouds located at different environments in the interstellar medium is very useful for our knowledge of the chemical evolution of the galaxy. Taking into account that the physical processes that take place in the molecular gas during the born and the evolution of massive stars have a deep influence in the chemistry, the study of molecular abundances towards massive star-forming regions is very important.

Nowadays there are several observational studies about the $^{13}CO/C^{18}O$ abundance ratio ($X^{13/18}$) towards nearby molecular clouds. These works give observational support to the $C^{18}O$ selective photodissociation due to the interaction between the far ultraviolet (FUV) radiation and the molecular gas. It is necessary to increase the sample of molecular clouds located at different distances and affected in different ways by nearby or embedded HII regions and OB associations to study the selective photodissociation. Using $^{12}CO$, $^{13}CO$, and $C^{18}O$ J = 1–0 data obtained from the FOREST unbiased galactic plane imaging survey performed with the Nobeyama 45 m telescope, the authors analyzed the filamentary infrared dark cloud IRDC 34.43 + 0.24 located at the distance of about 3.9 kpc. This IRDC is related to several HII regions and young stellar objects. Assuming local thermodynamic equilibrium, Areal et al. (2019) obtained: $0.8 \times 10_{16} < N(^{13}CO) < 4 \times 10^{17}$ cm$^{-2}$ (average value = $4.2 \times 10^{16}$ cm$^{-2}$), $0.6 \times 10^{15} < N(C^{18}O) < 4.4 \times 10^{16}$ cm$^{-2}$ (average value = $5.0 \times 10^{15}$ cm$^{-2}$), and $3 < X^{13/18} < 30$ (average = 8) across the whole IRDC. Larger values of $X^{13/18}$ were found towards portions of the cloud related to the HII regions associated with the N61 and N62 bubbles and with the photodissociation regions (PDRs), precisely the regions in which FUV photons are strongly interacting with the molecular gas. Their result represents an observational support to the $C^{18}O$ selectively photodissociation phenomenon occurring in a quite distant filamentary IRDC. Additionally, based on IR data from the Hi-GAL survey, the FUV radiation field was estimated in Habing units, and the dust temperature ($T_d$) and $H_2$ column density ($N(H_2)$) distribution was studied. Using the average of N(H2), values in close agreement with the canonical abundance ratios $[H_2]/[^{13}CO]$ and $[H_2]/[C^{18}O]$ were derived. However, the obtained ranges in the abundance ratios show that if an accurate analysis of the molecular gas is required, the use of the canonical values may introduce some bias. Thus, it is important to consider how the gas is irradiated by the far ultraviolet photons across the molecular cloud. The analysis of $X^{13/18}$ is a good tool to perform that. Effects of beam dilution and clumpiness were studied also (Areal et al. 2019).

The Large and Small Magellanic Clouds are two of the best-studied satellite galaxies around the Milky Way galaxy. They are both dwarf irregular galaxies with an average metallicity lower by about a factor of 2 and 5, respectively, compared to the Milky Way thin disk, and show ongoing star formation. Because of their proximity to the Milky Way galaxy, it is possible to perform detailed studies of the individual stars and the stellar populations within them. Both the LMC and SMC harbor populations of young massive stars as well as thousands of evolved asymptotic giant branch stars, including many carbon stars. The population of AGB in the Magellanic clouds is incredibly useful for studies of stellar evolution and nucleosynthesis. This is because the stars are at a known distance from us, so we can estimate their absolute brightness and luminosities. Furthermore, because the AGB population is so large, they span the entire range of AGB initial masses, which evolve from stars of 1 to about 8 $M_{sun}$, allowing us to place important constraints on the physics of this most uncertain phase of stellar evolution.

Karakas et al. (2018) presented new theoretical stellar yields and surface abundances for asymptotic giant branch models with a metallicity appropriate for stars in the Small Magellanic Cloud (Z = 0.0028, [Fe/H] ≈ −0.7). New evolutionary sequences and post-processing nucleosynthesis results are presented for initial masses between 1 and 7 $M_{sun}$, where the 7 $M_{sun}$ is a super-AGB star with an O–Ne core. Models above 1.15 $M_{sun}$ become carbon rich during the AGB, and hot bottom burning begins in models $M \geq 3.75$ $M_{sun}$. They presented stellar surface abundances as a function of thermal pulse number for elements between C to Bi and for a selection of isotopic ratios for elements up to Fe and Ni (e.g. $^{12}C/^{13}C$), which can be compared to observations. The integrated stellar yields are presented for each model in the grid for hydrogen, helium, and all stable elements from C to Bi. The authors present evolutionary sequences of intermediate-mass models between 4 and 7 $M_{sun}$ and nucleosynthesis results for three masses (M = 3.75, 5, and 7 $M_{sun}$) including s-process elements for two widely used AGB mass-loss prescriptions. They discuss these new models in the context of evolved AGB and post-AGB stars in the SMCs; barium stars in our galaxy; the composition of galactic globular clusters, including Mg isotopes with a similar metallicity to our models; and pre-solar grains, which may have an origin in metal-poor AGB stars.

The atomic phase of the interstellar medium plays a key role in the formation process of molecular clouds. Due to the line-of-sight confusion in the galactic plane that is associated with its ubiquity, atomic hydrogen emission has been challenging to study. Syed et al. (2022) investigated the physical properties of the Maggie filament, a large-scale filament identified in H I emission at line-of-sight velocities, 3LSR ~ −54 km s$^{-1}$. By employing the high-angular resolution data from The H I/OH Recombination line survey of the inner Milky Way (THOR), the authors have been able to study H I emission features at negative 3LSR velocities without any line-of-sight confusion due to the kinematic distance ambiguity in the first galactic quadrant. In order to investigate the kinematic structure, we decomposed the emission spectra using the automated Gaussian fitting algorithm GAUSSPY+. The authors identified one of the largest, coherent, mostly atomic H I filaments in the Milky Way. The giant atomic filament Maggie, with a total length of 1.2 ± 0.1 kpc, is not detected in most other tracers, and it does not show signs of active star formation. At a kinematic distance of 17 kpc, Maggie is situated below (by ≈500 pc), but parallel to, the galactic H I disk and is trailing the predicted location of the outer arm by 5 – 10 km $s^{-1}$ in longitude-velocity space. The centroid velocity exhibits a smooth gradient of less than ±3 km $s^{-1}$ (10 $pc)^{-1}$ and a coherent structure to within ±6 km $s^{-1}$. The line widths of ~10 km $s^{-1}$ along the spine of the filament are dominated by nonthermal effects. After correcting for optical depth effects, the mass of Maggie's dense spine is estimated to be $7.2^{+2.5}_{-1.9} \times 10^5$ $M_{sun}$.

The mean number density of the filament is ~4 $cm^{-3}$, which is best explained by the filament being a mix of cold and warm neutral gas. In contrast to molecular filaments, the turbulent Mach number and velocity structure function suggest that Maggie is driven by transonic to moderately supersonic velocities that are likely associated with the galactic potential rather than being subject to the effects of self-gravity or stellar feedback. The probability density function of the column density displays a log-normal shape around a mean of $\langle N_{H\,I}\rangle = 4.8 \times 10^{20}$ $cm^{-2}$, thus reflecting the absence of dominating effects of gravitational contraction. While Maggie's origin remains unclear, we hypothesize that Maggie could be the first in a class of atomic clouds that are the precursors of giant molecular filaments.

In order to investigate the impact of radio jets on the interstellar medium of galaxies hosting active galactic nuclei (AGN), Audibert et al. (2023) presented subarcsecond resolution Atacama Large Millimeter/Submillimeter Array (ALMA) CO(2–1) and CO(3–2) observations of the Teacup galaxy. This is a nearby ($D_L$ = 388 Mpc) radio-quiet type-2 quasar (QSO2) with a compact radio jet ($P_{jet} \approx 10^{43}$ erg $s^{-1}$) that subtends a small angle from the molecular gas disk. Enhanced emission line widths perpendicular to the jet orientation have been reported for several nearby AGN for the ionized gas. For the molecular gas in the Teacup, not only did they find this enhancement in the velocity dispersion but also a higher brightness temperature ratio (T32/T21) perpendicular to the radio jet compared to the ratios found in the galaxy disk. Their results and the comparison with simulations suggest that the radio jet is compressing and accelerating the molecular gas and driving a lateral outflow that shows enhanced velocity dispersion and higher gas excitation. These results provide further evidence that the coupling between the jet and the ISM is relevant to AGN feedback even in the case of radio-quiet galaxies.

Iglesias-Groth and Marin-Dobrincic (2023) reported a rich molecular chemistry in the gas of the IC 348 star cluster of the Perseus Molecular Cloud. They presented Spitzer 10–34 μm spectroscopic observations of the diffuse gas in the inner region of the star-forming region IC 348 of the Perseus Molecular Cloud. They found evidence for the strongest mid-IR bands of common molecules as $H_2$, OH, $H_2O$, $CO_2$, and $NH_3$ and of several carbonaceous molecules, which may play an important role in the production of more complex hydrocarbons: HCN, $C_2H_2$, $C_4H_2$, $HC_3N$, $HC_5N$, $C_2H_6$, $C_6H_2$, $C_6H_6$. The excitation diagram of $H_2$ reveals the presence of warm gas (270 ± 30 K) at the observed locations. Assuming this temperature, the derived abundances of $CO_2$ and $NH_3$ relative to $H_2$ are $10^{-8}$ and $10^{-7}$, respectively. From the water lines they obtained an abundance of order $10^{-6}$ and higher gas temperatures. The abundances derived for HCN and $C_2H_2$, key molecules in the development of prebiotic building blocks, are of order $10^{-7}$ and $10^{-9}$, respectively. More complex molecules such as PAHs and the fullerenes $C_{60}$ and $C_{70}$ are also present. IC 348 appears to be very rich and diverse in molecular content. The JWST spectroscopic capabilities may provide details on the spatial distribution of all these molecules and extend the present search to more complex hydrocarbons.

In unveiling the nature of the first stars, the main astronomical clue is the elemental compositions of the second generation of stars, observed as extremely metal-poor stars, in the Milky Way. However, no observational constraint was available on their multiplicity, which is crucial for understanding early phases of galaxy formation. Hartwig et al. (2023) developed a new data-driven method to classify observed EMP stars into mono- or multi-enriched stars with support vector machines. They also used their own nucleosynthesis yields of core-collapse supernovae with mixing fallback that can explain many of the observed EMP stars. Their method predicts, for the first time, that 31.8% ± 2.3% of 462 analyzed EMP stars are classified as mono-enriched. This means that the majority of EMP stars are likely multi-enriched, suggesting that the first stars were born in small clusters. Lower-metallicity stars are more likely to be enriched by a single supernova, most of which have high carbon enhancement. In addition, they found that Fe, Mg. Ca, and C are the most informative elements for this classification. Also, oxygen is very informative despite its low observability. Their data-driven method sheds a new light on solving the mystery of the first stars from the complex data set of galactic archeology surveys.

Recent observations have shown that filaments are ubiquitous in molecular clouds. Most dense clumps and cores are formed in filaments, and they may play a key role in the star formation process. The filaments in molecular clouds may overlap to form a web, creating junctions of filaments and the hub-filament systems. In a scenario of global hierarchical collapse of molecular clouds dominated by gravity, these filaments constitute channels for gas funneled from the extended cloud to the dense clumps. The clumps at the junction will accrete more materials from the surroundings and they become more massive and more likely to form high-mass star clusters (Ma et al. 2023). Compared to isotropic accretion, the accretion flow along the filament is a more efficient mechanism because it takes a longer time to gather mass for the high-mass star-forming cores before channels are destroyed. Several hub-filament systems show evidence of channeling gas feeding to the junctions with ongoing high-mass star formation. For clouds in gravitational collapse, the gas flows will be accelerated towards the center of the potential well because of gravitational attraction. It is expected to produce a characteristic V-shape on the gas velocity structure, which traces the accelerated gas motion around the central high mass clumps or cores. Such a feature has been detected within the OMC-1 cloud.

Ma et al. (2023) studied the hub-filament system G323.46−0.08 based on archival molecular line data from the SEDIGISM $^{13}CO$ survey and infrared data from the GLIMPSE, MIPS, and Hi-GAL surveys. G323.46−0.08 consists of three filaments, F-north, F-west, and F-south, that converge toward the central high-mass clump AGAL 323.459−0.079. F-west and Part 1 of the F-south show clear large-scale velocity gradients 0.28 and 0.44 km $s^{-1}$ $pc^{-1}$, respectively. They seem to be channeling materials into AGAL 323.459−0.079. The minimum accretion rate was estimated to be 1216 $M_\odot$ $Myr^{-1}$. A characteristic V-shape appears around AGAL 323.459−0.079 in the PV diagram, which traces the accelerated gas motions under gravitational collapse. This has also been supported by model fitting results. All three filaments are supercritical, and they have fragmented into many dense clumps. The seesaw patterns near most dense clumps in the PV diagram suggests that mass accretion also occurs along the filament toward the clumps. Their results show that filamentary accretion flows appear to be an important mechanism for supplying the materials necessary to form the central high-mass clump AGAL 323.459−0.079 and to propel the star forming activity taking place therein.

The Magellanic Stream (MS)—an enormous ribbon of gas spanning 140° of the southern sky trailing the Magellanic clouds—has been extensively mapped in the five decades since its discovery. However, despite concerted efforts, no stellar counterpart to the MS has been conclusively identified. This stellar stream would reveal the distance and 6D kinematics of the MS, constraining its formation and the past orbital history of the clouds. Chandra et al. (2023) have been conducting a spectroscopic survey of the most distant and luminous red giant stars in the galactic outskirts. From this data set, they have discovered a prominent population of 13 stars matching the extreme angular momentum of the clouds, spanning up to 100° along the MS at distances of 60–120 kpc. Furthermore, these kinematically selected stars lie along an [α/Fe]-deficient track in chemical space from $-2.5 <$ [Fe/H] $<- 0.5$, consistent with their formation in the clouds themselves. They identify these stars as high-confidence members of the Magellanic Stellar Stream. Half of these stars are metal-rich and closely follow the gaseous MS, whereas the other half are more scattered and metal-poor. They argue that the metal-rich stream is the recently formed tidal counterpart to the MS, and they speculate that the metal-poor population was thrown out of the SMC outskirts during an earlier interaction between the clouds. The Magellanic Stellar Stream provides a strong set of constraints—distances, 6D kinematics, and birth locations—that will guide future simulations toward unveiling the detailed history of the Clouds.

### 2.2.4 Solar Neighborhood

Bouy and Alves (2015) constructed a 3D map of the spatial density of OB stars within 500 pc from the Sun using the HIPPARCOS catalogue and find three large-scale stream-like structures that allow a new view on the solar neighborhood. The spatial coherence of these blue streams and the monotonic age sequence over hundreds of parsecs suggest that they are made of young stars, similar to the young streams that are conspicuous in nearby spiral galaxies. The three streams are

1. the Scorpius to Canis Majoris stream, covering 350 pc and 65 × 106 years of star formation history;
2. the Vela stream, encompassing at least 150 pc and 25 × 106 years of star formation history;
3. the Orion stream, including not only the well-known Orion OB1abcd associations but also a large previously unreported foreground stellar group lying only 200 pc from the Sun.

The map also reveals a remarkable and previously unknown nearby OB association between the Orion stream and the Taurus molecular clouds, which might be responsible for the observed structure and star formation activity in this cloud complex. This new association also appears to be the birthplace of Betelgeuse, as indicated by the proximity and velocity of the red giant. If this is confirmed, it would solve the long-standing puzzle of the origin of Betelgeuse. The well-known nearby star-forming low-mass clouds, including the nearby T and R associations Lupus, Cha, Oph, CrA, Taurus, Vela R1, and various low-mass cometary clouds in Vela and Orion, appear in this new view of the local neighborhood to be secondary star formation episodes that most likely were triggered by the feedback from the massive stars in the streams. They also recovered well-known star clusters of various ages that are currently cruising through the solar neighborhood; see Tables 2.2 and 2.3. Finally, they found no evidence of an elliptical structure such as the Gould belt, a structure we suggest is a 2D projection effect and not a physical ring.

The closest star to ours is 39,900,000,000,000 km (4.22 lightyears) away; some of the other nearby stars are as shown in Table 2.4.

Hou (2021) has discussed the spiral structure in the solar neighborhood. The spiral structure in the solar neighborhood is an important issue in astronomy. In the past few years, there has been significant progress in observation. The distances for a large number of good spiral tracers, that is, giant molecular clouds, high-mass star-formation region masers, HII regions, O-type stars, and young open clusters, have been accurately estimated, making it possible to depict the detailed properties of nearby spiral arms. In this work, we first give an overview of the research status for the galaxy's spiral structure based on different types of tracers. Then the objects with distance uncertainties better than 15% and <0.5 kpc are collected and combined to depict the spiral structure in the solar neighborhood. Five segments related to the Perseus, Local, Sagittarius-Carina, Scutum-Centaurus, and Norma arms are traced. With the large dataset, the parameters of the nearby arm segments are fitted and updated. Besides the dominant spiral arms, some substructures probably related to arm spurs or feathers are also noted and discussed.

For decades we have known that the Sun lies within the Local Bubble, a cavity of low-density, high-temperature plasma surrounded by a shell of cold, neutral gas and dust. However, the precise shape and extent of this shell, the impetus and timescale for its formation, and its relationship to nearby star formation have remained uncertain, largely due to low-resolution models of the local interstellar medium. Zucker et al. (2022) reported an analysis of the three-dimensional positions, shapes, and motions of dense gas and young stars within 200 pc of the Sun using new spatial and dynamical constraints. They found that nearly all of the star-forming complexes in the solar vicinity lie on the surface of the Local Bubble and that their young stars show outward expansion, mainly perpendicular to the bubble's surface. Tracebacks of these young stars' motions support a picture in which the origin of the Local Bubble was a burst of stellar birth and then death (supernovae) taking place near the bubble's center beginning approximately 14 Myr ago. The expansion of the Local Bubble created by the supernovae swept up the ambient interstellar medium into an extended shell that has now fragmented and collapsed into the most prominent nearby molecular clouds, in turn providing robust observational support for the theory of supernova-driven star formation.

Zhukovska et al. (2008) evaluated stardust in the solar neighborhood. They studied the population of dust grains in the interstellar medium of the Milky Way, its composition and evolution, and the input of these grains into the solar system. Part of these grains are formed in stellar ejecta or stellar winds of highly evolved stars from the refractory elements therein, and the resulting gas-dust mixture is ultimately mixed with the general interstellar matter. However, generally only some fraction of the refractory elements in the ejecta is really condensed into solid phases; a big mass fraction of the refractory elements returned to the ISM—or sometimes even most of it—stays in the gas phase. Turbulent mixing in the ISM rapidly intermingles the material that is ejected by the many different stellar sources. While all the dust particles from stellar sources, called stardust particles, retain their peculiar isotopic compositions of a number of elements indicative of their formation sites, in the gas phase the material from all sources is mixed together and the resulting isotopic composition of the mix is different from that of the stardust particles. The refractory elements in the ISM gas therefore have different isotopic compositions than the same elements found in stardust grains.

Zhukovska et al. (2008) calculated a chemical evolution model of the solar neighborhood in the Milky Way, which forms the basis for calculating a model of the evolution of the interstellar dust population at the galactocentric radius of the Milky Way. The model successfully passes all standard tests for the reliability of such models. In particular the abundance evolution of the important dust-forming elements is compared with observational results for the metallicity-dependent evolution of the abundances for G-type stars from the solar neighborhood. It is found that the tables of Nomoto et al. (2006) for the heavy element production give very good results for the abundance evolution of these important elements. The time evolution for the abundance of the following dust species is followed in the model: silicate, carbon, silicon carbide, and iron dust from AGB stars and from supernovae, as well as silicate, carbon, and iron dust grown in molecular clouds. It was shown that the interstellar dust population is dominated by dust accreted in molecular clouds; stardust only forms a minor fraction. Most of the dust material entering the solar system at its formation does not show isotopic abundance anomalies of the refractory elements; that is, inconspicuous isotopic abundances do not point to a solar system origin for dust grains. The observed abundance ratios of presolar dust grains formed in supernova ejecta and in AGB star outflows require that, for the ejecta from supernovae, the fraction of refractory elements condensed into dust be 0.15 for carbon dust and quite small (~$10^{-4}$) for other dust species.

TABLE 2.2 Ages Assumed for the Sco-CMa Clusters and Associations

| Group | Age | References |
|---|---|---|
| USco | 4–11 | Pecaut et al. (2012); Herczeg and Hillenbrand (2015) |
| UCL | 16 | Pecaut et al. (2012) |
| LCC | 17 | Pecaut et al. (2012) |
| IC2602 | 46 ± 6 | Dobbie et al. (2010) |
| IC 2391 | 50 ± 5 | Barrado y Navascués et al. (2004) |
| a Car | 60 | Platais et al. (1998) |
| Col135 | 40 | Claria and Kepler (1980); Rosario et al. (2015) |
| NGC2451A | 65 ± 15 | Hünsch et al. (2003) |

TABLE 2.3 Range of Age and Distance Measurements for the Clusters and Associations in Vela, after Bouy and Alves (2015)

| Group | Distance | Reference | Age | Reference |
|---|---|---|---|---|
| γ Vel | 336–424 | Dias et al. (2002); Anderson and Francis (2012) | 3–5 | Eldridge (2009) Jeffries et al. (2009) |
| NGC 451B | 302–489 | Dias et al. (2002); Kharchenko et al. (2013) | 44–170 | Dias et al. (2002) Kharchenko et al. (2013) |
| NGC 2547 | 361–457 | Dias et al. (2002); Anderson and Francis (2012) | 20–78 | Anderson and Francis (2012) Kharchenko et al. (2013) |
| Vela OB2 | 410–421 | Dias et al. (2002); de Zeeuw et al. (1999) | 15–20 | Dias et al. (2002) de Zeeuw et al. (1999) |
| Tr 10 | 366–424 | Dias et al. (2002); Anderson and Francis (2012) | 24–35 | Kharchenko et al. (2013) de Zeeuw et al. (1999) |
| NGC 2516 | 373–409 | Dias et al. (2002); Kharchenko et al. (2013), | 113–298 | Dias et al. (2002) Kharchenko et al. (2013) |

TABLE 2.4 Distance Information for Some Nearby Stars

| Star | Distance/Lightyears (ly) |
|---|---|
| Proxima Centauri | 4.22 |
| Barnard's Star | 5.97 |
| Wolf 359 | 7.7 |
| Sirius A & B | 8.6 |
| Epsilon Eridani | 10.5 |
| Ross 128 | 10.8 |
| Procyon | 11.4 |
| Tau Ceti | 11.9 |
| Altair | 16.3 |
| Eta Cassiopeia | 19 |

High-quality Hipparcos data for a complete sample of nearly 12,000 main-sequence and subgiant stars, together with Padua isochrones, are used to constrain the star formation history of the solar neighborhood and the processes that stochastically accelerate disk stars by Binney et al. (2000). The velocity dispersion of a coeval group of stars is found to increase with time from ~8 $kms^{-1}$ at birth as $t^{0.33}$. In the fits, the slope of the initial mass function (IMF) near 1 $M_{sun}$ proves to be degenerate with the rate at which the star formation rate declines. If the slope of the IMF is to lie near Salpeter's value, −2.35, the star formation rate has to be very nearly constant. The age of the solar neighborhood is found to be (11.2 ± 0.75) × $10^9$ years, with remarkably little sensitivity to variations in the assumed metallicity distribution of old disk stars. This age is only a gigayear younger than the age of the oldest globular clusters when the same isochrones and distance scale are employed. It is compatible with current indications of the redshift of luminous galaxy formation only if there is a large cosmological constant. A younger age is formally excluded because it provides a poor fit to the number density of red stars. Since this density is subject to a significantly uncertain selection function, ages as low as 9 × $10^9$ years are plausible even though they lie outside our formal error bars.

Isern (2019) has discussed the star formation history in the solar neighborhood as told by massive white dwarfs. White dwarfs (WDs) are the remnants of low- and intermediate-mass stars. Because of electron degeneracy, their evolution is just a simple gravothermal process of cooling. Recently, thanks to Gaia data, it has been possible to construct the luminosity function of massive ($0.9 \leq M/M_{sun} \leq 1.1$) WDs in the solar neighborhood (d < 100 pc). Because the lifetime of their progenitors is very short, the birth times of both parents and daughters are very close and facilitate the reconstruction of an (effective) star formation rate. This rate started growing from zero during the early galaxy and reached a maximum (6–7) × $10^9$ years ago. It declined and ~5 × $10^9$ years ago started to climb once more, reaching a maximum (2–3) × $10^9$ years ago; it has decreased since then. There are some traces of a recent star formation burst, but the method used here is not appropriate for recently born WDs.

### *2.2.4.1 History of the Solar Neighborhood*

Noguchi (2018) discussed the formation of solar-neighborhood stars in two generations separated by 5 billion years. The chemical compositions of stars encode those of the gas from which they formed, providing important clues regarding the formation histories of galaxies. A powerful diagnostic is the abundance of α elements (O, Mg, Si, S, Ca, and Ti) relative to iron, [α/Fe]. The α elements are synthesized and injected into the interstellar medium by type II supernovae, which occur about 10 million years after their originating stars form; by contrast, iron is returned to the interstellar medium by type Ia supernovae, which occur after a much longer timescale of roughly 1 billion years. Periods of rapid star formation therefore tend to produce high-[α/Fe] stellar populations (because only type II supernovae have time to contribute to interstellar-medium enrichment as the stellar population forms), whereas low-[α/Fe] stars require periods of star formation that last more than a few billion years (over which timescales type Ia supernovae begin to affect the elemental composition of the interstellar medium more strongly than type II supernovae). The existence of two distinct groups of stars in the solar neighborhood, one with high [α/Fe] and the other with low [α/Fe], therefore suggests two different origins, but the mechanism by which this bimodal distribution arose remains unknown.

Noguchi (2018) used a model of disk-galaxy evolution to show that the two episodes of star formation predicted by the "cold flow" theory of galactic gas accretion also explain the observed chemical bimodality. In this scenario, the high-[α/Fe] stars form early, during an initial phase of accretion that involves infalling streams of cold primordial gas. There is then a hiatus of around 2 billion years until the shock-heated gas in the galactic dark-matter halo has cooled because of radiation and can itself commence accretion. The low-[α/Fe] stars form during this second phase. The peaks in these two star-formation episodes are separated by around 5 billion years. In addition, the large-scale variation in the abundance patterns of these two stellar populations that has been observed for the Milky Way is partially explained by the spatial variation in this gas-accretion history.

Since the second half of the nineteenth century, it was recognized that the brightest stars are not distributed randomly in the sky but seem to form a belt (which afterwards became known as the Gould Belt) with an inclination of ~20o with respect to the plane of the Milky Way. Furthermore, O- and B-type stars clustered in loose groups that were named "associations". The Gould Belt was subsequently found to be associated with a significant amount of interstellar material, interpreted as an expanding ring of gas. Giant molecular clouds were also found to be related to the most prominent OB associations. This agrees well with the fact that OB associations are young, as supported by the ages derived from color-magnitude diagrams.

Zari et al. (2018) studied the three-dimensional arrangement of young stars in the solar neighborhood using the second release of the Gaia mission (Gaia DR2) and provided a new, original view of the spatial configuration of the star-forming regions within 500 pc of the Sun. By smoothing the star distribution through a Gaussian filter, they constructed three-dimensional density maps for early-type stars (upper-main sequence, UMS) and pre-main sequence (PMS) sources. The PMS and the UMS samples are selected through a combination of photometric and astrometric criteria. A side product of the analysis is a 3D, G-band extinction map, which they used to correct color-magnitude diagram for extinction and reddening. Both density maps show three prominent structures, Scorpius-Centaurus, Orion, and Vela. The PMS map shows a plethora of lower-mass star-forming regions, such as Taurus, Perseus, Cepheus, Cassiopeia, and Lacerta, which are less visible in the UMS map due to the lack of large numbers of bright, early-type stars. They reported the finding of a candidate new open cluster towards l, b ~ 218.5°, −2°, which could be related to the Orion star-forming complex. They estimated ages for the PMS sample and studied the distribution of PMS stars as a function of their age. They found that younger stars cluster in dense, compact clumps and are surrounded by older sources, whose distribution is instead more diffuse. The youngest groups found are mainly located in Scorpius-Centaurus, Orion, Vela, and Taurus. Cepheus, Cassiopeia, and Lacerta are instead more evolved and less numerous. In addition, they found that 3D density maps show no evidence for the existence of the ring-like structure which is usually referred to as the Gould Belt.

In conclusion, the 3D configuration of the star-forming regions in the solar neighborhood is far from being described by a ring-like structure such as the Gould Belt but is complex and filamentary. Additional analysis is required for precisely ordering all the star-forming regions according to their ages.

For the solar neighborhood, we have the most detailed observational data available for any galactic disk. In particular, the distributions within the solar neighborhood of stellar ages, metallicities, and space velocities are keys to deducing how the disk has evolved chemically and dynamically. Work directed at understanding how galaxies have formed and evolved is a major area of contemporary astronomy, and studies of the local disk have an important role to play in this effort. Despite much progress, many questions about the evolution of the local disk remain open. The Hipparcos catalogue provides uniquely useful data for studies of the solar neighborhood, because the data are homogeneous and of high quality. Dehnen and Binney (1998) defined a kinematically unbiased sample of Hipparcos stars and investigated its kinematics, while Binney et al. (2000) used the sample to model the history of star formation and the stochastic acceleration of stars in the disk.

Aumer and Binney (2009) redefined the sample and then re-worked these papers for several reasons. They used proper motions and parallaxes from the new reduction of Hipparcos data and Geneva–Copenhagen radial velocities for a complete sample of ~15,000 main-sequence and subgiant stars and new Padova isochrones to constrain the kinematics and star formation history of the solar neighborhood. They rederive the solar motion and the structure of the local velocity ellipsoids. When the principal velocity dispersions are assumed to increase with time as $t^{\beta}$, the index $\beta$ is larger for $\sigma_W$ ($\beta_W \approx 0.45$) than for $\sigma_U$ ($\beta_U \approx 0.31$). For the three-dimensional velocity dispersion, $\beta = 0.35$ is obtained. They exclude saturation of disk heating after ~3 Gyr, saturation after >4 Gyr combined with an abrupt increase in velocity dispersion for the oldest stars cannot be excluded. For all models, the star formation rate (SFR) is declining, being a factor of 2–7 lower now than it was at the beginning. Models in which the SFR declines exponentially favor very high disk ages between 11.5 and 13 Gyr and exclude ages below ~10.5 Gyr, as they yield worse fits to the number density and velocity dispersion of red stars. Models in which the SFR is the sum of two declining exponentials representing the thin and thick disks favor ages between 10.5 and 12 Gyr, with a lower limit of ~10.0 Gyr. Although in their models the SFR peaked surprisingly early, the mean formation time of solar-neighborhood stars is later than in ab initio models of galaxy formation, probably on account of weaknesses in such models.

Using 17 chemical elements as a proxy for stellar DNA, Jofré et al. (2017) presented a full phylogenetic study of stars in the solar neighborhood. This entails applying a clustering technique that is widely used in molecular biology to construct an evolutionary tree from which three branches emerge. These are interpreted as stellar populations that separate in age and kinematics and can be thus attributed to the thin disk, the thick disk, and an intermediate population of probable distinct origin. The authors found six lone stars of intermediate age that could not be assigned to any population with enough statistical significance. Combining the ages of the stars with their position on the tree, they were able to quantify the mean rate of chemical enrichment of each of the populations and thus show in a purely empirical way that the star formation rate in the thick disk is much higher than that in the thin disk. They were also able to estimate the relative contribution of dynamical processes such as radial migration and disk heating to the distribution of chemical elements in the solar neighborhood. Their method offers an alternative approach to chemical tagging methods with the advantage of visualizing the behavior of chemical elements in evolutionary trees. This offers a new way to search for "common ancestors" that can reveal the origin of solar neighborhood stars; see Figure 2.4.

#### *2.2.4.2 Magnetic Fields in the Solar Neighborhood*

Astronomical observations indicate that dynamically important magnetic fields are ubiquitous in the Universe, while their origin remains a mystery. Zhou et al. (2022a) provided a paradigm for understanding the origin of cosmic magnetism by taking into account the effects of the microphysics of collisionless plasmas on macroscopic astrophysical processes. They demonstrated that the first magnetic fields could be spontaneously generated in the Universe by generic motions of astrophysical turbulence through kinetic plasma physics, and cosmic plasmas were thereby ubiquitously magnetized. Their theoretical and numerical results set the stage for determining how these "seed" magnetic fields are further amplified by the turbulent dynamo (another central and long-standing question) and thus advance a fully self-consistent explanation of cosmic magneto-genesis.

Zhou et al. (2022a) show how an initially unmagnetized plasma may magnetize itself through kinetic instabilities arising self-consistently under the action of large-scale flows, which are ubiquitous and driven by a variety of large-scale processes in astrophysical environments. In unmagnetized, collisionless environments, the plasma flows are not of a purely fluid nature; instead, they are subject to phase mixing and Landau damping. These cause the plasma-distribution function to become anisotropic in velocity space, thereby providing free energy for microscopic instabilities, such as Weibel (1959), to grow rapidly on top of the slowly varying macroscopic flows. The Weibel instability produces fluctuations that extract free energy from the thermal anisotropy and generate kinetic-scale seed magnetic fields. As the Weibel magnetic field grows, the plasma becomes magnetized, leading to the saturation of the instability and modifying the macroscopic transport properties of the turbulence.

At plasma-kinetic scales, any macroscopic flow may be viewed locally as a shear flow and/or a compressional flow. Zhou et al. (2022a) focus on how an externally driven shear flow can spontaneously magnetize the plasma. They adopt a fully kinetic framework, in which the kinetic physics of both particle species is treated self-consistently. The sequence of events through which the plasma becomes magnetized involves multiple stages, each of which they consider in detail. They present an analytical model describing this phenomenon, which is supported by first-principles numerical simulations. Fundamentally, these calculations show that a large-scale shear flow provides an efficient mechanism for the self-magnetization of astrophysical plasmas.

High-mass star-forming molecular clouds in spiral galaxies primarily follow the spiral arms. As such, these molecular clouds and their young stellar objects are used to trace spiral structure within the Milky Way. Observations from Spitzer revealed that some of these star-forming clouds are dense ($\sim 10^4$ $cm^{-3}$), high mass, and exceptionally elongated (e.g., over 80 pc $\times$ 0.5 pc for the Nessie filament. These filamentary structures are called bones because they delineate the densest parts of arms in a spiral galaxy, just as bones delineate the densest parts of arms in a human skeleton. Zucker et al. (2015) and (2018) identified 18 bone candidates in the Milky Way using strict criteria: they must be velocity coherent along the structure, have aspect ratios of >50:1, lie within 20 pc of the galactic plane, and lie mostly parallel to the galactic plane. The physical properties of these bones are well characterized, including measurements of lengths, widths, aspect ratios, masses, column densities, dust temperatures, galactic altitudes, kinematic separation from arms in l–v space, and distances (Zucker et al. 2015, 2018). However, the magnetic field (henceforth B-field), which can potentially support the clouds against gravitational collapse or guide mass flow, has been mostly unconstrained for bones.

Star formation primarily occurs in filaments where magnetic fields are expected to be dynamically important. The largest and densest filaments trace spiral structure within galaxies. Over a dozen of these dense ($\sim 10^4$ cm$^{-3}$) and long (>10 pc) filaments have been found within the Milky Way, and they are often referred to as "bones". Until now, none of these bones have had their magnetic field resolved and mapped in their entirety. Stephens et al. (2022) introduced the SOFIA legacy project FIELDMAPS, which has begun mapping ~10 of these Milky Way bones using the HAWC+ instrument at 214 μm and 18.″2 resolution (or 0.58 pc resolution at a distance of 6.6 kpc). The authors presented a first result from this survey on the ~60 pc long bone G47. Contrary to some studies of dense filaments in the galactic plane, we find that the magnetic field is often not perpendicular to the spine (i.e., the center-line of the bone). Fields tend to be perpendicular in the densest areas of active star formation and more parallel or random in other areas. The average field is neither parallel nor perpendicular to the galactic plane nor the bone. The magnetic field strengths along the spine typically vary from ~20 to ~100 μG. Magnetic fields tend to be strong enough to suppress collapse along much of the bone, but for areas that are most active in star formation, the fields are notably less able to resist gravitational collapse.

The galactic global magnetic field is assumed to play a vital role in shaping galactic structures such as spiral arms and giant molecular clouds. However, the knowledge of magnetic field structures in the galactic plane at different distances is limited, as measurements used to map the magnetic field are the integrated effect along the line of sight. Doi et al. (2024) presented the first tomographic imaging of magnetic field structures in a galactic spiral arm. Using optical stellar polarimetry over a 17' × 10' field of view, the Sagittarius spiral arm was investigated. Combining data with stellar distances from the Gaia mission, they could isolate the contributions of five individual clouds along the line of sight by analyzing the polarimetry data as a function of distance. The observed clouds include a foreground cloud (d < 200 pc) and four clouds in the Sagittarius arm at 1.23, 1.47, 1.63, and 2.23 kpc. The column densities of these clouds

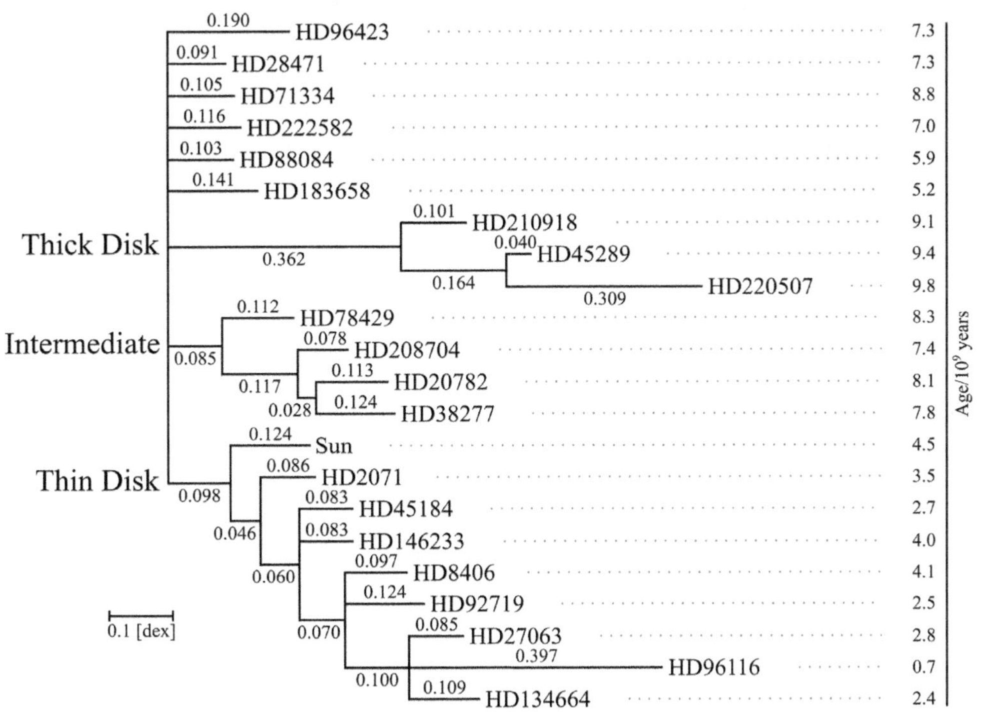


FIGURE 2.4 Unrooted phylogenetic tree of 22 solar twins in the solar neighborhood, created using 17 elemental abundances. Stellar populations are assigned considering the age and the dynamical properties of the stars and are indicated at the right. Branch lengths have units in dex, with the scale indicated at the left bottom, after Jofré et al. (2017).

range from 0.5 to $2.8 \times 10^{21}$ $cm^{-2}$. The magnetic fields associated with each cloud show smooth spatial distributions within their observed regions on scales smaller than 10 pc and display distinct orientations. The position angles projected on the plane of the sky, measured from the galactic north to the east, for the clouds in increasing order of distance are 135°, 46°, 58°, 150°, and 40°, with uncertainties of a few degrees. Notably, these position angles deviate significantly from the direction parallel to the galactic plane.

### 2.2.4.3 *Time Variation of Elemental Composition*

The Gaia data release 2 (DR2) catalogue is the best source of stellar astrometric and photometric data available today. The history of the Milky Way galaxy is written in stone in this data set. Parallaxes and photometry tell us where the stars are today, when were they formed, and with what chemical content, that is, their star formation history (SFH). Alzate et al. (2021) developed a Bayesian hierarchical model suited to reconstruct the SFH of a resolved stellar population. They studied the stars brighter than G = 15 within 100 pc of the Sun in Gaia DR2 and derived an SFH of the solar neighborhood in agreement with previous determinations and improving upon them because we detect chemical enrichment. Their results show a maximum of star formation activity about $10 \times 10^9$ years ago, producing large numbers of stars with slightly below solar metallicity (Z = 0.014), followed by a decrease in star formation up to a minimum level occurring around $8 \times 10^9$ years ago. After a quiet period, star formation rises to a maximum at about $5 \times 10^9$ years ago, forming stars of solar metallicity (Z = 0.017). Finally, star formation has been decreasing until the present, forming stars of Z = 0.03 at a residual level. They tested the effects introduced in the inferred SFH by ignoring the presence of unresolved binary stars in the sample, reducing the apparent limiting magnitude, and modifying the stellar initial mass function.

One of the primary tools to understanding the history of the solar neighborhood, and more generally the Milky Way, is the chemical composition of stars. Compilations of stellar abundances provide an overall picture of the chemical evolution of the solar neighborhood. Notable results obtained over the past few decades include correlations of metallicity with age and galactocentric distance and whether the Sun is suitably "average". Trends in the elemental abundances and a limited number of isotopic abundances relative to iron have also been observed over the whole metallicity range of the galactic disk. For example, oxygen and the other α-chain elements, relative to iron, vary systematically from being overabundant at [Fe/H] 1.0 dex to roughly solar at [Fe/H] ≈ 0.0. This decrease is widely taken to be caused by the contributions of supernovae Type Ia (SNe Ia) in an average, well-mixed interstellar medium.

Hinkel et al. (2014) compiled spectroscopic abundance data from 84 literature sources for 50 elements across 3058 stars in the solar neighborhood within 150 pc of the Sun to produce the Hypatia Catalog. They evaluated the variability of the spread in abundance measurements reported for the same star by different surveys. They also explored the likely association of the star within the galactic disk, the corresponding observation and abundance determination methods for all catalogs in Hypatia, the influence of specific catalogs on the overall abundance trends, and the effect of normalizing all abundances to the same solar scale. The resulting stellar abundance determinations in the Hypatia Catalog are analyzed only for thin-disk stars with observations that are consistent between literature sources. As a result of the large data set, they find that the stars in the solar neighborhood may reveal an asymmetric abundance distribution, such that a [Fe/H]-rich group near the midplane is deficient in Mg, Si, S, Ca, Sc ii, Cr ii, and Ni as compared to stars farther from the plane. The Hypatia Catalog has a wide number of applications, including exoplanet hosts, thick- and thin-disk stars, and stars with different kinematic properties.

Elemental abundances in the Milky Way provide stringent constraints not only on stellar astrophysics but also on the formation and evolutionary history of the Milky Way itself. Elements heavier than helium are synthesized inside and then ejected by dying stars. The next generation of stars form from gas clouds that include heavy elements from the previous stellar generations. Therefore, stars in the present-day galaxy are fossils that retain the information on the properties of stars from the past. From the elemental abundances of the present-day stars, it is possible to disentangle the star formation history of the host galaxy. This approach is called galactic archeology and can be applied not only to our Milky Way but also to other galaxies. For constraining the star formation histories of galaxies, the most important uncertainty is the set of nucleosynthesis yields (Kobayashi et al. 2020).

To reach a deeper understanding of the origin of elements in the periodic table, we construct galactic chemical evolution (GCE) models for all stable elements from C ($A = 12$) to U ($A = 238$) from first principles, that is, using theoretical nucleosynthesis yields and event rates of all chemical enrichment sources. This enables us to predict the origin of elements as a function of time and environment. In the solar neighborhood, we find that stars with initial masses of $M > 30M_{\odot}$ can become failed supernovae if there is a significant contribution from hypernovae (HNe) at $M \sim 20$–$50M_{\odot}$. The contribution to GCE from super-asymptotic giant branch stars (with $M \sim 8$–$10M_{\odot}$ at solar metallicity) is negligible, unless hybrid white dwarfs from low-mass super-AGB stars explode as so-called Type Iax supernovae or high-mass super-AGB stars explode as electron-capture supernovae (ECSNe). Among neutron-capture elements, the observed abundances of the second (Ba) and third (Pb) peak elements are well reproduced with our updated yields of the slow neutron-capture process (s-process) from AGB stars. The first peak elements (Sr, Y, Zr) are sufficiently produced by ECSNe together with AGB stars. Neutron star mergers can produce rapid neutron-capture process (r-process) elements up to Th and U, but the timescales are too long to explain observations at low metallicities. The observed evolutionary trends, such as for Eu, can well be explained if ~3% of 25–$50M_{\odot}$ HNe are magneto-rotational supernovae producing r-process elements. Along with the solar neighborhood, we also predict the evolutionary trends in the halo, bulge, and thick disk for future comparison with galactic archeology surveys (Kobayashi et al. 2020).

The prevalence of carbon and oxygen and the favorable thermodynamics of carbon monoxide bonds make CO a dominant molecular component of the gaseous protoplanetary disk. As solids condense out, the ratio of carbon to oxygen influences the planetesimal geology. If the C/O ratio is above a threshold value, then there is almost no free oxygen available to form silicates, and the geology will be dominated by carbonates; under high pressure, these are expected to form crystal diamond compositions. When the C/O ratio is below that threshold, planetesimal geology is primarily magnesium silicates. The threshold ratio is ~1 at the formation sites of the planetesimals, but due to the temperature structure and evolution of the protoplanetary disk, initial ratios as low as 0.65 can still lead to carbon rich conditions. The solar system C/O is ~ 0.54 and the rocky planets have magnesium silicate compositions despite carbon being almost ten times more abundant in the solar photosphere than silicon; see Kobayashi et al. (2020) and references therein.

The nucleosynthetic pathways for carbon and oxygen are slightly different. Although both are formed in the explosions of massive stars, the majority of carbon comes instead from lower-mass stars. The evolutionary pathway to carbon-rich dwarf is exceedingly narrow and involves mass transfer from an AGB companion. Oxygen is the most abundant heavy element produced by massive stars. Their more rapid evolution will result in an initially low C/O ratio that will increase as AGB winds contribute carbon to the interstellar medium. Galactic chemical evolution models predict that the ratio of C/O should peak at just slightly greater than solar. Photometric surveys and targeted searches for carbon dwarfs have found that fewer than 0.1% of main sequence stars are carbon rich (i.e. C/O > 1), from Brewer and Fischer (2016) and references therein.

The carbon to oxygen ratio in a protoplanetary disk can have a dramatic influence on the compositions of any terrestrial planets formed. In regions of high C/O, planets form primarily from carbonates, and in regions of low C/O, the ratio of magnesium to silicon determines the types of silicates which dominate the compositions. Brewer and Fischer (2016) present C/O and Mg/Si ratios for 849 F, G, and K dwarfs in the solar neighborhood. They find that the frequency of carbon-rich dwarfs in the solar neighborhood is < 0.13% and that 156 known planet hosts in the sample follow a similar distribution as all of the stars as a whole. The cosmic distribution of Mg/Si for these same stars is broader than the C/O distribution and peaks near 1.0 with ~ 60% of systems having $1 \leq$ Mg/Si < 2, leading to rocky planet compositions similar to the Earth. This leaves 40% of systems that can have planets that are silicate rich and may have very different compositions than our own.

Isotope ratios have opened a new window into the study of the details of stellar evolution, supernovae, and galactic chemical evolution. Kobayashi et al. (2011) presented the evolution of the isotope ratios of elemental abundances (from C to Zn) in the solar neighborhood, bulge, halo, and thick disk, using chemical evolution models with updated yields of asymptotic giant branch stars and core-collapse supernovae. The evolutionary history of each element is different owing to the effects of the initial progenitor mass and metallicity on element production. In the bulge and thick disk, the star formation timescale is shorter than in the solar neighborhood, leading to higher [α/Fe] ratios. Likewise, the smaller contribution from Type Ia supernovae in these regions leads to lower [Mn/Fe] ratios. Also in the bulge, the abundances of [(Na, Al, P, Cl, K, Sc, Cu, Zn)/Fe] are higher because of the effect of metallicity on element production from core-collapse supernovae. According to predictions by Kobayashi et al. (2011), it is possible to find metal-rich stars ([Fe/H] $\gtrsim -1$) that formed in the early Universe as a result of rapid star formation. The chemical enrichment time-scale of the halo is longer than in the solar neighborhood, and consequently the ratios of [(C, F)/Fe] and $^{12}C/^{13}C$ are higher owing to a significant contribution from low-mass AGB stars. While the [α/Fe] and [Mn/Fe] ratios are the same as in the solar neighborhood, the [(Na, Al, P, Cl, K, Sc, Cu, Zn)/Fe] ratios are predicted to be lower. Furthermore, they predict that isotope ratios such as $^{24}Mg/^{25,\,26}Mg$ are larger because of the contribution from low-metallicity supernovae. Using isotopic ratios, it is possible to select stars that formed in a system with a low chemical enrichment efficiency such as the satellite galaxies that were accreted on to our own Milky Way galaxy.

Roederer et al. (2022) presented a nearly complete rapid neutron-capture process (r-process) chemical inventory of the metal-poor ([Fe/H] = −1.46 +/− 0.10) r-process-enhanced ([Eu/Fe] = +1.32 +/−0.08) halo star HD 222925. This abundance set is the most complete for any object beyond the solar system, totaling 63 metals detected and 7 with upper limits. It comprises 42 elements from 31 <= Z <= 90, including elements rarely detected in r-process-enhanced stars, such as Ga, Ge, As, Se, Cd, In, Sn, Sb, Te, W, Re, Os, Ir, Pt, and Au. The authors derived these abundances from an analysis of 404 absorption lines in ultraviolet spectra collected using the Space Telescope Imaging Spectrograph on the Hubble Space Telescope and previously analyzed optical spectra. A series of appendices discusses the atomic data and quality of fits for these lines. The r-process elements from Ba to Pb, including all elements at the third r-process peak, exhibit remarkable agreement with the solar r-process residuals, with a standard deviation of the differences of only 0.08 dex (17%). In contrast, deviations among the lighter elements from Ga to Te span nearly 1.4 dex, and they show distinct trends from Ga to Se, Nb through Cd, and In through Te. The r-process contribution to Ga, Ge, and As is small, and Se is the lightest element whose production is dominated by the r-process. The lanthanide fraction, log ($X_{La}$) = −1.39 ± 0.09, is typical for r-process-enhanced stars and higher than that of the kilonova from the GW170817 neutron-star merger event. They advocate adopting this pattern as an alternative to the solar r-process-element residuals when confronting future theoretical models of heavy-element nucleosynthesis with observations.

## 2.3 PLANETS

In a new analysis based on the latest telescope data, University of Florida astronomers have discovered that a third of the planets around the most common stars in the galaxy could be in a goldilocks orbit close enough, and gentle enough, to hold onto liquid water—and possibly harbor life. The remaining two-thirds of the planets around these ubiquitous small stars are likely roasted by gravitational tides, sterilizing them.

Sagear and Ballard (2023) found that stars with multiple planets were the most likely to have the kind of circular orbits that allow them to retain liquid water. Stars with only one planet were the most likely to see tidal extremes that would sterilize the surface.

Since one-third of the planets in this small sample had gentle enough orbits to potentially host liquid water, that likely means that the Milky Way has hundreds of millions of promising targets to probe for signs of life outside our solar system.

Miyazaki and Masuda (2023) investigated how the occurrence rate of giant planets (minimum mass > $0.3M_{Jup}$) around Sun-like stars depends on the age, mass, and metallicity of their host stars. They developed a hierarchical Bayesian framework to infer the number of planets per star (NPPS) as a function of both planetary and stellar parameters. The framework fully takes into account the uncertainties in the latter by utilizing the posterior samples for the stellar parameters obtained by fitting stellar isochrone models to the spectroscopic parameters, Gaia DR3 parallaxes, and 2MASS $K_s$-band magnitudes adopting a certain bookkeeping prior. They applied the framework to 46 Doppler giants found around a sample of 382 Sun-like stars from the California Legacy Survey catalog that publishes spectroscopic parameters and search completeness for all the surveyed stars. They found evidence that the NPPS of hot Jupiters (orbital period P = 1–10 days) decreases roughly in the latter half of the main sequence over the timescale of O (Gyr), while that of cold Jupiters (P = 1–10 year) does not. Assuming that this decrease is real and caused by tidal orbital decay, the modified stellar tidal quality factor $Q'_\star$ is implied to be $O(10^6)$ for a Sun-like main-sequence star orbited by a Jupiter-mass planet with $P \approx 3$days.

### 2.3.1 Planet Formation Processes

Models of terrestrial planet formation predict that the final stages of planetary assembly—lasting tens of millions of years beyond the dispersal of young protoplanetary disks—are dominated by planetary collisions. It is through the giant impacts that planets like the young Earth grow to their final mass and achieve long-term stable orbital configurations. A key prediction is that these impacts produce debris. So far, the most compelling observational evidence for post-impact debris comes from the planetary system around the nearby 23-million-year-old A-type star HD 172555. This system shows large amounts of fine dust with an unusually steep size distribution and atypical dust composition, previously attributed to either a hypervelocity impact or a massive asteroid belt. Schnelderman et al. (2021) report the spectrally resolved detection of a carbon monoxide gas ring co-orbiting with dusty debris around HD 172555 between about 6 and 9 astronomical units—a region analogous to the outer terrestrial planet region of our solar system. Taken together, the dust and carbon monoxide detections favor a giant impact between large, volatile-rich bodies. This suggests that planetary-scale collisions, analogous to the Moon-forming impact, can release large amounts of gas as well as debris and that this gas is observable, providing a window into the composition of young planets.

Jergensen at al. (2022) have shown that binarity of protostar affects the evolution of the disk and planets. Nearly half of all stars similar to our Sun are in binary or multiple systems, which may affect the evolution of the stars and their protoplanetary disks during their earliest stages. The authors studied NGC 1333-IRAS2A, which is a young, Class 0, low-mass protostellar system located in the Perseus molecular cloud. It is known to drive two bipolar outflows that are almost perpendicular to each other on the sky and is resolved into binary components, VLA1 and VLA2, through long wavelength continuum observations.

Jergensen at al. (2022) reported spatially and spectrally resolved observations of a range of molecular species. They compared these to detailed magneto/hydrodynamic simulations: the comparisons showed that inhomogeneous accretion onto the circumstellar disks occurs in episodic bursts, driving a wobbling jet. The conclusion was that binarity and multiplicity in general strongly affect the properties of the emerging stars, as well as the physical and chemical structures of the protoplanetary disks and therefore potentially any emerging planetary systems.

Models of terrestrial planet formation predict that the final stages of planetary assembly, lasting tens of millions of years beyond the dispersal of young protoplanetary disks, are dominated by planetary collisions. It is through the giant impacts that planets like the young Earth grow to their final mass and achieve long-term stable orbital configurations. A key prediction is that these impacts produce debris. To date, the most compelling observational evidence for post-impact debris comes from the planetary system around the nearby 23 Myr-old A star HD 172555. This system shows large amounts of fine dust with an unusually steep size distribution and atypical dust composition, previously attributed to either a hypervelocity impact or a massive asteroid belt. Schneiderman et al. (2021) reported the spectrally resolved detection of a CO gas ring co-orbiting with dusty debris between ~6 and 9 au (astronomical units)—a region analogous to the outer terrestrial planet region of our solar system. Taken together, the dust and CO detections favor a giant impact between large, volatile-rich bodies. This suggests that planetary-scale collisions, analogous to the Moon-forming impact, can release large amounts of gas as well as debris, and that this gas is observable, providing a window into the composition of young planets.

The elemental compositions of planet-hosting stars serve as proxies for the primordial compositions of protoplanetary disks within which the planets form. The temperature profile of the disk governs the condensation fronts of various compounds, and although these chemically distinct regions migrate and mix during the disk lifetime, they can still leave an imprint on the compositions of the forming planets. Observable atmospheric compositions of hot Jupiters, when compared against their host stars, could potentially constrain their formation and migration processes. Brewer et al. (2017) compared the measured planetary and stellar abundances of carbon and oxygen for ten systems with hot Jupiters. If the planets formed by core accretion with significant planetesimal accretion and migrated through the disk, the hot Jupiter atmospheres should be super-stellar in O/H and substellar in C/O. On the contrary, however, they found that currently reported abundances of hot Jupiters have generally superstellar C/O ratios, although present uncertainties on the reported O/H and C/O ratios are too large to reach a firm conclusion. In one case, HD 209458b, however, the elevated C/O and depleted O/H of the planet compared to the host star are significant enough to suggest an origin far beyond the ice line, with predominantly gas accretion and subsequent disk-free migration. Improved measurements from the James Webb Space Telescope will enable more precise measurements for more hot Jupiters, and we predict, based on the current marginal trend, that a sizable fraction of hot Jupiters will show enrichment of C/O over and lower O/H than their hosts, similar to HD 209458b.

Understanding how binary companions affect the formation, evolution, and survival of exoplanets is also an important component to understanding planet formation overall. Theoretical studies show that a companion can truncate the protoplanetary disk, leaving less material for planets to form, causing the migration of gas giant planets, or dispersing the disk before planets are able to form.

Lester et al. (2021) presented high angular resolution imaging observations of 517 host stars of TESS exoplanet candidates using the 'Alopeke and Zorro speckle cameras at Gemini North and South. The sample consists mainly of bright F, G, and K stars at distances of less than 500 pc. Our speckle observations span angular resolutions of ~20 mas out to 1.2 arcsec, yielding spatial resolutions of <10 to 500 AU for most stars, and our contrast limits can detect companion stars 5–9 magnitudes fainter than the primary at optical wavelengths. They detect 102 close stellar companions and determine the separation, magnitude difference, mass ratio, and estimated orbital period for each system. Their observations of exoplanet host star binaries reveal that they have wider separations than field binaries, with a mean orbital semi-major axis near 100 AU. Other imaging studies have suggested this dearth of very closely separated binaries in systems which host exoplanets, but incompleteness at small separations makes it difficult to disentangle unobserved companions from a true lack of companions. With their improved angular resolution and sensitivity, they confirm that this lack of close exoplanet host binaries is indeed real. They also search for a correlation between planetary orbital radii vs. binary star separation, but given the very short orbital periods of the TESS planets, they do not find any clear trend. They do note that in exoplanet systems containing binary host stars, there is an observational bias against detecting Earth-size planet transits due to transit depth dilution caused by the companion star.

Hedges et al. (2021) reported the discovery of two planetary systems around comoving stars: TOI-2076 (TIC 27491137) and TOI-1807 (TIC 180695581). TOI-2076 is a nearby (41.9 pc) multiplanetary system orbiting a young, $(204 \pm 50) \times 10^6$ years, bright K-type star. TOI-1807 hosts a single transiting planet and is similarly nearby (42.58 pc), similarly young at $(180 \pm 40) \times 10^6$ years, and bright. Both targets exhibit significant, periodic variability due to starspots, characteristic of their young ages. Using photometric data collected by TESS, the researchers identify three transiting planets around TOI-2076 with radii of $R_b = 3.3 \pm 0.04\ R_\oplus$, $R_c = 4.4 \pm 0.05\ R_\oplus$, and $R_d = 4.1 \pm 0.07\ R_\oplus$. Planet TOI-2076b has a period of $P_b = 10.356$ days. For both TOI-2076c and d, TESS observed only two transits, separated by a 2-year interval in which no data were collected, preventing a unique period determination. A range of long periods (<17 days) are consistent with the data.

Hedges et al. (2021) identified a short-period planet around TOI-1807 with a radius of $R_b = 1.8 \pm 0.04\ R_\oplus$ and a period of $P_b = 0.549$ days. Their close proximity; bright, cool host stars; and young ages make these planets excellent candidates for follow-up. TOI-1807b is one of the best-known small ($R < 2\ R_\oplus$) planets for characterization via eclipse spectroscopy and phase curves with JWST. TOI-1807b is the youngest ultra-short-period planet discovered to date, providing valuable constraints on formation timescales of short-period planets. Given the rarity of young planets, particularly in multiple-planet systems, these planets present an unprecedented opportunity to study and compare exoplanet formation and young planet atmospheres at a crucial transition age for formation theory.

Placing the architecture of the solar system within the broader context of planetary architectures is one of the primary topics of interest within planetary science. Exoplanet discoveries have revealed a large range of system architectures, many of which differ substantially from the solar system's model. One particular feature of exoplanet demographics is the relative prevalence of super-Earth planets, for which the solar system lacks a suitable analog, presenting a challenge to modeling their interiors and atmospheres. Kane (2023) presented the results of a large suite of dynamical simulations that insert a hypothetical planet in the mass range 1–10 $M_{\oplus}$ within the semimajor axis range 2–4 au, between the orbits of Mars and Jupiter. He showed that, although the system dynamics remain largely unaffected when the additional planet is placed near 3 au, Mercury experiences substantial instability when the additional planet lies in the range 3.1–4.0 au, and perturbations to the Martian orbit primarily result when the additional planet lies in the range 2.0–2.7 au. Kane (2023) further shows that, although Jupiter and Saturn experience relatively small orbital perturbations, the angular momentum transferred to the ice giants can result in their ejection from the system at key resonance locations of the additional planet.

According to Hon et al. (2023), when main-sequence stars expand into red giants, they are expected to engulf nearby planets. Until now, the absence of planets with short orbital periods around post-expansion, core-helium-burning red giants has been interpreted as evidence that short-period planets around Sun-like stars do not survive the giant expansion phase of their host stars. Hon et al. (2023) presented the discovery that the giant planet 8 Ursae Minoris b orbits a core-helium-burning red giant. At a distance of only 0.5 au from its host star, the planet would have been engulfed by its host star, which is predicted by standard single-star evolution to have previously expanded to a radius of 0.7 au. Given the brief lifetime of helium-burning giants, the nearly circular orbit of the planet is challenging to reconcile with scenarios in which the planet survives by having a distant orbit initially. Instead, the planet may have avoided engulfment through a stellar merger that either altered the evolution of the host star or produced 8 Ursae Minoris b as a second-generation planet. This system shows that core-helium-burning red giants can harbor close planets and provides evidence for the role of non-canonical stellar evolution in the extended survival of late-stage exoplanetary systems.

#### *2.3.1.1 Protoplanetary Disks*

Intermediate-mass planets, from super-Earth to Neptune-sized bodies, are the most common types of planet in the galaxy. The prevailing theory of planet formation-core accretion predicts the existence of substantially fewer intermediate-mass giant planets than have been observed. The competing mechanism for planet formation—disk instability—can produce massive gas giant planets on wide orbits, such as HR 87995, by direct fragmentation of the protoplanetary disk. Previously, fragmentation in magnetized protoplanetary disks has been considered only when the magneto-rotational instability is the driving mechanism for magnetic field growth. However, this instability is naturally superseded by the spiral-driven dynamo when more realistic, non-ideal magneto-hydrodynamic conditions are considered. Deng et al. (2021) reported on magneto-hydrodynamic simulations of disk fragmentation in the presence of a spiral-driven dynamo. Fragmentation leads to the formation of long-lived bound protoplanets with masses that are at least one order of magnitude smaller than in conventional disk instability models. These light clumps survive shear and do not grow further owing to the shielding effect of the magnetic field, whereby magnetic pressure stifles the local inflow of matter. The outcome is a population of gaseous-rich planets with intermediate masses, while gas giants are found to be rarer, in qualitative agreement with the observed mass distribution of exoplanets.

To determine the origin of the spiral structure observed in the dust continuum emission of Elias 2–27, Paneque-Carreno et al. (2021) analyzed multi-wavelength continuum ALMA data with a resolution of ~0.2 arcsec (~23 au) at 0.89, 1.3, and 3.3 mm. Elias 2–27 is a young (0.8 Myr) M0 star located at a distance of $116^{+19}_{-10}$ pc in the ρ Oph star-forming region. It harbors an unusually massive protoplanetary disk; the disk-to-star mass ratio of Elias 2–27 is reported to be ~0.3. They also studied the kinematics of the disk with $^{13}$CO and C$^{18}$O ALMA observations in the J = 3–2 transition. The spiral arm morphology is recovered at all wavelengths in the dust continuum observations, where they measure contrast and spectral index variations along the spiral arms and detect subtle dust-trapping signatures. They determined that the emission from the midplane is cold and interpret the optical depth results as signatures of a higher disk mass than previous constraints. From the gas data, they searched for deviations from Keplerian motion and traced the morphology of the emitting surfaces and the velocity profiles. They found an azimuthally varying emission layer height in the system, large-scale emission surrounding the disk, and strong perturbations in the channel maps, co-located with the spirals. Additionally, they developed multigrain dust and gas SPH simulations of a gravitationally unstable disk and compared them to the observations. Given the large-scale emission and highly perturbed gas structure, together with the comparison of continuum observations to theoretical predictions, they proposed infall-triggered gravitational instabilities as an origin for the observed spiral structure. Overall, Elias 2–27 appears to be a strong candidate to be a gravitationally unstable protoplanetary disk, but there are many tests to be done in order to determine if this is in fact the origin of the observed spirals.

According to current understanding of stellar formation, the elemental ratios in a solar photosphere represent the elemental ratios of the nebula from which the central star formed (with very few exceptions, like deuterium). Since planetesimals and planets are formed within this nebula, their composition is expected to match that of their host star. This assumption is probably true for highly refractory species, since their high condensation temperature places their condensation line close to the star, but it may not apply to volatile species. Although the latter species will have no effect on the elemental ratios in the rocky part of planets (Fe/Si and Mg/Si), it is expected that they have an effect on the C/O ratio in the atmosphere as a result of differences in ice line positions during the early lifetime of the nebula while planets accrete. It is particularly difficult to estimate the amount of C and O in planetary bodies since they mostly occur in volatile species in a solar nebula. Observations and simulations show that the abundance of C in the bulk of the planets of the solar system is much lower than it was in the solar nebula, requiring element fractionation processes during the early stages of planet formation (Thiabaud et al. 2015).

Similarly, recent astronomical observations of other stellar systems have shown that the C/O ratio in atmospheres of planets can deviate significantly from the host star value and can also be different in the atmosphere of different planets in the same planetary system. Öberg et al. (2011) showed that these variations could be due to the different locations of condensation of $H_2O$, CO, and $CO_2$, resulting in an increase in the C/O ratios during condensation. The main shortcoming of the studies of Öberg et al. (2011) is that they did not take into account the planet migration during their formation or the gas phase evolution, which can both significantly change the C/O ratio in the atmosphere of growing planets.

Direct observations of gaseous exoplanets reveal that their gas envelope has a higher C/O ratio than that of the host star (e.g., Wasp 12-b). This has been explained by considering that the gas phase of the disk could be inhomogeneous, exceeding the stellar C/O ratio in regions where these planets formed, but few studies have considered the drift of the gas and planet migration. Thiaband et al. (2015) used their previous models for computing the chemical composition, together with a planet formation model, to which they added the composition and drift of the gas phase of the disk, which is composed of the main volatile species $H_2O$, CO, $CO_2$, $NH_3$, $N_2$, $CH_3OH$, $CH_4$, $H_2S$, $H_2$, and He. The study focuses on the region where ice lines are present and influence the C/O ratio of the planets. Their modeling shows that the condensation of volatile species as a function of radial distance allows for C/O enrichment in specific parts of the protoplanetary disk of up to four times the solar value. This leads to the formation of planets that can be enriched in C/O in their envelope up to three times the solar value. Planet migration, gas phase evolution, and disk irradiation enable the evolution of the initial C/O ratio that decreases in the outer part of the disk and increases in the inner part of the disk. The total C/O ratio of the planets is governed by the contribution of ices accreted, suggesting that high C/O ratios measured in planetary atmospheres are indicative of a lack of exchange of material between the core of a planet and its envelope or an observational bias. It also suggests that the observed C/O ratio is not representative of the total C/O ratio of the planet.

Modeling shows that the condensation of volatile species as a function of radial distance allows for C/O enrichment in specific parts of the protoplanetary disk of up to four times the solar value. This leads to the formation of planets that can be enriched in C/O in their envelope up to three times the solar value. Planet migration, gas phase evolution, and disk irradiation enable the evolution of the initial C/O ratio that decreases in the outer part of the disk and increases in the inner part of the disk. The total C/O ratio of the planets is governed by the contribution of ices accreted, suggesting that high C/O ratios measured in planetary atmospheres are indicative of a lack of exchange of material between the core of a planet and its envelope or an observational bias. It also suggests that the observed C/O ratio is not representative of the total C/O ratio of the planet; see Thiabaud et al. (2015).

The precursors to larger, biologically relevant molecules are detected throughout interstellar space, but determining the presence and properties of these molecules during planet formation requires observations of protoplanetary disks at high angular resolution and sensitivity. Ilee et al. (2021) presented 0.3" observations of $HC_3N$, $CH_3CN$, and c-$C_3H_2$ in five protoplanetary disks observed as part of the Molecules with ALMA at Planet-Forming Scales (MAPS) Large Program. They detected all molecules in four of the disks (GM Aur, AS 209, HD 163296, and MWC 480), with tentative detections of c-$C_3H_2$ and $CH_3CN$ in IM Lup. They observed a range of morphologies—central peaks, single or double rings—with no clear correlation in morphology between molecule or disk. Emission is generally compact and on scales comparable with the millimeter dust continuum. They performed both disk-integrated and radially resolved rotational diagram analysis to derive column densities and rotational temperatures. The latter reveals 5–10 times more column density in the inner 50–100 au of the disks when compared with the disk-integrated analysis. They demonstrated that $CH_3CN$ originates from lower relative heights in the disks when compared with $HC_3N$, in some cases directly tracing the disk midplane. Finally, they found good agreement between the ratio of small to large nitriles in the outer disks and comets. The results indicate that the protoplanetary disks studied here are host to significant reservoirs of large organic molecules and that this planet- and comet-building material can be chemically similar to that in our own solar system.

The complex organic molecules detected in star-forming regions are the precursors of the prebiotic molecules that can lead to the emergence of life. By studying COMs in more evolved protoplanetary disks, one can gain a better understanding of how they are incorporated into planets. Brunken et al. (2022) presented ALMA band 7 observations of the dust and ice trap in the protoplanetary disk around Oph IRS 48. They reported the first detection of dimethyl ether ($CH_3OCH_3$) in a planet-forming disk and a tentative detection of methyl formate ($CH_3OCHO$). They determined column densities for the detected molecules and upper limits on non-detected species using the CASSIS spectral analysis tool. The inferred column densities of $CH_3OCH_3$ and $CH_3OCHO$ with respect to methanol ($CH_3OH$) are of order unity, indicating unusually high abundances of these species compared to other environments. Alternatively, the $^{12}CH_3OH$ emission is optically thick and beam diluted, implying a higher $CH_3OH$ column density and a smaller emitting area than originally thought. The presence of these complex molecules can be explained by thermal ice sublimation, where the dust cavity edge is heated by irradiation and the full volatile ice content is observable in the gas phase. This work confirms the presence of oxygen-bearing molecules more complex than $CH_3OH$ in protoplanetary disks for the first time. It also shows that it is indeed possible to trace the full interstellar journey of COMs across the different evolutionary stages of star, disk, and planet formation.

Chemical evolution in the protoplanetary disk midplane can modify the composition of ices and gases. Eistrup et al. (2016) investigated if and how chemical evolution affects the abundances and distributions of key volatile species in the midplane of a protoplanetary disk in the 0.2–30 AU range. A full chemical network including gas-phase, gas-grain interactions, and grain-surface chemistry is used to evolve chemistry in time, for $1 \times 10^6$ years. Great diversity is observed in the relative abundance ratios of the main considered species: $H_2O$, CO, $CO_2$, $CH_4$, $O_2$, $NH_3$, and $N_2$. The choice of ionization level and initial abundances, as well as the extent of chemical reaction types included, are all factors that affect the chemical evolution. The only exception is the inheritance scenario with a low ionization level, which results in negligible changes compared with the initial abundances, regardless of whether grain-surface chemistry is included. The chemical processing changes the C/O ratios for gas and ice significantly with radius and model assumptions. For high ionization levels, chemical processing becomes significant after a few times $10^5$ years. It is concluded that chemical evolution in the disk midplane needs to be considered in the determination of the volatile composition of planetesimals.

Exoplanet atmospheres are thought be built up from accretion of gas as well as pebbles and planetesimals in the midplanes of planet-forming disks. The chemical composition of this material is usually assumed to be unchanged during the disk lifetime. However, chemistry can alter the relative abundances of molecules in this planet-building material. Eistrup et al. (2018) assessed the impact of disk chemistry during the era of planet formation. This is done by investigating the chemical changes to volatile gases and ices in a protoplanetary disk midplane out to 30 AU for up to $7 \times 10^6$ years, considering a variety of different conditions, including a physical midplane structure that is evolving in time, and considering two disks with different masses. An extensive kinetic chemistry gas-grain reaction network was utilized to evolve the abundances of chemical species over time. Two disk midplane ionization levels (low and high) were explored, as well as two different makeups of the initial abundances ("inheritance" or "reset"). Given a high level of ionization, chemical evolution in protoplanetary disk midplanes becomes significant after a few times $10^5$ years and is still ongoing by $7 \times 10^6$ years between the $H_2O$ and the $O_2$ icelines. Inside the $H_2O$ iceline, and in the outer, colder regions of the disk midplane outside the $O_2$ iceline, the relative abundances of the species reach (close to) steady state by $7 \times 10^6$ years. Importantly, the changes in the abundances of the major elemental carbon and oxygen-bearing molecules imply that the traditional "stepfunction" for the C/O ratios in gas and ice in the disk midplane (as defined by sharp changes at icelines of $H_2O$, $CO_2$ and CO) evolves over time and cannot be assumed to be fixed, with the C/O ratio in the gas even becoming smaller than the C/O ratio in the ice.

Understanding the driving forces behind spiral arms in protoplanetary disks remains a challenge due to the faintness of young giant planets. MWC 758 hosts such a protoplanetary disk with a two-armed spiral pattern that is suggested to be driven by an external giant planet. Wagner et al. (2023) presented observations in the thermal infrared that are uniquely sensitive to redder (that is, colder, or more attenuated) planets than past observations at shorter wavelengths. They detected a giant protoplanet, MWC 758c, at a projected separation of roughly 100 au from the star. The spectrum of MWC 758c is distinct from the rest of the disk and consistent with emission from a planetary atmosphere with $T_{eff} = 500 \pm 100$ K for a low level of extinction ($A_V \leq 30$) or a hotter object with a higher level of extinction. Both scenarios are commensurate with the predicted properties of the companion responsible for driving the spiral arms. MWC 758c provides evidence that spiral arms in protoplanetary disks can be caused by cold giant planets or by those whose optical emission is highly attenuated. MWC 758c stands out both as one of the youngest giant planets known and as one of the coldest and/or most attenuated. Furthermore, MWC 758c is among the first planets to be observed within a system hosting a protoplanetary disk.

Espaillat et al. (2023) reported the first detection of variability in the mid-infrared neon line emission of a protoplanetary disk by comparing a JWST Mid-InfraRed Instrument Medium Resolution Spectrometer spectrum of SZ Cha taken in 2023 with a Spitzer Infrared Spectrograph Short-High spectrum of this object from 2008. They measured the [Ne III]-to-[Ne II] line flux ratio, which is a diagnostic of the high-energy radiation field, to distinguish between the dominance of EUV- or X-ray-driven disk photo-evaporation. They found that the [Ne III]-to-[Ne II] line flux ratio changes significantly from ~1.4 in 2008 to ~0.2 in 2023. This points to a switch from EUV-dominated to X-ray-dominated photo-evaporation of the disk. They present contemporaneous ground-based optical spectra of the Hα emission line that show the presence of a strong wind in 2023. They propose that this strong wind prevents EUV radiation from reaching the disk surface while the X-rays permeate the wind and irradiate the disk. They speculate that at the time of the Spitzer observations, the wind was suppressed and EUV radiation reached the disk. These observations confirm that the MIR neon emission lines are sensitive to changes in high-energy radiation reaching the disk surface. This highlights the [Ne III]-to-[Ne II] line flux ratio as a tool to gauge the efficiency of disk photo-evaporation in order to provide constraints on the planet formation timescale. However, multi-wavelength observations are crucial to interpret the observations and properly consider the star–disk connection.

Planets form in protoplanetary disks of gas and dust around young stars that are undergoing their own formation process. The amount of material in the disk determines how big the planets can grow. Stefánsson et al. (2023) observed a nearby low-mass star using near-infrared spectroscopy. They used radial velocity observations to detect a Neptune-mass exoplanet orbiting LHS 3154, a star that is nine times less massive than the Sun. The exoplanet's orbital period is 3.7 days, and its minimum mass is 13.2 Earth masses, which is almost the mass of Neptune. Theoretical models do not predict the formation of such a massive planet around a low-mass star. The authors used simulations to show that its presence could be explained if the protoplanetary disk were 10 times more massive than expected for the host star.

Ramírez-Tannus et al. (2023) presented the first results of the eXtreme UV Environments (XUE) James Webb Space Telescope program, which focuses on the characterization of planet-forming disks in massive star-forming regions. These regions are likely representative of the environment in which most planetary systems formed. Understanding the impact of environment on planet formation is critical in order to gain insights into the diversity of the observed exoplanet populations. XUE targets 15 disks in three areas of NGC 6357, which hosts numerous massive OB stars, including some of the most massive stars in our galaxy. Thanks to the JWST, for the first time, study of the effect of external irradiation on the inner (<10 au), terrestrial-planet-forming regions of protoplanetary disks was possible. In this study, the authors report on the detection of abundant water, CO, $^{12}CO_2$, HCN, and $C_2H_2$ in the inner few au of XUE 1, a highly irradiated disk in NGC 6357. In addition, small, partially crystalline silicate dust is present at the disk surface. The derived column densities, the oxygen-dominated gas-phase chemistry, and the presence of silicate dust are surprisingly similar to those found in inner disks located in nearby, relatively isolated low-mass star-forming regions. Their findings imply that the inner regions of highly irradiated disks can retain similar physical and chemical conditions to disks in low-mass star-forming regions, thus broadening the range of environments with similar conditions for inner disk rocky planet formation to the most extreme star-forming regions in our galaxy.

#### *2.3.1.2 Exoplanets*

In 1992, the first ever exoplanets were discovered orbiting a pulsar called PSR B1257 + 12; see Wolszczan and Frail (1992). The planetary system is now known to host at least three planets similar in mass to the rocky planets in our solar system. Since then, a handful of pulsars have been found to host planets. However, the extremely violent conditions surrounding the births and lives of pulsars make "normal" planet formation unlikely, and many of these detected planets are exotic objects (such as planets made mostly of diamond), unlike those we know in our solar system. Since then there has been many discoveries, and today (7/9/2023) the http://exoplanet.eu/catalog/ is showing 5508 planets: 4067 planetary systems and 878 multiple planet systems.

Observations of exoplanets have shown that the solar system might not be a standard planetary system. This was initially implied after the first detections of hot Jupiters (although these turned out later to be rather rare but seemingly prevalent due to a detection bias) and further confirmed by the fact that super-Earths, absent in the solar system, are very common. To date, super-Earths have only been observed at short orbits, but planet-formation models indicate that they are also prevalent at large radial distances, where they could retain a non-negligible H–He envelope, which is the original atmosphere accreted from the protoplanetary disk. At several AU away from the host star, stellar radiation is insufficient to thermally evaporate a substantial amount of the atmospheric H–He (Lous et al. 2022).

So far, more than 5500 exoplanets have been detected in the Milky Way. The Hubble Space Telescope alone has been examining them for the last 20 years. The first non-questionable detection of water was reported in 2013 by measurement of water vapor on Jupiter's moon Europa. No evidence of the presence of life has been reported so far.

The majority have been detected using the transit method, which is based on looking for and measuring the dimming of light as a planet passes in front of the host star, as seen from Earth. The size of a planet can then be derived from the amplitude of the dimming and the size of the host star. A substantial number of planets have also been found using the radial velocity method, which involves detecting and measuring the "Doppler wobble" of a star—that is, the motion of a star caused by the gravitational pull of the planet that is orbiting around it.

It is striking to recognize that we have been aware of planets orbiting other stars for only the past quarter century. More than 5500 of those exoplanets have been cataloged by astronomers in that time. Statistical extrapolations from that population suggest that more than one planet orbits each star, on average. The big questions in astrophysics research have always been concerned with origins. The abundance of new worlds allows us to establish the context for our solar system and to refine ideas about the creation, evolution, and biological potential of planets. Of the many exciting discoveries made about exoplanets, perhaps the most profound is their diversity. No one knows how common a planetary configuration like our solar system may be, but it's clear that a much broader range of planetary properties exists than we can find among our immediate neighbors, Andrews (2021).

Instruments mounted on NASA's James Webb Space Telescope have improved capabilities of capturing the distinct signatures of water, evidence of clouds and haze in the atmosphere surrounding giant planets orbiting distant Sun-like stars. Recently, the Webb Telescope has studied properties of an exoplanet called WASP-96 in the constellation Phoenix in the southern sky roughly 1150 lightyears away. This exoplanet is not similar to any from our solar system; it is significantly hotter, and it orbits extremely close to its star, as reported by Garner (2022).

Ultra-hot Jupiters are currently the most accessible laboratories for the study of exoplanet atmospheres. Their size and large atmospheric scale heights, combined with the proximity to the host stars, make them appealing targets for the study of light transmitted through planetary atmospheres. Two of the best examples can be found with WASP-76b and WASP-121b. Both planets are inflated ultra-hot Jupiters on orbits with periods under 2 days and equilibrium temperatures close to 2500 K. Azevedo-Silva et al. (2022) added a novel detection of $Ba^+$ to the known atmospheric compositions of WASP-76b and WASP-121b, the heaviest species detected to date in any exoplanetary atmosphere, with additional new detections of Co and $Sr^+$ and tentatively $Ti^+$ for WASP121b. They also confirmed the presence of $Ca^+$, Cr, Fe, H, Li, Mg, Mn, Na, and V on both WASP-76b and WASP-121b, with the addition of Ca, $Fe^+$, and Ni for the latter.

Recently, observations of the exoplanet WASP-39b by JWST have been reported (see O'Callaghan 2022). WASP-39b is a "hot Jupiter" located 215 parsecs (700 lightyears) from Earth. The gas giant is about one-third the mass of Jupiter and more similar in composition to Saturn. It orbits its host star in just 4 Earth days and is eight times closer to it than Mercury is to our Sun, making it incredibly hot at nearly 900°C. This proximity and its consequent brightness make the planet inhospitable to life as we know it but made it the "perfect target" for JWST to observe early in its life and test its exoplanet capabilities.

By analyzing blight filtered through WASP-39b's atmosphere by transmission spectroscopy the research team was able to detect water, carbon monoxide, sodium, potassium, and more in the planet's atmosphere, in addition to the carbon dioxide. The gives the planet a similar composition to Saturn. They also detected sulfur dioxide produced by light splitting water in the WASP-39b's atmosphere into hydrogen and hydroxide, which reacts with hydrogen sulfide to produce sulfur dioxide. It looks that photochemistry is probably an important process on other potentially habitable planets.

The population of planets smaller than approximately 1.7 Earth radii ($R_\oplus$) is widely interpreted as consisting of rocky worlds, generally referred to as super-Earths. This picture is largely corroborated by radial velocity mass measurements for close-in super-Earths but lacks constraints at lower insolations. Piaulet et al. (2022) presented the results of a detailed study of the Kepler-138 system using 13 Hubble and Spitzer transit observations of the warm-temperate $1.51 \pm 0.04 R_\oplus$ planet Kepler-138 d combined with new radial velocity measurements of its host star obtained with the Keck/High Resolution Echelle Spectrometer. They found evidence for a volatile-rich "water world" nature of Kepler-138 d, with a large fraction of its mass contained in a thick volatile layer. This finding is independently supported by transit timing variations and radial velocity observations, as well as the flat optical/infrared transmission spectrum. Quantitatively, they inferred a composition of $11^{+3}_{-4}$% volatiles by mass or ~51% by volume, with a 2,000-km-deep water mantle and atmosphere on top of a core with an Earth-like silicates/iron ratio. Any hypothetical hydrogen layer consistent with the observations ($<0.003\,M_\oplus$) would have swiftly been lost on a ~10 Myr timescale. The bulk composition of Kepler-138 d therefore resembles those of the icy moons rather than the terrestrial planets in the solar system. They conclude that not all super-Earths are rocky worlds, but that volatile-rich water worlds exist in an overlapping size regime, especially at lower insolations. Their photodynamical analysis also reveals that Kepler-138 c is a slightly warmer twin of Kepler-138 d (that is, another water world in the same system) and infer the presence of Kepler-138 e, a likely non-transiting planet at the inner edge of the habitable zone (Piaulet et al. 2022).

Vissapragada et al. (2022) presented evidence of tidally driven inspiral in the Kepler-1658 (KOI-4) system, which consists of a giant planet ($1.1R_J$, $5.9M_J$) orbiting an evolved host star ($2.9R_{sun}$, $1.5M_{sun}$). Using transit timing measurements from Kepler, Palomar/WIRC, and TESS, they showed that the orbital period of Kepler-1658b appears to be decreasing at a rate $\dot{P} = 131_{-22}^{+20}$ ms yr$^{-1}$, corresponding to an infall timescale $P/\dot{P} \approx 2.5 \times 10^6$ years. The authors consider other explanations for the data, including line-of-sight acceleration and orbital precession, but find them implausible. The observed period derivative implies a tidal quality factor $Q = 2.50_{-0.62}^{+0.85} \times 10^4$, in good agreement with theoretical predictions for inertial wave dissipation in subgiant stars. Additionally, while it probably cannot explain the entire inspiral rate, a small amount of planetary dissipation could naturally explain the deep optical eclipse observed for the planet via enhanced thermal emission. As the first evolved system with detected inspiral, Kepler-1658 is a new benchmark for understanding tidal physics at the end of the planetary life cycle.

Many new planets orbiting evolved stars are being discovered with TESS; see Grunblatt et al. (2022a) and (2022b) and Saunders et al. (2022). If the tidal quality factor obtained here is applicable to other evolved planet-hosting stars, then most of their planets are nearing the ends of their lives.

An impressive 5500 exoplanets and counting have been detected thus far, largely thanks to the past and ongoing radial-velocity (RV) and transit surveys. On the hunt for an Earth analog, out of these thousands of planets, only ~50 have been found to sit in the so-called habitable zone (HZ) of their stellar host, which is defined to be the circumstellar region in which liquid water could potentially exist on the surface of the planet. Only 20 of these are considered Earth sized, defined by radii of $0.8\ R_\oplus < Rp < 1.6\ R_\oplus$ or by masses of $0.5\ M_\oplus < M \sin ip < 3\ M_\oplus$. Moreover, a majority of them have been discovered around M-dwarf stellar hosts, most likely due to the ease in detectability considering the higher planet-to-star mass and radius ratios.

The definition of the HZ is not a definite implication for a life-hosting world but rather acts as a good indicator for a planet to show further promising potential for markers of surface habitability. There are in fact a variety of factors that affect its habitability potential, such as X-ray/UV emission, age in regard to the stellar host, or processes due to the planet itself, including its atmospheric composition or its ability to retain certain elements in its atmosphere. For this reason, it is not only crucial to first gather planets that are situated within the HZ, but also, it is necessary to build a better understanding of the stellar effects on the planet's habitability.

Kossakowski et al. (2023) recently published the CARMENES search for exoplanets around M dwarfs, Wolf 1069 b: an Earth-mass planet in the habitable zone of a nearby very low-mass star. They presented the discovery of an Earth-mass planet ($M_b \sin i = 1.26 \pm 0.21\ M_\oplus$) on a 15.6 d orbit of a relatively nearby (d ~ 9.6 pc) and low-mass ($0.167 \pm 0.011\ M_\oplus$) M5.0 V star, Wolf 1069. Sitting at a separation of $0.0672 \pm 0.0014$ au away from the host star puts Wolf 1069 b in the habitable zone, receiving an incident flux of $S = 0.652 \pm 0.029\ S_\oplus$. The planetary signal was detected using telluric-corrected radial-velocity data from the CARMENES spectrograph, amounting to 262 spectroscopic observations covering almost 4 years. There are additional long-period signals in the RVs, one of which we attribute to the stellar rotation period. This is possible thanks to our photometric analysis including new, well-sampled monitoring campaigns undergone with the OSN and TJO facilities that supplement archival photometry (i.e., from MEarth and SuperWASP), and this yielded an updated rotational period range of $P_{rot} = 150 - 170$ d, with a likely value at $169.3^{+3.7}_{-3.6}$ d. The stellar activity indicators provided by the CARMENES spectra likewise demonstrate evidence for the slow rotation period, though not as accurately due to possible factors such as signal aliasing or spot evolution. Their detectability limits indicate that additional planets more massive than one Earth mass with orbital periods of less than 10 days can be ruled out, suggesting that perhaps Wolf 1069 b had a violent formation history. This planet is also the sixth closest Earth-mass planet situated in the conservative HZ, after Proxima Centauri b, GJ 1061 d, Teegarden's Star c, and GJ 1002 b and c. Despite not transiting, Wolf 1069 b is nonetheless a very promising target for future three-dimensional climate models to investigate various habitability cases as well as for sub-m $s^{-1}$ RV campaigns to search for potential inner sub-Earth-mass planets in order to test planet formation theories.

Stars and planets both form by accreting material from a surrounding disk. Because they grow from the same material, theory predicts that there should be a relationship between their compositions. In their study, Adibekyan et al. (2021) searched for a compositional link between rocky exoplanets and their host stars. They estimated the iron-mass fraction of rocky exoplanets from their masses and radii and compared it with the compositions of their host stars, which they assumed reflect the compositions of the protoplanetary disks. They found a correlation (but not a 1:1 relationship) between these two quantities, with a slope of >4, which was interpreted as being attributable to planet formation processes. Super-Earths and super-Mercuries appeared to be distinct populations with differing compositions, implying differences in their formation processes.

Gaidos et al. (2021) reported the discovery of a resolved (0".9) substellar companion to a member of the 1–5 Myr Taurus star-forming region. The host star (2M0437) is a single mid-M type ($T_{eff} \approx 3100$ K) dwarf with a position, space motion, and color-magnitude that support Taurus membership, and possible affiliation with a ~2.5 × $10^6$-year-old sub-group. A comparison with stellar models suggests a (2–5) × $10^6$ year age and a mass of 0.15–0.18 $M_{\odot}$. Although K2 detected quasi-periodic dimming from close-in circumstellar dust, the star lacks detectable excess infrared emission from a circumstellar disk, and its Hα emission is not commensurate with accretion. Astrometry based on three years of AO imaging shows that the companion (2M0437b) is co-moving, while photometry of two other sources at larger separation indicates they are likely heavily-reddened background stars. A comparison of the luminosity of 2M0437b with models suggests a mass of (3–5) × $M_{JUP}$, well below the deuterium-burning limit, and an effective temperature of 1400–1500K, characteristic of a late L spectral type. The H-K color is redder than the typical L dwarf but comparable to other directly detected young planets, such as those around HR 8799. The discovery of a super-Jupiter around a very young, very low-mass star challenges models of planet formation by either core accretion (which requires time) or disk instability (which requires mass). They also detected a second, co-moving, widely separated (75") object that appears to be a heavily extinct star. This is certainly a fellow member of this Taurus sub-group and statistically likely to be a bound companion.

The radius valley, a bifurcation in the size distribution of small, close-in exoplanets, is hypothesized to be a signature of planetary atmospheric loss. Such an evolutionary phenomenon should depend on the age of the star–planet system. David et al. (2021) studied the temporal evolution of the radius valley using two independent determinations of host star ages among the California–Kepler Survey (CKS) sample. They found evidence for a wide and nearly empty void of planets in the period–radius diagram at the youngest system ages (< 2–3 Gyr) represented in the CKS sample. They showed that the orbital period dependence of the radius valley among the younger CKS planets is consistent with that found among those planets with asteroseismically determined host star radii. Relative to previous studies of preferentially older planets, the radius valley determined among the younger planetary sample is shifted to smaller radii. This result is compatible with an atmospheric loss timescale on the order of gigayears for progenitors of the largest observed super-Earths. In support of this interpretation, they showed that the planet sizes that appear to be unrepresented at ages <~2–3 Gyr are likely to correspond to planets with rocky compositions. The results suggest that the size distribution of close-in exoplanets and the precise location of the radius valley evolve over gigayears.

Mishra et al. (2023a) presented a novel, model-independent framework for studying the architecture of an exoplanetary system at the system level. This framework allows us to characterize, quantify, and classify the architecture of an individual planetary system. Their aim in this endeavor is to generate a systematic method to study the arrangement and distribution of various planetary quantities within a single planetary system. They proposed that the space of planetary system architectures be partitioned into four classes: similar, mixed, anti-ordered, and ordered. They applied our framework to observed and synthetic multi-planetary systems, thereby studying their architectures of mass, radius, density, core mass, and the core water mass fraction. They explored the relationships between a system's (mass) architecture and other properties. Their work suggests that:

1. similar architectures are the most common outcome of planet formation;
2. internal structure and composition of planets shows a strong link with their system architecture;
3. most systems inherit their mass architecture from their core mass architecture;
4. most planets that started inside the ice line and formed in situ are found in systems with a similar architecture; and
5. most anti-ordered systems are expected to be rich in wet planets, while most observed mass ordered systems are expected to have many dry planets.

They found, in good agreement with theory, that observations are generally biased towards the discovery of systems whose density architectures are similar, mixed, or anti-ordered. This study probes novel questions and new parameter spaces for understanding theory and observations. Future studies may utilize the framework to not only constrain the knowledge of individual planets but also the multi-faceted architecture of an entire planetary system.

Since there are four classes of planetary system architecture: similar, mixed, anti-ordered, and ordered, Mishra et al. (2023b) investigated the formation pathways leading to these four architecture classes. To understand the role of nature versus nurture in sculpting the final (mass) architecture of a system, they applied their architecture framework to synthetic planetary systems—formed via core-accretion—using the Bern model. General patterns emerge in the formation pathways of the four architecture classes. Almost all planetary systems emerging from protoplanetary disks whose initial solid mass was less than one Jupiter mass are similar. Systems emerging from heavier disks may become mixed, anti-ordered, or ordered. Increasing dynamical interactions (planet–planet, planet–disk) tends to shift a system's architecture from mixed to anti-ordered to ordered. Their model predicts the existence of a new metallicity–architecture correlation. Similar systems have very high occurrence around low-metallicity stars. The occurrence of the anti-ordered and ordered classes increases with increasing metallicity. The occurrence of mixed architecture first increases and then decreases with increasing metallicity. In their synthetic planetary systems, the role of nature is disentangled from the role of nurture. Nature (or initial conditions) pre-determines whether the architecture of a system becomes similar; otherwise nurture influences whether a system becomes mixed, anti-ordered, or ordered. Exomoons represent a crucial missing puzzle piece in our efforts to understand extrasolar planetary systems. To address this deficiency, Kipping et al. (2021) described an exomoon survey of 70 cool, giant transiting exoplanet candidates found by Kepler. They identified only one exhibiting a moon-like signal that passes a battery of vetting tests: Kepler-1708 b. The authors showed that Kepler-1708 b is a statistically validated Jupiter-sized planet orbiting a Sun-like quiescent star at 1.6 au. The signal of the exomoon candidate, Kepler-1708 b-i, is a 4.8σ effect and is persistent across different instrumental detrending methods, with a 1% false-positive probability via injection–recovery. Kepler-1708 b-i is ~2.6 Earth radii and is located in an approximately coplanar orbit at ~12 planetary radii from its ~1.6 au Jupiter-sized host. Future observations will be necessary to validate or reject the candidate.

Despite their cool temperatures, M dwarfs have become hot targets for exoplanet surveys. The large planet-to-star ratios of these systems result in relatively deep transits and large radial velocity amplitudes. In addition, they emit more strongly in the infrared. This leads to favorable conditions for atmospheric characterization by the Hubble Space Telescope and James Webb Space Telescope.

Thanks to the relative ease of finding and characterizing small planets around M dwarf stars, these objects have become cornerstones in the field of exoplanet studies. The current paucity of planets in long-period orbits around M dwarfs make such objects particularly compelling, as they provide clues about the formation and evolution of these systems. In their study, Schanche et al. (2022) presented the discovery of TOI-2257 b (TIC 198485881), a long-period (35 d) sub-Neptune orbiting an M3 star at 57.8 pc. Its transit depth is about 0.4%, large enough to be detected with medium-size, ground-based telescopes. The long transit duration suggests the planet is in a highly eccentric orbit ($e \sim 0.5$), which would make it the most eccentric planet that is known to be transiting an M-dwarf star. They combined TESS and ground-based data obtained with the 1.0-m SAINT-EX, 0.60-m TRAPPIST-North, and 1.2-m FLWO telescopes to find a planetary size of 2.2 $R_\oplus$ and an orbital period of 35.19 days. In addition, they made use of archival data, high-resolution imaging, and vetting packages to support their planetary interpretation. With its long period and high eccentricity, TOI-2257 b falls in a novel slice of parameter space. Despite the planet's low equilibrium temperature (~ 256 K), its host star's small size ($R_* = 0.311 \pm 0.015$) and relative infrared brightness ($K_{mag} = 10.7$) make it a suitable candidate for atmospheric exploration via transmission spectroscopy.

The ultra-hot Jupiter WASP-189 b has a high equilibrium temperature of $T_{eq} = 2{,}641 \pm 34$K due to its close proximity to its hot A-type host star. It is one of the brightest transiting planet systems currently known, making it very amenable for spectroscopic studies of its atmosphere. The system is well characterized, thanks to extensive photometric observations with CHEOPS, including a precise measurement of the orbital parameters, which the authors adopted in this study. In recent years, similar systems have garnered considerable interest within studies that use high-resolution ground-based spectrographs. These have been used to reveal a myriad of absorbing atoms in their emission and transmission spectra (Prinoth et al. 2022).

The temperature of an atmosphere decreases with increasing altitude, unless a shortwave absorber that causes a temperature inversion exists. Ozone plays this role in the Earth's atmosphere. In the atmospheres of highly irradiated exoplanets, the shortwave absorbers are predicted to be titanium oxide (TiO) and vanadium oxide (VO). Detections of TiO and VO have been claimed using both low and high spectral-resolution observations, but subsequent observations have failed to confirm these claims or overturned them. Here we report the unambiguous detection of TiO in the ultra-hot Jupiter WASP-189 using high-resolution transmission spectroscopy. This detection is based on applying the cross-correlation technique to many spectral lines of TiO from 460 to 690 nm. Moreover, we report detections of metals, including neutral and singly ionized iron and titanium, as well as chromium, magnesium, vanadium and manganese (Fe, Fe+, Ti, Ti+, Cr, Mg, V, Mn). The line positions of the detected species differ, which we interpret as a consequence of spatial gradients in their chemical abundances, such that they exist in different regions or dynamical regimes. This is direct observational evidence for the three-dimensional thermochemical stratification of an exoplanet atmosphere derived from high-resolution ground-based spectroscopy (Prinoth et al. 2022).

Ultra-short period planets undergo strong tidal interactions with their host star which lead to planet deformation and orbital tidal decay. WASP-103b is the exoplanet with the highest expected deformation signature in its transit light curve and one of the shortest expected spiral-in times. Measuring the tidal deformation of the planet would allow us to estimate the second-degree fluid Love number and gain insight into the planet's internal structure. Moreover, measuring the tidal decay timescale would allow us to estimate the stellar tidal quality factor, which is key to constraining stellar physics. Barros et al. (2022) obtained 12 transit light curves of WASP-103b with the CHaracterising ExOplanet Satellite (CHEOPS) to estimate the tidal deformation and tidal decay of this extreme system. We modelled the high-precision CHEOPS transit light curves together with systematic instrumental noise using multi-dimensional Gaussian process regression informed by a set of instrumental parameters. To model the tidal deformation, they used a parametrization model which allowed them to determine the second-degree fluid Love number of the planet. They combined their light curves with previously observed transits of WASP-103b with the Hubble Space Telescope and Spitzer to increase the signal-to-noise of the light curve and better distinguish the minute signal expected from the planetary deformation. They estimated the radial Love number of WASP-103b to be hf = 1.59 + 0.45 −0.53. This is the first time that the tidal deformation was directly detected (at 3σ) from the transit light curve of an exoplanet. Combining the transit times derived from CHEOPS, HST, and Spitzer light curves with the other transit times available in the literature, they found no significant orbital period variation for WASP-103b. However, the data show a hint of an orbital period increase instead of a decrease, as is expected for tidal decay. This could be either due to a visual companion star if this star is bound, the Applegate effect, or a statistical artifact. In conclusion, the estimated Love number of WASP-103b is similar to Jupiter's. This will allow us to constrain the internal structure and composition of WASP-103b, which could provide clues on the inflation of hot Jupiters. Debris disks are a common outcome of the star and planet formation process and serve as a signpost of mature planetary systems. Bright debris disks are observed around some 25%30% of main-sequence stars. The true incidence is likely to be higher, since the sensitivity of current observations limits our ability to detect debris disks to those systems that are at least an order of magnitude more luminous than the disk generated by the solar system's Kuiper Belt. While the dust in debris disks is certainly worthy of study in its own right, it also presents an opportunity to trace the properties of planetary systems.

Debris disks are tenuous, dusty belts surrounding main-sequence stars generated by collisions between planetesimals. HD 206893 is one of only two stars known to host a directly imaged brown dwarf orbiting interior to its debris ring, in this case at a projected separation of 10.4 au. Nederlander et al. (2021) resolved structure in the debris disk around HD 206893 at an angular resolution of 24 au and wavelength of 1.3 mm with the Atacama Large Millimeter/Submillimeter Array. They observed a broad disk extending from a radius of <51 au to $194^{+13}{}_{-2}$ au. They modeled the disk with a continuous, gapped, and double power-low model of the surface density profile and found strong evidence for a local minimum in the surface density distribution near a radius of 70 au, consistent with a gap in the disk with an inner radius of $673^{+8}{}_{-16}$ au and with $31^{+11}{}_{-7}$ au. A gapped structure has been observed in four other debris disks—essentially every other radially resolved debris disk observed with sufficient angular resolution and sensitivity with ALMA—and could be suggestive of the presence of an additional planetary-mass companion.

The magnetosphere of an exoplanet has yet to be unambiguously detected. Investigations of star–planet interaction and neutral atomic hydrogen absorption during transit to detect magnetic fields in hot Jupiters have been inconclusive, and interpretations of the transit absorption non-unique. In contrast, ionized species escaping a magnetized exoplanet, particularly from the polar caps, should populate the magnetosphere, allowing detection of different regions from the plasmasphere to the extended magnetotail and characterization of the magnetic field producing them. Ben-Jaffel et al. (2021) reported ultraviolet observations of HAT-P-11 b, a low-mass (0.08 $M_J$) exoplanet showing strong, phase-extended transit absorption of neutral hydrogen (maximum and tail transit depths of 32 ± 4% and 27 ± 4%) and singly ionized carbon (15 ± 4% and 12.5 ± 4%). They showed that the atmosphere should have less than six times the solar metallicity (at 200 bar), and the exoplanet must also have an extended magnetotail (1.8–3.1 au). The HAT-P-11 b equatorial magnetic field strength should be about 1–5 G. Their panchromatic approach using ionized species to simultaneously derive metallicity and magnetic field strength can now constrain interior and dynamo models of exoplanets, with implications for formation and evolution scenarios.

Proxima Centauri is the closest star to the Sun. This small, low-mass, mid-M dwarf is known to host an Earth-mass exoplanet with an orbital period of 11.2 days within the habitable zone, as well as a long-period planet candidate with an orbital period of close to 5 yr. Faria et al. (2022) reported on the analysis of a large set of observations taken with the ESPRESSO spectrograph at the VLT aimed at a thorough evaluation of the presence of a third low-mass planetary companion, which started emerging during a previous campaign. Radial velocities were calculated using both a cross-correlation function (CCF) and a template matching approach. The RV analysis includes a component to model Proxima's activity using a Gaussian process (GP). They used the CCF's full width at half maximum to help constrain the GP and study other simultaneous observables as activity indicators in order to assess the nature of any potential RV signals. Faria et al. (2022) detected a signal at 5.12 ± 0.04 days with a semi-amplitude of 39 ± 7 cm $s^{-1}$. The analysis of subsets of the ESPRESSO data, the activity indicators, and chromatic RVs suggest that this signal is not caused by stellar variability but instead by a planetary companion with a minimum mass of 0.26 ± 0.05 $M_\oplus$ (about twice the mass of Mars) orbiting at 0.029 au from the star. The orbital eccentricity is well constrained and compatible with a circular orbit.

A possible planet candidate in an external galaxy detected through X-ray transit is reported by Di Stefano et al. (2021). Many lines of reasoning suggest that external galaxies should host planetary systems, but detecting them by methods typically used in our own galaxy is not possible. An alternative approach is to study the temporal behavior of X-rays emitted by bright extragalactic X-ray sources, where an orbiting planet would temporarily block the X-rays and cause a brief eclipse. They reported on such a potential event in the X-ray binary M51-ULS-1 in the galaxy M51. They examined a range of explanations for the observed X-ray dip, including a variety of transiting objects and enhancements in the density of gas and dust. The latter are ruled out by the absence of changes in X-ray colors, save any with sharp density gradients that cannot be probed with the data. Instead, the data are well fit by a planet transit model in which the eclipser is most likely to be the size of Saturn. They also found that the locations of possible orbits are consistent with the survival of a planet bound to a mass-transfer binary.

Prior studies have hypothesized that some polluted white dwarfs record a continent-like granitic crust—which is abundant on Earth and perhaps uniquely indicative of plate tectonics. But these inferences derive from only a few elements, none of which define rock type. Putirka and Xu (2021) presented the first estimates of rock types on exoplanets that once orbited polluted white dwarfs—stars whose atmospheric compositions record the infall of formerly orbiting planetary objects—examining cases where Mg, Si, Ca, and Fe are measured with precision. They found no evidence for continental crust, or other crust types, even after correcting for core formation. However, the silicate mantles of such exoplanets are discernable: one case is Earth-like, but most are exotic in composition and mineralogy. Because these exoplanets exceed the compositional spread of >4,000 nearby main sequence stars, their unique silicate compositions are unlikely to reflect variations in parent star compositions. Instead, polluted white dwarfs reveal greater planetary variety in our solar neighborhood than currently appreciated, with consequently unique planetary accretion and differentiation paths that have no direct counterparts in our solar system.

Ultrashort-period (USP) exoplanets have orbital periods shorter than 1 day. Precise masses and radii of USP exoplanets could provide constraints on their unknown formation and evolution processes. Lam et al. (2021) reported the detection and characterization of the USP planet GJ 367b using high-precision photometry and radial velocity observations. GJ 367b orbits a bright (*V*-band magnitude of 10.2), nearby, and red (M-type) dwarf star every 7.7 hours. GJ 367b has a radius of 0.718 ± 0.054 Earth-radii and a mass of 0.546 ± 0.078 Earth-masses, making it a sub-Earth planet. The corresponding bulk density is 8.106 ± 2.165 grams per cubic centimeter—close to that of iron. An interior structure model predicts that the planet has an iron core radius fraction of 86 ± 5%, similar to that of Mercury's interior.

Dalba et al. (2022) reported the discovery of TOI-2180 b, a 2.8 $M_J$ giant planet orbiting a slightly evolved G5 host star. This planet transited only once in Cycle 2 of the primary Transiting Exoplanet Survey Satellite mission. Citizen scientists identified the 24-hr single-transit event shortly after the data were released, allowing a Doppler monitoring campaign with the Automated Planet Finder telescope at Lick Observatory to begin promptly. The radial velocity observations refined the orbital period of TOI-2180 b to be 260.8 ± 0.6 days, revealed an orbital eccentricity of 0.368 ± 0.007, and discovered long-term acceleration from a more distant massive companion. They conducted ground-based photometry from 14 sites spread around the globe in an attempt to detect another transit. Although they did not make a clear transit detection, the nondetections improved the precision of the orbital period. They predict that TESS will likely detect another transit of TOI-2180 b in Sector 48 of its extended mission. They use giant planet structure models to retrieve the bulk heavy-element content of TOI-2180 b. When considered alongside other giant planets with orbital periods over 100 days, they find tentative evidence that the correlation between planet mass and metal enrichment relative to stellar is dependent on orbital properties. Single-transit discoveries like TOI-2180 b highlight the exciting potential of the TESS mission to find planets with long orbital periods and low irradiation fluxes despite the selection biases associated with the transit method.

Pineda and Villadsen (2023) reported coherent radio bursts from known M-dwarf planet-host YZ Ceti. Here is their report: observing magnetic star–planet interactions (SPIs) offers promise for determining the magnetic fields of exoplanets. Models of sub-Alfvénic SPIs predict that terrestrial planets in close-in orbits around M dwarfs can induce detectable stellar radio emission, manifesting as bursts of strongly polarized coherent radiation observable at specific planet orbital positions. They presented 2–4 GHz detections of coherent radio bursts on the slowly rotating M dwarf YZ Ceti, which hosts a compact system of terrestrial planets, the innermost of which orbits with a 2-day period. Two coherent bursts occur at similar orbital phases of YZ Ceti b, suggestive of an enhanced probability of bursts near that orbital phase. They modeled the system's magnetospheric environment in the context of sub-Alfvénic SPIs and determine that YZ Ceti b can plausibly power the observed flux densities of the radio detections. However, we cannot rule out stellar magnetic activity without a well-characterized rate of non-planet-induced coherent radio bursts on slow rotators. YZ Ceti is therefore a candidate radio SPI system, with unique promise as a target for long-term monitoring.Circumbinary planets, those that orbit around both stars of a central binary star system, challenge our understanding of planet formation. With only 12 binary systems known to host circumbinary planets, identifying more of these planets, along with their physical properties, could help to discern some of the physical processes that govern planet formation. Standing et al. (2023) analyzed radial-velocity data obtained by the HARPS and ESPRESSO spectrographs and report the detection of BEBOP-1 c, a gas giant planet with a mass of 65.2 ± 11.8 Earth masses ($M_{\oplus}$) orbiting around both stars of an eclipsing binary star system with a period of 215.5 ± 3.3 days. The system TOI-1338, hereafter referred to as BEBOP-1, which also hosts the smaller and inner transiting planet TOI-1338 b, is only the second confirmed multiplanetary circumbinary system. They do not detect TOI-1338 b with radial-velocity data alone, and we can place an upper limit on its mass of

$21.8\,M_{\oplus}$ with 99% confidence. TOI-1338 b is amenable to atmospheric characterization using JWST, so the BEBOP-1 system has the potential to act as a benchmark for circumbinary exo-atmospheric studies.

An early catalog of planet-hosting multiple-star systems of order three and higher has been published recently by Cuntz et al. (2022). They presented a catalog, status 1 July 2022, of triple and higher-order systems identified containing exoplanets based on data from the literature, including various analyses. The authors explored statistical properties of the systems with a focus on both the stars and the planets. So far, about 30 triple systems and one to three quadruple systems, including (mildly) controversial cases, have been found. The total number of planets is close to 40. All planet-hosting triplestar systems are highly hierarchic, consisting of a quasi-binary complemented by a distant stellar component, which is in orbit about the common center of mass. Furthermore, the quadruple systems are in fact pairs of close binaries ("double–doubles"), with one binary harboring a planet. For the different types of star–planet systems, the authors introduced a template for the classifications of planetary orbital configurations in correspondence to the hierarchy of the system and the planetary host. The data show that almost all stars are main-sequence stars. However, the stellar primaries tend to be more massive (i.e., corresponding to spectral types A, F, and G) than expected from single-star statistics, a finding also valid for stellar secondary's but less pronounced. Tertiary stellar components are almost exclusively low-mass stars of spectral type M. Almost all planets have been discovered based on either the radial velocity method or the transit method. Both gas giants (the dominant type) and terrestrial planets (including super-Earths) have been identified.

Seven rocky planets orbit the nearby dwarf star TRAPPIST-1, providing a unique opportunity to search for atmospheres on small planets outside the solar system. Thanks to the recent launch of the James Webb Space Telescope, possible atmospheric constituents such as carbon dioxide ($CO_2$) are now detectable. Recent JWST observations of the innermost planet TRAPPIST-1b showed that it is most probably a bare rock without any $CO_2$ in its atmosphere. Zieba et al. (2023) reported the detection of thermal emission from the dayside of TRAPPIST-1c with the Mid-Infrared Instrument (MIRI) on JWST at 15 μm. They measured a planet-to-star flux ratio of $f_p/f_* = 421 \pm 94$ ppm, which corresponds to an inferred dayside brightness temperature of $380 \pm 31$ K. This high dayside temperature disfavors a thick, $CO_2$-rich atmosphere on the planet. The data rule out cloud-free $O_2/CO_2$ mixtures with surface pressures ranging from 10 bar (with 10 ppm $CO_2$) to 0.1 bar (pure $CO_2$). A Venus-analog atmosphere with sulfuric acid clouds is also disfavored at 2.6σ confidence. Thinner atmospheres or bare-rock surfaces are consistent with their measured planet-to-star flux ratio. The absence of a thick, $CO_2$-rich atmosphere on TRAPPIST-1c suggests a relatively volatile-poor formation history, with less than $9.5_{-2.3}^{+7.5}$ Earth oceans of water. If all planets in the system formed in the same way, this would indicate a limited reservoir of volatiles for the potentially habitable planets in the system.

Trojan asteroids are common inhabitants of the solar system. They are minor bodies populating the L4 and L5 Lagrange regions of a planet, leading and trailing it 60° apart in the same orbital path. Laughlin and Chambers (2002) theoretically demonstrated that Trojans as massive as the main planet could be long-term stable, thus inspiring the concept of co-orbital planets. The fact that the stability condition for co-orbitals is met for pairs of bodies with similar masses opens the possibility of searching for exo-trojans using the same methods and instruments as for the currently confirmed exoplanets (radial velocities, transit timing variations, transits themselves, or the combination of transit and radial velocity data). However, the efforts in the search for exo-trojan planets have resulted in a select number of unconfirmed candidates so far. According to Balsalobre-Ruza et al. (2023) high-spatial resolution Atacama Large Millimeter/Submillimeter Array data have revealed a plethora of substructures in protoplanetary disks. Some of those features are thought to trace the formation of embedded planets. One example is the gas and dust that accumulated in the co-orbital Lagrangian regions L4/L5, which were tentatively detected in recent years and might be the pristine material for the formation of Trojan bodies. The work of Balsalobre-Ruza et al. (2023) is part of the TROY project, whose ultimate goal is to find robust evidence of exo-trojan bodies and study their implications in the exoplanet field. In their work they focus on the early stages of the formation of these bodies by inspecting the iconic system PDS 70, the only confirmed planetary system in formation. They found emission at a ~4σ level (~6σ when correcting from a cleaning effect) at the position of the L5 region of PDS 70 b. This emission corresponds to a dust mass in a range of 0.03–2 $M_{Moon}$, which potentially accumulated in this gravitational well.

A paper by Lee et al. (2023) is a follow-up observation to reinterpret the radial velocity of HD 36384, where the existence of planetary systems is known to be ambiguous. In giants, it is, in general, difficult to distinguish the signals of planetary companions from those of stellar activities. Thus, known exoplanetary giant hosts are relatively rare. This group, for many years, has obtained RV data in evolved stars using the high-resolution, fiber-fed Bohyunsan Observatory Echelle Spectrograph (BOES) at the Bohyunsan Optical Astronomy Observatory (BOAO). The authors reported the results of RV variations in the M giant HD 36384. They have found two significant periods of 586 d and 490 d. Considering the orbital stability, it is impossible to have two planets at such close orbits. To determine the nature of the RV variability variations, they analyzed the HIPPARCOS photometric data, some indicators of stellar activities, and line profiles. A significant period of 580 d was revealed in the HIPPARCOS photometry. H$\alpha$ EW variations also show a meaningful period of 582 d. Thus, the period of 586 d may be closely related to the rotational modulations and/or stellar pulsations. On the other hand, the other significant period of 490 d is interpreted as the result of the orbiting companion. Our orbital fit suggests that the companion was a planetary mass of 6.6 $M_{Jupiter}$ and is located at 1.3 AU from the host.

Mireles et al. (2023) reported the discovery and validation of two long-period giant exoplanets orbiting the early K dwarf TOI-4600 (V = 12.6, T = 11.9), first detected using observations from the Transiting Exoplanet Survey Satellite by the TESS Single Transit Planet Candidate Working Group. The inner planet, TOI-4600 b, has a radius of 6.80 ± 0.31 $R_\oplus$ and an orbital period of 82.69 days. The outer planet, TOI-4600 c, has a radius of 9.42 ± 0.42 $R_\oplus$ and an orbital period of 482.82 days, making it the longest-period confirmed or validated planet discovered by TESS to date. They combined TESS photometry and ground-based spectroscopy, photometry, and high-resolution imaging to validate the two planets. With equilibrium temperatures of 347 K and 191 K, respectively, TOI-4600 b and c add to the small but growing population of temperate giant exoplanets that bridge the gap between hot/warm Jupiters and the solar system's gas giants. TOI-4600 is a promising target for further transit and precise RV observations to measure the masses and orbits of the planets as well as search for additional nontransiting planets. Additionally, with Transit Spectroscopy Metric values of ~30, both planets are amenable for atmospheric characterization with JWST. Together, these will lend insight into the formation and evolution of planet systems with multiple giant exoplanets.

Due to their relatively small masses and sizes, M dwarfs are prime targets in the search for small (Rp ~ 1–4 $R_\oplus$) planets using current detection techniques. Furthermore, it is likely that the only rocky worlds around main sequence stars whose atmospheres we will be able to study over the next decade are transiting planets detected around late-type stars. Among the more than 5500 planets detected to date, fewer than 300 have been found to be transiting around M dwarfs. With such a small sample size, it is not surprising to see that there are close to 30 known exoplanets around M dwarfs that are in the habitable zone of their star—and only about ten of those planets are transiting. There is also the late-K-dwarf HZ desert that is of interest. It should be noted that although the true habitability of M dwarfs has been highly debated in great part due to their frequent and energetic flares, it has been shown that stellar activity and magnetic flaring is dramatically diminished for some stars.

The radius valley carries implications for how the atmospheres of small planets form and evolve, but this feature is visible only with highly precise characterizations of many small planets. Oddo et al. (2023) present the characterization of nine planets and one planet candidate with both NASA TESS and ESA CHEOPS observations, which adds to the overall population of planets bordering the radius valley. While five of the planets—TOI 118 b, TOI 262 b, TOI 455 b, TOI 560 b, and TOI 562 b—have already been published, they vet and validate transit signals as planetary using follow-up observations for four new TESS planets, including TOI 198 b, TOI 244 b, TOI 444 b, and TOI 470 b. While a three times increase in primary mirror size should mean that one CHEOPS transit yields an equivalent model uncertainty in transit depth as about nine TESS transits in the case that the star is equally as bright in both bands, they found that CHEOPS transits typically yield uncertainties equivalent to between two and 12 TESS transits, averaging 5.9 equivalent transits. Therefore, the fits to CHEOPS transits provide overall lower uncertainties on transit depth and better precision relative to fits to TESS transits; uncertainties for these fits do not always match expected predictions given photon-limited noise. They found no correlations between number of equivalent transits and any physical parameters, indicating that this behavior is not strictly systematic but rather might be due to other factors such as in-transit gaps during CHEOPS visits or nonhomogeneous detrending of CHEOPS light curves.

In the last decade, a dozen close-in giant planets have been discovered orbiting stars with spectral types ranging from M0 to M4, a mystery since known formation pathways do not predict the existence of such systems. Triaud et al. (2023) confirmed TOI-4860 b, a Jupiter-sized planet orbiting an M4.5 host, a star at the transition between fully and partially convective interiors. First identified with TESS data, they validated the transiting companion's planetary nature through multicolor photometry from the TRAPPIST-South/North, SPECULOOS, and MuSCAT3 facilities. Their analysis yields a radius of 0.76 ± 0.02 $R_{Jup}$ for the planet, a mass of 0.34 $M_{sun}$ for the star, and an orbital period of 1.52 d using the newly commissioned SPIRIT InGaAs camera at the SPECULOOS-South Observatory; they collect infrared photometry in *zYJ* that spans the time of secondary eclipse. These observations do not detect a secondary eclipse, placing an upper limit on the brightness of the companion. The planetary nature of the companion is further confirmed through high-resolution spectroscopy obtained with the IRD spectrograph at Subaru Telescope, from which they measured a mass of 0.67 ±0.14 $M_{Jup}$. Based on its overall density, TOI-4860 b appears to be rich in heavy elements, like its host star.

The TRAPPIST-1 system is host to seven transiting planets with masses and radii indicating a mostly rocky composition with, possibly, volatiles such as a surface water layer (Agol et al. 2021). The relatively small host star is a Jupiter-sized M8-type dwarf, boosting the amplitude of potential spectral signatures from planetary atmospheres seen in transmission. The TRAPPIST-1 planets orbit their star in a compact configuration, allowing more transit observations within a given amount of time relative to planets with longer orbital periods. All these features combined make TRAPPIST-1 ideally suited to study multiple Earth-sized, rocky exoplanets from the same environment with some in the habitable zone of the system (Gillon et al. 2020).

Each TRAPPIST-1 planet has been observed in transmission both from space and from the ground, confidently rejecting cloud-free, hydrogen-rich atmospheres. Secondary eclipse observations of TRAPPIST-1 b with JWST/MIRI are consistent with little to no atmosphere given the lack of heat redistribution. Lim et al. (2023) presented the first transmission spectra of TRAPPIST-1 b obtained with JWST/NIRISS over two visits. The two transmission spectra show moderate to strong evidence of contamination from unocculted stellar heterogeneities, which dominates the signal in both visits. The transmission spectrum of the first visit is consistent with unocculted starspots and the second visit exhibits signatures of unocculted faculae. Fitting the stellar contamination and planetary atmosphere either sequentially or simultaneously, they confirm the absence of cloud-free, hydrogen-rich atmospheres but cannot assess the presence of secondary atmospheres. They find that the uncertainties associated with the lack of stellar model fidelity are one order of magnitude above the observation precision of 89 ppm (combining the two visits). Without affecting the conclusion regarding the atmosphere of TRAPPIST-1 b, this highlights an important caveat for future explorations, which calls for additional observations to characterize stellar heterogeneities empirically and/or theoretical works to improve model fidelity for such cool stars. This need is all the more justified, as stellar contamination can affect the search for atmospheres around the outer, cooler TRAPPIST-1 planets for which transmission spectroscopy is currently the most efficient technique.

Kenworthy et al. (2023) reported recently on a planetary collision afterglow and transit of the resultant debris cloud. That is, planets grow in rotating disks of dust and gas around forming stars, some of which can subsequently collide in giant impacts after the gas component is removed from the disk. Monitoring programs with the warm Spitzer mission have recorded substantial and rapid changes in mid-infrared output for several stars, interpreted as variations in the surface area of warm, dusty material ejected by planetary-scale collisions and heated by the central star: for example, NGC 2354–ID8, HD 166191, and V488 Persei. In their paper, Kenworthy et al. (2023) reported combined observations of the young (about 300 million years old), solar-like star ASASSN-21qj: an infrared brightening consistent with a blackbody temperature of 1,000 Kelvin and a luminosity that is 4% that of the star, lasting for about 1,000 days, partially overlapping in time with a complex and deep, wavelength-dependent optical eclipse that lasted for about 500 days. The optical eclipse started 2.5 years after the infrared brightening, implying an orbital period of at least that duration. These observations are consistent with a collision between two exoplanets of several to tens of Earth masses at 2–16 astronomical units from the central star. Such an impact produces a hot, highly extended post-impact remnant with sufficient luminosity to explain the infrared observations. Transit of the impact debris, sheared by orbital motion into a long cloud, causes the subsequent complex eclipse of the host star.

Lissauer et al. (2023) recently published (November 2) a draft version of the Updated Catalog of Kepler Planet Candidates: Focus on Accuracy and Orbital Periods. Their catalog prioritizes accuracy of planetary dispositions and properties over uniformity. This catalog contains 4376 transiting planet candidates, including 1791 residing within 709 multi-planet systems, and provides the best parameters available for a large sample of Kepler planet candidates. They also provide a second set of stellar and planetary properties for transiting candidates that are uniformly derived for use in occurrence rates studies. Estimates of orbital periods have been improved, but, as in previous catalogs, their tabulated values for period uncertainties do not fully account for transit timing variations (TTVs). They show that many planets are likely to have TTVs with long periodicities caused by various processes, including orbital precession, and that such TTVs imply that ephemerides of Kepler planets are not as accurate on multi-decadal timescales as predicted by the small formal errors (typically 1 part in $10^6$ and rarely $> 10^{-5}$) in the planets' measured mean orbital periods during the Kepler epoch. Analysis of normalized transit durations implies that eccentricities of planets are anti-correlated with the number of companion transiting planets. Their primary catalog lists all known Kepler planet candidates that orbit and transit only one star; for completeness, they also provide an abbreviated listing of the properties of the two dozen non-transiting planets that have been identified around stars that host transiting planets discovered by Kepler.

Previous studies of the exoplanet LTT 1445Ac concluded that the light curve from the Transiting Exoplanet Survey Satellite was consistent with both grazing and nongrazing geometries. As a result, the radius and hence density of the planet remained unknown. To resolve this ambiguity, Pass et al. (2023) observed the LTT 1445 system for six spacecraft orbits of the Hubble Space Telescope using WFC3/UVIS imaging in spatial scan mode, including one partial transit of LTT 1445Ac. This imaging produces resolved light curves of each of the three stars in the LTT 1445 system. The authors confirmed that the planet transits LTT 1445A and that LTT 1445C is the source of the rotational modulation seen in the TESS light curve, and they refined the estimate of the dilution factor for the TESS data. They performed a joint fit to the TESS and HST observations, finding that the transit of LTT 1445Ac is not grazing with 97% confidence. They measured a planetary radius of $1.07^{+0.10}_{-0.07}$ $R_\oplus$. Combined with previous radial velocity observations, their analysis yields a planetary mass of $1.37 \pm 0.19$ $M_\oplus$ and a planetary density of $5.9^{+1.8}_{-1.5}$ g $cm^{-3}$; LTT 1445Ac is likely an Earth analog with respect to its mass and radius.

Planets with radii between that of the Earth and Neptune (hereafter referred to as "sub-Neptunes") are found in close-in orbits around more than half of all Sun-like stars. However, their composition, formation, and evolution remain poorly understood. According to Luque et al. (2023) the study of multiplanetary systems offers an opportunity to investigate the outcomes of planet formation and evolution while controlling for initial conditions and environment. Those in resonance (with their orbital periods related by a ratio of small integers) are particularly valuable because they imply a system architecture practically unchanged since its birth. The authors presented the observations of six transiting planets around the bright nearby star HD 110067. They found that the planets follow a chain of resonant orbits. A dynamical study of the innermost planet triplet allowed the prediction and later confirmation of the orbits of the rest of the planets in the system. The six planets are found to be sub-Neptunes with radii ranging from $1.94R_{\oplus}$ to $2.85R_{\oplus}$. Three of the planets have measured masses, yielding low bulk densities that suggest the presence of large hydrogen-dominated atmospheres.

### 2.3.2 Free-Floating Planets

Free-floating planets, floating freely without a parent star, may be more common than stars. Researchers discovered at least 70 free-floating planets in a region of the Milky Way known as the Upper Scorpius OB stellar association

Delorme et al. (2012) reported a 4–7 Jupiter-mass free-floating planet in the young moving group AB Doradus called CFBDSIR2149–0403. Using the CFBDSIR wide field survey for brown dwarfs, they identified CFBDSIRJ214947.2–040308.9, a late T dwarf with an atypically red $J–K_S$ color. They obtained an X-Shooter spectra, with signal detectable from 0.8 μm to 2.3 μm, which confirmed a T7 spectral type with an enhanced Ks-band flux indicative of a potentially low-gravity, young object. The comparison of their near infrared spectrum with atmosphere models for solar metallicity shows that CFBDSIRJ214947.2–040308.9 is probably a 650 – 750 K, $\log g = 3.75 - 4.0$ substellar object. Using evolution models, this translates into a planetary mass object with an age in the 20–200 Myr range. An independent Bayesian analysis from proper motion measurements results in a 87% probability that this free-floating planet is a member of the 50–120-Myr-old AB Doradus moving group, which strengthens the spectroscopic diagnosis of youth. By combining the atmospheric characterization with the age and metallicity constraints arising from the probable membership to the AB Doradus moving group, they found that CFBDSIRJ214947.2–040308.9 is probably a 4–7 Jupiter mass, free-floating planet with an effective temperature of ~700 K and a $\log g$ of ~4.0, typical of the late T-type exoplanets that are targeted by direct imaging. We stress that this object could be used as a benchmark for understanding the physics of the similar T-type exoplanets that will be discovered by the upcoming high-contrast imagers.

Some low-mass planets are expected to be ejected from their parent planetary systems during early stages of planetary system formation. According to planet-formation theories, such as the core accretion theory, typical masses of ejected planets should be between 0.3 and 1.0 $M_{Earth}$. Although in practice such objects do not emit any light, they may be detected using gravitational microlensing via their light-bending gravity. Microlensing events due to terrestrial-mass rogue planets are expected to have extremely small angular Einstein radii (<1 uas) and extremely short timescales (<0.1 day). Mróz et al. (2020) presented the discovery of the shortest-timescale microlensing event, OGLE-2016-BLG-1928, identified to date ($t_E \approx 0.0288$ day = 41.5 min). Thanks to the detection of finite-source effects in the light curve of the event, they were able to measure the angular Einstein radius of the lens $\theta_E = 0.842 \pm 0.064$ uas, making the event the most extreme short-timescale microlens discovered to date. Depending on its unknown distance, the lens may be a Mars- to Earth-mass object, with the former possibility favored by the Gaia proper motion measurement of the source. The planet may be orbiting a star, but they ruled out the presence of stellar companions up to the projected distance of 8.0 au from the planet. Their discovery demonstrates that terrestrial-mass free-floating planets can be detected and characterized using microlensing.

A free-floating planet (FFP) is a planetary-mass object that orbits a non-stellar massive object (e.g. a brown dwarf) or the galactic center. The presence of exomoons orbiting FFPs has been theoretically predicted by several models. Under specific conditions, these moons are able to retain an atmosphere capable of ensuring the long-term thermal stability of liquid water on their surface. Ávila et al. (2021) modeled this environment with a one-dimensional radiative-convective code coupled to a gas-phase chemical network including cosmic rays and ion-neutral reactions. They found that, under specific conditions and assuming stable orbital parameters over time, liquid water can be formed on the surface of the exomoon. The final amount of water for an Earth-mass exomoon is smaller than the amount of water in Earth oceans but enough to host the potential development of primordial life. The chemical equilibrium time-scale is controlled by cosmic rays, the main ionization driver in our model of the exomoon atmosphere.

The nature and origin of free-floating planets are still largely unconstrained because of a lack of large homogeneous samples to enable a statistical analysis of their properties. So far, most FFPs have been discovered using indirect methods; microlensing surveys have proved particularly successful to detect these objects down to a few Earth masses. However, the ephemeral nature of microlensing events prevents any follow-up observations and individual characterization. Several studies have identified FFPs in young stellar clusters and the galactic field, but their samples are small or heterogeneous in age and origin. Miret-Roig et al. (2021) reported the discovery of between 70 and 170 FFPs (depending on the assumed age) in the region encompassing Upper Scorpius and Ophiuchus, the closest young OB association to the Sun. They found an excess of FFPs by a factor of up to seven compared with core-collapse model predictions, demonstrating that other formation mechanisms may be at work. They estimate that ejection from planetary systems might have a contribution comparable to that of core collapse in the formation of FFPs. Therefore, ejections due to dynamical instabilities in giant exoplanet systems must be frequent within the first 10 Myr of a system's life.

Kondo et al. (2023) reported that the PRime-focus Infrared Microlensing Experiment (PRIME) will be the first to conduct a dedicated near-infrared microlensing survey by using a 1.8-m telescope with a wide field of view of 1.45 $deg^2$ at the South African Astronomical Observatory. The major goals of the PRIME microlensing survey are to measure the microlensing event rate in the inner galactic bulge to help design the observing strategy for the exoplanet microlensing survey by the Nancy Grace Roman Space Telescope and to make a first statistical measurement of exoplanet demographics in the central bulge fields where optical observations are very difficult owing to the high extinction in these fields. The authors conducted a simulation of the PRIME microlensing survey to estimate its planet yields and determine the optimal survey strategy, using a galactic model optimized for the inner galactic bulge. In order to maximize the number of planet detections and the range of planet mass, they compared the planet yields among four observation strategies and predicted that PRIME would detect planetary signals for 42–52 planets (1–2 planets with $Mp \leq 1M_{\oplus}$, 22 – 25 planets with mass $1M_{\oplus} < Mp \leq 100M_{\oplus}$, 19–25 planets $100M_{\oplus} < Mp \leq 10,000M_{\oplus}$), per year depending on the chosen observation strategy.

Free-floating planetary-mass objects may represent the end states of disrupted exoplanetary systems and "failed" stars. Except for the very young and most massive planets, such objects are too faint to be detected directly, and their identification relies on gravitational microlensing, whereby the planet traverses the observer's line of sight towards a background star.

McDonald et al. (2021) present the first short-duration candidate microlensing events from the Kepler K2 mission. From late April to early July 2016, Campaign 9 of K2 obtained high temporal cadence observations over a 3.7 $deg^2$ region of the galactic bulge. Its primary objectives were to look for evidence of a free-floating planet population using microlensing, and demonstrate the feasibility of space-based planetary microlensing surveys. Though Kepler K2 is far from optimal for microlensing, the recently developed MCPM photometric pipeline enables us to identify and model microlensing events. The authors describe their blind event-selection pipeline in detail and use it to recover 22 short-duration events with effective time-scales $t_{eff} <$ 10 d previously announced by the OGLE and KMTNet ground-based surveys. They also announce five new candidate events. One of these is a caustic-crossing binary event, modelled in a companion study. The other four have very short durations ($t_{eff} < 0.1$ d) typical of an Earth-mass FFP population. Whilst Kepler was not designed for crowded-field photometry, the K2C9 data set clearly demonstrates the feasibility of conducting blind space-based microlensing surveys towards the galactic bulge.

## 2.4 SOLAR SYSTEM

Our solar system was formed about $4.6 \times 10^9$ years ago through the gravitational collapse of an interstellar cloud. The small interstellar dust particles grew bigger and accreted more material, eventually forming planets or moons. The remaining material, not incorporated into planets, formed the populations of comets and asteroids that experienced a diverse history.

It is an established fact that solar system orbits around the center of the Milky Way galaxy with period of (225–250) × $10^6$ Earth's years, Stacy (2002). Since its birth, the Sun has made about 20 trips around the center of the galaxy; during this time, the Sun has made passages through the spiral arms of the disk. Gies and Helsel (2005) made a calculation of the Sun's motion through the Milky Way over the last 500 × $10^6$ years and found that the Sun has traversed four spiral arms in this period. The integration is based upon estimates of the Sun's current position and speed from measurements with Hipparcos and upon a realistic model for the galactic gravitational potential. They estimate the times of the Sun's past spiral arm crossings for a range in assumed values of the spiral pattern angular speed. They found that for a difference between the mean solar and pattern speed of $\Omega_\odot - \Omega_p = 11.9 \pm 0.7$ km s$^{-1}$ kpc$^{-1}$, the Sun has traversed four spiral arms at times that appear to correspond well with long-duration cold periods on Earth. This supports the idea that extended exposure to the higher cosmic ray flux associated with spiral arms can lead to increased cloud cover and long ice age epochs on Earth.

The Sun's astrophysical birth environment affected the formation and composition of the solar system. Primitive meteorites display mass-independent oxygen isotope anomalies that were likely caused by ultraviolet (UV) photochemistry of CO gas-phase molecules, either (1) in the outer solar nebula by light from the young Sun or (2) in the parent molecular cloud by light from nearby stars. However, measurements of oxygen isotopes alone cannot unambiguously constrain the UV spectrum of the source responsible for the photochemistry. Sulfur, with four stable isotopes, can be used as a more direct probe of the astrophysical environment of mass-independent photochemistry. Vacher et al. (2021) reported the in situ isotopic analysis of paired oxygen and sulfur isotope systematics in cosmic symplectite (COS), magnetite-pentlandite intergrowths, in the primitive ungrouped carbonaceous chondrite Acfer 094. They showed that COS grains contain mass-independent sulfur isotope anomalies (weighted means of $\Delta^{33}S = +3.84 \pm 0.72$‰ and $\Delta^{36}S = -6.05 \pm 2.25$‰, 2SE) consistent with $H_2S$ photochemistry by UV from massive O and B stars close to the solar system's parent molecular cloud, and inconsistent with UV from the protosun. The presence of coupled mass-independent sulfur and oxygen ($\Delta^{17}O = 86 \pm 6$‰, 2SE) isotope anomalies in COS imply that these anomalies originated in the same astrophysical environment. The authors proposed that this environment is the photo-dissociation region (PDR) of the solar system's parent molecular cloud, where nearby massive stars irradiated the edge of the cloud. They concluded that the Sun's stellar neighbors, likely O and B stars in a massive-star-forming region, affected the composition of the solar system's primordial building blocks.

The disk of the galaxy from the solar circle outwards appears to display four-arm spiral structure as seen in the emission of atomic hydrogen and molecular CO and in the distribution of star forming regions. Gies and Helsel (2005) show (see Figure 2.5) the appearance of the galactic spiral arm patterns. The dotted line through the center of the galaxy indicates the current location of the central bar according to Bissantz et al. (2003). The placement of the Sun's trajectory in this diagram depends critically on the relative angular pattern speeds of the Sun and the spiral arms. The mean advance in azimuth in the model of the Sun's motion corresponds to a solar angular motion of $\Omega_\odot = 26.3$ km s$^{-1}$ kpc$^{-1}$. If the difference in the solar and spiral arm pattern speeds, $\Omega_\odot - \Omega_p$, is greater than zero, then the Sun overtakes the spiral pattern and progresses in a clockwise direction in depiction of the galactic plane. Unfortunately, the spiral pattern speed is not well established and may in fact be different in the inner and outer parts of the galaxy. Several studies advocate a spiral pattern speed of $\Omega_p = 20 \pm 5$ km s$^{-1}$ kpc$^{-1}$; therefore, Figure 2.5 shows the Sun's trajectory projected onto the plane for this value ($\Omega_\odot - \Omega_p = 6.3$ km s$^{-1}$ kpc$^{-1}$). Diamonds along the Sun's track indicate its placement at intervals of 100 × $10^6$ years. It is evident that for this assumed pattern speed the Sun has passed through only two arms over the last 500 × $10^6$ years. However, if a lower but still acceptable pattern speed of $\Omega_p = 14.4$ km s$^{-1}$ kpc$^{-1}$ is assumed ($\Omega_\odot - \Omega_p = 11.9$ km s$^{-1}$ kpc$^{-1}$), then the Sun has crossed four spiral arms in the past 500 × $10^6$ years and has nearly completed a full rotation ahead of the spiral pattern.

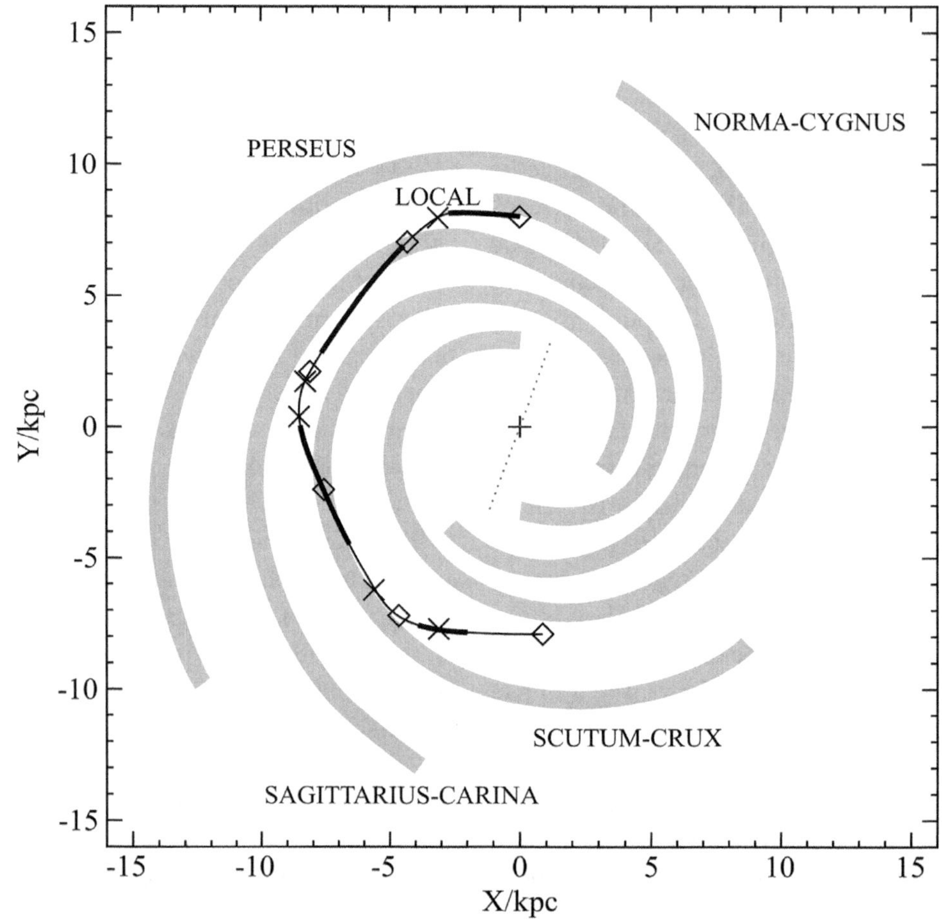


FIGURE 2.5 A depiction of the spiral arm pattern of the galaxy as viewed from above the plane. The plus sign marks the center of the galaxy while the main four arms plus the local (Orion) spur are indicated as gray shaded regions. The dotted line through the center of the galaxy indicates the location of the central bar (Bissantz et al. 2003). The Sun's path in the reference frame of the spiral arms is indicated with a solid line (for $\Omega_p = 20$ km s$^{-1}$ kpc$^{-1}$), and diamonds mark time intervals of $10^8$ years back in time from the present (top diamond). The thick portions correspond to icehouse times, and the X signs indicate times of large mass extinctions, after Gies and Helsel (2005).

Thus, the choice of the spiral pattern speed dramatically influences any conclusions about the number and timing of Sun's passages through the spiral arms over this time interval.

The oxygen isotopic composition of the Sun inferred from the measurements of the solar wind returned by the NASA Genesis spacecraft is $^{16}$O-enriched [$\Delta^{17}$O = −28.4 ± 3.6 per mil (‰)] relative to the whole-rock oxygen isotopic compositions of Mars, Moon, chondrites and achondrites, and oxygen isotopic compositions of chondrule phenocrysts, which, on a three-isotope oxygen diagram ($\delta^{17}$O versus $\delta^{18}$O), plot close to the terrestrial fractionation line (−7‰ < $\Delta^{17}$O < 3‰). The Sun is $^{16}$O-enriched ($\Delta^{17}$O = −28.4 ± 3.6‰) relative to the terrestrial planets, asteroids, and chondrules (−7‰ < $\Delta^{17}$O < 3‰). Ca, Al-rich inclusions (CAIs), the oldest solar system solids, approach the Sun's $\Delta^{17}$O. Ultraviolet CO self-shielding resulting in formation of $^{16}$O-rich CO and $^{17,18}$O-enriched water is the currently favored mechanism invoked to explain the observed range of $\Delta^{17}$O. However, the location of CO self-shielding (molecular cloud or protoplanetary disk) remains unknown. Krot et al. (2020) showed that CAIs with predominantly low ($^{26}$Al/$^{27}$Al)$_0$, <5 × 10$^{-6}$, exhibit a large inter-CAI range of $\Delta^{17}$O, from −40‰ to −5‰. In contrast, CAIs with the canonical ($^{26}$Al/$^{27}$Al)$_0$ of ~5 × 10$^{-5}$ from unmetamorphosed carbonaceous chondrites have a limited range of $\Delta^{17}$O, −24 ± 2‰. Because CAIs with low ($^{26}$Al/$^{27}$Al)$_0$ are thought to have predated the canonical CAIs and formed within first 10,000–20,000 years of the solar system evolution, these observations suggest oxygen isotopic heterogeneity in the early solar system was inherited from the protosolar molecular cloud.

The solar system formed from interstellar dust and gas in a molecular cloud. Astronomical observations show that typical interstellar dust consists of amorphous (*a*-) silicate and organic carbon. Bona fide physical samples for laboratory studies would yield unprecedented insight about solar system formation, but they were largely destroyed. The most likely repositories of surviving presolar dust are the least altered extraterrestrial materials, interplanetary dust particles (IDPs) with probable cometary origins. Cometary IDPs contain abundant submicron *a*-silicate grains called GEMS (glass with embedded metal and sulfides), believed to be carbon free. Some have detectable isotopically anomalous *a*-silicate components from other stars, proving they are preserved dust inherited from the interstellar medium. However, it is debated whether the majority of GEMS predate the solar system or formed in the solar nebula by condensation of high-temperature (>1,300 K) gas. Ishii et al. (2018) mapped IDP compositions with single nanometer-scale resolution and find that GEMS contain organic carbon. Mapping reveals two generations of grain aggregation, the key process in growth from dust grains to planetesimals, mediated by carbon. GEMS grains, some with *a*-silicate subgrains mantled by organic carbon, comprise the earliest generation of aggregates. These aggregates (and other grains) are encapsulated in lower-density organic carbon matrix, indicating a second generation of aggregation. Since this organic carbon thermally decomposes above ~450 K, GEMS cannot have accreted in the hot solar nebula and formed, instead, in the cold presolar molecular cloud and/or outer protoplanetary disk. This suggests that GEMS are consistent with surviving interstellar dust, condensed in situ, and cycled through multiple molecular clouds.

The initial isotopic composition of water in the solar system is of paramount importance to understanding the origin of water on planetary bodies but remains unknown, despite numerous studies. Aléon et al. (2022) used the isotopic composition of hydrogen in calcium–aluminum-rich inclusions from primitive meteorites, the oldest solar system rocks, to establish the hydrogen isotopic composition of water at the onset of solar system formation. They reported the hydrogen isotopic composition of nominally anhydrous minerals from CAI fragments trapped in a once-melted host CAI. Primary minerals have extremely low D/H ratios, with δD values down to −850‰, recording the trapping of nebular hydrogen. Minerals rich in oxidized iron formed before the capture of the fragments record the existence of a nebular gas reservoir with an oxygen fugacity substantially above the solar value and a D/H ratio within 20% of that of the Earth's oceans. Hydrogen isotopes also correlate with oxygen and nitrogen isotopes, indicating that planetary reservoirs of volatile elements formed within the first $2 \times 10^5$ years of the solar system, during the main CAI formation epoch. We propose that the isotopic composition of inner solar system water was established during the collapse of the protosolar cloud core owing to a massive admixture of interstellar water.

Meteorites and other hunks of space rock have been colliding with the Earth and Moon for their entire history. Counting impact craters on both bodies provides some estimate of the flux of material bombarding the inner solar system. In addition, if it can be measured, a meteorite's composition yields information about the geological processes that were responsible for its formation. A lunar micrometeorite preserves the solar system's early history (Lopatka 2022).

### 2.4.1 History

The solar system likely formed from the collapse of a molecular cloud core within a cluster of perhaps ~$10^3$–$10^4$ stars, Adams (2010). Dense cloud cores are supported against self-gravity by a combination of turbulent motion, magnetic fields, thermal (gas) pressure, and centrifugal forces. As turbulence within a cloud dissipates over time, magnetic fields delay the initial collapse until they are depleted by ambipolar diffusion and rapid collapse begins (Boss and Goswami 2006). Another possible trigger of collapse is a shock front from a nearby supernova (Cameron and Truran 1977), suggested by isotopic anomalies preserved in meteorites. The best chance of better constraining the events that triggered collapse is a combination of astronomical observations and modeling of disks and measurements of the isotopic record inherited from the molecular cloud and injected nucleosynthetic materials preserved in meteorites. Regardless of the collapse mechanism, the Sun started as a protosolar core supported by thermal pressure while still undergoing infall from the surrounding cloud material.

Arzoumanian et al. (2023) give insights on the Sun birth environment in the context of star cluster formation in hub–filament systems. Cylindrical molecular filaments are observed to be the main sites of Sunlike star formation, while massive stars form in dense hubs at the junction of multiple filaments. The role of hub–filament configurations has not been discussed yet in relation to the birth environment of the solar system and to infer the origin of isotopic ratios of short-lived radionuclides (SLRs, such as $^{26}Al$) of calcium–aluminum-rich inclusions observed in meteorites. Arzoumanian et al. (2023) presented simple analytical estimates of the impact of stellar feedback on the young solar system forming along a filament of a hub–filament system. They found that the host filament can shield the young solar system from stellar feedback, both during the formation and evolution of stars (stellar outflow, wind, and radiation) and at the end of their lives (supernovae). They also showed that a young solar system formed along a dense filament can be enriched with supernova ejecta (e.g., $^{26}Al$) during the formation timescale of CAIs. They also propose that the streamers recently observed around protostars may be channeling the SLR-rich material onto the young solar system. They claim that considering hub–filament configurations as the birth environment of the Sun is important when deriving theoretical models explaining the observed properties of the solar system.

In their paper, Bonsor et al. (2022) discuss the rapid formation of exoplanetesimals revealed by white drafts. The timing of formation of the first planetesimals determines the mode of planetary accretion and their geophysical and compositional evolution. Astronomical observations of circumstellar disks and solar system geochronology provide evidence for planetesimal formation during molecular cloud collapse, much earlier than previously estimated. They presented observational evidence from white dwarf planetary systems for planetesimal formation occurring during the first few hundred thousand years after cloud collapse in exoplanetary systems. A substantial fraction of white dwarfs have accreted planetary material rich in iron core or mantle material. For the exo-asteroids accreted by white dwarfs to form iron cores, substantial heating is required. By simulating planetesimal evolution and collisional evolution, they showed that the most likely heat source is short-lived radioactive nuclides such as $^{26}Al$ (which has a half-life of $\sim 0.7 \times 10^6$ years). Core-rich materials in the atmospheres of white dwarfs, therefore, provide independent evidence for rapid planetesimal formation, concurrent with star formation.

Spiral arm transits may affect Earth in other ways as well. Yeghikyan and Fahr (2004) suggest that during some spiral passages the Earth may encounter interstellar clouds of sufficient density to alter the chemistry of the upper atmosphere and trigger an ice age of relatively long duration. The higher stellar density in the arms may more effectively perturb the Oort cloud of comets and lead to a greater chance of large impacts on Earth, and this combined with the possible lethal effects of nearby supernova explosions could cause mass extinctions during during passages through the spiral arms (Leitch and Vasisht 1998). On the other hand, the record of terrestrial impact craters suggests a variation on a time scale shorter than the inter-arm crossing time, but possibly related to the Sun's oscillations above and below the disk plane (Stothers 1998).

The effects of the last supernova explosion before the formation of the solar system are considered by Ustinova (2007) using noble gases as examples. Acceleration of generated supernova matter in the explosive shock wave led to its initial fractionation and to the formation of small-scale isotopic heterogeneity of primordial matter. This is fixed as some isotopic anomalies in high-temperature phases of the earliest condensates of carbonaceous chondrites, as well as in the isotopic systems of noble gases, and is the basis of the supernova phenomenon. Two main manifestations of shock-wave acceleration in noble gases are investigated: the change in the isotopic ratios of their cosmogenic components due to the increasing hardness of the spectrum of nuclear-active particles and the fractionation of gases: the enrichment of their isotopic systems with heavier isotopes. The reality of the processes under consideration is demonstrated through the example of noble gases of solar corpuscular radiation in lunar ilmenites. The absence of r-process products among extinct radionuclides in Ca-and Al-rich inclusions of carbonaceous chondrites with a formation interval of less than or equal to $1 \times 10^6$ years supports the idea that the last supernova was an Ia-type supernova, which possibly played an important role in the origin of the solar system, see Table 2.5.

Isotope abundance ratios have an important role in astronomy and planetary sciences, providing insights into the origin and evolution of the solar system, interstellar chemistry, and stellar nucleosynthesis. In contrast to deuterium/hydrogen ratios, carbon isotope ratios are found to be roughly constant (around 89) in the solar system but do vary on galactic scales with a $^{12}C/^{13}C$ isotopologue ratio of around 68 in the current local interstellar medium. In molecular clouds and protoplanetary disks, $^{12}CO/^{13}CO$ ratios can be altered by ice and gas partitioning, low-temperature isotopic ion-exchange reactions, and isotope-selective photo dissociation. Zhang et al. (2021) reported observations of $^{13}CO$ in the atmosphere of the young, accreting super-Jupiter TYC 8998–760–1b, at a statistical significance of more than six sigma. Marginalizing over the planet's atmospheric temperature structure, chemical composition and spectral calibration uncertainties suggests a $^{12}CO/^{13}CO$ ratio of $31_{-10}^{+17}$ (90% confidence), a substantial enrichment in $^{13}C$ with respect to the terrestrial standard and the local interstellar value. As the current location of TYC 8998–760–1b at greater than or equal to 160 astronomical units is far beyond the CO snowline, they postulated that it accreted a substantial fraction of its carbon from ices enriched in $^{13}C$ through fractionation.

Astronomical observations reveal that protoplanetary disks around young stars commonly have ring- and gap-like structures in their dust distributions. These features are associated with pressure bumps trapping dust particles at specific locations, which simulations show are ideal sites for planetesimal formation. Izidoro et al. (2021) showed that our solar system may have formed from rings of planetesimals—created by pressure bumps—rather than a continuous disk. They modeled the gaseous disk phase assuming the existence of pressure bumps near the silicate sublimation line (at $T \sim 1{,}400\,K$), water snowline (at $T \sim 170\,K$) and CO snowline (at $T \sim 30\,K$). Their simulations show that dust piles up at the bumps and forms up to three rings of planetesimals: a narrow ring near 1 au, a wide ring between ~3–4 au and ~10–20 au and a distant ring between ~20 and ~45 au. They use a series of simulations to follow the evolution of the innermost ring and show how it can explain the orbital structure of the inner solar system and provides a framework to explain the origins of isotopic signatures of Earth, Mars, and different classes of meteorites. The central ring contains enough mass to explain the rapid growth of the giant planets'

TABLE 2.5 SOLAR SYSTEM TIMELINE

| $10^9$ Years Ago | Events |
|---|---|
| 13.8 | The Big Bang forms the universe |
| 4.6 | A group of protostars, one of which will become the Sun, form from a cloud of debris left by prior star explosions in the Milky Way |
| 4.59 | The giant planets *Jupiter*, *Saturn*, *Uranus*, and *Neptune* form around the protosun. At least *Uranus* and *Neptune* form closer to the Sun than where they are today. One or more ice giants may have also formed that were later ejected from the solar system |
| 4.55 | The Sun begins fusing hydrogen into helium |
| 4.5 | Mercury, Venus, Earth, and Mars form. A Mars-sized planet collides with Earth, and the debris forms the Moon |
| 4.5–4.1 | The Sun gravitationally separates from its protostar siblings |
| 4.1–3.8 | The giant planets' orbits shift, scattering small worlds throughout the solar system. Some bombard the inner planets and likely deliver water and organics to Earth |
| 4–3 | Small world bombardment causes widespread volcanism on the inner planets |
| 3.8–3.5 | Life begins on Earth |
| 3 | Mars loses most of its atmosphere and water |
| 2.5 | Photosynthetic organisms evolve on Earth and pump oxygen into our atmosphere, helping create the blend of gases we breathe today |
| 1 | Volcanism on the Moon stops |
| 0.1 | An impact on the Moon forms Tycho crater. Saturn gets its rings |
| 0.066 | A giant asteroid impacts Earth, triggering global climate change. Three quarters of life, including the dinosaurs, go extinct |
| Now | The solar system is a much calmer place now, though occasional asteroid impacts still threaten Earth |

cores. The outermost ring is consistent with dynamical models of solar system evolution proposing that the early solar system had a primordial planetesimal disk beyond the current orbit of Uranus.

The characterization of objects that have best preserved the mineralogical and molecular phases formed in the earliest stages of the solar system evolution is key to understanding the processes that led to the formation of the planets in their diversity. The Hayabusa mission of the Japan Aerospace Exploration Agency has returned for the first time samples collected at the surface of a C-type asteroid, Ryugu. They are now preserved at the Extraterrestrial Samples Curation Center of the Japan Aerospace Exploration Agency at the Institute of Space and Astronautical Science in Sagamihara, Japan, where they are submitted to a first round of purely non-destructive analyses, as reported by Pilorget et al. (2021). The MicrOmega hyperspectral microscope developed at the Institut d'Astrophysique Spatiale (Orsay, France), which operates in the near-infrared range (0.99–3.65 μm), is performing their mineralogical and molecular characterization down to the scale of a few tens of micrometers. Strong features at 2.7 μm (indicating their OH-rich content) and at 3.4 μm (diagnostic of the presence of organics) dominate at a global scale, but key distinctive signatures have been identified at a submillimeter scale. In particular, carbonates (a fraction of them enriched in iron) as well as NH-rich compounds have been detected. The occurrence of volatile-rich species, likely originating from the outer solar system, would support Ryugu having preserved both pristine material and altered phases, which are now available for refined laboratory analyses with the potential to draw new insights into the formation and evolution paths of planetary bodies in our solar system.

Short-lived radioactive isotopes with half-lives between 0.1 to 100 Myr can be used to probe the origin of the solar system. In their work, Lawson et al. (2021) examined the core-collapse supernovae production of the 15 SLRs produced: $^{26}$Al, $^{36}$Cl, $^{41}$Ca, $^{53}$Mn, $^{60}$Fe, $^{92}$Nb, $^{97}$Tc, $^{98}$Tc, $^{107}$Pd, $^{126}$Sn, $^{129}$I, $^{135}$Cs, $^{146}$Sm, $^{182}$Hf, and $^{205}$Pb. They probed the impact of the uncertainties of the core-collapse explosion mechanism by examining a collection of 62 core-collapse models with initial masses of 15, 20, and 25 $M_\odot$, explosion energies between $3.4 \times 10^{50}$ and $1.8 \times 10^{52}$ ergs and compact remnant masses between 1.5 $M_\odot$ and 4.89 $M_\odot$. They identified the impact of both explosion energy and remnant mass on the final yields of the SLRs. Isotopes produced within the innermost regions of the star, such as $^{92}$Nb and $^{97}$Tc, are the most affected by the remnant mass, 92Nb varying by five orders of magnitude. Isotopes synthesized primarily in explosive C-burning and explosive He-burning, such as $^{60}$Fe, are most affected by explosion energies. $^{60}$Fe increases by two orders of magnitude from the lowest to the highest explosion energy in the 15 $M_\odot$ model. The final yield of each examined SLR is used to compare to literature models.

Low-albedo asteroids preserve a record of the primordial solar system planetesimals and the conditions in which the solar nebula was active. However, the origin and evolution of these asteroids are not well constrained. Here we measured visible and near-infrared (about 0.5–4.0 μm) spectra of low-albedo asteroids in the mid-outer main belt. Takir et al. (2023) showed that numerous large (diameter > 100 km) and dark (geometric albedo < 0.09) asteroids exterior to the dwarf planet Ceres' orbit share the same spectral features, and presumably compositions, as Ceres. They also developed a thermal evolution model that demonstrates that these Ceres-like asteroids have highly porous interiors, accreted relatively late at $(1.5–3.5) \times 10^6$ years after the formation of calcium–aluminum-rich inclusions, and experienced maximum interior temperatures of <900 K. Ceres-like asteroids are localized in a confined heliocentric region between about 3.0 au and 3.4 au, but were probably implanted from more distant regions of the solar system during the giant planet's dynamical instability.

Our understanding of the evolution of organic molecules, and their voyage from molecular clouds to the early solar system and Earth, has changed dramatically. Incorporating recent observational results from the ground and space, as well as laboratory simulation experiments and new methods for theoretical modeling, this review recapitulates the inventory and distribution of organic molecules in different environments. The evolution, survival, transport, and transformation of organics is monitored, from molecular clouds and the diffuse interstellar medium to their incorporation into solar system material such as comets and meteorites. Ehrenfreund and Charnley (2000) constrained gas phase and grain surface formation pathways to organic molecules in dense interstellar clouds, using recent observations with the Infrared Space Observatory (ISO) and ground-based radio-telescopes. The main spectroscopic evidence for carbonaceous compounds in the diffuse interstellar medium is discussed (UV bump at 2200 Å, diffuse interstellar bands, extended red emission, and infrared absorption and emission bands). They critically reviewed the signatures and unsolved problems related to the main organic components suggested to be present in the diffuse gas, such as polycyclic aromatic hydrocarbons (PAHs), fullerenes, diamonds, and carbonaceous solids.

In the solar system, space missions to comet Halley and observations of the bright comets Hyakutake and Hale-Bopp have recently allowed a reexamination of the organic chemistry of dust and volatiles in long-period comets. Ehrenfreund and Charnley (2000) reviewed the advances in this area and also discuss progress being made in elucidating the complex organic inventory of carbonaceous meteorites. The knowledge of organic chemistry in molecular clouds, comets, and meteorites and their common link provides constraints for the processes that lead to the origin, evolution, and distribution of life in the galaxy.

A new class of Kuiper Belt objects (KBOs) that lie beyond Neptune with semi-major axes greater than 250 astronomical units show orbital anomalies that have been interpreted as evidence for an undiscovered ninth planet. Brown and Mathur (2023) showed that a modified gravity theory known as modified Newtonian dynamics (MOND) provides an alternative explanation for the anomalies using the well-established secular approximation. They predicted that the major axes of the orbits would be aligned with the direction toward the galactic center and that the orbits cluster in phase space, in agreement with observations of KBOs from the new class. Thus, MOND, which can explain galactic rotation without invoking dark matter, might also be observable in the outer solar system.

#### *2.4.1.1 Habitable Earth Formation*

Miyazaki and Korenaga (2022) described how a wet heterogeneous mantle created a habitable world in the Hadean. The Hadean Earth is often viewed as an uninhabitable world with a molten surface, active volcanism, and a hot steamy atmosphere. The picture of harsh surface conditions may be expected from the Moon-forming giant impact and the subsequent bombardment by leftover planetesimals, but ancient geological records suggest that by approximately $4.3 \times 10^9$ years ago, the Earth's surface had become similar to the present-day environment with oceans and granitic crust. Hadean detrital zircons, enriched in heavy oxygen and yielding low melting temperatures, indicate that liquid water was already present at the Earth's surface. Furthermore, by the end of the Hadean, the global carbon cycle is likely to have stabilized the climate and maintained a surface temperature similar to the present day. The first $500 \times 10^6$ years of our planet, therefore, should have experienced a dramatic change in the climate, evolving quickly from being uninhabitable to habitable.

The Hadean eon, following the global-scale melting of the mantle is expected to be a dynamic period, during which Earth experienced vastly different conditions. Geologic records, however, suggest that the surface environment of Earth was already similar to the present by the middle of the Hadean. Under what conditions a harsh surface environment could turn into a habitable one remains uncertain. Miyazaki and Korenaga (2022) showed that a hydrated mantle with small-scale chemical heterogeneity, created as a result of magma ocean solidification, is the key to ocean formation, the onset of plate tectonics and the rapid removal of greenhouse gases, which are all essential to create a habitable environment on terrestrial planets. When the mantle is wet and dominated by high-magnesium pyroxenites, the removal of carbon dioxide from the atmosphere is expected to be more than ten times faster than the case of a pyrolitic homogeneous mantle and could be completed within $160 \times 10^6$ years. Such a chemically heterogeneous mantle would also produce oceanic crust rich in olivine, which is reactive with ocean water and promotes serpentinization. Therefore, conditions similar to the Lost City hydrothermal field may have existed globally in the Hadean seafloor.

Earth is the only planet known to have continents, although how they formed and evolved is unclear. According to Hill (2022), the Pilbara region of Western Australia is the best-preserved remnant of Earth's ancient continental crust. Johnson et al. (2022) studied the oxygen isotope composition of dated magmatic zircon from this area and found that continental remnant was built in three stages. According to the authors a second low-$\delta^{18}O$ zircon cluster at around $3.4 \times 10^9$ years is contemporaneous with spherule beds that provide the oldest material evidence for giant impacts on Earth (Schmieder and Kring 2020).

The Kepler space telescope collected continuous photometric data for nearly 3.5 years from a small 0.25% patch of the celestial sphere. This uniform monitoring of 150,000 stars enabled the discovery of over 4700 transiting exoplanet candidates through an automated detection pipeline. Removing the human component from the signal vetting process enabled the first homogeneous transiting small planet sample suitable for exoplanet population studies. Through full automation, the sample completeness can be measured via transit injection/recovery procedures. Here, artificial transit signals are injected into the raw light curves, enabling quantification of the pipeline's detection efficiency as a function of, for instance, the injection signal strength. In addition, the final Kepler catalog (DR25) also provided a measure of the sample reliability (the rate of false alarm signals) by testing the software's ability to remove systematics from the final catalog (Coughlin 2017).

One of the main objectives of the Kepler mission was to provide a baseline measurement for the occurrence of Earth analogs ($\eta_{\oplus}$). Despite the scarce completeness in this area of parameter space, several studies have extrapolated from more populated regions, providing measurements ranging from 0.1–0.4 Earth-like planets per main-sequence star. Currently, it remains unclear what planet features are important for habitability, making precise measurements unattainable. Moreover, these extrapolations are model-dependent and require a more empirical sampling of the Earth-like region of parameter space to reduce our uncertainty. Unfortunately, the continuous photometric data collection of the Kepler field was terminated due to mechanical issues on board the spacecraft after 3.5 years, limiting the completeness of these long-period small planets.

Zink et al. (2021) provided the first full K2 transiting exoplanet sample, using photometry from Campaigns 1–8 and 10–18, derived through an entirely automated procedure. This homogeneous planet candidate catalog is crucial to perform a robust demographic analysis of transiting exoplanets with K2. They identified 747 unique planet candidates and 57 multiplanet systems. Of these candidates, 366 have not been previously identified, including one resonant multiplanet system and one system with two short-period gas giants. By automating the construction of this list, measurements of sample biases (completeness and reliability) can be quantified. They carried out a light-curve-level injection/recovery test of artificial transit signals and found a maximum completeness of 61%, a consequence of the significant detrending required for K2 data analysis. Through this operation, they attained measurements of the detection efficiency as a function of signal strength, enabling future population analysis using this sample. They assessed the reliability of our planet sample by testing the vetting software EDI-Vetter against inverted transit-free light curves. They estimated that 91% of the planet candidates are real astrophysical signals, increasing up to 94% when limited to the FGKM dwarf stellar population. They also constrained the contamination rate from background eclipsing binaries to less than 5%. The presented catalog, along with the completeness and reliability measurements, enabled robust exoplanet demographic studies to be carried out across the fields observed by the K2 mission for the first time.

Ertel and Fields (2023) reported on determination of the distances to recent near-Earth supernovae based on geological and lunar $^{60}Fe$ measurements. Near-Earth supernova blasts, which engulf the solar system, have left traces of their ejecta in the geological and lunar records. There is now a wealth of data on live radioactive $^{60}Fe$ pointing to a supernova at $3 \times 10^6$ years ago, as well as the recent discovery of an event at $7 \times 10^6$ years ago. The authors used the available measurements to evaluate the distances to these events. For the better-analyzed supernova at $3 \times 10^6$ years, samples include deep-sea sediments, ferromanganese crusts, and lunar regolith; they explored the consistency among and across these measurements, which depends sensitively on the uptake of iron in the samples as well as possible anisotropies in the $^{60}Fe$ fallout. There is also significant uncertainty in the astronomical parameters needed for these calculations. They performed a parameter study on the effects that the ejected 60Fe mass from a core-collapse supernova and the fraction of dust that survives the remnant have on the resulting distance. They found that with an ejected $^{60}Fe$ mass of $3 \times 10^{-5}$ $M_{\odot}$ and a dust fraction of 10%, the distance range for the supernova $3 \times 10^6$ years ago is $D \sim 20$–140 pc, with the most likely range between 50 and 65 pc. Using the same astrophysical parameters, the distance for the supernova at 7 Myr ago is $D \sim 110$ pc.

### 2.4.2 Present: Sun, Planets and Their Moons, Asteroids

Hunt et al. (2022) discussed the dissipation of the solar nebula constrained by impacts and core cooling in planetesimals. Rapid cooling of planetesimal cores has been inferred for several iron meteorite parent bodies on the basis of metallographic cooling rates and linked to the loss of their insulating mantles during impacts.

However, the timing of these disruptive events is poorly constrained. They used the short-lived $^{107}Pd$–$^{107}Ag$ decay system to date rapid core cooling by determining Pd–Ag ages for iron meteorites and showed that closure times for the iron meteorites equate to cooling in the time frame ~(7.8–11.7) × $10^6$ years after calcium–aluminum-rich inclusion formation, and that they indicate that an energetic inner solar system persisted at this time. This probably results from the dissipation of gas in the protoplanetary disk, after which the damping effect of gas drag ceases. An early giant planet instability between 5 and 14 Myr after calcium–aluminum-rich inclusion formation could have reinforced this effect. This correlates well with the timing of impacts recorded by the Pd–Ag system for iron meteorites (Hunt et al. 2022).

According to Cardenas and Lamb (2022). a major question in Mars's history is whether the planet had an ocean in its low-elevation northern hemisphere. They showed evidence that a substantial sedimentary accumulation, at least 900 m thick and covering hundreds of thousands of square km, developed at the margin of an ancient ocean. Aeolis Dorsa is the name of a trough-like region of Mars about 500 km wide and 900 km long, surrounded east and west by 2 km-high plains and to the south by steep slopes up to the southern hemisphere. There, they mapped over 6500 km of exposed fluvial ridges, landforms that show the locations of ancient river channels. They organized the ridges into 20 branching systems. Measurements of ancient flow directions and stratigraphic positions record a history of an ocean-scale shoreline that, over time, rose at least 900 m vertically while retreating from north to south. The modern topography is defined by differential erosion into this stratigraphy and is different from the ancient geography. Sea-level rise is consistent with an extended warm and wet climate, not a mostly frozen early Mars. In their consideration, the assumption was made that the landforms seen now are essentially frozen in place from the time of their formation, perhaps as long as 3.5 × $10^9$ years ago.

Cardenas and Lamb (2022) interpreted the region Aeolis Dorsa, Mars, to be an erosional window into a sedimentary basin along a northern ocean. The basin-filling strata exposed there are at least 900 m thick and span an area of at least 9 × $10^4$ $km^2$, though the total accumulation may have been as thick as 2 km and as extensive as 4 × $10^5$ $km^2$. Along this erosional window, 20 distinct systems of fluvial ridges have been exhumed, and record deposition in river deltas and a submarine channel during different intervals of Martian geologic history. They used the stratigraphic positions and paleoflow directions of the 20 ridge systems to track the evolution of this shoreline through 900 m of sea-level rise consistent with a warm and wet early Mars. The reconstruction indicates that the shoreline initially prograded, and was later pushed south toward the highlands during a time of net sea-level rise. Sea-level rise of this magnitude is consistent with a northern ocean.

The decay of formerly live $^{182}Hf$ with a half-life of 9 Myr results in variations in the abundance of $^{182}W$ in early solar system objects. Lee et al. (2002) demonstrated that major excesses in $^{182}W$ in some lunar samples are the results of cosmogenic additions. Apollo 17 high-Ti mare basalts yield high $^{182}W/^{184}W$ of up to $\varepsilon_W = +11 \pm 1$. Even more extreme variations of up to $\varepsilon_W = +22 \pm 1$ are found for mineral separates, although these lavas were erupted more than 500 Myr after the start of the solar system. The measured $^{182}W$ excess in the separated minerals is correlated with their Ta/W, confirming theoretical models that implicate the $^{181}Ta(n,\gamma)^{182}Ta(\beta^-)^{182}W$ reaction from cosmic irradiation as the most likely cause. In contrast, olivine basalt 15555, which has a low cosmic ray exposure age, displays no internal $^{182}W$ variations and defines an Ow of +1.3 ± 0.4. This is consistent with earlier conclusions that the Moon formed about 50 × $10^6$ years after the start of the solar system. The high-Ti mare basalt source, with very high Hf/W, has a W isotopic composition that is not grossly different, from which a time limit of ~ 70 × $10^6$ years after the start of the solar system can be inferred for the formation of ilmenite-rich layers in the final stages of the lunar magma ocean.

Due to its dynamic nature, it is inconceivable for any isotopic heterogeneities (radiogenic or stable) to have survived the Moon-forming giant impact and the subsequent lunar magma ocean. Therefore, the radiogenic W signature on the Moon most likely reflects in situ decay of $^{182}Hf$ within the lunar mantle after its Hf/W was sufficiently enriched by the extraction of a low Hf/W crust and a small lunar core towards the cessation of lunar magma ocean. Using the measured W and Hf/W ratios in these mare basalts, the formation of the Moon is estimated at ca. (50 ± 10) × $10^6$ years after the start of the solar system, identical to previous estimates. In addition, the duration of the lunar magma ocean is estimated to be no more than 30 Myr after the Moon formed (Lee et al. 2002).

The formation of Earth's Moon by the impact of a Mars-sized body with the Earth is widely accepted. While the high temperatures predicted by models of this process might be expected to have driven off volatile elements, such elements are preserved in lunar glasses. Scenarios for the origin of such elements include original accretion during the formation of Earth and subsequent accretion from exogenous sources. Each leaves distinct geochemical fingerprints in the form of the isotopic compositions of noble gases. Will et al. (2022) conclude that the Moon inherited indigenous noble gases from Earth's mantle by the Moon-forming impact and propose storage in the incompatible element-enriched reservoir.

Fractures on asteroids develop due to stress driven by diurnal temperature variations at spatial scales ranging from sub-millimeters to meters. This damage has consequences for material properties of these asteroids with implications for the preservation of primordial signature acquired during the accretional phases in the protoplanetary disk of our solar system, Delbo et al. (2022).

Widespread hydration was detected on the lunar surface through observations of a characteristic absorption feature at 3 µm by three independent spacecraft. Whether the hydration is molecular water ($H_2O$) or other hydroxyl (OH) compounds is unknown and there are no established methods to distinguish the two using the 3-µm band. However, a fundamental vibration of molecular water produces a spectral signature at 6-µm that is not shared by other hydroxyl compounds. Honniball et al. (2020) presented observations of the Moon at 6 µm using the NASA/DLR Stratospheric Observatory for Infrared Astronomy (SOFIA). Observations reveal a 6-µm emission feature at high lunar latitudes due to the presence of molecular water on the lunar surface. Based on the strength of the 6-µm band, they estimated abundances of about 100 to 400 $\mu g\,g^{-1}$ $H_2O$. They found that the distribution of water over the small latitude range is a result of local geology and is probably not a global phenomenon. They also suggested that a majority of the detected water must be stored within glasses or in voids between grains sheltered from the harsh lunar environment, allowing the water to remain on the lunar surface.

The Galileo probe measured the elemental abundances in Jupiter's atmosphere and found that several heavy elements are enriched by a factor of ~3 compared to a protosolar composition. Also, the recent measurement of Jupiter's water abundance by Juno implies that oxygen is enriched by a factor of a few. The origin of the heavy-element enrichment of Jupiter's atmosphere remains unknown, and several ideas have been suggested to explain this enrichment. One idea is that the atmospheric enrichment is caused by the erosion of a primordial heavy-element core. However, in this case, the materials dissolved into the deep interior must be mixed by convection and be delivered to the upper envelope.

Jupiter's atmosphere is enriched with heavy elements by a factor of about 3 compared to a protosolar composition. The origin of this enrichment and whether it represents the bulk composition of the planetary envelope remain unknown. Internal structure models of Jupiter suggest that its envelope is separated from the deep interior and that the planet is not fully mixed. This implies that Jupiter's atmosphere was enriched with heavy elements just before the end of its formation. Such enrichment can be a result of late planetesimal accretion. However, in situ Jupiter formation models suggest a decreasing accretion rate with increasing planetary mass, which cannot explain Jupiter's atmospheric enrichment. Shibata and Helled (2022) modeled Jupiter's formation and showed that the migration of proto-Jupiter from ~20 au to its current location can lead to late planetesimal accretion and atmospheric enrichment. Late planetesimal accretion does not occur if proto-Jupiter migrates only a few astronomical units. They suggested that if Jupiter's outermost layer is fully mixed and is relatively thin (up to ~20% of its mass), such late accretion can explain its measured atmospheric composition. It is therefore possible that Jupiter underwent significant orbital migration followed by late planetesimal accretion.

Alongside Earth and Mars, Titan is the third planetary body in the solar system to show evidence for wide-spread and diverse sedimentary environments, including lakes, rivers, canyons, fans, dissected plateaux, and sand dunes suggesting a strong control of climate on Titan's surface processes and landscape formation. Sediments on Saturn's moon are thought to largely consist of mechanically weak organic grains, prone to rapid abrasion into dust. Yet Titan's equatorial dunes have likely been active for around $10^4$–$10^5$ years. Sustaining Titan's dunes over geologic timescales requires a mechanism that produces sand-sized particles at equatorial latitudes. Lapôtre et al. (2022) explored the hypothesis that a combination of abrasion, when grains are transported by winds or methane rivers, and sintering, when they are at rest, could produce sand grains that maintain an equilibrium size. Their model demonstrates that seasonal sediment transport may produce sand under Titan's surface conditions and could explain the latitudinal zonation of Titan's landscapes. Their findings support the hypothesis of global, source-to-sink sedimentary pathways on Titan, driven by seasons, and mediated by episodic abrasion and sintering of organic sand by rivers and winds.

Jupiter's moon Europa is a prime candidate for extraterrestrial habitability in our solar system. The surface landforms of its ice shell express the subsurface structure, dynamics, and exchange governing this potential. Double ridges are the most common surface feature on Europa and occur across every sector of the moon, but their formation is poorly understood, with current hypotheses providing competing and incomplete mechanisms for the development of their distinct morphology. Culberg et al. (2022) presented the discovery and analysis of a double ridge in northwest Greenland with the same gravity-scaled geometry as those found on Europa. Using surface elevation and radar sounding data, they showed that this double ridge was formed by successive refreezing, pressurization, and fracture of a shallow water sill within the ice sheet. If the same process is responsible for Europa's double ridges, the results suggest that shallow liquid water is spatially and temporally ubiquitous across Europa's ice shell.

#### *2.4.2.1 Earth*

Terrestrial planet formation concludes with a number of giant impacts among Moon- and Mars-sized planetary embryos. The well-studied giant-impact scenario involves a protoplanet, Theia, colliding with proto-Earth. Earth's Moon subsequently formed from orbitally bound impact debris. Although it is challenging to reconcile the observed lunar isotopes and volatiles with the giant-impact hypothesis, this Moon-formation hypothesis accounts for several key features of the Earth–Moon system, including the current angular momentum, the Moon's small core and the high mass of the Moon compared with the Earth. However, direct evidence for the existence of Theia remains elusive. Yuan et al. (2023) demonstrated that a fraction of Theia mantle material (TMM) could have entered the largely solid lower layer of the post-impact Earth's mantle during the canonical Moon-forming impact. This mostly molten TMM could later have solidified and sunk to Earth's lowermost mantle and now constitute the seismically observed large low-velocity provinces (LLVPs) in the present day.

Rampino et al. (2021) performed spectral analyses on the ages of 89 well-dated major geological events of the last $260 \times 10^6$ from the recent geologic literature. These events include times of marine and non-marine extinctions, major ocean-anoxic events, continental flood-basalt eruptions, sea-level fluctuations, global pulses of intraplate magnetism, and times of changes in seafloor-spreading rates and plate reorganizations. The aggregate of all 89 events shows ten clusters in the last $260 \times 10^6$ years, spaced at an average interval of $\sim 26.9 \times 10^6$ years, and Fourier analysis of the data yields a spectral peak at $27.5 \times 10^6$ years at the $\geq 96\%$ confidence level. A shorter period of $\sim 8.9 \times 10^6$ years may also be significant in modulating the timing of geologic events. The results suggest that global geologic events are generally correlated, and seem to come in pulses with an underlying $\sim 27.5 \times 10^6$ year cycle. These cyclic pulses of tectonics and climate change may be the result of geophysical processes related to the dynamics of plate tectonics and mantle plumes, or might alternatively be paced by astronomical cycles associated with the Earth's motions in the solar system and the galaxy.

There are several major theories about how the Earth and Moon formed. The most widely accepted scenario for the formation of the Earth–Moon system involves a dramatic impact between the proto-Earth and some other cosmic body. In their article, Cattaneo and Hughes (2022) discuss the role magnetism played in this process. The authors argue that the Earth–Moon system was formed within constraints provided by the fact that the Earth is strongly magnetized. By now, it is accepted, that a dynamo operating in the outer liquid core maintains the Earth's magnetic field. However, because of the rapid rotation of the Earth, this dynamo has the peculiar property that it can maintain a strong field but cannot amplify a weak one. Therefore, the Earth must have been magnetized at a very early epoch, either pre-impact or as a result of the impact itself. Either way, any realistic model of the formation of the Earth–Moon system must include magnetic field evolution.

It was Shaviv (2002) who made a persuasive arguments that there are correlations between extended cold periods on Earth and Earth's exposure to a varying cosmic-ray flux. The short term variability of the galactic cosmic ray flux (CRF) reaching Earth has been previously associated with variations in the global low altitude cloud cover. This CRF variability arises from changes in the solar wind strength. However, cosmic ray variability also arises intrinsically from variable activity of and motion through the Milky Way. Thus, if indeed the CRF climate connection is real, the increased CRF witnessed while crossing the spiral arms could be responsible for a larger global cloud cover and a reduced temperature, thereby facilitating the occurrences of ice ages. In particular, the variable CRF recorded in iron meteorites appears to vary synchronously with the appearance ice ages. Marsh and Svensmark (2000) show that there is a close correlation between the CRF and low altitude cloud cover over a 15 year time span. Thus, one might expect that extended periods of high CRF lead to increased cloud cover and surface cooling that result in long-term ($10^6$ years) ice ages. Shaviv (2002) discussed the cosmic ray diffusion model which considers the motion of the galactic spiral arms. He also elaborated on the structure and dynamics of the Milky Way's spiral arms bringing forth new argumentation using HI observations which imply that the galactic spiral arm pattern speed appears to be that which fits the glaciation period and the cosmic-ray flux record extracted from iron meteorites. In addition, he showed that apparent peaks in the star formation rate history, as deduced by several authors, coincide with particularly icy epochs, while the long period of 1 to 2 × $10^9$ years before present, during which no glaciations are known to have occurred, coincides with a significant paucity in the past star formation rate; see Figure 2.6, modified after Shaviv (2002).

Svensmark (2006) studied a connection between climate and the solar system's motion perpendicular to the galactic plane during the last 200 × $10^6$ years. An imprint of galactic dynamics is found in a long-term record of the Earth's climate that is consistent with variations in the solar system oscillation around the galactic midplane. From small modulations in the oscillation frequency of Earth's climate the following features of the galaxy along the solar circle can be determined: (1) the mass distribution; (2) the timing of two spiral arm crossings (31 and 142 Myr); (3) spiral arm/interarm density ratio ($\rho_{arm}/\rho_{interarm} \approx 1.5–1.8$); and finally, (4) using current knowledge of spiral arm positions, a pattern speed of $\Omega_P = 13.6 \pm 1.4$ km $s^{-1}kpc^{-1}$ is determined.

A possible connection between climate and the solar system's motion perpendicular to the galactic plane

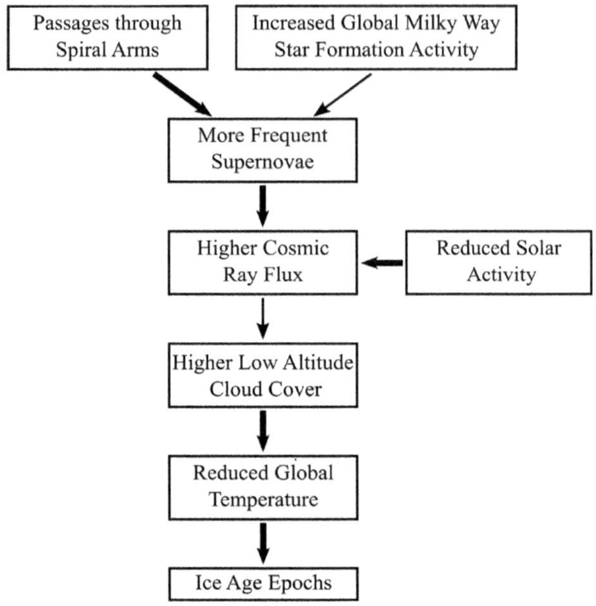


FIGURE 2.6 The simplified model for how the galaxy affects the ice ages. Solid arrows reflect physical cause and effect. Bold lines are hard-to-dispute relations, while thin lines are reasonable relations based on observational correlations. Modified after Shaviv (2002).

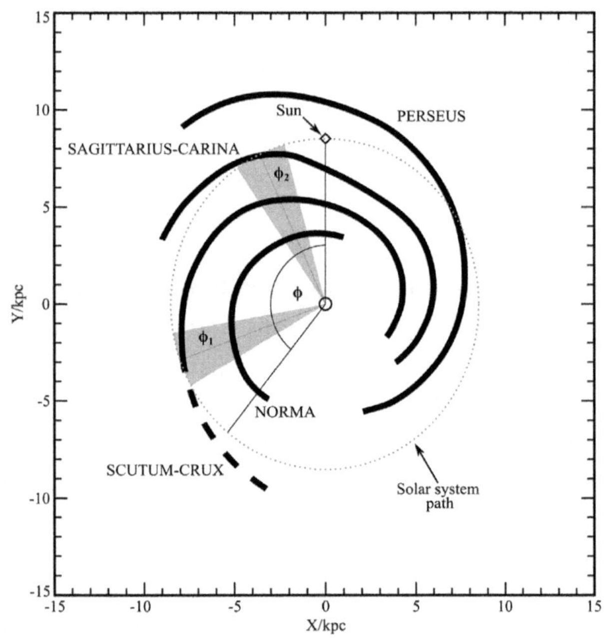


FIGURE 2.7 Overview of the solar system path through the Milky Way galaxy. The known parts of the four spiral arms are shown as thick lines together with symbols $\varphi_1$ and $\varphi_2$ being the angles at which the solar system crosses the spiral arms, together with the estimated uncertainty. The semi-circle is the angle the solar system has traveled relative to the spiral arms during the last 200 × $10^6$ years; see Svensmark (2006).

during the last 200 Myr has been found. In $\delta^{18}O$ proxy data of Earth's climate from the Phanerozoic database an approximately 30 Myr period is identified. From small frequency modulations of this period the following features of the galaxy along the solar circle can be determined:

1. the mass distribution;
2. the timing of two spiral arm crossings (31 and 142 Myr);
3. Spiral arm/interarm ratio (ρarm/ρinterarm ≈ 1.5–1.8); and
4. using current knowledge of spiral arm positions, a pattern speed of ΩP= 13.6 ± 1.4 km s−1kpc−1 is found (Svensmark 2006).

It is important to note that the present study is fundamentally different from the previous ones in one respect. It determines several features of two most recent spiral arm passages from climate data restricted to modulation of timescales between 20–60 Myr, much shorter than the characteristic time for spiral passage ≈140 Myr. The results obtained are consistent with previously reported properties of the Milky Way and give further confidence in the significance of cosmic ray variations and importance in climate changes.

Kirkland et al. (2022) proposed new thinking about the production of Earth's continental crust that is generally regarded as a function of planetary differentiation driven by internal processes. They report time series analysis of the Hf isotopic composition of zircon grains from the North Atlantic and Pilbara cratons, the archetypes of Archean plate tectonic and non-plate tectonic settings, respectively. A ~ (170–200) × $10^6$ years. frequency is recognized in both cratons that matches the transit of the solar system through the galactic spiral arms, where the density of stars is high. An increase in stellar density is consistent with an enhanced rate of Earth bombardment by comets, the larger of which would have initiated crustal nuclei production via impact-driven decompression melting of the mantle. Hence, the production and preservation of continental crust on the early Earth may have been fundamentally influenced by exogenous processes. A test of this model using oxygen isotopes in zircon from the Pilbara craton reveals correlations between crust with anomalously light isotopic signatures and exit from the Perseus spiral arm and entry into the Norma spiral arm, the latter of which matches the known age of terrestrial spherule beds. The data support bolide impact, which promoted the growth of crustal nuclei, on solar system transit into and out of the galactic spiral arms.

The origin and evolution of solar system bodies, including water on Earth, have been discussed based on the assumption that the relevant ingredients were simply silicates and ices. However, large amounts of organic matter have been found in cometary and interplanetary dust, which are recognized as remnants of interstellar/precometary grains. Precometary organic matter may therefore be a potential source of water; however, to date, there have been no experimental investigations into this possibility. Nakano et al. (2020) experimentally demonstrated that abundant water and oil are formed via the heating of a precometary-organic-matter analog under conditions appropriate for the parent bodies of meteorites inside the snow line. This implies that $H_2O$ ice is not required as the sole source of water on planetary bodies inside the snow line. Further, they could explain the change in the oxidation state of the Earth from an initially reduced state to a final oxidized state. Their study also suggests that petroleum was present in the asteroids and is present in icy satellites and dwarf planets.

The time of origin of the geodynamo has important implications for the thermal evolution of the planetary interior and the habitability of early Earth. It has been proposed that detrital zircon grains from Jack Hills, Western Australia, provide evidence for an active geodynamo as early as 4.2 × $10^9$ years ago. However, Borlina et al.'s (2020) combined paleomagnetic, geochemical, and mineralogical studies on Jack Hills zircons indicate that most have poor magnetic recording properties and secondary magnetization carriers that postdate the formation of the zircons. Therefore, the existence of the geodynamo before 3.5 × $10^9$ years ago remains unknown.

The origin of volatile species such as water in the Earth–Moon system is a subject of intense debate but is obfuscated by the potential for volatile loss during the Giant Impact that resulted in the formation of these bodies. One way to address these topics and place constraints on the temporal evolution of volatile components in planetary bodies is by using the observed decay of $^{87}Rb$ to $^{87}Sr$ because Rb is a moderately volatile element, whereas Sr is much more refractory. Borg et al. (2022) showed that lunar highland rocks that crystallized ~4.35 × $10^9$ years ago exhibit very limited ingrowth of $^{87}Sr$, indicating that prior to the Moon-forming impact, the impactor commonly referred to as "Theia" and the proto-Earth both must have already been strongly depleted in volatile elements relative to primitive meteorites. These results imply that (1) the volatile element depletion of the Moon did not arise from the Giant Impact, (2) volatile element distributions on the Moon and Earth were principally inherited from their precursors, (3) both Theia and the proto-Earth probably formed in the inner solar system, and (4) the Giant Impact occurred relatively late in solar system history.

Progressive crystallization of Earth's inner core drives convection in the outer core and magnetic field generation. Determining the rate and pattern of inner-core growth is thus crucial to understanding the evolution of the geodynamo. The growth history of the inner core is probably recorded in the distribution and strength of its seismic anisotropy, which arises from deformation texturing constrained by conditions at the inner-core solid–fluid boundary. Frost et al. (2021) showed from analysis of seismic body wave travel times that the strength of seismic anisotropy increases with depth within the inner core, and the strongest anisotropy is offset from Earth's rotation axis. Then, using geodynamic growth models and mineral physics calculations, they simulate the development of inner-core anisotropy in a self-consistent manner. From this, they found that an inner core composed of hexagonally close-packed iron–nickel alloy, deformed by a combination of preferential equatorial growth and slow translation, can match the seismic observations without requiring hemispheres with sharp boundaries. Their model of inner-core growth history is compatible with external constraints from outer-core dynamics and supports arguments for a relatively young inner core (~0.5–1.5) × $10^9$ years and a viscosity >$10^{18}$ Pa-s. The inner core underwent preferential equatorial growth and translation after nucleation ~0.5–1.5 billion years ago, according to an analysis of its seismic anisotropy and self-consistent geodynamic simulations.

Insights into magma ocean dynamics from the transport properties of basaltic melt is discussed by Bajgain et al. (2022). The viscosity of magma plays a crucial role in the dynamics of the Earth: from the crystallization of a magma ocean during its initial stages to modern-day volcanic processes. However, the pressure-dependence behavior of viscosity at high pressure remains controversial. In this study, they report the results of first-principles molecular dynamics simulations of basaltic melt to show that the melt viscosity increases upon compression along each isotherm for the entire lower mantle after showing minima at ~6 GPa. However, elevated temperatures of the magma ocean translate to a narrow range of viscosity, 0.01–0.03 P-s. This low viscosity implies that the crystallization of the magma ocean could be complete within a few million years. These results also suggest that the crystallization of the magma ocean is likely to be fractional, thus supporting the hypothesis that present-day mantle heterogeneities could have been generated during the early crystallization of the primitive mantle.

The isotopic composition of water in Earth's oceans is challenging to recreate using a plausible mixture of known extraterrestrial sources such as asteroids—an additional isotopically light reservoir is required. The Sun's solar wind could provide an answer to balance Earth's water budget. Daly et al. (2021) used atom probe tomography to directly observe an average ~1 mol% enrichment in water and hydroxyls in the solar-wind-irradiated rim of an olivine grain from the S-type asteroid Itokawa. They also experimentally confirm that H+ irradiation of silicate mineral surfaces produces water molecules. These results suggest that the Itokawa regolith could contain ~20 l $m^{-3}$ of solar-wind-derived water and that such water reservoirs are probably ubiquitous on airless worlds throughout our galaxy. The production of this isotopically light water reservoir by solar wind implantation into fine-grained silicates may have been a particularly important process in the early solar system, potentially providing a means to recreate Earth's current water isotope ratios.

The size and velocity distribution of cosmic dust particles entering the Earth's atmosphere is uncertain. Carrillo-Sánchez et al. (2015) showed that the relative concentrations of metal atoms in the upper mesosphere, and the surface accretion rate of cosmic spherules, provide sensitive probes of this distribution. Three cosmic dust models are selected as case studies: two are astronomical models, the first constrained by infrared observations of the Zodiacal Dust Cloud and the second by radar observations of meteor head echoes; the third model is based on measurements made with a spaceborne dust detector. For each model, a Monte Carlo sampling method combined with a chemical ablation model is used to predict the ablation rates of Na, K, Fe,

Mg, and Ca above 60 km and cosmic spherule production rate. It appears that a significant fraction of the cosmic dust consists of small (<5 μg) and slow (<15 km $s^{-1}$) particles.

The short-lived Hf-W isotope system has a wide range of important applications in cosmochemistry and geochemistry. The siderophile behavior of W, combined with the lithophile nature of Hf, makes the system uniquely useful as a chronometer of planetary accretion and differentiation; see Kleine and Walker (2017). Tungsten isotopic data for meteorites show that the parent bodies of some differentiated meteorites accreted within 1 million years after solar system formation. Melting and differentiation on these bodies took ~1–3 million years and was fueled by decay of $^{26}$Al. The timescale for accretion and core formation increases with planetary mass and is ~10 million years for Mars and >34 million years for Earth. The nearly identical $^{182}$W compositions for the mantles of the Moon and Earth are difficult to explain in current models for the formation of the Moon. Terrestrial samples with ages spanning ~4 billion years reveal small $^{182}$W variations within the silicate Earth, demonstrating that traces of Earth's earliest formative period have been preserved throughout Earth's history.

### 2.4.3 Chemical Element Abundances

Chemical abundances can only be determined from stellar spectra and, therefore, with the Sun set as the reference for any chemical diagnostics study, this effort requires self-consistent unbiased analyses of stellar and solar data. In addition, the chemical composition of the Sun is required in the context of various studies in astrophysics, among them in the calculation of standard solar models (SSMs) used to describe the evolution of the Sun from the pre-main-sequence to its present age. In their work, Magg et al. (2022) provided a new analysis of the solar chemical composition. The methodology was chosen to ensure that it is suitable for the analysis of any star (not only the Sun), with its very high-quality observations.

Table 2.6 lists solar chemical abundances on the astronomical scale A(i) = log (N(i)/$N_H$) + 12, as recommended by Magg et al. (2022).

Table 2.6 also lists the final meteoritic abundances transformed to the solar photospheric abundance scale and the associated measurement errors. Uncertainties listed for the meteoritic abundances are purely statistical, but a systematic and fully correlated 0.02 dex uncertainty should be included for all elements. The last row reports the total Z/X, where the abundances of volatile elements are always taken from the photospheric scale. The agreement between the photospheric and meteoritic scales is excellent, as shown in Figure 2.8.

Elements are of special interest for the origin of life issue are (H-C-N-O) based, relying on the number of bulk (Na-Mg-P-S-Cl-K-Ca) and trace elements (Cr-Mn-Fe-Co-Ni-Cu-Zn-Se-Mo-W, and possibly Li-B-F-Si-V-As), and they should be considered by special attention.

TABLE 2.6 Solar Chemical Abundances on the Astronomical Scale A(i) = log (n(i)/NH) + 12, as Recommended by Magg et al. (2022)

| Element | $A(El)_{ph\pm\sigma(tot)}$ | $A(Al)_{met\pm\sigma(stat)}$ | Reference |
|---|---|---|---|
| H | 12.0 | - | - |
| C | 8.56 ± 0.05 | - | Magg et al. (2022) |
| N | 7.98 ± 0.08 | - | Magg et al. (2022) |
| O | 8.77 ± 0.04 | - | Magg et al. (2022) |
| F | 4.40 ± 0.25 | 4.67 ± 0.09 | Maiorca et al. (2014) |
| Ne | 8.15 ± 0.11 | - | Magg et al. (2022) |
| Na | 6.29 ± 0.02 | 6.33 ± 0.04 | Zhao et al. (2016) |
| Mg | 7.55 ± 0.05 | 7.58 ± 0.02 | Magg et al. (2022) |
| Al | 6.43 ± 0.03 | 6.48 ± 0.03 | Nordlander and Lind (2017) |
| Si | 7.59 ± 0.07 | 7.57 ± 0.02 | Magg et al. (2022) |
| P | 5.41 ± 0.03 | 5.48 ± 0.03 | Scott et al. (2015a) |
| S | 7.16 ± 0.11 | 7.21 ± 0.03 | Caffau et al. (2011) |
| Cl | 5.25 ± 0.12 | 5.29 ± 0.07 | Lodders (2019) |
| Ar | 6.50 ± 0.10 | - | Lodders (2019) |
| K | 3.07 ± 0.10 | 5.12 ± 0.02 | Zhao et al. (2016) |
| Ca | 4.94 ± 0.04 | 6.32 ± 0.03 | Magg et al. (2022) |
| Sc | 3.07 ± 0.04 | 3.09 ± 0.03 | Zhao et al. (2016) |
| Ti | 4.94 ± 0.05 | 4.96 ± 0.03 | Bergemann (2011) |
| V | 3.89 ± 0.08 | 4.01 ± 0.03 | Scott et al. (2015b) |
| Cr | 5.74 ± 0.05 | 5.69 ± 0.02 | Bergemann and Cescutti (2010) |
| Mn | 5.52 ± 0.04 | 5.53 ± 0.03 | Bergemann et al. (2019) |
| Fe | 7.50 ± 0.06 | 7.51 ± 0.02 | Magg et al. (2022) |
| Co | 4.95 ± 0.04 | 4.92 ± 0.02 | Bergemann et al. (2010) |
| Ni | 6.24 ± 0.04 | 6.25 ± 0.0 | Magg et al. (2022) |
| Z/X | 0.0225 ± 0.0014 | 0.0226 ± 0.0014 | |

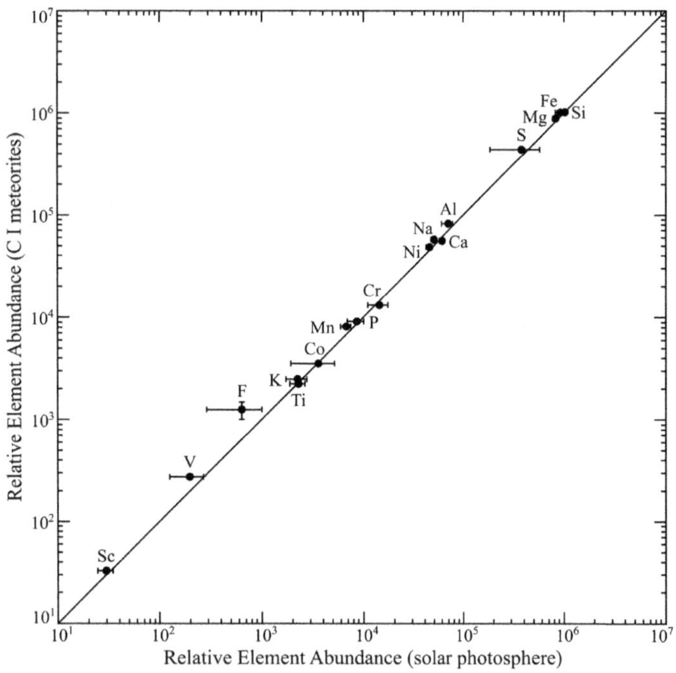


FIGURE 2.8 Comparison of the photospheric estimates with meteoritic values (after Magg et al. 2022).

The most abundant isotope of lithium, $^{7}Li$, is the heaviest element created during Big Bang nucleosynthesis (BBN), and its primordial abundance can be used to probe the standard model of cosmology. Also, Li abundance in the interstellar medium increases by a factor larger than 10 during the evolution of the galaxy, this suggests one or more sites of fresh Li production. Elements $^{6}$ Li, $^{9}$ Be, $^{10}B$, and $^{11}B$ are instead not produced in significant quantities during BBN or in stars. They are created by the interaction of energetic galactic cosmic rays (GCRs) with the interstellar medium that can happen in two different channels: a direct process in which protons and α-particles in GCRs collide with CNO nuclei in the ISM or an inverse process where CNO nuclei instead collide with protons and α-particles in the ISM.

Due to their production sites, as well as how they are processed and destroyed in stars, the light elements are excellent tools to investigate a number of crucial issues in modern astrophysics: from stellar structure and non-standard processes at work in stellar interiors to age dating of stars, from pre-main sequence evolution to the star formation histories of young clusters and associations and to multiple populations in globular clusters, from Big Bang nucleosynthesis to the formation and chemical enrichment history of the Milky Way galaxy and its populations, just to cite some relevant examples.

The composition of the early solar system can be inferred from meteorites. Many elements heavier than iron were formed by the rapid neutron capture process (r-process), but the astrophysical sources where this occurred remain poorly understood. Côté et al. (2021) demonstrated that the near-identical half-lives of the radioactive r-process nuclei iodine-129 and curium-247 preserve their ratio, irrespective of the time between production and incorporation into the solar system. They constrained the last r-process source by comparing the measured meteoritic ratio $^{129}I/^{247}Cm = 438 \pm 184$ with nucleosynthesis calculations based on neutron star merger and magneto-rotational supernova simulations. Moderately neutron-rich conditions, often found in merger disk ejecta simulations, are most consistent with the meteoritic value. Uncertain nuclear physics data limit confidence in this conclusion.

Stellar nucleosynthesis is a widely acknowledged theory for the formation of all elements in our universe; traditionally, we say that the highest-mass stars transmuted lighter elements into heavier elements lighter than iron. Fukuhara et al. (2021) proposed that the formation of 25 elements with smaller atomic numbers than iron resulted from an endothermic nuclear transformation of two nuclei confined in the natural compound lattice core of the Earth's lower mantle at high temperatures and pressures. This process is accompanied by the generation of neutrinos and is influenced by excited electrons generated by stick-sliding during supercontinent evolution, mantle convection triggered by major asteroid collisions, and nuclear fusion in the Earth's core. Therefore, their study suggests that the Earth itself has been able to create lighter elements by nuclear transmutation.

According to Haba et al. (2021), the niobium-92–zirconium-92 ($^{92}Nb$–$^{92}Zr$) decay system with a half-life of 37 Ma has great potential to date the evolution of planetary materials in the early solar system. Moreover, the initial abundance of the p-process isotope 92Nb in the solar system is important for quantifying the contribution of p-process nucleosynthesis in astrophysical models. Current estimates of the initial $^{92}Nb/^{93}Nb$ ratios have large uncertainties compromising the use of the 92Nb–92Zr cosmochronometer and leaving nucleosynthetic models poorly constrained. Here, the initial $^{92}Nb$ abundance is determined to high precision by combining the $^{92}Nb$–$^{92}Zr$ systematics of cogenetic rutiles and zircons from mesosiderites with U–Pb dating of the same zircons. The mineral pair indicates that the

$^{92}Nb/^{93}Nb$ ratio of the solar system started with (1.66 ± 0.10) × $10^{-5}$, and their $^{92}Zr/^{90}Zr$ ratios can be explained by a three-stage Nb–Zr evolution on the mesosiderite parent body. Because of the improvement by a factor of 6 of the precision of the initial solar system $^{92}Nb/^{93}Nb$, one can show that the presence of $^{92}Nb$ in the early solar system provides further evidence that both type Ia supernovae and core-collapse supernovae contributed to the light p-process nuclei.

The heaviest chemical elements are naturally produced by the rapid neutron-capture process (*r*-process) during neutron star mergers or supernovae. The *r*-process production of elements heavier than uranium (transuranic nuclei) is poorly understood and inaccessible to experiments so must be extrapolated by using nucleosynthesis models. Roederer et al. (2023) examined element abundances in a sample of stars that are enhanced in *r*-process elements. The abundances of elements ruthenium, rhodium, palladium, and silver (atomic numbers $Z = 44$ to 47; mass numbers $A = 99$ to 110) correlate with those of heavier elements ($63 \leq Z \leq 78$, $A > 150$). There is no correlation for neighboring elements ($34 \leq Z \leq 42$ and $48 \leq Z \leq 62$). They interpret this as evidence that fission fragments of transuranic nuclei contribute to the abundances. The results indicate that neutron-rich nuclei with mass numbers >260 are produced in *r*-process events.

## REFERENCES


Adams, F. C. 2010. The birth environment of the solar system. Annual Review of Astronomy and Astrophysics 48: 47–85.

Adibekyan, V., Dorn, C., Sousa, S. G., et al. 2021. A compositional link between rocky exoplanets and their host stars. Science 374: 330–332. https://doi.org/10.1126/science.abg8794.

Agol, E., Dorn, C., Grimm, S. L., et al. 2021. Refining the transit-timing and photometric analysis of TRAPPIST-1: Masses, radii, densities, dynamics, and ephemerides. The Planetary Science Journal 2(1): 38. https://hdl.handle.net/2268/256346.

Aléon, J., Lévy, D., Aléon-Toppani, A., et al. 2022. Determination of the initial hydrogen isotopic composition of the solar system. Nature Astronomy Letters. https://doi.org/10.1038/s41550-021-01595-7.

Ali, B., Paul, D., Eckart, A., et al. 2020. Kinematic structure of the galactic center S cluster. The Astrophysical Journal 896(2): article id.100 (19 pages). https://doi.org/10.3847/1538-4357/ab93ae.

Aller, L. H. 1942. The spectra of the emission nebulosities in messier 33. Astrophysical Journal 95: 52–57. https://doi.org/10.1086/144372.

Alzate, J. A., Bruzual, G., and Díaz-González. 2021. Star formation history of the solar neighborhood as told by *Gaia*. Monthly Notices of the Royal Astronomical Society 501(1): 302–328.

Amarsi, A. M., Lind, K., Osorio, Y., et al. 2020. The Galah survey: Non-LTE departure coefficients for large spectroscopic surveys. Astronomy & Astrophysics 642: A62 (18 pages). https://doi.org/10.1051/0004-6361/202038650.

Anderson, E., Francis, C. 2012. XHIP: An extended hipparcos compilation. Astron. Lett. 38: 331–346. https://doi.org/10.1134/S1063773712050015.

Andrews, S. M. 2021. The structures of protoplanetary disks. Physics Today August 2021: 37–41.

Andrews, S. M., and Williams, J. P. 2005. Circumstellar dust disks in Taurus-Auriga: The submillimeter perspective. The Astrophysical Journal 631: 1134–1160. https://doi.org/10.1086/432712.

Areal, M. B., Paron, S., Ortega, M. E., and Duvidovich, L. 2019. Study of the $^{13}CO/C^{18}O$ abundance ratio towards the filamentary infrared dark cloud IRDC 34.43 + 0.24. Publications of the Astronomical Society of Australia 36: e049 (8 pages). https://doi.org/10.1017/pasa.2019.44.

Arzoumanian, D., Arakawa, S., Kobayashi, M. I. N., et al. 2023. Insights on the sun birth environment in the context of star cluster formation in hub– filament systems. The Astrophysical Journal Letters 947: L29 (10 pages). https://doi.org/10.3847/2041-8213/acc849.

Audibert, A., Almeida, C. R., Garcia-Burillo, S., et al. 2023. Jet-induced molecular gas excitation and turbulence in the teacup. Astronomy & Astrophysics 671: L12 (9 pages). https://doi.org/10.1051/0004-6361/202345964.

Aumer, M., and Binney, J. J. 2009. Kinematics and history of the solar neighbourhood revisited. Monthly Notices of the Royal Astronomical Society 397: 1286–1301.

Ávila, P. J., Grassi, T., Bovino, S., et al. 2021. Presence of water on exomoons orbiting free-floating planets: A case study. International Journal of Astrobiology 20: 300–311.

Azevedo-Silva, T., Demangeon, O. D. S., Santos, N. C., et al. 2022. Detection of barium in the atmospheres of the ultra-hot gas giants WASP-76b and WASP-121b*–Together with new detections of Co and Sr+ on WASP-121b. Astronomy & Astrophysics 666: L10 (21 pages). https://doi.org/10.1051/0004-6361/202244489.

Bajgain, S. K., Ashley, A. W., Mookherjee, M., et al. 2022. Insights into magma ocean dynamics from the transport properties of basaltic melt. Nature Communication 13: 7590 (10 pages). https://doi.org/10.1038/s41467-022-35171-y.

Balsalobre-Ruza, O., de Gregorio-Monsalvo, I. J., Lillo-Box, I. J., et al. 2023. Tentative co-orbital submillimeter emission within the Lagrangian region L5 of the protoplanet PDS 70 b. Astronomy & Astrophysics 675: A172 (8 pages). https://doi.org/10.1051/0004-6361/202346493.

Barrado y Navascués, D., Stauffer, J. R., and Jayawardhana, R. 2004. Spectroscopy of very low mass stars and brown dwarfs in IC 2391: Lithium depletion and Hα emission. The Astrophysical Journal 614: 386–397. https://doi.org/10.1086/423485.

Barros, S. C. C., Akinsanmi, B., Boué, G., et al. 2022. Detection of the tidal deformation of WASP-103b at 3σ with CHEOPS. Astronomy & Astrophysics 657: A52 (20 pages). https://doi.org/10.1051/0004-6361/202142196.

Battaglia, G., Helmi, A., and Breddels, M. 2013. Internal kinematics and dynamical models of dwarf spheroidal Galaxies around the Milky Way. New Astronomy Reviews 57(3–4): 52–79. https://doi.org/10.1016/j.newar.2013.05.003.

Belokurov, V. 2013. Galactic archaeology: The dwarfs that survived and perished. New Astronomy Reviews 57(3–4): 100–121. https://doi.org/10.1016/j.newar.2013.07.001.

Ben-Jaffel, L., Ballester, G. E., Muñoz, A, G., et al. 2021. Signatures of strong magnetization and a metal-poor atmosphere for a Neptune-sized exoplanet. Nature Astronomy. https://doi.org/10.1038/s41550-021-01505-x.

Bergemann, M. 2011. Ionization balance of Ti in the photospheres of the sun and four late-type stars. Monthly Notices of the Royal Astronomical Society 413(3): 2184–2198. https://doi.org/10.1111/j.1365-2966.2011.18295.x.

Bergemann, M., and Cescutti, G. 2010. Chromium: NLTE abundances in metal-poor stars and nucleosynthesis in the galaxy. Astronomy and Astrophysics 522: A9 (13 pages). https://doi.org/10.1051/0004-6361/201014250.

Bergemann, M., Gallagher, A. J., Eitner, P., et al. 2019. Observational constraints on the origin of the elements. I. 3D NLTE formation of Mn lines in late-type stars. Astronomy and Astrophysics 631: A80 (28 pages). https://doi.org/10.1051/0004-6361/201935811.

Bergemann, M., Pickering, J. C., and Gehren, T. 2010. NNLTE analysis of Co I /Co II lines in spectra of cool stars with new laboratory hyperfine splitting constants. Mon. Not. R. Astron. Soc. 401: 1334–1346. doi:10.1111/j.1365-2966.2009.15736.x.

Binney, J. 2013. Dynamics for galactic archaeology. New Astronomy Reviews 57(3–4): 29–122. http://dx.doi.org/10.1016/j.newar.2013.08.001.

Binney, J. J., Dehnen, W., and Bertelli, G. 2000. The age of the solar neighborhood. Monthly Notices of the Royal Astronomical Society 318(3): 658–664. https://doi.org/10.1046/j.1365-8711.2000.03720.x MNRAS.

Bissantz, N., Englmaier, P., and Gerhard, O. 2003. Gas dynamics in the Milky Way: Second pattern speed and large-scale morphology. Monthly Notices of the Royal Astronomical Society 340(3): 949–968. https://doi.org/10.1046/j.1365-8711.2003.06358.x.

Bland-Hawthorn, J., and Gerhard, O. 2016. The galaxy in context: Structural, kinematic, and integrated properties. Annual Review of Astronomy and Astrophysics 54: 529–596. https://doi.org/10.1146/annurev-astro-081915-023441.

Blum, J. Levasseur-Regourd, A. C., Muñoz, O., et al. 2008. Dust in space. Euro Physics News 39(3) 27–29. https://doi.org/10.1051/epn:2008303.

Bonsor, A., Lichtenberg, T., Drążkowska, J., and Buchan, A. 2022. Rapid formation of exoplanetesimals revealed by white dwarfs. Nature Astronomy 7(1): 1–10. https://doi.org/10.1038/s41550-022-01815-8.

Borg, L. E., Brennecka, G. A., and Kruijer, T. S. 2022. The origin of volatile elements in the Earth-moon system. PNAS 119(8): e2115726119 (5 pages). https://doi.org/10.1073/pnas.2115726119.

Borlina, C. S., Weiss, B. P., Lima, E. A., et al. 2020. Reevaluating the evidence for a Hadean-eoarchean dynamo. Science Advances 6(15): eaav9634. https://doi.org/10.1126&sciadv.aav9634.

Boss, A. P., and Goswami, J. N. 2006. Presolar Cloud Collapse and the Formation and Early Evolution of the solar system. In D. S. Lauretta and H. Y. McSween Jr., (eds.), Meteorites and the Early solar system II (pp.171–186). Tucson: University Arizona Press.

Bouy, H., and Alves, J. 2015. Cosmography of OB stars in the solar neighbourhood. Astronomy & Astrophysics 584: A26 (13 pages). https://doi.org/10.1051/0004-6361/201527058.

Brewer, J. M., and Fischer, D. A. 2016. C/O and Mg/Si ratios of stars in the solar neighborhood. The Astrophysical Journal 831: 20 (7 pages). https://doi.org/10.3847/0004-637X/831/1/20.

Brewer, J. M., Fischer, D. A., and Madhusudhan, N. 2017. C/O and O/H ratios suggest some hot Jupiters originate beyond the snow line. The Astronomical Journal 153: 83 (6 pages). https://doi.org/10.3847/1538-3881/153/2/83.

Brown, K., and Mathur, H. 2023. Modified Newtonian dynamics as an alternative to the planet nine hypothesis. The Astronomical Journal 166: 168 (21 pages). https://doi.org/10.3847/1538-3881/acef1e.

Brunken, G. C., Booth, A. S., Leeker, M., et al. 2022. A major asymmetric ice trap in a planet-forming disk III. First detection of dimethyl ether. Astronomy & Astrophysics 59: A29 (11 pages). https://doi.org/10.1051/0004-6361/202142981.

Burbidge, E. M., Burbidge, G. R., Fowler, W. A., and Hoyle, F. 1957. Synthesis of the elements in stars. Reviews of Modern Physics 29(4): 547 (108 pages). https://doi.org/10.1103/RevModPhys.29.547.

Cabrera, T., and Rodriguez, C. L. 2023. Runaway and hypervelocity stars from compact object encounters in globular clusters. The Astrophysical Journal 953: 19 (13 pages). 10 https://doi.org/10.3847/1538–4357/acdc22.

Caffau, E. Ludwig, H. G., Steffen, M., et al. 2011. Solar chemical abundances determined with a CO5BOLD 3D model atmosphere. Solar Physics 268: 255–269. https://doi.org/10.1007/s11207-010-9541–4.

Cameron, A. G. W., and Truran, J. W. 1977. The Supernova trigger for formation of the solar system. Icarus 30: 447–461.

Cardenas, B. T., and Lamb, M. P. 2022. Paleogeographic reconstructions of an ocean margin on mars based deltaic sedimentology at Aeolis Dorsa. Journal of Geophysical Research 127: e2022JE007390 (25 pages). https://doi.org/10.1029/2022JE007390.

Carretero-Castrillo, Ribó, M., and Paredes, J. M. 2023. Galactic runaway O and be stars found using Gaia DR3. Astronomy & Astrophysics. https://doi.org/10.1051/0004-6361/202346613.

Carrillo-Sánchez, J. D., Plane, J. M. C., Feng, W., et al. 2015. On the size and velocity distribution of cosmic dust particles entering the atmosphere. Geophysical Research Letters 42: (8 pages). https://doi.org/10.1002/2015GLO65149. On arXiv: https://doi.org/10.48550/arxiv.2311.01827.

Cashman, F. H., Fox, A. J., Wakker, B. P., et al. 2023. Caught in the act: A metal-rich high-velocity cloud in the inner galaxy. The Astrophysical Journal 944: 65 (17pages). https://doi.org/10.3847/1538-4357/acaaa5.

Castelvecchi, D. 2022. A supernova could light up the Milky Way at any time. Astronomers will be watching. Nature 602(7898): 562–565. https://doi.org/10.1038/d41586-022-00425-8.

Cattaneo, F., and Hughes, D. W. 2022. How was the Earth-moon system formed? New insights from the geodynamo. Proceedings of the National Academy of Sciences 119(44): e2120682119. https://doi.org/10.1073/pnas.2120682119.

Chamberlain, J. W., and Aller, L. H. 1951. The atmospheres of A-type subdwarfs and 95 Leonis. Astrophysical Journal 114: 52 (21 pages). https://doi.org/10.1086/145451.

Chandra, V., Naidu, R. P., Conroy, C., et al. 2023. Discovery of the Magellanic stellar stream out to 100 kpc. The Astrophysical Journal 956: 110 (15 pages). https://doi.org/10.3847/1538-4357/acf7bf.

Chevance, M., Kruijssen, J. M. D., Vazquez-Semadeni, E., et al. 2020. The molecular cloud lifecycle. Space Science Reviews 216: 50 (42 pages). https://doi.org/10.1007/s11214-020-00674-x.

Claria, J. J., and Kepler, S. O. 1980. Wide-and narrow-band photometry of stars in a field around Collinder 135. Publications of the Astronomical Society of the Pacific 92(548): 501–513.

Clavel, M., Terrier, R., Goldwurm, A., et al. 2013. Echoes of multiple outburst of Sagittarius A* revealed by Chandra. Astronomy and Astrophysics 558: A32 (15 pages). https://doi.org/10.1051/0004-6361/201321667.

Conroy, C., Naidu, R. P., Garavito-Camargo, N., et al. 2021. All-sky dynamical response of the galactic halo to the large Magellanic cloud. Nature 592: 534–536. https://doi.org/10.1038/s41586-021-03385-7.

Côté B., Eichler, M., López, A. Y., et al. 2022. $^{129}$I and $^{247}$Cm in meteorites constrain the last astrophysical source of solar r-process elements. Science 371: 945–948. https://doi.org/10.1126/science.aba1111.

Coughlin, J. L. 2017. Planet Detection Metrics: Robovetter Completeness and Effectiveness for Data Release 25. Kepler Science Document, KSCI19114–002, CA: NASA Ames Research Center Moffett Field.

Crutcher, R. M. 2003. Observations of Magnetic Fields in Molecular Clouds. In B. Uyaniker, W. Reich, and R. Wielebinski (eds.), The Magnetized Interstellar Medium (pp. 123–132), Proceedings of the Conference, Held in Antalya, Turkey, 8–12 September, Katlenburg-Lindau: Copernicus GmbH.

Culberg, R., Scroeder, D. M., and Steibrügge, G. 2022. Double ridge formation over shallow water sills on Jupiter's moon Europa. Nature Communication 13: 2007 (10 pages). https://doi.org/10.1038/s41467-022-29458-3.

Cuntz, M., Luke, G. E., Millard, M. J., et al. 2022. An early catalog of planet-hosting multiple-star systems of order three and higher. The Astrophysical Journal Supplement Series 263: 33 (23 pages). https://doi.org/10.3847/1538-4365/ac9302.

Dalba, P. A., Kane, S. R., Dragomir, D., et al. 2022. The TESS-keck survey. VIII. Confirmation of a transiting giant planet on an eccentric 261 day orbit with the automated planet finder telescope. The. Astronomical Journal 163: 61 (17 pages). https://doi.org/10.3847/1538-3881/ac415b.

Daly, L., Lee, M. R., Hallis, L. J., et al. 2021. Solar wind contributions to Earth's oceans. Nature Astronomy 5: 1275–1285. https://doi.org/10.1038/s41550-021-01487-w.

Daly, R. A., Donahue, M., O'Dea, M. C., et al. 2023. New black hole spin values for Sagittarius a* obtained with the outflow method. Monthly Notices of the Royal Astronomical Society 527(1): 428–436 (10 pages). arXiv:2310.12108v1 [astro-ph.GA]. https://doi.org/10.1093/mnras/stad3228.

David, T. J., Contardo, G., Sandoval, A., et al. 2021. Evolution of the exoplanet size distribution: Forming Large super-Earths over billions of years. The Astronomical Journal 161: 265 (29 pages). https://doi.org/10.3847/1538-3881/abf439.

De Cia, A., Jenkins, E. B., Fox, A. J., et al. 2021. Large metallicity variations in the galactic interstellar medium. Nature 597: 206–208.

De Jong, T., and Kamijo, F. 1973. Growth and destruction of interstellar grains in the presence of low energy cosmic rays. Astronomy and Astrophysics 25: 363–370.

de Zeeuw, P. T., Hoogerwerf, R., de Bruijne, J. H. J., et al. 1999. The Astronomical Journal 117(1): 354–399 NLTE analysis of Co I /Co II lines in spectra of cool stars with new laboratory hyperfine splitting constants. doi:10.1086/300682.

Deason, A. J., Fattahi, A., Frenk, C. S., et al. 2020. The edge of the galaxy. Monthly Notices of the Royal Astronomical Society 496(3): 3929– 3942. https://doi.org/10.1093/mnras/staa1711.

Dehnen, W., and Binney, J. J. 1998. Local stellar kinematics from HIPPARCOS data. Monthly Notices of the Royal Astronomical Society 298(2): 387–394. https://doi.org/10.1046/j.1365-8711.1998.01600.x.

Delbo, M., Walsh, K. J., Matonti, C., et al. 2022. Aligment of fractures on Bennu's boulders indicative of rapid asteroid surface evolution. Nature Geoscience 15: 453–457.

Delorme, P., Gagné, J., Malo, L., et al. 2012. CFBDSIR2149–0403: A 4–7 Jupiter-mass free-floating planet in the young moving group AB Doradus? Astronomy & Astrophysics 548: A26 (14 pages). https://doi.org/10.1051/0004-6361/201219984.

Deng, H., Mayer, L., and Helled, R. 2021. Author correction: Formation of intermediate-mass planets via magnetically controlled disk fragmentation. Nature Astronomy 5: 967 (11 pages). https://doi.org/10.1038/s41550-021-01400-5.

Di Stefano, R., Berndtsson, J., Urquhart, R., et al. 2021. A possible planet candidate in an external galaxy detected through X-ray transit. Nature Astronomy 5: 1297–1307. https://doi.org/10.1038/s41550-021-01495-w.

Dias, W. S., Alessi, B. S., Moitinho, A., Lépine, J. R. D. 2002. New catalogue of optically visible open clusters and candidates. Astronomy and Astrophysics 389: 871–873. doi:10.1051/0004-6361:20020668.

Dobbie, P. D., Lodieu, N., and Sharp, R. G. 2010. IC 2602: A lithium depletion boundary age and new candidate low-mass stellar members. Monthly Notices of the Royal Astronomical Society 409(3): 1002–1012. https://doi.org/10.1111/j.1365-2966.2010.17355.xMNRAS.409.1002.

Doi, Y., Nakamura, K., Kawabata, K. S., et al. 2024. Tomographic imaging of the Sagittarius spiral arm's magnetic field structure. The Astrophysical Journal 961: 13 (20 pages) https://doi.org/10.3847/1538-4357/ad0fe2.

Dominik, C., Blum, J., Cuzzi, J., and Wurm, G. 2007. Aggregation and transport of dust in disks as initial steps toward planet formation. In B. Reipurth, D. Jewitt, and K. Keil (eds.), Protostars and Planets V (pp. 783–800). Tucson: University. of Arizona Press.

Donlon; II, T., Newberg, H. J., Sanderson, R., et al. 2020. the Milky Way's shell structure reveals the time of a radial collision. The Astrophysical Journal 902: 119 (27 pages). https://doi.org/10.3847/1538-4357/abb5f6.

Draine, B. T. 2010. Observed Properties of Interstellar Dust. IPMU 2010.04.20. Lecture Notes: Lectures on Interstellar Dust at the Institute for Physics and Mathematics of the Universe (IPMU) at the University of Tokyo. www.astro.princeton.edu/~draine/dust/Draine_IPMU_Lectures.pdf.

Dwek E. 2016. Iron: A key element for understanding the origin and evolution of interstellar dust. The Astrophysical Journal 825(2): 136. https://doi.org/10.3847/0004-637x/825/2/136.

Eckart, A., and Genzel, R. 1996. Observations of stellar proper motions near the galactic centre. Nature 383: 415–417.

Eddington, A. S. 1920. The internal constitution of the stars. The Observatory 43: 341–358.

Ehrenfreund, P., and Charnley, S. B. 2000. Organic molecules in the interstellar medium, comets, and meteorites: A voyage from dark clouds to the early Earth. Annual Review of Astronomy and Astrophysics 38: 427–483. https://doi.org/10.1146/annurev.astro.38.1.427.

Eistrup, C., Walsh, C., and van Dishoeck, E. F. 2016. Setting the volatile composition of (exo)planet-building material. Does Chemical Evolution in Disk Midplanes Matter? arXiv:1607.06710v1 [astro-ph.EP], 22 Jul.

Eistrup, C., Walsh, C., and van Dishoeck, E. F. 2018. Molecular abundances and C/O ratios in chemically evolving planet-forming disk midplanes. Astronomy & Astrophysics 613: A14 (16 pages). https://doi.org/10.1051/0004-6361/201731302.

Eldridge, J. J. 2009. A new-age determination for γ2 Velorum from binary stellar evolution models. Mon. Not. R. Astron. Soc. 400: L20–L23. doi:10.1111/j.1745-3933.2009.00753.x.

Englmaier, P., and Gerhard, O. 1999. Gas dynamics and large-scale morphology of the Milky Way galaxy. MNRAS 304: 512–534.

Ertel, A. F., and Fields, B. D. 2023. Distances to recent near-Earth supernovae from geological and lunar $^{60}$Fe. arXiv:2309.11604v1 [astro-ph.SR], 20 Sep.

Espaillat, C. C., Thanathibodee, T., Pittman, C. V., et al. 2023. JWST detects neon line variability in a protoplanetary disk. The Astrophysical Journal Letters 958: L4 (7 pages). https://doi.org/10.3847/2041-8213/ad023d.

Faria, J. P., Suárez Mascareño, A., Figueira, P., et al. 2022. A candidate short-period sub-Earth orbiting Proxima Centauri. Astronomy & Astrophysics 658: A115 (16 pages). https://doi.org/10.1051/0004-6361/202142337.

Franchini, M., Morossi, C., Di Marcantonio, P., et al. 2020. The Gaia-ESO survey: Carbon abundance in the galactic thin and thick disks. The Astrophysical Journal 888: 55. https://doi.org/10.3847/1538-4357/ab5dc4.

Franchini, M., Morossi, C., Di Marcantonio, P., et al. 2021. The Gaia-eSO survey: Oxygen abundance in the galactic thin and thick disks. The Astrophysical Journal 161: 9. https://doi.org/10.3847/1538-3881/abc69b.

Frost, D. A., Lasbleis, M., Chandler, B., and Romanowicz, B. 2021. Dynamic history of the inner core constrained by seismic anisotropy. Nature Geoscience. https://doi.org/10.1038/s41561'021-00761-w.

Fukuhara, M., Yoshino, A., and Fujima, N. 2021. Earth factories: Creation of the elements from nuclear transmutation in Earth's lower mantle. AIP Advances 11: 105113. https://doiorg/10.1063/5.0061584.

Gaia Collaboration. 2021. Gaia early data release 3: Acceleration of the solar system from Gaia astrometry. Astronom. Astrophys 649: A9 (19 pages). https://doi.org/10.1051/0004-6361/202039734.

Gaidos, E., Hirano, T., Kraus, A. L., et al. 2021. Zodiacal Exoplanets in Time (ZEIT) XII: A Directly-Imaged Planetary-Mass Companion to a Young Taurus M Dwarf Star. arXiv:2110.08655v1 [astro-ph.EP], 16 Oct.

Garner, R., ed. 2022. NASA's Webb Reveals Steamy Atmosphere of Distant Planet in Detail. NASA, Page Last Updated: Jul. 13. www.nasa.gov/webbfirstimages.

Gillon, M., Meadows, V., Agol, E., et al. 2020. Bulletin of the American Astronomical Society 52: 0208. https://doi.org/10.3847/25c2cfeb.afbf0205.

Ginsburg, A., Barnes, A. T., Battersby, C. D., et al. 2023. JWST reveals widespread CO ice and Gas absorption in the galactic center cloud G0.253 + 0.016. The Astrophysical Journal 959(1): 36 (13 pages). (13pp). https://doi.org/10.3847/1538-4357/acfc34.

Goharji, A., and Adams, S. 1984. The C/O abundance ratio in the planetary nebula IC 2501. Monthly Notices of the Royal Astronomical Society 210: 683–691.

Gough, E. 2023. Astronomers find dozens of massive stars fleeing the Milky Way. Physical Organic, 10 Nov.

GRAVITY Collaboration. 2017, First light for GRAVITY: Phase referencing optical interferometry for the Very Large Telescope Interferometer. Astronomy & Astrophysics 602: A94.

GRAVITY Collaboration. 2018. Detection of orbital motions near the last stable circular orbit of the massive black hole SgrA*. Astronomy & Astrophysics 618: L10.

GRAVITY Collaboration. 2019. A geometric distance measurement to the galactic center black hole with 0.3% uncertainty. Astronomy & Astrophysics 625: L10. https://doi.org/10.1051/0004-6361/201935656.

GRAVITY Collaboration. 2022a. Deep images of the galactic center with gravity. Astronomy & Astrophysics 657: A82 (24 pages). https://doi.org/10.1051/0004-6361/202142459.

GRAVITY Collaboration. 2022b. Mass distribution in the galactic center based on interferometric astrometry of multiple stellar orbits. Astronomy & Astrophysics 657: L12 (17 pages). https://doi.org/10.1051/0004-6361/202142465.

Griffith, E., Johnson, J. A., and Weinberg, D. H. 2019. Abundance ratios in GALAH DR2 and their implications for nucleosynthesis. The Astrophysical Journal 886: 84 (21 pages). https://doi.org/10.3847/1538-4357/ab4b5d.

Grün, E., Krüger, H., and Srama, R. 2019. The dawn of dust astronomy. Space Science Reviews 215: 46 (51 pages). https://doi.org/10.1007/s11214-019-0610-1.

Grunblatt, S. K., Saunders, N., Chontos, A., et al. 2022a. TESS Giants transiting giants III: An eccentric warm Jupiter supports a period-eccentricity relation for giant planets transiting evolved stars. arXiv:2210.17062 [astro-ph.EP], 31 Oct.

Grunblatt, S. K., Saunders, N., Sun, M., et al. 2022b. TESS giants transiting giants. II. The hottest Jupiters orbiting evolved stars. The Astronomical Journal 163: 120 (16 pages). https://doi.org/10.3847/1538-3881/ac4972.

Haba, M. K., Lai, Y.-J., Wotzlaw, J.-F., et al. 2021. Precise initial abundance of niobium-92 in the solar system and implications for p-process nucleosynthesis. PNAS 118(8): e2017750118 (7 pages). https://doi.org/10.1073/pnas.2017750118.

Han, J. J., Conroy, C., Johnson, B. D., et al. 2022. The stellar halo of the galaxy is tilted and doubly broken. The Astronomical Journal 164: 249 (16 pages). https://doi.org/10.3847/1538-3881/ac97e9.

Hartwig, T., Ishigaki, M. N., Kobayashi, C., et al. 2023. Machine learning detects multiplicity of the first stars in stellar archaeology data. The Astrophysical Journal 946(1): 20 (18 pages). https://doi.org/10.3847/1538-4357/acbcc6.

Hassani, H., Rosolowsky, E., Leroy, A. K., et al. 2023. Phangs–JWST first results: The 21 μm compact source population. The Astrophysical Journal Letters 944: L21 (11pages). https://doi.org/10.3847/2041-8213/aca8ab.

Hayes, C. R., Majewski, S. R., Shetrone, M., et al. 2018. Disentangling the galactic halo with APOGEE. I. Chemical and kinematical investigation of distinct metal-poor populations. The Astrophysical Journal 852(1): 49 (18 pages). https://doi.org/10.3847/1538-4357/aa9cec.

Hedges, C., Hughes, A., Zhou, G., et al. 2021. TOI-2026 and TOI-1807: Two young, comoving planetary systems with 50 pc identified by TESS that are ideal candidates for further follow up. The Astronomical Journal 162: 54 (27 pages). https://doi.org/10.3847/1538-3881/ac06cd.

Herczeg, G. J., and Hillenbrand, L. A. 2015. Empirical isochrones for low mass stars in nearby young associations. The Astrophysical Journal 808: 23 (18 pages). https://doi.org/10.1088/0004-637X/808/1/23.

Heywood, I., Rammala, I., Camilo, F., et al. 2022. The 1.28 GHz MeerKAT galactic center mosaic (29 pages). arXiv:2201.10541v3 [astro-ph.GA], 27 Jan.

Hill, H. M. 2022. Giant meteorites could be responsible for Earth's continents. Physics Today 75(10): 16–18.

Hinkel, N. R., Timmes, F. X., Young, P. A., et al. 2014. Stellar abundances in the solar neighborhood: The Hypatia catalog. The Astronomical Journal 148: 54 (33 pages). https://doi.org/10.1088/0004-6256/148/3/54.

Hon, M., Huber, D., Rui, N. Z., et al. 2023. A close-in giant planet escapes engulfment by its star. Nature 618: 917 (18 pages). https://doi.org/10.1038/s41586-023-06029-0.

Honniball, C. I., Lucey, P. G., Li, S., et al. 2020. Molecular water detected on the sunlit moon by SOFIA. Nature Astronomy 5: 121–127. https://doi.org/10.1038/s41550-020-01222-x.

Hou, L. G. 2021. The spiral structure in the solar neighborhood. Frontiers in Astronomy and Space Science 8: 671670 (16 pages). https://doi.org/10.3389/fspas.2021.671670.

Huang, Y., Beers, T. C., Wolf, C., et al. 2022. Beyond spectroscopy. I. Metallicities, distances, and age estimates for over 20 million stars from SMSS DR2 and gaia EDR3. The Astrophysical Journal 925: 164 (22 pages). https://doi.org/10.3847/1538-4357/ac21cb.

Hünsch, M., Weidner, C., and Schmitt, J. H. M. M. 2003 An X-ray study of the open clusters NGC 2451 A and B. Astronomy & Astrophysics 402(2): 571–586. https://doi.org/10.1051/0004-6361:20030268.

Hunt, A. C., Theis, K. J., Rehkämper, M., et al. 2022. The dissipation of the solar nebula constrained by impacts and core cooling in planetesimals. Nature Astronomy 6: 812–818. https://doi.org/10.1038/s41550-022-01675-2.

Iglesias-Groth, S., and Marin-Dobrincic, M. 2023. A rich molecular chemistry in the gas of the IC 348 star cluster of the Perseus Molecular Cloud. Monthly Notices of the Royal Astronomical Society, Volume 521(2): 2248–2269. https://doi.org/10.1093/mnras/stad495.

Ilee, J. D., Walsh, C., Booth, A. S., et al. 2021. Molecules with ALMA at planet-forming scales (MAPS). IX. Distribution and Properties of the Large Organic Molecules $HC_3N$, $CH_3CN$, and c-$C_3H_2$. arXiv:2109.06319v2 [astro-ph.EP].

Isern, J. 2019. The star formation history in the solar neighborhood as told by massive white dwarfs. The Astrophysical Journal Letters 878: L11 (4 pages). https://doi.org/10.3847/2041-8213/ab238e.

Ishii, H. A., Bradley, J. P., Bechtel, H. A., et al. 2018. Multiple generations of grain aggregation in different environments preceded solar system body formation. PNAS 115(26): 6608-6613.

Izidoro, A., Dasgupta, R., Raymond, S, N., et al. 2021. Planetesimal rings as the cause of the solar system's planetary architecture. Nature Astronomy. https://doi.org/10.1038/s41550-021-01557-z.

Jeffries, R. D., Naylor, T., Walter, F. M., et al. 2009. The stellar association around Gamma Velorum and its relationship with Vela OB2. Mon. Not. R. Astron. Soc. 393: 538–556. doi:10.1111/j.1365-2966.2008.14162.x.

Jergensen, J. K., Kuruwita, R. L., Harsono, D., et al. 2022. Binarity of protostar affects the evolution of the disk and planets. Nature. Article (4 pages). https://doi.org/10.1038/s41586-022-04659-4.

Jiao, Y., Hammer, F., Wang, H., et al. 2023. Detection of the Keplerian decline in the Milky Way rotation curve. Astronomy & Astrophysics. Manuscript no. AA:2023: 47513 ©ESO 2023. 23 Sep.

Jofré, P., Das, P., Bertranpetit, J., and Foley, R. 2017. Cosmic phylogeny: Reconstructing the chemical history of the solar neighbourhood with an evolutionary tree. Monthly Notices of the Royal Astronomical Society 467: 1140–1153. https://doi.org/10.1093/mnras/stx075.

Johnson, T. E., Kirkland, C. L., Lu, Y., et al. 2022. Giant impacts and the origin and evolution of continents. Nature 608: 330–335. https://doi.org/10.1038/s41586-022-04956-y.

Kalvāns, J. 2018. Temperature spectra of interstellar dust grains heated by cosmic rays: Dark cloud cores. arXiv: 1810.00708v1 [astro-ph.GA], 1 Oct.

Kane, S. R. 2023. The dynamical consequences of a super-Earth in the solar system. The Planetary Science Journal 4: 38 (17pages). https://doi.org/10.3847/PSJ/acbb6b.

Karakas, A. I., Lugaro, M., Carlos, M., et al. 2018. Heavy-element yields and abundances of asymptotic giant branch models with a small Magellanic cloud metallicity. Monthly Notices of the Royal Astronomical Society 477: 421–437.

Kenworthy, M., Lock, S., Kennedy, G., et al. 2023. A planetary collision afterglow and transit of the resultant debris cloud. Nature 622: 251–254. https://doi.org/10.1038/s41586-023-06573–9.

Kharchenko, N. V., Piskunov, A. E., Schilbach, E., et al. 2013. Global survey of star clusters in the Milky Way II. The catalogue of basic parameters. Astronomy & Astrophysics 558: A53. doi:10.1051/0004-6361/201322302.

Kipping, D., Bryson, S., Burke, C., et al. 2021. An exomoon survey of 70 cool giant exoplanets and the new candidate Kepler-1708 b-i. Nature Astronomy. https://doi.org/10.1038/s41550-021-01539-1.

Kirkland, C. L., Sutton, P. J., Erickson, T., et al. 2022. Did transit through the galactic spiral arms seed crust production on the early Earth? Geology. Published online 23 Aug. https://doi.org/10.1130/G50513.1.

Kleine, T., and Walker, R. J. 2017. Tungsten isotopes in planets. Annual Review of Earth and Planetary Sciences 45: 389–417. https://doi.org/10.1146/annurev-earth-063016-020037.

Kobayashi, C., Karakas, A. I., and Lugaro, M. 2020. The origin of elements from carbon to uranium. The Astrophysical Journal 900(2): 179 (33 pages).

Kobayashi, C., Karakas, A. I., and Umeda, H. 2011. The evolution of isotope ratios in the Milky Way galaxy. Monthly Notices of the Royal Astronomical Society 414(4): 3231–3250. https://doi.org/10.1111/j.1365-2966.2011.18621.x.

Kondo, I., Sumi, T., Koshimoto, N., et al. 2023. Prediction of planet yields by the prime-focus infrared microlensing experiment microlensing survey. The Astronomical Journal 165(6): 254 (18 pages). https://doi.org/10.3847/1538–3881/acccf9.

Kossakowski, D., Kürster, M., Trifonov, T. 2023. The CARMENES search for exoplanets around M dwarfs. Wolf 1069 b: Earth-Mass Planet in the Habitable Zone of a Nearby, Very Low-Mass Star. (26 pages). arXiv:2301.02477v2 [astro-ph.EP], 2 Feb.

Krot, A. N., Nagashima, K., Lyons, J. R., et al. 2020. Oxygen isitopicnheterogeneity in the early solar system inherited from the protostar molecular cloud. Science Advances 6: eaay2724 (7 pages).

Lam, K. W. F., Csizmadia, S., Astudillo-Defru, N., et al. 2021. GJ 367b: A dense, ultrashort-period sub-Earth planet transiting a nearby red dwarf star. Science 374: 1271–1275.

Laporte, C. F. P., Koposov, S. E., Belokurov, V. 2022. Kinematics beats dust: Unveiling nested substructure in the perturbed outer disc of the Milky Way. Monthly Notices of the Royal Astronomical Society: Letters 510(1): L13–L17. https://doi.org/10.1093/mnrasl/slab109.

Lapôtre, M. G. A., Maslaska, M. J., and Cable, M. L. 2022. The role of seasonal sediment transport and sintering in shaping titan's landscapes: A Hypothesis. Geophysical Research Letters 49: e202; https://doi.org/10.1029/202/GL097605.

Laughlin, G., and Chambers, J. E. 2002. Extrasolar trojans: The viability and detectability of planets in the 1: 1 Resonance. The Astronomical Journal 124: 592–600.

Lazarian, A. 2003. Magnetic fields via polarimetry: Progress of Grain Alignment Theory. Journal of Quantitative Spectroscopy and Radiative Transfer 79: 881 (30 pages). https://doi.org/10.1016/S0022-4073%2802%2900326-6.

Lawson, T. V., Pignatari, M., Stancliffe, R. J., et al. 2021. Radioactive nuclei in the early solar system: Analysis of the 15 isotopes produced by core-collapse supernovae. (30 pages). arXiv:2303.08918v1 [astro-ph.GA], 15 Mar. https://doi.org/10.1093/mnras/stab3684.

Lee, B.-C., Jeong, G., Koo, J.-R., et al. 2023. A search for exoplanets around northern circumpolar stars. VIII. Filtering out a planet cycle from the multi-period radial velocity variations in M giant HD 36384. Journal of the Korean Astronomical Society, JKAS 56(2): 195–199. https://doi.org/10.5303/JKAS.2023.56.2.195.

Lee, D.-C., Halliday, A. N., Leya, I., et al. 2002. Cosmogenic tungsten and the origin and earliest differentiation of the moon. Earth and Planetary Science Letters 198: 267–274.

Leitch, E. M., and Vasisht, G. 1998. Mass extinctions and the sun's encounters with spiral arms. New Astronomy 3: 51–56. https://doi.org/10.1016/S1384-1076(97)00044-4.

Lester, K. V., Matson, R. A., Howell, S. B., et al. 2021. Speckle observations of TESS exoplanet host stars. II. Stellar Companions at 1–1000 AU and Implications for Small Planet Detection (30 pages). arXiv:2106.13354v1 [astro-ph.SR], 24 Jun.

Levan, A. Gompertz, B. P., Salafia, O. S., et al. 2023. Heavy element production in a compact object merger observed by JWST. Nature (23 pages). https://doi.org/10.1038/s41586-023-06759–1.

Li, T. S., Ji, A. P., Pace, A. B., et al. 2022. $S^5$: The orbital and chemical properties of one dozen stellar streams. arXiv: 2110.06950v2 [astro-ph.GA], 2 Jan.

Lim, O., Benneke, B., Doyon, R., et al. 2023. Atmospheric Reconnaissance of TRAPPIST-1 b with JWST/NIRISS: Evidence for strong stellar contamination in the transmission spectra. The Astrophysical Journal Letters 955: L22 (20 pages). https://doi.org/10.3847/2041-8213/acf7c4.

Lissauer, J. J., Rowe, J. F., Jontof-Hutter, D., et al. 2023. Updated Catalog of Kepler planet candidates: Focus on accuracy and orbital periods. (62 pages). arXiv:2311.00238v1 [astro-ph.EP], 1 Nov.

Lodders, K. 2019. Solar elemental abundances. arXiv:1912.00844v1[astro-ph.SR].https://doi.org/10.48550/arXiv.1912.00844.

Lopatka, A. 2022. A lunar micrometeorite preserves the solar system-s early history. Physics Today: 16–18.

Lous, M. M., Helled, R., and Mordasini, C. 2022. Potential long-term habitable conditions on planets with primordial H–He atmospheres. Nature astronomy. https://doi.org/10.1038/s41550-022-01699-8.

Love, S. G., Pettit, D. R., and Messenger, S. R. 2014. Particle aggregation in microgravity: Informal experiments on the international space station. Meteoritic & Planetary Science 49: 732–739.

Luque, R., Osborn, H. P., Leleu, A., et al. 2023. A resonant sextuplet of sub-Neptunes transiting the bright star HD 110067. Nature 623: 932–937. https://doi.org/10.1038/s41586-023-06692–3.

Ma, Y., Zhou, J., Esimbek, J., et al. 2023. Gravitational collapse and accretion flows in the hub filament system G323.46–0.08. Astronomy &Astrophysics 676: A15 (11 pages). https://doi.org/10.1051/0004-6361/202346248.

Magg, E., Bergemann, M., Serenelli, A., et al. 2022. Observational constraints on the origin of the elements. IV. Standard composition of the sun. Astronomy & Astrophysics 661: A140 (18 pages). https://doi.org/10.1051/0004-6361/202142971.

Maiolino, R., and Mannucci, F. 2019. De re metallica: The cosmic chemical evolution of galaxies. The Astronomy and Astrophysics Review 27: 3 (187 pages). https://doi.org/10.1007/s00159-018-0112-2.

Maiorca, E., Uitenbroek, H., Uttenthaler, S., et al. 2014. A new solar fluorine abundance and a fluorine determination in the two open clusters M67 and NGC 6404. The Astrophysical Journal 788: 149 (10 pages). https://doi.org/10.1088/0004-637X/788/2/149.

Mannel, T., Bentley, M. S., Schmied, R., et al. 2016. Fractal cometary dust—a window into the early solar system. Monthly Notices of the Royal Astronomical Society 462: S304–S311.

Marsh, N., and Svensmark, H. 2000. Cosmic rays, clouds, and climate. Space Science Reviews 94: 215–230.

Martin, N. F., Ibata, R. A., Starkenburg, E., et al. 2022b. The pristine survey—XVI. The Metallicity of 26 stellar streams around the Milky Way detected with the STREAMFINDER in Gaia EDR3. arXiv:2201.01310v2 [astro-ph.GA]. https://doi.org/10.48550/arXiv.2201.01310.

Martin, N. F., Venn, K. A., Aguado, D. S., et al. 2022a. A stellar stream remnant of a globular cluster below the metallicity floor. Nature 601: 45–48. https://doi.org/10.1038/s41586-021-04162-2.

Mathis, J. S. 1990. Interstellar dust and extinction. Annual Review of Astronomy and Astrophysics 28: 37–70.

Matthews, L. S., Shotorban, B., and Hyde, T. W. 2013. Cosmic dust aggregation with stochastic charging. The Astrophysical Journal 776: 103 (9 pages). https://doi.org/10.1088/0004-637X/776/2/103.

Miyazaki, S., and Masuda, K. 2023. Evidence that the occurrence rate of hot Jupiters around sun-like stars decreases with stellar age. arXiv:2309.14605v1 [astro-ph.EP]. https://doi.org/10.48550/arXiv.2309.14605.

McDonald, I., Kerins, E., Poleski, R., et al. 2021. Kepler K2 campaign 9–I. Candidate short-duration events from the first space-based survey for planetary microlensing. MNRAS 505: 5584–5602.

Meingast, S., Alves, J., and Rottensteiner, A. 2021. Extended stellar systems in the solar neighborhood V. Discovery of coronae of nearby star clusters. Astronomy/Astrophysics 645: A84 (23 pages). https://doi.org/10.1051/0004-6361/202038610.

Mireles, I., Dragomir, D., Osborn, H. P., et al. 2023. TOI-4600 b and c: Two long-period giant planets orbiting an early K dwarf. The Astrophysical Journal Letters 954: L15 (13 pages). https://doi.org/10.3847/2041-8213/aceb69.

Miret-Roig, N., Bouy, H., Raymond, S. N., et al. 2021. A rich population of free-floating planets in the Upper Scorpius Young Stellar Association. Nature Astronomy Letters. https://doi.org/10.1038/s41550–021–01513-x.

Mishra, L., Albert, Y., Udry, S., and Mordrasini, C. 2023a. Framework for the architecture of exoplanetary systems I. Four classes of planetary system architecture. Astronomy & Astrophysics 670: A68 (28 pages). https://doi.org/10.1051/0004-6361/202243751.

Mishra, L., Albert, Y., Udry, S., and Mordrasini, C. 2023b. Framework for the architecture of exoplanetary systems II. Nature versus nurture: Emergent formation pathways of architecture classes. Astronomy & Astrophysics 670: A69 (11 pages). https://doi.org/10.1051/0004-6361/202244705.

Miyazaki, Y., and Korenaga, J. 2022. A wet heterogeneous mantle creates a habitable world in the Hadean. Nature 603: 86–90. https://doi.org/10.1038/s41586-021-04371-9.

Montalbán, J., Mackereth, J. T., Miglio, A., et al. 2021. Chronologically dating the early assembly of the Milky Way. Nature Astronomy Letters. https://doi.org/10.1038/s41550-021-01347-7.

Mróz, P, Poleski, R., Gould, A., et al. 2020. A terrestrial-mass rogue planet candidate detected in the shorteat-timescale microlensing event. arXiv:2009.12377v2 [astro-ph.EP], 20 Oct.

Naidu, R. P., Conroy, C., Bonaca, a., et al. 2021. Reconstructing the last major merger of the Milky Way with the H3 survey. The Astrophysical Journal 923: 92 (24 pages). https://doi.org/10.3847/1538-4357/ac2d2d.

Nakano, H., Hirakawa, N., Matsubara, Y., et al. 2020. Precometary organic matter: A hidden reservoir of water inside the snow line. Nature Scientific Reports 10: 7755 (13 pages). https://doi.org/10.1038/s41598-020-64815-6.

Nederlander, A., Hughes, A. M., Fehr, A. J., et al. 2021. Resolving structure in the debris disk around HD 206893 with ALMA. The Astrophysical Journal 917: 5 (17 pages). https://doi.org/10.3847/1538-4357/abdd32.

Noguchi, M. 2018. The formation of solar-neighbourhood stars in two generations separated by 5 billion years. Nature 559(7715): 585–588.

Nomoto, K., Tominaga, N., Umeda, H., et al. 2006. Nucleosynthesis yields of core-collapse supernovae and hypernovae, and galactic chemical evolution. Nuclear Physics A 777: 424-458. https://doi.org/10.1016/j.nuclphysa.2006.05.008.

Nordlander, T., and Lind, K. 2017. Non-LTE aluminium abundances in late-type stars. Astronomy and Astrophysics 607: A75 (20 pages). https://doi.org/10.1051/0004-6361/201730427.

Öberg, K. I., Murray-Clay, R., and Bergin, E. A. 2011. The effects of snowlines on C/O in planetary atmospheres. The Astrophysical Journal Letters 743(1): L16 (5 pages). https://doi.org/10.1088/2041-8205/743/1/L16.

O'Callaghan, J. 2022. First active chemistry on an exoplanet revealed by Webb telescope. Nature News. 22 Nov. https://doi.org/10.1038/d41586-022-03820-3.

Oddo, D., Dragomir, D., Brandeker, A., et al. 2023. Characterization of a set of small planets with TESS and CHEOPS and an analysis of photometric performance. The Astronomical Journal 165: 134 (36 pages). https://doi.org/10.3847/1538-3881/acb4e3.

Ou, X., Eilers, A.-C., Necib, L., and Frebel, A. 2023. The dark matter profile of the Milky Way inferred from its circular velocity curve. arXiv:2303.12838v2 [astro-ph.GA], 26 May.

Palme, H., Lodders, K., and Jones, A. 2014 Solar System Abundances of the Elements. In K. Turekian and H. Holland (eds.), Treatise on Geochemistry 2nd Edition (pp. 15–36). Pub. Elsevier Ltd.

Paneque-Carreno, T., Pérez, L. M., Benisty, M., et al. 2021. Spiral arms and a massive dust disk with non-Keplerian kinematics: Possible evidence for gravitational instability in the disk of Elias 2–27. arXiv:2103.14048v2 [astro-ph.EP], 24 May.

Pass, E. K., Winters, J. G., Charbonneau, D., et al. 2023. HST/WFC3 light curve supports a terrestrial composition for the closest exoplanet to transit an M dwarf. The Astronomical Journal 166: 171 (8 pages). https://doi.org/10.3847/1538-3881/acf561.

Pecaut, M. J., Mamajek, E. E., and Bubar, E. J. 2012. A revised age for Upper Scorpius and the star formation history among the F-type members of the Scorpius-Centaurus OB Association. The Astrophysical Journal 746(2): Article 154 (22 pages). https://doi.org/10.1088/0004-637X/746/2/154.

Peißker, F., Zajaček, M., Sabha, N. B., et al. 2023. X3: A high-mass young stellar object close to the supermassive black hole Sgr A*. The Astrophysical Journal 944: 231 (28 pages). https://doi.org/10.3847/1538-4357/aca977.

Piaulet, C., Benneke, B., Almenara, J. M., et al. 2022. Evidence for the volatile-rich composition of a 1.5-Earth-radius planet. Nature Astronomy. https://doi.org/10.1038/s41550-022-01835-4.

Pilorget, C., Okada, T., Hamm, V., et al. 2021. First compositional analysis of Ryugu samples by the MicrOmega hyperspectral microscope. Nature Astronomy. https://doi.org/10.1038/s41550-021-01549-z.

Pineda, J. S., and Villadsen, J. 2023. Coherent radio bursts from known M-dwarf planet-host YZ Ceti. Nature Astronomy. https://doi.org/10.1038/s41550-023-01914-0.

Planck Collaboration. 2020. Planck 2018 results VI. Cosmological parameters. Astronomy & Astrophysics 641: 15, A6 (67 pages). https://doi.org/10.1051/0004-6361/201833910.

Platais, I., Kozhurina-Platais, V., and van Leeuwen, F. 1998. A search for star clusters from the HIPPARCOS data. The Astronomical Journal 116(5): 2423–2430.

Potapov, A., and McCoustra, M. 2021. Physics and chemistry on the surface of cosmic dust grains: A laboratory view. International Reviews in Physical Chemistry 40(2): 299–364.

Prinoth, B., Hoeijmakers, H. J., Kitzmann, D., et al. 2022. Titanium oxide and chemical inhomogeneity in the atmosphere of the exoplanet WASP-189 b. Nature Astronomy Letters. https://doi.org/10.1038/s41550-021-01581-z.

Putirka, K. D., and Xu, S. 2021. Polluted white dwarfs reveal exotic mantle rock types on exoplanets in our solar neighborhood. Nature Communications 12: 6168 (7 pages). https://doi.org/10.1038/s41467-021-26403-8.

Ramírez-Tannus, M. C., Bik, A., Cuijpers, L., et al. 2023. XUE: Molecular inventory in the inner region of an extremely irradiated protoplanetary disk. The Astrophysical Journal Letters 958: L30 (14 pages). https://doi.org/10.3847/2041-8213/ad03f8.

Rampino, M. R., Caldeira, K., and Zhu, Y. 2021. A pulse of the Earth: A 27.5-Myr underlying cycle in coordinated geological events over the last 260 Myr. Geoscience Frontiers 12: 101245 (7 pages). https://doi.org/10.1016/j.gsf.2021.101245.

Randich, S., and Magrini, L. 2021. Light elements in the universe. Frontiers in Astronomy and Space Science 8: 616201 (28 pages). https://doi.org/10.3389/fspas.2021.616201.

Roederer, I. U., Vassh, N., Holmbeck, E. M., et al. 2023. Element abundance patterns in stars indicate fission of nuclei heavier than uranium. Science 382(6675): 1177–1180. HTTPS://DOI.ORG/10.1126/science.adf1341.

Romano, D., Franchini, M., Grisoni, V., et al. 2020. On the variation of carbon abundance in galaxies and its implications. arXiv e-printsarXiv:2005.05717.

Rosario, M., Tijerina, R. S., Cargile, P., and James, D. 2015. Cosmography of OB stars in the solar neighbourhood. Astronomy Astrophysics 584: A26 (13 pages). https://doi.org/10.1051/0004-6361/201527058.

Russeil, D. 2003. Star-forming complexes and the spiral structure of our galaxy. Astronomy and Astrophysics 397(1): 133–146. https://doi.org/10.1051/0004-6361:20021504.

Sagear, S., and Ballard, S. 2023. The orbital eccentricity distribution of planets orbiting M dwarfs. Proceedings of the National Academy of Sciences 120(23): e2217398120. https://doi.org/10.1073/pnas.2217398120.

Salaris, M., and Cassisi, S. 2005. Evolution of Stars and Stellar Populations. Chichester: John Wiley & Sons Ltd.

Sawala, T., Cautun, M., Frenk, C., et al. 2022. The Milky Way's plane of satellites is consistent with $\Lambda$CDM. Nature Astronomy. https://doi.org/10.1038/s41550-022-01856-z.

Saunders, N., Grunblatt, S. K., Huber, D., et al. 2022. TESS giants transiting giants. I.: A noninflated hot Jupiter orbiting a massive subgiant. Astronomical Journal 163: 53 (15 pages). https://doi.org/10.3847/1538–3881/ac38a1.

Schanche, N., Pozuelos, F. J., Günther, M. N., et al. 2022. TOI-2257 b: A highly eccentric long-period sub-Neptune transiting a nearby M dwarf. Astronomy/Astrophysics 657: mA45 (18 pages). https://doi.org/10.1051/0004-6361/2021142280.

Schmieder, M., and Kring, D. A. 2020. Earth's impact events through geologic time: A list of recommended ages for terrestrial impact structures and deposits. Astrobiology 20: 91–141.

Schnelderman, T., Matrà, L., Jackson, A. P., et al. 2021. Carbon monoxide gas produced by a giant impact in the inner region of a young system. Nature 598: 425–428.

Scibelli, S., and Shirley, Y. 2020. Prevalence of complex organic molecules in starless and prestellar cores within the Taurus molecular cloud. The Astrophysical Journal 891: 73 (17 pages). https://doi.org/10.3847/1538-4357/ab7375.

Scott, P., Grevesse, N., Asplund, M., et al. 2015a. The elemental composition of the Sun I. The intermediate mass elements Na to Ca. Astronomy and Astrophysics 573: A25 (19 pages). https://doi.org/10.1051/0004-6361/201424109.

Scott, P., Asplund, M., Grevesse, N., et al. 2015b. Astronomy and Astrophysics 573: A26 (33 pages). https://doi.org/10.1051/0004-6361/201424110.

Scott, N., van de Sande, J., Sharma, S., et al. 2021. Identification of an [α/Fe]-enhanced thick disk component in an edge-on Milky Way analog. The Astrophysical Journal Letters 913: L11 (6 pages). https://doi.org/10.3847/2041-8213/abfc57.

Shaviv, N. J. 2002. The spiral structure of the Milky Way, cosmic rays, and ice age epochs on Earth. arXiv:astro-ph/0209252v1, 12 Sep.

Shibata, S., and Helled, R. 2022. Enrichment of Jupiter's atmosphere by late planetesimal bombardment. The Astrophysical Journal Letters 926: L37 (9 pages). https://doi.org/10.3847/2041-8213/ac54b1.

Shirazi, A. R., Haghi, H., Khalaj, P., et al. 2022. The escape of globular clusters from the satellite dwarf galaxies of the Milky Way. Monthly Notices of the Royal Astronomical Society 513(3): 3526–3540. https://doi.org/10.1093/mnras/stac1070.

Stacy, L. 2002. Period of the Sun's Orbit Around the Galaxy (Cosmic Year). In G. Elert (ed.), The Physics Factbook. https://hypertextbook.com/facts/2002/StacyLeong.shtml.

Standing, M. R., Sairam, L., Martin, D. V., et al. 2023. Radial-velocity discovery of a second planet in the TOI-1338/BEBOP-1 circumbinary system. Nature Astronomy 7: 702–714. https://doi.org/10.1038/s41550-023-01948-4.

Stefánsson, G., Mahadevan, S., Miguel, Y., et al. 2023. A Neptune-mass exoplanet in close orbit around a very low-mass star challenges formation models. Science 382(6674): 1031–1035. https://doi.org/10.1126/science.abo0233.

Stephens, I. W., Myers, P. C., Zucker, C., et al. 2022. The magnetic field in the Milky Way filamentary bone G47. arXiv:2201.11933v2 [astro-ph.GA], 08 Feb.

Stothers, R. B. 1998. Galactic disc dark matter, terrestrial impact cratering and the law of large numbers. Monthly Notices of the Royal Astronomical Society 300: 1098–1104. https://doi.org/10.1046/j.1365-8711.1998.02001.x.

Svensmark, H. 2006. Imprint of galactic dynamics on Earth's climate. Astronomische Nachrichten 327(9): 866–870.

Syed, J., Soler, J. D., Beuther, H., et al. 2022. The "Maggie" filament: Physical properties of a giant atomic cloud. Astronomy & Astrophysics 657: A1 (19 pages). https://doi.org/10.105/0004-6361/202141265.

Takir, D., Neumann, W., Raymond, S. N., et al. 2023. Late accretion of Ceres-like asteroids and their implantation into the outer main belt. Nature Astronomy. https://doi.org/10.1038/s41550-023-01898-x.

Thiabaud, A., Marboeuf, U., Alibert, Y., et al. 2015. Gas composition of the main volatile elements in protoplanetary discs and its implication for planet formation. Astronomy & Astrophysics 574: A138 (11 pages). https://doi.org/10.1051/0004-6361/201424868.

Tinsley B. M. 1980. Evolution of the stars and gas in galaxies. Fundamentals of Cosmic Physics 5: 287–388. https://doi.org/10.48550/arXiv.2203.02041.

Tominaga, R. T., Inutsuka, S.-I., and Kobayashi, H. 2021. Coagulation instability in protoplanetary disks: A novel mechanism connecting collisional growth and hydrodynamical clumping of dust particles. The Astrophysical Journal 923: 34 (24 pages). https://doi.org/10.3847/1538-4357/ac173a.

Triaud, A. H. M., Dransfield, G., Kagetani, T., et al. 2023. An M dwarf accompanied by a close-in giant orbiter with SPECULOOS. Monthly Notices of the Royal Astronomical Society 525: L98–L104. https://doi.org/10.1093/mnrasl/slad097.

Tsuge, M., Molpeceres, G., Aikawa, Y., et al. 2023. Surface diffusion of carbon atoms as a driver of interstellar organic chemistry. Nature Astronomy. https://doi.org/10.1038/s41550-023-02071–0.

Ustinova, G. 2007. On the problem of solar system origin: The regularities of noble gas fractionation in shock waves. Solar System Research 41(3): 231–255.

Vacher, L. G., Ogliore, R. C., Liu, N., and Fike, D. A. 2021. Cosmic symplectite recorded irradiation by nearby massive stars in the solar system's parent molecular cloud. Geochimica et Cosmochimica Acta 309: 135–150.

Vissapragada, S., Chontos, A., Greklek-McKeon, M., et al. 2022. The possible tidal demise of Kepler's first planetary system. The Astrophysical Journal Letters 941L31 (8 pages). https://doi.org/10.3847/2041-8213/aca47e.

Wagner, K., Stone, J., Skemer, A., et al. 2023. Direct images and spectroscopy of a giant protoplanet driving spiral arms in MWC 758. Nature Astronomy. https://doi.org/10.1038/s41550-023-02028-3.

Wang, Q. D. 2021. Chandra large-scale mapping of the galactic centre: Probing high-energy structures around the central molecular zone. Monthly Notices of the Royal Astronomical Society 504: 1609–1618. https://doi.org/10.1093/mnras/stab801.

Weibel, E. S. 1959. Spontaneously growing transverse waves in a plasma due to an anisotropic velocity distribution. Physical Review Letters 2: 83–84. https://doi.org/10.1103/PhysRevLett.2.83.

Wielgus, M., Moscibrodzka, M., Vos, J., et al. 2022. Orbital motion near Sagittarius A*. Constraints from polarimetric ALMA observations. A&A 665: L6 (pages 11). https://doi.org/10.1051/0004-6361/202244493.

Will, P., Busemann, H., Riebe, M. E. I., and Maden, C. 2022. Indigenous noble gases in the Moon's interior. Science Advances 8: eabl4920 (8 pages). https://doi.org/10.1126/sciadv.abl4920.

Wolszczan, A., and Frail, D. 1992. A planetary system around the millisecond pulsar PSR1257 + 12. Nature 355: 145–147. https://doi.org/10.1038/355145a0.

Wurm, G., and Blum, J. 2000. An experimental study on the structure of cosmic dust aggregates and their alignment by motion relative to gas. The Astrophysical Journal 529: L57-L60.

Xu, Y., Newberg, H. J., Carlin, J. L., et al. 2015. Rings and radial waves in the disk of the Milky Way. The Astrophysical Journal 801: 105 (25 pages). https://doi.org/10.1088/0004-637X/801/2/105.

Yang, H.-Y. K., Ruszkowski, M., and Zweibel, E. G. 2022. Fermi and eROSITA bubbles as relics of the past activity of the galaxy's central black hole. Nature Astronomy Articles https://doi.org/10.1038/s41550-022-01618-x.

Yates, R. M., Henriques, B., Thomas, P. A., et al. 2013. Modelling element abundances in semi-analytic models of galaxy formation. Monthly Notices of the Royal Astronomical Society 435: 3500–3520.

Yeghikyan, A., and Fahr, H. 2004. Terrestrial atmospheric effects induced by counterstreaming dense interstellar cloud material. Astronomy & Astrophysics 425(3): 1113–1118. https://doi.org/10.1051/0004-6361:20041100.

Yong, D., Kobayashi, C., Da Costa, G. S., et al. 2021. R-process elements from magnetorotational hypernovae. Nature 595: 223–226.

Yuan, Q., Li, M., Desch, S. J., et al. 2023. Moon-forming impactor as a source of Earth's basal mantle anomalies. Nature 623: 95–99. https://doi.org/10.1038/s41586-023-06589-1.

Yusef-Zadeh, F., Arendt, R. G., Wardle, M., et al. 2022. Statistical properties of the population of the galactic center filaments: The spectral index and equipartition magnetic field. arXiv:2201.10552v1 [astro-ph.GA], 25 Jan.

Zari, E., Hashemi, H., Brown, A. G. A., et al. 2018. 3D mapping of young stars in the solar neighbourhood with *Gaia* DR2. Astronomy & Astrophysics 620: A172 (16 pages). https://doi.org/10.1051/0004-6361/201834150.

Zhang, Y., Snellen, I. A. G., Bohn, A. J., et al. 2021. The $^{13}CO$-rich atmosphere of a young accreting super-Jupiter. Nature 595: 370–372.

Zhao, G., Mashonkina, L., Yan, H. L., et al. 2016. Systematic Non-LTE Study of the -.6 ≤ [Fe/H] ≤ 0.2 F and G dwarfs in the solar neighborhood. II. Abundance patterns from Li to Eu. The Astrophysical Journal 833(2): 225, (26 pages). https://doi.org/10.3847/1538-4357/833/2/225.

Zhukovska, S., Gail, H.-P., and Trieloff, M. 2008. Evolution of interstellar dust and stardust in the solar neighbourhood. Astronomy & Astrophysics 479(2): 453–480.

Zhou, S., Aragón-Salamanca, A., Merrifield, M., et al. 2022a. Are Milky-Way-like galaxies like the Milky Way? A view from SDSS-IV/MaNGA. arXiv:2212.09127v1 [astro-ph.GA]. https://doi.org/10.48550/arXiv.2212.09127.

Zhou, M., Zhdankin, V., Kunz, M. V., et al. 2022b. Spontaneous magnetization of collisionless plasma through the action of a shear flow. arXiv:2110.01134v1 [physics.plasm-ph], 4 Oct.

Zieba, S., Kreidberg, L., Ducrot, E., et al. 2023. No thick carbon dioxide atmosphere on the rocky exoplanet TRAPPIST-1 c. Nature Article 2023 (21 pages). https://doi.org/10.1038/s41586-023–06232-z.

Zink, J. K., Hardegree-Ullman, K. K., Christiansen, J. L., et al. 2021. Scaling K2. IV. A uniform planet sample for campaigns 1–8 and 10–18. The Astronomical Journal 162: 259 (20 pages). https://doi.org/10.3847/1538-3881/ac2309.

Zucker, C., Battersby, C., and Goodman, A. 2015. The skeleton of the Milky Way. The Astrophysical Journal 815: 23 (25 pages). https://doi.org/10.1088/0004-637X/815/1/23.

Zucker, C., Battersby, C., and Goodman, A. 2018. Physical properties of large-scale galactic filaments. The Astrophysical Journal 864: 153 (34 pages). https://doi.org/10.3847/1538-4357/aacc66.

Zucker, C., Goodman, A. A., Alves, J., et al. 2022. Star formation near the sun driven by expansion of the local bubble. Nature 601: 334–337.

## ADDITIONAL READING

Belbruno, E., and Green, J. 2022. When leaving the solar system: Dark matter makes a difference. Monthly Notices of the Royal Astronomical Society 510: 5154–5163.

Bernard, E. J. 2017. Reconstructing the star formation history of the solar neighbourhood with Gaia. Rediscovery our Galaxy, Proceedings IAU Symposium No. 33. https://doi.org/10.1017/S1743921317008341.

Boone, R. H., King, J. R., and Soderblom, D. R. 2006. Metallicity in the solar neighbourhood out to 60 pc. New Astronomy Reviews 50: 526–529.

Brown, T., Wilson, C. D., Zabel, N., et al. 2021. VERTICO: The Virgo environment traced in CO survey. The Astrophysical Journal 257: 21 (24 pages). https://doi.org/10.3847/1538-4365/ac28f5.

Cooke, G. J., Marsh, D. R., Walsh, C., et al. 2022. A revised lower estimate of ozone columns Earth's oxygenated history. Royal Society Open Science 9: 211165 (20 pages). https://doi.org/10.1098/rsos211165.

De Pater, I., Fletcher, L. N., Reach, W. T., et al. 2021. SOFIA observations of variability in Jupiter's para-H2 distributions and subsurface emission characteristics of the Galilean satellites. The Planetary Science Journal 2: 226 (13 pages). https://doi.org/10.3847/PSJ/ac2d24.

Farihi, J., Hermes, J. J., Marsh, T. R., et al. 2022. Relentless and complex transits from a planetosimal debris disc. Monthly Notices of the Royal Astronomical Society 511: 1647-1666.

Feltzing, S., and Chiba, M. 2013. Elemental abundances in the Milky Way stellar disk(s), bulge, and halo. New Astronomy Reviews 57(3–4): 80–99. https://doi.org/10.1016/j.newar.2013.06.001.

Gautam, A., Crocker, R. M., Ferrario, L., et al. 2022. Millisecond pulsars from accretion-induced collapse as the origin of the galactic centre gamma-ray excess signal. Nature Astronomy Letters. https://doi.org/10.1038/s41550-022-01658-3.

Gies, D. R., and Helsel, J. W. 2005. Ice age epochs and the sun's path through the galaxy. The Astrophysical Journal 626(2): 844–848.

Guzik, P., and Drahus, M. 2021. Gaseous atomic nickel in the coma of interstellar comet 2I/Borisov. Nature 593: 375–378.

Hammouda, T., Boyet, M., Frossard, P., Cartier, C. 2022. The message of oldhamites from enstatite chondrites. Progress in Earth and Planetary Science 9 (1): article no. 13 (20 pages). https://doi.org/10.1186/s40645-022-00471-w.

Kluska, J., Van Winckel, H., Coppée, Q., et al. 2022. A population of transition disks around evolved stars: Fingerprints of planets. Catalog of disks surrounding galactic post-AGB binaries. Astronomy & Astrophysics 658: A36 (24 pages). https://doi.org/10.1051/0004-6361/202141690.

Leane, R. K., and Smirnov, J. 2021. Exoplanets as sub-GeV dark matter detectors. Physical Review letters 126: 161101 (11 pages).

Manfroid, J. Hutsemékers, D., and Jehin, E. 2021. Iron and nickel atoms in cometary atmospheres even far from the Sun. Nature 593(7859): 372–374.

Pacucci, F., and Loeb, A. 2022 Accretion from winds of red giant branch stars may reveal the supermassive black hole in Leo I. The Astrophysical Journal Letters 940: L33 (4 pages). https://doi.org/10.3847/2041–8213/ac9b21.

Paquet, M., Moynier, F., Yokoyama, T., et al. 2022. Contribution of Ryugu-like material to Earth's volatile inventory by U and Zn isotopic analysis. Nature Astronomy. https://doi.org/10.1038/s41550-022-01846-1.

Rocha-Pinto, H. J., Scalo, J., Maciel, W. J., and Flynn, C. 2000 An intermittent star formation history in a "normal" disk galaxy: The Milky Way. The Astrophysical Journal 531(2): L115 (4 pages). https://doi.org/10.1086/312531.

Roederer, I. U., Lawler, J. E., Den Hartog, E. A., et al. 2022. The R-process alliance: A nearly complete R-process abundance template derived from ultraviolet spectroscopy of the R-process-enhanced metal-poor star HD 222925. arXiv:2205.03426v1 [astro-ph.SR], 6 May.

Romanelli, N., and DiBraccio, G. A. 2021. Occurrence rate of ultra-low frequency waves in the foreshock of Mercury increases with heliocentric distances. Nature Communications 12: 6748. https://doi.org/10.1038/s41467-021-26344-2.

Rosenberg, J., and Ebrahimi, F. 2021. Onset of plasmoid reconnection during magnetorotational instability.

The Astrophysical Journal Letters 920: L29 (10 pages). https://doi.org/10.3847/2041-8213/ac2b2e.+
Rowther, S., Nealon, R., and Meru, F. 2022. Wrapping away gravitational instabilities in protoplanetary discs. The Astrophysical Journal 925: 163 (10 pages). https://doi.org/10.3847/1538-4357/ac3975.
Siraj, A., and Loeb, A. 2022. The mass budget necessary to explain 'Oumuamua as nitrogen iceberg. New Astronomy 92: 101730. https://doi.org/10.1016/j.newast.2021.101730.
Su, K. Y. L., Kennedy, G. M., Schlawin, E., et al. 2022. A star-sized impact-produced dust clump in the terrestrial zone of the HD 166191 system. The Astrophysical Journal 927: 135 (13 pages). https://doi.org/10.3847/1538-4357/ac4bbb.
Svensmark, H. 2021. Supernova rates and burial of organic matter. Geophysical Research Letters 48: e2021GL096376. https://doi.org/10.1029/2021GL096376.
Williams, D. A. 2005. Gas and dust in the interstellar medium. Journal of Physics: Conference Series 6: 1–17.

CHAPTER 3

# Life

## 3.1 INTRODUCTION

Different viewpoints, many with deep philosophical and historical roots, have shaped the scientific study of the origin of life. Some of these argue that primeval life was based on simple anaerobic microorganisms able to use a wide inventory of abiotic organic materials (i.e. a heterotrophic origin), whereas others invoke a more sophisticated organization, one that thrived on simple inorganic molecules (i.e. an autotrophic origin). While many scientists assume that life started as a self-replicative molecule, the first gene, a primitive self-catalytic metabolic network has also been proposed as a starting point. Even the emergence of the cell itself is a contentious issue: did boundaries and compartments appear early or late during life's origin? Starting with a recent definition of life, based on concepts of autonomy and open-ended evolution, it is proposed here that, first, organic molecules self-organized in a primordial metabolism located inside protocells. The flow of matter and energy across those early molecular systems allowed the generation of more ordered states, forming the cradle of the first genetic records. Thus, the origin of life was a process initiated within ecologically interconnected autonomous compartments that evolved into cells with hereditary and true Darwinian evolutionary capabilities. In other words, the individual existence of life preceded its historical-collective dimension (Peretó 2005). Table 3.1 shows controversial research topics in the origin of life problem.

TABLE 3.1 Research Topics in the Origin of Life Area

| | |
|---|---|
| **Place of origin**<br>Earth | **Place of origin**<br>Elsewhere in Universe: Other planets, molecular clouds |
| **Heterotrophic origin**<br>Complex prebiotic chemistry and simple metabolism | **Autotrophic origin**<br>Simple chemical ambient and complex metabolism |
| **Genetics first**<br>Early, spontaneous, self-replicative polymers | **Metabolism first**<br>Primitive bioenergetic mechanism |
| **Cells as latecomers**<br>Cells as mere compartments for replicators | **Cells as early invents**<br>Cells as necessary elements for bioenergetics |

The three bottom controversies have been identified already by Peretó (2005); today, up to the present day, the first entry in the Table 3.1 has been subject to discussion. For example, Valković and Obhođaš (2022) have proposed origins of chiral life in interstellar molecular clouds as early as $(4 \pm 1) \times 10^9$ years after Big Bang. Such early life had to live through many turbulences in the Universe, including the last one, to survive the planet formation process.

Photosynthesis is the ultimate source of nearly all energy used in all living organisms. This process is carried out by plants, algae, and bacteria to transform the energy of sunlight into chemical energy that can be stored and used by them and other organisms. Without photosynthesis, life on Earth as we know it would not be possible!

Photosynthesis is the process by which plants use sunlight, water, and carbon dioxide to create oxygen and energy in the form of simple sugars such as glucose and sucrose. During photosynthesis, plants take in carbon dioxide ($CO_2$) and water ($H_2O$) from the air and soil. Inside the plant, the water is converted into oxygen and $CO_2$ into glucose by a chemical process. The plant then releases the oxygen into the air and stores energy in glucose molecules. Later, this simple sugar is converted into more complex molecules like cellulose, lipids, or amino acids, needed for plants to develop and grow.

DOI: 10.1201/9781003366607-3

Photosynthesis is generally assumed to be initiated by a single photon from the Sun, which, as a weak light source, delivers at most a few tens of photons per $nm^2sec$ within a chlorophyll absorption band. Yet much experimental and theoretical work over the past 40 years has explored the events during photosynthesis subsequent to absorption of light from intense, ultrashort laser pulses. Li et al. (2023) used single photons to excite under ambient conditions the light-harvesting 2 (LH2) complex of the purple bacterium *Rhodobacter sphaeroides*, comprising B800 and B850 rings that contain 9 and 18 bacteriochlorophyll molecules, respectively. Excitation of the B800 ring leads to electronic energy transfer to the B850 ring in approximately 0.7 ps, followed by rapid B850-to-B850 energy transfer on an approximately 100-fs timescale and light emission at 850–875 nm. Using a heralded single-photon source along with coincidence counting, they established time correlation functions for B800 excitation and B850 fluorescence emission and demonstrated that both events involve single photons. They also found that the probability distribution of the number of heralds per detected fluorescence photon supports the view that a single photon can upon absorption drive the subsequent energy transfer and fluorescence emission and hence, by extension, the primary charge separation of photosynthesis. An analytical stochastic model and a Monte Carlo numerical model capture the data, further confirming that absorption of single photons is correlated with emission of single photons in a natural light-harvesting complex.

## 3.2 WHAT IS LIFE?

Information on the first signs of life on Earth appears as carbon isotope signatures in rocks 3.95 billion years of age (Tashiro et al. 2017). They presented the total organic carbon contents and carbon isotope values of graphite ($\delta^{13}C_{org}$) and carbonate ($\delta^{13}C_{carb}$) in the oldest metasedimentary rocks from northern Labrador. A clear correlation between the $\delta^{13}C_{org}$ values and metamorphic grade indicates that variations in the $\delta^{13}C_{org}$ values are due to metamorphism and that the pre-metamorphic value was lower than the minimum value. They concluded that the large fractionation between the $\delta^{13}C_{carb}$ and $\delta^{13}C_{org}$ values, up to 25‰, indicates the oldest evidence of organisms greater than $3.95 \times 10^9$ years.

All living things are built from cells: small, membrane-enclosed units filled with a concentrated aqueous solution of chemicals and possessing the capability to create copies of themselves by growing and then dividing in two. This makes cells the fundamental unit of life (Alberts et al. 2014).

- Cells come in a variety of shapes and sizes, typically about 5–20 μm in diameter. They are diverse in their chemical requirements. Some need oxygen for living; for others, this gas is deadly. In addition to differences in size, shape, and chemical requirements, cells have enormous differences in their functions. In spite of all of this variety, cells resemble one another to a high degree in their chemistry. They are composed of the same sorts of molecules participating in the same type of chemical reactions.
- In all organisms, DNA molecules carry out genetic information. In every cell, DNA polymer chains are made from same set of four monomers (nucleotides) bound together in different sequences conveying information. In every cell, the information encoded in the DNA is transcribed into a chemically related set of polymers called RNA. A subset of these RNA molecules is translated into another type of polymer called protein. This flow of information, from DNA to RNA to protein, is fundamental to life and often referred to as the life central dogma, as illustrated in Figure 3.1, following Alberts et al. (2014).
- The presence or absence of a nucleus in the cell is used as the basis for a simple but fundamental classification of all living things. Organisms whose cells have a nucleus are called eukaryotes, while those whose cells do not have a nucleus are called prokaryotes (bacteria and archaea). Prokaryotes are the most diverse and numerous cells on Earth. Most of them live as single-celled organisms, and some join together to form chains, clusters, or other structures. Mitochondria is the name for organelles that generate energy in eukaryotic cells. It is assumed that they have evolved from aerobic bacteria that lived inside the anaerobic ancestors of today's eukaryotic cells.
- Mariscal (2021) recently updated the entry in the Stanford Encyclopedia of Philosophy about the phenomenon of life. The author presents definitions of life from antiquity to Darwin to contemporary definitions, including some definitional skepticism. The question of life's nature cannot be separated

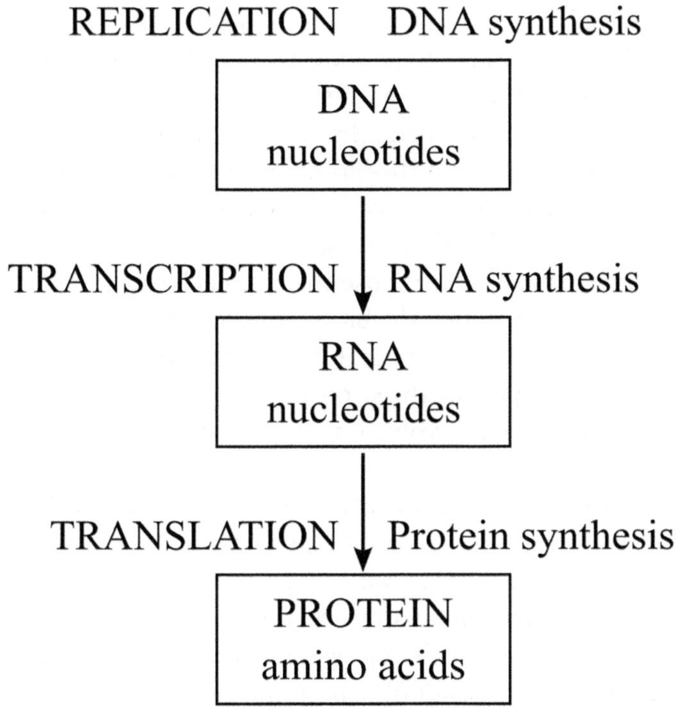


FIGURE 3.1 Schematic presentation of genetic flows from DNA to RNA (transcription) and from RNA to protein (translation), representing a sequence known as the central dogma.

> from the question of its origin. Very often, it is assumed that one speaks about life on planet Earth. However, evidence shows that as soon as it was not molten, Earth was filled with life. How life started and why it started so quickly remain some of the most intriguing open questions for advocates of an Earthly origin of life.

Proteins are built from amino acids, and all organisms use the same set of 20 amino acids to make their proteins. A cell's genome is the name for the entire sequence of nucleotides in an organism's DNA. It provides a genetic program that instructs the cell how to behave. Each cell is capable of carrying out variety of biological tasks.

A cell reproduces its DNA and then divides in two, passing a copy of the genetic instructions encoded in its DNA to each of its daughter cells. Occasionally, the pattern is modified when two cells of the same species fuse, pooling their DNA. The genetic cards are therefore shuffled, re-dealt, and redistributed in new combinations to the next generation, making the process of evolution. This process probably lasted for (3.5–3.8) $\times$ $10^9$ years. This is why present-day cells are so similar; they have inherited their genetic instructions from the same common ancestor.

Gómez-Márquez (2021) presents some interesting thoughts on what a life problem is. According to him, life is a process that takes place in highly organized organic structures and is characterized by being preprogrammed, interactive, adaptive, and evolutionary. If life is the process, living beings are the system in which this process takes place. He also wonders whether viruses can be considered living things. His conclusion is that viruses satisfy all the vital characteristics common to all living things. In addition, they played an important role in the evolution of species. He also argues that if there were life elsewhere in the universe, it would be very similar to what we know on this planet because the laws of physics and the composition of matter are universal and because of the principle of the inexorability of life.

Another interesting approach is taken by Dupré and O'Malley (2009). They address three fundamental questions: What does it mean for an entity to be living? What is the role of inter-organismic collaboration in evolution? What is a biological individual? Their central argument is that life arises when lineage-forming entities collaborate in metabolism. By conceiving of metabolism as a collaborative process performed by functional wholes, which are associations of a variety of lineage-forming entities, they avoid the standard tension between reproduction and metabolism in discussions of life—a tension particularly evident in discussions of whether viruses are alive. Their perspective assumes no sharp distinction between life and non-life and does not equate life exclusively with cellular or organismal status. They reach this conclusion through an analysis of the capabilities of a spectrum of biological entities in which they include the pivotal case of viruses as well as prions, plasmids, organelles, intracellular and extracellular symbionts, and unicellular and multicellular life-forms. The usual criterion for classifying many of the entities of their continuum as non-living is autonomy. This emphasis on autonomy is problematic, however, because even paradigmatic biological individuals, such as large animals, are dependent on symbiotic associations with many other organisms. These composite individuals constitute the metabolic wholes on which selection acts. Finally, their account treats cooperation and competition not as polar opposites but as points on a continuum of collaboration. They suggest that competitive relations are a transitional state, with multi-lineage metabolic wholes eventually outcompeting selfish competitors, and that this process sometimes leads to the emergence of new types or levels of wholes. Their view of life as a continuum of variably structured collaborative systems leaves open the possibility that a variety of forms of organized matter—from chemical systems to ecosystems—might be usefully understood as living entities.

According to Witzany (2020), in searching for life in extraterrestrial space, it is essential to act based on an unequivocal definition of life. In the 20th century, life was defined as cells that self-replicate, metabolize, and are open for mutations, without which genetic information would remain unchangeable, and evolution would be impossible. Current definitions of life derive from statistical mechanics, physics, and chemistry of the 20th century in which life is considered to function like a machine, ignoring a central role of communication. Recent observations show that context-dependent meaningful communication and network formation (and control) are central to all life forms. Evolutionary relevant new nucleotide sequences now appear to have originated from social agents such as viruses, their parasitic relatives, and related RNA networks, not from errors. By applying the known features of natural languages and communication, a new 21st-century definition of life can be reached in which communicative interactions are central to all processes of life. A new definition of life must integrate the current empirical knowledge about interactions between cells, viruses, and RNA networks to provide better explanatory power than the 20th-century narrative.

We have modified the approach suggested by Witzany (2020) that the bio-communication approach is identified four levels into which cellular organisms are involved since the start of their life until death. Instead, we propose the approach illustrated in Figure 3.2, in which the bio-communication approach could be identified with two inputs (environmental and memory) and three outputs into which cellular organisms have involved since the start of their life until death.

Primarily, life is a process. The main characteristic of this process is the coordinated organization of complex interactions that we see as protein-based organisms of three domains of life, their reproduction, and metabolism, all mediated by complex interwoven gene regulation as a result of communication. Living nature is structured and organized by language and communication within and among organisms, viruses, and RNA networks. If communication is damaged or disturbed, coordination and organization may be incomplete, and normal function becomes dysregulated, leading to a broad variety of diseases. Without RNA world agents, no cellular gene regulation could take place. Without viruses and related infectious agents, these capabilities of RNA stem-loop group behavior as gene inventors and

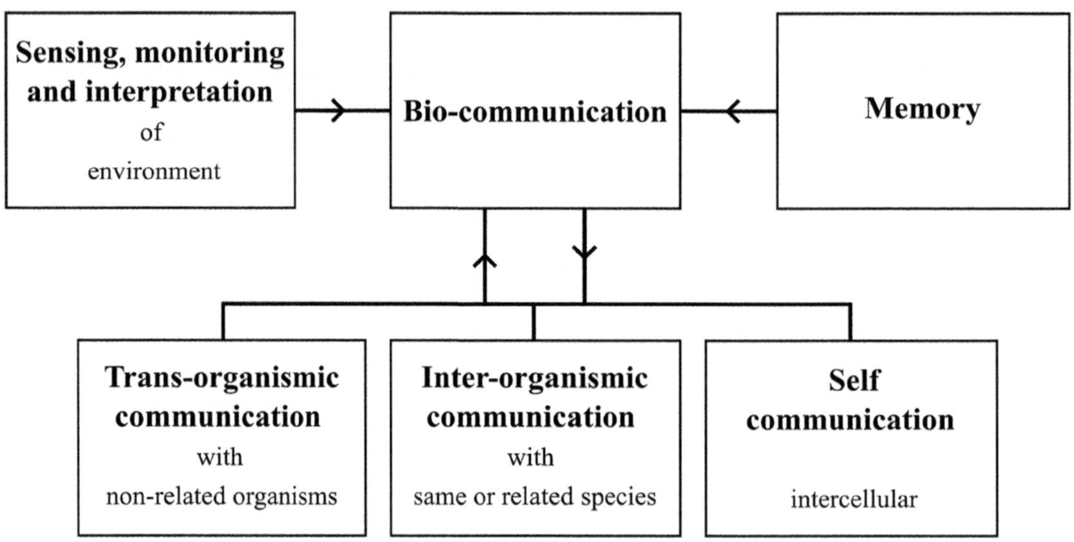


FIGURE 3.2 The bio-communication approach identified with two inputs and three outputs into which cellular organisms have involved since the start of their life until death; modification of Witzany (2020).

regulators would not have been integrated into cellular host genomes.

### 3.2.1 Thermodynamical Considerations

This section provides a short glossary of terms used for easier reading and understanding of the formulas and statements.

#### *3.2.1.1 Glossary*

*Enthalpy*, normally denoted *H*, is a thermodynamic property; it is equal to the sum of the internal energy plus the product of the pressure and the volume:

$$H = E + pV$$

where E = internal energy; p = pressure; V = volume.

Changes in enthalpy during heating and cooling operations are important; the change in enthalpy $\Delta H$ is given by

$$\Delta H = \Delta E + \Delta pV$$

$$\Delta H = \Delta E + p\Delta V + V\Delta$$

where Δ indicates the final minus the initial value. If a process takes place at constant pressure (i.e. $\Delta p = 0$), then

$$\Delta H = \Delta E + p\Delta V$$

From the first law of thermodynamics, $q = \Delta E + p\Delta V$, then $q = \Delta H$, where q is the heat exchanged between a system and its surroundings.

*Thermodynamic potentials*: Five common thermodynamic potentials are shown in Table 3.2.

TABLE 3.2 Five Common Thermodynamic Potentials

| Name | Symbol | Formula | Definition |
|---|---|---|---|
| Internal energy | U | $\int(TdS - pdV + \Sigma\mu_i dN_i$ | Energy contained within the system. |
| Helmholtz free energy | F | U – TS | Energy available in the system. |
| Enthalpy | G | U + pV | The capacity to do non-mechanical work plus the capacity to release heat. |
| Gibbs free energy | H | U + pV – TS | The maximum amount of non-expansion work that can be extracted from the closed system. |
| Landau potential | $\Omega$, $\Phi_G$ | $U - TS - \Sigma\,\mu_i N_i$ | A quantity used in statistical mechanics, especially for irreversible processes in open systems. |

where: $T$ = temperature, $S$ = entropy, $p$ = pressure, $V$ = volume. $N_i$ is the number of particles of type $i$ in the system, and $\mu_i$ is the chemical potential for an $i$-type particle.

The second law of thermodynamics makes no distinction between living and non-living things. Indeed, the concepts of thermodynamics constitute the unifying principles of physics, chemistry, and biology. The second law of thermodynamics implies the principle that the total entropy, which is a measure of disorder, must increase steadily. Even though thermodynamics itself does not describe processes as a function of time, the second law defines a unique direction of time (time's arrow) as the direction in which total entropy increases. Nevertheless, thermodynamics does not exclude local exceptions. Living things are local exceptions (Udgaonkar 2001).

Life on earth is possible because the temperature is such that it is high enough to permit structure-creating reactions and yet low enough to ensure that vital biochemical structures are not destroyed by the chaotic agitation from random thermal motion. Life can cease if a relatively small change in temperature occurs in either direction. At lower temperatures, metabolism slows down and effectively stops, while at higher temperatures, the specific structural organization in the cell that facilitates metabolic reactions disappears. It is truly a delicate balance. During the process of ageing, the balance shifts in the direction of anarchy. Death is the thermodynamically favored state: it represents a large increase in entropy as molecular structure yields to chaos. Although human beings expend a lot of energy to avert death, it is a state of too high a probability to be evaded.

Progress in nature is nearly always achieved by the development of more complicated biological structures, usually when these structures are better adapted to the environment. This development occurs by accumulation of small random changes in the DNA master plan, not by starting again from scratch. In other words, the mechanism of evolution makes the whole process irreversible. Thus, the analogy between the theory of the evolution of life and the laws of statistical physics and entropy, both of which apply to large populations (of organisms and molecules, respectively), extends to the concept of the irreversibility of time.

One manifestation of the high-order and low-entropy characteristic of living systems is evident in the intricate and hierarchical but yet very precise structures of proteins, into which their polypeptide chains comprising amino acid residues fold. It is the structure that determines the function of any protein, whether in catalysis, structural support, locomotion, transport, or communication. The functional or native state of a protein has very low entropy because its conformation is highly restricted. On the other hand, the unfolded form can exist in many different conformations: even if each amino acid can adopt only three positions, a polypeptide chain of 100 amino acid residues can adopt $3^{100}$ or $10^{47}$ different conformations.

Although life is driven by spontaneous reactions, it is only by the intervention of large, very elaborately structured molecules in cells, usually proteins called enzymes, which the rates of these reactions are brought into biologically relevant time scales. Entropy plays a big role in enzyme catalysis. Reactions in solution are usually slow because of the entropic cost in bringing the reactants or reactant and catalyst together.

Molecular motors carry out the basic functions that keep cells living: these protein molecules build or destroy other proteins, transport materials, pump ions, or propel the cell. They do their jobs by converting chemical energy into mechanical energy with near 100% efficiency. The directed motion of molecular motors, overcoming entropy, is crucial for processing information in biology. It appears that biological nanomotors might be able to perform their tasks because they use the energy from biased Brownian motion to convert the molecular disorder around them into order; see Udgaonkar (2001).

For a process taking place at constant pressure, the enthalpy change is equal to the heat absorbed or evolved. If the enthalpy change is positive, heat is absorbed and the reaction is endothermic. If the enthalpy change is negative, heat is evolved and the reaction is termed exothermic. For this reason, because many processes take place at a constant pressure, enthalpy is often equated with heat content. For example, when methane is burnt in air, the equation is as follows:

$$CH_4 + 2O_2 \rightarrow CO_2 + 2H_2O, \Delta H = -890\ \mathrm{kJmol^{-1}}$$

Thus, when 16 g of methane burns at a constant pressure, 890 kJ of energy is released. Usually, we are not interested in absolute enthalpy values, and in many situations, the enthalpy is expressed as a value above an arbitrarily chosen zero value. Often, in refrigeration, the enthalpy of a saturated liquid at −40°C is given a value of zero. With thermodynamic fluids it is common to talk about the specific enthalpy, the enthalpy associated with unit mass of the fluid (in kJ kg−1). Processes which take place at constant enthalpy ($\Delta H = 0$) are known as isenthalpic.

Animate matter represents a highly organized, self-assembled amount of substance clearly separated by a semipermeable membrane from its environment (inanimate matter) (Alberts et al. 2014). In biological terms, an organism is characterized by cellular structure that corresponds to a thermodynamic system.

Popovic (2019) discussed the thermodynamic properties of microorganisms by determination and analysis of enthalpy; entropy; and Gibbs free energy of biomass, cells, and colonies of 32 microorganism species. The author reviewed the methods for estimation of thermodynamic properties of biomass, including standard enthalpy of combustion $H_C$, enthalpy of formation from chemical elements $H_f$, entropy S, and Gibbs free energy of formation from chemical elements $G_f$. These parameters were calculated on molar and mass specific basis for 32 microorganism species, including 14 bacteria, 7 yeast, and 11 algae species (see Figure 3.3). It was found that $H_f$, S, and $G_f$ are, respectively, similar for all the analyzed species, due to the fact that all living organisms have a common ancestor and thus a similar chemical composition.

The standard molar enthalpy of combustion $H_C$ of organic matter is proportional to the number of electrons that it transfers to oxygen during combustion.

Growth of microorganisms occurs in five phases during time and culture ageing: (1) lag, (2) exponential, (3) declining growth rate, (4) stationary, and (5) death phase. The duration of the phases depends on availability of nutrients in the environment. The exponential phase is the period when growth is the most intense. The number of microorganisms is described as a function of time by the equation

$$N_{cells} = N_{cells,0} \times 2^{t/t_d}$$

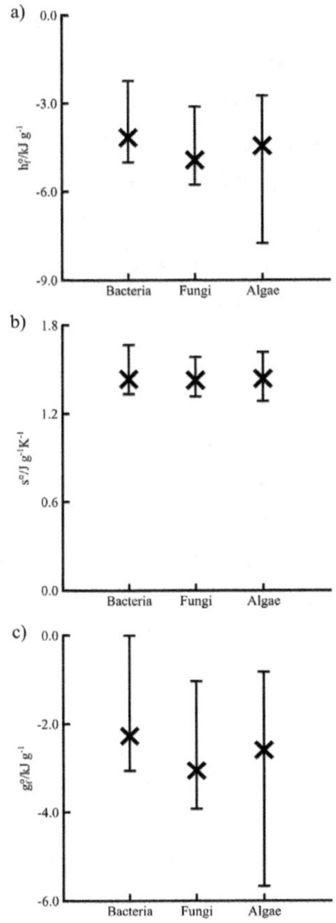


FIGURE 3.3 A comparison between average (a) enthalpy of formation from elements, (b) entropy, and (c) Gibbs free energy of formation of bacteria, fungi, and algae. The range of values (maximal, average, and minimal values) for each group are shown as reported by Popovic (2019).

where $N_{cells,0}$ is the initial cell number and $t_d$ is division time. A cell or a colony, as growing open systems, are characterized by thermodynamic parameters, including entropy, which can be calculated knowing that a cell consists of dry biomass and water.

Each thermodynamic parameter has very similar values for all the analyzed microorganisms, according to Popovic (2019). This is because all living organisms share a universal common ancestor and thus have similar chemical compositions. Standard enthalpies of formation from elements of all analyzed microorganisms are negative, because during hypothetical formation of biomass from elements, oxygen and nitrogen partially oxidize other elements in the biomass. All microorganisms have positive standard specific entropy, due to the third law of thermodynamics. Gibbs energies of formation from elements of all the microorganisms are negative.

Popovic and Minceva (2021) applied this approach for the evaluation of standard thermodynamic properties, biosynthesis rates, and the driving force of growth of some agricultural plants (see Figure 3.4). In their work, the elemental compositions of the analyzed plants were used to calculate empirical formulas, thermodynamic properties of plant tissues, and their growth, as well as to formulate growth reactions. The calculated thermodynamic properties were compared with those of viruses, bacteria, fungi, archaea, and human tissues, as shown in Figure 3.4.

Though all theories for the origin of life require a source of energy to promote primordial chemical

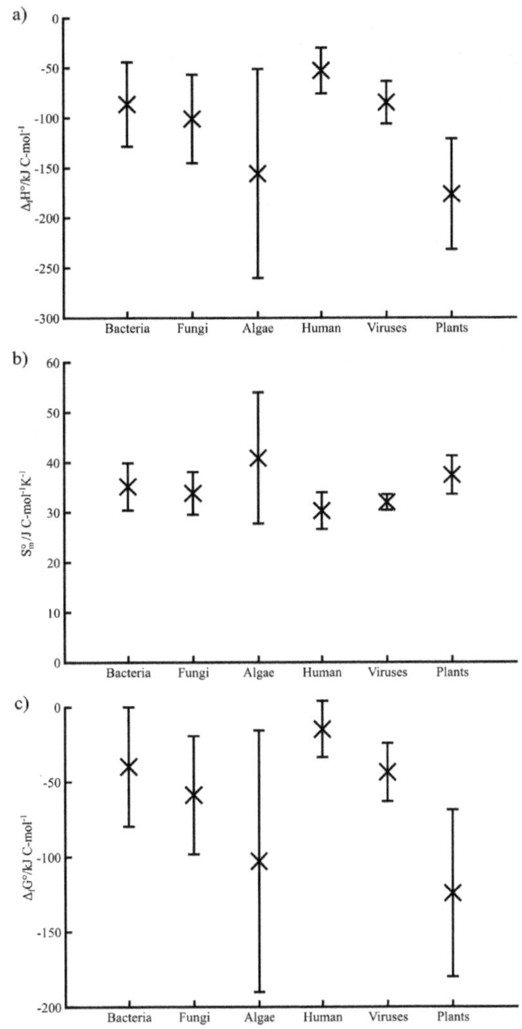


FIGURE 3.4 Comparison of thermodynamic properties of formation of live matter: (a) standard enthalpy of formation, $\Delta fH^0$; (b) standard molar entropy, $Sm^0$; (c) standard Gibbs energy of formation $\Delta fG^0$. The error bars show the spread in thermodynamic properties within each group, following Popovic and Minceva (2021).

reactions, the nature of energy that drove the emergence of metabolism at origins is still debated. Wimmer et al. (2021) reasoned that evidence for the nature of energy at origins should be preserved in the biochemical reactions of life itself, whereby changes in free energy, G, which determine whether a reaction can go forward, should help specify the source. By calculating values of ΔG across the conserved and universal core of 402 individual reactions that synthesize amino acids, nucleotides, and cofactors from $H_2$, $CO_2$, $NH_3$, $H_2S$, and phosphate in modern cells, they found that 95–97% of these reactions are exergonic (ΔG ≤ 0 kJ·mol$^{-1}$) at pH 7–10 and 80–100°C under non-equilibrium conditions, with $H_2$ replacing biochemical reductants. While 23% of the core's reactions involve ATP hydrolysis, 77% are ATP independent, thermodynamically driven by 1G of reactions involving carbon bonds. They identified 174 reactions that are exergonic by −20 to −300 kJ·mol$^{-1}$ at pH 9 and 80°C and that fall into only ten reaction types. The 402 reactions of the biosynthetic core trace to the last universal common ancestor (LUCA), and reveal that synthesis of LUCA's chemical constituents required no external energy inputs such as electric discharge, UV light, or phosphide minerals. The biosynthetic reactions of LUCA uncover a natural thermodynamic tendency of metabolism to unfold from energy released by reactions of $H_2$, $CO_2$, $NH_3$, $H_2S$, and phosphate.

The present data uncover a unique thermodynamic link between core biochemistry as a whole and the conditions of a geochemical environment known to have existed on the early Earth. The reactions of the core require neither membrane proteins, cytochromes, quinones, nor light. Their thermodynamics indicate that the core biosynthetic reactions of microbial metabolism could have arisen from soluble and surface-catalyzed reactions in the dark, under hot, aqueous, $H_2$-bearing geochemical environments, independent of exposed land masses (light) or the existence of water with a low ionic strength. Though ATP provides energy for roughly one fourth of the core's reactions, a three fourth's majority of reactions derive their energy release from reactions of carbon compounds germane to metabolism itself, sources of chemical energy that, with the exception of thioesters and acyl phosphates, have escaped the focus of previous investigations into early metabolic evolution. While estimates of ΔG are, of course, silent on reaction rates, activation energy, and catalysts, the crucial energetic role of hydrogen and the exergonic biochemical reactions of carbon reported here uncover a natural thermodynamic tendency for the individual reactions of metabolism to arise from $H_2$, $CO_2$, $NH_3$, and $H_2S$ in the presence of phosphate, following Wimmer et al. (2021).

Macroscopically, the entropy (S) for a thermodynamically reversible process is defined by the following equation as the heat (Q) into the system divided by the uniform temperature (T).

$$S = Q/T$$

In the modern microscopic interpretation of entropy in statistical mechanics, entropy is a logarithmic measure of the number of states ($W_i$) with significant probability of being occupied:

$$S = k_B \ln W_i$$

where $k_B$ is the Boltzmann constant, $k_B = 1.38 \times 10^{-23}\,J\,K^{-1}$.

Nevertheless, it is necessary to understand that the proper definition of the second law of thermodynamics says that the entropy of an adiabatically isolated system never decreases. In this context a living cell or organism is not an isolated system, since it gets the nutrients from the exterior; that is, there is an exchange of heat or matter with the environment, and by doing so, we have to consider the system an open one, together with its environment, restoring balance to the universe in an increase of entropy, or, in other words, an increase of disorder; see Venegas-Gomez (2017).

The main question now is to understand the main differences between a cell and the standard Carnot engine. Following Himeoka and Kaneko (2017), there are three main differences: (1) cells contain catalysts; (2) the catalysts are synthetized within the cell as a result of catalytic reactions; and (3) cell volume growth results from membrane synthesis from nutrient components, facilitated by the catalyst. One can try to analyze these three points by assuming the analogy of a cell to the Carnot cycle with the substitution of particle number for heat and chemical potential for temperature in the Carnot diagram.

Several ideas have been proposed trying to understand the production entropy inside a cell, even arguing going against the tendency of the second law, which is invalid for a living system, because the principle that entropy can only increase or remain constant applies only to a closed system which is adiabatically isolated, meaning no heat can enter or leave. Whenever a system can exchange either heat or matter with its environment, an entropy decrease of that system is entirely compatible with the second law.

In his landmark book. Schrödinger asked, “What is life?”, and stated that life feeds on negative entropy. In von Stockar and Liu (1999), the validity of this statement is discussed through a careful thermodynamic analysis of microbial growth processes. In principle, both feeding on negative entropy, that is, yielding products of higher entropy than the substrates, and generating heat can be used by microorganisms to rid themselves of internal entropy production resulting from maintenance and growth processes. The authors reviewed the literature data in order to compare these two mechanisms. It was shown that entropy-neutral, entropy-driven, and entropy-retarded growth exist. The analysis of some particularly interesting microorganisms shows that enthalpy-retarded microbial growth may also exist, which would signify a net uptake of heat during growth. However, the existence of endothermic life has never been demonstrated in a calorimeter. The internal entropy production in live cells also reflects itself in Gibbs energy dissipation accompanying growth, which is related quantitatively to the biomass yield. An empirical correlation of Gibbs energy dissipation in terms of the physico-chemical nature of the growth substrate has been proposed in the literature and can be used to predict the biomass yield approximately. The ratio of enthalpy change and Gibbs energy change can also be predicted since it is shown to be approximately equal to the same ratio of the relevant catabolic process alone.

Microbial metabolism consists of irreversible processes and therefore generates entropy continuously within the cell. The living cell maintains itself, however, in a state of constant entropy by exporting the internally generated entropy to the environment. This may occur by generating metabolic products of higher entropies than the nutrients, a process that was described as “life feeds on negative entropy” by Schrödinger. With this statement, however, Schrödinger disregarded heat generation as a mechanism for entropy export and “was misleading by interpreting the feeding too literally by emphasizing the ordered structures of nutrients as the source of negative entropy”.

The export of the internally generated entropy in these forms of both “negative entropy” and heat reflects itself in a decrease, or dissipation, of Gibbs energy in the environment. The amount of Gibbs energy dissipated per C-mol of new biomass grown is the key parameter for understanding the thermodynamics of microbial growth. On the one hand, it is linked to the rate of metabolism and therefore regarded as the driving force for growth.

Life is considered something different from non-living things, but no single driving force can account for all the different aspects of life, which consist of different levels of hierarchy, such as metabolism, cell physiology, multi-cellular development and organization, population dynamics, ecosystem, and evolution. Although free energy is evidently the driving force in biochemical reactions, there is no established relationship between metabolic energy and spatiotemporal organization of living organisms or between metabolic energy and genetic information. Since Schrödinger pointed out the importance of exporting entropy in maintaining life, misunderstandings of entropy have been obstacles in constructing a unified view on the driving forces of life. Sato (2012) presented a simplified conceptual framework for unifying driving forces of life at various different levels of hierarchy. The key concept is “entropy deficit”, or simply “inhomogeneity”, which is defined as the difference of maximal possible entropy and actual entropy. This is equivalent to information content in genetic information and protein structure and is also defined similarly for non-homogeneous structures in ecosystems and evolution. Entropy deficit or inhomogeneity is a unified measure of all driving forces of life, which could be considered a scientific equivalent to the “élan vital” of Bergson.

It may be generally accepted that all life on the Earth, except some chemo-lithotrophs in deep ocean, depends on photosynthesis, but it is not easy to envisage the importance and necessity of photosynthesis for the maintenance and development of all living beings. Schrödinger’s notion of negative entropy is famous, but in his time, the theory of entropy in photosynthesis was not fully understood, and this was the reason various confusions were invoked later. Some people attacked him because entropy is always positive. The correct meaning of “negative entropy” should be “decrease in entropy” in a dynamic system. Another confusion was that light had negative value of entropy, which was not correct at all. There are several books (at least in Japanese) written by physicists stating that photosynthesis is driven by entropy dissipation through transpiration. An active enzyme is just a fraction of all possible polypeptides of identical amino acid composition and therefore has a lower entropy than the random mixture of polypeptides. This increases the activity (in the sense of thermodynamics, meaning effective concentration) of the enzyme. This is the source of stabilization energy of native enzyme and enzyme/substrate complexes and consequently lowered activation energy.

Most biologists may agree that metabolism can be viewed as processes of non-equilibrium thermodynamics, even though it is extremely complex and organized within a cell. One can understand metabolism in the words of generalized entropy deficit or free energy as described in the present chapter. Self-organization or development can be the most prominent characteristics of both unicellular rand multi-cellular organisms, but all the processes of intracellular and individual organization can be explained by networks of signals that, as a whole, appear as emergent systems, which can be understood as self-organizing dissipative structures and therefore driven by entropy deficit.

All irreversible processes arise and persist to produce entropy. Entropy production is not incidental to such processes but rather the very reason for their origin and persistence. Michaelian (2011) took such a thermodynamic perspective on the origin of life, recognizing that entropy production is not only the vital force of life but the fundamental link between life in the biosphere today and its origin in the Archean. Today the greatest entropy production in the biosphere is due to visible photon absorption and dissipation into heat by organic material in liquid water and the subsequent degradation of the established heat gradient through the water cycle. Following this link back in time to the Archean environment leads to a suggestion for a mechanism for the origin of life based on UV photon absorption and dissipation by RNA and DNA.

Martin et al. (2001) considered the early evolution of biological energy conservation. Life is a persistent, self-specified set of chemical reactions far from equilibrium. In modern microbes, core carbon and energy metabolism are what keep cells alive. In very early chemical evolution, the forerunners of carbon and energy metabolism were the processes of generating reduced carbon compounds from $CO_2$ and the mechanisms of harnessing energy as compounds capable of doing some chemical work. The process of serpentinization at

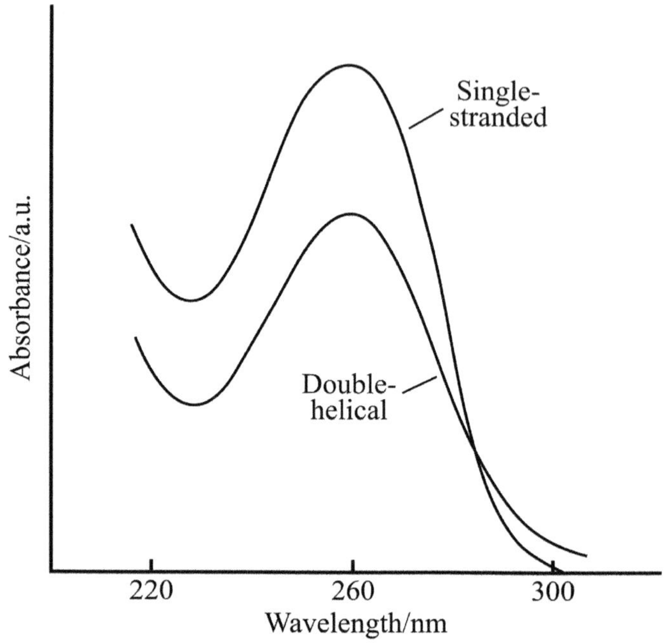


FIGURE 3.5 Absorption maximum in the ultraviolet at 260 nm of single-strand and double-helix DNA. The difference, marked by the double arrow, is known as hypochroism and is due to the random orientation of the bases in single-strand DNA (or RNA).

alkaline hydrothermal vents holds promise as a model for the origin of early reducing power, because $Fe^{2+}$ in the Earth's crust reduces water to $H_2$ and inorganic carbon to methane. The overall geochemical process of serpentinization is similar to the biochemical process of methanogenesis, and methanogenesis is similar to acetogenesis in that both physiologies allow energy conservation from the reduction of $CO_2$ with electrons from $H_2$. Electron bifurcation is a newly recognized cytosolic process that anaerobes use to generate low-potential electrons; it plays an important role in some forms of methanogenesis and, via speculation, possibly in acetogenesis. Electron bifurcation likely figures into the early evolution of biological energy conservation.

Some modern cells overcome the energetically uphill reaction of generating low potential FeS clusters (ferredoxin) with electrons from $H_2$ via electron bifurcation; it seem possible that early biochemical systems did as well. While modern electron bifurcation entails flavin adenine dinucleotide (FAD) (or quinones), there is room for (testable) speculation that simpler inorganic catalysts (transition metals) might have been able to do the job. From thermodynamics, the synthesis of amino acids is the main energy-releasing reaction when it comes to the synthesis of cell mass under reducing hydrothermal conditions; their synthesis compensates for the endergonic synthesis of bases. In metabolism, the bases come from amino acids and carbamates, and in early evolution a similar sequence can be envisaged. However, in the beginning there is just no way that base synthesis could have been highly specific. The early real RNA world (assuming it existed) must have consisted of messy bases synthesized spontaneously without the help of genes. The chemical environment at the heart of the present considerations is full of reactive C1 and methyl groups (on their way to becoming methane, acetyl groups, and acetate), and it is noteworthy that ribosome–tRNA interactions at the heart of the genetic code will not work properly without the myriad base modifications in tRNA and rRNA, the vast majority of which are methylations. That a chemical imprint of the environment in which genes and proteins arose is preserved in the chemical structure of modern tRNA and rRNA bases is something to consider (Martin et al. 2001).

Recently, Michaelian (2022) discussed non-equilibrium thermodynamic foundations of the origin of life. There is little doubt that life's origin followed from the known physical and chemical laws of nature. The most general scientific framework incorporating the laws of nature and applicable to most known processes to good approximation is that of thermodynamics and its extensions to treat out-of-equilibrium phenomena. The event of the origin of life should therefore also be amenable to such an analysis.

If the thermodynamic dissipation theory for the origin of life proves to be the correct theory, then, given particular, but probably common, environmental conditions on planets of hot K-type, G-type, and hotter stars (having an important UVC component), there will be, in fact, a thermodynamic imperative for an origin of life similar to our own based on carbon. In this case, the origin of life would not have been a single fortuitous chemical event constrained to perhaps a few "lucky" planets but rather a thermodynamic imperative and would appear on many, if not most, planets of these types of stars. In this case we would certainly not be alone in the universe, although, because of the specific physical requirements for carbon-based molecular dissipative structuring, developed complex lifeforms based on carbon, such as we know them on Earth, may be relatively scarce. However, the first steps of life, corresponding to the "self-organization" of organic pigments, should be occurring on the surface of every planet circling any star which has at least some continuous light in the 205 to 285 nm UVC region. Michaelian (2022) proposed that the fundamental molecules of life, including nucleobases and their stacking into DNA and RNA polymers, were, at life's origin, "self-organized" dissipative structures arising under the UVC photons that reached Earth's surface at the beginning of the Archean.

The theory presented previously views the origin of life as a thermodynamic imperative, the "self-organization" of a new irreversible process of organic pigment formation for the dissipation of photons and the coupling of the resulting heat of dissipation to other abiotic irreversible processes, in particular the water cycle. This irreversible process, which we call "life", appeared once the environmental conditions became appropriate about 3.85 billion years ago (e.g. surface UVC light and high temperatures) under the thermodynamic imperative of increasing the Earth's entropy production in its interaction with its solar environment. Such conditions have never again occurred on Earth's surface, and this fact could explain the apparent emergence of life at only a single instant of time in Earth's history. Other theories would have to allow for the possibility that life could have arisen at numerous times throughout Earth's history, evidence for which is lacking (Michaelian 2022).

The perspective taken by the author is that the origin of life was not a scenario of organic material organization driven by natural selection leading to "better adapted" organisms or to greater chemically stability (e.g. UV-resistant organisms) but rather a scenario of the dissipative structuring of material under the imposed UVC solar photon potential leading to an organization of material in space and time (biosynthetic pathways) providing more efficient routes to the dissipation of the externally imposed photon potential. The dissipative synthesis of an ever-larger array of photochemical catalysts and cofactors would mean that ever more complex biosynthetic pathways would emerge through thermodynamic selection on dissipation to promote the synthesis of novel pigments for dissipating not only the fundamental UVC and other UV regions, but eventually the entire short wavelength region of the solar photon spectrum, eventually reaching the red-edge (~700 nm), which is the approximate limit of biological photon dissipation on Earth today.

Liquid-liquid phase separation in living cells can lead to the formation of biomolecular condensates that aid intracellular organization (Choi et al. 2020 and references therein). These condensates have different functions such as compartmentalization of reactions, buffering of molecules, and midcell localization during cell division. However, in a thermal equilibrium system, the liquids will completely segregate through a coarsening process. To arrest this process, the system must be brought out of equilibrium by supplying energy, such as via fuel-driven chemical reactions. This has been shown to lead to "active droplet" systems that exhibit a wealth of novel phenomena not encountered in thermal equilibrium (Weber et al. 2019).

Enzyme-enriched condensates can organize the spatial distribution of their substrates by catalyzing non-equilibrium reactions. Conversely, an inhomogeneous substrate distribution induces enzyme fluxes through substrate-enzyme interactions. Demarchi et al. (2023) found that condensates move toward the center of a confining domain when this feedback is weak. Above a feedback threshold, they exhibit self-propulsion, leading to oscillatory dynamics. Moreover, catalysis-driven enzyme fluxes can lead to interrupted coarsening, resulting in equidistant condensate positioning, and to condensate division.

### 3.2.2 Cells: The Fundamental Units of Life

A living cell is a self-reproducing system of molecules separated and protected from outside environment by the membrane. The membrane is made of a two-ply sheet of lipid molecules, about 5 nm (approx. 50 atoms) thick, into which proteins have been inserted. Cells must also have the ability to let nutrients inside, and waste products must be let out.

Eukaryotic cells, in general, are bigger and more elaborate than bacteria and archaea. Some live independent lives as single-celled organisms, such as amoebae and yeasts; others live in multicellular assemblies. All of the more complex multicellular organisms—including plants, animals, and fungi—are formed from eukaryotic cells. By definition, all eukaryotic cells have a nucleus.

The nucleus is usually the most prominent organelle in a eukaryotic cell. It is enclosed within two concentric membranes that form the nuclear envelope, and it contains molecules of DNA—extremely long polymers that encode the genetic information of the organism. In the light microscope, these giant DNA molecules become visible as individual chromosomes when they become more compact before a cell divides into two daughter cells. DNA also carries the genetic information in prokaryotic cells; these cells lack a distinct nucleus not because they lack DNA, but because they do not keep their DNA inside a nuclear envelope, segregated from the rest of the cell contents.

Mitochondria are generators of chemical energy for the cell. They harness the energy from the oxidation of food molecules, such as sugars, to produce adenosine triphosphate, or ATP—the basic chemical fuel that powers most of the cell's activities. Because the mitochondrion consumes oxygen and releases carbon dioxide in the course of this activity, the entire process is called cellular respiration—essentially, breathing on a cellular level. Without mitochondria, animals, fungi, and plants would be unable to use oxygen to extract the energy they need from the food molecules that nourish them.

Mitochondria contain their own DNA and reproduce by dividing in two. Because they resemble bacteria in so many ways, they are thought to have been derived from bacteria that were engulfed by some ancestor of present-day eukaryotic cells. This evidently created a symbiotic relationship in which the host eukaryote and the engulfed bacterium helped one another to survive and reproduce. Like mitochondria, chloroplasts contain their own DNA, reproduce by dividing in two, and are thought to have evolved from bacteria—in this case, from photosynthetic bacteria that were engulfed by an early eukaryotic cell.

The cell interior is in constant motion. The cytoskeleton is a dynamic jungle of protein ropes that are continually being strung together and taken apart; its filaments can assemble and then disappear in a matter of minutes. Motor proteins use the energy stored in molecules of ATP to trundle along these tracks and cables, carrying organelles and proteins throughout the cytoplasm, and racing across the width of the cell in seconds. In addition, the large and small molecules that fill every free space in the cell are swept back and forth by random thermal motion, constantly colliding with one another and with other structures in the cell's crowded cytoplasm.

Eukaryotic cells are typically 10 times the length and 1000 times the volume of prokaryotic cells, although there is huge size variation within each category. They also possess a whole collection of features, a cytoskeleton, mitochondria, and other organelles, that set them apart from bacteria and archaea; see Figure 3.6.

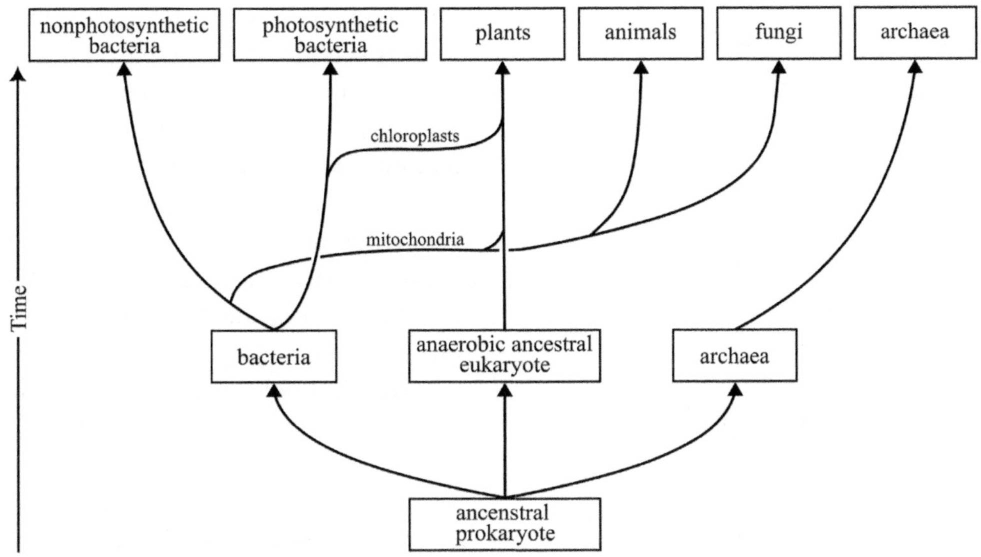


FIGURE 3.6 Where did eukaryotes come from? The eukaryotic, bacterial, and archaean lineages diverged from one another very early in the evolution of life on Earth. Some time later, eukaryotes are thought to have acquired mitochondria; later still, a subset of eukaryotes acquired chloroplasts. Mitochondria are essentially the same in plants, animals, and fungi, and therefore were presumably acquired before these lines diverged (Alberts et al. 2014).

To be able to quantify the entropy production in a living cell, we need to consider simple models, which allow us to understand these reactions in the simplest possible way. In a cell, a huge number of chemical is synthetized by mutual catalyzation, leading to replication of molecules allowing the cell to grow until the cell is large enough to divide into two. A protocell is a model of a cell created in order to investigate how chemical compositions are synthetized by mutual catalyzation, leading to replication of molecules that allow the growth of the cell until its division (Mavelli et al. 2014).

Motility is a fundamental feature of living matter, encompassing single cells and collective behavior. Such living systems are characterized by non-conservativity of energy and a large diversity of spatiotemporal patterns. Thus, fundamental physical principles to formulate their behavior are not yet fully understood. Ishimoto et al. (2023) explored a violation of Newton's third law in motile active agents by considering non-reciprocal mechanical interactions known as odd elasticity. By extending the description of odd elasticity to a nonlinear regime, they presented a general framework for the swimming dynamics of active elastic materials in low-Reynolds-number fluids, such as wavelike patterns observed in eukaryotic cilia and flagella. They also investigated the nonlocal interactions within a swimmer using generalized material elasticity and applied these concepts to biological flagellar motion. Through simple solvable models and the analysis of Chlamydomonas flagella waveforms and experimental data for human sperm, they demonstrated the wide applicability of a nonlocal and non-reciprocal description of internal interactions within living materials in viscous fluids, offering a unified framework for active and living matter physics.

The outer membrane of a gram-negative bacteria encapsulates the plasma membrane thereby protecting it from the harsh external environment. This membrane acts as a sieving barrier due to the presence of special membrane-spanning proteins called "pores". These porins are β-barrel channel proteins that allow the passive transport of hydrophilic molecules and are impermeable to large and charged molecules. Many porins form trimers in the outer membrane. They are abundantly present on the bacterial surface and therefore play various significant roles in the host–bacteria interactions. These include the roles of porins in the adhesion and virulence mechanisms necessary for the pathogenesis, along with providing resistance to the bacteria against the antimicrobial substances. They also act as the receptors for phage and complement proteins and are involved in modulating the host cellular responses.

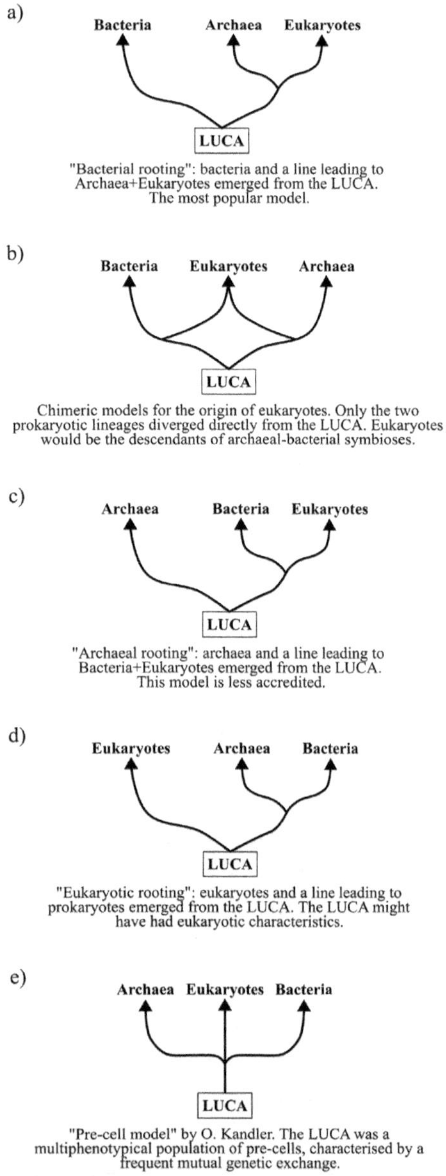


FIGURE 3.7 Current hypotheses for the evolution of three domains of life from LUCA, indicated by blue sphere. Models A to D show that all the properties attributed to the LUCA co-existed in the same cell, whether as in E they were present collectively in a population of primitive cell, after Moreira and Lopez-Garcia (2007).

In addition, the potential use of porins as adjuvants, vaccine candidates, therapeutic targets, and biomarkers is now being exploited. In this review, we focus briefly on the structure of the porins, along with their important functions and roles in the host-bacteria interactions; see Sharma et al. (2022). Porins are also present in the cell wall of gram-positive bacteria and perform a similar function but are less abundant compared to gram-negative bacteria. The pore size of the gram-positive porins is also comparably larger.

In eukaryotes, the mitochondrial outer membrane also possesses porins that help in the transport of metabolites across the mitochondrial outer membrane, regulation of mitochondrial metabolism, and apoptosis. Open and closed conformations of some mitochondrial porins are regulated by the voltage across the outer membrane and they show anion selectivity. Thus, they are also known as voltage-gated anion-selective channels. Porins are also found in chloroplasts. They help in the transport of many metabolites, as the chloroplast is a center for metabolic reactions like photosynthetic $CO_2$ fixation, amino acid synthesis, and ammonia assimilation. The ubiquitous distribution of porins in almost all the major kingdoms of life and acquisition of more diverse functions shows that they are one of the important biomolecules, after Sharma et al. (2022).

### 3.2.3 Life, Time, and Memory

All living things show behavior—they respond to stimuli from the environment. The environment is dynamic, and the behavior is adaptive. Because of this, the behavior has a temporal component—the organism anticipates its future in relation to the present. The success of this behavioral anticipation is essential to survival. In this sense, all living organisms are aware of time. This is the origin of memory which could be carried on by many means, including the building block arrangements and similar.

Eddington (1928) coined the phrase "arrow of time" to describe an implicit directionality in time. This was based on Boltzmann's recognition that, for an isolated system, the spontaneous direction of any process was towards increased entropy or disorder. Although several other arrows of time have been proposed, this thermodynamic arrow seems the least dependent on prior assumptions. However, in the context of our universe, it does depend for directionality on recognition of an initially constrained, dense, high-temperature, high-entropy state, whose evolution was triggered by the Big Bang. With respect to the subsequent expansion into free space, and the evolving universe as a whole, this starting condition is considered to have been a low entropy state. from Crofts (2007).

Memory is usually associated with higher organisms rather than bacteria. However, evidence is mounting that many regulatory networks within bacteria are capable of complex dynamics and multi-stable behaviors that have been linked to memory in other systems; see Wolf et al. (2008). It is well established that bacteria that have experienced different environmental histories may respond differently to current conditions. These memory effects may be more than incidental to the regulatory mechanisms controlling acclimation or to the status of the metabolic stores. They may be regulated by the cell and confer fitness to the organism in the evolutionary game it participates in.

Wolf et al. (2008) proposed that history-dependent behavior is a potentially important manifestation of memory, worth classifying and quantifying. Guided by this idea, they developed an information-theory based conceptual framework for measuring both the persistence of memory in microbes and the amount of information about the past encoded in history-dependent dynamics. This method produces a phenomenological measure of cellular memory without regard to the specific cellular mechanisms encoding it. They applied this framework to a strain of *Bacillus subtilis* engineered to report on commitment to sporulation and degradative enzyme (AprE) synthesis and estimate the capacity of these systems and growth dynamics to remember ten distinct cell histories prior to application of a common stressor. The analysis performed suggested that *B. subtilis* remembers, both in short and long term, aspects of its cell history and that this memory is distributed differently among the observables. While this study does not examine the mechanistic bases for memory, it presents a framework for quantifying memory in cellular behaviors and is thus a starting point for studying new questions about cellular regulation and evolutionary strategy.

It has long been known that bacterial cells that have experienced different environmental histories may respond differently to current conditions. History-dependent behavior, whether short or long term, controlled or incidental, reflects a form of memory. In experiments performed by Wolf et al. (2008), they quantified the ability of three *B. subtilis* stress response systems—sporulation, degradative enzyme synthesis, and growth—to remember ten distinct cell histories prior to application of a common stressor. They chose to observe commitment to sporulation because the sporulation decision is bistable, and bistability is associated with memory. They observed growth because it is the most accessible measure of cellular health and fitness and is an integrator of many other aspects of cell function; thus it may show interesting differences depending on cell history. If cells could use a memory of past conditions to predict future conditions and delay sporulation, an expensive process, if the environment is likely to improve or accelerate sporulation if the starvation period is likely to be long, they might improve their odds for long-term survival, as illustrated in Figure 3.8. Among the many ingenious genetic and biochemical programs employed by *B. subtilis* to cope with environmentally adverse conditions are its ability to take up extracellular DNA, competence; differentiate into an inert heat, chemical, and UV-resistant spore; secrete

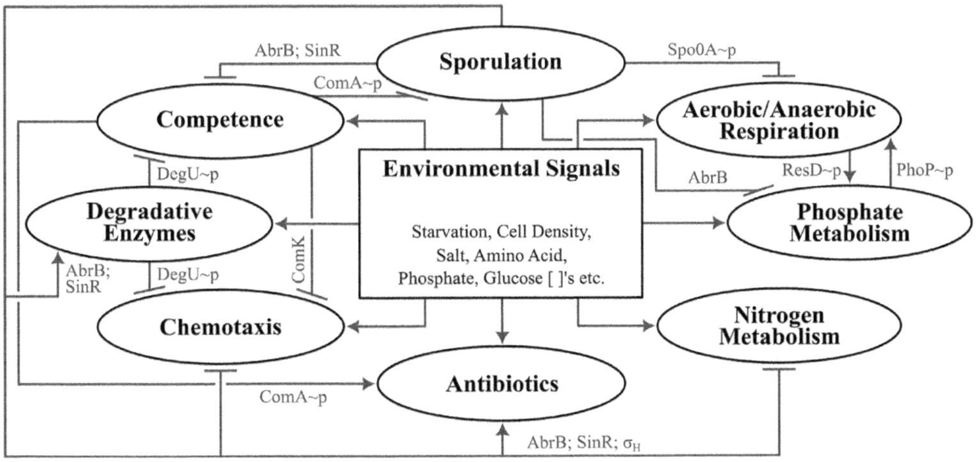


FIGURE 3.8 The *B. subtilis* stress response meta-network, where each oval represents both a stress response and the regulatory network of 100 or so interacting molecular species that regulates it; see Wolf et al. (2008).

degradative enzymes to identify and digest new food sources; become motile and chemotax toward possibly better surroundings; synthesize antibiotics to eliminate competitors in the same ecosystem; turn on alternative metabolic pathways; and form biofilms, just to name a few. The cross-repressive feedback between sporulation and competence, and the many positive feedback loops within each large "individual" stress response pathway, are suggestive of switches and other elements that could potentially encode memory.

Ulrich and Zhulin (2010) published the MiST2 database, which identifies and catalogs the repertoire of signal transduction proteins in microbial genomes. Signal transduction systems regulate the majority of cellular activities, including the metabolism, development, host-recognition, biofilm production, virulence, and antibiotic resistance of human pathogens. MiST2 currently contains 966 complete and 157 draft bacterial and archaeal genomes, which collectively contain more than 245,000 signal transduction proteins. The majority (66%) of these are one-component systems, followed by two-component proteins (26%), chemotaxis (6%), and finally ECF factors (2%).

Bacteria are capable of sensing and responding to an astonishing variety of environmental signals: amino acids, sugars, oxygen, pH, osmolality, temperature, light, secondary metabolites, molecular waste products (e.g., ammonia), environmental DNA, and other microbes (both conspecifics and other species)—even physical objects such as tiny, chemically inert beads. Signal transduction systems in bacteria come in several basic forms that perform a wide range of behavioral and physiological roles (Ulrich and Zhulin 2010): one-component systems (1CSs); 2CSs, which also come in hybrid forms; CT systems, a special sub-class of 2CSs; extra-cytoplasmic function sigma factor proteins, the so-called "third pillar of bacterial ST" Staroń et al. (2009); and non-canonical systems that involve three and more components; see Marijuán et al. (2010).

Memory, or history-dependent behavior, can manifest across multiple time scales. Short-term, or transient, memory is stored by the system for some time and then "forgotten". Systems may also have either "effective" long-term memory if the transient dynamics are long compared to environmental fluctuations or "true" asymptotic memory if the stationary state of the system depends on initial conditions, as occurs in nonlinear systems with multiple attractors. For an example of the latter, the state of a bi-stable switch encodes an asymptotic memory of the last switching event.

Lyon (2015) reviewed the evidence in bacteria for capacities encompassed by the concept of cognition and presented the historical look into the idea that microbes might have some knowledge of their environment. According to the author, cognition is a biological function unlike any other. We agree with the author's statements about memory. "Memory, needless to say, is critical to cognition. Without memory, present circumstances have no context; the detection of change is impossible. Without the ability to detect change, the decision to alter behavior can only be random, haphazard. Without memory, learning of any kind is impossible". A similar signal transduction (ST) mechanism appear to be at work in all species, albeit with different leading biochemical actors and different degrees of regulatory complexity. For example, ST in bacterial two-component systems relies principally on histidine kinases for protein phospho-transfer and DNA methylation, while genomic analysis shows that serine/threonine kinases, their analog in eukaryotes, make up roughly 1 in 4 of the more than 500 protein kinases involved in ST in humans (see Table 3.3).

Biological cognition is defined as "the mechanisms by which animals acquire, process, store, and act on information from the environment. These include perception, learning, memory, and decision making". Cognition thus is composed of the total suite of mechanisms that underwrite information acquisition, storage, processing, and use. As do animals and plants, bacteria monitor the biotic and abiotic features of their surrounding milieu using a wide variety of sensory systems and adapt their behavior and physiology in response to what they perceive in order to stay alive and reproduce. While the ecological niche of some bacteria is fairly stable, if sometimes extreme, many inhabit highly mutable environments and conduct highly complex lifestyles within them, which presents considerable challenges, some of which are capable of being anticipated and others not. Like all organisms, bacteria do not gather information in a vacuum but are able to assess the significance of the signals they receive relative to their own functioning, their internal milieu.

> Memory, typically the capacity to retain information acquired during the organism's lifetime for a period sufficient to affect behavior, is part

TABLE 3.3 Genome Size and Signal Transduction, by Millions of Base Pairs (Mbp) and Type of System

| Organism | Genome ($10^6$ bp) | 1CS | 2CS | CT | ECF | OTHER | TOTAL Types |
|---|---|---|---|---|---|---|---|
| *Vibrio parahaemolyticus* | 39.3 | 2,624 | 497 | 149 | 20 | 31 | 3,321 |
| *Burkholderia xenovorans* | 9.7 | 719 | 152 | 46 | 12 | 8 | 937 |
| *Myxococcus xanthus* | 9.1 | 315 | 263 | 60 | 41 | 8 | 687 |
| *Escherichia coli* 0157 | 7.6 | 362 | 78 | 13 | 1 | 5 | 459 |
| *Pseudomonas aeruginosa* | 6.4 | 447 | 131 | 46 | 21 | 8 | 655 |
| *Vibrio harveyii* | 6.1 | 417 | 90 | 35 | 6 | 7 | 555 |
| *Vibrio cholerae* | 4.2 | 230 | 89 | 66 | 3 | 6 | 394 |
| *Bacillus subtilis* | 4.2 | 254 | 71 | 18 | 7 | 2 | 352 |
| *Staphlococcus aureus* | 3.1 | 119 | 34 | - | - | 2 | 156 |
| *Mycoplasma genitalium* | 0.6 | 5 | - | - | 1 | - | 6 |
| *Buchnera aphidocola* | 0.4 | 3 | - | - | - | - | 3 |

*Source:* MiST2 database: a comprehensive genomics resources on microbial signal transduction, Ulrich and Zhulin (2010).

> of the bedrock of cognition. For centuries memory, especially in its long-term modes, has been intrinsic to the anglophone concept of "mind". As stated earlier, the idea that bacteria exhibit some form of memory, or "memory effects," is now four decades old.

As mentioned earlier, Wolf et al. (2008) selected *B. subtilis* for its sensitivity to environmental conditions and the well-known mechanisms governing sporulation. These mechanisms, triggered as part of a stress response, exhibit switch-like bistability, the basis of memory effects in computers, some physical compounds (i.e., magnetized iron), and neuronal activity in animals. Bistability is believed to have a role in human memory, particularly working memory, but it has also been implicated in memory-associated processes in cellular plasticity and differentiation, such as in bone marrow.

Sporulation is an irreversible, energetically costly process of morphological and functional differentiation initiated as a last resort under starvation conditions or a "bet-hedging" strategy when nutrient supplies deteriorate. The developmental sequence unfolds over 8 h, and results in the formation of specialized cells resistant to heat, chemical, and UV stress. This involves the death of a significant proportion of the population, so the existential ramifications of movement toward sporulation are profound (for the survival of individual cells, at least), even before commitment becomes irreversible (González-Pastor et al. 2003).

Bacteria prudently regulate their metabolic phenotypes by sensing the availability of specific nutrients, expressing the required genes for their metabolism, and repressing them after specific metabolites are depleted. It is unclear, however, how genetic networks maintain and transmit phenotypic states between generations under rapidly fluctuating environments. By subjecting bacteria to fluctuating carbon sources (glucose and lactose) using microfluidics, Lambert and Kussell (2014) discovered two types of non-genetic memory in *Escherichia coli* and analyzed their benefits. First, phenotypic memory conferred by transmission of stable intracellular lac proteins dramatically reduces lag phases under cyclical fluctuations with intermediate timescales (1–10 generations). Second, response memory, a hysteretic behavior in which gene expression persists after removal of its external inducer, enhances adaptation when environments fluctuate over short timescales (<1 generation).

Using a mathematical model, the authors analyzed the benefits of memory across environmental fluctuation timescales. They showed that memory mechanisms provided an important class of survival strategies in biology that improve long-term fitness under fluctuating environments. These results could be used to understand how organisms adapt to fluctuating levels of nutrients, antibiotics, and other environmental stresses.

Cellular membrane potential plays a key role in the formation and retrieval of memories in the metazoan brain, but it remains unclear whether such memory can also be encoded in simpler organisms like bacteria. Yang et al. (2020) have shown that single-cell-level memory patterns could be imprinted in bacterial biofilms by light-induced changes in the membrane potential. They demonstrated that transient optical perturbations generate a persistent and robust potassium-channel-mediated change in the membrane potential of bacteria within the bio-film. The light-exposed cells respond in an anti-phase manner, relative to unexposed cells, to both natural and induced oscillations in extracellular ion concentrations. This anti-phase response, which persists for hours following the transient optical stimulus, enables a direct single-cell resolution visualization of spatial memory patterns within the biofilm. The ability to encode robust and persistent membrane-potential-based memory patterns could enable computations within prokaryotic communities and suggests a parallel between neurons and bacteria.

Motion is a key characteristic of every form of life. Even at the microscale, it has been reported that colonies of bacteria can generate nano-motion on mechanical cantilevers, but the origin of these nanoscale vibrations has remained unresolved. Rosłoń et al. (2022) presented a new technique using drums made of ultrathin bilayer graphene, where the nano-motion of single bacteria can be measured in its aqueous growth environment. A single *Escherichia coli* cell is found to generate random oscillations with amplitudes of up to 60 nm, exerting forces of up to 6 nN to its environment. Using mutant strains that differ by single gene deletions that affect motility, they were able to pinpoint the bacterial flagella as the main source of nano-motion. By real-time tracing of changes in nano-motion on administering antibiotics, they demonstrated that graphene drums could perform antibiotic susceptibility testing with single-cell sensitivity. These findings deepen our understanding of processes underlying cellular dynamics and pave the way towards high-throughput and parallelized rapid screening of the effectiveness of antibiotics in bacterial infections with graphene devices.

Although life on Earth is unified by its use of chemical compounds and reactions, providing a detailed model for universal biochemistry, it does not allow quantitative predictions about examples not yet observed. Gagler et al. (2022) introduced a more generalizable concept of biochemical universality that is more akin to the kind of universality found in physics. Using annotated genomic datasets including an ensemble of 11,955 metagenomes, 1,282 archaea, 11,759 bacteria, and 200 eukaryotic taxa, they showed how enzyme functions form universality classes with common scaling behavior in their relative abundances across the datasets. They verified that these scaling laws were not explained by the presence of compounds, reactions, and enzyme functions shared across known examples of life. They also demonstrated how these scaling laws can be used as a tool for inferring properties of ancient life by comparing their predictions with a consensus model for LUCA as well as illustrating how network analyses shed light on the functional principles underlying the observed scaling behaviors. Together, their results established the existence of a new kind of biochemical universality, independent of the details of life on Earth's component chemistry, with implications for guiding the search for missing biochemical diversity on Earth or for biochemistries that might deviate from the exact chemical makeup of life as we know it, such as at the origins of life, in alien environments, or in the design of synthetic life.

Recently Bhattacharyya et al. (2023) have discovered that a heritable iron memory enables decision-making in *Escherichia coli*. In humans, repeated encounters with a wide variety of stimuli can reactivate specific neurons to produce long-term memories. They report here a multigenerational memory in *Escherichia coli* swarming motility, where bacteria "remember" their swarming experience for several generations. They show unambiguously that the molecular basis of this memory is the levels of available cellular iron. The act of swarming "conditions" the cells with this memory. Given the central role of iron in cellular metabolism, an iron-based memory might offer the advantage of providing a hub connecting various stress responses such as antibiotic survival and biofilms.

The importance of memory in bacterial decision-making is relatively unexplored. Bhattacharyya et al. (2023) show that a prior experience of swarming is remembered when *Escherichia coli* encounters a new surface, improving its future swarming efficiency. They conducted >10,000 single-cell swarm assays to discover that cells store memory in the form of cellular iron levels. This "iron" memory preexists in planktonic cells, but the act of swarming reinforces it. A cell with low iron initiates swarming early and is a better swarmer, while the opposite is true for a cell with high iron. The swarming potential of a mother cell, which tracks with its iron memory, is passed down to its fourth-generation daughter cells. This memory is naturally lost by the seventh generation, but artificially manipulating iron levels allows it to persist much longer. A mathematical model with a time-delay component faithfully recreates the observed dynamic interconversions between different swarming potentials. They demonstrate that cellular iron levels also track with biofilm formation and antibiotic tolerance, suggesting that iron memory may influence other physiologies.

Let us finish this section with reference to the new book by Ball (2024), whose motto is that we cannot assume that cells are computers and genes are their code. According to him, most genes do not have a preset function that can be determined from their DNA sequence. Instead, genes' activity, for instance, or the length of protein that they encode, depends on myriad external factors, from the diet to the environment in which the organism develops. In addition, many genes can influence each trait. According to Ball (2024), it is a huge oversimplification to say that genes cause this trait or that disease. The reality is that organisms are extremely robust, and a particular function can often be performed even when key genes are removed. The author also criticizes the thinking by which a protein with a fixed shape binding to its target is similar to how a key fits into a lock. Many proteins, he points out, have disordered domains—sections whose shape is not fixed, but changes constantly. This "fuzziness and imprecision" is not sloppy design but an essential feature of protein interactions. Being disordered makes proteins "versatile communicators", able to respond rapidly to changes in the cell, binding to different partners and transmitting different signals depending on the circumstance. Almost 70% of protein domains might be disordered.

## 3.3 ELEMENT CONCENTRATION FACTORS FOR PRIMITIVE EUKARYOTIC, BACTERIAL, AND ARCHAEAL TAXA

The molar concentrations of 34 elements in marine organisms were compared with the availability of these elements in their natural environment and some physicochemical properties of the elements, that is, the charge/radius ratio or ionic potential and the "noble gas deviation" or NGD-index. The NGD-index is defined in this chapter as a value that expresses the deviation of the electron configuration of an element with that of the previous and subsequent noble gas configurations. The value for the NGD-index can be obtained by multiplying the number of electrons which must be lost or gained to obtain the configuration of, respectively, the previous and the subsequent noble gas, divided by their squared sum. The concentration factor (ratio of element concentration in biological material and in the environment) appears to be a periodic function of the element number, very similar to the NGD-index; see Figure 3.9. The relation is most obvious when the composition of biological material is compared with that of sea water and river water and becomes less clear in the comparison with the composition of crustal rocks. These results do not exclude a freshwater origin of marine organisms. The close fit between the concentration factor and the NGD-index offers, with some limitations, a possibility to assess the concentration factors of elements for which up to now no data have been available, after Spaargaren (1991).

### 3.3.1 Overview of Knowledge on Elemental Composition of Earth's LUCA

Around 4 billion years ago there lived a microbe called LUCA—the last universal common ancestor. There is evidence that it could have lived a somewhat "alien" lifestyle, hidden away deep underground in iron-sulfur rich hydrothermal vents. Anaerobic and autotrophic, it didn't breath air and made its own food from the dark, metal-rich environment around it. Its metabolism depended upon hydrogen, carbon dioxide, and nitrogen, turning them into organic compounds such as ammonia. Most remarkable of all, this little microbe was the beginning of a long lineage that encapsulates all life on Earth, after Cooper (2017).

If we trace the tree of life far enough back in time, we come to find that we're all related to LUCA. If the war cry for our exploration of Mars is "follow the water",

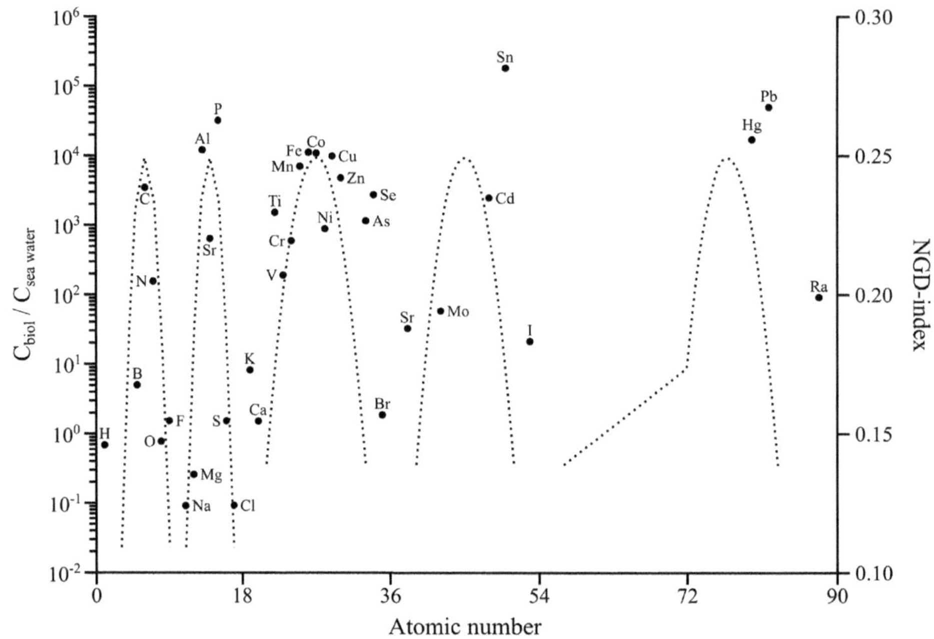


FIGURE 3.9 Concentration factors ($C_{biol}/C_{sea}$ water: symbols referring to the left-hand ordinate) and NGD-index (dotted line, right-hand ordinate) as a function of atomic number, after Spaargaren (1991).

then in the search for LUCA it's "follow the genes". The study of the genetic tree of life, which reveals the genetic relationships and evolutionary history of organisms, is called phylogenetics. Over the last 20 years, our technological ability to fully sequence genomes and build up vast genetic libraries has enabled phylogenetics to truly come of age and has taught us some profound lessons about life's early history.

For a long time it was thought that the tree of life formed three main branches, or domains, with LUCA at the base—eukarya, bacteria, and archaea. The latter two—the prokaryotes—share similarities in being unicellular and lack a nucleus and are differentiated from one another by subtle chemical and metabolic differences. Eukarya, on the other hand, are the complex, multicellular life forms composed of membrane-encased cells, each incorporating a nucleus containing the genetic code as well as the mitochondria "organelles" powering the cell's metabolism. The eukarya are considered so radically different from the other two branches as to necessarily occupy their own domain.

However, a new picture has emerged that places eukarya as an offshoot of bacteria and archaea. This "two-domain tree" was first hypothesized by Lake at al. (1984) but only got a foothold in the last decade, in particular due to the work of Embley and Martin (2006). In their paper, Weiss et al. (2016) describe the new perspective on LUCA and the two-domain tree with phylogenetics.

Previous studies of LUCA looked for common, universal genes that are found in all genomes, based on the assumption that if all life has these genes, then these genes must have come from LUCA. This approach has identified about 30 genes that belonged to LUCA, but they're not enough to tell us how or where it lived. Another tactic involves searching for genes that are present in at least one member of each of the two prokaryote domains, archaea and bacteria. This method has identified 11,000 common genes that could potentially have belonged to LUCA, but it seems far fetched that they all did: with so many genes, LUCA would have been able to do more than any modern cell can.

Martin et al. (2001) realized that a phenomenon known as lateral gene transfer (LGT) was muddying the waters by being responsible for the presence of most of these 11,000 genes. LGT involves the transfer of genes between species and even across domains via a variety of processes such as the spreading of viruses or homologous recombination that can take place when a cell is placed under some kind of stress.

A schematic of the two-domain tree, with eukaryotes evolving from endosymbiosis between members of the two original trunks of the tree, archaea and bacteria, after Weiss et al. (2016).

A growing bacteria or archaea can take in genes from the environment around them by "recombining" new genes into their DNA strand. Often this newly adopted DNA is closely related to the DNA already there, but sometimes the new DNA can originate from a more distant relation. Over the course of 4 billion years, genes can move around quite a bit, overwriting much of LUCA's original genetic signal. Genes found in both archaea and bacteria could have been shared through LGT and hence would not necessarily have originated in LUCA.

When considering life on Earth, it must be noted that LUCA is not the origin of life. The earliest evidence of life on Earth dates to 3.7 billion years ago in the form of stromatolites, which are layers of sediment laid down by microbes. Presumably, life on Earth may have existed even before that. Yet LUCA's arrival and its evolution into archaea and bacteria could have occurred at any point between 2 and 4 billion years ago.

## 3.4 BIOLOGICAL EFFECTS OF MAGNETIC FIELDS

HMF is an extremely weak magnetic field with its total magnetic flux intensity of less than one-tenth of the GMF, while GMF is a geocentric axial dipole field with the greatest intensity at the magnetic poles and the smallest at the magnetic equator. The field intensity has a mean value of 50 μT. The interstellar space magnetic field is much smaller than that of the Earth. Zero magnetic field, near-zero (null) magnetic field, hypo-geomagnetic field, low magnetic field, low-level magnetic field, and extremely low magnetic field are synonymous for describing the sub-geomagnetic nature of the outer space. Mo et al. (2012) proposed intensity values of $0 < |B| \leq 5$ μT for the definition of HMF. The term "lunar magnetic field" has been adopted for a magnetic field strength that lies in the range of 0–300 nT, while "Mars magnetic field" has been used to describe range values of 300 nT to 5 μT (Xue et al. 2021).

Some species are capable to respond to magnetic fields in their environment. For example, magnetotactic bacteria (MTB) produce membrane-enveloped single-domain magnetite ($Fe_3O_4$), greigite ($Fe_3S_4$), or both, which are called magnetosomes. Although the morphological features of magnetosomes are different among taxonomic groups, all magnetosomes function as a compass needle for moving along the Earth's geomagnetic field.

MTB are a diverse group of microorganisms with the ability to orient and migrate along geomagnetic field lines. This unique feat is based on specific intracellular organelles, the magnetosomes, which, in most MTB, comprise nanometer-sized, membrane-bound crystals of magnetic iron minerals organized into chains via a dedicated cytoskeleton. Because of the special properties of magnetosomes, MTB are of great interest for paleomagnetism, environmental magnetism, biomarkers in rocks, magnetic materials, and biomineralization in organisms, and bacterial magnetites have been exploited for a variety of applications in modern biological and medical sciences. Yan et al. (2012) described general characteristics of MTB and their magnetic mineral inclusions but focus mainly on the magnetosome formation and the magnetisms of MTB and bacterial magnetosomes, as well as on the significances and applications of MTB and their intracellular magnetic mineral crystals.

Magnetosome-producing microorganisms can sense and move toward the redox gradient and have been extensively studied in terrestrial and shallow marine sediment environments. However, given the difficulty of sampling, MTB are poorly explored in deep-sea hydrothermal fields. In the study by Nakano et al. (2023), a deep-sea hydrothermal vent chimney from the Southern Mariana Trough was collected using a remotely operated submersible. The mineralogical and geochemical characterization of the vent chimney sample showed an internal iron redox gradient. Additionally, the electron microscopy of particles collected by magnetic separation from the chimney sample revealed MTB cells with bullet-shaped magnetosomes, and there were minor occurrences of cuboctahedral and hexagonal prismatic magnetosomes. Genome-resolved metagenomic analysis was performed to identify microorganisms that formed magnetosomes. A metagenome-assembled genome (MAG) affiliated with Nitrospinae had magnetosome genes such as mamA, mamI, mamM, mamP, and mamQ. Furthermore, a diagnostic feature of MTB genomes, such as magnetosome gene clusters (MGCs), including mamA, mamP, and mamQ, was also confirmed in the Nitrospinae-affiliated MAG. Two lines of evidence support the occurrence of MTB in a deep-sea, inactive hydrothermal vent environment.

In the past 50 years, the interplanetary magnetic field varies between 2 and 8 nT, with a mean value of 6.6 nT, corresponding to an increase of the solar magnetic flux,

Steinhilber et al. (2010). The field strengths of individual planets are very different from each other. For example, it is about 350–700 nT on Mercury, and the magnetic field at the Venus surface is much smaller than that at the Earth. At the surface of Jupiter and Saturn, it is more than a dozen times that of the Earth. The main destinations for human interplanetary exploration are the Moon and Mars. The Moon does not possess a global dipole magnetic field. Data obtained from several lunar orbiting spacecrafts indicate that the Moon's minimal magnetic field intensity is usually several nanotesla and unevenly distributed, Shkuratov et al. (1999). The maximum field intensity is about 300 nT, mostly located in the highlands on the far side of the Moon. "Magnetic refuges" are lunar regions of high residual magnetic field that can shield cosmic rays to some extent, where a human lunar base can be set up for lunar exploration projects, according to Wieser et al. (2010). Mars is a multi-pole planet with many local magnetic fields where the southern hemisphere has a higher magnetic field. Cain et al. (2003) predicted that there might be some regions of high residual magnetic field at the surface of Mars with intensity up to several microtesla. These data indicate that the maximum value of interplanetary HMF intensity would not exceed 5 μT. The field strength of each planet is positively correlated with its mass, core radius, and rotational angular velocity. Solar wind activity can interfere with the planet's magnetic field. As the core of the planet cools, its internal "magneto" slowly shuts down and fades away. In addition, celestial events such as asteroid strike may also induce the disappearance and uneven distribution of the magnetic field. (Xue et al. 2021).

Schematics of the light-dependent chemical magnetoreception model (Ritz et al. 2000) are shown in Figure 3.10. CRY proteins (D) exist in an inactive ground state. Absorption of UV or blue light promotes electron transfer to an acceptor atom (A), resulting in a radical pair (D+ and A-) that further undergoes magnetically singlet-triplet interconversion

Verkin et al. (1976) described effect of a weak magnetic field. A stationary device was constructed for conducting experiments in a weak magnetic field. A ferromagnetic screen was used to produce in the working volume a magnetic field of 26 nT [tesla]. The effect of a weak magnetic field on the following bacteria was studied: *Escherichia coli* var. communior, *Bacterium prodigiosum* [*Serratia marcescens*], *Staphylococcus aureus* 209,

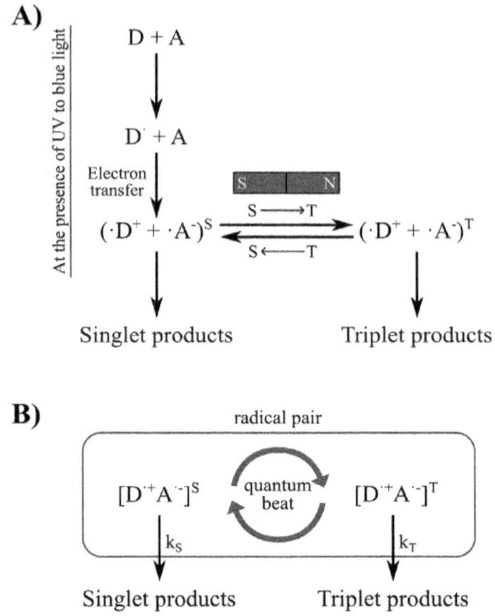


FIGURE 3.10 Schematics of the light-dependent chemical magnetoreception model, modified from Ritz et al. (2000).

and *Bacillus anthracoides*. Cultivation of the bacteria in these conditions for a long time resulted in changes of their pigment and morphological, cultural, and biochemical properties.

Binhi and Prato (2017a) published an early review of experiments and theories of the biological effects of hypomagnetic field. They concentrated on nT magnetic field intensities. This review contains, currently, the most extensive bibliography on the biological effects of hypomagnetic field. This includes both a review of known experimental results and the putative mechanisms of magneto-reception and their explanatory power with respect to the hypomagnetic field effects. They show that the measured correlations of the HMF effect with HMF magnitude and inhomogeneity and type and duration of exposure are statistically absent. This suggests that there is no general biophysical MF target similar for different organisms. This also suggests that magneto-reception is not necessarily associated with evolutionary developed specific magneto-receptors in migrating animals and magnetotactic bacteria. Independently, there is nonspecific magneto-reception that is common for all organisms, manifests itself in very different biological observables as mostly random reactions, and is a result of MF interaction with magnetic moments at a physical level-moments that are present everywhere in macromolecules and proteins and can sometimes transfer the magnetic signal at the level of downstream biochemical events. The corresponding universal mechanism of magneto-reception that has been given further theoretical analysis allows one to determine the parameters of magnetic moments involved in magneto-reception-their gyromagnetic ratio and thermal relaxation time—and so to better understand the nature of MF targets in organisms.

It is known that the weak magnetic field (MF) can trigger a variety of biological responses. MF action can change the concentration of different substances and signaling molecules—this is the area of research primarily in laboratory magnetobiology. The geomagnetic field affects the everyday behavior of some animals. It is also used as a navigational cue by many animals in long seasonal migrations. Background MFs of commercial frequencies sometimes correlate with the incidence of some forms of cancer and other pathologies. The electromagnetic (EM) environment is a subject of research in the epidemiology of electromagnetic fields. It is standardized by safety standards and is accounted for in animal behavior studies. Finally, heliobiology investigates the correlation of the geomagnetic disturbance to the health of humans and other biospheric processes, the MF being regarded as an immediate cause of the correlations. All these effects form a variety of the so-called non-thermal magnetic biological effects.

Despite the abundance and variety of empirical data, the mechanisms underlying magnetoreception, although speculated on extensively, are still not identified.

Binhi and Prato (2017a) use the term "MF target" rather than "MF receptor". The term "receptor" implies that there is a specific receptor—a biophysical construct that is capable of reacting to MF and transferring the signal to the biological level. An example is the magnetosome chain in magnetotactic bacteria, or the magnetosensitive retina in the eyes of some birds, for example. This is not the case in nonspecific magnetoreception, where biological magnetic effects exist, while specific magnetoreceptors do not, for example, in *E. coli* cells and other species that have not developed through evolutionary pressures specific magnetic receptors. In this case we are talking about MF targets that are potentially different over different organisms and even over individuals. Further, known proteins and parts of macromolecules can presumably be MF targets depending on local conditions. As well, we have been careful to distinguish between a primary physical target, that is, magnetic moments, and a biophysical MF target that contains the moments and can produce a biophysical or biochemical signal depending on the state of these moments. These signals may not have a definite transduction path and elicit a biological reaction mostly by chance. A clear example of the MF biophysical target is a flavin–tryptophan radical pair, while the pair of electron spin magnetic moments inside is a physical primary MF target. However it is not always the case that an MF-induced change in the state of physical MF targets results in the biophysical construct producing a potentially observable signal.

Biological effects of HMFs are interesting in that organisms on Earth might have adapted to the presence of the geomagnetic field in a billion-year evolution. Elimination of the geomagnetic field may affect organisms. A hypomagnetic field has been suggested to influence negatively on the early development of organisms and functioning of the central nervous system, although positive effects are also possible. Besides the practical importance, hypomagnetic effects are also of fundamental importance, as their physical origin remains unclear. Hence, here a significant theoretical effort has been directed toward a general understanding of these effects.

It should be pointed out that of 137 considered experimental data, 54 are for magnetic field intensity less or equal to 100 nT. Each reported and reviewed work is unique and not replicated by an independent research group. Over 95% of considered studies report significant effect of HMF; therefore, it can be concluded that the biological effect is well established although the mechanism remains unknown. According to the authors, the closest to the primary physical level are studies that report on the involvement of gene expression in magneto-reception, with different genes expressing in different situations.

A wide range of exposure times required for the appearance of the magnetic effect—from minutes to months—is another fact, indicating that there is no a single type of specialized biomolecules responsible for magneto-reception. Different biomolecules with very different characteristic speeds of functioning may become MF targets—the mediators of a magnetic signal transduction from the objects of the physical level to downstream systems and reactions; see Binhi and Prato (2017a). However, the conclusion that there cannot be a biophysical magneto-receptor common to all organisms, is not completely obvious, and is a supposition rather than a logical inference.

The biological effect often occurs in EMFs that have only vanishingly small inductive or thermal action. This fact, known as the "kT problem", can be expressed as:

$$mH << kT$$

H—MF magnitude
m—the magnetic moment of the putative MF target
k—Boltzmann constant
T—temperature of the target

It is not clear how a change in magnetic energy that is much smaller than the scale of the thermal fluctuations influences the rate of a chemical reaction, whilst this influence is necessary to excite a biological response. One could say that a correct explanation of magneto-reception should transform the previous inequality to at least an approximate equality.

Some chemical reactions are known to change their rate in the MF due to formation of the intermediary magnetically sensitive state—a spin-correlated pair of radicals; see Buchachenko et al. (1978) and Salikhov et al. (1984). This is one of the most actively discussed magneto-reception mechanisms and is called radical pair mechanism (RPM). The total spin of the radicals or the relative orientation of the spins affects the probability of the product formation. For example, probability of recombination of the newly formed radicals in a homolytic reaction depends on the relative orientation of their spins. A constant MF changes the likelihood of a favorable orientation and, thereby, able to shift the chemical balance. This mechanism does not possess a frequency selectivity in the low-frequency range, since the evolution of the magneto-sensitive spin state occurs within a very short lifetime of the pair, usually on the order of $10^{-9}$–$10^{-7}$ s.

In another paper, Binhi and Prato (2017b) considered a nonuniform precession of magnetic moment under the action of the MF, whose direction is unchanged while the value changes. This MF does not cause quantum transitions; hence the classical model of the Larmor precession is sufficient. It turns out that this precession mode has interesting properties, similar to those of non-specific magneto-reception. The precession of magnetic moments in MF precedes any biophysical or biochemical mechanism of magneto-reception and largely determines the spectral and non-linear characteristics of the biological response. Objects in biological cells that possess a precessing magnetic moment are unpaired electrons, paramagnetic ions, protons and other magnetic nuclei. Protein-bound ions and the rotations of molecular groups with a distributed electric charge may also have a virtual magnetic moment.

The mechanism considered is based on the following points: the external MF acts on a magnetic moment; the magnetic moment precesses and undergoes thermal relaxation. A biological effect occurs if, during the relaxation time or faster, the MF imparts a significant disturbance to the dynamics of the magnetic moment. A measure of the disturbance is a deviation from the state of the unperturbed uniform precession in the GMF. Idealizations of the model:

1. uniform precession is a natural background for the microscopic events in organisms;
2. the events of the next level—biophysical or biochemical events—form a nonhomogeneous Poisson process, the rate of which varies periodically with the precession phase in the local coordinate system of the target; and
3. a biological effect is associated with the disturbance of precession of the magnetic moments of the same type—the disturbance being averaged over time and over the realizations of random precession phase.

Of course, the biological effect is observed only when the biophysical-level changes go through the stages of transformation at the biochemical, physiological, and biological levels of the system (Binhi and Prato 2017b).

In this scenario, the average probability of biophysical events P(H, h, O, γ, τ, β) depends on six values. Three of these are MF variables: the constant MF component H and the amplitude h and frequency O of the variable component. Three others are MF target parameters: the gyromagnetic ratio γ and the parameters τ (relaxation time) and β, describing respectively the thermal and the "signal" interaction of the magnetic moment with its immediate environment.

Binhi and Prato (2017a) also described the so-called molecular gyroscope mechanism. Interestingly, the magnetoreception mechanisms presented previously involve, in one way or another, a variety of rotations: the precession of magnetic moments of different natures and the rotations of nanoparticles that have their own intrinsic magnetic moment. The presence of rotations in the models of magnetoreception is not by chance. This is a consequence of the specificity of the MF interaction with matter. This link is carried out only through the interaction with magnetic moments. The magnetic moments, on the one hand, experience a torque in the MF and, on the other hand, are bound with spins or can be generated by rotations of electric charges.

The essence of the mechanism is a rotation of large fragments of macromolecules or amino acid residues with distributed electric charge. Due to this rotation, a magnetic moment appears that interacts with an external MF. In the quantum description, the rotation of fragments is an interference of their angular states with a nonzero magnetic quantum number. How can molecular gyroscopes arise? The folding of long protein chains that are synthesized by ribosomes results in the formation of protein globules. In the process of folding/maturation, there occurs an evolutionarily determined and workable conformation of a protein molecule. Incorrect folding can break or make impossible a specific function of the protein.

The characteristic time of folding is highly dependent on the length of the protein chain, ranging from microseconds to seconds. At some stages of the folding, virtual cavities free of water molecules may occur within the protein. Indeed, hydrophobic cavities of the order of 1 nm or less push out water molecules. In these cavities, the amino acid residues (molecular gyroscopes) rotate within milliseconds whilst looking for the best position. That would be enough for the action of ELF MFs on such rotations and, therefore, on the path and the result of the protein folding. The magnetoreception mechanism that utilizes the interference of quantum states of a molecular gyroscope is based on the fact that the main rotational degree of freedom of a gyroscope can long remain nonthermalized, or "cold". Rotations of small molecules that are fixed inside virtual cavities are well protected from the thermal vibrations of the cavity walls and, therefore, have a relatively long lifetime. In combination with quantum interference effects, these coherent long-lived rotations allow one to explain magnetoreception. In this way, the molecular gyroscope mechanism resolves the kT problem (Binhi et al. 2006).

Binhi and Prato (2017a) concluded their paper with the statement "The critical level of constant MF, or $1/\gamma\tau$, is inversely proportional to the product of the gyromagnetic factor and the thermal relaxation time of the magnetic moment of the target. In this or in smaller fields, an effect of the MF on molecular targets is possible". Recently, Binhi and Rubin (2022) presented the main historical facts and listed the general observed properties of magnetobiological phenomena. The presence of nonspecific magnetobiological effects in the everyday is also shown. Particular attention is paid to the formation of theoretical concepts in magnetobiology and the state of the art in this area of research. Some details are provided on the molecular mechanisms of the nonspecific action of a magnetic field on organisms.

Zadeh-Haghighi and Simon (2022a) recently published a review article on magnetic field effects in biology from the perspective of the radical pair mechanism. They reported some hundreds of studies reporting that weak magnetic fields can significantly influence various biological systems. However, the underlying mechanisms behind these phenomena remain elusive. Remarkably, the magnetic energies implicated in these effects are much smaller than thermal energies. The authors reviewed these observations and suggested an explanation based on the radical pair mechanism, which involves the quantum dynamics of the electron and nuclear spins of transient radical molecules. While the radical pair mechanism has been studied in detail in the context of avian magneto-reception, the studies reviewed showed that magneto-sensitivity is widespread throughout biology. Zadeh-Haghighi and Simon (2022a) reviewed magnetic field effects on various physiological functions, discussing static, hypo-magnetic, and oscillating magnetic fields, as well as isotope effects. Afterwards, they reviewed the radical pair mechanism as a potential unifying model for the described magnetic field effects and discussed plausible candidate molecules for the radical pairs. The authors reviewed recent studies proposing that the radical pair mechanism provides explanations for isotope effects in xenon anesthesia and lithium treatment of hyperactivity, magnetic field effects on the circadian clock, and hypomagnetic field effects on neurogenesis and microtubule assembly.

Sensitivity to weak magnetic fields is abundant throughout biology, as discussed in numerous review articles. Effects of either static or oscillating weak magnetic fields have been reported on the circadian clock, electron transfer in cryptochrome, stem cells, calcium concentration, the brain's functions such as action potentials, reactive oxygen species (ROS), development, neuronal activities, DNA, memory, anxiety, analgesia, genetics, and many other functions. Despite the wealth of observations, thus far, there is no clear explanation for the mechanism behind these phenomena. This is mainly due to the fact that the corresponding energies for such effects are far smaller than thermal energies. However, there is a promising quantum physics (or spin chemistry) concept that can account for the effects of such weak fields: the radical pair mechanism. This mechanism, which is an example of the emerging field of quantum biology, has been studied in significant detail in the comparatively narrow context of bird magnetoreception, where it is accepted as one of the leading potential explanations for how birds sense magnetic fields, and in particular the Earth's magnetic field, for the purpose of navigation. It is known that birds and amphibians, and in all likelihood other vertebrates, have not one but two magnetoreception mechanisms, a magnetite-based detector that provides the high sensitivity necessary for sensing weak spatial gradients in the magnetic field and a light-dependent magnetic compass that underlies a magnetic map sense. The latter is thought to be based on the radical pair mechanism (Zadeh-Haghighi and Simon 2022b).

The radical pair mechanism involves magnetically sensitive intermediate molecules, so-called radical pairs; see, for example, Hore and Mouritsen (2016). The key ingredient is the spin correlation between two unpaired electrons, one on the donor molecule and the other on the acceptor molecule. Depending on the initial spin configuration of the donor and acceptor molecules, this initial spin correlation of the radical pair will be either a singlet (S) or a triplet (T) state, which are, respectively, spin-0 and spin-1 states. Due to the spin interactions with its environment (in particular with external magnetic fields and with nearby nuclear spins), the state of the radical pair will oscillate between S and T states (Timmel et al. 1998). Each spin state, S and T, can lead to different reaction products, providing an example of spin chemistry. The energies induced by these magnetic fields are hundreds of thousands of times smaller than thermal energies, kBT (kB is the Boltzmann constant and T is temperature), which are associated with motions, rotation, and vibrations in biological environments. In thermal equilibrium, the energies required to alter the rate or yield of a chemical transformation should be at least comparable to kBT. Due to this, the radical pair mechanism was originally ignored in the context of physiology. However, the situation differs in systems far from thermal equilibrium, which is the case for radical pairs (Hore and Mouritsen 2016). Sensitivity to weak magnetic fields is one of the key properties of radical pair reactions. Nowadays, many research laboratories study the role of radical pairs in (bio)chemical reactions; see, for example, Beretta et al. (2019).

Recent studies have proposed roles for radical pairs beyond avian magnetoreception, in particular in xenon-induced anesthesia, lithium effects on mania, magnetic field and lithium effects on the circadian clock, and hypomagnetic field effects on microtubule reorganization and neurogenesis, where hypomagnetic fields are fields much weaker than that of the Earth. Zadeh-Haghighi and Simon (2022b) suggest that the radical pair mechanism is in fact quite common in biology and that it may provide an explanation for many of the weak magnetic field effects on physiological functions that have been observed.

Spin is an inherently quantum property that emerges from Dirac's relativistic quantum mechanics and is described by two numbers, S and $m_s$, respectively, the spin quantum number and the spin projection quantum number. Electrons, protons, and neutrons have spins of S = 1/2. Having an angular momentum characteristic, spin can be coupled not only with external magnetic fields but also with other spin in its vicinity. For instance, coupling of two electron spins, $S_A$ and $S_B$, results in a total spin of $S_T$, which has a quantum number of either S = 1 or S = 0. The latter case is called a singlet state, with $m_s = 0$, and the former is called a triplet state, with $m_s = 0, \pm 1$. Radicals are molecules with an odd number of electrons in the outer shell. A pair of radicals can be formed by breaking a chemical bond or electron transfer between two molecules. It is important to note that in reactions of organic molecules, spin is usually a conserved quantity, which is essential for magnetic field effect in biochemical reactions. For example, a radical pair can be created if a bond between a pair of molecules [A · · · D] breaks or an electron is transferred from D to A, [$A^{-}$. ··· $D^{+}$] (D and A denote donor and acceptor molecules). A radical pair may be in a superposition of singlet and triplet states, depending on the parent molecule's spin configuration. Assuming that the initial state of the electron pairs before separation was a singlet (triplet), the recombination of unpaired electrons can only happen if they stayed in a singlet.

Discussion involves, however, only mT and μT magnetic field intensities (data available and of some practical relevance). The interactions between the unpaired electron spins on each radical and external magnetic field is essential for generating MFEs. This interaction is called the Zeeman effect. The nuclear spins of radical molecules also experience applied magnetic fields; however, as nuclear magnetogyric ratios are much smaller than that of the electrons, these interactions are negligible.

In addition, the authors consider hyperfine interaction, exchange interaction, and dipolar interaction. The sensitivity of certain reactions to weak magnetic fields relies on the oscillations between singlet and triplet states of radical pairs, also known as "quantum beats" (Steiner and Ulrich 1989). If the radicals are separated enough spatially, having the same energies, singlet and triplet will undergo a coherent interconversion process, quantum beating. The interconversion is tuned by the magnetic fields experienced by the electrons, including Zeeman and hyperfine interactions. At low magnetic fields, the main drive for S–T interconversion is due to the hyperfine interactions. Obeying selection rules, the singlet and triplet yields will follow different chemical pathways, which depend on the timing of the coherent spin dynamics (Hore 2021). These quantum beats have just recently been observed directly (Mims et al. 2021).

Although hypo-magnetic fields belong to the static magnetic field category, the effects due to extremely low magnetic fields are often particularly significant compared to other magnetic field effects. It can be shown that for different relaxation and reaction rates, the hypo-magnetic field effects are significant.

It should be kept in mind that the curve in Figure 3.11 is calculated trying to fit available data. However, data with very small field intensities are scarce or nonexistent. The authors used in their analyses only the following data obtained with field intensity in the μT region:

1. 74 μT from Ermakov et al. (2022),
2. 0.4 mT from Wu et al. (2014), and
3. 100 μT from Höytö et al. (2017) and Luukkonen et al. (2014).

In other words, they do not cover the nT region at all, which is our prime interest (intensity of galactic magnetic fields and fields in molecular clouds). Therefore, Figure 3.11 can be misleading, since effects are seen and expected to be of relevance for the problem of origin of life.

It should be noted that the radical pair models used in the studies are simplified, partly because the exact radical pair molecules involved in these systems are still unknown. This is the case even in the context of avian magnetoreception, where the proposed radical pairs include flavin–tryptophan, flavin–tyrosin, and flavin–superoxide, among others; see Hong et al. (2020). More

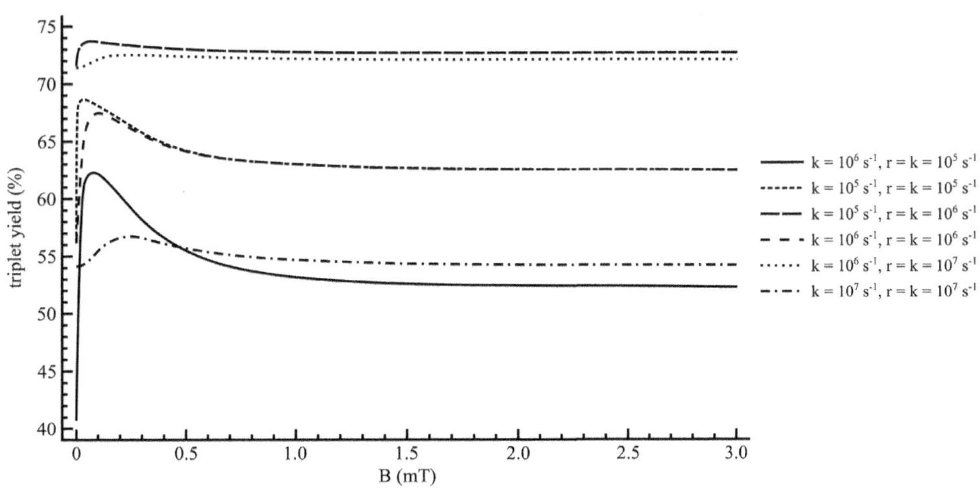


FIGURE 3.11 Triplet yield vs applied magnetic field for different reaction and spin relaxation rates for a simple model of a radical pair. In this model, one of the radicals is coupled to a nucleus with a hyperfine coupling constant of 1 mT. For different values of the rates, one can see a pronounced dip near the zero field, together with a maximum close to the value of the geomagnetic field (~ 0.05 mT).

realistic models of the radical pairs may provide further insight into the underlying mechanism behind these phenomena. This might involve including multiple nuclei, dipolar, and exchange interactions in the models. It should also be pointed out that including these interactions can reduce the predicted effect size; see Hogben et al. (2009). However, this may be balanced by potential amplification effects in the biological systems.

It should be noted that reproducibility of weak magnetic field effects in biology has been a challenge. There are several studies reporting failed attempts at independent replications of magnetic field effects in biological systems. However, this problem is not confined to this particular area of the life sciences. For example, a recent analysis of high-impact cancer studies concluded that only 5 out of 53 papers could be fully reproduced, after Kaiser (2021). A lot of these issues are likely due to the complexity of biological systems. Despite these challenges, it seems unlikely that all of the hundreds of magnetic field effects on biological

Accumulating evidence has revealed that gut microbiota is affected by both genetic and environmental factors. As one of the important environmental factors, the diminishing or disappearing GMF could affect living organisms at different levels of biological organization; see review article by Erdmann et al. (2021). However, it is unclear whether and how long-term HMF exposure affects gut microbiota in mammals.

Zhan et al. (2022) showed:

1. long-term HMF exposure affects the diversity and composition of mouse gut bacterial microbiota, and returning to the GMF environment could recover the changes in gut microbiota caused by HMF;
2. long-term HMF exposure may affect the metabolic potential of gut bacterial microbiota and reduce the content of SCFAs in the colon; and
3. long-term HMF exposure increases the ROS levels in epithelial layers and the number of cell proliferation in the colon.

The possible associations among the change of gut microbiota, high ROS levels in the colon, and increased colonic cell proliferation can provide some valuable hints to better understanding the effects of HMF on the gut microbiota in mice.

The biological effects of electromagnetic fields are probably mediated by plasma membrane. Intramembrane proteins (IMPs) play an important role, which represent a class of proteins located in the lipid bilayer of the cell membrane which function as ion channels, enzymes, or receptors of chemical signal conclusions. The exposure to the 50 Hz, peak amplitude 1.8 mT EMR can cause the formation of ion channels without cell control and can induce in this way important metabolic modifications.

Zablotskii et al. (2022) studied the effects of high magnetic fields on the diffusion of biologically active molecules. The diffusion of biologically active molecules is a ubiquitous process, controlling many mechanisms and the characteristic time scales for pivotal processes in living cells. The authors showed how a high static magnetic field (MF) affects the diffusion of paramagnetic and diamagnetic species including oxygen, hemoglobin, and drugs. They derived and solved the equation describing diffusion of such biologically active molecules in the presence of an MF as well as reveal the underlying mechanism of the MF's effect on diffusion. They found that a high MF accelerates diffusion of diamagnetic species while slowing the diffusion of paramagnetic molecules in cell cytoplasm. When applied to oxygen and hemoglobin diffusion in red blood cells, their results suggest that an MF may significantly alter the gas exchange in an erythrocyte and cause swelling.

### 3.4.1 Concentration Factors as Functions of Magnetic Field Intensity

The mechanism through which the geomagnetic field might have influenced the extinction of the species was proposed some time ago; see Valkovic (1977). It has been assumed that the concentration factor for essential trace elements is also dependent on the magnetic field intensity. The concentration factor for element $E_i$ is then defined as:

$$CF(E_i, H) = C_{org.}\left(E_i, H, P_{j,env}\right) / C_{env.}\left(E_i, H, P_{j,env}\right)$$

where
Ei is for life-essential elements,
H is magnetic field intensity,
Pi,env are other environmental parameters.

The organisms concentrate all elements present in their environment. However, not all of the elements are essential for life. Criteria which an element must satisfy to be essential for life are well established. The relation of the uptake of an essential element to yield or growth may be considered a definition of essentiality. There is a rather narrow range of adequacy of element concentration in organisms. Smaller concentrations result in different abnormalities induced by deficiencies, which are accompanied by pertinent, specific biochemical changes. Higher concentrations result in toxicity. As a curiosity, let us mention that in plants it is possible to have, under severe deficiency conditions, a decrease in concentration of an element which results in a small increase in growth. This phenomenon is known as the Piper–Steenbjerg effect; see Piper (1942) and Steenbjerg (1951).

The terrestrial magnetic field is a factor in biological evolution. Approximately every 200,000 years, the Earth's magnetic polarity changes its sign, and during the reversal, living organisms find themselves in a significantly reduced MF for periods of millennia. In discussion of the biological role of this exposure to the HMF, one should also consider a synchronous factor: ionizing solar radiation, which normally is repelled away by the geomagnetic field.

According to Baxter (2009), the accumulation of a given element is a complex process controlled by a network of gene products critical for uptake, binding, transportation, and sequestration. Many of these genes and physiological processes affect more than one element. Therefore, to understand how the essential elements are regulated, it is necessary to measure as many of the essential elements contained in a cell, tissue, or organism (the ionome) as possible. The elements, which share components of their network, vary depending on the species and genotype of the plants, in this case, that are studied and environment they are grown in. Several recent papers describe high-throughput elemental profiling studies of how the ionome responds to the environment or explore the genetics that control the ionome. When combined with new genotyping technologies, ionomics provides a rapid way to identify genes that control elemental accumulation in plants. This type of consideration should be extended to all living species and all essential elements.

Extremely low-frequency magnetic fields are known to affect biological systems. In many cases, biological effects display "windows" in biologically effective parameters of the magnetic fields: most dramatic is the fact that the relatively intense magnetic fields sometimes do not cause appreciable effect, while smaller fields of the order of 10–100 μT do. Linear resonant physical processes do not explain the frequency windows in this case. Amplitude window phenomena suggest a nonlinear physical mechanism. Such a nonlinear mechanism has been proposed recently to explain those "windows". It considers the quantum-interference effects on the protein-bound substrate ions. Magnetic fields cause an interference of ion quantum states and change the probability of ion-protein dissociation. This ion-interference mechanism predicts specific magnetic-field frequency and amplitude windows within which the biological effects occur. It agrees with a lot of experiments. However, according to the mechanism, the lifetime Gamma(–1) of ion quantum states within a protein cavity should be an unrealistic value, more than 0.01 s for frequency band 10–100 Hz. Binhi and Savin (2002) proposed a biophysical mechanism which

1. retains the attractive features of the ion interference mechanism, that is, predicts physical characteristics that might be experimentally examined, and
2. uses the principles of gyroscopic motion and removes the necessity to postulate large lifetimes.

The mechanism considers the dynamics of the density matrix of the molecular groups, which are attached to the walls of protein cavities by two covalent bonds, that is, molecular gyroscopes. Numerical computations have shown almost free rotations of the molecular gyroscopes. The relaxation time due to van der Waals forces was about 0.01 s for the cavity size of 28 Å.

### 3.4.2 Mechanical Generation of Spin

Spin current, a flow of spins, can be generated by using couplings between an electron spin and other physical quantities such as a magnetic field, magnetization, and orbital degrees of freedom. These conventional methods, however, force us to use magnetic materials or rare metals with strong spin-orbit coupling. In this study, we present a new route for spin-current generation by considering the coupling between spin and mechanical rotation. In their paper, Matsuo et al. (2013) define spin-rotation coupling (SRC) as

$$\mathrm{H} = -\mathbf{S} \times \Omega$$

where S is the electron spin angular momentum and $\Omega$ is the mechanical rotation frequency. If we rewrite the SRC as $H = -\mathbf{S} - \mathbf{B}_\Omega$, where $\mathbf{B}_\Omega = \mathbf{\Omega}\gamma^{-1}$ is an effective magnetic field due to mechanical rotation with gyromagnetic ratio γ (gyromagnetic ratio the ratio of the magnetic momentum in a particle to its angular momentum). The SI unit is the radian per second per tesla, $rad{\cdot}s^{-1}{\cdot}T^{-1}$), the SRC can be regarded as the Zeeman coupling. Historically, the SRC is known as a fundamental origin of the Barnett and Einstein–de Haas effects and plays a crucial role in tests of general relativity by the use of the gyroscopic effect in gravitational fields due to Earth.

In their paper, the focus is on induced surface acoustic waves (SAWs), rotational deformations that vary in space and time. In this case, the gradient of rotational modes acts on an electron spin as the gradient of a magnetic field. For an electron spin, spin prefers to align to the effective magnetic field and to flow from the bottom to the surface for lower Zeeman energy. As a result, the stripe pattern of the spin accumulation arises at the surface.

Later, Matsuo et al. (2015) provided the theory of the coupling between mechanical angular momentum and spin current and showed that spin current can be generated from mechanical motion, including rigid rotation and elastic deformations. Spin transport theory in accelerating frames is constructed by using the low-energy expansion of the generally covariant Dirac equation. Barnett (1915) discovered the effect of magnetization induced by rotation. He measured magnetization of iron, which is rotated with the angular velocity Ω, and showed that the magnetization M is given by

$$M = \chi\Omega/\gamma$$

where χ is the magnetic susceptibility and γ is the gyromagnetic ratio. The result means that an effective magnetic field,

$$B_\Omega = \Omega/\gamma$$

is induced in the rotating body. Both Einstein–de Haas (1915) and Barnett's experiments established conversion between magnetic and mechanical angular momentum. Progress in nanotechnology has allowed investigation of the coupling of nano-magnetism and mechanical rotation; see Zolfagharkhani et al. (2008) and Tejada et al. (2010).

Danner et al. (2020) reported the spin-rotation coupling observed in neutron interferometry. They observed phase shifts arising as a consequence of the spin's coupling with the angular velocity of a rotating magnetic field. This coupling is a purely quantum mechanical extension of the Sagnac (1913) effect, a phase shift induced between two counter-rotating light waves in a rotating interferometer. The resulting phase shifts linearly depend on the frequency of the rotation of the magnetic field. Their results agree with the predictions derived from the Pauli–Schrödinger equation. Their interferometer experiment confirms the phase shift on the neutrons' wave function, which is a purely quantum mechanical effect and cannot be explained with the semi-classical Bloch equations. The observed linear dependence of the phase shift on the rotation frequency validates the spin-rotation coupling of neutrons as a purely quantum mechanical extension of the Sagnac effect.

Vijayakumar and Gopinathan (1996) presented a calculation of spin-orbit coupling constants of transition metal atoms and ions in density functional theory. Reliable calculations of atomic and molecular properties of d- and f-series elements and molecules containing these elements present a formidable problem in theoretical chemistry. In such systems, relativistic and electron correlation effects are important. The calculated values of spin-orbit coupling constants and radial moments can be used to compute spin-orbit splitting in molecules.

Quantum control of chemical reactions is an important goal in chemistry and physics. Ultra-cold chemical reactions are often controlled by preparing the reactants in specific quantum states. Sikorsky et al. (2018) demonstrate spin-controlled atom–ion inelastic (spin-exchange) processes and chemical (charge-exchange) reactions in an ultracold Rb-$Sr^+$ mixture. The ion's spin state is controlled by the atomic hyperfine spin state via spin-exchange collisions, which polarize the ion's spin parallel to the atomic spin. They achieved ~90% spin polarization due to the absence of strong spin-relaxation channel. Charge-exchange collisions involving electron transfer are only allowed for $(RbSr)^+$ colliding in the singlet manifold. Initializing the atoms in various spin states affects the overlap of the collision wave function with the singlet molecular manifold and therefore also the reaction rate. Their observations agree with theoretical predictions.

## 3.5 SPORULATION

Some bacteria, such as *Bacillus subtilis*, withstand starvation by forming dormant spores that revive when nutrients become available. Although sporulation and spore revival jointly determine survival in fluctuating environments, the relationship between them has been unclear. Here we show that these two processes are linked by a phenotypic "memory" that arises from a carry-over of molecules from the vegetative cell into the spore. By imaging life histories of individual *B. subtilis* cells using fluorescent reporters, Mutlu et al. (2017) demonstrated that sporulation timing controls nutrient-induced spore revival. Alanine dehydrogenase contributes to spore memory and controls alanine-induced outgrowth, thereby coupling a spore's revival capacity to the gene expression and growth history of its progenitors. A theoretical analysis, and experiments with signaling mutants exhibiting altered sporulation timing, support the hypothesis that such an intrinsically generated memory leads to a tradeoff between spore quantity and spore quality, which could drive the emergence of complex microbial traits.

Several population-level studies have found that environmental conditions, during and even after sporulation, modulate the average resistance of spores, as well as their capacity for revival, by influencing their molecular composition. Furthermore, a considerable portion of the spore's proteinaceous cargo derives from its progenitor cell, presumably through carry-over of stable proteins, although little is known about their functional relevance for spore revival. According to Mutlu et al. (2017), one may thus consider the history-dependent portion of a spore's composition a "memory" of the past that could influence a spore's future revival response. Indeed, memory effects may couple the heterogeneous dynamics of sporulation and spore revival at the level of the life histories of individual cells. However, experimental evidence for this hypothesis has been lacking, and the potential consequences of such memory for linking of sporulation to spore revival are unclear.

In addition, Mutlu et al. (2017) addressed these questions by imaging life histories of individual *B. subtilis* cells under changing nutrient conditions. They found that heterochronic sporulation resulted in the formation of spore subpopulations with a long-term memory that controls their responses to nutrient-induced spore revival. This memory stems from the growth and gene-expression history of individual cells prior to the onset of sporulation, and they identified alanine dehydrogenase (Ald) as an important molecular determinant of this memory. High levels of Ald in the spore core support outgrowth following alanine-dependent germination, provided sufficient alanine is in the medium. Moreover, this intrinsically generated phenotypic memory leads to the emergence of a tradeoff between spore quantity and spore quality.

The basis of the mathematical model by Mutlu et al. (2017) are following: They consider cells that contain an enzyme e, which is required for spore revival. Upon nutrient downshift at time $t = 0$, cells stop enzyme production but continue to grow at rate $\lambda$ so that the normalized enzyme concentration in every cell decreases as $e(t) = \exp(-\lambda t)$. Cells also progress at rate $\eta$ from their naïve state $c_0$, through intermediate states $c_1, \ldots c_k$, toward the starved state $c_s$, where they may die with rate $\mu$. Cell may sporulate from all states $c_1 \ldots c_k, c_s$ with rate $\delta$. Sporulation at time t generates a spore denoted as $s^{e(t)}$, which conserves the current enzyme level e(t). The dynamical equations for the state populations are:

$$\partial_t c_0(t) = (\lambda - \eta) c_0(t);\ \partial_t c_i(t) = \eta c_{i-1}(t) + (\lambda - \eta) c_i(t) \text{ for } i > 0;$$

$$\partial_t c_s(t) = \eta c_k(t) + (\lambda - \mu) c_s(t);$$

$$\partial_t s^e(t) = \delta\left[e - e(t)\right] \times \delta \times \left\{\textstyle\sum_{i=1}^{k} c_i(t) + c_s(t)\right\}.$$

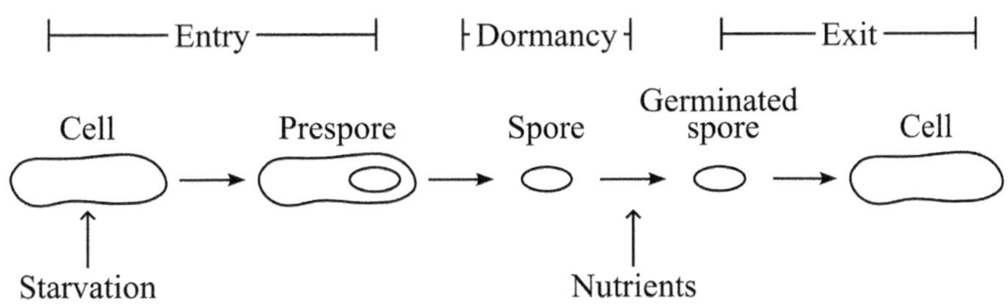


FIGURE 3.12 Tracking the life histories of individual bacteria to study the role of memory in the *B. subtilis* life cycle.

These equations should be solved for $k = 2$ with $c_0(0) = 1$ and all other populations 0 as the initial condition. The spore quantity or spore yield is a number of spores generated per initial cell, $s = \lim_{t\to\infty} \int_{e=0}^{1} s^e(t)de$. The spore quality is defined by the average enzime level per spore, $e_{av} = \int_{e=0}^{1} e\rho(e)de$, where $\rho(e) = \lim_{t\to\infty} s^e(t)/s$ is the distribution of enzyme levels in the generated spore population.

Veening et al. (2005) described how phosphatases modulate the bistable sporulation gene expression pattern in *Bacillus subtilis.* That is, the spore formation in the gram-positive bacterium *Bacillus subtilis* is a last resort adaptive response to starvation. To initiate sporulation, the key regulator in this process, Spo0A, needs to be activated by the so-called phosphorelay. Within a sporulating culture of *B. subtilis,* some cells initiate this developmental program, while other cells do not. Therefore, initiation of sporulation appears to be a regulatory process with a bistable outcome. Using a single cell analytical approach, Veening et al. (2005) showed that the auto-stimulatory loop of spo0A is responsible for generating a bistable response resulting in phenotypic variation within the sporulating culture. It is demonstrated that the main function of RapA, a phosphorelay phosphatase, is to maintain the bistable sporulation gene expression. As rapA expression is quorum regulated, it follows that quorum sensing influences sporulation bistability. Deletion of spo0E, a phosphatase directly acting on Spo0A~P, resulted in abolishment of the bistable expression pattern. Artificial induction of a heterologous Rap phosphatase restored heterogeneity in a rapA or spo0E mutant. These results demonstrate that with external phosphatases, *B. subtilis* can use the phosphorelay as a tuner to modulate the bistable outcome of the sporulating culture. This shows that *B. subtilis* employs multiple pathways to maintain the bistable nature of a sporulating culture, stressing the physiological importance of this phenomenon.

Spore formation in *Bacillus subtilis* is a complex adaptive response to starvation. The process of sporulation is governed by a multicomponent phosphor-relay which consists of five histidine kinases (KinA, KinB, KinC, KinD, and KinE) and two phosphorelay proteins (Spo0F and Spo0B); see Perego and Hoch (2002).

Multiple environmental and physiological signals are fed into this system, and under appropriate conditions, this leads to phosphorylation of Spo0A, the key sporulation transcription factor. Within an isogenic culture of sporulating *B. subtilis,* some cells initiate the developmental program of sporulation, whereas others do not. Therefore, initiation of sporulation appears to be a regulatory process with a bistable outcome. Phenotypic variation could benefit the fitness of the species, because the heterogeneous population is able to quickly react to changing environments. As sporulation is an energy-intensive process, and irreversible after its earliest stage, cells that are delayed to commit to sporulate could have an advantage over sporulating cells if food resources were to become plentiful again, Veening et al. (2005). It is believed that stochastic fluctuations causes some cells to reach the threshold level to activate the feedforward loop, and these cells will end up in the "high expressing" population, whereas others do not reach this threshold and remain in the "low expressing" state.

When *B. subtilis* cells reach the end of exponential growth, and various environmental signals promote the activation of the sporulation response regulator Spo0A, this protein in turn activates more than 40 genes directly, including the sigma-H dependent spoIIA operon comprising the early sporulation genes spoIIAA, spoIIAB, and sigF.

## 3.6 CHIRALITY

Here is an old quote which well describes the opinion of the author of this book:

> The first is the fact that although most compounds involved in fundamental life processes, such as sugars and amino acids, are chiral and although the energy of both enantiomers and the probability of their formation in an achiral environment are equal, only one enantiomer occurs in nature; the enantiomers involved in life processes are the same in men, animals, plants and microorganisms, independent on their place and time on Earth. Many hypotheses have been conceived about this subject, which can be regarded as one of the first problems of molecular theology. One possible explanation is that the creation of living matter was an extremely improbable event, which occurred only once.
>
> Prelog (1976)

Chiral molecules have non-planar structures (usually tetrahedral), and they always have one or more chiral centers (usually around an asymmetric carbon atom). The number of chiral centers within a specific chiral molecule determines whether there is only an enantiomeric pair of this molecule or more possibilities of enantiomers and diastereomers (non-superimposable mirror images of one another). For instance, 17 out of the 20 amino acids possess one chiral center. For each of them, there are two possibilities: an enantiomeric pair. Threonine and isoleucine have two chiral centers, and they have two enantiomers and two diastereomers.

The vast preponderance of biologically formed chiral compounds are synthesized exclusively as one or other enantiomer. Enantiomers are isomers for which the left- and right-handed forms have the same physical behavior but respond differently in chemical reactions with other chiral molecules. In chemical processes, left- and right-handed molecules are generally formed in equal amounts, unless chiral reagents, catalysts, or surfaces are present to introduce a bias. Also, some organisms may synthesize the same chiral compound in different enantiomeric forms. For the formation of regular structures when these monomers are assembled, however, it is generally more efficient to have the same handedness in all molecules. Thus, 100% of amino acids in Terran (a model of life that is based on the life that we know on Earth) proteins have the same absolute spatial relationship of the amino group, the acid group, and the side chain distinctive for the amino acid and are all designated "L". Likewise, all of the sugars in nucleic acids have the opposite spatial relationship and are designated "D". Enantiomeric excess (also called chirality, or handedness), the predominance of one enantiomer over the other for a given assembly of atoms, is therefore believed to be a universal feature of chemistry characteristic of living systems. Once an organism dies and its biochemicals are released into the environment, their chiral purity (and optical activity) may or may not persist, depending on the relative chemical stability of the bonds in the vicinity of the chiral center. Various natural chemical processes can lead to racemization, the formation of mixtures of the two enantiomers. While racemization may result in loss or corruption of a biological signature, the rate at which it happens can also have a practical application. The best known example is the dating of fossil organic matter on the basis of the degree of amino acid racemization.

Molecules are three-dimensional. A carbon atom with four single bonds lies at the center of a tetrahedron. The atoms to which the carbon is bonded are at the vertices of the tetrahedron. Those atoms are in turn likely to be bonded to other atoms. If the chemical structures at the four corners all differ, however slightly, the mirror images of the tetrahedron will not be superimposable. Mirror-image stereoisomers that are not superimposable are called enantiomers; stereoisomers that are not enantiomers are called diastereomers. Enantiomers possess chirality, or handedness, and when dissolved rotate the plane of polarized light when it is passed through the solution. Life on Earth makes use of only a limited number of diastereomers of all those that are possible. Moreover, biotic processes display an enantiomeric excess; for example, left-handed amino acids and right-handed sugars almost exclusively predominate in living systems. Carbon atoms bearing four different substituents are said to be chiral centers. If a molecule has n chiral centers, it will, in most cases, have 2n stereoisomers. There will, for example, be 256 stereoisomers of a compound with eight chiral centers. Each will have exactly the same chemical formula and pattern of connectivity among its atoms (A is connected to B is connected to C and D, and so on). Only the arrangements of those atoms in space will differ, and there will be 256 variations. Life functions by using only a small subset of all possible stereoisomers.

Enantiomeric excess can be detected in a number of ways. Direct observation of optical activity, that is, determination of the specific rotation of a compound, is cumbersome. Biochemical detection is also possible, although methodologies are generally specific to individual compounds or compound types. Indirect measurement through a chromatographic procedure, for example, gas chromatography or gas chromatography-mass spectrometry on a chiral stationary phase, has wide applicability and is very sensitive. Chirality is often viewed as the unique province of biological systems. For this reason, researchers were surprised to find that several of the amino acids found in the Murchison meteorite were slightly enantiomerically enriched. Especially notable were enantiomeric enrichments of 7 to 9% in three specific amino acids not known in forms of life currently on Earth, thus making contamination from terrestrial sources unlikely. If the amino acids in Murchison were not generated by living processes, it is possible that enantiomeric enrichment can be achieved by processes that are independent of life (by preferential destruction process like bombardment by polarized protons). This weighs against the usual thinking that enantiomeric enrichment is a unique signature of life. However, the data on enantiomeric excess are also consistent with a biotic origin, The larger asteroids may have stayed clement inside for several hundreds of millions of years, raising the possibility that life, or some borderline form, emerged there. This would be the case if the Murchison meteorite originated in an asteroid (or a comet) that supported life based on a-methyl amino acids that preferred one enantiomer to the other. While few would argue so, this result is consistent with life having emerged in the asteroid belt (National Academy of Sciences 2007).

The electron's spin is essential to the stability of matter, and control over the spin opens up avenues for manipulating the properties of molecules and materials. The Pauli exclusion principle requires that two electrons in a single spatial eigenstate have opposite spins, and this fact dictates basic features of atomic states and chemical bond formation. Because of its spin, the electron has a magnetic moment, and a magnetic field is usually required to separate (or split) the energies of its two possible spin states (quantum numbers of ½ or −½). This splitting energy is typically small and comparable to ambient thermal energies (~0.025 eV). When chiral molecules approach a substrate or another molecule, charge redistribution occurs. This charge polarization is accompanied by spin redistribution and a transient spin accumulation at the electric poles. Upon chemical adsorption, the spin of the electron on the surface that is opposite to the spin on the electric pole of the molecule has the more stable spatial orbital (nascent bonding), meaning that adsorbate interactions with antiparallel spins (one on the surface and one on the molecule) should be more stable, see Naaman et al. (2019). The electric field induces a spin-selective electron displacement that results in a transient spin polarization at the electric poles ($\delta^{(+)}$ and $\delta^{(-)}$ at the ends of the helices). The spin associated with each pole depends on the handedness of the molecule.

Figure 3.13 shows a schematic description of the electron transmission through a chiral potential of pitch p. The electron moves within the potential, while a constant force, like the classical centripetal force, acts on it in the direction perpendicular to its velocity. This force is like a Lorentz force, F, that results from a magnetic field, B, applied along the axis of the molecule. The effective spin–orbit coupling is then given by $\neg\mu \cdot B$, where $\neg\mu$ is the magnetic moment of the spin.

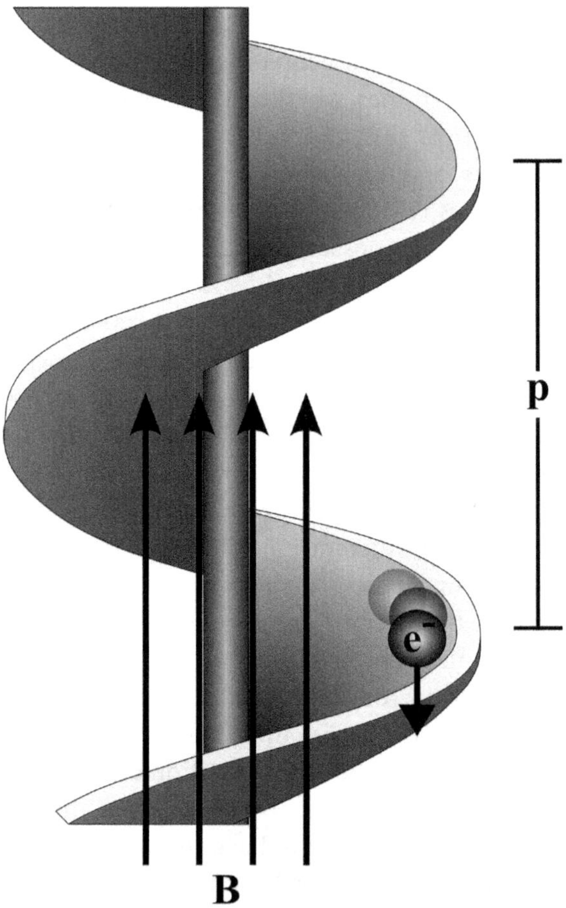


FIGURE 3.13 A schematic description of the electron transmission through a chiral potential of pitch p.

Spin alignment and the associated exchange energy linked to molecular chirality can play a role inefficient charge transfer, charge separation, bio-recognition, and protein processes, all important for biological functions. As an example, let us mention the work by Zwang et al. (2016) describing experiments on magnetized DNA-modified electrodes to explore spin-selective electron transport through hydrated duplex DNA. Their results show that the two spins migrate through duplex DNA with a different yield and that spin selectivity requires charge transport through the DNA duplex. Significantly, shifting the same duplex DNA between right-handed B- and left-handed Z-forms leads to a diode-like switch in spin selectivity; which spin moves more efficiently through the duplex depends upon the DNA helicity. With DNA, the supramolecular organization of chiral moieties, rather than the chirality of the individual monomers, determines the selectivity in spin, and thus a conformational change can switch the spin selectivity

The recently discovered chiral-induced spin selectivity (CISS) effect gives rise to a spin selective electron transmission through biomolecules first reported by Ray et al. (1999). Depending on the handedness of the molecule, electrons of a certain spin can traverse the molecule more easily in one direction than in the other. These directions are reversed for electrons of opposite spin. A number of different chiral organic molecules and biomolecules have been shown to act as spin filters, of which some of the most selective spin filters are nucleic acids; see Xie et al. (2011) and Rosenberg et al. (2015), and peptides; see Kettner et al. (2015). These observations are surprising, because spin filtering is commonly associated with magnetic materials or with substances that possess large spin orbit couplings (SOCs), rather than organic molecules which are typically neither magnetic nor have large SOC. Michaeli et al. (2016) described three processes arising from the chirality of molecules and specifically by the chiral-induced spin selectivity effect: long-range electron transfer, spin effects on the oxidation of water, and enantio-selectivity in bio-recognition events. They discussed studies which have been used to show that electron transfer through biomolecules can be highly efficient and spin selective, and they discussed how multi-electron reactions can have particular reaction pathways preferred by the spin filtering effect. In addition, they conjectured about a beneficial role that chiral-induced spin selectivity might play in enantioselective bio-recognition events. Each of these fundamental processes is important to biology, and the role played by spin and chirality is significant.

Naaman et al. (2020) discusses recent experiments that bear on the chiral-induced spin selectivity mechanism and its manifestation in electronic and magnetic properties of chiral molecules and materials. Although the discussion emphasizes newer experiments, such as the magnetization dependence of chiral molecule interactions with ferromagnetic surfaces, early experiments, which reveal the nonlinear scaling of the spin filtering with applied potential, are also described. In many of the theoretical studies, one has had to invoke unusually large spin–orbit couplings in order to reproduce the large spin filtering observed in experiments. Experiments imply that exchange interactions and Pauli exclusion constraints are an important aspect of CISS. They also demonstrate the spin-dependent charge flow between a ferromagnetic substrate and chiral molecules. With these insights in mind, a simplified model is described in which the chiral molecule's spin polarization is enhanced by a spin blockade effect to generate large spin filtering.

In recent theoretical work by Dalum and Hedegård (2019) one possible solution for the issue of small spin–orbit coupling was provided for a helical chain of atoms. In this treatment, the spin–orbit coupling on each atom in the chiral molecule is given by $\lambda L\sigma$, where $\lambda$ is the spin–orbit coupling observed for carbon atoms (5 meV), L the angular momentum, and $\sigma$ the Pauli spin matrix. The total spin–orbit coupling of the chiral molecules is then expressed as the vector sum of the atomic terms:

$$H_{SOC} = \sum \lambda_i \mathbf{L}_i \cdot \sigma_i$$

The SOC for the chiral chain can be significantly larger (order of magnitude) than the atomic value. This SOC, along with accidental degeneracies (curve crossings) that appear in the spectrum of chiral molecules, yields a sizable spin polarization. Although quantitative comparisons are not available, this sort of description may account for the large spin polarization seen in transport and in photoemission experiments. The approaches described previously are one-electron (orbital) models, whereas CISS may require a many-body description.

One can consider another possible mechanism, which is based on a spin blockade idea. First, consider a process for generating a spin polarization in a chiral molecule. When an electric field (or gradient in electrochemical potential) is applied across a molecule, or a monolayer of molecules, charge polarization occurs. This charge displacement current generates opposite spin polarizations at the negative and the positive electric poles of the molecule; which spin density exists in excess on the pole with electron density excess depends on the molecule's handedness. It is important to appreciate that the total spin of the molecule is still zero because of the total spin conservation. The spin polarization is a transient local property; namely, it means that in the molecule there are sites in which the net spin density is not zero. However, the sum over all those sites results in net zero spin. For chiral molecules with a relatively small SOC of about 5 meV, a spin polarization, $\Delta P(V)$, of a few percent accompanies this charge polarization. Please note that the spin polarization will depend on the electric field applied on the molecule (hence on the voltage applied), because it requires charge polarization, which is field dependent.

Now we consider how a molecule's small spin polarization causes strong spin filtering. As an electron moves from a donor (or negatively biased electrode) into the chiral molecule, it can have, in principle, two possible spin orientations with regard to its velocity. Because of the spin polarization at the chiral molecule's site near the donor/electrode, a difference in energy, ΔeV, is associated with the injection into the molecule of one spin over the other. This difference in energy is

$$\Delta eV = \Delta P(V) E_{ST}$$

in which $E_{ST}$ is the energy gap between the singlet and triplet states of the molecule. To motivate this choice, consider a thought experiment in which a molecular radical is being reduced by an incident electron. For a chiral molecule, the unpaired electron has a preferred spin direction in the molecular frame. Depending on the relative orientation of the reducing electron's spin and that of the molecular radical, a molecular anion can form in either a singlet state or a triplet state, as a result of the Pauli exclusion principle. In this way, the spin constraints applied in this thought experiment/reaction result in large energy differences for the two possible reactions (singlet–triplet splitting energy, EST). In many spin-filtering measurements, the excess electron is transiting through the molecule, and the molecule's spin polarization is not 100%. In the previous equation, the factor ΔP scales the energy splitting to account for the imperfect degree of spin polarization.

As an example, consider the interaction between two helical molecules as represented by the schematic diagram in Figure 3.14. When two molecules interact, charge polarization occurs. In the case of chiral molecules this charge polarization is accompanied by spin polarization, because of the CISS effect. Figure 3.14A shows the case of two molecules with the same helicity, for which the dynamic charge polarization (indicated by the asymmetric blue shading and the charge displacement q) also generates a spin polarization (indicated by the black arrows on the red balls). Because the molecules are of the same helicity (chirality), the spin polarization in the regions of enhanced electron density point in the same direction, and the regions of depleted electron density point in the same, but opposite, directions. For the displacement shown in Figure 3.14A along the helical axis, the spin polarizations point outward along the helical axis. If the molecules are close enough together for sufficient electronic cloud overlap, then one must consider the exchange interaction between the molecules. For the case shown in Figure 3.14A the spin polarizations are aligned anti-parallel (paired like a singlet state), whereas for the case shown in Figure 3.14B for two interacting molecules of opposite chirality (represented by helices of opposite handedness), the spin polarizations have the opposite directions (pointed inward for the left-handed helix, whereas it is outward for the right-handed helix). In this latter case, the exchange interaction between the molecules is characterized by two spins parallel to each other, analogous to a triplet state, and this heterochiral interaction (Figure 3.14B) is more destabilizing

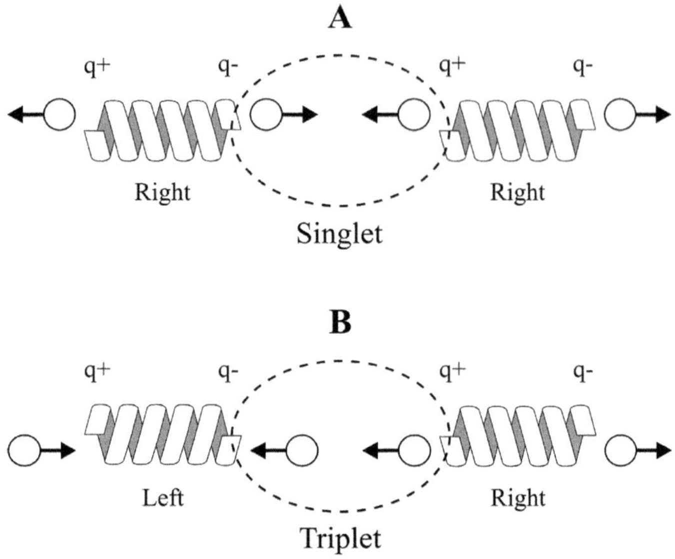


FIGURE 3.14 A) The diagram illustrates the interaction of two helices with the same handedness. Because of the induced dipole interaction charge q is transferred from one side of the helix to the other, and because of the CISS effect a spin polarization is generated (represented by a black arrow on the red ball). The side of the helix with excess electron density (q-) has a spin polarization parallel to the helical axis, and the side of the helix with a decrement of electron density (q+) has a spin polarization anti-parallel to the helical axis. In the region of electron density overlap (dotted circle-singlet surface), the two electron densities have a total spin polarization like that of a singlet state. B) When the helical charge distribution is of the opposite chirality, its spin polarization points in the opposite direction, hence the arrows are pointed inward for the left-handed helix. For the two helices of opposite chirality the overlap region (in the dotted circle-triplet surface) is characterized by two spins parallel to each other, like a triplet state (after Michaeli et al. 2016).

than the homochiral interaction (Figure 3.14A). Another presentation of the same phenomenon is to say that a hole with one spin and an electron with opposite spin interact at the end of the chiral molecule and that the hole attracts the electron of the opposite spin more strongly than it does an electron of the same spin. In summary, this interaction between the spins is enantioselective; irrespective of its magnitude, it gives rise to an enantioselective force in nature that is underappreciated.

Figure 3.15 presents a scheme of the model; upon applying an electric field on the chiral molecule, either by two leads or by having contacts with different electrochemical potentials at the two sides (represented by the displaced yellow lines), charge moves in response to the field (excess charge indicated by q+ and q− near the helical coil, which represents the molecule). The charge distribution, following the application of the applied voltage, is presented by the green and purple wave-like curves. Because of the chiral molecule's SOC, the charge reorganization is spin dependent. Hence, each curve represents a different spin alignment, spin-aligned parallel (green), or antiparallel (purple) relative to the electron displacement direction. The dotted line represents the field inside the molecule, assuming that the molecule has a very low dielectric constant. For a SOC of a few meV, the charge reorganization generates a spin polarization, ΔP, of a few percent.

Multiheme cytochromes, located on the bacterial cell surface, function as long-distance (>10 nm) electron conduits linking intracellular reactions to external surfaces. This extracellular electron transfer process, which allows microorganisms to gain energy by respiring solid redox-active minerals, also facilitates the wiring of cells to electrodes. While recent studies have suggested that achiral-induced spin selectivity effect is linked to efficient electron transmission through biomolecules, this phenomenon has not been investigated in extracellular electron conduits. Using magnetic conductive probe atomic force microscopy, Hall voltage measurements, and spin-dependent electrochemistry of the decaheme cytochromes MtrF and OmcA from the metal-reducing bacterium *Shewanella oneidensis* MR-1, Mishra et al. (2019) showed that electron transport through these extracellular conduits is spin-selective. This study has implications for understanding how spin-dependent interactions and magnetic fields may control electron transport across biotic–abiotic interfaces in both natural and biotechnological systems.

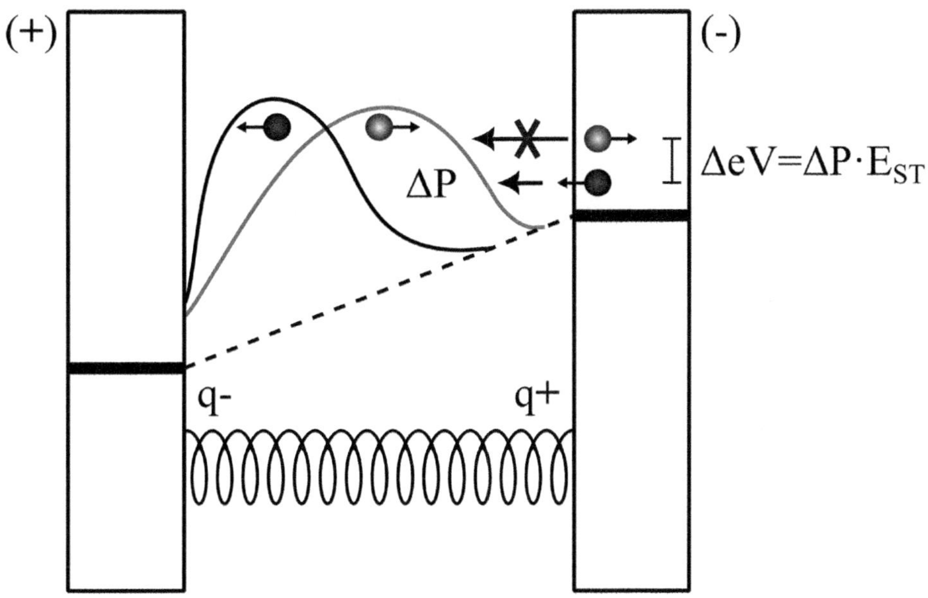


FIGURE 3.15 Scheme describing the mechanism of the CISS effect in terms of a small spin polarization, ΔP, that arises from the SOC. This small spin polarization causes a spin blockade, because of the Pauli principle, that is proportional to the singlet–triplet energy gap, EST, in the molecule. The purple and green curves represent the charge distribution occurring upon applying the field across the molecule, for electrons with spin aligned parallel (green) or antiparallel (purple) to their velocities. The molecule is presented schematically as a coil. The yellow lines indicate the Fermi energy at each electrode, and the dotted line shows the electric field across the molecule, assuming a molecule with a very low dielectric constant.

The functionality of many biological systems depends on reliable electron transfer with minimal heating. Interestingly, nature realizes electron transport via insulating molecules, in contrast to human-made electronic devices, which are based on metals and semiconductors. The high efficiency of electron transfer through these organic molecules is unexpected for tunneling-based transport, and it is one of the most compelling questions in the field. Furthermore, it has been shown that the electron tunneling probability is strongly spin dependent. Michaeli and Naaman (2019) demonstrated that the chiral structure of these molecules gives rise to robust coherent electron transfer. They introduced spin into the analysis of tunneling through organic helical molecules and showed that they support strong spin filtering accompanied by enhanced transmission. Thus, their study resolves two key questions posed by transport measurements through organic molecules.

Chiral organic molecules (like DNA) exhibit surprisingly rich spin physics, such as efficient spin filtering. Chirality-induced spin selectivity is now an umbrella term that defines a wide range of phenomena in which the chirality of molecular species imparts significant spin selectivity to various electron processes. The interplay between molecular handedness and electron spin suggests spin-orbit coupling as a natural mechanism for CISS, given that it is easy to show that "current through a coil" arguments yield an effect that is lower by many orders of magnitude. Early theoretical efforts have indeed confirmed that SOC may provide a qualitative explanation for some aspects of the experimental findings. Quantitatively, however, such calculations have consistently predicted effects that were smaller by up to several orders of magnitude than those observed experimentally. Additional theoretical research efforts (see Evers et al. 2022) have shed more light on CISS, but a complete quantitative theory of the effect remains elusive and its microscopic origins are insufficiently understood.

### 3.6.1 Origins of Chirality

The fact that sugars, amino acids, and the biological polymers they construct exist exclusively in one of two possible mirror-image forms has fascinated scientists and laymen alike since Pasteur's first painstaking separation of the enantiomorphic crystals of a tartrate salt. Yet, it was only in the late 20th century that experimental studies began to probe how biological homochirality, a signature of life, arose from a prebiotic world that presumably contained equal amounts of both mirror-image forms of these molecules. A review by Blackmond (2019) discusses experimental studies aimed at understanding how chemical reactions, physical processes, or a combination of both may provide prebiotically relevant mechanisms for the enrichment of one form of a chiral molecule over the other to allow for the emergence of biological homochirality.

How homochirality concerning biopolymers (DNA/RNA/proteins) could have originally occurred (i.e., arisen from a non-life chemical world, which tended to be chirality-symmetric) is a long-standing scientific puzzle. For many years, people have focused on exploring plausible physic-chemical mechanisms that may have led to prebiotic environments biased to one chiral type of monomers (e.g., D-nucleotides against L-nucleotides; L-amino-acids against D-amino-acids)—which should have then assembled into corresponding polymers with homochirality. Chen and Ma (2020) showed, by computer simulation—with a model based on the RNA world scenario—that the biased-chirality may have been established at polymer level instead, just deriving from a racemic mixture of monomers (i.e., equally with the two chiral types). In other words, the results suggested that the homochirality may have originated along with the advent of biopolymers during the origin of life, rather than somehow at the level of monomers before the origin of life.

Stereogenic $sp^3$-hybridized carbon centers are fundamental building blocks of chiral molecules. Unlike dynamic stereogenic motifs, such as $sp^3$-nitrogen centers or atropisomeric biaryls, $sp^3$-carbon centers are usually fixed, requiring intermolecular reactions to undergo configurational changes. Bismillah et al. (2023) report the internal enantiomerization of fluxional carbon cages and the consequences of their adaptive configurations for the transmission of stereo chemical information. The $sp^3$-carbon stereochemistry of the rigid tricyclic cages is inverted through strain-assisted Cope rearrangements, emulating the low-barrier configurational dynamics typical for sp3-nitrogen inversion or conformational isomerism. This dynamic enantiomerization can be stopped, restarted or slowed by external reagents, while the configuration of the cage is controlled by neighboring, fixed stereogenic centers. As part of a phosphoramidite–olefin ligand, the fluxional cage acts as a conduit to transmit stereo-chemical information from the ligand while also transferring its dynamic properties to chiral-at-metal coordination environments, influencing catalysis, ion pairing and ligand exchange energetics.

Cope rearrangements of chiral 9-BB cages simultaneously invert every stereogenic $sp^3$-carbon center of their structures. These configurational rearrangements occur rapidly and reversibly, achieving the uncommon property of dynamic $sp^3$-carbon stereochemistry—one that has remained surprisingly rare since Le Bel (1874) and van't Hoff (1874) first identified tetrahedral carbon as a source of molecular chirality in 1874. Both the rate of $sp^3$-carbon inversion and the equilibrium distribution of isomers are sensitive to changes in the 9-BB structure. On the one hand, the dynamics of the rearrangement processes are controlled through manipulation of covalent bonding or metal coordination of the 9-BB olefin groups, providing convenient functional handles. On the other hand, the cage adapts its configuration to minimize steric interactions with nearby fixed stereogenic elements and, in so doing, is able to transmit the stereo-chemical information across its rigid, tricyclic backbone. When interfaced with transition-metal complexes, the dynamic cage conveys a stereo-chemical preference to the chiral-at-metal center. Controllable and adaptable $sp^3$-carbon stereochemistry of this kind can be exploited in enantioselective synthesis and chiral functional materials.

Biological molecules chose one of two structurally chiral systems, which are related by reflection in a mirror. Globus and Blandford (2020) proposed that this choice was made, causally, by cosmic rays, which are known to play a major role in mutagenesis. The authors have shown that magnetically polarized cosmic rays that dominate at ground level today can impose a small but persistent chiral bias in the rate at which they induce structural changes in simple, chiral monomers that are the building blocks of biopolymers. A much larger effect should be present with helical biopolymers, in particular those that may have been the progenitors of ribonucleic acid and deoxyribonucleic acid. It is shown that the interaction can be both electrostatic, just involving the molecular electric field, and electromagnetic, also involving a magnetic field. It is argued that this bias can lead to the emergence of a single, chiral life form over an evolutionary timescale. If this mechanism dominates, then the handedness of living systems should be universal. Experiments are proposed to assess the efficacy of this process. A prediction of their model is that the mutation rate is dependent upon the spin polarization of the radiation. A possible experiment would be to measure the mutation rate of two cultures of bacteria under spin-polarized radiation (either $e\pm$ or $\mu\pm$) of different lodacity with energy above the threshold necessary to induce double strand breaks in DNA (~50 eV). If the coupling between lodacity and molecular chirality is efficient in introducing a chiral bias, one of the two cultures should exhibit a lower mutation rate. We emphasize that much can be learned experimentally from the comparison of chiral molecules involved in biology and using both signs of lodacity that can be created at accelerators. It is not necessary to create "mirror life" to proceed. Once the dominant processes are identified, we can have confidence in our understanding of particle physics and quantum chemistry to draw the necessary conclusions. Additional experiments, relevant to THIS electromagnetic models, involve measuring the magnetic structure and properties of biopolymers. If these experiments show that the evolution of bacteria is sensitive to polarization, this will be a good indication that magnetically polarized cosmic rays are an important piece of the chiral puzzle of life.

We proposed a new mechanism that might result in chirality excess, assuming that amino acids were synthesized on dust particles in interstellar space. This is the bombardment by cosmic rays of high-energy protons polarized in magnetic fields. Polarized proton cosmic rays preferentially destroy one isomer because of significant asymmetry in proton (in cosmic rays)—proton scattering (in amino acid/sugar) on the surface of ISM dust particles aligned by the magnetic fields; see Valković and Obhođaš (2022). Such dust particle alignment has been observed for solar magnetic field, and it is assumed to be the reason the light scattered by cometary dust becomes circularly polarized (Kolokolova et al. 2016). Considering the chemical composition of dust in interstellar molecular clouds (Zhukovska et al. 2008) and recent findings in the field of prebiotic solid-state chemistry (Oba et al. 2019; Stolar et al. 2020), which show that the evolution from molecular clouds to stars and planets provides a suitable environment for nucleobase synthesis in space, we can hypothesize that life arose chiral because originated in the IMCs with the critical role of dust particles, magnetic fields, and exposure to cosmic rays. The arguments supporting the hypothesis are put forward based on numerous astrophysical observations and physics laws.

Together with the phenomenon of elements' essentiality, we should consider a closely related phenomenon of chirality. A living organism is an organized system of molecules with specific handedness, called chirality. Chiral molecules are designated D (dextrorotatory) or L (levorotatory) according to the right or left direction, respectively, in which the crystalline forms rotate polarized light. Biological polymers (e.g., nucleic acids and proteins), whose function is determined by their shape, use almost exclusively D sugars and L amino acids. The exception is glycine, as it has two (indistinguishable) hydrogen atoms attached to its alpha (central) carbon. Laboratory synthesis made from optically inactive starting materials yields racemic (1:1) mixtures of L and D isomers. Nineteen of the 20 amino acids used in the synthesis of proteins can exist as L or D enantiomorphs. As early as 1953, Frank proposed a model of enantioselective autocatalysis where molecules maintain their own production while inhibiting the formation of the opposite enantiomer. This important model showed that spontaneous asymmetric synthesis is a natural property of life. Joyce et al. (1984) found that template-directed reactions with a racemic mixture autocatalyze the replication of a polynucleotide of the same chirality, while the reactions with the opposite template were far less efficient. Moreover, it was noted that the incorporation of the monomers of the opposite chirality terminates the polymerization. Yin et al. (2015) showed that chiral macroanions demonstrate chirality recognition behavior by forming a homogeneous blackberry structure via long-range electrostatic interactions between the individual enantiomers in their racemic solutions. Adding chiral co-anions suppresses the self-assembly of one enantiomer while maintaining the assembly of the other one. However, for enantiomeric amplification to start, previously some physical or chemical process or their combination had to crack the symmetry between left-handed and right-handed molecules. There have been several suggestions, including (1) polarized light (see De Marcellus et al. 2011; Garcia et al. 2019; MacDermott 2012); (2) optically active quartz (see Bonner 2020); (3) natural radioactivity (see MacDermott 2012); (4) cosmic rays (see Boyd et al. 2018; Globus and Blandford 2020); and (5) a nonenergetic mechanism (atom addition surface reaction; Ioppolo et al. 2020). The small chiral bias produced in proposed processes is unlikely to lead to a homochiral state, and some prebiotic mechanism is still required for enantiomeric amplification; see Blackmond (2010).

### 3.6.2 Preferential Destruction

Numerical simulations suggest that unpolarized light with a twisted phase could help sort left- and right-handed molecules into separate ring-shaped traps (Forbes and Green 2022).

Valković and Obhođaš (2022) proposed a new mechanism that might result in chirality excess, assuming that amino acids were synthesized on dust particles in interstellar space. This is bombardment by cosmic rays of high-energy protons polarized in magnetic fields. Polarized proton cosmic rays preferentially destroy one isomer because of significant asymmetry in proton (in cosmic rays)—proton scattering (in amino acid/sugar) on the surface of ISM dust particles aligned by the magnetic fields. Such dust particle alignment has been observed for solar magnetic fields, and it is assumed to be the reason the light scattered by cometary dust becomes circularly polarized; see Kolokolova et al. (2016). Actually, the most popular explanation of CP formation is the scattering of light on aligned/irregular dust particles or on the particles that contain homochiral molecules.

### 3.6.3 Preferential Destruction of Enantiomers by Spin-Polarized Cosmic-Ray Protons

The sources of accelerated protons that might collide with individual atoms or dust particles in IMCs are the primary cosmic rays irradiated by stars, black holes, or the BB itself. The primary cosmic rays consist of 95% protons, 4% of helium, and 1% of heavier elements up to iron. Their energies could be enormous, from $10^9$ eV (1GeV) up to $10^{20}$ eV ($10^8$ TeV). Since they are charged (otherwise they cannot be accelerated), they can be polarized by magnetic fields on their path through the IMCs. Polarized protons colliding with dust particles might be elastically scattered in spin-polarized proton–proton (p-p) reaction, either proton (beam or target) can be polarized. One of the main characteristics of the p-p elastic scattering reaction in the energy range of primary cosmic-ray protons is the asymmetry between the number of scatterings on the left versus on the right, which can be above 5%; see Akchurin et al. (1993), Bazilevsky et al. (2011), Bravar et al. (2005), Okada et al. (2008), and Adamczyk et al. (2013). Polarization-induced asymmetry produced in p-p scattering has been observed even at low energies of 30 and 50 MeV; see Batty et al. (1963). The spin-polarized p-p reaction has been overlooked in the scientific literature as the possible cause of prebiotic molecular chirality. Sugars and amino acids involved in life fundamental processes of all taxa in the tree of life are exclusively single enantiomers, although both enantiomers have a similar probability of their formation in an achiral environment (Prelog 2006). Enantiomer's selection is critical for molecular interaction and replication processes; thus, it is crucial for understanding the origin of life.

Preferential destruction of one enantiomer leads to a disbalance of molecules available from the primordial reservoir of IMCs, which would almost certainly result in an extreme bias of enantiomers selection by emerging life; see Blackmond (2010) and McGuire and Carroll 2016). Analysis of amino acids in meteoritic samples, such as the one that hit Murchison, Australia, shows an excess of left-handed enantiomers of up to 10% suggesting that such an enantiomeric excess has been generated before the Earth's formation, McGuire and Carroll (2016). The process discussed by the authors could have generated an enantiomeric excess in left-handed amino acids and right-handed sugars, thus linking the origins of life's enantiomeric bias to IMCs. Since the life in the universe emerged probably long before the occurrence of our solar system, as suggested by a comparison of life and stellar elements abundance curves, it is reasonable to assume that dust that formed our solar system already contained life in some primitive form.

#### *3.6.3.1 Proposed Experimental Verification Preferential Destruction of One Isomer Due to Asymmetry in p-p Scattering*

We proposed the spin-polarized proton–proton scattering as a potential physical process that takes place in IMCs environments and could lead to enrichment of L-type amino acids and D-type sugars. Analyzing power AN(t) for p-p elastic scattering in the energy range of energetic cosmic ray protons has been studied by several groups; see Akchurin et al. (1993), Bravar et al. (2005), Okada et al. (2008). Assuming that amino acids, and some sugars, could be synthesized on dust particles in interstellar space, the observed optical activity may be a result of cosmic ray bombardment. High-energy polarized protons in cosmic rays may be able to preferentially destroy one isomer on the dust particle surface because of significant asymmetry in proton (in cosmic rays)–proton (chirality center in amino acid or sugar) scattering.

Things to be done:

- Model calculations of preferential destruction of enantiomers.
- Accelerator-based experiment on preferential destruction of one isomer (amino acids and sugars) due to asymmetry in proton (in cosmic rays)-proton (in amino acid) scattering.

### 3.6.4 Learning Chemical Networks

Chirality preference in living matter may arise spontaneously to optimize energy harvesting. The model developed by Piñeros and Tlusty (2022) showed that highly dissipating systems and large energy differences are more prone to inducing chiral symmetry breaking. Furthermore, the calculations revealed that such transitions are almost inevitable, so it is reasonable to believe they may generically occur in random chemical reaction systems. Thus, the energy harvesting optimization-based model demonstrated by the group explains how homochirality could have spontaneously arisen from the harsh, energy-rich environment of the early planet Earth.

Living systems have evolved to efficiently consume available energy sources using an elaborate circuitry of chemical reactions which, puzzlingly, bear a strict restriction for asymmetric chiral configurations. While autocatalysis is known to promote such chiral symmetry breaking, whether a similar phenomenon may also be induced in a more general class of configurable chemical systems—via energy exploitation—is a sensible yet underappreciated possibility. This work examines this question within a model of randomly generated complex chemical networks. We show that chiral symmetry breaking may occur spontaneously and generically by harnessing energy sources from external environmental drives. Key to this transition are intrinsic fluctuations of achiral-to-chiral reactions and tight matching of system configurations to the environmental drives, which together amplify and sustain diverged enantiomer distributions. These asymmetric states emerge through steep energetic transitions from the corresponding symmetric states and sharply cluster as highly-dissipating states. The results thus demonstrate a generic mechanism in which energetic drives may give rise to homochirality in an otherwise totally symmetrical environment, and from an early-life perspective, might emerge as a competitive, energy-harvesting advantage.

## 3.7 IMPORTANCE OF METALS AND HEAVY ELEMENTS FOR LIFE PROCESSES

Recently, attention have attracted heavy elements (i.e. Mo, W) essential for living species. One of the topic recently debated is the origin of $^{180}W$ excesses in iron meteorites. Cook et al. (2018) studied a number of IIAB iron meteorites in order to determine the contribution from galactic cosmic rays (GCRs) and from potential decay of $^{184}Os$ in minor isotope $^{180}W$. Their measurements represent the first clear evidence for p-process variability in W isotopes.

The $^{180}W$ excesses observed in IIAB irons do not have a single origin. These excesses consist of three components, which include a small initial nucleosynthetic anomaly upon which ingrowth from $^{184}Os$ decay has occurred and by GCR induced effects. The same approach could be used to elucidate if heterogeneity in p-process process produced W was widespread in the early solar system or if it is restricted to the IIAB parent body.

Valković and Obhođaš (2022) have presented numerous examples of observations, experiments, and theoretical considerations, which we have used to synthesize our hypothesis concerning the origin of life on interstellar dust particles. The universe was born only with plenty of hydrogen isotopes, some helium, and lithium, while all other elements were formed as the universe aged through star formation processes and their lives and deaths, which resulted in dust clouds as a birthplace of a new generation of stars. During this process of the universe aging, the chemical element abundance curve has been changing in such a manner that at the time $T_{origin}$ coincided with the abundance curve of living matter. Our hypothesis defines $T_{origin}$ as a time when conditions were right for life to originate in the primitive form. Exposure to cosmic rays and magnetic fields as well as physical, chemical, and biological processes made life chiral. Our preliminary considerations of the solar neighborhood as a part of the Milky Way galaxy indicate period of $(4 \pm 1) \times 10^9$ years after BB for the origin of life on the carbonaceous dust particles in the IMCs of the solar neighborhood. The same could be expected for similar zones in other spiral galaxies. Emerged life could have survived the planet formation processes to evolve in habitable zones.

According to Bromberg et al. (2022), quantifying structural relationships of metal-binding sites suggests origins of biological electron transfer. Biological redox reactions drive planetary biogeochemical cycles. Using a novel, structure-guided sequence analysis of proteins, they explored the patterns of evolution of enzymes responsible for these reactions. Their analysis reveals that the folds that bind transition metal–containing ligands have similar structural geometry and amino acid sequences across the full diversity of proteins. Similarity across folds reflects the availability of key transition metals over geological time and strongly suggests that transition metal–ligand binding had a small number of common peptide origins. They observe that structures central to the similarity network come primarily from oxidoreductases, suggesting that ancestral peptides may have also facilitated electron transfer reactions. Last, their results reveal that the earliest biologically functional peptides were likely available before the assembly of fully functional protein domains over 3.8 billion years ago.

The origin(s) of biologically catalyzed redox reactions have been obscured by the marked expansion of protein folds following the Great Oxidation Event approximately 2.35 billion years ago. Hence, a phylogeny of these proteins rooted in sequence space is nearly impossible. The arrangement of multiple independent peptides into catalytically active structures further complicates any sequence-based evolutionary analysis. Such an analysis would require accounting for the coevolution of non-sequential amino acid sequences that describe structural domains. Therefore, Bromberg et al. (2022) focus on elucidating the evolution of these peptides based on their structures. Evolution of protein structures entails understanding how new folds arose from previously existing ones. Using network analysis, they trace distant relationships between metal-binding proteins. They observed that existing transition metal–binding folds are similar structurally and carry similar sequences within the overlapping structures. This observation suggests that they might have had a single or small number of common ancestors. Moreover, while metal binding folds in both current redox and non-redox proteins are similar, the central structures are most often derived from redox proteins. These central structures are enriched in prebiotically available amino acids. Their analysis suggests that "simple" folds, found in extant oxidoreductases, may have been incorporated into the complete contemporary range of metal binding and molecular functionality carried out by many enzyme families. They also identified small structural motifs that are frequently repeated within network-central structures. These motifs describe the appearance of central folds and are thus likely at the root of life.

Three conclusions from our work are most salient. First, the peptides at the origins of life were likely used for facilitating electron transfer. They have since been widely repurposed in the complete variety of biological functions and, hence, have adapted their sequences. However, the likely ancient folds are still enriched in amino acids that must have been prebiotically available. Second, the interchangeability of metals used in the biological activity reflects the likely contemporaneous availability of these metals on geological time scales, with folds binding these metals possibly adapted from a common ancestor. This observation suggests that the more robust geological chronometers could be useful in exploring the timing of biological evolution. Last, they find that the earliest biologically functional peptides were likely available before the assembly of fully functional protein folds over 3.8 billion years ago. These may have been used as cofactors for RNA-driven catalytic activity together with the available metal ligands, as evidenced by both their nucleotide and metal binding abilities. This scaffolding could then have allowed for more complicated protein fold assembly facilitated by bound metals and led to the structural diversity we observe today (Bromberg et al. 2022).

## 3.8 PREFERENTIAL DESTRUCTION AS A FUNDAMENTAL BIO-PROCESS

Famiano et al. (2018) consider the model involving asymmetries in the way certain particles interact with biomolecules at the nuclear level. It works through selective destruction: If there are initially equal numbers of left- and right-handed molecules and one kind is destroyed preferentially, then there is an imbalance. A nucleus within an amino acid could be destroyed in a spin-polarized way via the weak force, a parity-violating interaction. One can look at destructive interactions involving spin-polarized leptons—specifically neutrinos, which are inherently chiral—and nitrogen atoms in amino acids. The destructive mechanism explored is inverse beta decay. In that decay, the nitrogen atom becomes carbon-14 or oxygen-14, a process that destroys the whole amino acid. There is also a nuclear recoil, which can literally rip the molecule apart.

However, this process is chirality selective only if the amino-acid molecules all align with each other. If that happens, only amino acids of one chirality will be pointing in a direction that allows their nitrogen atoms to interact with the neutrinos. This alignment could happen through crystallization, or it could happen in an electric field. This process should work for any polarized lepton, not just neutrinos.

Chirality of the amino acids could be established in the magnetic field of a nascent neutron star from a core-collapse supernova or massive collapsar. The magnetic field would orient the $^{14}N$ nuclei, and the alignment of its nuclear spin with respect to those of the electron antineutrinos emitted from the collapsing star would determine the probability of destruction of the $^{14}N$ nuclei by interactions with the antineutrinos. It is estimated that the bulk polarization of the $^{14}N$ nuclei in large rotating meteoroids in such an environment. The present work adds a crucial piece of this model by describing the details by which the selective $^{14}N$ nuclear destruction would produce molecular chiral selectivity by the neutrino-induced interactions with the $^{14}N$ nuclei bound in amino acids polarized in strong magnetic fields.

According to Famiano et al. (2018), the chirality of the amino acids could be established in the magnetic field of a nascent neutron star from a core-collapse supernova or massive collapsar. The magnetic field would orient the $^{14}N$ nuclei, and the alignment of its nuclear spin with respect to those of the electron antineutrinos emitted from the collapsing star would determine the probability of destruction of the $^{14}N$ nuclei by interactions with the antineutrinos. Subsequent work estimated the bulk polarization of the $^{14}N$ nuclei in large rotating meteoroids in such an environment. The present work adds a crucial piece of this model by describing the details by which the selective $^{14}N$ nuclear destruction would produce molecular chiral selectivity. The effects of the neutrino-induced interactions on the $^{14}N$ nuclei bound in amino acids polarized in strong magnetic fields are studied. It is shown that electric fields in the reference frame of the nuclei modify the magnetic field at the nucleus, creating nuclear magnetizations that are asymmetric in chirality. The antineutrino cross sections depend on this magnetization, creating a selective destructive effect. The environmental conditions and sites in which such a selection mechanism could occur are discussed. Selective destruction of D-enantiomers results in enantiomeric excesses which may be sufficient to drive subsequent autocatalysis necessary to produce the few-percent enantiomeric excesses found in meteorites and subsequent homochirality. Molecular quantum chemical calculations were performed for alanine, and the chirality-dependent effects studied were included. A preference for left-handed molecules was found, and enantiomeric excesses as high as 0.02% were estimated for molecules in the electromagnetic conditions expected from a core-collapse supernova

The origin of life, based on the homochirality of biomolecules, is a persistent mystery. Did life begin by using both forms of chirality, and then one of the forms disappeared? Or did the choice of homochirality precede the formation of biomolecules that could ensure replication and information transfer? Is the natural choice of L-amino acids and D-sugars on which life is based deterministic or random? Is the handedness present in/of the Universe from its beginning? The whole biosystem on Earth and all living creatures are chiral. Many theories try to explain the origin of life and chirality on the Earth, such as the panspermia hypothesis, primordial soup hypothesis, and theory of parity violation in weak interactions. Additionally, heavy neutrinos and the impact of the fact that only left-handed particles decay, and even dark matter, all have to be considered, after Devinsky (2021).

## 3.9 MOLECULES ESSENTIAL FOR LIFE

Rimola et al. (2022) published a review article on tracing the primordial chemical life of glycine. Glycine (Gly), $NH_2CH_2COOH$, is the simplest amino acid. Although it has not been directly detected in the interstellar gas-phase medium, it has been identified in comets and meteorites, and its synthesis in these environments has been simulated in terrestrial laboratory experiments. Likewise, condensation of Gly to form peptides in scenarios resembling those present in a primordial Earth has been demonstrated experimentally. Thus, Gly is a paradigmatic system for biomolecular building blocks to investigate how they can be synthesized in astrophysical environments, transported and delivered by fragments of asteroids (meteorites, once they land on Earth) and comets (interplanetary dust particles that land on Earth) to the primitive Earth, and there react to form biopolymers as a step towards the emergence of life. Quantum chemical investigations addressing these Gly-related events have been performed, providing fundamental atomic-scale information and quantitative energetic data. However, they are spread in the literature and difficult to harmonize in a consistent way due to different computational chemistry methodologies and model systems. This review aims to collect the work done so far to characterize, at a quantum mechanical level, the chemical life of Gly, that is, from its synthesis in the interstellar medium up to its polymerization on Earth.

The first step, Gly formation, could take place in the early phases of the solar-type planetary system formation. In principle, Gly can be synthesized in the gas phase and/or on the surfaces of the dust grains; see Aponte et al. (2017). Glycine and methylamine are meteoritic water-soluble organic compounds that provide insights into the processes that occurred before, during, and after the formation of the solar system. Both glycine and methylamine and many of their potential synthetic precursors have been studied in astrophysical environments via observations, laboratory experiments, and modeling. In spite of these studies, the synthetic mechanisms for their formation leading to their occurrence in meteorites remain poorly understood. Typical $^{13}C$-isotopic values ($\delta^{13}C$) of meteoritic glycine and methylamine are $^{13}C$-enriched relative to their terrestrial counterparts; thus, analyses of their stable carbon isotopic compositions ($^{13}C/^{12}C$) may be used not only to assess terrestrial contamination in meteorites but also to provide information about their synthetic routes inside the parent body. Here, we examine potential synthetic routes of glycine and methylamine from a common set of precursors present in carbonaceous chondrite meteorites, using data from laboratory analyses of the well-studied CM2 meteorite Murchison. Several synthetic mechanisms for the origins of glycine and methylamine found in carbonaceous chondrites may be possible, and the prevalence of these mechanisms will largely depend on (1) the molecular abundance of the precursor molecules and (2) the levels of processing (aqueous and thermal) that occurred inside the parent body.

Rimola et al. (2022) studied the grain surface routes in synthesis of Gly. Grains of the dense molecular clouds consist of a refractory material core surrounded by icy mantles. Because of the watery nature of the ices, a reasonable path towards interstellar Gly is that occurring in water solution, that is, the Strecker synthesis. Strecker reactions comprise the synthesis of amino acids under acidic conditions by the reactivity of aldehydes/ketones with ammonia and hydrogen cyanide. For the particular case of Gly, the reaction involves three steps:

1. $H_2C{=}O + NH_3 \rightarrow NH_2CH_2OH \rightarrow NH{=}CH_2 + H_2O$;
2. $NH{=}CH_2 + HCN \rightarrow NH_2CH_2CN$;
3. $NH_2CH_2CN + 2H_2O \rightarrow NH_2CH_2COOH + NH_3$.

This reaction is particularly attractive from an interstellar perspective because the reactants (i.e., $H_2C{=}O$, $NH_3$, and HCN) are compounds usually identified as minor species in the ice mantles, and the $NH{=}CH_2$ and $NH_2CH_2CN$ intermediates are compounds observationally detected in different interstellar environments. Because of that, different works have simulated the Strecker synthesis (partly, totally, or modifications of it) for the formation of Gly in the presence of water molecules mimicking the watery ice mantles.

Rimola et al. (2022) consider only Gly transportation and delivery to primitive Earth. They do not consider eventual survival in the Earth formation process! Irrespective of the way through which Gly forms in space, the next step to evolve into a more biochemically complex form in terrestrial environments is its transportation and delivery to the primitive Earth. A feasible way was that Gly, once synthesized, was adsorbed and entrapped on asteroidal dust grains (either of cometary, meteoritic or any other origin) and then released to the early Earth's surface. Although the chances of Gly delivery were highly likely during the epoch of intense meteoritic bombardments, it is worth mentioning that even today, great amounts of these dust grains continuously penetrate the atmosphere, enriching our planet with organic matter; see Chyba and Sagan (1992).

Ancestral metabolic processes involve the reversible oxidation of molecular hydrogen by hydrogenase. Extant hydrogenase enzymes are complex, comprising hundreds of amino acids and multiple cofactors. Timm et al. (2023) designed a 13-amino acid nickel-binding peptide capable of robustly producing molecular hydrogen from protons under a wide variety of conditions. The peptide forms a di-nickel cluster structurally analogous to a Ni-Fe cluster in [NiFe] hydrogenase and the Ni-Ni cluster in acetyl-CoA synthase, two ancient, extant proteins central to metabolism. These experimental results demonstrate that modern enzymes, despite their enormous complexity, likely evolved from simple peptide precursors on early Earth.

Glucose is a sugar with the molecular formula $C_6H_{12}O_6$. Glucose is overall the most abundant monosaccharide, a subcategory of carbohydrates. Glucose is mainly made by plants and most algae during photosynthesis from water and carbon dioxide, using energy from sunlight, where it is used to make cellulose in cell walls, the most abundant carbohydrate in the world.

In energy metabolism, glucose is the most important source of energy in all organisms. Glucose for metabolism is stored as a polymer, in plants mainly as starch and amylopectin and in animals as glycogen. Glucose circulates in the blood of animals as blood sugar. The naturally occurring form of glucose is d-glucose, while its stereoisomer l-glucose is produced synthetically in comparatively small amounts and is less biologically active. Glucose is a monosaccharide containing six carbon atoms and an aldehyde group and is therefore an aldohexose. The glucose molecule can exist in an open-chain (acyclic) as well as ring (cyclic) form. Glucose is naturally occurring and is found in its free state in fruits and other parts of plants. In animals, glucose is released from the breakdown of glycogen in a process known as glycogenolysis.

Glucose, as intravenous sugar solution, is on the World Health Organization's List of Essential Medicines. It is also on the list in combination with sodium chloride.

### 3.9.1 Evolution

The next step is evolution, the formation of more complex molecules, that is, proteins, made up of building blocks: amino acids. The common property of all proteins is that they consist of long chains of α-amino (alpha amino) acids. Although more than 100 amino acids occur in nature, particularly in plants, only 20 types are commonly found in most proteins. In protein molecules, the α-amino acids are linked to each other by peptide bonds between the amino group of one amino acid and the carboxyl group of its neighbor. Sanger and Thompson (1953) were the first to determine the amino acid sequence in the glycyl chain of insulin, indicating that all amino-acids in this protein are L-type amino acids.

The complete tree of life is now available, but methods to visualize it are still needed to meet needs in research, teaching, and science communication. Dynamic visualization of million-tip trees requires many challenges in data synthesis, data handling, and computer graphics to be overcome; see Wong and Rosindell (2022). Their tree of life explorer is now broadly complete; it has attracted large number of online users and is backed by a novel long-term sustainability plan.

An origins-of-life scenario depicts Darwinian evolution from self-replicating molecules, such as RNA, toward complex living systems. How molecular replicators could develop complexity by continuously expanding information and functions is a central issue in prebiotic evolution. An expected route for complexification is that novel RNA replicators successively emerged and co-replicated so that increased genetic information can be stored at a population level before their assembly into a long genome. Although several theoretical studies investigated the possibility of complexification and stable coexistence of molecular replicators, empirical demonstration has been challenging

In prebiotic evolution, self-replicating molecules are believed to have evolved into complex living systems by expanding their information and functions openendedly. Theoretically, such evolutionary complexification could occur through successive appearance of novel replicators that interact with one another to form replication networks. Mizuuchi et al. (2022) performed long-term evolution experiments on RNA that replicates using a self-encoded RNA replicase. The RNA diversifies into multiple coexisting host and parasite lineages, whose frequencies in the population initially fluctuate and gradually stabilize. The final population, comprising five RNA lineages, forms a replicator network with diverse interactions, including cooperation to help the replication of all other members. These results support the capability of molecular replicators to spontaneously develop complexity through Darwinian evolution, a critical step for the emergence of life.

Our understanding of the bacterial cell cycle is framed largely by population-based experiments that focus on the behavior of idealized average cells. Most famously, the contributions of Cooper and Helmstetter (1968) help to contextualize the phenomenon of overlapping replication cycles observed in rapidly growing bacteria. Despite the undeniable value of these approaches, their necessary reliance on the behavior of idealized average cells masks the stochasticity inherent in single-cell growth and physiology and limits their mechanistic value. To bridge this gap, Sanders et al. (2023) proposed an updated and agnostic framework, informed by extant single-cell data, which quantitatively accounts for stochastic variations in single-cell dynamics and the impact of medium composition on cell growth and cell cycle progression. In this framework, stochastic timers sensitive to medium composition affect the relationship between cell cycle events, accounting for observed differences in the relationship between cell cycle events in slow- and fast-growing cells. They conclude with a roadmap for potential application of this framework to long-standing open questions in the bacterial cell cycle field.

Proliferation of organisms across the tree of life requires effective coordination of cell growth, DNA replication, and division. Coordination is challenging in bacteria for which population mass doubling times can vary as much as fivefold with nutrient availability. In many bacteria, including the model organism *Escherichia coli*, the time required to complete a round of DNA replication can be longer than the mass doubling time, particularly under nutrient-rich conditions, resulting in multiple ongoing cycles of DNA replication on the same chromosomal template.

Early work on the *E. coli* cell cycle focused on the relationships between growth rate and parameters: cell size, RNA content, DNA content, and nutrient composition. Schaechter et al. (1958) observed that the average mass of *Salmonella typhimurium* increases exponentially with nutrient-imposed increases in population mass doubling time. Protein, RNA, and DNA content similarly increase, indicating that their overall concentration (mass/mass) remains constant.

Recognizing the value of single-cell data as a framework from which to understand the molecular mechanisms underlying cell cycle progression in bacterial cells is just the first step. Larger, more comprehensive single-cell datasets spanning a wide range of MDTs and media compositions are essential to determine the relationship between nutrient availability and cell cycle progression at high resolution not only in *E. coli* but also in other bacteria, model, and non-model alike.

A planet's suitability to sustain life is the combination of a number of chemical and physical characteristics, including planetary size and stellar metallicity. Among these is the requirement for the presence of liquid water and hence, the planet's position in the circumstellar habitable zone. The initial development of life on Earth is also linked to iron. Cells utilize iron in numerous and highly conserved processes, including DNA replication, intermediate metabolism, and gene expression, exploiting the chemical properties of iron that facilitate redox reactions and bond formation in multiple orientations. That iron plays such a fundamental role across the gamut of biology suggests that its utilization is ancestral and hence may reflect environmental conditions at the initiation of life.

Iron is an irreplaceable component of proteins and enzyme systems required for life. This need for iron is a well-characterized evolutionary mechanism for genetic selection. However, there is limited consideration of how iron bioavailability, initially determined by planetary accretion but fluctuating considerably at global scale over geological time frames, has shaped the biosphere. Wade et al. (2021) described influences of iron on planetary habitability from formation events $>4 \times 10^9$ years and initiation of biochemistry from geochemistry through oxygenation of the atmosphere to current host–pathogen dynamics. By determining the iron and transition element distribution within the terrestrial planets, planetary core formation is a constraint on both the crustal composition and the longevity of surface water, hence a planet's habitability. As such, stellar compositions, combined with metallic core-mass fraction, may be an observable characteristic of exoplanets that relates to their ability to support life. On Earth, the stepwise rise of atmospheric oxygen effectively removed gigatons of soluble ferrous iron from habitats, generating evolutionary pressures. Phagocytic, infectious, and symbiotic behaviors, dating from around the Great Oxygenation Event, refocused iron acquisition onto biotic sources, while eukaryotic multicellularity allows iron recycling within an organism. These developments allow life to more efficiently utilize a scarce but vital nutrient. Initiation of terrestrial life benefitted from the biochemical properties of abundant mantle/crustal iron, but the subsequent loss of iron bioavailability may have been an equally important driver of compensatory diversity. This latter concept may have relevance for the predicted future increase in iron deficiency across the food chain caused by elevated atmospheric $CO_2$.

The capacity for biological systems to adapt to variable and unpredictable conditions, and to maintain certain key survival-requisite properties within tight tolerances, is fundamental to life itself. According to Araujo and Liotta (2023), the evolution of life, at the molecular level, is driven by the generation and diversification of adaptation mechanisms. A universal description of adaptation-capable chemical reaction network (CRN) structures has remained elusive until now, since currently known criteria for adaptation apply only to a tiny subset of possible CRNs. They identified the definitive structural requirements that characterize all adaptation-capable collections of interacting molecules, however large or complex. They showed that these network structures implement a form of integral control in which multiple independent integrals can collaborate to confer the capacity for adaptation on specific molecules. Using an algebraic algorithm informed by these findings, they demonstrated the existence of embedded integrals in a variety of biologically important CRNs that have eluded previous methods, and for which adaptation has been observed experimentally. This definitive picture of biological adaptation at the level of intermolecular interactions represents a blueprint for adaptation-capable signaling networks across all domains of life, and for the design of synthetic biosystems.

Physical laws, such as the laws of motion, gravity, electromagnetism, and thermodynamics, codify the general behavior of varied macroscopic natural systems across space and time. Wong et al. (2023) propose that an additional, hitherto-unarticulated law is required to characterize familiar macroscopic phenomena of our complex, evolving universe. An important feature of the classical laws of physics is the conceptual equivalence of specific characteristics shared by an extensive, seemingly diverse body of natural phenomena. Identifying potential equivalencies among disparate phenomena—for example, falling apples and orbiting moons or hot objects and compressed springs—has been instrumental in advancing the scientific understanding of our world through the articulation of laws of nature. A pervasive wonder of the natural world is the evolution of varied systems, including stars, minerals, atmospheres, and life. These evolving systems appear to be conceptually equivalent in that they display three notable attributes:

1. They form from numerous components that have the potential to adopt combinatorially vast numbers of different configurations;
2. processes exist that generate numerous different configurations; and
3. configurations are preferentially selected based on function.

The authors identify universal concepts of selection—static persistence, dynamic persistence, and novelty generation—that underpin function and drive systems to evolve through the exchange of information between the environment and the system. Accordingly, they propose a "law of increasing functional information": The functional information of a system will increase (i.e., the system will evolve) if many different configurations of the system undergo selection for one or more functions.

Many plants employ osmotic and hydrostatic pressure to generate movement for survival, but little is known about the cellular mechanisms involved. Even though plants are essentially sessile in nature, they usually conduct reversible shape-shifting activities of organ movement over a wide range of sizes and timescales. The primary force driving plant movement is hydrostatic pressure, that is, a force that pushes. Plant movement is often associated with motor cells that expand (or contract) upon osmotic changes to generate movement. Wang et al. (2023) reported a new cell type in angiosperms termed "contractile cells" in the stigmas of the flowering plant *Chirita pumila* with a much-expanded rough endoplasmic reticulum (RER). Cryo-scanning electron microscopy and transmission electron microscopy analyses revealed that the RER is continuously distributed throughout the entirety of cells, confirmed by endoplasmic reticulum (ER)–specific fluorescent labeling, and is distinct from the common feature of plant ER. The RER is water sensitive and extremely elongated with water absorption. The authors showed that the contractile cells drive circadian stigma closing–bending movements in response to day-to-night moisture changes. RNA-seq analyses demonstrated that contractile cells have distinct molecular components. Furthermore, multiple microstructural changes in stigma movements convert an anti-selfing structure into a device promoting selfing—a unique cellular mechanism of reproductive adaptation for uncertain pollination environments.

The Great Oxidation Event (GOE), ~$2.4 \times 10^9$ years ago, represents an irreversible shift in atmospheric composition that dramatically altered weathering and nutrient cycles on the early Earth. The GOE caused fundamental changes to the chemistry of Earth's surface environments. However, the effect of these changes on the biosphere is unknown, due to a worldwide lack of well-preserved fossils from this time. Barlow et al. (2023) investigated exceptionally preserved, large spherical aggregate (SA) microfossils permineralized in chert from the c. $2.4 \times 10^9$ years old Turee Creek Group in Western Australia. Field and petrographic observations, Raman spectroscopic mapping, and in situ carbon isotopic analyses uncover insights into the morphology, habitat, reproduction, and metabolism of this unusual form, whose distinctive SA morphology has no known counterpart in the fossil record. Comparative analysis with microfossils from before the GOE reveals the large SA microfossils represent a step up in cellular organization. Morphological comparison to extant microorganisms indicates the SAs have more in common with coenobial algae than coccoidal bacteria, emphasizing the complexity of this microfossil form. The remarkable preservation here provides a unique window into the biosphere, revealing an increase in the complexity of life coinciding with the GOE.

The long-term diversification of the biosphere responds to changes in the physical environment, see Salles et al. (2023). Yet, over the continents, the nearly monotonic expansion of life started later in the early part of the Phanerozoic eon than the expansion in the marine realm, where instead the number of genera waxed and waned over time. A comprehensive evaluation of the changes in the geodynamic and climatic forcing fails to provide a unified theory for the long-term pattern of evolution of life on Earth. Salles et al. (2023) coupled climate and plate tectonics models to numerically reconstruct the evolution of the Earth's landscape over the entire Phanerozoic eon and then compared to paleodiversity datasets from marine animal and land plant genera. Their results indicate that biodiversity is strongly reliant on landscape dynamics, which at all times determine the carrying capacity of both the continental domain and the oceanic domain. In the oceans, diversity closely adjusted to the riverine sedimentary flux that provides nutrients for primary production. On land, plant expansion was hampered by poor edaphic conditions until widespread endorheic basins resurfaced continents with a sedimentary cover that facilitated the development of soil-dependent rooted fora, and the increasing variety of the landscape additionally promoted their development.

Ball (2024) questioned classic views of evolution which is often regarded as a slow affair of letting random mutations change one amino acid for another and seeing what effect it produces. However, in fact, proteins are typically made up of several sections called modules; by reshuffling, duplicating, and tinkering with these modules, a useful new protein is produced. The often-presented view of biology is oversimplified and out of date. Scientists must set the record straight, argues the author; see the book review by Noble (2024).

## 3.10 MOLECULES IN SPACE

In order to better understand the galactic power sources associated with galaxy evolution, it is important to investigate the physical/chemical properties of a nucleus in galaxies using molecular lines, which are likely to be interstellar gas tracers free from dust extinction. Several extreme and large-scale phenomena, such as starbursts, active galactic nuclei (AGNs), and merging occur in external galaxies, unlike local objects in the Milky Way. Thus far, systematic unbiased spectral scans (i.e., molecular line surveys) have been performed for typical nuclear starbursts, AGNs, and (ultra-)luminous infrared galaxies [(U)LIRGs] using large single-dish telescopes. Recently, an important and interesting issue about galaxies has been the investigation of power sources in active galaxies, focusing especially on the obscured galactic nuclei, which are the central engines of the galaxy starburst or AGN. Observations revealing the power sources may provide key information regarding the evolution of galaxies.

For example, Nakajima et al. (2023) recently presented an imaging molecular line survey in the 3-mm band (85–114 GHz) focused on one of the nearest galaxies with an active galactic nucleus, NGC 1068, based on observations taken with the Atacama Large Millimeter/Submillimeter Array. Distributions of 23 molecular transitions were obtained in the central ~3 kpc region, including both the circumnuclear disk (CND) and starburst ring (SBR) with 60 and 350 pc resolution. The column densities and relative abundances of all the detected molecules are estimated under the assumption of local thermodynamic equilibrium in the CND and SBR. Then, they discuss the physical and chemical effects of the AGN on molecular abundance corresponding to the observation scale. They found that $H^{13}CN$, SiO, HCN, and $H^{13}CO^+$ are abundant in the CND relative to the SBR. In contrast, $^{13}CO$ is more abundant in the SBR. Based on the calculated column density ratios of N(HCN)/N($HCO^+$), N(HCN)/N(CN), and other molecular distributions, they conclude that the enhancement of HCN in the CND may be due to high-temperature environments resulting from strong shocks, which are traced by the SiO emission. Moreover, they found that the abundance of CN in the CND is significantly lower than the expected value of the model calculations in the region affected by strong radiation. The expected strong X-ray irradiation from the AGN has a relatively low impact on the molecular abundance in the CND than mechanical feedback.

The presence of prebiotic molecules in planetary systems is a common result of star formation rather than a unique feature of our own solar system. Recently Berné et al. (2023) used NASA's James Webb Space Telescope to detect a new carbon compound in space for the first time. Known as methyl cation, $CH_3^+$, the molecule is important because it aids the formation of more complex carbon-based molecules. Methyl cation was detected in a young star system, with a protoplanetary disk, known as d203–506, which is located about 1,350 lightyears away in the Orion Nebula. In the last decades, the formation of organic molecules in space has been considered to happen mostly at the surface of grains. The detection of $CH_3^+$ indicates that alternative gas-phase routes are available to activate the organic chemistry, when UV radiation is present. External UV irradiation is expected to occur during the early life of most protoplanetary disks, making UV-driven organic chemistry common for the chemical evolution of most protoplanetary disks and of the early the solar system.

Establishing the reservoirs of nitriles (an organic compound containing a cyanide group—CN bound to an alkyl group) in the interstellar medium is of importance for on the Earth based RNA-origin of life. In this theory (Gilbert 1986), RNA could have performed both metabolic and genetic roles. The process by which inert matter generated first the building blocks of RNA, ribonucleotides, and ultimately RNA itself, remains a mystery (Rivilla et al. 2022). A plausible origin of this prebiotic material was assumed to be extraterrestrial delivery during the heavy bombardment of meteorites and comets that occurred around $3.9 \times 10^9$ years ago. These basic molecular precursors have been already formed prior to the formation of the solar system, in its parental molecular cloud, through the chemistry that takes place in the interstellar medium.

McGeoch and McGeoch (2021) discussed the structural organization of space polymers. Extraterrestrial polymers of glycine with iron have been characterized by mass spectrometry to have a core mass of 1494 Da, with dominant rod-like variants at mass to charge ratios of 1567 and 1639. Several principal macro-structural morphologies are observed in solvent extracts from a chondritic Vigarano class alteration type 3 meteoritic material. The first is an extended sheet of linked (three-legged) triskelia containing the 1494 Da core entity that encloses gas bubbles in the solvent. A second is of fiber-like crystals found here, via X-ray diffraction, to be multiple-walled nanotubes made from a square lattice of the 1494 Da polymer. A third is a dispersion of floating phantom-like short tubes of up to 100 $\mu$m length with characteristic angled bends that suggest the influence of a specific underlying protein structure. They proposed that the angled tubes are the observable result of a space-filling super-polymerization of 1638 Da polymer subunits guided by the tetragonal symmetry of linking silicon bonds. Distorted hexagonal sheets are linked by perpendicular subunits in a three-dimensional hexagonal diamond structure to fill the largest possible volume. This extended very low-density structure is conjectured to have dominated in a process of chemical selection because it captured a maximum amount of molecular raw material in the ultra-low density of molecular clouds or of the proto-solar nebula. This could have led ultimately to the accretion of the earliest planetary bodies.

Brunken et al. (2022) reported a first detection of dimethyl ether in a major asymmetric ice trap in a planet-forming disk. The complex organic molecules (COMs) detected in star-forming regions are the precursors of the prebiotic molecules that can lead to the emergence of life. By studying COMs in more evolved protoplanetary disks, we can gain a better understanding of how they are incorporated into planets. This paper presents ALMA band 7 observations of the dust and ice trap in the protoplanetary disk around Oph IRS 48. They report the first detection of dimethyl ether ($CH_3OCH_3$) in a planet-forming disk and a tentative detection of methyl formate ($CH_3OCHO$). They determined column densities for the detected molecules and upper limits on non-detected species using the CASSIS spectral analysis tool. The inferred column densities of $CH_3OCH_3$ and $CH_3OCHO$ with respect to methanol ($CH_3OH$) are of order unity, indicating unusually high abundances of these species compared to other environments. Alternatively, the $^{12}CH_3OH$ emission is optically thick and beam diluted, implying a higher $CH_3OH$ column density and a smaller emitting area than originally thought. The presence of these complex molecules can be explained by thermal ice sublimation, where the dust cavity edge is heated by irradiation and the full volatile ice content is observable in the gas phase. This work confirms the presence of oxygen-bearing molecules more complex than $CH_3OH$ in protoplanetary disks for the first time. It also shows that it is indeed possible to trace the full interstellar journey of COMs across the different evolutionary stages of star, disk, and planet formation.

Icy grain mantles are the main reservoir of the volatile elements that link chemical processes in dark, interstellar clouds with the formation of planets and the composition of their atmospheres. The initial ice composition is set in the cold, dense parts of molecular clouds, before the onset of star formation. With the exquisite sensitivity of the James Webb Space Telescope, this critical stage of ice evolution is now accessible for detailed study. Here we show initial results of the Early Release Science program Ice Age that reveal the rich composition of these dense cloud ices. Weak ice features, including $^{13}CO_2$, $OCN^-$, $^{13}CO$, OCS, and complex organic molecule functional groups, are now detected along two pre-stellar lines of sight. The $^{12}CO_2$ ice profile indicates modest growth of the icy grains. Column densities of the major and minor ice species indicate that ices contribute between 2% and 19% of the bulk budgets of the key C, O, N, and S elements. Our results suggest that the formation of simple and complex molecules could begin early in a water-ice-rich environment; see McClure et al. (2023).

Krasnokutski (2021) described the biological molecules delivered to Earth on board meteorites and comets as being one of the possible causes of the origin of life. Therefore, our understanding of the routes of formation of biomolecules in space should shed light on the possibility of the existence of habitable extrasolar planets. The large abundance of organic molecules is found in the space regions with the lowest temperature. Different routes of organics formation in these areas were suggested. The most probable route of organic molecule formation seems to be the surface reactions on the cold dust particles present in the molecular clouds or in the protoplanetary disks. The two main routes are currently considered for their formation in astrophysical environments. The first one involves the energetic processing of the ice mantels covering the refractory dust grains, caused most commonly by UV photons and/or cosmic rays. The second non-energetic route includes an addition of atomic species to the molecules present in ice mantels. The second route could potentially lead to a much bigger amount of organics, as the energetic processing not only creates the radicals required for the molecular growth but also leads to the destruction of the formed molecules. In this article, the authors demonstrate that complex organic molecules as well as important biological molecules can be formed due to the reaction of C atoms with the mantels of molecular ices covering refractory dust grains present in the interstellar medium. With four valence electrons, C atoms act as glue joining simple non-organic molecules and converting them into organic matter. The formation of many molecules is barrierless and thus can happen at low temperatures. The barrierless reaction $C + NH_3 + CO \rightarrow NH_2CHCO$ attracts particular interest. The product of this reaction is an isomer of the glycine residue of a polypeptide and expected to be efficiently formed in the translucent molecular clouds. The polymerization of these molecules leads to the formation of proteins that, according to some theories, are life's first molecules. Considering a high abundance of atomic carbon in the ISM, they expect a high efficiency of the formation of a large variety of different organic molecules and show why the amount of organic material formed by condensation of atomic carbon can be underestimated.

Life appeared on Earth about $4 \times 10^9$ years ago, but we do not know the processes that made it possible. One of the proposed scenarios is the so-called RNA world, which suggests that early forms of life relied solely on RNA to store genetic information and to catalyze chemical reactions. The RNA-world hypothesis was originally not very popular among the astrobiology community because it was far from trivial to form RNA from its main constituents: a nitrogenous base, a ribose sugar, and a phosphate. Even the prebiotic formation of these basic building blocks of the RNA is still, so far, unknown. In recent years, however, it has become clear that ribonucleotides could have formed instead from simpler molecules such as cyanamide ($NH_2CN$), cyanoacetylene, or glycolaldehyde [$CH_2(OH)CHO$], all derivatives of hydrogen cyanide (HCN) in an aqueous medium.

In the past decade, astrochemistry has witnessed an impressive increase in the number of detections of complex organic molecules. Some of these species are of prebiotic interest, such as glycolaldehyde, the simplest sugar, and oraminoacetonitrile, a possible precursor of glycine. Recently, we have reported the detection of two new nitrogen-bearing complex organics, glycolonitrile and Z-cyanomethanimine, known to be intermediate species in the formation process of ribonucleotides within theories of a primordial RNA world for the origin of life. In this study, the authors present deep and high-sensitivity observations toward two of the most chemically rich sources in the galaxy: a giant molecular cloud in the center of the Milky Way (G + 0.693–0.027) and a proto-Sun (IRAS16293–2422 B). Jiménez-Serra et al. (2020) explored whether the key precursors considered to drive the primordial RNA-world chemistry are also found in space. High-sensitivity observations reveal that urea is present in G + 0.693–0.027 with an abundance of $\sim 5 \times 10^{-11}$. This is the first detection of this prebiotic species outside a star-forming region. Urea remains undetected toward the proto-Sun IRAS16293–2422 B (upper limit to its abundance of $\leq 2 \times 10^{-11}$). Other precursors of the RNA-world chemical scheme such as glycolaldehyde or cyanamide are abundant in space, but key prebiotic species such as 2-amino-oxazole, glyceraldehyde, or dihydroxyacetone are not detected in either source. Future more sensitive observations targeting the brightest transitions of these species will be needed to disentangle whether these large prebiotic organics are definitely present in space; see Jiménez-Serra et al. (2020).

Peptide bonds, as the molecular bridges that connect amino acids, are crucial to the formation of proteins. Searches and studies of molecules with embedded peptide-like bonds are thus important for the understanding of protein formation in space. Li et al. (2021) reported the first tentative detection of propionamide ($C_2H_5CONH_2$), the largest peptide-like molecule detected in space toward Sagittarius B2(N1) at a position called N1E that is slightly offset from the continuum peak. New laboratory measurements of the propionamide spectrum were carried out in the 9–461 GHz range, which provide good opportunity to check directly for the transition frequencies of detected interstellar lines of propionamide. The observed result indicates that propionamide emission comes from the warm, compact cores in Sagittarius B2, in which massive protostellars are forming. The column density of propionamide toward Sgr B2(N1E) was derived to be $1.5 \times 10^{16}$ $cm^{-2}$, which is three-fifths of that of acetamide and one-nineteenth of that of formamide. This detection suggests that large peptide-like molecules can form and survive during star-forming process and may form more complex molecules in the interstellar medium. The detection of propionamide bodes well for the presence of polypeptides, as well as other complex prebiotic molecules in the ISM.

Krasnokutski et al. (2022) proved experimentally that the condensation of carbon atoms on the surface of cold solid particles (cosmic dust) leads to the formation of isomeric polyglycine monomers (aminoketene molecules). Following encounters between aminoketene molecules, they polymerize to produce peptides of different lengths. The chemistry involves three of the most abundant species (CO, C, and $NH_3$) present in star-forming molecular clouds and proceeds via a novel pathway that skips the stage of amino acid formation in protein synthesis. The process is efficient, even at low temperatures, without irradiation or the presence of water. The delivery of biopolymers formed by this chemistry to rocky planets in the habitable zone might be an important element in the origins of life.

A recent suggestion by Kerkeni and Simmie (2023) that acetamide, $CH_3C(O)NH_2$, could be readily formed on water–ice grains by the acid induced addition of water across the C≡N bond has now been shown to be credible. Computational modeling of the reaction between R–CN (R = H, $CH_3$) and a cluster of 32 molecules of water and one $H_3O^+$ proceeds catalytically to form first a hydroxy imine R–C(OH)=NH and second an amide R–C(O) $NH_2$. Quantum mechanical tunneling, computed from small-curvature estimates, plays a key role in the rates of these reactions. This work represents the first reasonable effort to show, in general, how amides can be formed from nitriles and water, which are abundant substrates, reacting on a water–ice cluster containing catalytic amounts of hydrons in the interstellar medium with consequential implications toward the origins of life.

Tryptophan is among the 20 vital amino acids necessary for protein synthesis, crucial for life's evolution on Earth. Iglesias-Groth (2023) used spectra of the Spitzer Space Telescope to conduct a search for the aromatic amino acid tryptophan in the interstellar gas of the young star cluster IC 348. For all the strongest mid-infrared (mid-IR) laboratory bands of tryptophan, we have found counterpart emission lines in the observed spectrum which are consistent in wavelength and strength with the laboratory measurements. Assuming that the detected emission lines are due to tryptophan and using the measured fluxes, we estimate a tryptophan column density in the line of sight of the core of IC 348 in the range $10^9$–$10^{11}$ $cm^{-2}$. The observed emission lines are also found in the combined spectrum of >30 interstellar locations obtained in diverse unrelated star-forming regions observed by Spitzer. This could be an indication that the molecule causing the emission is widespread in interstellar space. Future high spectral resolution mid-IR searches for proteinogenic amino acids in protostars, protoplanetary disks, and the interstellar medium will be key to study an exogenous origin of meteoritic amino acids and to understand how the pre-biotic conditions for life were set in the early Earth.

## REFERENCES


Adamczyk, L., Agakishiev, G., Aggarwal, M. M., et al. 2013. Single spin asymmetry an in polarized proton-proton elastic scattering at $\sqrt{s}$=200 GeV. Physics Letters B 719: 62–69.

Akchurin, N., Langland, J., Onel, Y., et al. 1993. Analyzing power measurement of pp elastic scattering in the Coulomb-nuclear interference region with the 200-GeV/c polarized-proton beam at Fermilab. Physical Review D 48(7): 3026. https://doi.org/10.1103/PhysRevD.48.3026.

Alberts, B., Bray, D., Hopkin, K., et al. 2014. Essential Cell Biology 4th Edition. New York: Garland Science, Taylor and Francis Group.

Aponte, J. C., Elsila, J. E., Glavin, D. P., et al. 2017. Pathways to meteoritic glycine and methylamine. ACS Earth Space Chem 1: 3–13.

Araujo, R. P., and Liotta, L. A. 2023. Universal structures for adaptation in biochemical reaction networks. Nature Communications 14: 2251 (12 pages). https://doi.org/10.1038/s41467-023-38011-9.

Ball, P. 2024. How Life Works: A User's Guide to the New Biology (560 pages). Stuttgart: Pan Macmillan. Pan Macmillan is part of Holtzbrinck Publishing Group.

Barlow, E. V., House, C. H., Liu, M.-C., et al. 2023. Distinctive microfossil supports early Paleoproterozoic rise in complex cellular organization. Geobiolog 22(1): 1–23. https://doi.org/10.1111/gbi.12576.

Barnett S. J. 1915. Magnetization by rotation. Physical Review 6: 239–270. https://doi.org/10.1103/PhysRev.6.239.

Batty, C. J., Gilmore, R. S., Stafford, G. H. 1963. Measurement of the polarization in proton-proton scattering at 30 and 50 MeV. Nuclear Physics 45: 481–491.

Baxter, I. 2009. Ionomics: Studying the social network of mineral nutrients. Current Opinion in Plant Biology 12(3): 381–386. https://doi.org/10.1016/j.pbi.2009.05.002.

Bazilevsky, A., Alekseev, I., Aschenauer, E., al. 2011. Measurements of the energy dependence of the analyzing power in pp elastic scattering in the CNI region. Journal of Physics: Conference Series 295: 012096. https://doi.org/10.1088/1742-6596/295/1/012096.

Beretta, G., Mastorgio, A. F., Pedrali, L., et al. 2019. The effects of electric, magnetic and electromagnetic fields on microorganisms in the perspective of bioremediation. Reviews in Environmental Science and Bio/Technology 18: 29–75. https://doi.org/10.1007/s11157-018-09491-9.

Berné, O., Martin-Drumel, M. A., Schroetter, I., et al. 2023. Formation of the methyl cation by photochemistry in a protoplanetary disk. Nature (41 pages). https://doi.org/10.1038/s41586-023-06307-x.

Bhattacharyya, S., Bhattarai, N., Pfannenstiel, D. M., et al. 2023. A heritable iron memory enables decision-making in Escherichia coli. PNAS 120(48): e2309082120 https://doi.org/10.1073/pnas.2309082120.

Binhi, V. N., Milyaev, V. A., Chernavskii, D. S., and Rubin, A. B. 2006. The paradox of magnetobiology: Analysis and prospects for solution. Biophysics 51(3): 497–503. https://doi.org/10.1134/S000635090603025.

Binhi, V. N., and Prato, F. S. 2017a. Biological effects of the hypomagnetic field: An analytical review of experiments and theories. PLoS One 12(6): e0179340. https://doi.org/10.1371/journal. pone.0179340.

Binhi, V. N., and Prato, F. S. 2017b. A physical mechanism of magnetoreception: Extension and analysis. Bioelectromagnetics 38(1): 41–52. https://doi.org/10.1002/bem.22011.

Binhi, V. N., and Rubin, A. B. 2022. Theoretical concepts in magnetobiology after 40 years of research. Cells 11(2): 274 (15 pages). https://doi.org/10.3390/cells11020274.

Binhi, V. N., and Savin, A. V. 2002. Molecular gyroscope and biological effects of weak ELF magnetic fields. Physical Review E 65(5). 051912.

Bismillah, A. N., Johnson, T. G., Hussein, B. A., et al. 2023. Control of dynamic $sp^3$-C stereochemistry. Nature Chemistry. https://doi.org/10.1038/s41557-023-01156-7.

Blackmond, D. G. 2010. The origin of biological homochirality. Cold Spring Harb Perspect Biol 2: a002147. https://doi.org/10.1101/cshperspect.a002147.

Blackmond, D. G. 2019. The origin of biological homochirality. Cold Spring Harb Perspect. Biol 11: a032540. https://doi.org/10.1101/cshperspect.a032540.

Bonner, W. A. 2020. Parity violation and the evolution of biomolecular homochirality. Chirality 12(3): 114–126.

Boyd, R. N., Famiano, M., Onaka, T., and Kajino, T. 2018. Sites that can produce left-handed amino acids in the supernova neutrino amino acid processing model. The Astrophysical Journal 856(1): article id. 26, 5.

Bravar, A., Alekseev, I., Bunce, G., et al. 2005. Spin Dependence in Elastic Scattering in the CNI Region. Brookhaven National Laboratory (8 pages). Upton, NY, USA: Report BNL-75128–2005-CP.

Bromberg, Y., Aptekmann, A. A., Mahlich, Y., et al. 2022. Quantifying structural relationships of metal-binding sites suggests origins of biological electron transfer. Science Advances 8: eabj3984 (13 pages). https://doi.org/10.1126/sciadv.abj3984.

Brunken, N. G., Booth, A. S., Leemker, M., et al. 2022. A major asymmetric ice trap in a planet-forming disk III. First detection of dimethyl ether. Astronomy & Astrophysics 659: A29 (11 pages). https://doi.org/10.1051/0004-6361/202142981.

Buchachenko, A. L., Sagdeev, R. Z., and Salikhov, K. M. 1978. Magnetic and Spin Effects in Chemical Reactions. Novosibirsk: Nauka.

Cain, J. C., Ferguson, B. B., and Mozzoni, D. 2003. An n=90 internal potential function of the martian crustal magnetic field. Journal of Geophysical Research (Planets) 108(E2): 5008 (19 pages). https://doi.org/10.1029/2000JE001487.

Chen, Y., Ma, W. 2020. The origin of biological homochirality along with the origin of life. PLoS Computational Biology 16(1): e1007592. https://doi.org/10.1371/journal.pcbi.1007592.

Choi, J. M., Holehouse, A. S., and Pappu, R. V. 2020. Physical principles underlying the complex biology of intracellular phase transitions. Annual Review of Biophysics 49: 107–133. https://doi.org/10.1146/annurev-biophys-121219–081629.

Chyba, C., and Sagan, C. 1992. Endogenous production, exogenous delivery and impact-shock synthesis of organic molecules: An inventory for the origins of life. Nature 355: 125–132. https://doi.org/10.1038/355125a0.

Cook, D. L., Smith, T., Leya, I., et al. 2018. Excess $^{180}W$ in IIAB iron meteorites: Identification of cosmogenic, radiogenic, and nucleosynthetic components. Earth and Planetary Science Letters 503: 29-36.

Cooper, K. 2017. Looking for LUCA, the Last Universal Common Ancestor. https://astrobiology.nasa.gov/news/looking-for-luca-the-last-universal-common-ancestor/.

Cooper, S., and Helmstetter, C. E. 1968 Chromosome replication and the division cycle of Escherichia coli B/r. Journal of Molecular Biology 31: 519–540. https://doi.org/10.1016/0022-2836(68)90425-7.

Crofts, A. R. 2007. Life, information, entropy and time. Complexity 13(1): 14–50. https://doi.org/10.1002/cplx.20180.

Dalum, S., and Hedegård, P. 2019. Theory of chiral induced spin selectivity. Nano Letters 19: 5253–5259. https://doi.org/10.1021/acs.nanolett.9b01707.

Danner, A., Demirel, B., Kersten, W., et al. 2020. Spin-rotation coupling observed in neutron Interferometry. NPJ Quantum Information 6: 23 (6 pages). https://doi.org/10.1038/s41534020-0254-8.

De Marcellus, P., Meinert, C., Michel Nuevo, M., et al. 2011. Non-racemic amino acid production by ultraviolet irradiation of achiral interstellar ice analogs with circularly polarized light. The Astrophysical Journal Letters 727(2): L27 (6 pages).

Demarchi, L., Goychuk, A., Maryshev, I., and Frey, E. 2023. Enzyme-enriched condensates show self-propulsion, positioning, and coexistence. Physical Review Letters 130: 128401 (6 pages). https://doi.org/10.1103/PhysRevLett.130.128401.

Devinsky, F. 2021. Chirality and the origin of Life. Symmetry 13: 2277. https://doi.org/10.3390/sym13122277.

Dupré, J., O'Malley, M. A. 2009. Varieties of living things: Life at the intersection of lineage and metabolism. Philosophy and Theory in Biology 1: e003 (25 pages). https://doi.org/10.3998/ptb.6959004.0001.003.

Eddington, A. S. 1928. The Nature of the Physical World. Cambridge: Cambridge University Press (182 pages).

Einstein, A., and de Haas, W. J. 1915. Experimental proof of the existence of Ampere's molecular currents. Verhandlungen der Deutschen Physikalischen Gesellschaft 17: 152.

Embley, T. M., and Martin, W. 2006. Eukaryotic evolution, changes and challenges. Nature 440: 623–630.

Erdmann, W., Kmita, H., Kosicki, J. Z., Kaczmarek, L. 2021. How the geomagnetic field influences life on earth—an integrated approach to geomagnetobiology. Origins of Life and Evolution of Biospheres 51: 231–257. https://doi.org/10.1007/s11084-021-09612-5.

Ermakov, A., Afanasyeva, V., Ermakova, O., et al. 2022. Effect of weak alternating magnetic fields on planarian regeneration. Biochemical and Biophysical Research Communications 592: 7–12. https://doi.org/10.1016/j.bbrc.2021.12.096.

Evers, F., Aharony, A., Bar-Gill, N., et. al. 2022. Theory of chirality induced spin selectivity: Progress and challenges. Advanced Materials 34(13): 2106629 (11 pages). https://doi.org/10.1002/adma.202106629.

Famiano, M. A., Boyd, R. N., Kajino, T., and Onaka, T. 2018. Selection of amino acid chirality via neutrino interactions with 14N in crossed electric and magnetic fields. Astrobiology 18(2): 190–206. https://doi.org/10.1089/ast.2017.1686.

Forbes, K. A., and Green, D. 2022. Enantioselective optical gradient forces using 3D structured vortex light. Optics Communications 515(2): 128197. https://doi.org/10.1016/j.optcom.2022.128197.

Frank, C. 1953. On spontaneous asymmetric synthesis. BiochimicaetBiophysicaActa11:459-463.https://doi.org/10.1016/0006-3002(53)90082-1.

Gagler, D. C., Karas, B., Kempes, C. P., et al. 2022. Scalling laws in enzime function reveal a new kind of biochemical universality. PNAS 119(9): e2106655119 (12 pages). https://doi.org/10.1073/pnas.2106655119.

Gilbert, W. 1986. Origin of life: The RNA world. Nature 319: 618. https://doi.org/10.1038/319618a0.

Globus, N., and Blandford, R. D. 2020. The chiral puzzle of life. The Astrophysical Journal Letters 895: L11 (14 pages). https://doi.org/10.3847/2041-8213/ab8dc6.

González-Pastor, J. E., Hobbs, E. C., and Losick, R. 2003. Cannibalism by sporulating bacteria. Science 301(5632): 510–513. https://doi.org/10.1126/science.1086462.

Himeoka, Y., and Kaneko, K. 2017. Theory for transitions between exponential and stationary phases: Universal laws for lag time. Physical Review X 7(2): 021049 (16 pages). https://link.aps.org/doi/10.1103/PhysRevX.7.021049.

Hogben, H. J., Efimova, O., Wagner-Rundell, N., et al. 2009. Possible involvement of superoxide anddioxygen with cryptochrome in avianmagnetoreception: Origin of Zeeman resonances observed by in vivo EPR spectroscopy. Chemical Physics Letters 480: 118–122. https://doi.org/10.1016/j.cplett.2009.08.051.

Hong, G., Pachter, R., Essen, L.-O., and Ritz, T. 2020. Electron transfer and spin dynamics of the radical-pair in the cryptochrome from Chlamydomonas reinhardtii by computational analysis. The Journal of Chemical Physics 152(6): 065101. https://doi.org/10.1063/1.5133019.

Hore, P. J. 2021. Radical quantum oscillations. Science 374: 1447–1448. https://doi.org/10.1126/science.abm9261.

Hore, P. J., and Mouritsen, H. 2016. The radical-pair mechanism of magnetoreception. Annual Review of Biophysics 45: 299–344. https://doi.org/10.1146/annurevbiophys-032116-094545.

Höytö, A., Herrala, M., Luukkonen, J., et al. 2017. Cellular detection of 50 Hz magneticfields and weak blue light: Effects on superoxide levels and genotoxicity. International Journal of Radiation Biology 93(6): 646–652. https://doi.org/10.1080/09553002.2017.1294275.

Iglesias-Groth, S. 2023. A search for tryptophan in the gas of the IC 348 star cluster of the Perseus molecular cloud. Monthly Notices of the Royal Astronomical Society 523(2): 2876–2886. https://doi.org/10.1093/mnras/stad1535.

Ioppolo, S., Fedoseev, G., Chuang, K.-J., et al. 2020. A non-energetic mechanism for glycine formation in the interstellar medium. Nature Astronomy 5: 197–205. https://doi.org/10.1038/s41550-020-01249-0.

Ishimoto, K., Moreau, C., and Yasuda, K. 2023. Odd elastohydrodynamics: Non-reciprocal living. Material in a Viscous Fluid. PRX Life 1: 023002 (16 pages). https://doi.org/10.1103/PRXLife.1.023002.

Jelen, B. I., Giovannelli, D., and Falkowski, P. G. 2016. The role of microbial electron transfer in the coevolution of the biosphere and geosphere. Annual Review of Microbiology 70: 45–62.

Jiménez-Serra, I., Martin-Pintado, J. Rivilla, V. M., et al. 2020. Toward the RNA-world in the interstellar medium—detection of urea and search of 2-amino-oxazole and simple sugars. Astrobiology 20(9): 1048–1066. https://doi.org/10.1089/ast.2019.2125.

Johnson, F. 1972. Spectroscopy of tetrabenzporphin molecules and possible astrophysical implications. Mémoires Société Royale des Sciences de Liège (6e série, tome III, pp. 391-407).

Joyce, G., Visser, G., van Boeckel, C., et al. 1984. Chiral selection in poly(C)-directed synthesis of oligo(G). Nature 310: 602–604. https://doi.org/10.1038/310602a0.

Kaiser, J. 2021. More than half of high-impact cancer lab studies could not be replicated in controversial analysis. Science 374(6573): 1311. https://doi.org/10.1126/science.acx9788.

Kerkeni, B., and Simmie, J. M. 2023. Peptide bonds in the interstellar medium: Facile catalytic formation from nitriles on water–ice grains. The Journal of Physical Chemistry A 127: 5382–5389. https://doi.org/10.1021/acs.jpca.3c00719.

Kettner, M., Göhler, B., and Zacharias, H. 2015. Spin filtering in electron transport through chiral oligopeptides. The Journal of Physical Chemistry C 119: 14542–14547. http://pubs.acs.org/doi/abs/10.1021/jp509974z.

Kolokolova, L., Koenders, C., Goetz, C., et al. 2016. Clues to cometary circular polarization from studying the magnetic field in the vicinity of the nucleus of comet 67P/Churyumov–Gerasimenko. Monthly Notices of the

Royal Academy of Science 462: S422–S431. https://doi.org/10.1093/mnras/stw2843.

Krasnokutski, S. A. 2021. Did life originate from low-temperature areas of the Universe? Fizika Nizkikh Temperatur 47: 219–226, translated by AIP author services. Low Temperature Physics 47: 199–205. https://doi.org/10.1063/10.0003519.

Krasnokutski, S. A., Chuang, K.-J., Jäger, C., et al. 2022. A pathway to peptides in space through the condesation of atomic carbon. Nature Astronomy Articles. https://doi.org/101038/s41550-021-01577-9.

Lake, J. A., Henderson, E., Oakes, M., and Clark, M. W. 1984. Eocytes: A new ribosome structure indicates a kingdom with a close relationship to eukaryotes. Proceedings of the National Academy of Sciences of the United States of America 81: 3786–3790.

Lambert, G., and Kussell, E. 2014. Memory and fitness optimization of bacteria under fluctuating environments. PLoS Genetics 10(9): e1004556 (10 pages). https://doi.org/10.1371/journal.pgen.1004556.

Le Bel, J. A. 1874. Sur les relations qui existent entre les formules atomiques des corps organiques, et le pouvoir rotatoire de leurs dissolutions. Bulletin de la Société Chimique de France 22: 337–347.

Li, J., Wang, J., Lu, X., et al. 2021. Propionamide ($C_2H_5CONH_2$): The largest peptide-like molecule in space. The Astrophysical Journal 919: 4 (18 pages). https://doi.org/10.3847/1538–4357/ac091c.

Li, Q., Orcutt, K., Cook, R. L., et al. 2023. Single-photon absorption and emission from a natural photosynthetic complex. Nature 619: 300–304. https://doi.org/10.1038/s41586-023-06121–5.

Luukkonen, J., Liimatainen, A., Juutilainen., J, and Naarala, J. 2014. Induction of genomic instability, Oxidative processes, and mitochondrial activity by 50 Hz magnetic fields in human SH-SY5Y neuroblastoma cells. Mutation Research/Fundamental and Molecular Mechanisms and Mutagenesis 760: 33–41. https://doi.org/10.1016/j.mrfmmm.2013.12.002.

Lyon, P. 2015. The cognitive cell: Bacterial behavior reconsidered. Frontiers in Microbiology 6: 264 (18 pages). https://doi.org/10.3389/fmicb.2015.00264.

MacDermott, A. J. 2012. Perspective and Concepts: Biomolecular Significance of Homochirality: The Origin of Homochiral Signature of Life (Comprehensive Chirality). In E. M. Carreira and H. Yamamoto (eds.), (vol. 8, pp. 11–38). Amsterdam: Elsevier. https://api.semanticscholar.org/CorpusID:92524852;https://doi.org/10.1016/b978-0-08-095167–6.00804–1.

Marijuán, P. C., Navarro, J., and del Moral, R. 2010. On prokaryotic intelligence: Strategies for sensing the environment. BioSystems 99: 94–103. https://doi.org/10.1016/j.biosystems.2009.09.004.

Mariscal, C. 2021. Life. In Entry in The Stanford Encyclopedia of Philosophy (pp. 1095–5054). Copyright 2021 by The Metaphysics Research Lab, Department of Philosophy, Stanford University. Stanford, CA: Library of Congress Catalog Dana.

Martin, W., Hoffmeister, M., Rotte, C., and Henze, K. 2001. An overview of endosymbiotic models for the origins of eukaryotes, their ATP-producing organelles (mitochondria and hydrogenosomes), and their heterotrophic lifestyle. Biological Chemistry 382: 1521–1539.

Matsuo, M., Leda, J., Harli, K., et al. 2013. Mechanical generation of spin current bz spin-rotation coupling. Physical Review B 87: 180402(R). https://doi.org/10.1103/PhysRevB.87.180402.

Matsuo, M., Leda, J., and Maekawa, S. 2015. Mechanical generation of spin current. Frontiers in Physics 3: 54 (9 pages). https://doi.org/10.3389/fphy.2015.00054.

Mavelli, F., Altamura, E., Cassidei, L., and Stano, P. 2014. Recent theoretical approaches to minimal artificial cells. Entropy 16: 2488–2511.

McClure, M. K., Rocha, W. R. M., Pontoppida, K. M., et al. 2023. An ice age JWST inventory of dense molecular cloud ices. Nature Astronomy. https://doi.org/10.1038/s41550-022-01875-w.

McGeoch, J. E. M., and McGeoch, M. W. 2021. Structural organization of space polymers. Physics of Fluids 33: 067118. https://doi.org/10.1063/5.0054860.

McGuire, B. A., and Carroll, P. B. 2016. Mirror asymmetry in life and in space. Physics Today 69(11): 86–87. https://doi.org/10.1063/PT.3.3375.

Michaeli, K., Kantor-Uriel, N., Naaman, R., David, H., and Waldeck, D. H. 2016. The electron's spin and molecular chirality-how are they related and how do they affect life processes? Chemical Society Review 45(23): 6478–6487.

Michaeli, K., and Naaman, R. 2019. Origin of spin dependent tunneling through chiral molecules. The Journal of Physical Chemistry C 123: 17043.

Michaelian, K. 2011. Entropy production and the origin of life. Journal of Modern Physics 2(6A): 595–601. https://doi.org/10.4236/jmp.2011.226069.

Mims, D., Herpich, J., Lukzen, N. N., et al. 2021. Readout of spin quantum beats in a charge-separated radical pair by pump-push spectroscopy. Science 374: 1470–1474. https://doi.org/10. 1126/science.abl4254.

Mishra, S., Pirbadian, S., and Mondal, A. K. 2019. Spin-dependent electron transport through bacterial cell surfacemultiheme electron conduits. Journal of the American Chemical Society 141: 19198–19202. https://doi.org/10.1021/jacs.9b09262J.

Mizuuchi, R., Furubayashi, T., and Ichihashi, N. 2022. Evolutionary transition from a single RNA replicator to a multiple replicator network. Nature Communications 13: 1460 (10 pages). https://doi.org/10.1038/s41467-022-29113-x.

Mo, W. C., Liu, Y., and He, R. Q. 2012. A biological perspective of the hypomagnetic field: From definition towards mechanism. Progress in Biochemistry and Biophysics 39: 835–842.

Moreira, D., and Lopez-Garcia, P. 2007. The Last Common Ancestor of Modern Cells. In M. Gargaud, H. Martin, and P. Claeys (eds.), Lectures in Astrobiology (vol II, pp. 305–317). Berlin: Springer.

Mutlu, A., Trauth, S., Ziesack, M., et al. 2017. Phenotypic memory in Bacillus subtilis links dormancy entry and exit by a spore quantity-quality tradeoff. Nature Communication 9(1): 69. https://doi.org/10.1038/s41467-017-02477-1.

Naaman, R., Paltiel, Y., and Waldeck, D. H. 2019. Chiral molecules and the electron spin. Nature Reviews Chemistry 3: 250–260. https://doi.org/10.1038/s41570-019-0087-1.

Naaman, R., Paltiel, Y., and Waldeck, D. H. 2020. Chiral molecules and the spin selectivity effect. Physical Chemistry Letters 11: 3660–3666.

Nakajima, T., Takano, S., Tosaki, T., et al. 2023. Molecular abundance of the circumnuclear region surrounding an active galactic nucleus in NGC 1068 based on an imaging line survey in the 3 mm band with ALMA. The Astrophysical Journal 955: 27 (26 pages). https://doi.org/10.3847/1538-4357/ace4c7.

Nakano, S., Furutani, H., Kato, S., et al. 2023. Bullet-shaped magnetosomes and metagenomic-based magnetosome gene profiles in a deep-sea hydrothermal vent chimney. Frontiers in Microbiology 14: 1174899 (11 pages). https://doi.org/10.3389/fmicb.2023.1174899.

National Academy of Sciences. 2007. The Limits of Organic Life in Planetary Systems. Washington, DC: The National Academies Press. (116 pages). ISBN 978-0-309-10484-5. https://doi.org/10.17226/11919. Available at http://nap.edu/11919.

Noble, D. 2024. It's time to admit that genes are not the blueprint for life. Nature 626: 254–255: https://doi.org/10.1038/d41586-024-00327-x.

Oba, Y., Takano, Y., Naraoka, H., et al. 2019. Nucleobase synthesis in interstellar ices. Nature Communication 10: 4413. https://doi.org/10.1038/s41467-019-12404-1.

Okada, H., Alekseev, I., Bravar, A., et al. 2008. Absolute polarimetry at RHIC. arXiv:0712.1389v2 [nucl-ex], 9 Jan.

Perego, M., and Hoch, J. A. 2002. Two-component systems, phosphorelays, and regulation of their activities by phosphatases. In A. L. Sonenshein, J. A. Hoch, and R. Losick (eds.), *Bacillus subtilis* and its Closest Relatives: From Genes to Cell (pp. 473–481). Washington, DC: American Society for Microbiology Press.

Peretó, J. 2005. Controversies on the origin of life. International Microbiology 8: 23–31. www.im.microbios.org.

Piñeros, W. D., and Tlusty, T. 2022 Spontaneous chiral symmetry breaking in a random driven chemical system. Nature Communication 13: 2244 (8 pages). https://doi.org/10.1038/s41467-022-29952-8.

Piper, C. S. 1942. Investigations on copper deficiency in plants. The Journal of Agricultural Science 32: 143–178.

Popovic, M. 2019. Thermodynamic properties of microorganisms: Determination and analysis of enthalpy, entropy, and Gibbs free energy of biomass, cells and colonies of 32 microorganism species. Heliyon 5(6): e01950. https://doi.org/10.1016/j.heliyon.2019.e01950.

Popovic, M., and Minceva, M. 2021. Standard thermodynamic properties, biosynthesis rates, and the driving force of growth of five agricultural plants. Frontiers in Plant Science 12: 671868 (17 pages). https://doi.org/103389/fpls.2021.671868.

Prelog, V. 1976. Chirality in chemistry. Journal of Molecular Catalysis 1(3): 159–172.

Prelog, V. 2006. Chirality in chemistry–Nobel lecture, December 12. The Nobel Foundation 1975 & Croatica Chemica Acta 79 (3): 49–57.

Ray, K., Ananthavel, S. P., Waldeck, D. H., and Naaman, R. 1999. Asymmetric scattering of polarized electrons by organized organic films made of chiral molecules. Science 283: 814–816. https://doi.org/10.1126/science.283.5403.814.

Rimola, A., Balucani, N., Ceccarelli, C., and Ugliengo, P. 2022. Tracing the chemical life of glycine: A review from quantum chemical simulations. International Journal of Molecular Sciences 23: 4252 (36 pages). https://doi.org/10.3390/ijms23084252.

Ritz T., Adem, S., and Schulten, K. 2000. A model for photoreceptor-based magnetoreception in birds. Biophysical Journal 78: 707–718. https://doi.org/10.1016/S0006–3495(00)76629-X.

Rivilla, V. M., Jiménez-Serra, I., Martin-Pintado, J., et al. 2022. Molecular precursors of the RNA-world in space: New nitriles in the G+0.693–0.027 molecular cloud. Frontiers in Astronomy and Space Science 9: 876870 (18 pages).

Rosenberg, R. A., Mishra, D., and Naaman, R., Prof. 2015. Chiral selective chemistry induced by natural selection of spin-polarized electrons. Communication. https://onlinelibrary.https://doi.org/10.1002/anie.201501678.

Rosłoń, I. E., Japaridze, A., Steeneken, P. G., et al. 2022. Probing nanomotion of single bacteria with graphene drums. Nature Nanotechnology. https://doi.org/10.1038/s41565-022-01111-6.

Sagnac, G. 1913. The demonstration of the luminiferous aether by an interferometer in uniform rotation. Comptes Rendus 157: 708–710.

Salikhov, K. M., Molin, Yu. N., Sagdeev, R. Z., and Buchachenko, A. L. 1984. Spin Polarization And Magnetic Effects In Radical Reactions. In Studies in Physical and Theoretical Chemistry (vol. 22). Amsterdam: Elsevier. https://doi.org/10.1002/bbpc.198400037.

Salles, T., Husson, L., Lorcery, M., et al. 2023. Landscape dynamics and the phanerozoic diversification of the biosphere. Nature (25 pages). https://doi.org/10.1038/s41586-023-06777-z.

Sanders, S., Joshi, K., Levin, P. A., and Iyer-Biswas, S. 2023. Beyond the average: An updated framework for understanding the relationship between cell growth, DNA replication, and division in a bacterial system. PLoS Genetics 19(1): e1010505 (18 pages). https://doi.org/10.1371/journal.pgen.1010505.

Sanger, F., and Thompson, E. O. P. 1953. The amino-acid sequence in the glycyl chain of insulin. 1. The investigation of peptides from enzymatic hydrolysates. Biochemistry Journal 53(3): 353–366. https://doi.org/10.1042/bj0530353.

Sato, N. 2012. Scientific élan vital: Entropy deficit or inhomogeneity as a unified concept of driving forces of life in hierarchical biosphere driven by photosynthesis. Entropy 14: 233–251. https://doi.org/10.3390/e14020233.

Schaechter, M., Maaløe, O., and Kjeldgaard, N. O. 1958. Dependency on medium and temperature of cell size and chemical composition during balanced grown of salmonella typhimurium. The Journal of General Microbiology 19: 592–606. https://doi.org/10.1099/00221287-19-3-592.

Sharma, A., Yadav, S. P., Sarma, D., and Arunika Mukhopadhaya, A. 2022. Modulation of host cellular responses by gram-negative bacterial porins. Chapter 2 in Advances in Protein Chemistry and Structural Biology 128. Elsevier Inc. https://doi.org/10.1016/bs.apcsb.2021.09.004.

Shkuratov, Y. G., Kaydash, V. G., and Opanasenko, N. V. 1999. Iron and titanium abundance and maturity degree distribution on the lunar nearside. Icarus 137: 222–234. doi/10.1006/icar.1999.6046.

Sikorsky, T., Meir, Z., Ben-shlomi, R., et al. 2018. Spin-controlled atom–ion chemistry. Nature Communication 9: 920 (5 pages). https://doi.org/10.1038/s41467-018-03373-y.

Spaargaren, D. H. 1991. The biological use of chemical elements: Selection on environmental availability and electron configuration. Oceanologica Acta 14(6): 569–574.

Staroń, A., Sofia, H. J., Dietrich, S., et al. 2009. The third pillar of bacterial signal transduction: Classification of the extracytoplasmic function (ECF) sigma factor protein family. Molecular Microbiology 74(3): 557–581. https://doi.org/10.1111/j.1365-2958.2009.06870.x.

Steenbjerg, F. 1951. Yield curves and chemical plant analyses. Plant and Soil 3: 97–109.

Steiner, U. E., and Ulrich, T. 1989. Magnetic field effects in chemical kinetics and related phenomena. Chemical Reviews 89: 51–147. https://doi.org/10.1021/cr00091a003.

Steinhilber, F., Abreu, J. A., Beer, J., and McCracken, K. G. 2010. Interplanetary magnetic field during the past 9300 years inferred from cosmogenic radionuclides. Journal of Geophysical Research 115: a01104 (14 pages). https://doi.org/10.1029/2009JA014193.

Stolar, T., Lukin, S., Etter, M., et al. 2020. DNA-specific selectivity in pairing of model nucleobases in the solid state. Chemical Communication 56: 13524–13527. The Royal Society of Chemistry. https://doi.org/10.1039/D0CC03491F.

Tashiro, T., Ishida, A., Hori, M., et al. 2017. Early trace of life from 3.95 Ga sedimentary rocks in Labrador, Canada. Nature 549: 516–518. https://doi.org/10.1038/nature24019.

Tejada, J., Zysler, R. D., Molins, E., and Chudnovsky, E. M. 2010. Evidence for quantization of mechanical rotation of magnetic nanoparticles. Physical Review Letters 104(2): 027202. https://doi.org/10.1103/PhysRevLett.104.027202.

Timm, J., Pike, D. H., Mancini, J. A., et al. 2023. Design of a minimal di-nickel hydrogenase peptide. Science Advances 9(10): eabq1990. https://doi.org/10.1126/sciadv.abq1990.

Timmel, C., Till, U., Brocklehurst, B., et al. 1998. Effects of weak magnetic fields on free radical recombination reactions. Molecular Physics 95: 71–89. https://doi.org/10.1080/00268979809483134.

Udgaonkar, J. B. 2001. Entropy in biology. Resonance 6: 61–66. https://doi.org/10.1007/BF02837738.

Ulrich, L. E., and Zhulin, I. B. 2010. The MiST2 database: A comprehensive genomics resource on microbial signal transduction. Nucleic Acids Research 38 (Database issue): D401–407. https://doi.org/10.1093/nar/gkp940.

Valkovic, V. 1977. A possible mechanism for the influence of geomagnetic field on the evolution of life. Origins Life Evol B 8: 7–11. https://doi.org/10.1007/bf009 30933.

Valković, V., and Obhođaš, J. 2022. Origins of chiral life in interstellar molecular clouds. The Astronomical Journal 163: 270 (14 pages). https://doi.org/10.3847/1538-3881/ac63d0.

van't Hof, J. H. 1874. Sur les formules de structure dans l'espace. Archives Neerlandaises 9: 445–454.

Veening, J.-W., Hamoen, L. W., and Kuipers, O. P. 2005. Phosphatases modulate the bistable sporulation gene expression pattern in bacillus subtilis. Molecular Microbiology 56(6): 1481–1494. https://doi.org/10.1111/j.1365-2958.2005.04659.x.

Venegas-Gomez, A. 2017. The thermodynamics of the living organisms: Entropy. Production in the Cell. arXiv: 1410.8820v.2 [physics.bio-ph], 12. Dec.

Verkin, B. I., Bondarenko, S. I., Sheremet, V. I., et al. 1976. The effect of a weak magnetic field on some bacterial species. Mikrobiologiya 45(6): 1067–1070.

Vijayakumar, M., and Gopinathan, M. S. 1996. Spin-orbit coupling constants of transition metal atoms and ions in density functional theory. Journal of Molecular Structure (Theochem) 361: 15–19.

von Stockar, U., and Liu, J.-S. 1999. Does microbial life always feed on negative entropy? Thermodynamic analysis of microbial growth. Biochimica et Biophysica Acta (BBA)—Bioenergetics 1412 (3): 191–211. https://doi.org/10.1016/S0005-2728(99)00065-1.

Wade, J., Byrne, D. J., Ballentine, C. J., and Drakesmith, H. 2021. Temporal variation of panetary iron as a driver of evolution. PNAS 118(51): e2109865118, (8 pages). https://doi.org/10.1073/PNAS.2109865118.

Wang, Y.-Z., Lin, Y.-X., Qi Liu, Q., et al. 2023. A new type of cell related to organ movement for selfing in plants. National Science Review 10: nwad208 (14 pages). https://doi.org/10.1093/nsr/nwad208.

Weber, C. A., Zwicker, D., Jülicher, F., and Lee, C. F. 2019. Physics of active emulsions. Reports on Progress in Physics 82(6): 064601 (40 pages). https://doi.org/10.1088/1361-6633/ab052b.

Weiss, M., Sousa, F., Mrnjavac, N., et al. 2016. The physiology and habitat of the last universal common ancestor. Nature Microbiology 1: 16116 (12 pages). https://doi.org/10.1038/nmicrobiol.2016.116.

Wieser, M., Barabash, S., Futaana, Y., et al. 2010. First observation of a mini-magnetosphere above a lunar magnetic anomaly using energetic neutral atoms. Geophysical Research Letters 37: L05103 (4 pages). https://doi.org/10.1029/2009GL041721.

Wimmer, J. L. E., Xavier, J. C., Vieira, A. d. N., et al. 2021. Energy at origins: Favorable thermodynamics of biosynthetic reactions in the last universal common ancestor (LUCA). Frontiers in Microbiology 12: 793664 (18 pages). https://doi.org/10.3389/fmicb.2021.793664.

Witzany, G. 2020. What is life? Frontiers in Astronomy and Space Sciences 7: 7 (13 pages). https://doi.org/10.3389/fspas.2020.00007.

Wolf, D. M., Fontaine-Bodin, L., Bischofs, I., et al.2008. Memory in microbes: Quantifying history-dependent behavior in a bacterium. PLoS One 3(2): e1700. https://doi.org/10.1371/journal.pone.0001700.

Wong, M. L., Cleland, C. E., Arend Jr. D., et al. 2023. On the roles of function and selection in evolving systems. PNAS 120(43): e2310223120 (11 pages). https://doi.org/10.1073/pnas.2310223120.

Wong, Y., and Rosindell, J. 2022. Dynamic visualisation of million-tip trees: The OneZoom project. Methods in Ecology and Evolution 13: 303–313. https://doi.org/10.1111/2041-210X.13766.

Wu, X., Cao, M.-P., Chu, K.-P., et al. 2014 Weak power frequency magneticfield acting similarly to EGF stimulation, inducesacute activations of the EGFR sensitive actincytoskeleton motility in human amniotic cells. PLOS One 9(2): e87626 (17 pages). https://doi.org/10.1371/journal.pone.0087626.

Xie, Z., Markus, T. Z., Cohen, S. R., et al. 2011. Spin specific electron conduction through DNA oligomers. Nano Letters 11(11): 4652–4655. https://doi.org/10.1021/nl2021637.

Xue, X., Ali, Y. F., Luo, W., et al. 2021. Biological effects of space hypomagnetic environment on circadian rhythm. Frontiers in Physiology 12: 643943 (7 pages). https://doi.org/10.3389/fphys.2021.643943.

Yan, L., Zhang, S., Chen, P., et al. 2012. Magnetotactic bacteria, magnetosomes and their application. Microbiological Research 167: 507–519. https://doi.org/10.1016/j.micres.2012.04.002.

Yang, C.-Y., Bialecka-Fornal, M., Weatherwax, C., et al. 2020. Encoding membrane-potential-based memory within a microbial community. Cell Systems 10: 417–423.

Yin, P., Zhang, Z., Lv, H., et al. 2015. Chiral recognition and selection during the self-assembly process of protein-mimic macroanions. Nature Communication 6: 6475. https://doi.org/10.1038/ncomms7475.

Zablotskii, V., Polyakova, T., and Dejneka, A. 2022. Effects of high magnetic fields on the diffusion of biologically active molecules. Cells 11: 81 (13 pages). https://doi.org/10.3390/cells11010081.

Zadeh-Haghighi, H., and Simon, C. 2022a. Magnetic field effects in biology from the perspective of the radical pair mechanism. Journal of the Royal Society–Interface 19: 20220325 (39 pages). https://doi.org/10.1098/rsif.2022.0325.

Zadeh-Haghighi, H., and Simon, C. 2022b. Radical pairs may play a role in microtubule reorganization. Scientific Reports 12: 1–11. https://doi.org/10.1038/s41598-022-10068-4.

Zhan, A., Luo, Y., Qin, H., et al. 2022. Hypomagnetic field exposure affecting gut microbiota, reactive oxigen species levels, and colonic cell proliferation in mice. Bioelectromagnetics 43: 462–475.

Zhukovska, S., Gail, H.-P., and Trieloff, M. 2008. Evolution of interstellar dust and stardust in the solar neighbourhood. Astronomy & Astrophysics 479: 453–480. https://doi.org/10.1051/0004-6361:20077789.

Zolfagharkhani, G., Gaidarzhy, A., Degiovanni, P., et al. 2008. Nanomechanical detection of itinerant electron spin flip. Nature Nanotechnology 3: 720–723. https://doi.org/10.1038/nnano.2008.311.

Zwang, T. J., Hurlimann, S., Hill, M. G., and Barton, J. K. 2016. Helix-dependent spin filtering through the DNA duplex. Journal of the American Chemical Society 138: 15551–15554. https://doi.org/10.1021/jacs.6b10538.

## ADDITIONAL READINGS

Adawy, A. 2022. Functional chirality: From small molecules to supramolecular assemblies. Symmetry 14: 292 (27 pages). https://doi.org/10.3390/sym14020292.

Anbar, A. D. 2008. Elements and evolution. Science 322: 1481–1483. https://doi.org/10.1126/science.1163100.

Aristov, V. V., Buchelnikov, A. S., and Nechipurenko, Y. D. 2022. The use of the statistical entropy in some new approaches for the description of biosystems. Entropy 24: 172(17 pages). https://doi.org/10.3390/e24020172.

Bejan, A. 2016. Life and evolution as physics. Communicative & Integrative Biology 9(3): e1172159 (13 pages). https://doi.org/10.1080/19420889.2016.1172159.

Bray, M. S., Lenz, T. K., Haynes, J. W., et al. 2018. Multiple prebiotic metals mediate translation. PNAS 115(48): 12164–12169. www.pnas.org/cgi/doi/10.1073/PNAS.1803636115.

Dorta-Urra, A., and Bargueño, P. 2019. Homochirality: A perspective from fundamental physics. Symmetry 11: 661 (9 pages). https://doi.org/10.3390/sym11150661.

Egel, R. 2012. Life's Order, complexity, organization, and its thermodynamic-holistic imperatives. Life 3: 323–363. https://doi.org/10.3390/life2040323.

Escobar, J. F., Vaca-González, J. J., Guevara, J. M., and Garzón-Alvarado, D. A. 2020. Effect of magnetic and electric fields on plasma membrane of single cells: A computational approach. Engineering Reports 2: e12125 (14 pages). https://doi.org/10.1002/eng2.12125.

Feng, L. 2022. Nebula-relay hypothesis: The chirality of biological molecules in molecular clouds. Frontiers in Astronomy and Space Sciences 9: article 794067. https://doi.org/10.3389/fspas,2022.794067.

Funk, R. H. W., and Monsees, T. K. 2006. Effects of electromagnetic fields on cells: Physiological and therapeutical approaches and molecular mechanisms of interaction. Cells Tissues Organs 182: 59–78. https://doi.org/10.1159/000093061.

Garcia, A. D., Meinert, C., Sugahara, H., et al. 2019. The astrophysical formation of asymmetric molecules and the emergence of a chiral bias. Life 9(1): 29. https://doi.org/10.3390/life9010029.

Gomez-Marquez, J. 2021. What is life? Molecular Biology Reports. published online 27 July. https://doi.org/10.1007/s11033-021-06594-5.

He, Y., Gu, Z., Lu, W., et al. 2018. Atmospheric humidity and particle charging state on agglomeration of aerosol particles. Atmospheric Environment 197: 141–149.

Honniball, C. I., Lucey, P. G., and Li, S. 2020. Molecular water detected on the sulit Moon by SOFIA: Nature Astronomy Letters. https://doi.org/10.1038/s41550-020-01222-x.

Hu, R., Damiano, M., Scheucher, M., et al. 2021. Unveiling shrouded oceans on temperature sub-neptunes via transit signatures of solubility equilibria vs. Gas therochemistry. arXiv:2108.04745 [astro-ph.EP], 10 Aug.

Kudella, P. W., Tkachenko, A. V., Salditt, A., et al. 2021. Structured sequences emerge from random pool when replicated by templated ligation. PNAS 118(8): e2018830118 (8 pages). https://doi.org/10.1073/pnas.2018830118.

Mac Dermott, A. J. 2012. Perspective and Concepts: Biomolecular Significance of Homochirality: The Origin of the Homochiral Signature of Life. In E. M. Carreira and H. Yamamoto (eds.), Comprehensive Chirality 8 (pp. 11–38). Elsevier, Amsterdam.

Marboeuf, U., Thiabaud, A., Alibert, Y., et al. 2014. From planetesimals to planets: Volatile molecules. Astronomy & Astrophysics 570: A36 (16 pages). https://doi.org/10.1051/0004-6361/201423431.

Martin, W. F. 2012. Hydrogen, metals, bifurcating electrons, and proton gradients: The early evolution of biological energy conservation. FEBS Letters 586: 485–493. https://doi.org/101016/j.febslet.2011.09.031.

Masood, S., Saleem, I., Smith, D., and Chu, W.-K. 2020. Growth pattern of magnetic field treated bacteria. Current Microbiology 77(2): 194–203. https://doi.org/10.1007/s00284-019-01820-7.

Michaelian, K. 2022. Non-equilibrium thermodynamic foundations of the origin of life. Foundations 2: 308–337. https://doi.org/10.3390/foundations2010022.

National Research Council. 2007. Exploring Organic Environments in the Solar System. Washington, DC: The National Academies Press.

Preiner, M., Xavier, J. C., Viera, do N. A., et al. 2019. Catalysts, autocatalysis and the origin of metabolism. Interface Focus 9: 20190072. http://dx.doi.org/10.1098/rsfs2019.0072.

Price, M. E. 2017. Entropy and selection: Life as an adaption for universe replication. Complexity 2017: article ID 4745379 (4 pages). https://doi.org/10.1155/2017/4745379.

Roushan, P., Seo, J., Parker, C. V., et al. 2009. Topological surface states protected from backscattering by chiral spin texture. Nature 460(7259): 1106–1109. https://doi.org/10.1038/nature08308.

Valkovic, V. 1990. Origin of Trace Element Requirement by Living Matter. In B. Gruber and J. H. Yopp (eds.), Symmetries in Sciences IV (pp. 213–242). New York, NY: Plenum Press.

Westsson, E., Stephen, P., and Ger, K. 2020. The effect of magnetic field on catalytic properties in core-shell type particles. Frontiers in Chemistry 8: 163 (9 pages). www.frontiersin.org/articles/10.3389/fchem.2020.00163.

Yada, T., Abe, M., Okada, T., et al. 2022. Preliminary analysis of the Hayabusa2 samples returned from C-type asteroid Ryugu. Nature Astronomyx Letters. https://doi.org/10.1038/s41550-021-01550-6.

CHAPTER 4

# How the Universe Accommodated Life—Origin of Life

## 4.1 INTRODUCTION

The hints answering the question of how Universe accommodated life are spread throughout the book. Assuming that the origins of life seeds are everywhere in the Universe implies that

- Everywhere in the Universe, there are four forces governing physical processes which also define laws obeyed by living matter;
- The nuclear force predominates in element formation, capable of producing all elements essential for life;
- Molecules (life building blocks) are available everywhere in the Universe, with an important role of cosmic rays in their formation and electromagnetic force in molecule formation and interaction.

Intergalactic clouds, as well as galactic ones, are birthplace of solar systems as well as primitive life forms. Once formed, life forms have to survive the planet formation process and find a planet convenient for the evolution of cellular life.

The chapter starts with a discussion of chemical element abundances in the universe. It is a vital question when life is considered. Life as we know it is (H-C-N-O) based and relies on a number of bulk elements (Na-Mg-P-S-Cl-K-Ca) for its existence. In addition, life needs a group of elements called trace elements that includes some (or all) of the following chemical elements: Li-B-F-Si-V-Cr-Mn-Fe-Co-Ni-Cu-Zn-As-Se-Mo-I-W. The universe had to go through the formation of a variety of bodies and processes to produce this group of elements in adequate concentrations and concentration ratios to satisfy life requirements.

To study these processes and to characterize the observed objects, astronomers have introduced a variable called metallicity and studied its behavior in various objects and different locations at variable times. Metallicity is one of the crucial factors that determine stellar evolution. To characterize the properties of stellar populations, one needs to know the fraction of stars forming at different metallicities.

Understanding processes leading to the formation of elements required by life is of the utmost importance. Before they are all formed, one cannot speak about the existence of life, since they are needed for providing proteins with unique coordination, catalytic, and electron transfer properties. Of special interest are processes influencing C/O ratios and processes responsible for production of heavy life-required elements (Mo, I, W). About half of the stable nuclei heavier than iron are believed to be synthesized during the late stages of evolution of stars with masses in the range $0.8 \pm 8$ solar masses. These elements are then expelled into the interstellar medium through stellar winds after being “dredged up” towards the surface of the stars. These processes generated interest in the study of heavy metal stars.

The emergence of life is necessary stage in molecular cloud evolution, as documented by hundreds of different types of molecules already detected in these clouds, including sugars and amino acids. More than 70 amino acids and several DNA bases were found in meteorites, which represent the original composition of the molecular cloud in which our solar system was formed. To answer the question of where and when life originated,

DOI: 10.1201/9781003366607-4

we can hypothesize that life originated when the element abundance curve of living matter and the cosmic environment in which life originated coincided. This coincidence occurring at a particular space/time region of the universe can indicate when life originated, $T_{origin}$. At $T_{origin}$, the chemical abundance ratios of selected elements for a particular cosmic environment, or the whole universe, and living matter, corrected by the values of concentration factors, should be close to 1. The life abundance curve that we have considered is of the only life we know: that on the planet Earth. It could be universal since it resulted from the physics and chemistry laws, which are probably valid everywhere in the universe. This is the way we discuss the Big Bang and origin of life, life on rogue planets, life in molecular clouds, and $T_{origin}$ for different parts of Universe.

When the relative abundances of the life-forming elements evaluated for the last universal common ancestor (LUCA) are compared with the solar neighborhood stellar abundances, a striking similarity is found. We tested the hypothesis that in some particular regions and at some particular time, the abundance curve of the first living matter and the universe coincided. Indeed, the best agreement between the two curves was obtained for $(4 \pm 1) \times 10^9$ years after the Big Bang, indicating the time of the origin of life. All organisms evolved on the Earth independently of place and time are leading to the LUCA and involve chiral molecules such as L amino acids and D sugars in fundamental life processes. The growing evidence from carbonaceous meteorite analysis shows an excess of L-type amino acids and D-type sugars, suggesting that the increase in L-type or D-type molecular chirality is the process that takes place in planetary and stellar forming systems; thus the life emerging from interstellar molecular clouds had to be chiral. Spin-polarized proton–proton scattering is a potential physical process that takes place in Interstellar Molecular Cloud (IMC) environments and leads to enrichment of L-type amino acids and D-type sugars.

However, in such an environment, the first living thing was formed (born), not LUCA but older; we call it the first universal stardust ancestor—FUSDA. The birthday of FUSDA is not fixed in time but rather in elemental composition of the dust environment. In this sense it is universal; it happened only once since the universe began aging, and once the concentration of life-essential elements was able to satisfy life requirements, FUSDA was born. This could happen at different times in different parts of the universe. FUSDA into Earth's LUCA evolution probably happened after Earth formation, since, it was FUSDA who had capability of surviving Earth formation process. LUCA had to adjust to Earth's magnetic field and other environmental parameters.

## 4.2 CHEMICAL ABUNDANCES

Hydrogen is by far the most abundant element, accounting for about 92% of the atoms in the universe. The next most abundant element is helium, accounting for 7.1% of the universe's atoms. In general, the universe contains more atoms of elements with lighter atomic masses than atoms of heavier elements. More commonly, a table of abundance describes elements in terms of mass percent. Figure 4.1 shows the distribution of element abundances in the Universe. Of interest to us are the abundances of elements essential for life: life, as we know it, is (H-C-N-O) based and relies on a number of bulk elements (Na-Mg-P-S-Cl-K-Ca) for its existence. All existing organisms use a number of elements for providing proteins with unique coordination, catalytic, and electron transfer properties. This group of elements is called trace elements, and it includes some (or all) of the following chemical elements: Li-B-F-Si-V-Cr-Mn-Fe-Co-Ni-Cu-Zn-As-Se-Mo-I-W.

Table 4.1 shows their relative abundances in the Universe (by mass) as well as their electron configuration.

The outermost orbital shell of an atom is called its valence shell, and the electrons in the valence shell are valence electrons. Valence electrons are the highest-energy electrons in an atom and are therefore most reactive. While inner electrons do not participate in chemical bonding and reactions, valence electrons can be gained, lost, or shared to form chemical bonds.

Half of the chemical elements heavier than iron are produced by the rapid neutron capture process (r-process). The sites and yields of this process are disputed, with candidates including some types of supernovae (SNe) and mergers of neutron stars. For example, Wallner et al. (2021) searched for two isotopic signatures in a sample of Pacific Ocean crust—iron-60 ($^{60}Fe$) (half-life, $2.6 \times 10^6$ years), which is predominantly produced

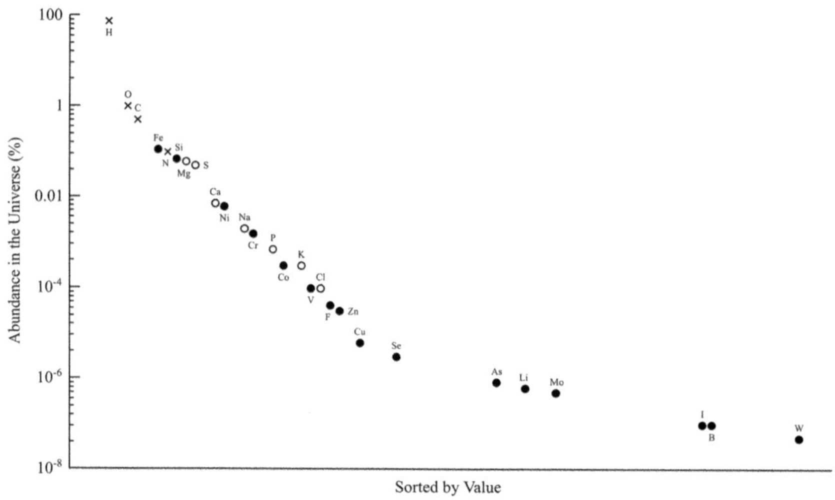


FIGURE 4.1 Relative abundance in the Universe of elements essential for life.

TABLE 4.1 Universe Abundances of Elements Essential for Life

| Life | Atomic Number, Z | Symbol | Element Name | Relative Abundance | Electron Configuration | Abundance by Mass % |
|---|---|---|---|---|---|---|
| BASE | 1 | H | Hydrogen | 1 | $1s^1$ | 75 |
| | 8 | O | Oxygen | 3 | $1s^2 2s^2 2p^4$ | 1 |
| | 6 | C | Carbon | 4 | $1s^2 2s^2 2p^2$ | 0.5 |
| | 7 | N | Nitrogen | 7 | $1s^2 2s^2 2p^3$ | 0.10 |
| BULK | 12 | Mg | Magnesium | 9 | $1s^2 2s^2 2p^6 3s^2$ | 0.06 |
| | 16 | S | Sulfur | 10 | $1s^2 2s^2 2p^6 3s^2 3p^5$ | 0.05 |
| | 20 | Ca | Calcium | 12 | $1s^2 2s^2 2p^6 3s^2 3p^6 4s^2$ | 0.007 |
| | 11 | Na | Sodium | 15 | $1s^2 2s^2 2p^6 3s^1$ | 0.002 |
| | 15 | P | Phosphorus | 18 | $1s^2 2s^2 2p^6 3s^2 3p^3$ | $7 \times 10^{-4}$ |
| | 19 | K | Potassium | 19 | $1s^2 2s^2 2p^6 3s^2 3p^6 4s^1$ | $3 \times 10^{-4}$ |
| | 17 | Cl | Chlorine | 22 | $1s^2 2s^2 2p^6 3s^2 3p^5$ | $1 \times 10^{-4}$ |
| ESSENTIAL TRACE ELEMENTS | 26 | Fe | Iron | 6 | $1s^2 2s^2 2p^6 3s^2 3p^6\ 4s^2 3d^6$ | 0.11 |
| | 14 | Si | Silicon | 8 | $1s^2 2s^2 2p^6 3s^2 3p^2$ | 0.07 |
| | 28 | Ni | Nickel | 13 | $1s^2 2s^2 2p^6 3s^2 3p^6\ 4s^2 3d^8$ | 0.006 |
| | 24 | Cr | Chromium | 16 | $1s^2 2s^2 2p^6 3s^2 3p^6\ 4s^2 3d^5$ | 0.015 |
| | 25 | Mn | Manganese | 17 | $1s^2 2s^2 2p^6 3s^2 3p^6\ 4s^2 3d^3$ | $8 \times 10^{-4}$ |
| | 27 | Co | Cobalt | 21 | $1s^2 2s^2 2p^6 3s^2 3p^6\ 4s^2 3d^7$ | $3 \times 10^{-4}$ |
| | 23 | V | Vanadium | 23 | $1s^2 2s^2 2p^6 3s^2 3p^6\ 4s^2 3d^3$ | $1 \times 10^{-4}$ |
| | 9 | F | Fluorine | 24 | $1s^2 2s^2 2p^5$ | $4 \times 10^{-5}$ |
| | 30 | Zn | Zinc | 25 | $1s^2 2s^2 2p^6 3s^2 3p^6\ 4s^2 3d^{10}$ | $3 \times 10^{-5}$ |
| | 29 | Cu | Copper | 27 | $1s^2 2s^2 2p^6 3s^2 3p^6\ 4s^1 3d^{10}$ | $6 \times 10^{-6}$ |
| | 34 | Se | Selenium | 32 | $1s^2 2s^2 2p^6 3s^2 3p^6\ 4s^2 3d^{10} 4p^4$ | $3 \times 10^{-6}$ |
| | 33 | As | Arsenic | 41 | $1s^2 2s^2 2p^6 3s^2 3p^6\ 4s^2 3d^{10} 4p^3$ | $8 \times 10^{-7}$ |
| | 3 | Li | Lithium | 44 | $1s^2 2s^1$ | $6 \times 10^{-7}$ |
| | 42 | Mo | Molybdenum | 45 | $1s^2 2s^2 2p^6 3s^2 3p^6\ 4s^2 3d^{10} 4p^6 5s^1 4d^5$ | $5 \times 10^{-7}$ |
| | 5 | B | Boron | 62 | $1s^2 2s^2 2p^1$ | $1 \times 10^{-7}$ |
| | 53 | I | Iodine | 63 | $1s^2 2s^2 2p^6 3s^2 3p^6\ 4s^2 3d^{10} 4p^6 5s^1 4d^{10} 5p^5$ | $1 \times 10^{-7}$ |
| | 74 | W | Tungsten | 74 | $1s^2 2s^2 2p^6 3s^2 3p^6\ 4s^2 3d^{10} 4p^6 5s^1 4d^{10} 5p^6\ 6s^2 4f^{14} 5d^4$ | $5 \times 10^{-8}$ |

in massive stars and ejected in supernova explosions, and $^{244}Pu$ (half-life, $80.6 \times 10^6$ years), which is produced solely in r-process events. They detected two distinct influxes of $^{60}Fe$ to Earth in the last 10 million years and accompanying lower quantities of $^{244}Pu$. The $^{244}Pu/^{60}Fe$ influx ratios are similar for both events. The $^{244}Pu$ influx is lower than expected if SNe dominate r-process nucleosynthesis, which implies some contribution from other sources.

Dauphas et al. (2022) discussed the extent, nature, and origin of potassium and rubidium depletion and isotopic fractionations in Earth, the Moon, and other planetary bodies. They found that moderately volatile elements (MVEs) are depleted and isotopically fractionated in the Moon relative to Earth. The data reported by the authors has shaken the paradigm of formation of the Moon (so-called canonical model of Moon formation) by giant impact involving the collision of the proto-Earth with a Mars-size impactor at a velocity just above the mutual escape velocity and an impact angle of 45°. The model cannot explain the large depletions in moderately volatile elements in lunar rocks. Dauphas et al. (2022) examined the origin of the depletions of the Moon in K, Cu, Rb, Zn, Ga, and Sn. For example, K and Rb are depleted by factors of $3.96 \pm 0.87$ and $4.71 \pm 1.07$, respectively, while the K/U weight ratios of Earth and Moon are estimated to be 9704 and 2448, respectively, and the $^{87}Rb/^{86}Sr$ atomic ratios of Earth and the Moon are estimated to be 0.0725 and 0.0154, respectively. A quantitative understanding of the origin of the depletion in MVE of the Moon is challenging; however, the measured isotopic fractionations for MVEs in lunar rocks provide new insights into the formation mechanisms or the conditions that prevailed when these elements were depleted. It is estimated that the loss of elements in the aftermath of the Moon-forming impact was 74.8% (±5.5%) of K, 78.8% (±4.8%) of Rb, 95.3% of Cu, 99.5% of Zn, 89.4% of Ga, and 96.9% of Sn.

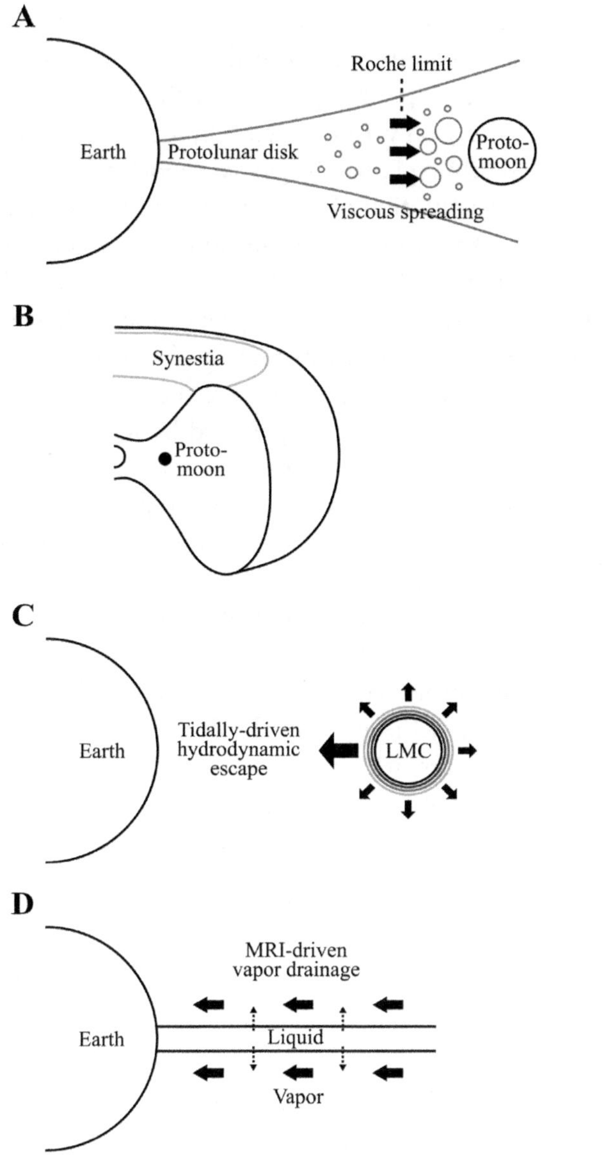


FIGURE 4.2 A. Incomplete condensation model of Canup et al. (2015).

B. Incomplete condensation synestia model of Lock et al. (2018)

C. MRI-driven vapor drainage model of Charnoz and Michaut (2015) and Nie and Dauphas (2019).

D. Tidally driven hydrodynamic escape of MVEs from the lunar magma ocean (Sossi et al. 2019; Charnoz et al. 2021).

Figure 4.2 shows the four scenarios proposed to explain MVE depletion of the Moon. Several giant impact models can explain the dynamical and physical properties of the Earth–Moon system, but these models cannot readily explain why the Moon is depleted in MVEs relative to Earth. Several recent studies have shown that lunar rocks were enriched in the heavy isotopes of several MVEs, notably K and Rb. Because the equilibrium isotopic fractionation factors between condensate and vapor were unknown, the message that these heavy isotope enrichments conveyed about the conditions of volatile element depletion in the Moon was uncertain.

Dauphas et al. (2022) used an ab initio approach to calculate the reduced partition function ratios (β-factors) of various minerals and gas species and found that condensation or evaporation where each parcel of vapor is in equilibrium with the liquid cannot explain the heavy isotopic composition of the Moon relative to Earth. The heavy isotope enrichment of the Moon is more readily explained by partial vaporization in a gas at an under-saturation of ~99% for most MVEs, regardless of element volatility. Constant under-saturation during evaporation can be achieved if the vapor is advected away and elements are removed in proportion to their partial vapor pressures. A possible setting for such evaporation is viscous drainage onto Earth of the vapor layer in the protolunar disk, possibly powered by MRI. The timescale needed to deplete Na and K in the protolunar disk to levels encountered in lunar rocks is typically a month to a year, which is longer than the initial disk heating and vaporization phase that can last just a few days and is shorter than the lifetime of the disk of ~100 years, meaning that it is a viable scenario. Loss of MVEs from the LMO by hydrodynamic escape is not favored because the timescale to achieve lunar-like depletion levels is ~0.1–$10^3 \times 10^6$ years, which is much longer than the ~1000-year lifetime of an exposed magma ocean free of flotation crust. The stark difference in timescales between the protolunar disk and LMO settings stems from differences in temperature and evaporation geometries.

### 4.2.1 Metallicity

Astronomers call metallicity the abundance of elements present in an object that are heavier than hydrogen and helium, that is, the ones made in stars or processes involving them, that is, collisions, supernovae explosions, and so on. Most of the normal currently detectable (i.e. non-dark) matter in the universe is either hydrogen or helium, and astronomers use the word "metals" as a convenient short term for "all elements except hydrogen and helium". This word use is distinct from the conventional chemical or physical definition of a metal as an electrically conducting solid. Stars and nebulae with relatively high abundances of heavier elements are called "metal-rich" in astrophysical terms, even though many of those elements are nonmetals in chemistry.

All elements essential for life (except hydrogen) are in this category. Astronomers use several different methods to describe and approximate metal abundances, depending on the available tools and the object of interest. Some methods include determining the fraction of mass attributed to gas versus metals or measuring the ratios of the number of atoms of two different elements as compared to the ratios found in the Sun. Stellar composition is often simply defined by the parameters X, Y, and Z. Here X represents the mass fraction of hydrogen, Y is the mass fraction of helium, and Z is the mass fraction of all the remaining chemical elements; therefore $X + Y + Z = 1$. In most stars, nebulae, $H_{II}$ regions, and other astronomical sources, hydrogen and helium are the two dominant elements. The hydrogen mass fraction is generally expressed as $X = m_H/M$, where M is the total mass of the object and $m_H$ the mass of the hydrogen it contains. Similarly, the helium mass fraction is $Y = m_{He}/M$. The remainder of the elements are collectively called "metals", and the mass fraction of elements heavier than He is $Z = \sum m_e/M = 1 - X - Y$, where the summation is over all elements heavier than He. It should be noted that the metallicity of our sun is $Z_{sun} = 0.0134$. Due to the effects of stellar evolution, neither the initial composition nor the present-day bulk composition of the Sun is the same as its present-day surface composition.

The overall stellar metallicity is conventionally defined using the total hydrogen content, since its abundance is considered relatively constant in the Universe, or the iron content of the star, which has an abundance that is generally linearly increasing in time in the Universe. Hence, iron can be used as a chronological indicator of nucleosynthesis. Iron is relatively easy to measure with spectral observations in the star's spectrum given the large number of iron lines in the star's spectra (even though oxygen is the most abundant heavy element). The abundance ratio is the common logarithm of the ratio of a star's iron abundance compared to that of the Sun and is calculated thus:

$$[\mathrm{Fe}/\mathrm{H}] = \log_{10}\left(N_{Fe}/N_H\right)_{star} - \log_{10}\left(N_{Fe}/N_H\right)_{sun}$$

where $N_{Fe}$ and $N_H$ are the number of iron and hydrogen atoms per unit volume, respectively. The unit often used for metallicity is the dex, short for decimal exponent. By this formulation, stars with a higher metallicity than the Sun have a positive common logarithm, whereas those more dominated by hydrogen have a corresponding negative value. Young population I stars have significantly higher iron-to-hydrogen ratios than older population II stars. Primordial population III stars are estimated to have metallicity less than −6, a millionth of the abundance of iron in the Sun.

The fraction of heavier elements, called metallicity, provides one method of measurement when a particular star or nebula formed. A low-metallicity star must have formed a long time ago, while a higher-metallicity star is of more recent vintage. The first stars would have had effectively zero metallicity. Stellar metallicity is also related to whether a star has planets. Higher-metallicity stars like the Sun are more likely to have planets than lower-metallicity stars, based on data from NASA's Kepler telescope and other observatories. Using that information, astronomers took the observed metallicity of stars throughout the Milky Way and estimated the total number of potential planets.

Stars tend to form in clusters, and many remain in those clusters for their entire life cycle. As a result, the metallicities of stars reveal information about a cluster's age and environment where it formed. For instance, in the Milky Way, old low-metallicity star clusters are found in the halo, the region of the galaxy surrounding the bright inner regions where the Sun lives.

This is important because similar low-metallicity clusters are common in small galaxies. According to the theory of galaxy formation, big galaxies like the Milky Way form out of smaller galaxies, even eating them whole. The various metallicities of star clusters give a partial history of our galaxy's cannibalism, providing some details of how it formed and grew.

The same principles apply to other galaxies. Individual stars are usually too faint to see within distant galaxies, but astronomers can measure the collective metallicity of many stars together. High-metallicity galaxies tend to be brighter than their low-metallicity cousins, and many galaxies have higher metallicities in their inner regions than out toward the edges. As with the Milky Way, this is a reflection of the individual history of the galaxy, measured in part by where the various chemical elements live.

To calibrate automatic pipelines that determine atmospheric parameters of stars, one needs a sample of stars, or "benchmark stars", with well-defined parameters to be used as a reference. Jofré et al. (2014) provided detailed documentation of the iron abundance determination of the 34 FGK-type benchmark stars that are selected to be the pillars for calibration of the 1 billion Gaia stars. They cover a wide range of temperatures, surface gravities, and metallicities. Up to seven different methods were used to analyze an observed spectral library of high resolutions and high signal-to-noise ratios. The metallicity was determined by assuming a value of effective temperature and surface gravity obtained from fundamental relations; that is, these parameters were known a priori and independently from the spectra. The authors present a set of metallicity values obtained in a homogeneous way for the sample of benchmark stars. In addition to this value, they provide detailed documentation of the associated uncertainties. Finally, they report a value of the metallicity of the cool giant $\psi$ Phe for the first time.

Metallicity is one of the crucial factors that determine stellar evolution. To characterize the properties of stellar populations, one needs to know the fraction of stars forming at different metallicities. Knowing how this fraction evolves over time is necessary, for example, to estimate the rates of occurrence of any stellar evolution related phenomena (e.g. double compact object mergers, gamma-ray bursts). Such theoretical estimates can be confronted with observational limits to validate the assumptions about the evolution of the progenitor system leading to a certain transient. However, to perform the comparison correctly, one needs to know the uncertainties related to the assumed star formation history and chemical evolution of the Universe. Chruslinska and Nelemans (2019) combined the empirical scaling relations and other observational properties of star-forming galaxies to construct the distribution of the cosmic star formation rate density at different metallicities and redshifts. They addressed the question of uncertainty of this distribution due to currently unresolved questions, such as the absolute metallicity scale, the flattening in the star formation–mass relation, or the low-mass end of the galaxy mass function. They found that the fraction of stellar mass formed at metallicities <10% solar (>solar) since $z = 3$ varies by ~18% (~26%) between the extreme cases considered in our study. This uncertainty stems primarily from the differences in the mass–metallicity relations obtained with different methods.

Çelebi et al. (2019) derived photometric metallicity and absolute magnitude calibrations using F-G spectral type main-sequence stars in the solar neighborhood with precise spectroscopic, photometric and Gaia astrometric data for UBV photometry. The sample consists of 504 main-sequence stars covering the temperature, surface gravity, and color index intervals $5300 < T_{eff} < 7300$ K, $\log g > 4$ (cgs), and $0.3 < (B - V)_0 < 0.8$ mag, respectively. Stars with relative trigonometric parallax errors $\sigma\pi/\pi \leq 0.01$ were preferred from Gaia DR2 data for the estimation of their MV absolute magnitudes. In order to obtain calibrations, $(U - B)_0$ and $(B - V)_0$ color indices of stars were preferred, and a multi-variable second order equation was used. Calibrations are valid for main-sequence stars in the metallicity and absolute magnitude ranges $-2 < [Fe/H] < 0.5$ dex and $2.5 < M_V < 6$ mag, respectively. The mean value and standard deviation of the differences between original and estimated values for the metal abundance and absolute magnitude are $(\Delta[Fe/H]) = 0.00 \pm 0.11$ dex and $(\Delta MV) = 0.00 \pm 0.22$ mag, respectively. In this work, it was shown that more precise iron abundance and absolute magnitude values were obtained with the new calibrations, compared to previous calibrations in the literature.

## 4.3 MASS–METALLICITY TIME DEPENDENCE

Galaxies show a tight luminosity–metallicity relation in which those with higher luminosities also have higher metallicities. This relation has been attributed to a more generalized relation, the mass–metallicity relation. The mass–metallicity relation, $Z(M_*)$, has been observed in the nearest galaxies, most of them located in the Local Group, using gas-phase metallicities but also using spectroscopic metallicities of individual stars. Hidalgo (2017) presented a new approach to study the mass–metallicity relation and its dependency on time. He used the star formation history (SFH) derived from color-magnitude diagram fitting techniques of a sample of Local Group (LG) dwarfs to obtain stellar masses, metallicities, and star-formation rates (SFRs) to analyze the mass–metallicity relation as a function of the ages of their stellar populations. The accurate SFHs allow a time resolution of about $2 \times 10^9$ years at the oldest ages for a total redshift range of $0 \le z \le 3$. The mass–metallicity relation retrieved for the sample of LG dwarfs was compared with a large dataset of literature data obtained in a wide redshift range. Neither of the two independent datasets shows a clear evolution of the mass–metallicity relation slope with redshift. However, when the star-formation rate is added as an additional parameter in the relation, it shows a dependence on the redshift in the sense that the coefficient of the mass decreases with increasing redshift, while the coefficient for the SFR is almost constant with time. This result suggests an increasing contribution with time of the galaxy stellar mass to the metallicity of the stars that formed most recently, but it also shows that the SFR can play a fundamental role in shaping the mass–metallicity relation.

Most structural and evolutionary properties of galaxies strongly rely on the stellar initial mass function (IMF), the distribution of the stellar mass formed in each episode of star formation. The IMF shapes the stellar population in all stellar systems and so has become one of the most fundamental concepts of modern astronomy. Both constant and variable IMFs across different environments have been claimed despite a large number of theoretical and observational efforts. However, the measurement of the IMF in galactic stellar populations has been limited by the relatively small number of photometrically observed stars, leading to high uncertainties. Li et al. (2023) reported a star-counting result based on approximately 93,000 spectroscopically observed M-dwarf stars, an order of magnitude more than previous studies, in the 100–300-parsec solar neighborhood. They found unambiguous evidence of a variable IMF that depends on both metallicity and stellar age. Specifically, the stellar population formed at early times contains fewer low-mass stars compared with the canonical IMF, independent of stellar metallicities. In more recent times, however, the proportion of low-mass stars increases with stellar metallicity. The variable abundance of low-mass stars in our Milky Way establishes a powerful benchmark for models of star formation and can heavily affect results in galactic chemical-enrichment modeling, mass estimation of galaxies, and planet-formation efficiency

Nakajima et al. (2023) published a JWST census for the mass–metallicity star-formation relations at $z = 4–10$ with the self-consistent flux calibration and the proper metallicity calibrators. They presented the evolution of the mass–metallicity (MZ) relations at $z = 4–10$ derived with 135 galaxies identified in the JWST/NIRSpec data taken from the three major public spectroscopy programs of ERO, GLASS, and CEERS. Because there are many discrepancies between flux measurements reported by early ERO studies, they first established the NIRSpec data reduction procedure for reliable emission-line flux measurements and errors successfully explaining Balmer decrements with no statistical tensions via thorough comparisons of the early ERO studies. Applying the reduction procedure to the 135 galaxies, they obtained emission-line fluxes for physical property measurements. They confirmed that 10 out of the 135 galaxies with [O iii]λ4363-lines have electron temperatures of $\simeq(1.1–2.3) \times 10^4$ K, similar to lower-z star-forming galaxies, that can be explained by heating of young massive stars. They derived metallicities of the ten galaxies by the direct method and the rest of the galaxies with strong lines by the metallicity calibrations of Nakajima et al. (2022) applicable for these low-mass metal-poor galaxies, anchoring the metallicities with the direct-method measurements. They obtained MZ relations and star-formation rate–MZ relations over z = 4–10 and found that there is a small evolution of the MZ relation from $z \sim 2–3$ to z = 4–10, while, interestingly, the SFR-MZ relation shows no evolution up to $z \sim 8$ but a significant decrease at $z > 8$ beyond the error. This SFR-MZ relation decrease at $z > 8$ may suggest a break of the metallicity equilibrium state via star-formation, inflow, and outflow, while further statistical and local-baseline studies are needed for a conclusion.

Li et al. (2022) presented the mass–metallicity relation (MZR) at z = 2–3 in the stellar mass range of $M_{star} \approx 10^{6.5}–10^{9.5} M_{sun}$ using 55 dwarf galaxies in the Abell 2744 and SMACS J0723–3732 galaxy cluster fields. These dwarf galaxies are identified and confirmed by deep JWST/NIRISS imaging and slitless grism spectroscopic observations. Taking advantage of the gravitational lensing effect, they extended the previous MZR relation at z = 2–3 to a much lower mass regime by more than 2.5 orders of magnitude compared with previous studies. They found that the MZR has a shallower slope at the low-mass end ($M_{star} < 10^9 M_{sun}$) compared to that at the high-mass end ($M_{star} > 10^9 M_{sun}$), with a slope turnover point at around the stellar mass of $10^9 M_{sun}$. This implies that dominating feedback processes in dwarf galaxies may be different from that in galaxies with higher mass. From z = 3 to z = 2, the metallicity of the dwarf galaxies is enhanced by ≈0.1 dex for a given stellar mass, consistent with the mild evolution found in galaxies with higher mass. Further, they confirmed the existence of a 3D relation between the gas phase metallicity, stellar mass, and star formation rate, that is, a fundamental metallicity relation (FMR), in dwarf galaxies at $z = 2 - 3$. Their derived FMR, which has no significant redshift evolution, can be used as a benchmark to understand the origin of the anti-correlation between SFR and metallicity of dwarf galaxies in the high-redshift Universe.

About 380,000 years after the Big Bang, the universe saw the creation of the first neutral atoms, mainly H and He from the recombination of electrons onto heavier nuclei of protons and neutrons. This phase of the evolution of the universe transformed the previously opaque universe where the mean free path of photons of photons was short compared to cosmological distance scales to the transparent universe we live in today. This change in the phase happened at an estimated redshift z = 1100 when the universe temperature was about 4000 K. The universe kept on cooling, and molecules formed that eventually led to the formation of the first stars. This major event in the universe's lifetime, the cosmic dawn, probably happened 200–400 million years after the Big Bang at about z = 20–16. It marked the development of the main factories that produced the rest of the heavy elements (metals) contained in the universe until today. Metallicity traces the building of the elements. Single-element abundances are much more difficult to estimate because one obviously needs an excellent-quality spectroscopy. Unfortunately, when trying to detect and analyze galaxies before the end of reionization, this detailed objective might be out of reach, even for the largest telescopes. The metallic content in galaxies provides information not only on integrated star formation rate in galaxies (through element synthesis) but also on the fraction of metals exchanged with the inter-galactic medium (IGM). Metals are unique tracers of galaxy evolution and of the past history of feedback because the physical processes that shape galaxies and determine their evolution leave imprints on metal distribution and relative abundance of different elements. The mass–metallicity relation is a strong correlation between these two physical parameters. The original interpretation of the shape of this relation is that the saturation appears because galactic outflows regulate the metal content, and an equilibrium metallicity is reached when the metals ejected by outflows are balanced by those produced by star formation. The crucial parameters are therefore the metallicity and the stellar mass (and the SFR). Burgarella et al. (2020) showed how to best measure them in the young universe when galaxies were small (low stellar mass) and at low metallicity.

Peña and Flores-Durán (2019) studied metallicity gradients derived from planetary nebulae (PNe) using O, Ne, and Ar abundances are compared to those from H II regions in the galaxies M 31, M 33, NGC 300, and the Milky Way. Galactocentric rad II and chemical abundances were collected from the literature, carefully selecting a homogeneous sample for each galaxy. Metallicity gradients shown by PNe are flatter than those of H II regions in all cases. The extreme case is M 31, where PN abundances are not related to galactocentric distances and the gradients are consistent with zero. To analyze the evolution of gradients with time, they built gradients for Peimbert Type I and non-Type I PNe, finding that Type I PNe show steeper gradients than non-Type I PNe, more similar to those of H II regions, indicating that the chemical gradients might steepen with time. Alternatively, the flat gradients for old PNe show that radial migration could have an important role in the evolution of galaxies.

Torrey et al. (2019) presented the evolution of the mass–metallicity relation and its scatter in IllustrisTNG. The coevolution of galaxies and their metal content serves as an important test for galaxy feedback models. They analyzed the distribution and evolution of metals within the IllustrisTNG simulation suite with a focus on the gas-phase mass–metallicity relation. They found that the IllustrisTNG model broadly reproduces the slope and normalization evolution of the MZR across the redshift range $0 < z < 2$ and mass range $10^9 < M_*/M_\odot < 10^{10.5}$. They made predictions for the high-redshift ($2 < z < 10$) metal content of galaxies, which is described by a gradual decline in the normalization of the metallicity with an average high-redshift ($z > 2$) evolution fit by $d\log(Z)/dz \approx -0.064$. Their simulations indicate that the metal retention efficiency of the interstellar medium (ISM) is low: a majority of gas-phase metals (~85% at $z = 0$) live outside of the ISM, either in an extended gas disk, the circumgalactic medium, or outside the halo. Nevertheless, the redshift evolution in the simulated MZR normalization is driven by the higher gas fractions of high-redshift galaxies, not by changes to the metal retention efficiency. The scatter in the simulated MZR contains a clear correlation with the gas-mass or star formation rate of the system, in agreement with the observed fundamental metallicity relation. The scatter in the MZR is driven by a competition between periods of enrichment- and accretion-dominated metallicity evolution. The authors expect that while the normalization of the MZR declines with redshift, the slope of the correlation between metallicity and gas-mass at fixed stellar mass is not a strong function of redshift. The results indicate that the gas fraction dependence of "regulator"-style models allows them to simultaneously explain the shape, redshift evolution, and existence of correlated scatter with gas fraction about the MZR.

Henry et al. (2021) studied the mass–metallicity relation at z ~ 1–2 and its dependence on the star formation rate. They presented a new measurement of the gas-phase mass–metallicity relation and its dependence on star formation rates at $1.3 < z < 2.3$. Their sample comprised 1056 galaxies with a mean redshift of $z = 1.9$, identified from the Hubble Space Telescope Wide Field Camera 3 (WFC3) grism spectroscopy in the Cosmic Assembly Near-Infrared Deep Extragalactic Survey and the WFC3 Infrared Spectroscopic Parallel Survey. This sample is four times larger than previous metallicity surveys at z ~ 2 and reaches an order of magnitude lower in stellar mass ($10^8$ $M_{sun}$). Using stacked spectra, they found that the MZR evolves by 0.3 dex relative to z ~ 0.1. Additionally, they identify a subset of 49 galaxies with high signal-to-noise (S/N) spectra and redshifts between $1.3 < z < 1.5$, where Hα emission is observed along with [O III] and [O II]. With accurate measurements of SFR in these objects, they confirm the existence of a mass–metallicity–SFR (M–Z–SFR) relation at high redshifts. These galaxies show systematic differences from the local M–Z–SFR relation, which vary depending on the adopted measurement of the local relation. However, it remains difficult to ascertain whether these differences could be due to redshift evolution, as the local M–Z–SFR relation is poorly constrained at the masses and SFRs of the sample. Last, they reproduced the sample selection in the IllustrisTNG hydro dynamical simulation, demonstrating that the line flux limit lowers the normalization of the simulated MZR by 0.2 dex. They showed that the M–Z–SFR relation in IllustrisTNG has an SFR dependence that is too steep by a factor of around 3.

## 4.4 PROCESSES LEADING TO FORMATION OF ELEMENTS REQUIRED BY LIFE

The organic cycle includes the eleven elements: H, C, N, O, Na, Mg, P, S, Cl, K, and Ca, which form the bulk of the living matter. They all have very low atomic weights and they belong to the lowest 20 elements of the periodic table. These elements are also among the most abundant elements in the Universe. The only nonessential light element seems to be Be and maybe Al. In addition, a number of elements have been recognized as essential trace elements. For example, essential trace elements for some complex species are: F, Si, V, Cr, Mo, Fe, Co, Ni, Cu, Zn, Se, Mo, Sn, I, and W. For the most part, the essential trace elements are transition metals with unfilled d-orbitals. None of the 39 elements beyond tungsten ($Z = 74$) has ever been shown to have any physiological significance. Table 4.2 shows the list of stable isotopes and formation process elements essential for life.

TABLE 4.2 Stable Isotopes and Formation Process Elements Essential for Life

| Life | Atomic Number, Z | Symbol | Element Name | Relative Abundance | Stable Isotopes | Formation Processes |
|---|---|---|---|---|---|---|
| Base | 1 | H | Hydrogen | 1 | $^{1}$**H**(99.99%), $^{2}$H | Primordial |
| | 8 | O | Oxygen | 3 | $^{16}$**O**(99.78%), $^{17}$O, $^{18}$O | α capture<br>core collapse SNe |
| | 6 | C | Carbon | 4 | $^{12}$**C**(98.93%), $^{13}$C | α capture<br>AGB stars |
| | 7 | N | Nitrogen | 7 | $^{14}$**N**(99.63%), $^{15}$N | AGB stars |
| Bulk | 12 | Mg | Magnesium | 9 | $^{24}$**Mg**(78.99%), $^{25}$Mg, $^{26}$Mg | α capture<br>core collapse SNe |
| | 16 | S | Sulfur | 10 | $^{32}$**S**(94.93%), $^{33}$S, $^{34}$S, $^{36}$S | α capture<br>core collapse SNe |
| | 20 | Ca | Calcium | 12 | $^{40}$**Ca**(96.94%), $^{42}$Ca, $^{43}$Ca, $^{44}$Ca, $^{46}$Ca | α capture<br>core collapse SNe |
| | 11 | Na | Sodium | 15 | $^{23}$**Na** | Carbon burning |
| | 15 | P | Phosphorus | 18 | $^{31}$**P** | Core-collapse SNe |
| | 19 | K | Potassium | 19 | $^{39}$**K**(93.26%), $^{41}$K | Type II Sne |
| | 17 | Cl | Chlorine | 22 | $^{35}$**Cl**(75.78%), $^{37}$Cl | Explosive nucleosynthesis |
| Essential trace elements | 26 | Fe | Iron | 6 | $^{54}$Fe, 56**Fe**(91.75%), $^{57}$Fe, $^{58}$Fe | Produced by SNe Ia |
| | 14 | Si | Silicon | 8 | $^{28}$**Si**(92.23%), $^{29}$Si, $^{30}$Si | α capture<br>core collapse SNe |
| | 28 | Ni | Nickel | 13 | $^{58}$**Ni**(68.08%), $^{60}$Ni, $^{61}$Ni, $^{62}$Ni, $^{64}$Ni | Produced by SNe Ia |
| | 24 | Cr | Chromium | 16 | $^{50}$Cr, 52**Cr**(83.79%), $^{53}$Cr, $^{54}$Cr | Produced by SNe Ia |
| | 25 | Mn | Manganese | 17 | $^{55}$**Mn** | Produced by SNe Ia |
| | 27 | Co | Cobalt | 21 | $^{59}$**Co** | |
| | 23 | V | Vanadium | 23 | $^{51}$**V** | |
| | 9 | F | Fluorine | 24 | $^{19}$**F** | AGB stars |
| | 30 | Zn | Zinc | 25 | $^{64}$**Zn**(48.63%), $^{66}$Zn, $^{67}$Zn, $^{68}$Zn, $^{70}$Zn | Largely by HNe |
| | 29 | Cu | Copper | 27 | $^{63}$**Cu**(69.17%), $^{65}$Cu | Largely by HNe |
| | 34 | Se | Selenium | 32 | $^{74}$Se, $^{76}$Se, $^{77}$Se, $^{78}$Se,$^{82}$Se $^{80}$**Se**(49.61%) | In part through the r-process |
| | 33 | As | Arsenic | 41 | $^{75}$**As** | In part r-process |
| | 3 | Li | Lithium | 44 | $^{6}$Li, 7**Li**(92.41%) | Big Bang<br>Classical Novae |
| | 42 | Mo | Molybdenum | 45 | $^{92}$Mo, $^{94}$Mo, $^{95}$Mo, $^{96}$Mo, $^{97}$Mo, $^{98}$**Mo**(24.13%) | In part r-process |
| | 5 | B | Boron | 62 | $^{10}$B, 11**B**(80.1%) | Cosmic rays |
| | 53 | I | Iodine | 63 | $^{127}$**I** | In part r-process |
| | 74 | W | Tungsten | 74 | $^{180}$W, $^{182}$W, $^{183}$W, $^{184}$**W**(30.64%), $^{186}$W | All isotopes both s- and r-processes, $^{180}$W only p-process |

Camprubi et al. (2017) studied the possible role of iron catalysis at the origin of life. Iron–sulfur proteins are ancient and drive fundamental processes in cells, notably electron transfer and $CO_2$ fixation. Iron–sulfur minerals with equivalent structures could have played a key role in the origin of life. However, the "iron–sulfur world" hypothesis has had a mixed reception, with questions raised especially about the feasibility of a pyrites-pulled reverse Krebs cycle. Phylogenetics suggests that the earliest cells drove carbon and energy metabolism via the acetyl CoA pathway, which is also replete in Fe(Ni)S proteins. Deep differences between bacteria and archaea in this pathway obscure the ancestral state. These differences make sense if early cells depended on natural proton gradients in alkaline hydrothermal vents. If so, the acetyl CoA pathway diverged with the origins of active ion pumping, and ancestral $CO_2$ fixation might have been equivalent to methanogens, which depend on a membrane-bound NiFe hydrogenase, energy-converting hydrogenase. This uses the proton-motive force to reduce ferredoxin, thence $CO_2$. The mechanism suggests that pH could modulate reduction potential at the active site of the enzyme, facilitating the difficult reduction of $CO_2$ by $H_2$. This mechanism could be generalized under abiotic conditions so that steep pH differences across semi-conducting Fe(Ni)S barriers drive not just the first steps of $CO_2$ fixation to C1 and C2 organics such as CO, $CH_3SH$, and $CH_3COSH$ but a series of similar carbonylation and hydrogenation reactions to form longer-chain carboxylic acids such as pyruvate, oxaloacetate, and α-ketoglutarate, as in the incomplete reverse Krebs cycle found in methanogens. They suggested that the closure of a complete reverse Krebs cycle, by regenerating acetyl CoA directly, displaced the acetyl CoA pathway from many modern groups. A later reliance on acetyl CoA and ATP eliminated the need for the proton-motive force to drive most steps of the reverse Krebs cycle; for details, see also Biologydictionary.net Editors (2017).

Ever since its discovery in 1966, the reverse Krebs cycle has been proposed as an ancient pathway of $CO_2$ fixation, potentially even prebiotic. However, the conception of primordial metabolic cycles was challenged by Orgel on the basis that a steep decline in yield at each step would preclude non-enzymatically catalyzed flux through the full cycle. We see this as a valid criticism. The reverse Krebs cycle is also phylogenetically restricted almost entirely to bacteria and so is unlikely to date back to the common ancestor of bacteria and archaea (LUCA). The short, linear acetyl CoA pathway, which uses Fe(Ni)S proteins and electrochemical ion gradients across membranes to drive both carbon and energy metabolism from $H_2$ and $CO_2$, solves many of these problems. This pathway is found in both methanogenic archaea and acetogenic bacteria, but deep differences in the mechanism of electron bifurcation and methyl synthesis obscure the ancestral pathway of $CO_2$ fixation. Camprubi et al. (2017) proposed that these differences arose with the origin of active pumping and that ancestral carbon fixation in the first cells resembled that of methanogens, which use electrochemical ion gradients to drive ferredoxin reduction via the membrane-bound NiFe hydrogenase Ech.

These factors map onto the topology of pores in submarine alkaline hydrothermal vents, in which natural proton gradients across Fe(Ni)S barriers could drive the reduction of $CO_2$ by $H_2$. It is suggested that the initial reduction of $CO_2$ to a methyl group proceeded via a Fischer Tropsch–type mechanism on the acid side of an Fe(Ni)S surface, followed by carbonylation via a Koch-type mechanism. These endergonic reactions ought to be facilitated by pH differences between phases, which modulate the reduction potential of $H_2$, $CO_2$, and Fe(Ni)S minerals such as mackinawite. Similar hydrogenations and carbonylations, facilitated by natural proton gradients across Fe(Ni)S barriers, could generate longer-chain carboxylic acids (up to six carbons) via a prebiotic equivalent to the reverse, incomplete Krebs cycle in methanogens. Finally, after the evolution of cells with genes and proteins, closure of the complete reverse Krebs cycle should displace the acetyl CoA pathway from bacteria and archaea, except under anoxic conditions and at high partial pressures of $H_2$. This framework helps explain the origins of autotrophic $CO_2$ fixation in alkaline vents, the early divergence of bacteria and archaea, and the later evolution of the reductive Krebs cycle in bacteria.

### 4.4.1 Processes Influencing C/O Ratios

A minimal concentration of $^8$Be can start undergoing reactions with other $^4$He nuclei in to produce an unstable excited state of the mass-12 carbon. Some of these excited $^{12}$C nuclei emit a gamma-ray fast enough and become stable before they disintegrate. This sequence of nuclear processes is called the triple-alpha process. Its result is a combination of three alpha particles (that is, $3 \times {}^4$He nuclei) into a $^{12}$C nucleus. The triple-alpha reaction and the $^{12}C(\alpha,\gamma)^{16}O$ reaction operate side by side, and their relative rates determine the relative abundances of carbon and oxygen created; see Hoyle (1954), Kirsebom (2013), and Freer and Fynbo (2014).

Carbon is mainly produced in the core of a red giant star. It is stored there until the end of the star's life. It will still be present when the star explodes as a supernova and spreads its material into interstellar space. The sum of $^{12}C+\alpha$ masses is 7.1695 MeV. Therefore, the reaction $^{12}C(\alpha,\gamma)^{16}O$ should occur through a known nonresonant level at 7.1169 MeV. However, the 7.1169 level is just below $^{12}C + \alpha = 7.1695$ MeV, and therefore the resonance cannot happen.

However, with the introduction of 7.65 MeV resonance, the carbon yield has increased by a factor of about $10^7$ compared to $3\alpha \rightarrow {}^{12}C$. At high densities and temperatures, this reaction proceeds through the excited states of $^{12}C$ at 7.65 and 9.638 MeV; actually, it should be noted that at higher temperatures (a few times $10^9$ K) the $3 \times {}^4He \rightarrow {}^{12}C$ reaction proceeds primarily through the 9.638 MeV state. The reaction rates depend on the widths for radiative deexcitation of the states involved. Under these conditions, the possibility exists that inelastic collisions of the excited nuclei with n, p, α, and other hadrons can enhance the deexcitation process. We are mainly interested in neutrons because of the nonexistence of the Coulomb barrier, that is, in the reactions $^{12}C^*(n,n')^{12}C_{g.s.,4.43}$. Experimentally we can investigate the inverse reaction $^{12}C_{g.s}(n',n)^{12}C^*$ for kinematically allowed $^{12}C$ excitations. In this chapter, we report our study of this reaction at $E_n = 14.16$ MeV.

From the inspection of $^{12}C$ (Table 4.3) and $^{16}O$ energy levels, it could be concluded that the nuclei "remember" formation process. This "memory" explains why the cluster model describes so well the low-lying excited states in light nuclei (Valkovic 2021).

The rate of the triple-α reaction also affects the synthesis of heavy elements in the Ga–Cd range in proton-rich neutrino-driven outflows of core-collapse supernovae; see Wanajo (2006), Fröhlich et al. (2006), and Wanajo et al. (2011). At the beginning, outflows contain only protons and neutrons that subsequently combine into α-particles, then $^{12}C$ and some heavier elements. Although the most dominant is 3α→ 12C reaction involving the 7.65 MeV state, very high nucleon densities may cause the importance of proton and neutron scattering processes, altering the effective width of 7.65 MeV state (Turan and Kozlowsky 1969, Beard et al. 2017). The results of these calculations indicate that, for very high temperatures and particle densities > 108 g cm-3, the rates for inelastic scattering to the ground state of $^{12}C$ can exceed the rates of spontaneous radiative transitions by a significant factor.

TABLE 4.3 Low-Lying Energy Levels of $^{12}C$

| E level (keV) | $J^\pi$ | Γ |
|---|---|---|
| 0 | $0^+$ | STABLE |
| 4439.82 | $2^+$ | $10.8 \times 10^{-3}$ eV |
| 654.07 | $0^+$ | 9.3 eV |
| 9641 | $3^-$ | 46 keV |
| 9870 | $2^+$ | 850 keV |
| 9930 | $0^+$ | 2710 keV |
| 10300 | $(0^+)$ | 3.0 MeV |
| 10847 | $1^-$ | 273 keV |

The triple alpha process that converts $^4He$ into $^{12}C$ is one of the fundamental reactions in astrophysics; its rate influences the stellar productions of many elements. For the intermediate-temperature range (0.1–2) × $10^9$ K, the triple-alpha reaction proceeds through two low-lying s-wave resonances with the reaction rate R

$$R \infty \left(\Gamma_{\pm o}\Gamma_{rad} / \Gamma\right) T^{-3/2} \exp\left(-Q / kT\right) \quad (1)$$

where

Q—the energy of 7.65 MeV 12C excited state (Hoyle state) relative to the α + α + α threshold;

$\Gamma_{\alpha o}$—the partial α decay width to the ground state of 8Be;

$\Gamma_{rad}$—the radiative width, $\Gamma_{rad} = \Gamma_\gamma + \Gamma_\pi$, where $\Gamma_\pi$ is the e± pair decay width of the Hoyle state;

Γ—total decay width of the excited state, $\Gamma = \Gamma_\alpha + \Gamma_\gamma + \Gamma_\pi$.

It is well established that $\Gamma_{rad/}\Gamma_\alpha \ll 1$, so it follows $\left(\Gamma_{\alpha o}\Gamma_{rad} / \Gamma\right) = \left(\Gamma_{\alpha o}\Gamma_{rad} / \Gamma_\alpha + \Gamma_{rad}\right) \approx \Gamma_{rad}$. Accordingly, the reaction rate at given T depends only on two quantities, Q and $\Gamma_{rad}$.

Oxygen is primarily produced by process of α-capture by $^{12}C_{g.s}$. However, the α-induced reaction could be initiated on $^{12}C$-excited states produced by either by neutron or proton inelastic scattering or (n,n') or (p,p') reactions, leading to the reaction sequence:

$$^{12}C_{g.s.} + n(p) \rightarrow {}^{12}C^* + n'(p')$$
$$\rightarrow {}^{12}C^* + Q \rightarrow {}^{16}O^* \rightarrow {}^{16}O_{g.s.} + \gamma$$

In such a way, the quantity of 12C available to 16O production is effectively increased, consequently further reducing the C/O ratio.

The nucleosynthesis pathways of carbon and oxygen are different, so the majority of carbon comes from lower-mass star explosions, while oxygen is the more abundant in massive star explosions. Galactic chemical evolution models predict that the ratio of C/O should peak at a value just slightly greater than the solar one. The prevalence of C and O and the favorable thermodynamics of CO bonds make it a dominant molecular component of protoplanetary disks.

The carbon-to-oxygen ratio of a planet provides critical information about its primordial origins and subsequent evolution. A primordial C/O more significant than 0.8 results in an interior dominated by carbide, contrary to the silicate-dominated composition of Earth. The atmosphere can also be different from those in the solar system. The solar C/O is 0.54. Brewer and Fischer (2016) reported that the C/O ratio in the solar neighborhood peaked just below the solar value at 0.47, with a steep drop-off at super solar values. There are no stars in their 849-star sample with C/O > 0.7, implying that the formation of carbon-rich rocky planets is rare among stars in their sample.

Condensation of volatile species as a function of radial distance of C/O ratios has been calculated; see, for example, Thiabaud et al. (2015), showing enrichment in specific parts of the protoplanetary disk of up to four times the solar value. This leads to the formation of planets that can be enriched in C/O in their envelope up to three times the solar value. We should keep in mind that the total C/O ratio of the planets is governed by the contribution of the ices accreted (solid component); therefore the observed C/O ratio (in the atmosphere) is not representative of the total C/O ratio of the planet. The trends in dust composition as presented for the solar neighborhood by Zhukovska et al. (2008) indicate the C/O ratio for ISM dust particles in the solar neighborhood, starting from $3 \times 10^9$ years after Big Bang, to be > 1.

The reversed reaction $^{12}C^{*}(n',n)^{12}C$ is of interest from the astrophysical point of view since it enhances the triple-α reaction rate to the ground state of $^{12}C$. The triple-α reaction rate affects the oxygen primarily produced by process of α-capture by $^{12}C_{g.s}$. However, the α-induced reaction could be initiated on $^{12}C$-excited states, produced by either by neutron or proton inelastic scatterings or (n,n') or (p,p') reactions, leading to the reaction sequence described previously. In such a way, the quantity of $^{12}C$ available to $^{16}O$ production is effectively increased, consequently further reducing the C/O ratio. The carbon-to-oxygen ratio of a planet provides critical information about its primordial origins and subsequent evolution.

The rate of the triple-α reaction that forms $^{12}C$ affects the synthesis of heavy elements in the Ga–Cd range in proton-rich neutrino-driven outflows of core-collapse supernovae. Initially, these outflows contain only protons and neutrons; these later combine to form α particles, then $^{12}C$ nuclei via the triple-α reaction, and eventually heavier nuclei as the material expands and cools. Previous experimental work demonstrated that despite the high temperatures encountered in these environments, the reaction is dominated by the well-characterized Hoyle state resonance in $^{12}C$ nuclei. At sufficiently high nucleon densities, however, proton- and neutron-scattering processes may alter the effective width of the Hoyle state. This raises the questions of what the reaction rate in supernova outflows is and how changes affect nucleosynthesis predictions.

Jin et al. (2020) reported that in proton-rich core-collapse supernova outflows, these hitherto neglected processes enhance the triple-α reaction rate by up to an order of magnitude. The larger reaction rate suppresses the production of heavy proton-rich isotopes that are formed by the νp process (where ν is the neutrino and p is the proton) in the innermost ejected material of supernovae. Previous work on the rate enhancement mechanism did not anticipate the importance of this enhancement for proton-rich nucleosynthesis. Because the in-medium contribution to the triple-α reaction rate must be present at high densities, this effect needs to be included in supernova nucleosynthesis models. This enhancement also differs from earlier sensitivity studies that explored variations of the unenhanced rate by a constant factor, because the enhancement depends on the evolving thermodynamic conditions. The resulting suppression of heavy-element nucleosynthesis for realistic conditions casts doubt on the νp process being the explanation for the anomalously high abundances of $^{92,94}Mo$ and $^{96,98}Ru$ isotopes in the solar system and for the signatures of early Universe element synthesis in the Ga–Cd range found in the spectra of ancient metal-poor stars.

The C/O ratio of an exoplanet atmosphere (usually derived from relative abundances of the simple molecules listed previously) has been proposed to be a possible tool to link an exoplanet to its formation site in the natal protoplanetary disk. This is because disk midplanes have a radial gradient of C/O ratios in gas and ice (Öberg et al. 2011). This approach, to couple observed C/O ratios in exoplanets with their possible formation sites, typically assumes that the radial dependence of the midplane C/O ratio is defined solely by the iceline (snowline) positions of the main volatile species in a disk of fixed chemical composition; see, for example, Marboeuf et al. (2014). The position of the iceline for a particular species depends on the temperature structure of the disk midplane and the binding energy (or desorption energy) of the molecule to ice mantles on dust grains. Icelines thus affect the relative ratios of heavy elements in the gas and ice. For example, beyond the $H_2O$ iceline, the gas is depleted in oxygen, whilst the ice is enriched in oxygen. Similar changes occur across the $CO_2$, $CH_4$, and CO ice lines, thereby also altering the local C/O ratio in both the ice and gas. Due to the step-like features of that figure, caused by ice lines, the figure will be referred to as the traditional step function. Protoplanetary disks are not static objects: over the disk lifetime, from formation to dispersal, typically ~0.1–10 Myr, the disk generally cools, spreads, loses material (via accretion onto the star or dissipation), and thus becomes less dense over time.

Exoplanet atmospheres are thought be built up from accretion of gas as well as pebbles and planetesimals in the midplanes of planet-forming disks. The chemical composition of this material is usually assumed to be unchanged during the disk lifetime. However, chemistry can alter the relative abundances of molecules in this planet-building material. Eistrup et al. (2018) aim to assess the impact of disk chemistry during the era of planet formation. This was done by investigating the chemical changes to volatile gases and ices in a protoplanetary disk midplane out to 30 AU for up to 7 Myr, considering a variety of different conditions, including a physical midplane structure that is evolving in time, and also considering two disks with different masses. Methods. An extensive kinetic chemistry gas-grain reaction network was utilized to evolve the abundances of chemical species over time. Two disk midplane ionization levels (low and high) were explored, as well as two different makeups of the initial abundances ("inheritance" or "reset"). Results. Given a high level of ionization, chemical evolution in protoplanetary disk midplanes becomes significant after a few times $10^5$ years and is still ongoing by 7 Myr between the $H_2O$ and the $O_2$ ice lines. Inside the $H_2O$ ice line, and in the outer, colder regions of the disk midplane outside the $O_2$ ice line, the relative abundances of the species reach (close to) steady state by 7 Myr. Importantly, the changes in the abundances of the major elemental carbon and oxygen-bearing molecules imply that the traditional "step function" for the C/O ratios in gas and ice in the disk midplane (as defined by sharp changes at icelines of $H_2O$, $CO_2$, and CO) evolves over time and cannot be assumed fixed, with the C/O ratio in the gas even becoming smaller than the C/O ratio in the ice. In addition, at lower temperatures ($<29$ K), gaseous CO colliding with the grains gets converted into $CO_2$ and other more complex ices, lowering the CO gas abundance between the $O_2$ and CO thermal icelines. This effect can mimic a CO ice-line at a higher temperature than suggested by its binding energy.

Chemistry in the disk midplane is ionization driven and evolves over time. This affects which molecules go into forming planets and their atmospheres. In order to reliably predict the atmospheric compositions of forming planets, as well as to relate observed atmospheric C/O ratios of exoplanets to where and how the atmospheres have formed in a disk midplane, chemical evolution needs to be considered and implemented in planet formation models.

### 4.4.2 Processes Responsible for the Presence of Heavy Life-Required Elements (Mo, I, W)

One of the greatest questions for modern physics to address is how elements heavier than iron are created in extreme astrophysical environments. A particularly challenging part of that question is the creation of the so-called p-nuclei, which are believed to be mainly produced in some types of supernovae. The lack of needed nuclear data presents an obstacle in nailing down the precise site and astrophysical conditions. The observed distribution of heavy element abundances in our solar system provides a fingerprint of a complex interplay between nuclear properties and extreme astrophysical environments. Our understanding and identification of the stellar forges creating elements heavier than iron has improved significantly since the first attempts at understanding stellar nucleosynthesis in the 1950s. However, there are still mysteries regarding the astrophysical sites as well as the nuclear data needed to describe heavy-element nucleosynthesis.

Perhaps one of the most intriguing remaining mysteries concerns the 35 stable isotopes that cannot be explained by slow or rapid neutron-capture processes. The so-called p-process was suggested as an explanation for the existence of these isotopes (Burbidge et al. 1957). As of today, γ-induced photodisintegration of preexisting seed nuclei is understood to be the main production mechanism of the p-process, also known as the γ-process for this reason.

Favorable conditions for the p-process are found in the O-Ne layer of type II supernovae and in type Ia supernovae. Astrophysical model calculations are able to reproduce abundance patterns of most p-isotopes reasonably well, with some pivotal exceptions (Tveten et al. 2016). In particular, p-isotopes of mass $92 \leq A \leq 98$ are under-produced in calculations compared to the actual abundance of these isotopes. It has been suggested that the reason is related to the p-process seed nuclei. The underproduction could also be related to the details of the astrophysical site description. Experimental constraints on nuclear reaction rates are important to rule out the anomaly being related to the nuclear physics input.

Tveten et al. (2016) presented for the first time measurements on the nuclear-level density and average γ strength function of $^{92}$Mo. State-of-the-art p-process calculations systematically underestimate the observed solar abundance of this isotope. Their data provide stringent constraints on the $^{91}$Nb(p, γ)$^{92}$Mo reaction rate, which is the last unmeasured reaction in the nucleosynthesis puzzle of $^{92}$Mo. Based on their results, they concluded that the $^{92}$Mo abundance anomaly is not due to the nuclear physics input to astrophysical model calculations. The rapid neutron capture process (r-process) produces many of the heavy chemical elements, but the astrophysical settings where it occurs remain unclear. Leading candidates are neutron star mergers and some types of supernovae.

Molybdenum enzymes are described as ubiquitous because they have been isolated from organisms throughout the entire kingdom of life from simple bacteria to humans. Conservation of molybdenum enzymes through a billion years of evolution underscores their importance to the health of evolving organisms. It is now believed that molybdenum enzymes are the chemical descendants of tungsten enzymes. This hypothesis is based on the discovery of tungsten enzymes in archaebacteria, the most ancient organisms, where these tungsten enzymes have structures and functions very similar to those of the molybdenum enzymes found in higher organisms, including humans. The selection of tungsten over molybdenum by the archaebacteria is, in hindsight, consistent with the biogeochemistry of the anaerobic prehistoric environment. Molybdenum is an essential trace element for all higher organisms, where molybdenum enzymes have a number of important roles in the health of these organisms. Molybdenum in nitrate reductase is required by all plants for proper nitrogen assimilation. Molybdenum in xanthine oxidase, sulfite oxidase, and aldehyde oxidase is the functional center for these enzymes involved in the human diseases of gout, combined oxidase deficiency, and radical damage following cardiac failure. Combined oxidase deficiency is a rare genetic disease responsible for such severe neurological disorder that infant children with this defect rarely survive. Because of research directed at understanding the basis for this fatal disease, the tools of molecular biology and genetics have uncovered a link between one of the many proteins required to synthesize the molybdenum cofactor of these enzymes and a protein used in neuronal synapses. This illustrates how the biological impact of the molybdenum enzymes may reach beyond the limited context of their unique catalytic function.

Molybdenum is an essential trace element for all higher organisms because molybdenum enzymes have a number of important roles for maintaining the health of these organisms. All plants require molybdenum in nitrate reductase for proper nitrogen assimilation. In humans, the molybdenum enzymes xanthine oxidase, sulfite oxidase, and aldehyde oxidase are involved in the human diseases of gout, combined oxidase deficiency, and radical damage following cardiac failure. In industrial catalysis, molybdenum is used by the petroleum refining industry in multiple hydro-treating processes. In both industrial and biological catalysts, molybdenum is found in an environment including at least two and typically more sulfur atoms. Both molybdenum and sulfur exhibit an unusually wide range of formal oxidation states (+6 to −2) in their compounds, and the reduction potentials needed to access these oxidation states have considerable overlap.

Tungsten has five stable isotopes ($^{180}$W, $^{182}$W, $^{183}$W, $^{184}$W, $^{186}$W) with different nucleosynthetic origins: the rare $^{180}$W is a pure p-process nuclide, whereas $^{182}$W, $^{183}$W, $^{184}$W, and $^{186}$W are synthesized during both the s- and r-processes.

According to Kleine and Walker (2017), tungsten isotope measurements on extraterrestrial and terrestrial samples have led to the following discoveries:

> The parent bodies of differentiated meteorites accreted earlier than those of chondrites and underwent melting and differentiation due to internal heating by 26Al decay.

Complementary nucleosynthetic W isotope anomalies in chondrules and matrix indicate a solar nebular origin of chondrules and rule out that chondrules are the result of protoplanetary impacts.

The timescales of accretion and core formation increase with increasing planetary mass: Meteorite parent bodies accreted within the first few Ma, accretion of Mars took ~10 Ma, and the earliest time growth of Earth and segregation of its core can have been completed is within ~34 Ma of solar system formation.

The Moon exhibits a small $^{182}$W excess over the present-day silicate Earth, most likely resulting from the proportionally much lower mass of primitive material added to the Moon compared to Earth by late accretion.

Immediately following the giant impact, the Moon's and Earth's mantles had indistinguishable $^{182}$W compositions. This $^{182}$W homogeneity is not predicted in current formation models of the Moon and presents a key constraint for any successful model of lunar origin.

Terrestrial samples with ages spanning ~4 Ga reveal widespread $^{182}$W variations within the silicate Earth. The $^{182}$W anomalies either reflect derivation from mantle sources that lack some portion of the late veneer or are the product of very early differentiation processes in the mantle, predating the putative Moon-forming impact. Either way, the $^{182}$W heterogeneities in the silicate Earth demonstrate that signatures of Earth's earliest formative period have been preserved throughout most of Earth's history.

The discovery of a kilonova produced by the collision of two neutron stars brings researchers a big step closer to the explanation of one of the great puzzles of science: how the heaviest elements in the universe were formed.

Like other heavy elements, iodine is thought to have been formed through the rapid neutron capture process, formerly believed to have occurred primarily in supernovas but now mostly in collisions of neutron stars with other neutrons stars or black holes. Such events occurring long before the formation of the solar system seeded the primordial nebula with heavy elements.

Lattimer and Schramm (1974) examined the tidal breakup of a neutron star near a black hole. A simple model for the interaction is calculated, and the results show that the amount of neutron-star material ejected into the interstellar medium may be significant. Using reasonable stellar statistics, the estimated quantity of ejected material is found to be roughly comparable to abundance of r-process material.

## 4.5 HEAVY-METAL STARS

Low- and intermediate-mass stars with super solar metallicities constitute a known portion of the universe. Yet yields for asymptotic giant branch (AGB) stars with metallicities greater than $Z = 0.04$ do not exist in the literature. This contributes a significant uncertainty to galactic chemical evolution simulations. Cinquegrana and Karakas (2022) presented stellar yields of AGB stars for $M = 1–8M_{\odot}$ and $Z = 0.04–0.10$ and weighted these yields to represent the chemical contribution of a metal-rich stellar population. They found that as metallicity increases, the efficiency of the mixing episodes (known as the third dredge-up) on the thermally pulsing AGB (TP-AGB) decrease significantly. Consequently, much of the nucleosynthesis that occurs on the TP-AGB is not represented on the surface of very metal-rich stars. It instead remains locked inside the white dwarf remnant. The temperatures at the base of the convective envelope also decrease with increasing metallicity. For the intermediate-mass models, this results in the occurrence of only partial hydrogen burning at this location, if there is any burning at all. They also investigated heavy element production via the slow neutron capture process (*s*-process) for three 6-$M_{\odot}$ models: $Z = 0.04$, 0.05, and 0.06. There is minor production at the first *s*-process peak at strontium, which decreases sharply with increasing metallicity. They found the chemical contributions of the models are dominated by proton capture nucleosynthesis, mixed to the surface during the first and second dredge-up events. This conclusion is mirrored in stellar population yields, weighted towards the lower mass regime to reflect the mass distribution within a respective galaxy.

About half of the stable nuclei heavier than iron are believed to be synthesized during the late stages of evolution of stars with masses in the range $0.8 \pm 8$ solar masses. These elements are then expelled into the interstellar medium through stellar winds after being "dredged up" towards the surface of the stars. These processes occur when the star is in the asymptotic giant branch phase of its life. Nuclei (mainly iron) deep inside the star slowly capture neutrons and progressively build up heavier elements (the s-process). For AGB stars that formed early in the history of the galaxy, and that therefore have very low abundances of elements heavier than helium ("metals"), models predict that the s-process will accumulate synthesized material with atomic weights in the Pb ± Bi region. Such stars will therefore have large overabundances of lead relative to other heavy elements. Van Eck et al. (2001) reported the discovery of large amounts of lead in three metal-poor stars (HD187861, HD196944, and HD224959). Their analysis shows that these stars are more enriched in lead than in any other element heavier than iron. The excellent agreement between the observed and predicted abundances reinforces our current understanding of the detailed operation of the s-process deep in the interiors of AGB stars.

Karakas et al. (2022) presented new stellar evolutionary sequences of very metal-rich stars evolved with the Monash Stellar Structure code and with MESA. The Monash models include masses of 1–8 $M_\odot$ with metallicities $Z = 0.04$ to $Z = 0.1$ and are evolved from the main sequence to the thermally pulsing asymptotic giant branch. These are the first $Z = 0.1$ AGB models in the literature. The MESA models include intermediate-mass models with $Z = 0.06$ to $Z = 0.09$ evolved to the onset of the TP phase. The third dredge-up only occurs in intermediate-mass models $Z \leq 0.08$. Hot bottom burning shows a weaker dependence on metallicity, with the minimum mass increasing from 4.5 $M_\odot$ for $Z = 0.014$ to ≈5.5 $M_\odot$ for $Z = 0.04$, 6 $M_\odot$ for $0.05 \leq Z \leq 0.07$ and above 6.5 $M_\odot$ for $Z \geq 0.08$. The behavior of the $Z = 0.1$ models is unusual; most do not experience He-shell instabilities owing to rapid mass loss on the early part of the AGB. Turning off mass loss produces He-shell instabilities; however, thermal pulses are weak and result in no TDU. The minimum mass for carbon ignition is reduced from 8 $M_\odot$ for $Z = 0.04$ to 7 $M_\odot$ for $Z = 0.1$, which implies a reduction in the minimum mass for core-collapse supernovae. MESA models of similarly high metallicity ($Z = 0.06$–$0.09$) show the same lowering of the minimum mass for carbon ignition: carbon burning is detected in a 6 $M_\odot$ model at the highest metallicity ($Z = 0.09$) and in all 7 $M_\odot$ models with $Z \geq 0.06$. This demonstrates robustness of the lowered carbon-burning threshold across codes.

## 4.6 WHEN AND WHERE

The emergence of life is necessary stage in molecular cloud evolution. Hundreds of different types of molecules have been already detected in these clouds, including sugars and amino acids. More than 70 amino acids and several DNA bases were found in meteorites, which represent the original composition of the molecular cloud from which our solar system was formed.

Nitriles, a class of organic molecules with cyano group (a carbon atom bound with a triple unsaturated bond to a nitrogen atom), are a key precursor for molecules essential for life, such as ribonucleotides, a component of RNA. It is true that RNA can fulfill the functions of catalyzing reactions like enzymes and storing and copying information like DNA. Recent research (see Rivilla et al. 2022) shows that a wide range of nitriles occurs in interstellar space within the molecular cloud G+0.693–0.027 located near the center of the Milky Way galaxy. It appears that nitriles are among the most abundant chemical families in the universe. According to the authors, the chemical content of G+0.693–0.027 is similar to that of other star-forming regions in our galaxy and also to that of solar system objects like comets.

The discovery of extremophilic microbes has radically changed our conception of where life might be able to exist in the Universe—it causes us to reassess the concept of habitable zones. Scientists now realize that habitats suitable for microbial life are far more widely distributed in our solar system, and surely in the Universe as well, than was considered possible even in the most optimistic views of the 1980s and before. On the other hand, these same studies are showing that complex life—such as higher animals and plants—may have fewer suitable habitats than was previously thought. But just because life could exist in a place doesn't mean it is actually there; see Brownlee and Ward (2000).

Organic molecules are widely present in the dense interstellar medium, and many have been synthesized in the laboratory on Earth under the conditions typical for an interstellar environment. Until now, however, only relatively small molecules of biological interest have been demonstrated to form experimentally under typical space conditions. Here we prove experimentally that the condensation of carbon atoms on the surface of cold solid particles (cosmic dust) leads to the formation of isomeric polyglycine monomers (aminoketene molecules). Following encounters between aminoketene molecules, they polymerize to produce peptides of different lengths. The chemistry involves three of the most abundant species (CO, C, and NH3) present in star-forming molecular clouds and proceeds via a novel pathway that skips the stage of amino acid formation in protein synthesis. The process is efficient, even at low temperatures, without irradiation or the presence of water. The delivery of biopolymers formed by this chemistry to rocky planets in the habitable zone might be an important element in the origins of life (Krasnokutski et al. 2022).

Whereas modern proteins rely on a quasi-universal repertoire of 20 canonical amino acids (AAs), numerous lines of evidence suggest that ancient proteins relied on a limited alphabet of 10 "early" AAs and that the 10 "late" AAs were products of biosynthetic pathways. However, many nonproteinogenic AAs were also prebiotically available, which begs two fundamental questions: Why do we have the current modern amino acid alphabet and would proteins be able to fold into globular structures as well if different amino acids made up the genetic code? Makarov et al. (2023) experimentally evaluated the solubility and secondary structure propensities of several prebiotically relevant amino acids in the context of synthetic combinatorial 25-mer peptide libraries. The most prebiotically abundant linear aliphatic and basic residues were incorporated along with or in place of other early amino acids to explore these alternative sequence spaces. The results show that foldability was likely a critical factor in the selection of the canonical alphabet. Unbranched aliphatic amino acids were purged from the proteinogenic alphabet despite their high prebiotic abundance because they generate polypeptides that are oversolubilized and have low packing efficiency. Surprisingly, they found that the inclusion of a short-chain basic amino acid also decreases polypeptides' secondary structure potential, for which they suggest a biophysical model. Their results support the view that, despite lacking basic residues, the early canonical alphabet was remarkably adaptive at supporting protein folding and explain why basic residues were only incorporated at a later stage of protein evolution, according to Makarov et al. (2023).

In summary, their study on the solubility and secondary structure propensity of several prebiotically relevant amino acid alphabets supports the assertion that foldability played a critical early role in governing which amino acids ultimately became part of the canonical alphabet. Unbranched amino acids and short-chain basic amino acids were excluded, despite their prebiotic abundance because they "over-solubilized" polypeptides by stabilizing their unfolded conformations. Polypeptides with the propensity to cooperatively fold via hydrophobic packing would have then been better adapted for binding cofactors, recognizing RNA, and performing enzymatic catalysis. More broadly, this study supports the view that the early canonical alphabet, despite its deficiencies in relation to the modern canonical alphabet, was remarkably adaptive at supporting folding for the earliest proteins.

To shed some light on the problems of where and when life originated, we can hypothesize that life originated when the element abundance curve of the living matter and of the cosmic environment in which life originated coincided. This coincidence occurring at a particular space/time region of the universe can indicate when life originated, $T_{origin}$. In $T_{origin}$, the chemical abundance ratios of selected elements for a particular cosmic environment and living matter, corrected by the values of concentration factors, should be close to 1. The life abundance curve that we have considered is of the only life we know: that on the planet Earth. In our considerations, we have taken the estimated essential elements abundances of the last universal common ancestor; see Chopra and Lineweaver (2015), ignoring the fact that we do not know the essential elements concentration factors for LUCA, as we do not know its place of origin. The interstellar medium is a place of complex molecular synthesis. Many molecular species found on surfaces of comets, asteroids, meteorites, and interplanetary dust particles are used as life-forming molecules on the Earth (Ehrenfreund and Charnley 2000; Kwok 2004; Van Dishoeck 2014). When looking into the ISM as the possible environment of the origin of primitive life forms, our attention is on interstellar molecular clouds. IMCs contain gas and small micron-sized dust particles, which play a critical role as a catalyst in the formation of complex interstellar molecules. The molecular species forming on the dust particle are easily desorbed into the gas phase of IMCs upon formation (Greenberg 2002; Cazaux et al. 2016). These molecules are crucial for the initiation of life, and both life-forming molecules and a primitive life form can participate in planetary condensation from the ISM, initiating planetary life and its subsequent evolution. It is reasonable to suppose that life will not depend on resources that are scarce in its environment. The living matter needs only some chemical elements for its existence.

Life, as we know, is (H–C–N–O) based and relies on the number of bulk elements (Na–Mg–P–S–Cl–K–Ca) for its existence. All existing organisms use a number of elements for providing proteins with unique coordination, catalytic, and electron transfer properties. This group of elements is called trace elements, and it includes some (or all) of the following chemical elements: Li–B–F–Si–V–Cr–Mn–Fe–Co–Ni–Cu–Zn–As–Se–Mo–I–W. One can question the essentiality of some of these trace elements for life emerging. For example, boron, which is depleted in the solar system, played an essential role in the process of life forming, since its primary purpose has been to provide thermal and chemical stability in hostile environments (Scorei 2012). The fact that boron is not depleted in the ISM is yet another indication that life probably originated at a place different from Earth and the solar system.

There have been many attempts to elucidate the place and time of transition from abiotic to biotic forms that eventually led to our existence. It is currently accepted that this transition occurred on the Earth, probably after substantial delivery of the organic life building material during the heavy meteorite bombardment of the Earth some 3.9 Gy ago (Ehrenfreund and Menten 2001). The emerging of life in places different from Earth is yet to be confirmed. Many studies envisage such scenarios, but they all presume the existence of an Earth-like habitable zone for life to emerge. Our approach is that environments that are capable of producing life-forming building material, such as IMCs, are capable of producing life itself, whereas the existence of a habitable zone such as Earth is necessary for life to evolve into more complex forms. Since there are no other life forms on the Earth that could lead to roots different from LUCA, and our research has shown that LUCA is apparently much older than the solar system, we can assume that life originated only once. Otherwise, there should be different types of organisms available on Earth governed by different physical, chemical, and biological processes. Anyway, the processes such as photosynthesis and chemosynthesis that enable primitive organisms to be to some extent independent from their environmental resources are expected to be similar in extra-terrestrial and terrestrial bacteria (Sharov and Gordon 2017), thus resembling LUCA. These assumptions will be confirmed or disapproved after retrieving organisms from places other than Earth. The chemical composition of the universe is changing because of its aging, while LUCA, assumed to be the child of its stellar environment, has passed the information of required essential elements, as it was written in its genes at the time of its origin. By calculating the best agreement between essential elemental abundances in LUCA and corresponding elements of the solar neighborhood stars over their galactic evolution, we were able to find the time of LUCA's origin, presuming that the best agreement between the two curves is the most probable time of its origin, according to Valković and Obhođaš (2022).

Interactions between RNA and proteins are the cornerstone of many important biological processes from transcription and translation to gene regulation, yet little is known about the ancient origin of said interactions. Rout et al. (2023) hypothesized that peptide amyloids played a role in the origin of life and that their repetitive structure lends itself to building interfaces with other polymers through avidity. They reported that short RNA with a minimum length of three nucleotides binds in a sequence-dependent manner to peptide amyloids. The 3′–5′ linked RNA backbone appears to be well suited to support these interactions, with the phosphodiester backbone and nucleobases both contributing to the affinity. Sequence-specific RNA–peptide interactions of the kind identified here may provide a path to understanding one of the great mysteries rooted in the origin of life: the origin of the genetic code.

## 4.7 THE BIG BANG AND ORIGIN OF LIFE

Trevors (2006) examined the origin of life on Earth and its connection to superstring theory, which attempts to explain all phenomena in the universe (theory of everything) and unify the four known forces and relativity and quantum theory. The four forces of gravity, electromagnetism, and strong and weak nuclear were all present and necessary for the origin of life on the Earth. It was the separation of the unified force into four singular forces that allowed the origin of life.

The result of this work may also give an explanation for the homochirality of life. Even if activated monomers supplied to the polymerization cycle are a racemic mixture, life emerging from them would be homochiral if homochirality is a necessary requirement for an RNA polymer to show biological activities. It simply requires more time or volume for a homochiral polymer to be assembled by random polymerization, with $N_{nb}$ twice as large as when ignoring chirality. Changing $N_{nb}$ by a factor of two does not seriously affect the expected number of abiogenesis events in an inflationary universe. On the other hand, the expected number of abiogenesis events is much smaller than unity when we observe a star, a galaxy, or even the whole observable universe. This gives an explanation to Fermi's paradox. The observable universe is just a tiny part, whose volume is likely smaller than 1/1078 of the whole universe created by an inflation, and there is no strong reason to expect more than one abiogenesis event in such a small region. Even if Earth is the only planet that harbors life inside the observable universe, it does not contradict the Copernican principle, because life would have emerged on countless planets in the whole inflationary universe in which we exist.

Living organisms are the least entropic (i.e., the most complexly ordered and improbable) entities known to exist. Biological natural selection (BNS) (intended as shorthand for all types of biological selection, including natural selection proper, sexual selection, and kin selection), then, is the strongest known antientropic process, because it creates organisms. Organisms have extraordinarily low entropy because BNS endows them with adaptations that ultimately enable genetic replication. Achieving replication is a complex and highly ordered process and requires organisms that are themselves complex and highly ordered. That is, it requires organisms composed of myriad adaptations that are functionally specialized for solving diverse kinds of adaptive problems (such as breathing, seeing, digesting, and mating). By enabling the organism to succeed in the competition with conspecifics to survive and reproduce, the genes which encode these adaptations are ultimately able to replicate. Genes are the replicators in this process, and the organism—which is essentially a bundle of adaptations—is their vehicle of selection. In summary, then, the selective process works by favoring those genes which, by encoding adaptive designs that outcompete other designs, most successfully enable their own replication.

According to the theory of cosmological natural selection (CNS), we live in a multiverse in which the most-represented universe designs are those best at replication. Compared to BNS, CNS is a vastly less thoroughly documented and well understood process. Price (2017) considers what the consequences would be if CNS were true. He took an adaptationist perspective on the CNS theory that has already been developed by Smolin (1992). According to this theory, black holes are the mechanism by which universes replicate, and universe designs are selected to produce greater numbers of black holes. However, if we apply the adaptationist logic outlined above to Smolin's model, then life appears more likely than black holes to be a mechanism of universe replication. According to Smolin's model, black holes are adaptations designed by CNS. Other theorists, however, have proposed that intelligent life may constitute such an adaptation, an idea known as "cosmological natural selection with intelligence" or CNS-I, or alternatively, "cosmological artificial selection". CNS-I models propose that intelligence functions ultimately (after evolving to a sufficiently sophisticated state) to create new universes that replicate the physical laws and parameters of its home universe.

So which is more likely to be an adaptation for the replication of universes, intelligent life or black holes? Based on adaptationist logic, life would seem to be the more promising candidate of the two, simply because it is far more complexly ordered and less likely to arise by chance. As the most improbable known thing in the universe, designed by the strongest known antientropic process, life seems more likely than black holes—or any other known entity—to be a CNS-designed adaptation. That is not to say, however, that black holes could not also be such adaptations. Just as eyes and reduced skin pigmentation could both be BNS-selected adaptations, life and black holes could both be CNS-selected adaptations. But just as eyes are more likely to be an adaptation than reduced melanin, life is more likely to be an adaptation than black holes (Price 2017).

About $15 \times 10^6$ years after the big bang, the entire universe had cooled to the point where the electromagnetic radiation left over from its hot beginning was at about room temperature. Loeb (2014) described a possible habitable epoch of the early universe. In the redshift range $100 \leq (1 + z) \leq 137$, the cosmic microwave background (CMB) had a temperature of 273–373 K (0–100°C), allowing early rocky planets (if any existed) to have liquid water chemistry on their surface and be habitable, irrespective of their distance from a star. In the standard ΛCDM cosmology, the first star-forming halos within our Hubble volume started collapsing at these redshifts, allowing the chemistry of life to possibly begin when the Universe was merely $(10–17) \times 10^6$

years old. The possibility of life starting when the average matter density was a million times bigger than it is today argues against the anthropic explanation for the low value of the cosmological constant

However, the initial bottleneck for life was not a suitable temperature, as it is today, but rather the production of the essential elements. Loeb et al. (2016) discussed the relative likelihood for life as a function of cosmic time. According to the standard model of cosmology, the first stars in the observable Universe formed ~30 Myr after the Big Bang at a redshift, $z \sim 70$ (Naoz et al. 2006). Within a few Myr, the first supernovae dispersed heavy elements into the surrounding gas, enriching the second-generation stars with heavy elements. Remnants from the second generation of stars are found in the halo of the Milky Way galaxy and may have planetary systems in the habitable zone around them (Mashian and Loeb 2016). The related planets are likely made of carbon, and water could have been delivered to their surface by icy comets, in a similar manner to the solar system. The formation of water is expected to consume most of the oxygen in the metal poor interstellar medium of the first galaxies.

Loeb et al. (2016) tried to answer a question of is it most likely for life to emerge at the present cosmic time near a star like the Sun. They address this question by calculating the relative formation probability per unit time of habitable Earth-like planets within a fixed comoving volume of the Universe, dP(t)/dt, starting from the first stars and continuing to the distant cosmic future. They conservatively restricted their attention to the context of "life as we know it" and the standard cosmological model, ΛCDM. They find that unless habitability around low mass stars is suppressed, life is most likely to exist near ~0.1 Msun stars 10 trillion years from now. Spectroscopic searches for bio-signatures in the atmospheres of transiting Earth-mass planets around low-mass stars will determine whether present-day life is indeed premature or typical from a cosmic perspective.

Life requires stars for two reasons. Stars are needed to produce the heavy elements (carbon, oxygen, and so on, up to iron) out of which rocky planets and the molecules of life are made. Stars also provide a heat source for powering the chemistry of life on the surface of their planets. Each star is surrounded by a habitable zone where the surface temperature of a planet allows liquid water to exist. The approximate distance of the habitable zone, $r_{HZ}$, is obtained by equating the heating rate per unit area from the stellar luminosity, L, to the cooling rate per unit area at a surface temperature of $T_{HZ} \sim 300$ K, that is, $(L/4\pi r^2_{HZ}) \sim \sigma T^4 \times 10^6$, where σ is the Stefan-Boltzmann constant. According to the standard model of cosmology, the first stars in the observable Universe formed $\sim 30 \times 10^6$ years after the Big Bang at a redshift $z \sim 70$. Within a few $10^6$ years, the first supernovae dispersed heavy elements into the surrounding gas, enriching the second-generation stars with heavy elements. Remnants from the second generation of stars are found in the halo of the Milky Way galaxy and may have planetary systems in the habitable zone around them. The related planets are likely made of carbon, and water could have been delivered to their surface by icy comets, in a similar manner to the solar system. The formation of water is expected to consume most of the oxygen in the metal poor interstellar medium of the first galaxies.

According to Sutter (2023), life might have been possible just seconds after the Big Bang. For such a statement, the author uses an extremely broad but very useful definition: Life is everything that's subject to Darwinian evolution. This definition is handy because we'll be exploring the origins of life itself, which, by definition, will blur the boundaries between life and nonlife. Some theories speculate that the first self-replicating molecules, and hence the simplest possible form of life on Earth, could have arisen as soon as the oceans cooled, well over 4 billion years ago. Life as we know it requires a few key elements: hydrogen, oxygen, carbon, nitrogen, and phosphorus. With the exception of hydrogen, which appeared in the first few minutes after the Big Bang, all of these elements are created in the hearts of stars during their life cycles. So, as long as you have at least one or two generations of stars living and dying, and thereby spreading their elements out into the wider galaxy, one can have Earth-like life appearing in the universe.

So, life as we know it—built on chains of carbon, using oxygen to transport energy, and submersed in a bath of liquid water—may be much, much older than Earth. Even other hypothesized forms of life based on exotic biochemistries require a similar mixture of elements. For example, some alien life may use silicon instead of carbon as a basic building block or use methane instead of water as a solvent. No matter what, those elements have to come from somewhere, and that somewhere is in the cores of stars. Without stars you can't have chemical-based life. So statement by Valković and Obhođaš (2022) claiming that life was in the solar neighborhood already at $(4 \pm 1) \times 10^9$ years seems easily acceptable.

Sutter (2023) speculates that it might be possible to have life without chemistry. Although chemistry is a convenient way to store information, extract energy, and interact with the environment, there are other hypothetical pathways. For example, 95% of the energy contents of the universe are unknown to physics, literally sitting outside the known elements. Scientists aren't sure what these mysterious components of the universe, known as dark matter and dark energy, are made of. Perhaps there are additional forces of nature that work only on dark matter and dark energy. Maybe there are multiple "species" of dark matter—an entire "dark matter periodic table". Who knows what interactions and what dark chemistry play out in the vast expanses between the stars? Hypothetical "dark life" may have appeared in the extremely early universe, well before the emergence of the first stars, powered and mediated by forces we do not yet understand.

Some physicists have hypothesized that in the earliest moments of the Big Bang, the forces of nature were so extreme and so exotic that they could have supported the growth of complex structures. For example, these structures could have been cosmic strings, which are folds in space-time, anchored by magnetic monopoles. With sufficient complexity, these structures could have stored information. There would have been plenty of energy to go around, and those structures could have self-replicated, enabling Darwinian evolution. Any creatures existing in those conditions would have lived and died in the blink of an eye, their entire history lasting less than a second—but to them, it would have been a lifetime (Sutter 2023).

### 4.7.1 Life on Rogue Planets

We start this section with a discussion of Mehta (2022), contributing editor for the Planetary Society, who asked a question: Is life possible on rogue planets and moons? His starting point is that starless free-floating worlds might represent the most common habitable real estate of the universe. The search for planets around other stars in the MW galaxy has resulted in more than 5000 such objects. Quite a few of these exoplanets seem to be Earth-like, where surface conditions could sustain liquid water and some form of life. However, Earth-like planets are not the only places where life could form. Keeping for the moment our proposal that molecular clouds are an obvious location, we should look into other objects. It is known that in our solar system, icy moons orbiting giant planets far away from the Sun (Europa, Ganymede, and Enceladus, for example) can have underground, habitable oceans too. Their liquid water isn't due to the Sun's heat but rather warmed by friction between parts of their interiors being tugged by their planets' gravity. Obviously, sunlight, a surface, and an atmosphere are not necessary to make a world habitable.

It is assumed that planets that don't orbit any star, called free-floating planets or rogue planets, can harbor life too. These planets originally form around stars like any other but get kicked out of their system at some point due to gravitational effects of giant planets within. Planets may have been ejected from our solar system, too, over 4 billion years ago and now orbit our galaxy as dark worlds. Without a star, how can these dark worlds conceivably host life as we know it? Our exploration of the solar system combined with two decades of exoplanet research tells us there are several possibilities.

According to Stevenson (1999), during planet formation, rock and ice embryos on the order of Earth's mass may be formed, some of which may be ejected from the solar system as they scatter gravitationally from protogiant planets. These bodies can retain atmospheres rich in molecular hydrogen, which, upon cooling, can have basal pressures of $10^2$ to $10^4$ bars. Pressure-induced far-infrared opacity of $H_2$ may prevent these bodies from eliminating internal radioactive heat except by developing an extensive adiabatic (with no loss or gain of heat) convective atmosphere. This means that, although the effective temperature of the body is around 30 K, its surface temperature can exceed the melting point of water. Such bodies may therefore have water oceans whose surface pressure and temperature are like those found at the base of Earth's oceans. Such potential homes for life will be difficult to detect.

Life on free-floating worlds would still have to miraculously emerge using the planet's minuscule internal energy, compared to over 99% of Earth's energy coming from sunlight (Mehta 2022). Hypothetically, if a free-floating planet has a large enough moon, it could further heat the planet using tidal mechanisms, similar to our Moon and Earth. When the Moon formed more than 4.4 billion years ago, it was about 15 times closer to us than it is today. It induced such a strong tidal heating that scientists think the Moon may have played a key role in making the early Earth habitable. Even if such heating lasts only a few hundred million years, it could provide a richer source of energy than the free-floating planet's own heat to keep an ocean warm, initiate complex geology, and possibly develop microbial life.

Free-floating planets can have moons; see the paper by Rabago and Steffen (2019) discussing the survivability of moon systems around ejected gas giants. Orbital simulations show that more than 47% of moons can remain bound to exiled gas giant planets. Simulations with ejected Earth-mass planets show that more than 4% of them retain their Moon-sized satellite.

Rabago and Steffen (2019) examined the effects that planetary encounters have on the moon systems of ejected gas giant planets by conducting a suite of numerical simulations of planetary systems containing three Jupiter-mass planets (with the innermost planet at 3 au) up to the point where a planet is ejected from the system. The ejected planet has an initial system of 100 test-particle moons. They determined the survival probability of moons at different distances from their host planet, measure the final distribution of orbital elements, examine the stability of resonant configurations, and characterize the properties of moons that are stripped from the planets. They found that moons are likely to survive in orbits with semi-major axes out beyond 200 planetary radii (0.1 au in our case). The orbital inclinations and eccentricities of the surviving moons are broadly distributed and include nearly hyperbolic orbits and retrograde orbits. They found that a large fraction of moons in two-body and three-body mean-motion resonances also survive planetary ejection with the resonance intact. Moon–planet interactions, especially in the presence of mean-motion resonance, can keep the interior of the moons molten for billions of years via tidal flexing, as is seen in the moons of the gas giant planets in the solar system. Given the possibility that life may exist in the subsurface ocean of the Galilean satellite Europa, these results have implications for life on the moons of rogue planets—planets that drift through our galaxy with no host star.

### 4.7.2 Life in the Molecular Clouds

Valković and Obhođaš (2022), in their paper, discuss the origins of chiral life in interstellar molecular clouds and suggest some possible future activities which might prove these assumptions. They start with the fact that the living matter life needs only some chemical elements for its existence. Life, as we know, is (H-C-N-O) based and relies on the number of bulk elements (Na-Mg-P-S-Cl-K-Ca) for its existence. All existing organisms use a number of elements for providing proteins with unique coordination and catalytic and electron transfer properties. This group of elements is called trace elements, and it includes some (or all) of the following chemical elements: Li-B-F-Si-V-Cr-Mn-Fe-Co-Ni-Cu-Zn-As-Se-Mo-I-W.

On the other hand, experimental evidence shows that the chemical evolution of galaxies results in the changes of chemical composition (relative abundances of chemical elements) of stars, interstellar gas, and dust. The increase of chemical element abundances with time (elements heavier than Li) provides a clock for galactic aging. The older the galaxy, the more heavy elements it contains.

Therefore, to shed some light on the problems of where and when life originated, one could assume that life originated when the element abundance curve of the living matter and the one of the environment in which it originated coincided. This coincidence occurring at a particular time could indicate the phase of the Universe when life originated, $T_{origin}$. To estimate the closeness of different abundance curves, one should evaluate relations between elements and groups for life-essential elements and various media. The chemical elements abundance ratios for selected groups of elements for a particular medium and living matter, corrected by the values of concentration factors, should be close to one (1) in space/time region when and where life originated.

The remaining crucial question is which two-element abundance curves should be considered. The environmental abundance curve should be either the abundance curve of the part of the Universe or the individual galaxy where life supposedly originated at that time. The life abundance curve could be of the only life we know: that on the planet Earth. Therefore, we should take that of last universal common ancestor or some other primitive organism that can survive the planet forming process, ignoring the fact that we do not know their element concentration factors.

When looking into Universe, our attention is on interstellar molecular clouds and circumstellar envelopes, which are places of complex molecular synthesis (Ehrenfreund and Charnley 2000; Kwok 2004; van Dishoeck 2014). In addition to gas, interstellar material also contains small micron-sized particles. Gas–phase and gas–grain interactions result in the formation of complex molecules. Surface catalysis on solid particles enables molecule formation and chemical pathways that cannot proceed in the gas phase because of reaction barriers.

Interstellar dust appears to play a critical role in the formation of interstellar molecules. Molecules may be formed on or in grain surfaces. Many molecules have been observed in interstellar space. As early as 1972 Johnson (1972) has reported even the observation of interstellar porphyrins (molecule $MgC_{46}H_{30}N_6$). The existence of interstellar molecules suggests that such molecules support or are the metabolic products of an interstellar biota. Such molecules, participating in planetary condensation from the interstellar medium, can make a significant contribution to the origin of planetary life. A high number of molecules that are used in contemporary biochemistry on the Earth are present in the interstellar medium on surfaces of comets, asteroids, meteorites, and interplanetary dust particles. Therefore, we can assume that life originated in an interstellar molecular clouds with the critical role of dust particles, magnetic fields, and exposure to cosmic rays in their environment.

The solar system we live in is the result of the gravitational collapse of a small part of a giant molecular cloud. Assuming life originated before the origin of Earth, it could be hypothesized that life was brought to Earth in the process of its formation. Here we must consider the role of cosmic dust mineral grains in this process. It is reasonable to assume that life will not depend on resources that are scarce in its environment. Yet the chemical element boron, which is depleted in the solar system but not in the interstellar medium, played an essential role in the process of life forming, since its primary purpose has been to provide thermal and chemical stability in hostile environments (Cazaux et al. 2016).

In further discussion, we have to consider only well-characterized cosmic environments, ones for which the elemental abundance curves for elements of our interest could be constructed from the measured data or from existing models. We will consider two cosmic environments: the solar neighborhood only and also Milky Way galaxy as a whole. When more data are available, we should also look into DLA systems.

Possibilities for methanogenic and acetogenic life in molecular clouds have also been discussed by Feng (2023). He is strong advocate of life existing in molecular clouds. He discusses the possibility of methanogenic life in molecular clouds with methane as the final metabolic product. According to his calculations, it is easy to see that the chemical reaction through methanogenesis can release sufficient free energy. If methanogenic life exists in the presolar nebula, then it may be the ancestor of Earth's life, and there is some tentative evidence for several molecular studies. Methanogens on Earth maintain their activity through methanogenesis, a form of anaerobic respiration. Carbon dioxide is the terminal electron acceptor in methanogenesis through the following chemical reaction

$$CO_2 + 4H_2 = CH_4 + 2H_2O$$

With the same electron acceptor and donor, acetogenic processes are also possible through the following reaction

$$2CO_2 + 4H_2 = CH_3COOH + 2H_2O$$

As the $CO_2$ in molecular clouds is more abundant in the solid phase than in the gas phase, solid $CO_2$ provide abundant carbon sources and energy for life activities. Therefore, there might be life forms gathered around these solid carbon dioxide.

If there were methanogenic and/or acetogenic life in the pre-solar nebula and then on the early Earth, LUCA would be one type of methanogenic life. If so, the descendant of this ancient life would exist in all places in the solar system that are suitable for its survival, such as Europa and Titan. Coincidentally, the primitive atmosphere of the earth was rich in carbon monoxide and carbon dioxide. This could ensure the continued survival of methane life on Earth. The most important prediction of this model is that methanogenic and/or acetogenic life is the ancestor of life on Earth or, more generally, LUCA is one type of life form that uses $H_2CO$ or $H_2CO_2$ as an electron donor (acceptor) in the chemical reactions of generating energy. In general, this model provides a perfect origin for LUCA. There is already some tentative evidence about the probability that LUCA is one type of methanogenic and/or acetogenic life. Mat et al. (2008) has identified the minimum genome of LUCA and found that early life forms invented methanogenesis in the cooler temperature zones that led to the rise of LUCA after acclimatization to the hydrothermal vent environment.

Weiss et al. (2016) investigated all clusters and phylogenetic trees for 6.1 million protein coding genes from sequenced prokaryotic genomes in order to reconstruct the microbial ecology of LUCA. Among 286,514 protein clusters, they identified 355 protein families (~0.1%) that trace to LUCA by phylogenetic criteria. Because these proteins are not universally distributed, they can shed light on LUCA's physiology. Their functions, properties, and prosthetic groups depict LUCA as anaerobic, $CO_2$-fixing, $H_2$-dependent with a Wood–Ljungdahl pathway, $N_2$-fixing, and thermophilic. LUCA's biochemistry was replete with FeS clusters and radical reaction mechanisms. Its cofactors reveal dependence upon transition metals, flavins, S-adenosyl methionine, coenzyme A, ferredoxin, molybdopterin, corrins, and selenium. Its genetic code required nucleoside modifications and S-adenosyl methionine-dependent methylations. The 355 phylogenies identify clostridia and methanogens, whose modern lifestyles resemble that of LUCA, as basal among their respective domains. LUCA inhabited a geochemically active environment rich in $H_2$, $CO_2$, and iron. The data support the theory of an autotrophic origin of life involving the Wood–Ljungdahl pathway in a hydrothermal setting. The authors found that LUCA's genes point to methanogenic and acetogenic roots by surveying nearly 2000 genomes of modern microbes.

Reactions in interstellar ices are shown to be capable of producing key prebiotic molecules without energetic radiation that are necessary for the origins of life. When present in interstellar ices, carbamic acid ($H_2NCOOH$) can serve as a condensed-phase source of the molecular building blocks for more complex proteinogenic amino acids. In the work by Marks et al. (2023) Fourier transform infrared spectroscopy, applied during heating of analog interstellar ices composed of carbon dioxide and ammonia, identifies the lower limit for thermal synthesis being 62 ± 3 K for carbamic acid and 39 ± 4 K for its salt ammonium carbamate ($[H_2NCOO-][NH_4^+]$). While solvation increases the rates of formation and decomposition of carbamic acid in ice, the absence of solvent effects after sublimation results in a significant barrier to dissociation and a stable gas-phase molecule. Photoionization reflectron time-of-flight mass spectrometry permits an unprecedented degree of sensitivity toward gaseous carbamic acid and demonstrates sublimation of carbamic acid from decomposition of ammonium carbamate and again at higher temperatures from carbamic acid dimers. Since the dimer is observed at temperatures up to 290 K, similar to the environment of a protoplanetary disk, this dimer is a promising reservoir of amino acids during the formation of stars and planets.

## 4.8 $T_{ORIGIN}$ FOR DIFFERENT PARTS OF THE UNIVERSE

Totani (2020) discussed the emergence of life in an inflationary universe. Abiotic emergence of ordered information stored in the form of RNA is an important unresolved problem concerning the origin of life. A polymer longer than 40–100 nucleotides is necessary to expect a self-replicating activity, but the formation of such a long polymer with a correct nucleotide sequence by random reactions seems statistically unlikely. However, our universe, created by a single inflation event, likely includes more than $10^{100}$ Sun-like stars. If life can emerge at least once in such a large volume, it is not in contradiction with our observations of life on Earth, even if the expected number of abiogenesis events is negligibly small within the observable universe that contains only $10^{22}$ stars. Here, a quantitative relation is derived between the minimum RNA length lmin required to be the first biological polymer and the universe size necessary to expect the formation of such a long and active RNA by randomly adding monomers. It is then shown that active RNA can indeed be produced somewhere in an inflationary universe, giving a solution to the abiotic polymerization problem. On the other hand, lmin must be shorter than ~20 nucleotides for the abiogenesis probability close to unity on a terrestrial planet, but a self-replicating activity is not expected for such short RNA. Therefore, if extraterrestrial organisms of a different origin from those on Earth are discovered in the future, it would imply an unknown mechanism at work to polymerize nucleotides much faster than random statistical processes.

Carbon-enhanced metal-poor (CEMP) stars are a unique resource for galactic archaeology because they probe the properties of the first stars, early chemical evolution, and binary interactions at very low metallicity. Comparing the fractions and properties of CEMP stars in different galactic environments can provide us with unique insights into the formation and evolution of the Milky Way halo and its building blocks. In this work, we investigate whether directly comparing fractions of CEMP stars from different literature samples of very metal-poor ($[Fe/H] < -2.0$) stars is valid. Arensten et al. (2022) compiled published CEMP fractions and samples of galactic halo stars from the past 25 years and find that they are not all consistent with each other. Focusing on giant stars, they find significant differences between various surveys when comparing their trends of [Fe/H] versus [C/Fe] and their distributions of CEMP stars. To test the role of the analysis pipelines for low-resolution spectroscopic samples, they reanalyzed giant stars from various surveys with the SSPP and FERRE pipelines. They found systematic differences in [C/Fe] of ~0.1–0.4 dex, partly independent of degeneracies with the stellar atmospheric parameters. These systematics are likely due to the different pipeline approaches, different assumptions in the employed synthetic grids, and/or the comparison of different evolutionary phases. They conclude that current biases in (the analysis of) very metal-poor samples limit the conclusions one can draw from comparing different surveys.

## 4.9 THE BEST ESTIMATE FOR $T_{ORIGIN}$ IN THE SOLAR NEIGHBORHOOD

The exploration of galactic chemical composition across the Milky Way, and specifically across the solar neighborhood, provides insights into the chemical evolution of the universe. Since the formation of the first stars some 100 million years after the big bang (BB), heavier elements are synthesized in different stellar production processes at the expense of lighter elements. When the relative abundances of the life-forming elements evaluated for the last universal common ancestor are compared with the solar neighborhood stellar abundances, a striking similarity occurs. In this study, we tested the hypothesis that in some particular regions and at some particular time, the abundance curve of the first living matter and the universe coincided. Indeed, the best agreement between the two curves was obtained for $(4 \pm 1) \times 10^9$ years after the BB, indicating the time of the origin of life. All organisms that evolved on the Earth independently of place and time lead to the LUCA and involve chiral molecules such as L amino acids and D sugars in fundamental life processes. The growing evidence from carbonaceous meteorites analysis shows an excess of L-type amino acids and D-type sugars, suggesting that the increase in L-type or D-type molecular chirality is the process that takes place in planetary and stellar forming systems, thus the life emerging from interstellar molecular clouds had to be chiral. Here we propose spin-polarized proton–proton scattering as a potential physical process that takes place in IMC environments and could lead to enrichment of L-type amino acids and D-type sugars.

The elemental abundances for the solar neighborhood were obtained from the experimental values summarized in a paper by Kobayashi et al. (2020). The elemental abundances for LUCA were taken from a paper by Chopra and Lineweaver (2015) and expressed relative to iron. The agreement between life essential bulk and trace elements in LUCA and the solar neighborhood as a function of metallicity [Fe/H] were calculated for different [Fe/H] values corresponding to different times, T values, according to the age-[Fe/H] model presented in Kobayashi et al. (2020). Only those essential elements available for the evaluation of the solar neighborhood chemical evolution presented in Kobayashi et al. (2020) were selected (C, N, O, Na, Mg, Al, Si, P, S, K, Ca, V, Cr, Mn, Co, Ni, Cu, Zn, and Mo). Two approaches were used to estimate the time of the life origin, $T_{origin}$. First, we calculated the minimum of $\chi^2$ (best agreement) as defined by

$$\chi^2 = 1/N \sum_{i=1}^{N} \left(C_{i,LUCA} - C_{i,SN}\right)^2 / C_{i,SN}$$

where $C_i$ is the abundance of essential element i in LUCA and in the solar neighborhood (SN) where LUCA originated. In the second approach, we calculated the maximum coefficient of determination ($R^2$), that is, the percentage of the variance explained by the linear regression model between logarithmic values of abundances of essential elements in LUCA and the environment of its origin. $\chi^2$ and $R^2$ have been calculated for 19 essential elements (as mentioned) and also $\chi^2$ for the subgroup of trace elements (V, Cr, Mn, Co, Ni, Cu, and Zn).

For the determination of the time of the life origin, $T_{origin}$, we have to consider only well-characterized cosmic environments such as the solar neighborhood, for which the elemental abundance curves for elements of our interest could be constructed from the measured data or existing models. Recently, Kobayashi et al. (2020) constructed galactic chemical evolution models for the solar neighborhood. Their calculations result in concentration values for all stable elements from $^{12}C$ to $^{238}U$, based on theoretical nucleosynthesis yields and event rates of all nucleosynthesis processes involved in elemental synthesis as a function of time and environment. The basic equations of chemical evolution are described by Kobayashi et al. (2000). Considering the life abundance curve, it is the common understanding that all organisms on the tree of life have a common ancestor; that is, the tree is rooted in the LUCA.

Therefore, to establish an average elemental composition of life, one needs to estimate the relative abundance of elements in LUCA. One such effort was undertaken by Chopra et al. (2009) and Chopra and Lineweaver (2015). They considered the elemental composition of eukaryotic, bacterial, and archaeal species, as well as the phylogenetic relationship between them. Although the idea of LUCA or the progenitor is the most important for the study of early evolution and life's origin, the information about when, where, and how LUCA originated is still lacking. That is, considerable time and extinct organisms might have existed between the origin of life and the root of the tree, as defined by extant organisms (Cornish-Bowden and Cárdenas 2017). All these considerations are done by assuming that Earth's present-day three existing domains of life define LUCA's capabilities and characteristics. However, as pointed out by Cockell (2015), we cannot be sure that there were not domains that went extinct early in the history of life and took with them crucial information.

### 4.9.1 Evaluation of Possible Errors

In all cases, a $T_{origin}$ of (3–5) × $10^9$ years after the big bang is observed (or $(4 \pm 1) \times 10^9$ after BB). However, the large uncertainty of age determination derived from the age–[Fe/H] metallicity model, as well as uncertainties derived by extrapolation and interpolation of solar neighborhood abundances of elements from graphical diagrams presented in Kobayashi et al. (2020), should be taken into account. On the other hand, the corrections for concentration factors in LUCA would shift the $\chi^2$ and $R^2$ curves presumably toward the better agreement between two data sets (those for LUCA and the abundances of elements in stars of the solar neighborhood), but it would not change the shape of the $\chi^2$ and $R^2$ curves. For example, the correction for potassium in LUCA for a factor of 5.5, Obhođaš et al. (2021) shifts up the $R^2$ estimates for 0.035 (i.e., 3.5% better agreement was observed between abundance curves for LUCA and stars of the solar neighborhood throughout their chemical evolution).

A different approach to the estimation of the origin of life based on extrapolation of the genetic complexity of organisms back to just one base pair yielded $T_{origin} = (9.7 \pm 2.5) \times 10^9$ ago (or $(4.1 \pm 2.5) \times 10^9$ after BB; Sharov and Gordon (2013), which almost exactly match our estimation of $T_{origin} = (9.8 \pm 1) \times 10^9$ ago (or $(4.0 \pm 1.0) \times 10^9$ after BB). Our estimation of a $T_{origin}$ lower boundary of $3 \times 10^9$ years after BB can be further supported by the trends in dust composition as presented for the solar neighborhood by Zhukovska et al. (2008). The evolution of the dust composition is closely related to the stellar injection of the elements in ISM and therefore to the galactic chemical evolution, although slight differences can be observed with respect to volatile elements. In particular, one should pay attention to the C/O ratios found in ISM dust particles, stars of the solar neighborhood, and our planet, compared to the C/O ratio inferred for LUCA. Throughout the galactic evolution, the C/O ratio is <1 for most stars of the solar neighborhood. The C/O ratio for the Earth's crust is also less than 1 (Demayo 1983). Yet the C/O ratio for LUCA is above 1 and also for ISM dust particles in the solar neighborhood starting from $3 \times 10^9$ years after BB (Zhukovska et al. 2008), suggesting that life most probably originated on the carbonaceous dust particles of IMCs. Furthermore, the huge star formation rate in our galaxy for the first $3 \times 10^9$ years after BB in the bulge and thick disk regions (Kobayashi et al. 2020) probably excludes these regions as possible places for life to emerge, although they might have reached the right distribution of elemental abundances for the emergence of life earlier than the solar neighborhood and halo. However, if life was not able to be sustained in such restless systems, it may have been propagated into ISM during the supernova explosion to potentially seed the more friendly habitable zones such as our solar system.

## 4.10 FIRST UNIVERSAL STARDUST ANCESTOR

ISM is composed primarily of hydrogen in its atomic, ionic, or molecular form and dust particles. Two forms of ISM are distinguished: interstellar diffuse clouds and IMCs. Interstellar diffuse clouds consist of atomic hydrogen. They can have kinetic temperatures of 50–100 K and a density of hydrogen atoms below 300 per $cm^3$. Molecular hydrogen is the main component of IMCs, which are of interest to this study because they are the place where the complex organic molecules are formed on the dust particles which they contain. IMCs are at kinetic energies as low as 20–10 K, can have any density above 300 per $cm^3$, and at critical densities collapse to form stars and planetary systems. However, the dust in ISM is at a lower temperature than gas. In diffuse interstellar clouds, the ultraviolet radiation heats the dust to 15 K, while IMC dust is shielded, and the temperature may be as low as 5–10 K (Greenberg 1971, 2002; Greenberg and Li 1997). Recent laboratory achievements suggest that solid-state chemistry was crucial in the synthesis of RNA and DNA building blocks and their polymerization both on early Earth (Forsythe et al. 2015; Becker et al. 2018; Bolm et al. 2018; Lamour et al. 2019) and in ISM (Oba et al. 2020).

In such an environment, the first living being was formed (born); we call it the first universal stardust ancestor. The birthday of FUSDA is not fixed in time but rather in elemental composition of the dust environment. In this sense it is universal: it happened only once, since the universe is ageing, and once the concentration of life-essential elements satisfied the requirements for life, the FUSDA was born. This could happen at different times in different parts of the universe.

Oba et al. (2020) reported the simultaneous detection of all three pyrimidine (cytosine, uracil, and thymine) and three purine nucleobases (adenine, xanthine, and hypoxanthine) in interstellar ice analogs examined in environmental conditions simulating those in IMCs. Stolar et al. (2020) achieved self-assembly of DNA by heating solid nucleobase mixtures. The findings of Obhođaš et al. (2021) demonstrated that magnetic fields below 400 nT significantly enhance the growth of *Bacillus subtilis* compared to controls growing in the Earth's magnetic field. This might suggest that microorganisms favor extremely low magnetic fields. The Earth's magnetic field, of approximately the same intensity as presently (between 33 μT at the magnetic equator, and 67 μT at the magnetic poles), formed $4.2 \times 109$ yr ago, almost immediately after the lunar-forming giant impact (Tarduno et al. 2020). Moreover, it has been shown that microorganisms in a very similar way favor microgravity (Horneck et al. 2010). These findings of microorganisms favoring low magnetic and gravitational fields, evident as a 10-fold increase in their multiplication rate, strengthen the assumption that life originated in IMCs, where maximum magnetic field intensities are ~100 nT (Crutcher 2012) and strong gravitational fields are not yet formed.

## 4.11 FUSDA'S TRANSFORMATION INTO EARTH'S LUCA

Goldman et al. (2023) considered ancient electron transport chains (ETCs) a potential bridge between early evolutionary history and a protocellular stage that preceded it. Current phylogenetic evidence suggests that ancestors of several extant ETC components were present at least as late as the last universal common ancestor of life. In addition, recent experiments have shown that some aspects of modern ETCs can be replicated by minerals, protocells, or organic cofactors in the absence of biological proteins.

In their paper, Goldman et al. (2023) discussed the diversity of ETCs and other forms of emiosmotic energy conservation, described current work on the early evolution of membrane bioenergetics, and advocated for several lines of research to enhance this understanding by pairing top–down and bottom–up approaches.

By now it is well established that that a wide range of biomolecules can be formed abiotically, such as amino acids, carbohydrates, nucleobases, nucleotides, and membrane-forming amphiphilic compounds, as well as more complex molecules such as depsipeptides and nucleotide oligomers and supramolecular assemblies). The field as a whole has produced a substantial abiotic chemical inventory that can be used to develop plausible models for the origin of life.

A small number of protein families allow researchers to use standard phylogenetic methods to reconstruct evolutionary events prior to the LUCA with a high degree of confidence. These protein families resulted from duplications that occurred after the origin of life but before the time of the LUCA, yielding two ancestral nodes that can both be associated with the LUCA, or "universal paralogs". The branch connecting these LUCA nodes, therefore, represents an evolutionary event prior to the LUCA; see Kollman and Doolittle (2000) and Zhaxybayeva et al. (2005).

Tracing organismal histories on the timescale of the tree of life remains one of the challenging tasks in evolutionary biology, see for example Zhaxybayeva et al. 2005. The hotly debated questions include the evolutionary relationship between the three domains of life (e.g., which of the three domains are sister domains, are the domains para-, poly-, or monophyletic) and the location of the root within the universal tree of life. For the latter, many different points of view have been considered, but so far no consensus has been reached. The only widely accepted rationale to root the universal tree of life is to use anciently duplicated paralogous genes that are present in all three domains of life. To date only few anciently duplicated gene families useful for phylogenetic reconstruction have been identified. Zhaxybayeva et al. (2005)

presented results from a systematic search for ancient gene duplications using 12 representative, completely sequenced, archaeal and bacterial genomes. Phylogenetic analyses of identified cases show that the majority of datasets support a root between Archaea and Bacteria; however, some datasets support alternative hypotheses, and all of them suffer from a lack of strong phylogenetic signal. The results were discussed with respect to the impact of horizontal gene transfer on the ability to reconstruct organismal evolution. The exchange of genetic information between divergent organisms gives rise to mosaic genomes, where different genes in a genome have different histories. Simulations show that even low rates of horizontal gene transfer dramatically complicate the reconstruction of organismal evolution, and that the different most recent common molecular ancestors likely existed at different times and in different lineages.

According to Goldman et al. (2023) electron transport chains and the ATP synthase motor complex that they support, represent a ubiquitous and diverse strategy for converting redox potential into ATP in extant life. The ATP synthase enzyme is a membrane-spanning multiprotein complex that uses the motive force of proton gradients, or in some cases sodium gradients, to synthesize ATP from adenosine diphosphate (ADP) and inorganic phosphate. The ATP synthase enzyme is ubiquitous across the tree of life and its evolutionary history predates the LUCA.

Earth's LUCA is probably positioned at the evolutionary separation of the Archaea from the Eubacteria and before the symbiotic event believed to have led to the Eukarya. LUCA is probably not close to the origin of life, since it already had the capacity to code for many proteins and had some of the same bioenergetic capacities as modern organisms. An organism at the origin of life must have been vastly simpler. Even if acceptance of the giant viruses as living organisms forces the definition of LUCA to be revised, it will not alter the essential point that LUCA should be regarded as a recent player in the evolution of life; see Figure 4.3, following Cornish-Bowden and Cárdenas (2017). LUCA and the origin of life, which can be estimated from isotopic and fossil data (Joyce 1991; Arndt and Nisbet 2012) to date from 3.8 billion years ago. It is assumed that LUCA is much more recent, situated at the bifurcation of the Eubacteria and the Archaea, as suggested by Tuller et al. (2010). This bifurcation is marked with a filled circle. However, if the giant viruses are accepted as living organisms or as descendants of living organisms, or if free-living relatives of these viruses are discovered, LUCA will need to be redefined as an earlier entity (open circle). Regardless of any such redefinition, any LUCA with hundreds of genes coding for hundreds of proteins must be vastly more complicated than any self-sustaining system that existed at the origin of life. According to Glansdorff et al. (2008), "the first diagnostically identifiable Cyanobacteria are approximately 2.1 billion years old", and it is suggested that LUCA existed earlier.

No suggestion that there were no living organisms before LUCA has been made. On the contrary, there were probably many, as indicated in the Figure 4.4, but all of these are now extinct, as well as others that have become extinct more recently.

Evolving from a simple self-sustaining system to an organism with a coding capacity of hundreds of proteins is a huge step: even *Thermoplasma acidophilum*, with the smallest genome for a free-living organism, has about 1500 protein-coding genes; see Giovannoni et al. (2005). We cannot at present know how much this number could be decreased while still having a viable free-living organism. Understanding how the transition to an organism with a large coding capacity can have happened is a more challenging problem than understanding how LUCA could have evolved to *Homo sapiens*.

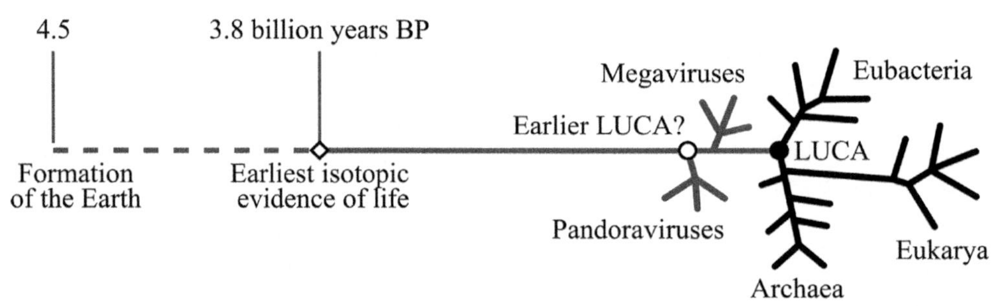


FIGURE 4.3 LUCA and the origin of life.

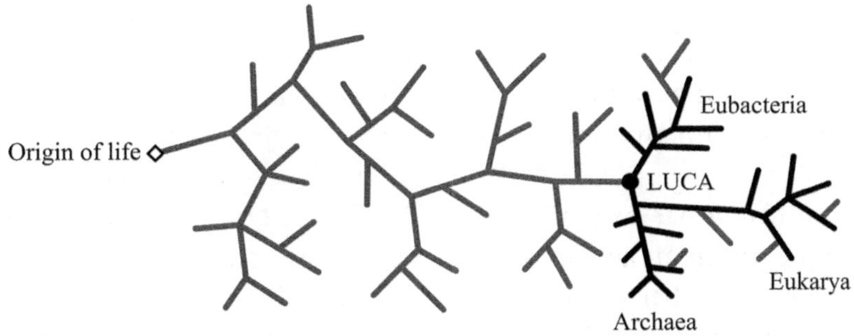


FIGURE 4.4 Life before LUCA.

For the post-LUCA evolution, we have at least a rough idea of how it happened and what mechanisms were involved, but pre-LUCA evolution is a black box, probably more difficult to understand than to understand how self-sustaining systems came to exist in the first place. If the theoretical ideas of Kauffman (1986) or Friston (2013) are valid, then self-sustaining systems may be inevitable, but getting from these to LUCA still presents enormous difficulties, and some crucial points, such as the origin of the genetic code, remain speculative. It is the view of Cornish-Bowden and Cárdenas (2017) that studying the pre-LUCA period and finding a plausible series of steps to LUCA is an urgent task, but it will not be solved as long as LUCA continues to be discussed as if it were an entity close to the origin of life.

### 4.11.1 LUCA

LUCA is a theoretical construct—it might or might not have been something we today would call an organism. It helps to bridge the conceptual gap between rocks and water on the early Earth and ideas about the nature of the first cells. Thoughts about LUCA span decades. Various ideas exist in the literature about how LUCA was physically organized and what properties it possessed. These ideas are traditionally linked to our ideas about the overall tree of life and where its root might lie. Phylogenetic trees are, however, ephemeral. It is their inescapable fate to undergo change as new data and new methods of phylogenetic inference emerge. Accordingly, the tree of life has been undergoing a great deal of change in time (Weiss et al. 2018).

The familiar three-domain tree of life presented by ribosomal RNA depicted LUCA as the last common ancestor of archaea, bacteria, and eukaryotes. In that framework, efforts to infer the gene content, and

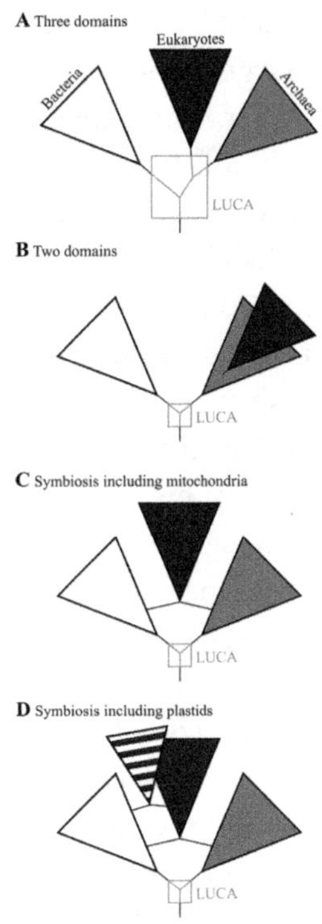


FIGURE 4.5 Different views on domain relationships in the tree of life.

(A) The three-domain tree: based on rRNA phylogeny, the three domains were of equal rank.

(B) The two-domain tree: modern trees show eukaryote cytosolic ribosomes branching within the diversity of archaeal ribosomes.

(C) As eukaryotes are not just grownup archaea, the eukaryote ancestor possessed mitochondria. If mitochondrial-derived genes are taken into account, the tree is no longer a bifurcating graph.

(D) If plastids are included, the tree becomes even less tree-like because the photosynthetic lineages of eukaryotes also acquired many genes from the plastid ancestor (see https://journals.plos.org/plosgenetics/article/figure?id=10.1371/journal.pgen.1007518.g001).

hence the properties of LUCA, boiled down to identifying genes that were present in eukaryotes, archaea, and bacteria. Phylogenetic advances and new metagenomic data are changing the three-domain tree into a two-domain tree.

A traditional approach to LUCA has been to simply look for the genes that are present in all genomes. That is easy enough, but the results are sobering. What one finds is a collection of about 30 genes, mostly for ribosomal proteins, telling us that LUCA had a ribosome and had the genetic code, which we already knew. That collection of about 30 genes has been in use for about 20 years as concatenated alignments to make trees of lineages based on larger amounts of data than rRNA sequences have to offer. The genes that are present in all lineages (or nearly all) inform us about how LUCA translated mRNA into protein, but they do not tell us about how or where LUCA lived (Weiss et al. 2018).

LUCA was an anaerobe, as long predicted by microbiologists (Decker et al. 1970). Its metabolism was replete with $O_2$-sensitive enzymes. These include proteins rich in $O_2$-sensitive iron–sulfur (FeS) clusters and enzymes that entail the generation of radicals (unpaired electrons) via S-adenosyl methionine (SAM) in their reaction mechanisms. That fits well with the view that FeS clusters represent very ancient cofactors in metabolism. It also fits with newer insights about the ancient and spontaneous (nonenzymatic) chemistry underlying SAM synthesis. LUCA lived from gases. For carbon assimilation, LUCA used the simplest and most ancient of the six known pathways of $CO_2$ fixation, called the acetyl–CoA (or Wood–Ljungdahl) pathway, which is increasingly central for our concepts on early evolution because of its chemical simplicity and exergonic nature.

Though the organization of inanimate matter into living cells with genetics can be charted in mathematical terms, the biochemical details remain elusive. For example, it is controversial whether LUCA had DNA. Several DNA-binding proteins trace to LUCA, so it would appear that LUCA possessed DNA, but it is unresolved whether LUCA could actually replicate DNA. For LUCA, DNA might just have been a chemically stable repository for RNA-based replication. Most earlier depictions of LUCA focused on what it was like; for example, whether it was like RNA, like a virus, whether it was like prokaryotes in terms of its genetic code or like eukaryotes in terms of its cellular organization. But traditional approaches lacked information about how and from what LUCA lived. A phylogenetic approach to LUCA by Weiss et al. (2018) uncovered information about what LUCA was doing: its physiology, its ecology, and its environment.

Work in search of the properties LUCA had is rooted in Charles Darwin's closing paragraph in *On the Origin of Species*: "that probably all the organic beings which have ever lived on this earth have descended from some one primordial form, into which life was first breathed". Since then, there have been numerous attempts to describe what LUCA is all about. For example, Krupovic et al. (2020) considered the reconstruction of the genome and phenotype of the LUCA a major challenge in evolutionary biology. According to them, all life forms are associated with viruses and/or other mobile genetic elements; therefore there should not be any doubt that the LUCA was a host to viruses. By projecting back in time using the extant distribution of viruses across the two primary domains of life, bacteria and archaea, and tracing the evolutionary histories of some key virus genes, they attempted a reconstruction of the LUCA virome. They concluded that it appears likely that the LUCA was a prokaryote-like organism (that is, like bacteria or archaea) of considerable genomic and organizational complexity. However, the nature of the replication and membrane machineries of LUCA remains unclear due to the drastic differences between the respective systems of bacteria and archaea, the two primary domains of life.

My thinking is very close to that of Koskela and Annila (2012) that in spite the fact that genomic sequences across diverse species seem to align towards a common ancestry. However, when evolution is understood not only as a biological process but as a general thermodynamic process, it becomes apparent that the quest for the last universal common ancestor is unattainable. Ambiguities in alignments are unavoidable because the driving forces and paths of evolution cannot be separated from each other. Thus tracking down life's origin is by its nature a non-computable task. The thermodynamic tenet clarifies that evolution is a path-dependent process of least-time consumption of free energy. The natural process is without a demarcation line between animate and inanimate.

According to Koskela and Annila (2012), when evolution is understood as a thermodynamic process of least-time free energy consumption, it follows that phylogenetic analyses, as done using genetic information, will not be able to determine common ancestors.

Moreover, when looking for LUCA, there is an inadvertently in-built impression that animate and inanimate would somehow be distinct from each other. This obscures comprehension of the principle of abiogenesis. Also the idea of aligning for a common ancestor is unintentionally a reductionist approach, but contemporary characters cannot be reduced one-to-one to primordial qualities because energy is not a constant of evolution. The modern evolutionary synthesis, by its gene-centered paradigm, narrows the view away from the general physical principle of evolution and thereby prevents obtaining a holistic understanding of nature as a unity.

Weiss et al. (2018) discuss LUCA as being between ancient Earth chemistry and the onset of genetics. They have a stand that the first cells came into existence on early Earth covered by ocean in a narrow window of time of only about $2 \times 10^8$ years. Since the genetic code and amino acid chirality are universal, all modern life forms ultimately trace back to that phase of evolution, back to LUCA who lived during this time. Therefore, it was LUCA that witnessed the onset of Darwinian evolution.

Lineweaver (2020), in his paper, found it useful to introduce the idea of multiple LUCAs. In Figure 4.6, the branch length to an extant organism is the distance from LUCA to the end of the branch. However, since we only have a very approximate idea of where LUCA is located (somewhere in the red oval), we cannot say which branch lengths are the shortest. Potential candidates for having the shortest branch lengths (and therefore being the best extant representative of LUCA) are Wirthbacteria within CPR and Riflebacteria just outside the CPR region (Brown et al. 2015).

The idea of multiple LUCAs seems problematic. How can there be more than one LUCA? Figure 4.6 introduces and helps explain the potential epistemological recession of LUCA deeper and deeper into the past as we discover more deeply rooted life forms. Our understanding of the microbial diversity on Earth is incomplete. The discovery of CPR and DPANN is an example of the ongoing discovery of organisms that are deeper branching and have shorter branches in the tree of life. There is an important distinction to be made between deep branches and short branches. Extant organisms with short branches are more representative of LUCA than extant organisms with long branches. However, branch length depends on where LUCA is.

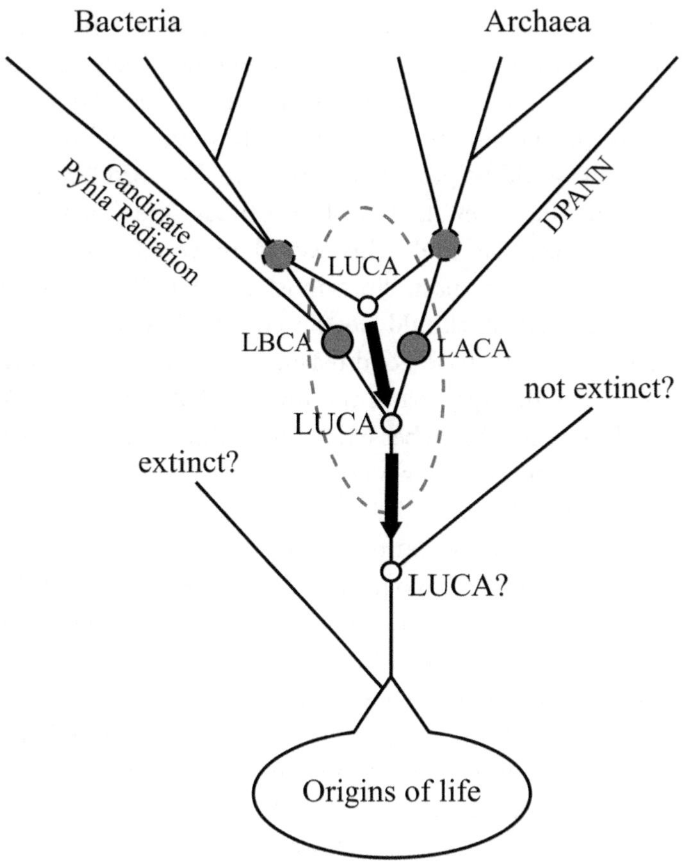


FIGURE 4.6 If all extant cellular life on Earth were known to us, the last universal common ancestor of all extant cellular life would be fixed in time and would have fixed properties. It would have lived about 3 or 4 billion years ago, and its properties could be inferred by comparing the most distantly related extant cellular life forms (Weiss et al. 2016). However, all extant cellular life on Earth is not known. Discoveries are still being made. The discovery of CPR (Hug et al. 2016; Castelle and Banfield 2018) has pushed the last bacterial common ancestor (LBCA) back in time. The discovery of a new group of archaea known as DPANN has pushed the last archaeal common ancestor (LACA) back in time. The existence of these two deeper ancestors can then push the last universal common ancestor of all extant cellular life deeper into the past. If metagenomics reveals an even deeper branch, then LUCA recedes even deeper into the past and becomes a better representative of the origins of life. Figure modified from Lineweaver and Chopra (2019).

An important goal of phylogenetic trees is to identify the last universal common ancestor of all cellular life and to find its metabolism (Weiss et al. 2016; Weiss et al. 2018). This estimate of LUCA's metabolism then becomes our best proxy for the first metabolisms of life. LUCA is normally understood as the last universal common ancestor of all known extant cellular life. Thus, as we learn more—as we more comprehensively sample extant life—LUCA can recede deeper into the past. If we were able to sample the DNA of billion-year-old fossils, then it would be profitable to remove the word "extant" and define a new LUCA of all cellular life (including extinct cellular life). This would move LUCA much deeper into the past, since most cellular life has gone extinct. LUCA should not be confused with the origin of life. We have good reasons to believe that many branches diverged from the lineage that led to LUCA and that these branches have gone extinct. If they hadn't gone extinct, LUCA would be even deeper and closer to the origin of life.

In his paper, Koonin (2009) proposed that the precellular stage of biological evolution unraveled within networks of inorganic compartments that harbored a diverse mix of virus-like genetic elements. This stage of evolution might make up the last universal cellular ancestor that more appropriately could be denoted last universal cellular ancestral state (LUCAS). Such a scenario recapitulates the ideas J. B. S. Haldane sketched in his classic 1928 essay. However, unlike in Haldane's day, considerable support for this scenario exits today: lack of homology between core DNA replication system components in archaea and bacteria, distinct membrane chemistries and enzymes of lipid biosynthesis in archaea and bacteria, spread of several viral hallmark genes among diverse groups of viruses, and the extant archaeal and bacterial chromosomes appear to be shaped by accretion of diverse, smaller replicons. Under the viral model of precellular evolution, the key components of cells originated as components of virus-like entities. The two surviving types of cellular life forms, archaea and bacteria, might have emerged from the LUCAS independently, along with, probably, numerous forms now extinct.

The concept of a precellular stage of biological evolution outlined here posits that the precellular stage of life's evolution took place within networks of inorganic compartments that hosted a diverse mix of virus-like genetic elements. It is further proposed that these ensembles of genetic elements were the ancestral state from which cells emerged, probably, in multiple, independent escapes, only two or three of which (the ancestors of bacteria and archaea, and possibly, eukarya) yielded stable cellular lineages that enjoyed a long-term evolutionary success. Considering this hypothetical consortial state of primordial life forms that eventually gave rise to cells, it seems reasonable to replace the acronym LUCA with LUCAS.

The primordial virus world model is, at least in parts, refutable and, potentially, testable. A discovery of an organism with an archaeal replication system but a

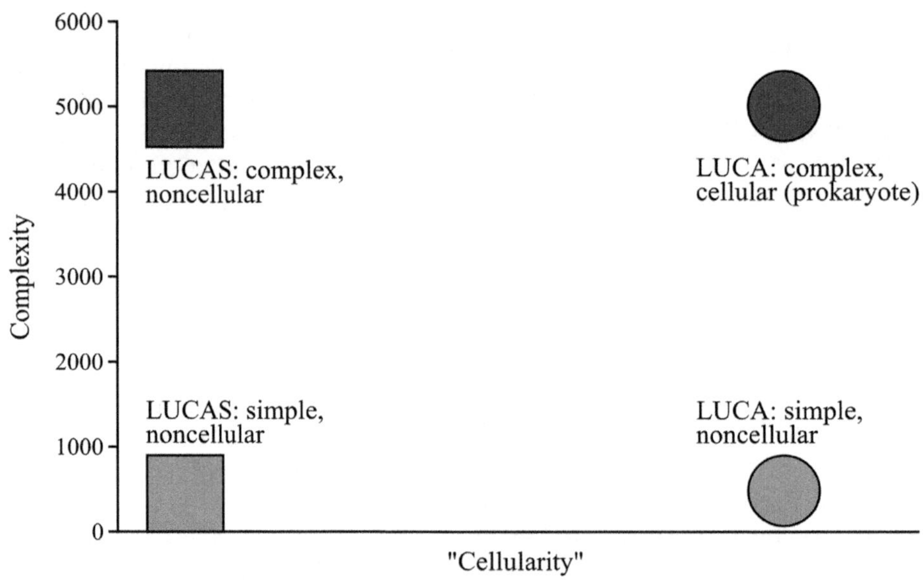


FIGURE 4.7 Genetic complexity and "cellularity" of LUCA(S): the space of logical possibilities. "Cellularity" is the degree of similarity to the organization of modern cells. The notion of "cellularity" is qualitative; there are no specific units. The complexity scale also could be considered arbitrary, but the units of complexity can be assumed to (roughly) represent the number of genes. The primordial virus world model delineated in this article implies a complex, noncellular LUCAS.

bacterial membrane (or vice versa) would come close to a refutation. Further study of the diversity of viruses might reveal new membrane translocation devices, for instance, packaging machines homologous to the H+-ATPases of cellular organisms. Such evidence would provide support for a role of viruses in the evolution of cellular membranes. Direct biochemical experiments on early evolution are inherently hard. However, this model might make them easier by splitting the gargantuan feat of evolving a cell into more manageable steps of evolution of virus-like agents (Koonin 2009).

According to Mahendrarajah et al. (2023) the timing of early cellular evolution, from the divergence of archaea and bacteria to the origin of eukaryotes, is poorly constrained. The ATP synthase complex is thought to have originated prior to LUCA, and analyses of ATP synthase genes, together with ribosomes, have played a key role in inferring and rooting the tree of life. The authors reconstructed the evolutionary history of ATP synthases using an expanded taxon sampling set and developed a phylogenetic cross-bracing approach, constraining equivalent speciation nodes to be contemporaneous, based on the phylogenetic imprint of endosymbioses and ancient gene duplications. This approach resulted in a highly resolved, dated species tree and establishes an absolute timeline for ATP synthase evolution. Their analyses show that the divergence of ATP synthase into F- and A/V-type lineages was a very early event in cellular evolution dating back to more than $4 \times 10^9$ years, potentially predating the diversification of archaea and bacteria. Their cross-braced, dated tree of life also provides insight into more recent evolutionary transitions including eukaryogenesis, showing that the eukaryotic nuclear and mitochondrial lineages diverged from their closest archaeal $(2.67–2.19) \times 10^9$ years and bacterial $(2.58–2.12) \times 10^9$ years relatives at approximately the same time, with a slightly longer nuclear stem-lineage.

### 4.11.2 Effects of Earth's Magnetic Field and Other Environmental Parameters

All organisms exhibit cyclical modulations in their levels of activity that are deemed to be of adaptive value. Long-term and short-term cycles are thus ubiquitous and can be regarded as "embodied rhythms of life", a temporally organized homeostatic activity dictated by or even exploiting the cyclic variations of environmental variables. Such variations are diverse and well known; examples are variations in day and night; the passing of the seasons and their associated periods of cold, dark, or wet, or any combination thereof; and also the abundance or lack of resources in ecological niches. Such cyclic variations are ubiquitously found from microorganisms to unicellular and multicellular organisms, including human beings and their socio-economic life, which also crucially depends on natural daily and seasonal rhythms. The 2017 Nobel Prize in Physiology or Medicine was awarded to those who discovered some of the molecular mechanisms underpinning circadian rhythms, providing the first mechanistic insights into how organisms physiologically organize their cyclic activities, in particular to the ~24-h period of the Earth's rotation; see Nobel Assembly (2017).

The celestial mechanics of the Sun, Moon, and Earth dominate the variations in gravitational force that all matter, live or inert, experiences on Earth. Expressed as gravimetric tides, these variations are pervasive and have forever been part of the physical ecology with which organisms evolved. De Mello Gallep and Robert (2021) give a brief review of previously proposed explanations that gravimetric tides constitute a tangible and potent force shaping the rhythmic activities of organisms. Through meta-analysis, they interrogated data from three study cases and show the close association between the omnipresent gravimetric tides and cyclic activity. As exemplified by free-running cyclic locomotor activity in isopods, reproductive effort in coral, and modulation of growth in seedlings, biological rhythms coincide with temporal patterns of the local gravimetric tide. These data reveal that, in the presumed absence of rhythmic cues such as light and temperature, local gravimetric tide is sufficient to entrain cyclic behavior. The present evidence thus questions the phenomenological significance of so-called free-run experiments.

It has been recently shown by Obhođaš et al. (2021) that preconcentration factors can be different in very low magnetic fields. *Bacillus subtilis* grown in magnetic fields of 100 nT, which are intensities found in IMCs (Crutcher 2012), required 5.5 times less potassium compared to the control culture growing in the Earth's magnetic field. Magnetic as well as gravitational fields are orders of magnitude weaker in IMCs compared to the Earth. Thus, the abundances derived for LUCA may be severely misleading. If potassium abundance in LUCA is corrected by factor of 5.5 inferred from the experiment with the *Bacillus subtilis* grown in the magnetic field of 100 nT (Obhođaš et al. 2021), the LUCA abundance curve resembles the Sun abundance curve more closely. The essential element concentration factors of primitive life forms for intensities of magnetic and gravitational fields found in IMCs will be systematically studied in our future research.

## REFERENCES


Amarsi, A. M., Nissen, P. E., and Skúladóttir, Á. 2019. Carbon, oxygen, and iron abundances in disk and halo stars: Implications of 3D non-LTE spectral line formation. Astronomy & Astrophysics 630: A104 (19 pages). https://doi.org/10.1051/0004-6361/201936265.

Arensten, A., Placco, V. M., Lee, Y. S., et al. 2022. On the inconsistency of [C/Fe] abundances and the fractions of carbon-enhanced metal-poor stars among various stellar surveys. Monthly Notices of the Royal Astronomical Society 515(3): 1–18. https://doi.org/10.1093/mnras/stac2062.

Arndt, N. T., and Nisbet, E. G. 2012. Processes on the young earth and the habitats of early life. Annual Review of Earth and Planetary Sciences 40: 521–549. https://doi.org/10.1146/annurev-earth-042711-105316.

Beard, M., Austin, S. M., and Cyburt, R. 2017. Enhancement of the triple alpha ratre in hot dense medium. arXiv:1708.07204v1 [nucl-th], 22 Aug.

Becker, S., Schneider, C., Okamura, H., et al. 2018. Wet-dry cycles enable the parallel origin of canonical and non-canonical nucleosides by continuous synthesis Nature Communications 9: 163. https://doi.org/10.1038/s41467-017-02639-1.

Biologydictionary.net Editors. 2017. Krebs cycle. Biology Dictionary, Biologydictionary.net, 12 Jan. https://biologydictionary.net/krebs-cycle/.

Bolm, C., Mocci, R., Schumacher, C., et al. 2018. Mechanochemical Activation of Iron Cyano Complexes: A Prebiotic Impact Scenario for the Synthesis of α-Amino Acid Derivatives. Angewandte Chemie International Edition 57(9): 2423–2426. https://doi.org/10.1002/anie.201713109.

Brewer, J. M., and Fischer, D. A. 2016. C/O and Mg/Si ratios of stars in the solar neighborhood. The Astrophysical Journal 831: 20 (7 pages). https://doi.org/10.3847/0004-637X/831/1/20.

Brown, C. T., Hug, L. A., Thomas, B. C., et al. 2015. Unusual biology across a group comprising more than 15% of domain Bacteria. Nature 523: 208–211.

Brownlee, D., and Ward, P. D. 2000. Rare Earth: Why Complex Life is Uncommon in the Universe. New York, NY: Copernicus Books; Springer Verlag.

Burgarella, D., Sakon, I., Buat, V., et al. 2020. Astro2020 Science White Paper Measuring the Metallicity of Low-Mass, Low-Metallicity Galaxies in the Early Universe and the Galactic Habitability. https://roman.gsfc.nasa.gov/science/Astro2020.

Camprubi, E., Jordan, S. F., Vasiliadou, R., and Lane, N. 2017. Hypothesis: Iron catalysis at the origin of life. International Union of Biochemistry and Molecular Biology, IUBMB Life 69(6): 373–381 (Wiley Online Library). https://doi.org/10.1002/iub.1632.

Canup, R. M., Visscher, C., Salmon, J., and Fegley, Jr, B. 2015. Lunar volatile depletion due to incomplete accretion within an impact-generated disk. Nature Geoscience 8(12): 918–921.

Castelle, C. J. and Banfield, J. F. 2018. Major new microbial groups expand diversity and alter our understanding of the tree of life. Cell 172: 1181–1197. https://doi.org/10.1016/j.cell.2018.02.016.

Cazaux, S., Minissale, M., Dulieu, F., and Hocuk, S. 2016. Dust as interstellar catalyst II. How chemical desorption impacts the gas. Astronomy and Astrophysics 585: A55 (9 pages). https://doi.org/10.1051/0004-6361/201527187.

Çelebi, M., Bilir, S., Ak, S. et al. 2019. Metallicity and absolute magnitude calibrations for F-G type main-sequence stars in the Gaia era. Astrophysics and Space Science 364: 172 (12 pages). https://doi.org/10.1007/s10509-019-3659-0.

Charnoz, S., and Michaut, C. 2015. Evolution of the protolunar disk: Dynamics, cooling timescale and implantation of volatiles onto the earth. Icarus 260: 440–463. https://doi.org/10.1016/j.icarus.2015.07.018.

Charnoz, S., Sossi, P. A., Lee, Y.-N., et al. 2021. Tidal pull of the Earth strips the proto-moon of its volatiles. Icarus 364: 114451 (55 pages). https://doi.org/10.1016/j.icarus.2021.114451.

Chopra, A. and Lineweaver, C. H. 2015, An Estimate of the Elemental Composition of LUCA. Astrobiology Science Conference. Chicago, Illinois, 7328.

Chopra, A., Lineweaver, C. H., Brocks, J. J., and Ireland, T. R. 2009. Palaeoecophylostoichiometrics: Searching for the elemental composition of the last universal common ancestor. Conference paper, Australian Space science Conference Series: 9th Conference Proceedings, Sidney 28–30 September. National Space Society of Australia Ltd. ISBN 13::978–0–9775740.

Chruslinska, M., and Nelemans, G. 2019. Metallicity of stars formed throughout the cosmic history based on the observational properties of star-forming galaxies. Monthly Notices of the Royal Astronomical Society 488(4): 5300–5326. https://doi.org/10.1093/mnras/stz2057.

Cinquegrana, G. C., and Karakas, A. I. 2022. The most metal-rich stars in the universe: Chemical contributions of low- and intermediate-mass asymptotic giant branch stars with metallicities within $0.04 \leq Z \leq 0.10$. Monthly Notices of the Royal Astronomical Society 510(2): 1557–1576. https://doi.org/10.1093/mnras/stab3379.

Cockell, C. 2015. Astrobiology – Understanding Life in the Universe. John Wiley & Sons Ltd. www.wiley.com/go/cockell/astrology.

Cornish-Bowden, A., and Cárdenas, M. L. 2017. Life before LUCA. Journal of Theoretical Biology 434(7): 68–74.

Crutcher, R. M. 2012. Magnetic fields in molecular clouds. Annual Review of Astronomy & Astrophysics 50: 29–63.

Dauphas, N., Nie, N. X., Blanchard, M., et al. 2022. The extent, nature, and origin of K and Rb depletions and isotopic fractionations in earth, the moon, and other planetary bodies. The Planetary Science Journal 3: 29 (39 pages). https://doi.org/10.3847/PSJ/ac2e09.

Decker, K., Jungerman, K., and Thauer, R. K. 1970. Energy production in anaerobic organisms. Angewandte Chemie International Edition 9: 138–158.

Demayo, A. 1983. In R. C. Weast (ed.), Handbook of Chemistry and Physics 64th edition (F–151). Boca Raton, FL: CRC Press.

de Mello Gallep, C., and Robert, D. 2021. Are cyclic plant and animal behaviours driven by gravimetric mechanical forces? Journal of Experimental Botany (11 pages). https://doi.org/10.1093/jxb/erab462.

Ehrenfreund, P., and Charnley, S. B. 2000. Organic molecules in the interstellar medium, comets, and meteorites: A voyage from dark clouds to the early earth. Annual Review of Astronomy and Astrophysics 38(1): 427–483.

Ehrenfreund, P., and Menten, K. M. 2001. From Molecular Clouds to the Origin of Life. In G. Horneck and C. Baumstark-Kahn (eds.), Astrobiology: The Quest for the Conditions of Life (pp. 7–24). Berlin: Springer Verlag; New York: Heidelberg; Hong Kong, London, Milan, Paris, Tokyo: Barcelona.

Eistrup, C., Walsh, C., and Dishoeck, E. F. 2018. Molecular abundances and C/O ratios in chemically evolving planet-forming disk midplanes. Astronomy & Astrophysics 613: A14 (16 pages). https://doi.org/10.1051/0004-6361/201731302.

Feng, L. 2023. Possibilities for methanogenic and acetogenic life in molecular cloud. arXiv:2311.14291v1 [physics.pop-ph], 24 Nov.

Forsythe, J. G., Yu, S. S., Mamajanov, I., et al. 2015. Ester-Mediated Amide Bond Formation Driven by Wet–Dry Cycles: A Possible Path to Polypeptides on the Prebiotic Earth. Angewandte Chemie International Edition 54(34): 9871–9875. https://doi.org/10.1002/anie.201503792.

Freer, M., and Fynbo, H. O. U. 2014. The Hoyle state in $^{12}C$. Progress in Particle and Nuclear Physics 78: 1–23. https://doi.org/10.1016/j.ppnp.2014.06.001.

Friston, K. 2013. Life as we know it. Journal of the Royal Society Interface 10(86): 20130475. https://doi.org/10.1098/rsif.2013.0475.

Fröhlich, C., Martinez-Pinedo, G., Liebendörfer, M. et al. 2006. Neutrino-induced nucleosynthesis of A>64 nuclei: The νp-process. Physical Review Letters 96(14): 142502. https://doi.org/10.1103/PhysRevLett.96.142502.

Gandhi, S. 2015. Origin of chirality in the universe. Berkeley Scientific Journal Symmetry 20(1): 1–2.

Giovannoni, S. J., Tripp, H. J., Givan, S., et al. 2005. Genome streamlining in a cosmopolitan oceanic bacterium. Science 309(5738): 1242–1245. https://doi.org/10.1126/science.1114057.

Glansdorff, N., Xu, Y., Labedan, B. 2008. The last universal common ancestor: Emergence, constitution and genetic legacy of an elusive forerunner. Biology Direct 3: 29. https://doi.org/10.1186/1745-6150-3-29.

Goldman, A. D., Weber, J. M., LaRowe, D. E., and Barge, L. M. 2023. Electron transport chains as a window into the earliest stages of evolution. PNAS 120(34): e2210924120 (9 pages). https://doi.org/10.1073/pnas.2210924120.

Greenberg, J. M. 2002. Cosmic dust and our origins. Surface Science 500: 793–822.

Greenberg, J. M., & Li, A. 1997, in ASP Conf. Ser. 124, Grain temperatures and emission in diffuse interstellar clouds. Diffuse Infrared radiation and the IRTS, ed. H. Okuda, T. Mataumoto, & T. L. Roelling (San Francisco, CA: ASP).

Henry, A., Rafelski, M., Sunnquist, B., et al. 2021. The mass–metallicity relation at z~1–2 and its dependence on the star formation rate. The Astrophysical Journal 919: 143 (29 pages). https://doi.org/10.3847/1538-4357/ac1105.

Hidalgo, S. L: 2017. Mass–metallicity relation of dwarf galaxies and its dependency on time: Clues from resolved systems and comparison with massive galaxies. Astronomy and Astrophysics 606: A115 (7 pages). https://doi.org/10.1051/0004-6361/201630264.

Horneck, G., Klaus, D. M., & Mancinelli, R. L. 2010. Space microbiology. Microbiology and Molecular Biology Reviews 74(1): 121–156. doi: 10.1128/MMBR.00016-09.

Hoyle, F. 1954. On nuclear reactions occurring in very hot stars. 1. The synthesis of elements from carbon to nickel. Astrophysical Journal Supplement Series 1: 121–146.

Hug, L. A., Baker, B. J., Anantharaman, K., et al. 2016. A new view of the tree of life. Nature Microbiology 1: 16048. https://doi.org/10.1038/nmicrobiol.2016.48.

Jin, S., Roberts, L. F., Austin, S. M., and Schatz, H. 2020. Enhanced triple-α reaction reduces proton-rich nucleosynthesis in supernovae. Nature 588: 57–60.

Jofré, P., Heiter, U., Soubiran, C., et al. 2014. Gaia FGK benchmark stars: Metallicity. Astronomy and Astrophysics 564: A133 (27 pages). https://doi.org/10.1051/0004-6361/201322440.

Johnson, F. M. 1972. Interstellar matter II: Diffuse interstellar lines and porphyrins. Annals of the New York Academy of Sciences 187: 186–206.

Joyce, G. F. 1991. The rise and fall of the RNA world. New Biologist 3: 399–407.

Karakas, A. I., Cinquegrana, G., Joyce, M. 2022. The most metal-rich asymptotic giant branch stars. Monthly Notices of the Royal Astronomical Society 509(3): 4430–4447. https://doi.org/10.1093/mnras/stab3205.

Kirsebom, O. S. 2013. $^{12}C$ and the triple-α reaction rate. Journal of Physics: Conference Series 436: 012072. https://doi.org/10.1088/1742-6596/436/1/012072.

Kleine, T., and Walker, R. J. 2017. Tungsten isotopes in planets. Annual Review of Earth and Planetary Sciences 45: 389–417. https://doi.org/10.1146/annurev-earth-063016-020037.

Kobayashi, C., Karakas, A. I., & Lugaro, M. 2020. The Origin of Elements from Carbon to Uranium. The Astrophysical Journal 900(2): 179 (33 pages). doi:10.3847/1538-4357/abae65.

Kollman, J. M., and Doolittle, R. F. 2000. Determining the relative rates of change for prokaryotic and eukaryotic proteins with anciently duplicated paralogs. Journal of Molecular Evolution 51: 173–181. https://doi.org/10.1007/s002390010078.

Koonin, E. V. 2009. On the origin of cells and viruses, primordial virus world scenario. Annals of the New York Academy of Sciences. 1178(1): 47–64. https://doi.org/10.1111/j.1749-6632.2009.04992.x.

Koskela, M., and Annila, A. 2012. Looking for the last universal common ancestor (LUCA). Genes 3: 81–87.

Krasnokutski, S. A., Chuang, K.-J., Jäger, C., Ueberschaar, N., and Henning, T. 2022. A pathway to peptides in space through the condensation of atomic carbon. Nature Astronomy. https://doi.org/10.1038/s41550-021-01577-9.

Krupovic, M., Dolja, V. V., and Koonin, E. V. 2020. The LUCA and its complex virome. Nature Reviews Microbiology 18: 661–670.

Kwok, S. 2004. The synthesis of organic and inorganic compounds in evolved stars. Nature 430: 985–99l.

Lamour, S., Pallmann, S., Hass, M., & Trapp, O. 2019. Prebiotic Sugar Formation Under Nonaqueous Conditions and Mechanochemical Acceleration Life, 9(2): 52. https://doi.org/10.3390/life9020052

Lattimer, J. M., and Schramm, D. N. 1974. Black-hole-neutron-star collisions. Astrophysical Journal 192: L145–L147. https://doi.org/10.1086/181612ć.

Li, J., Liu, C., Zhang, Z. Y., et al. 2023. Stellar initial mass function varies with metallicity and time. Nature 613: 460–462. https://doi.org/10.1038/s41586-022-05488-1.

Li, M., Cai, Z., Bian, F., et al. 2022. The Mass–Metallicity Relation of Dwarf Galaxies at the Cosmic Noon in the JWST Era. arXiv:2211.01382v1 [astro-ph.GA], 2 Nov.

Lineweaver, C. 2020. What Do the DPANN Archea and the CPR Bacteria Tell Us about the Last Universal Common Ancestors? In J. Seckbach and H. Stan-Lotter (eds.), Book: Extremophiles as Astrobiological Models (pp. 359–367). Beverly, MA: Wiley on line library, Scrivener Publishing LLC.

Lineweaver, C. H., and Chopra, A. 2019. The biological overview effect: Our place in nature. Journal of Big History 3(3): 109–122. https://doi.org/10.22339/jbh.v3i3.3360.

Lock, S. J., Stewart, S. T., Petaev, M. I., et al. 2018. The origin of the moon within a terrestrial synestia. Journal of Geophysical Research: Planets 123: 910–951.

Loeb, A. 2014. The habitable epoch of the early universe. arXiv:1312.0613v3 [astro-ph.CO]. https://doi.org/10.48550/arXiv.1312.0613.

Loeb, A., Batista, R. A., and Sloan, D. 2016. Relative Likelihood for Life as a Function of Cosmic Time (11 pages). arXiv:1606.08448v2 [astro-ph.CO] 26 Jul. Journal of Cosmology and Astroparticle Physics 2016: (11 pages). https://doi.org/10.1088/1475-7516/2016/08/040.

Mahendrarajah, T. A., Moody, E. R. R., Schrempf, D. et al. 2023. ATP synthase evolution on a cross-braced dated tree of life. Nature Communication 14: 7456. https://doi.org/10.1038/s41467-023-42924-w.

Makarov, M., Sanchez Rocha, A. C., Krystufek, R., et al. 2023. Early selection of the amino acid alphabet was adaptively shaped by biophysical constraints of foldability. Journal of the American Chemical Society 145: 5320–5329. https://doi.org/10.1021/jacs.2c12987.

Marboeuf, U., Thiabaud, A., Alibert, Y., et al. 2014. From planetesimals to planets: Volatile molecules. Astronomy and Astrophysics 570: A36 (16 pages). https://doi.org/10.1051/0004-6361/201423431.

Marks, J. H., Wang, J., Sun, B.-J., et al. 2023. Thermal synthesis of carbamic acid and its dimer in interstellar ices: A reservoir of interstellar amino acids. ACS Central Science (10 pages). https://doi.org/10.1021/acscentsci.3c01108.

Mashian, N. and Loeb, A. 2016. CEMP stars: Possible hosts to carbon planets in the early universe. Monthly Notices of the Royal Astronomical Society. [1603.06943].

Mat, W. K., Xue, H., Wong, J. T. 2008. The genomics of LUCA. Frontiers in Bioscience 13: 5605–5613. https://doi.org/10.2741/3103.

Mehta, J. 2022. Is life possible on rogue planets and moons? Starless free-floating worlds might represent the most common habitable real estate of the universe. The Planetary Society, Jan 18. www.planetary.org/articles/is-life-possible-on-worlds-without-stars.

Nakajima, K., Ouchi, M., Isobe, Y., et al. 2023. JWST census for the mass-metallicity star-formation relations at z = 4–10 with the self-consistent flux calibration and the proper metallicity calibrators. arXiv:2301.12825v2 [astro-ph.GA]. https://doi.org/10.48550/arXiv.2301.12825.

Nakajima, K., Ouchi, M., Xu, Y., et al. 2022. Empress. V. Metallicity diagnostics of Galaxies over 12 + log(O/H); 6.9–8.9 established by a local galaxy census: Preparing for JWST spectroscopy. Astrophysical Journal Supplement Series 262: 3 (27 pages). https://doi.org/10.3847/1538–4365/ac7710.

Naoz, S., Noter, S., and Barkana, R. 2006. The first stars in the universe. Monthly Notices of the Royal Astronomical Society: Letters 373(1): L98–L102. https://doi.org/10.1111/j.1745-3933.2006.00251.x.

Nie, N. X., and Dauphas, N. 2019. Vapor drainage in the protolunar disk as the cause for the depletion in volatile elements of the moon. The Astrophysical Journal Letters 884: L48 (10 pages). https://doi.org/10.3847/2041-8213/ab4a16.

Nobel Assembly. 2017. The Nobel Foundation. In A. Pontikis and J. Petterson (eds.), Annual Review. P.O. Box 5232, 102 45 Stockholm: The Nobel Foundation.

Oba, Y., Takano, Y., Naraoka, H., et al. 2020. Extraterrestrial hexamethylenetetramine in meteorites—a precursor of prebiotic chemistry in the inner solar system.

Nature Communications: article (8 pages). https://doi.org/10.1038/s41467-020-20038-x.

Öberg, K. I., Murray-Clay, R., and Bergin, E. A. 2011. The effects of snowlines on c/o in planetary atmospheres. The Astrophysical Journal Letters 743(1): L16 (5 pages). https://doi.org/10.1088/2041-8205/743/1/L16.

Obhođaš, J., Valković, V., Kollar, R., et al. 2021. The Growth and Sporulation of Bacillus subtilis in Nanotesla Magnetic Fields. Astrobiology 21(3): 323–331. https://doi.org/10.1089/ast.2020.2288.

Peña, M., and Flores-Durán, S. N. 2019. Metallicity gradients in M31, M 33, NGC 300 and the Milk Way using abundances of different elements. Revista Mexicana de Astronomia y Astrofisica 55(2). https://doi.org/10.22201/ia.01851101p.2019.55.02.13.

Price, M. E. 2017. Entropy and selection: Life as an adaptation for universe replication. Complexity 2017: Article ID 4745379 (4 pages). https://doi.org/10.1155/2017/4745379.

Rabago, I., and Steffen, J. H. 2019. Survivability of moon systems around ejected gas giants. Monthly Notices of the Royal Astronomical Society 489(2): 2323–2329. https://doi.org/10.1093/mnras/sty2552.

Rivilla, V. M., Jiménez-Serra, I., Martin-Pintado, J., et al. 2022. Molecular precursors of the RNA-world in space: New nitriles in the G+0.693–0.027 molecular cloud. Frontiers in Astronomy and Space Science 9: 876870 (18 pages).

Rout, S. K., Cadalbert, R., Schröder, N., et al. 2023. An analysis of nucleotide-amyloid interactions reveals selective binding to codon-sized RNA. Journal of the American Chemical Society 145(40): 21915–21924. https://doi.org/10.1021/jacs.3c06287.

Scorei, R. 2012. Is boron a prebiotic element? A mini review of the essentiality of boron for the appearance of life on earth. Origins of Life and Evolution of Biospheres 42: 440–453.

Sharov, A. A., and Gordon, R. 2013. Life Before Earth. In Cornell University Library (pp. 233–251). arXiv:1304.3381 [physics.gen-ph].

Smolin, L. 1992. Did the universe evolve? Classical and Quantum Gravity 9(1): 173–191. https://doi.org/10.1088/0264-9381/9/1/016.

Sossi, P., Charnoz, S., Lee, Y.-N., et al. 2019. Tidal pull of the earth strips the moon of its volatiles from a surface magma ocean. EPSC-DPS Joint Meeting 2019, Held 15–20 September 2019 in Geneva, Switzerland, id. EPSC-DPS2019–1522.

Stevenson, D. 1999. Life-sustaining planets in interstellar space? Nature 400: 32. https://doi.org/10.1038/21811.

Sutter, P. 2023. Life might have been possible just seconds after the Big Bang. The Universe. December 11. www.space.com/life-possible-seconds-after-big-bang.

Thiabaud, A., Marboeuf, U., Alibert, Y., et al. 2015. Gas composition of the main volatile elements in protoplanetary discs and its implications for planet formation. Astronomy and Astrophysics 574: A138 (11 pages). https://doi.org/10.1051/0004-6361/201424868.

Torrey, P., Vogelsberger, M., Marinacci, F., et al. 2019. The evolution of the mass–metallicity relation and its scatter in illustris TNG. Monthly Notices of the Royal Astronomical Society 484(4): 5587–5607. https://doi.org/10.1093/mnras/stz243.

Totani, T. 2020. Emergence of life in an inflationary universe. Nature Scientific Reports 10: 1671 (7 pages). https://doi.org/10.1038/s41598-020-58060-0.

Trevors, J. T. 2006. The Big Bang, superstring theory and the origin of life on the earth. Theory in Biosciences 124(3–4): 403–412. https://doi.org/10.1016/j.thbio.2005.04.002.

Truran, J. W., and Kozlovsky, B.-Z. 1969. The enhancement of the $3^4He \rightarrow ^{12}C$ reaction rate in dense matter by inelastic scattering processes. The Astrophysical Journal 158: 1021–1031. https://doi.org/10.1086/150260,

Tuller, T., Birin, H., Gophna, U., Kupiec, M., and Ruppin, E. 2010. Reconstructing ancestral gene content by coevolution. Genome Research 20: 122–132. https://doi.org/10.1101/gr.096115.109.

Tveten, G. M., Spyrou, A., Schwengner, R., et al. 2016. Completing the nuclear reaction puzzle of the nucleosynthesis of $^{92}Mo$. Physics Review C 94: 025804 (9 pages). https://doi.org/10.1103/PhysRevC.94.025804.

Valkovic, V. 2021. Origins of Life: Musings from Nuclear Physics. In Astrophysics and Astrobiology. Boca Raton, FL: CRC Press, Taylor and Francis Group.

Valković, V., and Obhođaš, J. 2022. Origins of chiral life in interstellar molecular clouds. The Astronomical Journal 163: 270 (14 pages). https://doi.org/10.3847/1538-3881/ac63d0.

Van Dishoeck, E. F. 2014. Astrochemistry of dust, ice and gas: Introduction and overview. Faraday Discussions 168: 9–47.

Van Eck, S., Goriely, S., Jorissen, A., and Plez, B. 2001. Discovery of three lead-rich stars. Nature 412: 793–795.

Wallner, A., Froehlich, M. B., Hotchkis, M. A. C., et al. 2021. $^{60}Fe$ and $^{244}Pu$ deposited on earth constrain the r-process yields of recent nearby supernovae. Science 372(6543): 742–745. https://doi.org/10.1126/science.aax3972.

Wanajo, S. 2006. The rp-process in neutrino-driven winds. Astrophysics Journal 647(2): 1323-1340. https://doi.org/10.1086/505483.

Wanajo, S., Janka, H.-T., and Kubono, S. 2011. Uncertainties in the νp-process: Supernova dynamics versus nuclear physics. Astrophysics Journal 729(1): 46 (18 pages). https://doi.org/10.1088/0004-637X/729/1/46.

Weiss, M. C., Preiner, M., Xavier, J. C. Zimorski, V., and Martin, W. F. 2018. The last universal common ancestor between ancient Earth chemistry and the onset of genetics. PLOS Genetics 14(8): e1007518 (19 pages). https://doi.org/10.1371/journal.pgen.1007518.

Weiss, M. C., Sousa, F., Mrnjavac, N., et al. 2016. The physiology and habitat of the last universal common ancestor. Nature Microbiology 1: 16116. https://doi.org/10.1038/nmicrobiol.2016.116.

Zhaxybayeva, O., Lapierre, P., and Gogarten, J. P. 2005. Ancient gene duplications and the root(s) of the tree of life. Protoplasma 227: 53–64. https://doi.org/10.1007/s00709-005-0135-1.

Zhukovska, S., Gail, H.-P., and Trieloff, M. 2008. Evolution of interstellar dust and stardust in the solar neighbourhood. Astronomy and Astrophysics 479: 453–480. https://doi.org/10.1051/0004-6361:20077789.

## ADDITIONAL READING

Adawy, A. 2022. Functional chirality: From small molecules to supramolecular assemblies. Symmetry 14: 292 (27 pages). https://doi.org/10.3390/sym14020292.

Adibekyan, V., Dorn, C., Sousa, S. G., et al. 2021. A compositional link between rocky exoplanets and their host stars. Science 374: 330–332.

Aristov, V. V., Buchelnikov, A. S., and Nechipurenko, Y. D. 2022. The use of the statistical entropy in some new approaches for the description of biosystems. Entropy 24: 172 (17 pages). https://doi.org/10.3390/e24020172.

Avnir, D. 2021. Critical review of chirality indicators of extraterrestrial life. New Astronomy Reviews 92: 101596 (8 pages). https://doi.org/10.1016/j.newar,2020.101596.

Bazan, E., Dokládal, P., and Dokladalova, E. 2019. Quantitative analysis of similarity measures of distributions. British Machine Vision Conference 2019, BMVC 2019, Cardiff, UK. https://hal.archives-ouvertes.fr/hal-01984970.

Bray, M. S., Lenz, T. K., Haynes, J. W., et al. 2018. Multiple prebiotic metals mediate translation. PNAS 115(48): 12164–12169. https://doi.org/10.1073/pnas.1803636115.

Burbidge, E. M., Burbidge, G. R., Fowler, W. A. and Hoyle, F. 1957. Synthesis of the elements in stars. Reviews of Modern Physics 29: 547–650. https://link.aps.org/doi/10.1103/RevModPhys.29.547.

Cescutti, G., Matteucci, F., Caffau, E., and François, P. 2012. Chemical evolution of the Milky Way: The origin of phosphorus (Research Note). Astronomy & Astrophysics 540: A33 (4 pages) https://doi.org/10.1051/0004-6361/201118188.

Chen, Y., and Ma, W. 2020. The origin of biological homochirality along with the origin of life. PLoS Computational Biology (20 pages). https://doi.org/10.1371/journal.pcbi.1007592.

Chernykh, M., Feldmeier, H., Neff, T., et al. 2010. Pair decay width of the Hoyle state and its role for stellar carbon production. arXiv:1004.3877v2 [nucl-ex], 16 Jun.

Crofts, A. R. 2007. Life, information, entropy and time: Vehicles for semantic inheritance. Complexity 13(1): 14–50.

Darwin, C. 1861. The Origin of Species by Means of Natural Selection. New York, NY: Appleton, D. and Company.

Davids, C. N., and Bonner, T. I. 1971. Enhancement of the 3 $^{4}He \rightarrow {}^{12}C$ reaction rate by inelastic proton scattering. The Astrophysical Journal 166: 405–410.

de Farias, S. T., Rê, T. G., and José, M. V. 2016. A proposal of the proteome before the last universal common ancestor (LUCA). International Journal of Astrobiology 15(1): 21–31.

Dorta-Urra, A., and Bargueño P. 2019. Homochirality: A perspective from fundamental physics. Symmetry 11(5): 661 (9 pages). https://doi.org/10.3390/sym11050661.

Glasgow, D. W., Purser, F. O., Hogue, H., et al. Differential elastic and inelastic scattering of 9- to 15-MeV neutrons from carbon. Nuclear Science and Engineering 64(4): 521–533.

Gómez, A. V. 2014. The thermodynamics of the living organisms: Entropy production in the cell. arXiv:1410.8820v2 [physics.bi-ph], 12 Dec.

Hansen, L. F., Dietrich, F. S., Pohl, B. A., Poppe, C. H., and Wong, C. 1985. Test of microscopic optical model potentials for neutron elastic scattering at 14.6 MeV over a wide mass range. Physical Review C 31(1): 111–119.

Haouat, G., Lachkar, J., Sigaud, J., Tatin, Y., and Cocu, F. 1978. Measurement of the differential elastic and inelastic neutron scattering cross section of carbon from 8.0 to 14.5 MeV. Nuclear Science and Engineering 65: 331–346.

Jiménez-Serra, I., Martin-Pintado, J., Rivilla, V. M., et al. 2020. Toward the RNA-world in the interstellar medium—detection of urea and search of 2-amino-oxazole and simple sugars. Astrobiology 20(9): 1048–1066.

Kauffman, S. A. 1986. Autocatalytic sets of proteins. Journal of Theoretical Biology 119(1): 1–24. https://doi.org/10.1016/s0022-5193(86)80047-9.

Kloprogge, J. T., and Hartman, H. 2022. Clays and the origin of life: The experiments. Life 12: 259 (54 pages). https://doi.org/10.3390/life12020259.

Kühnlein, A., Lanzmich, S. A., and Braun, D. 2021. tRNA sequences can assemble into a replicator. eLife 10: e63431 (22 pages). https://doi.org/10.7554/eLife.63431.

Lamzin, V. S., Dauter, Z., and Wilson, K. S. 1995. How nature deals with stereisomers. Current Opinion in Structural Biology 5: 830–836.

Laurent, G., Lacoste, D., and Gaspard, P. 2021. Emergence of homochirality in large molecular systems. PNAS 118(3): e2012741118 (6 pages). https://doi.org/10.1073/pnas.201241118.

Lei, F. 2021. Nebula-relay hypothesis: Primitive life in nebula and origin of life on earth. Acta Astronomica Sinica 62(3): 28 (8 pages).

Mack, G., Mertens, G., and Tornow, W. 1985. Fast neutron scattering from hydrogen and carbon. Nuclear Instruments and Methods in Physics Research B 10/11: 387–391.

Martin, N. F., Venn, K. A., Aguado, D. S., et al. 2022. A stellar stream remnant of a globular cluster below the metallicity floor. Nature 601. 45–48. https://doi.org/10.1038/s41586-021-04162-2.

Michaelian, K. 2022. Non-equilibrium thermodynamic foundations of the origin of life. Foundations 2: 308–337.

Moreira, D., and López-García, A. 2007. The Last Common Ancestor of Modern Cells. In M. Gatgaud, H. Martin, and P. Claeys (eds.), Lectures in Astrobiology, vol. II, Advances in Astrobiology Biophysical (pp. 305–318). Berlin: Springer, Heidelberg.

Morowitz, H. J., Srinivasan, V., and Smith, E. 2016. Ligand field theory and the origin of life as an emergent feature of the periodic table of elements. The Biological Bulletin 219: 1–6.

Nissden, P. E. 2013. The carbon-to-oxygen ratio in stars with planets. Astronomy & Astrophysics 552: A73 (10 pages). https://doi.org/10.1051/0004-6361/201321234.

Olson, N., Trostell, B., and Ramström, E. 1989. Neutron elastic and inelastic scattering from carbon in the energy range 16.5–22.0 MeV. Nuclear Physics A496: 505–529.

Papoular, R. 2008. On the formation and abundances of CO in envelopes of AGB stars. arXiv:0808.2790v1 [astro-ph], 20 Aug 20.

Potapov, A., Bouwman, J., Jäger, C., et al. 2020. Dust/ice mixing in cold regions and solid-state water in diffuse interstellar medium. Nature Astronomy. https://doi.org/10.1038/s41550-020-01214-x.

Romano, D., Matteucci, F., Zhang, Z. Y., Ivison, R. J., and Ventura, P. 2019. The evolution of CNO isotopes: The impact of massive stellar rotators. Monthly Notices of the Royal Astronomical Society 490: 2838–2854.

Sato, N. 2012. Scientific élan vital: Entropy deficit or inhomogeneity as a unified concept of driving forces of life in hierarchical biosphere driven by photosynthesis. Entropy 14.

Schoepp-Cothenet, B., von Lis, R., Philippot, P., et al. 2012. The ineluctable requirement for the trans-iron elements molybdenum and/or tungsten in the origin of life. Scientific Reports 2: 263 (5 pages). https://doi.org/10.1038/srep00263.

Smith, H. H., Hyde, A. S., Simkus, D. N., et al. 2021. The grayness of the origin of life. Life 11: 498 (18 pages). https://doi.org/10.3390/life11060498.

Stolte, D. 2020. Way before stars form, space clouds hold building blocks of life. University of Arizona, Original Study. https://doi.org/10.3847/1538-4357/ab7375.

Sudac, D., and Valkovic, V. 2022. Contribution to the understanding of the role of carbon in nucleosynthesis. Unpublished.

Thaler, D. S. 2021. Is global microbial biodiversity increasing, decreasing, or staying the same? Frontiers in Energy and Evolution 9: article 565649 (9 pages). https://doi.org/10.3389/fevo.2021.565649.

Udgaonkar, J. B. 2001. Entropy in biology. Resonance 6(9): 61–66.

Ulanowicz, R. E., and Hannon, B. M. 1987. Life and the production of entropy. Proceedings of the Royal Society of London. Series B, Biological Sciences 232(1267): 181–192.

Von Stockar, U., and Liu, J. S. 1999. Does microbial life always feed on negative entropy? Thermodynamic analysis of microbial growth. Biochimica et Biophysica Acta 1412: 191–211.

CHAPTER 5

# Present Life in the Universe

## 5.1 THE ANTHROPIC PRINCIPLE

It was Dyson who, in his book *Disturbing the Universe*, claimed that the architecture of the Universe is consistent with the hypothesis that the mind plays an essential role in its functioning (1981). Actually, the question, which remains open, is: what is the origin of the rules which govern the behavior of the Universe once it is created? Here we come to the anthropic principle. A critical examination is presented in detail in a nice paper by Earman (1987). The author starts by trying to understand the motivation for the meaning of the anthropic principle as it was first introduced by Dicke (1961) and Carter (1974).

Earman's (1987) critical examination of the anthropic principle concludes:

> Insofar as the various anthropic principles are directed at the evidentiary evaluation of cosmological theories they are usually interpretable in terms of wholly sensible ideas, but the ideas embody nothing new, being corollaries of any adequate account of confirmation. In addition, insofar as anthropic principles are directed at promoting Man or Consciousness to a starring role in the functioning of the universe, they fail; for either the promotion turns out to be an empty tease or else it rests on woolly and ill-founded speculations. There remains the potentially legitimate use of anthropic reasoning to alleviate the state of puzzlement into which some people have managed to work themselves over various features of the observable portion of our universe. The alleviation takes place by extending Dicke's selection effect notion, as originally applied to temporal stages of a standard big bang model, to spatio-temporal stages of a worlds within-worlds model. But to be legitimate, the anthropic reasoning must be backed by substantive reasons for believing in the required worlds-within worlds structure. Assertions by anthropic theorizers to the effect that modern science provides many such reasons should be taken with a large dose of salt. Neither classical general relativity theory nor quantum mechanics provide any firm grounds for taking worlds-within-worlds models seriously, and while various speculative versions of the new inflationary cosmology may eventually provide such grounds, the verdict is at present very much in doubt.

The basic features of galaxies, stars, planets, and the everyday world are essentially determined by a few microphysical constants and by the effects of gravitation. Many interrelations between different scales that at first sight seem surprising are straightforward consequences of simple physical arguments. But several aspects of our Universe—some of which seem to be prerequisites for the evolution of any form of life—depend rather delicately on apparent "coincidences" among the physical constants Carr and Rees (1979).

### 5.1.1 Cosmic Fine-Tuning

It is a fact that the universe supporting life depends on several of its fundamental characteristics, notably on the form of the laws of nature, on the values of some constants of nature, and on aspects of the universe's conditions in its very early stages. Friederich (2019) shows

DOI: 10.1201/9781003366607-5

an example of such fine-tuning for life in the difference between the masses of the two lightest quarks: the up and down quarks. Small changes in this difference would undermine the stability properties of the proton and neutron, which are the bound states of these quarks, and lead to a much less complex universe where bound states of quarks other than the proton and neutron dominate. Life would apparently also have been impossible if the mass of the electron, which is roughly ten times smaller than the mass difference between the down and up quark, had been somewhat larger in relation to that difference. Fine-tuning of the lightest quark masses with respect to the strength of the weak nuclear force has been found as well. The suggested instances of fine-tuning concern include:

- the strength of gravity,
- the strengths of the strong and weak nuclear forces,
- the mass of the Higgs boson,
- the vacuum energy density,
- the total energy density of the universe in its very early stages,
- the relative amplitude of energy density fluctuations in the very early universe,
- the initial entropy of the universe, and
- aspects of the very form of the known laws of nature.

Friederich (2019) assumes that at least some laws and constants are indeed fine-tuned for life in the sense just sketched. One of the most popular and most-discussed responses to the universe's fine-tuning for life is to suggest that the laws and constants might be environmental, that is, that they might differ in different regions of space-time and/or in different spatio-temporally disconnected universes. A collection of multiple universes with different constants and laws is referred to as a "multiverse". The version of the idea that the laws and constants are environmental that is most often discussed is the multiverse idea. The most intensely discussed concrete multiverse scenario is the landscape multiverse (Susskind 2005), which is taken to be suggested by a combination of string theory and eternal inflation.

The idea that underlies the suggested inference from the finding that life requires fine-tuned laws and constants to a multiverse is that, if there is a sufficiently vast and diverse multiverse, it is only to be expected that it contains at least one universe where the laws and constants are right for life. The famous (weak) anthropic principle states that "we must be prepared to take account of the fact that our location . . . is necessarily privileged to the extent of being compatible with our existence as observers" (Carter 1974). As this principle highlights, given that we are observers, the universe in which we find ourselves has to be one where the conditions are compatible with the existence of observers. This suggests that, on the assumption that there is a sufficiently diverse multiverse, it is neither surprising that there is at least one universe that is hospitable to life nor—since we could not have found ourselves in a life-hostile universe—that we find ourselves in a life-friendly one. Many physicists and philosophers regard this line of thought as suggesting the inference of a multiverse as a rational response to the finding that the conditions are right for life in our universe despite the required fine-tuning.

## 5.2 LIFE IN THE MOLECULAR CLOUDS

The small solid particles in the space between the stars provide the surfaces for the production of many simple and complex molecules. Processes involving the effects of ultraviolet irradiation of the thin (hundredth-micron) mantles are shown to produce a wide range of molecules and ions also seen in comets. Some of the more complex ones inferred from laboratory experiments are expected to play an important role in the origin of life. An outline of the chemical evolution of interstellar dust as observed and as studied in the laboratory is presented. Observations of comets are shown to provide substantial evidence for their being fluffy aggregates of interstellar dust as it was in the protosolar nebula, that is, the interstellar cloud which collapsed to form the solar system. The theory that comets may have brought the progenitors of life to the earth is summarized by Greenberg (2002). Among other relevant issues, the author presented the brief history of a dust grain and the basic steps in its evolution from formation to destruction. At the beginning, small silicate particles condense in the atmospheres of cool stars and are ejected into space. A pattern followed inside molecular clouds is described by the author as:

1. Silicate particles cool to T = 5–15K and act as condensation nuclei for accretion of gas atoms and molecules as a mantle of frost.

2. Complex chemical reactions between gas and solids leads to an H2O- and CO-dominated grain mantle.
3. Ultraviolet radiation of mantle breaks simple molecules, leading to new combinations and to complex organic molecules (photoprocessing).
4. Star formation occurs, leading to dissipation of portions of the collapsed cloud into surrounding low-density regions (diffuse clouds).
5. In the low-density regions (diffuse clouds), volatile icy mantles are either photoprocessed, evaporated, or destroyed, leaving only complex refractory mantles. These mantles are further subjected to photoprocessing and to destructive processes, which can break off small pieces.

Steps 1 to 5 are repeated many times (~50) until the grain is consumed by star formation or becomes part of the comet. The total mean lifetime of a dust grain is about 5 × 109 years (turnover time for the interstellar medium into and out of star), after Greenberg (2002).

Formation of hydrated molecules, and in particular $H_2O$, by sequential surface reactions between atoms and radicals with incoming H atoms is shown in Figure 5.1, after Greenberg (2002).

A schematic of the sequence of bond breaking and recombination of radicals occurring in ultraviolet irradiated interstellarice mantles is shown in Figure 5.2. Initially the silicate core accretes (by surface formation or condensation) a mantle of water, methane, ammonia, and carbon monoxide. This mantle is subjected to bombardment by ultraviolet photons whose energy is sufficient to break molecular bonds so that H atoms are detached and leave the radicals OH and others frozen (10 K) in the ice matrix. It is also shown how the detached H atoms may combine with CO to produce the HCO radical and the CO may combine with the O from a doubly photoprocessed water to produce carbon dioxide. All the molecules in the lower panel are seen in both laboratory and interstellar spectra, after Greenberg (2002).

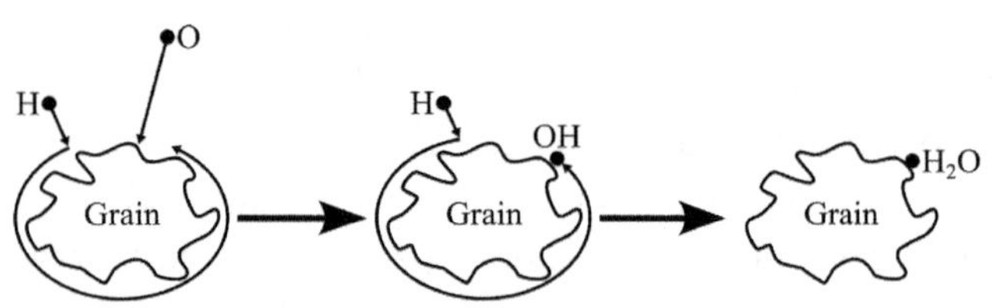


FIGURE 5.1 A schematic illustration of the formation of hydrated molecules, and in particular $H_2O$, by sequential surface reactions between atoms and radicals with incoming H atoms, after Greenberg (2002).

There is an interesting coincidence between when life first appeared and when the massive bombardment of the earth by comets and other objects was just beginning to tail off. It can be seen from the cratering record on other solar system objects, where the record has not been erased by weathering and crustal evolution (see Chyba 1990 for a review), that during the first 700 $\times 10^6$ years after the earth formed, it had undergone about 20 $\times 10^6$ impacts by comets—at least one 5-km comet every 10 years—and that it was not until 3.8 $\times$ $10^9$ years ago that this rate dropped quickly. It appears from the fossil evidence that life was already extensive 3.6 billion years ago and possibly even 200 million years before that based on carbon isotopic evidence. One may even speculate that the evidence for the earliest living organisms could be extended further back in time by noting that the lack of evidence is possibly because the amount of datable rocks decreases with time, similarly to the amount of identifiable fossils, so that the percentage of fossil material may not be zero 3.8 $\times 10^9$ years ago. In any case, if we assume that all that heavy bombardment was delivering millions of death blows to the earth similar to the single one which took care of the dinosaurs 65 $\times 10^6$ years ago, there was not a great window in time for life to have emerged. The question is whether comets were the source of life as well as death (Greenberg 2002).

While accounting for only a small fraction (–1%) of the galaxy's mass, the interstellar medium (ISM) is nevertheless an important part of the galactic ecosystem. Gravitational collapse of dense interstellar clouds leads to the formation of new stars, which produce heavier elements in their interiors by nucleosynthesis. The main reaction in stellar interiors is the nuclear fusion of H into He. In a later stage of stellar evolution C, N, and O are formed. Further nucleosynthesis occurs in massive stars, leading to elements as heavy as iron. At the end of their lifetime, which is mainly determined by their initial mass, stars return material to the interstellar environment by mass outflows, forming expanding shells and envelopes, as well as by violent explosions. Thus, the ISM represents an environment in which atoms, molecules, and solid matter undergo strong evolution and recycling (Ehrenfreund and Menten 2001).

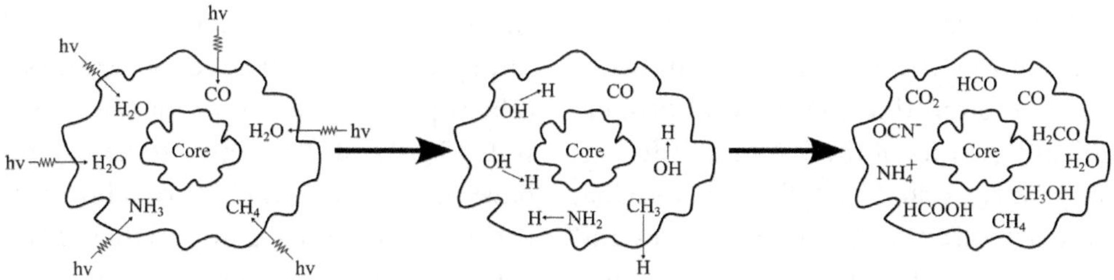


FIGURE 5.2 Schematic of the sequence of bond breaking and recombination of radicals occurring in ultraviolet irradiated interstellar ice mantles, after Greenberg (2002).

Nitriles play a key role as molecular precursors in any RNA-based scenario for the origin of life, therefore, the establishment of their reservoir in the interstellar medium is of great importance. A step in this direction is the work by Rivilla et al. (2022), with their report on the detection of several nitriles, including cyanic acid (HOCN) and three $C_4H_3N$ isomers (cyanoallene, $CH_2CCHCN$; propargyl cyanide, $HCCCH_2CN$; and cyanopropyne ($CH_3CCCN$) and the tentative detections of cyanoformaldehyde (HCOCN) and glycolonitrile ($HOCH_2CN$) in the molecular cloud G+0.693 – 0.027. This cloud, located in the Sgr B2 region of the center of our galaxy, is one of the most chemically rich sources in the ISM. Based on the derived molecular abundances of the different nitriles in this cloud and other interstellar sources, the authors have discussed their formation mechanisms in the ISM. They proposed that the observed HOCN abundance in this molecular cloud was mainly due to surface chemistry and subsequent shock-induced desorption, while HCOCN might be mainly formed through gas-phase chemistry. In the case of $HOCH_2CN$, several grain-surface routes from abundant precursors could produce it.

The derived abundances of the three $C_4H_3N$ isomers in G+0.693 – 0.027 are very similar and also similar to those previously reported in the dark cold cloud TMC-1. This suggests that the three isomers are likely formed through gas-phase chemistry from common precursors, possibly unsaturated hydrocarbons ($CH_3CCH$ and $CH_2CCH_2$) that react with the cyanide radical (CN). The rich nitrile feedstock found towards G+0.693 – 0.027 confirms that interstellar chemistry is able to synthesize in space molecular species that could drive the prebiotic chemistry of the RNA-world (Rivilla et al. 2022).

Cosmic dust grains with carbonaceous and siliceous composition represent the most pristine starting material for planetary systems, influence the thermodynamic properties of the medium by absorption and emission of stellar light, and provide a surface for key astrochemical reactions. Dust grains in cold dense astrophysical environments, such as dense interstellar clouds, protostellar envelopes, and planet-forming disks beyond the snow line, are typically considered mixtures of dust particles with molecular ices. Water is the main constituent of these ices, accounting for more than 60% of the ice in most lines of sight. Ices are believed to cover the surface of a dust core and/or to be physically mixed with dust.

Yang et al. (2023) published a rich molecular inventory of high-redshift dusty galaxies revealed by broadband spectral line surveys. Understanding the nature of high-redshift dusty galaxies requires a comprehensive view of their interstellar medium and molecular complexity. However, the molecular ISM at high redshifts is commonly studied using only a few species beyond $^{12}C\ ^{16}O$, limiting our understanding. In this paper, the authors present the results of deep 3-mm spectral line surveys using the NOrthern Extended Millimeter Array (NOEMA) targeting two strongly lensed dusty galaxies observed when the Universe was less than $1.8 \times 10^9$ years old: APM 08279 + 5255, a quasar at redshift z = 3.911, and NCv1.143 (H-ATLAS J125632.7 + 233625), a z = 3.565 starburst galaxy. The spectral line surveys cover rest-frame frequencies from about 330 to 550 GHz for both galaxies. They report the detection of 38 and 25 emission lines in APM 08279 + 5255 and NCv1.143, respectively. These lines originate from 17 species, namely CO, $^{13}CO$, $C^{18}O$, CN, CCH, HCN, $HCO^+$, HNC, CS, $C^{34}S$, $H_2O$, $H_3O^+$, NO, $N_2H^+$, CH, $c\text{-}C_3H_2$, and the vibrationally excited HCN and neutral carbon. The spectra reveal the chemical richness and the complexity of the physical properties of the ISM. By comparing the spectra of the two sources and combining the analysis of the molecular gas excitation, they found that the physical properties and the chemical imprints of the ISM are different: the molecular gas is more excited in APM 08279 + 5255, which exhibits higher molecular gas temperatures and densities compared to NCv1.143; the molecular abundances in APM 08279 + 5255 are akin to the values of local active galactic nuclei (AGN), showing boosted relative abundances of the dense gas tracers that might be related to high-temperature chemistry and/or the X-ray-dominated regions, while NCv1.143 more closely resembles local starburst galaxies. The most significant differences between the two sources are found in $H_2O$: the 448 GHz ortho-$H_2O(4_{23}–3_{30})$ line is significantly brighter in APM 08279 + 5255, which is likely linked to the intense far-infrared radiation from the dust powered by AGN. Their astrochemical model suggests that, at such high column densities, far-ultraviolet radiation is less important in regulating the ISM, while cosmic rays (and/or X-rays and shocks) are the key players in shaping the molecular abundances and the initial conditions of star formation. Both their observed CO isotopologs line ratios and the derived extreme ISM conditions (high gas temperatures, densities, and cosmic-ray ionization rates) suggest the presence of a top-heavy stellar initial mass function. From the ~330–550 GHz continuum, they also find evidence of nonthermal millimeter flux excess in APM 08279 + 5255 that might be related to the central supermassive black hole. Such deep spectral line surveys open a new window into the physics and chemistry of the ISM and the radiation field of galaxies in the early Universe.

Potapov et al. (2022) discussed dust/ice mixing in cold regions and solid-state water in the diffuse interstellar medium. Whether ice in cold cosmic environments is physically separated from silicate dust or mixed with individual silicate moieties is not known. However, different grain models give very different compositions and temperatures of grains. The aim of this study was to compare the mid-infrared spectra of laboratory silicate grain/water ice mixtures with astronomical observations to evaluate the presence of dust/ice mixtures in interstellar and circumstellar environments. The laboratory data can explain the observations, assuming reasonable mass-averaged temperatures for the protostellar envelopes and protoplanetary disks, demonstrating that a substantial fraction of water ice may be mixed with silicate grains. On the basis of the combination of laboratory data and infrared observations, Potapov et al. (2022) provided evidence of the presence of solid-state water in the diffuse interstellar medium. Their results have implications for future laboratory studies investigating cosmic dust grain analogs and for future observations trying to identify the structure, composition, and temperature of grains in different astrophysical environments.

Krasnokutski et al. (2022) considered a pathway to peptides in space through the condensation of atomic carbon. Organic molecules are widely present in the dense interstellar medium, and many have been synthesized in the laboratory on Earth under the conditions typical for an interstellar environment. Until now, however, only relatively small molecules of biological interest have been demonstrated to form experimentally under typical space conditions. They proved experimentally that the condensation of carbon atoms on the surface of cold solid particles (cosmic dust) leads to the formation of isomeric polyglycine monomers (aminoketene molecules). Following encounters between aminoketene molecules, they polymerize to produce peptides of different lengths. The chemistry involves three of the most abundant species (CO, C, and $NH_3$) present in star-forming molecular clouds and proceeds via a novel pathway that skips the stage of amino acid formation in protein synthesis. The process is efficient, even at low temperatures, without irradiation or the presence of water. The delivery of biopolymers formed by this chemistry to rocky planets in the habitable zone might be an important element in the origins of life.

### 5.2.1 Is Life Universal?

Scientists have detected key building blocks for RNA in a cloud of gas and dust at the heart of the Milky Way. Is this the key to how life arrived on Earth?

- Some of the key building blocks of life have been detected at the heart of the Milky Way.
- Known as nitriles, scientists spotted them in a molecular cloud of gas and dust.
- Nitriles are building blocks for RNA—a DNA-like nucleic acid in all living cells.
- Supports the RNA world theory of origin of life, which is that it arrived from space.

RNA can fulfil two functions: storing and copying information like DNA and catalyzing reactions like enzymes. According to the RNA world theory, nitriles and other building blocks for life need not necessarily all have arisen on Earth itself.

Astronomers have discovered a prebiotic molecule in molecular clouds in space, a finding that could be important in our understanding of how life evolved on Earth. The molecule, ethanolamine ($NH_2CH_2CH_2OH$), is known as a prebiotic compound that precedes the advent of life on Earth and possibly plays a role in it. In living organisms, such a molecule is part of the cell membrane, the barrier between different cells. Ethanolamine has previously been discovered in meteorites, and researchers suspected that it was something in the meteorites that must have facilitated the formation of the molecule. The new research, published by Rivilla et al. (2021) changes this picture. The team observed a molecular cloud called G+0.693–0.027 located near the center of the Milky Way. This is one of the most chemically rich clouds in the galaxy. Using a radio telescope, they spotted the faint emissions of ethanolamine in the cloud in a surprising abundance.

Rivilla et al. (2021) reported the detection towards the molecular cloud G+0.693 – 0.027 of several nitriles, including cyanic acid (HOCN), three $C_4H_3N$ isomers (cyanoallene, $CH_2CCHCN$; propargyl cyanide, $HCCCH_2CN$; and cyanopropyne ($CH_3CCCN$), and the tentative detections of cyanoformaldehyde (HCOCN) and glycolonitrile ($HOCH_2CN$). They also performed the first interstellar search for cyanoacetaldehyde ($HCOCH_2CN$), which was not detected. Based on the derived molecular abundances of the different nitriles in G+0.693 – 0.027 and other interstellar sources, they discussed their formation mechanisms in the ISM. They proposed that the observed HOCN abundance in G+0.693 – 0.027 is mainly due to surface chemistry and subsequent shock-induced desorption, while HCOCN might be mainly formed through gas-phase chemistry. In the case of $HOCH_2CN$, several grain-surface routes from abundant precursors could produce it. The derived abundances of the three $C_4H_3N$ isomers in G+0.693 – 0.027 are very similar and also similar to those previously reported in the dark cold cloud TMC-1. This suggests that the three isomers are likely formed through gas-phase chemistry from common precursors, possibly unsaturated hydrocarbons ($CH_3CCH$ and $CH_2CCH_2$) that react with the cyanide radical (CN). The rich nitrile feedstock found towards G+0.693 – 0.027 confirms that interstellar chemistry is able to synthesize in space molecular species that could drive the prebiotic chemistry of the RNA-world.

Unfortunately, they started their paper with

> Nitriles play a key role as molecular precursors in prebiotic experiments based on the RNA world scenario for the origin of life. These chemical compounds could have been partially delivered to the young Earth from extraterrestrial objects, stressing the importance of establishing the reservoir of nitriles in the interstellar medium.

Oba et al. (2022) reported the detection of nucleobases in three carbonaceous meteorites using state-of-the-art analytical techniques optimized for small-scale quantification of nucleobases down to the range of parts per trillion (ppt). In addition to previously detected purine nucleobases in meteorites such as guanine and adenine, they identified various pyrimidine nucleobases, such as cytosine, uracil, and thymine, and their structural isomers such as isocytosine, imidazole-4-carboxylic acid, and 6-methyluracil, respectively. Given the similarity in the molecular distribution of pyrimidines in meteorites and those in photon-processed interstellar ice analogs, some of these derivatives could have been generated by photochemical reactions prevailing in the interstellar medium and later incorporated into asteroids during solar system formation. This study demonstrates that all of the bases serving as building blocks of DNA and RNA have now been found in meteorites.

Cleaves II. et al. (2023) reported a significant advance in one of the most important problems in astrobiology—the development of a simple, reliable, and practical method for determining the biogenicity of organic materials in planetary samples, both on other worlds and for the earliest traces of life on Earth. They developed a robust method that combines pyrolysis GC-MS measurements of a wide variety of terrestrial and extraterrestrial carbonaceous materials with machine-learning-based classification to achieve ~90% accuracy in the differentiation between samples of abiotic origins vs. biotic specimens, including highly degraded, ancient, biologically derived samples. Such discrimination points to underlying "rules of biochemistry" that reflect the Darwinian imperative of biomolecular selection for function. These machine-learning methods do not rely on precise compound identification; rather, the relational aspects of chromatographic and mass peaks provide the needed information, which underscores this method's utility for detecting alien biology.

## 5.3 LIFE IN THE SOLAR SYSTEM

The sun is thought to be a third-generation star in the universe. The sun formed from a molecular cloud produced by the supernova explosion of the sun's predecessor star. During the process of star formation, interstellar molecules and dust become the building blocks for protostellar disk from which planets, comets, asteroids, and other small bodies are formed. A number of protoplanetary disks have been imaged by the Hubble Space Telescope, indicating that the formation of solar systems similar to our own is a rather common process.

The fate of organic molecules present in the cloud from which the solar system originated is tied to the knowledge of the chemistry and physics of protoplanetary disk. The fate of organic molecules during solar system formation is strongly tied to the local environment conditions and the radial distance from the protosun. Organic molecules in the inner solar nebula were destroyed, but organic molecules from the outer solar system, regions of low temperature and pressure, may have been integrated into solar system bodies in rather pristine forms (Ehrenfreund and Menten 2001).

Early Mars habitability is often discussed in the literature. For example, it is believed that during the Noachian period, $(4.1–3.7) \times 10^9$ years ago, Mars' crust may have provided a favorable environment for microbial life; see Cockell et al. (2016), Michalski et al. (2018). The porous brine material on the Mars surface would have created a physical space sheltered from ultraviolet and cosmic radiation and provided a habitat for simple microbial organisms that consumed $H_2$ and $CO_2$ as energy and carbon sources and produced methane as waste. Sauterey et al. (2022) presented a probabilistic assessment of Mars' habitability in in this period and quantified their biological feedback on Mars' atmosphere and climate. They found that subsurface habitability was very likely, limited by the extent of surface ice coverage. Michalski et al. (2022) also recently discussed the geological diversity and microbiological potential of lakes on Mars.

A review of habitable zones—for animals as well as microbes and in the galaxy and Universe as well as around our sun—leads to an inescapable conclusion: Earth is a rare place indeed. Perhaps the most intriguing finding of this line of research is that Earth is rare as much for its abundant metal content as for its location relative to the sun.

### 5.3.1 Origin of Life on Planet Earth

All life on Earth is dependent on biologically mediated electron transfer (i.e., redox) reactions that are far from thermodynamic equilibrium. Biological redox reactions originally evolved in prokaryotes and ultimately, over the first $\sim 2.5 \times 10^9$ years of Earth's history, formed a global electronic circuit. To maintain the circuit on a global scale requires that oxidants and reductants be transported; the two major planetary wires that connect global metabolism are geophysical fluids—the atmosphere and the oceans. Because all organisms exchange gases with the environment, the evolution of redox reactions has been a major force in modifying the chemistry at Earth's surface. Here we briefly review the discovery and consequences of redox reactions in microbes with a specific focus on the coevolution of life and geochemical phenomena; see Jelen et al. (2016).

During the period of early Earth's skies, ocean, and crust evolution, the planet was being bombarded by large meteorites. Early collisions likely destroyed near-surface environments, but they may also have delivered key prebiotic compounds to the planet, such as amino acids, sugars, purines (nitrogen-containing organic compounds that form the building blocks of modern DNA and RNA), and reactive phosphorus; see Furukawa et al. (2019). Most models of prebiotic chemistry suggest reducing conditions as the most probable path to generating essential prebiotic compounds. However, these considerations ignore the possibility of life forms of some sort surviving the Earth formation process from the molecular cloud.

Beyond the delivery and production of essential organic molecules through impacts, such events may have helped fuel life's beginnings in other ways, specifically by stimulating hydrothermal activity. The full range of consequences from early impacts, both conducive to and challenging for the beginnings of life, remains an important topic of research (Trail et al. 2022).

The new Prebiotic Chemistry and Early Earth Environments ($PCE_3$) Research Coordination Network (RCN) within NASA's Astrobiology Program was designed to bridge this gap. This consortium of researchers aims to transform origins-of-life research by enhancing communication across the disciplinary divide between early Earth geoscientists and prebiotic chemists. In this new collaborative scheme, possible prebiotic chemical scenarios can be filtered by minimizing assumptions and by environmental plausibility. The principal goal of $PCE_3$ is to cultivate a new research culture in which potential prebiotic pathways to life are tested within realistic planetary conditions and the dynamics and constraints of early Earth environments are fully integrated into origins hypotheses.

According to Sephus et al. (2022), decoding the complex relationships between life and the environments it inhabits is central for reconstruction of the factors that determine planetary habitability over geologic timescales. For billions of years, life has continuously adapted to dynamic physical conditions near the Earth's surface. Fossils and other preserved bio-signatures in the paleontological record are the most direct evidence for reconstructing the broad historical contours of this adaptive interplay. The problem is that bio-signatures dating to Earth's earliest history are exceedingly rare.

In their work, Sephus et al. (2022) combined phylogenetic inference of primordial rhodopsin proteins with modeled spectral features of the Precambrian Earth environment to reconstruct the paleo-biological history of this essential family of photoactive transmembrane proteins. Their results suggest that ancestral microbial rhodopsin likely acted as light-driven proton pumps and were spectrally tuned toward the absorption of green light, which would have enabled their hosts to occupy depths in a water column or biofilm where UV wavelengths were attenuated. Subsequent diversification of rhodopsin functions and peak absorption frequencies was made possible by the expansion of surface ecological niches induced by the accumulation of atmospheric oxygen. Inferred ancestors retain distinct associations between extant functions and peak absorption frequencies. Their findings suggest that novel information encoded by biomolecules can be used as "paleo-sensors" for conditions of ancient, inhabited niches of host organisms not represented elsewhere in the paleontological record. The coupling of functional diversification and spectral tuning of this taxonomically diverse protein family underscores the utility of rhodopsins as universal testbeds for inferring remotely detectable bio-signatures on inhabited planetary bodies.

There are several models of the process called origin of life on Earth. The most popular for decades was abiogenesis, also known as chemical origin theory, advocated by great many scientists ever since Oparin (1924) and Haldane (1929) proposed similar theories independently. In his work, Haldane (1929) also introduced the idea that the emergence of life was associated with the origin of viruses as well as the concept of prebiotic soup. However, it is well established that the planet Earth formed some $4.5 \times 10^9$ years ago. The first evidence for life on planet Earth is provided by microfossils; see Schopf (1993): he and his colleagues found worm-like patterns preserved in the Apex Chert, a rock formation in Australia dating back about $3.465 \times 10^9$ years. However, this finding has been questioned by a number of scientists; see, for example, Brasier et al. (2004). Schopf et al. (2017) presented carbon isotope analyses of 11 microbial fossils from the same Western Australian Apex chert, from which they inferred that two of the five species studied were primitive photosynthesizers, one was an Archaeal methane producer, and two others were methane consumers. This discovery of Archaea in the Archean is consistent with the rRNA "tree of life", and it confirmed the earlier disputed biogenicity of the Apex fossils. That would imply that only $1.035 \times 10^9$ years were needed for life to develop to this stage if it originated on the planet. This is rather short time interval considering that the planet provided very hostile conditions immediately after Earth formation. In addition, the heavy bombardment phase ended $3.8 \times 10^9$ before the present, as scaled from the lunar record.

In another paper, Mojzsis et al. (1996) reported a search for geochemical evidence of past biotic activity that has been preserved within minerals that are resistant to metamorphism. They present the ion-microprobe measurements of the carbon-isotope composition of carbonaceous inclusions within grains of apatite (basic calcium phosphate) from the oldest known sediment sequences, a ~$3.8 \times 10^9$-year-old banded iron formation from the Isua supra-crustal belt, West Greenland, and a similar formation from the nearby Akilia island that is possibly older than $3{,}85 \times 10^9$ years, Nutman et al. (1996). The carbon in the carbonaceous inclusions is isotopically light, indicative of biological activity; no known abiotic process can explain these data. Unless some unknown abiotic process exists which is able both to create such isotopically light carbon and then selectively incorporate it into apatite grains, the authors' results provide evidence for the emergence of life on

Earth by at least $3.8 \times 10^9$ years before present. This will leave an even shorter time period ($\sim 0.7 \times 10^9$ years) for the emergence of life after the formation of planet Earth.

Tashiro et al. (2017) reported early traces of life from $3.95 \times 10^9$-year-old sedimentary rocks in Labrador, Canada. The authors measured the total organic carbon contents and carbon isotope values of graphite ($\delta^{13}C_{org}$) and carbonate ($\delta^{13}C_{.}$) in the oldest metasedimentary rocks from northern Labrador (Komiya et al. 2015). Some pelitic rocks have low $\delta^{13}C_{org}$ values of –28.2, comparable to the lowest value in younger rocks. The consistency between crystallization temperatures of the graphite and metamorphic temperature of the host rocks establishes that the graphite does not originate from later contamination. A clear correlation between the $\delta^{13}C_{org}$ values and metamorphic grade indicates that variations in the $\delta^{13}C_{org}$ values are due to metamorphism and that the pre-metamorphic value was lower than the minimum value. The authors concluded that the large fractionation between the $\delta^{13}C_{carb}$ and $\delta^{13}C_{org}$ values, up to 25‰, indicates evidence of the oldest organisms, approximately $3.95 \times 10^9$ years before present. Consequently, life on planet Earth originated only $<550 \times 10^6$ years after planet formation! This may itself imply that the formation of life did not happen on Earth.

Obviously, this research has been disputed by those who believe that geological processes may have caused the anomalies observed.

The other line of thinking is panspermia theory (see Davies 1988), first suggested by Lord Kelvin in 1971. This theory is based on the assumption that primitive life exists in the solar system, Milky Way, and even entire universe. This primitive life is brought to Earth by meteorites in the case of the solar system or by comets from planet systems somewhere in the Milky Way. These seeds must have the ability to survive Earth's harsh environment. There are some drawbacks in this theory. If life came from different parts of the Milky Way, it would not have the same genetic expression system, so we would be forced to abandon the idea of LUCA, the last universal common ancestor. Not to mention that this theory just moves the solution to the question of how life originated to another location.

Feng (2021) and Feng (2022) introduced a modified version of panspermia theory, the so-called nebula-relay hypothesis, in an effort to explain the origin of life on Earth. According to Feng (2021), primitive life, which acted as the seeds of life on Earth, originated during the pre-solar era through physicochemical processes and then spread through the pre-solar nebula after the death of the pre-solar star. The history of life on Earth is then divided into three epochs: formation of primitive life in the pre-solar epoch, the pre-solar nebula epoch, and the formation of the solar system and the Earth age of life.

Brasier et al. (2004) question the biogenecity of putative bacterial and cyanobacterial "microfossils" from 3465 Ma Apex cherts of the Warrawoona Groop in Western Australia based on integrated multidisciplinary evidence obtained from the field and fabric mapping and high-resolution research into their context, sedimentology, and filament morphology. They cannot be distinguished from (and are reinterpreted as) secondary artifacts of amorphous carbon that formed during devitrification of successive generations of carbonaceous hydrothermal dyke vein quartz. Similar structures occur within associated carbonaceous volcanic glass. The null hypothesis of an abiotic or prebiotic origin for such ancient carbonaceous matter is sustained until mutually supporting contextual, morphological, and geochemical evidence for a bacterial rather than abiotic origin is forthcoming.

Sandberg et al. (2018) presented detailed estimates of both literature and scientific uncertainties in Drake equation parameters and came to the conclusion that we are alone in our galaxy (53–96%) and perhaps even in our observable universe (39–85%). In other words, their proper treatment of scientific uncertainties dissolves the Fermi paradox by showing that it is not unlikely for us to be alone in the Milky Way. Actually, their updated probabilities suggest that there is a substantial probability that we are indeed alone.

We shall elaborate this point in some more detail. Scientists here study the spectra from G+0.693–0.027, a molecular cloud near the center of the Milky Way. They detect a range of nitriles, key building blocks for RNA, including cyanoallene, propargyl cyanide, cyanopropyne, and possibly cyanoformaldehyde and glycolonitrile, none of which had previously been found in G+0.693–0.027. They conclude that nitriles are among the most abundant chemical families in the universe, which lends support to the RNA world theory of the origin of life on Earth, and possibly elsewhere in our galaxy. According to this scenario, life on Earth was originally based on RNA only, and DNA and protein enzymes evolved later. RNA can fulfill both functions: storing and copying information like DNA and catalyzing reactions like enzymes. According to the RNA world theory, nitriles and other building blocks for life needn't necessarily all have arisen on Earth itself: they might also have originated in space and "hitchhiked" to the young Earth inside meteorites and comets during the Late Heavy Bombardment period, between $4.1 \times 10^9$ and $3.8 \times 10^9$ years ago. In support, nitriles and other precursor molecules for nucleotides, lipids, and amino acids have been found inside contemporary comets and meteors.

The transition from abiotic chemistry to the earliest forms of cellular life represents a critical link in the chain of events that connect us to our cosmic origins. Ditzler et al. (2022) discuss the progression from abiotic precursors, to the simplest forms of life, to the emergence of complex cellular processes present in the last universal common ancestor of life on Earth. They consider progress in understanding which abiotic molecules served as the raw materials for the assembly of the earliest forms of life and how these prebiotic building blocks may have come together to support the basic functions necessary for life. They explore leading theories on the origin of life, the nature of the earliest cell-like entities (protocells), and the subsequent evolution of the genetic code and core metabolic processes present in LUCA.

It seems likely that the emergence of life from abiotic precursors required the recombination of modular molecular building blocks, but the identity of these building blocks and the manner in which they were combined could have been quite different from those used by life on Earth today. Understanding the emergence of life requires us to identify which materials were available, how they could have come together, and what functions they needed to support in order to initiate and sustain life. When considering which building blocks are available in prebiotic environments, there are two primary sources of information that we can examine, astronomical sources of organic molecules (e.g., meteorites, interplanetary dust particles, comets, asteroids) and laboratory experiments that attempt to simulate either astronomical processes or processes at play on the early Earth. The organic molecules present in extraterrestrial materials and laboratory simulations include many small molecules used by modern biology, along with a broad suite of additional molecules that may have served as building blocks in earlier forms of life on Earth or in life elsewhere in the universe. In addition, laboratory simulations of early Earth conditions have identified potential routes to the formation of macromolecular building blocks of life. Still, many of the important questions about the connection between abiotic chemistry and cellular life are far from being answered. Which building blocks did emerging life use, and from which prebiotic reservoirs were they drawn? Which of these molecules were delivered to the early Earth, which were the results of geological processes, and which were only introduced to biology through pre-existing biological systems? There are several possible paths life may have taken in transitioning from abiotic building blocks to cell. Three illustrated paths in this picture serve as

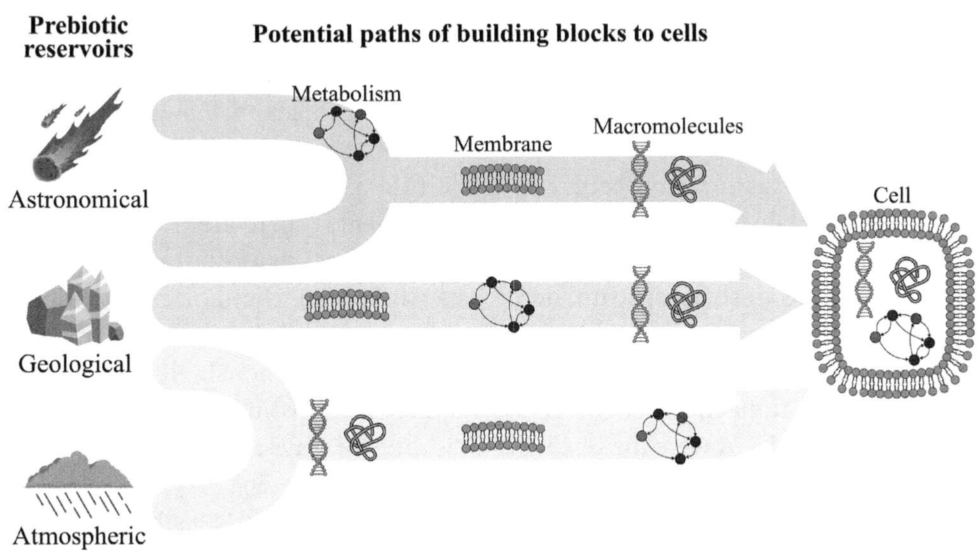


FIGURE 5.3 Potential paths from building blocks to cell, after Ditzler et al. (2022).

representatives of the many different proposed paths from building blocks to cells; see Figure 5.3, after Ditzler et al. (2022).

All known forms of life can be traced back through billions of years of evolution to a shared ancestral community, a community generally referred to as the last universal common ancestor. It may seem an impossible task to uncover the nature of this ancestral community and the communities that preceded it, but the sequences and structures of modern macromolecules are a surprisingly rich source of information that can be used to deduce some of the basic features of LUCA and the microbial communities that preceded it. There is considerable disagreement about what life might have been like up to and including the time of LUCA, especially with respect to the metabolic processes used by these communities, but several characteristics are broadly agreed upon. In general, features universally conserved in modern organisms were likely present in LUCA. On this basis, it can be safely inferred that the LUCA community had a sophisticated system of coded protein synthesis; that is, it had some form of lipid.

What about life before LUCA; can modern biology tell us anything specific about these even older communities? It appears that the answer is yes. Several modern macromolecules retain substructures within their conserved structural cores that likely supported early cellular life prior to lineage-specific divergence among these macromolecules. This allows us to use information embedded within modern biological systems to explore the nature of pre-LUCA life. Many of the details of the metabolic processes at play during the time of LUCA remain hidden, but some features of metabolism at that time can be clearly established by examining modern cells. Notably, we can safely infer that LUCA employed a complex, electrochemically driven membrane protein (ATP synthase) to synthesize ATP as an energy source for the cell. The ATP synthases are key players in extracting energy from the environment; are present in all cellular life; and are highly conserved in structure, function, and sequence and therefore were undoubtedly present at the time of LUCA. The complexity of their conserved structure and function clearly indicates that the ATP synthases present at the time of LUCA were the result of a prior multistep evolutionary process, with at least one clear instance of gene duplication in which both copies were incorporated into the ATP synthase. In addition to ATP synthesis, this protein assembly can operate in the reverse direction by consuming ATP to drive ion movement in the thermodynamically unfavorable direction, and it is possible that ATP-driven ion transport was its primary function in the pre-LUCA era. Regardless of how the ATP synthases evolved, their presence in LUCA has important implications for the nature of early cells. Specifically, because the ATP synthases rely on a thermodynamically favorable flow of ions across a membrane to support ATP synthesis, we can safely infer that organic cellular membranes that were relatively impermeable to ions existed prior to LUCA, from Ditzler et al. (2022)

They might also have originated in space and "hitchhiked" to the young Earth inside meteorites and comets during the "Late Heavy Bombardment" period. In support, nitriles and other precursor molecules for nucleotides, lipids, and amino acids have been found inside recent comets and meteors. The question is, where in space could these molecules have come from? Prime candidates are molecular clouds, which are dense and cold regions of the interstellar medium and are suitable for the formation of complex molecules.

One of the most intriguing questions in science has been the origin of life on Earth. Several hypotheses have been proposed to explain this. Since the time of Darwin's "warm little pond", a large number of different theories have been published. Currently, there is still a large gap in our understanding of how life on Earth was formed. The early history of Earth, roughly between 4.5 and 4.1 × $10^9$ years ago, is still largely shrouded in mystery, since there are no rocks from this era that we can use to reconstruct the early environments in which life was formed and evolved. The very first fossils that we know of from about 3.5 × $10^9$ years ago already show amazing complexity. How did this life come to be?

Kloprogge and Hartman (2022) discuss extensively the role of clays in the origin of life in particular the experiments. There are three groups of scientists dominating the search for the origin of life: the organic chemists (the Soup), the molecular biologists (RNA world), and the inorganic chemists (metabolism and transient-state metal ions), all of which have experimental adjuncts. According to Kloprogge and Hartman (2022) it is time for clays and the origin of life to have its experimental adjunct. The clay data coming from Mars and carbonaceous chondrites have necessitated a review of the role that clays played in the origin of life on Earth. The data from Mars have suggested that Fe-clays such as nontronite, ferrous saponites, and several other clays were formed on early Mars when it had sufficient water. This raised the question of the possible role that these clays may have played in the origin of life on Mars. This has put clays front and center in studies on the origin of life not only on Mars but also here on Earth. One of the major questions is: What was the catalytic role of Fe-clays in the origin and development of metabolism here on Earth? First, there is the recent finding of a chiral amino acid (isovaline) that formed on the surface of a clay mineral on several carbonaceous chondrites. This points to the formation of amino acids on the surface of clay minerals on carbonaceous chondrites from simpler molecules, such as $CO_2$, $NH_3$, and HCN. Additionally, there is the catalytic role of small organic molecules, such as dicarboxylic acids and amino acids found on carbonaceous chondrites, in the formation of Fe-clays themselves. Amino acids and nucleotides adsorb on clay surfaces on Earth and subsequently polymerize. All of these observations and more must be subjected to strict experimental analysis. This review provides an overview of what has happened and is now happening in the experimental clay world related to the origin of life. The emphasis is on smectite-group clay minerals, such as montmorillonite and nontronite (Kloprogge and Hartman 2022).

According to Preiner et al. (2019), if life on Earth started out in geochemical environments like hydrothermal vents, then it started out from gasses like $CO_2$, $N_2$, and $H_2$. Anaerobic autotrophs still live from these gases today, and they still inhabit the Earth's crust. In the search for connections between abiotic processes in ancient geological systems and biotic processes in biological systems, it becomes evident that chemical activation (catalysis) of these gasses and a constant source of energy are key. The $H_2$–$CO_2$ redox reaction provides a constant source of energy and anabolic inputs, because the equilibrium lies on the side of reduced carbon compounds. Identifying geochemical catalysts that activate these gasses on route to nitrogenous organic compounds and small autocatalytic networks will be an important step towards understanding prebiotic chemistry that operates only on the basis of chemical energy, without input from solar radiation. So, if life arose in the dark depths of hydrothermal vents, then understanding reactions and catalysts that operate under such conditions is crucial for understanding origins. This is presented in Figure 5.4; that is, purely geochemical reactions such as $CO_2$ fixation with $H_2$ can give rise to autocatalytic networks and proto/metabolism, as long as energy is released. Kinetically controlled reactions build up a specific set of products, which interact further to form an autocatalytic network that serves as a basis for higher complexity. C1, C2, and C3 represent carbon compounds with 1, 2, or 3 carbon atoms such as formyl groups, acetyl groups, and pyruvate.

Kobayashi et al. (2023) discussed the formation of amino acids and carboxylic acids in weakly reducing planetary atmospheres by solar energetic particles from the young Sun. According to them, life most likely started during the Hadean Eon; however, the environmental conditions which contributed to the complexity

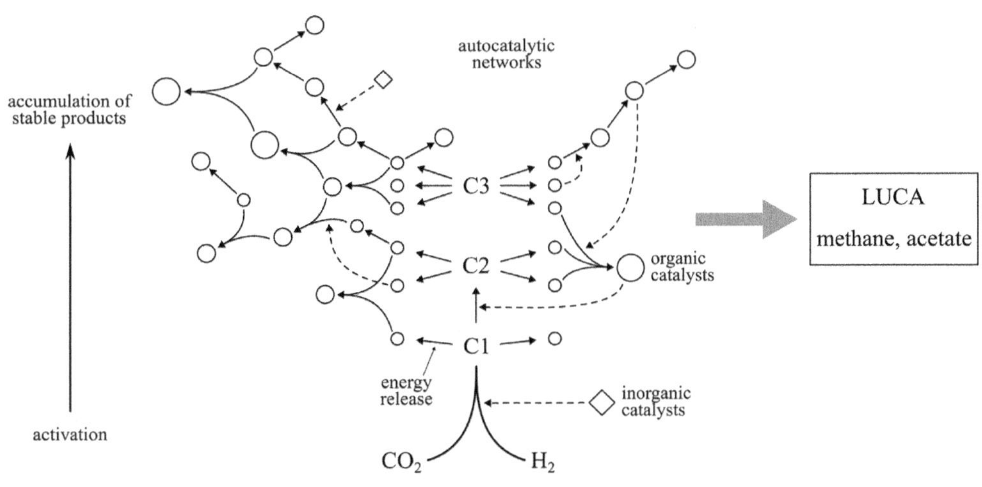


FIGURE 5.4 Purely geochemical reactions such as $CO_2$ fixation with $H_2$ give rise to autocatalytic networks and proto-metabolism.

of its chemistry are poorly known. A better understanding of various environmental conditions, including global (heliospheric) and local (atmospheric, surface, and oceanic), along with the internal dynamic conditions of the early Earth, are required to understand the onset of abiogenesis. Herein, we examine the contributions of galactic cosmic rays (GCRs) and solar energetic particles (SEPs) associated with superflares from the young Sun to the formation of amino acids and carboxylic acids in weakly reduced gas mixtures representing the early Earth's atmosphere. They also compare the products with those introduced by lightning events and solar ultraviolet light. In a series of laboratory experiments, they detected and characterized the formation of amino acids and carboxylic acids via proton irradiation of a mixture of carbon dioxide, methane, nitrogen, and water in various mixing ratios. These experiments show the detection of amino acids after acid hydrolysis when 0.5% (v/v) of initial methane was introduced to the gas mixture. In the set of experiments with spark discharges (simulation of lightning flashes) performed for the same gas mixture, they found that at least 15% methane was required to detect the formation of amino acids, and no amino acids were detected in experiments via UV irradiation, even when 50% methane was used. Carboxylic acids were formed in non-reducing gas mixtures (0% methane) by proton irradiation and spark discharges. Hence, they suggest that GCRs and SEP events from the young Sun represent the most effective energy sources for the prebiotic formation of biologically important organic compounds from weakly reducing atmospheres. Since the energy flux of space weather, which generated frequent SEPs from the young Sun in the first 600 million years after the birth of the solar system, was expected to be much greater than that of GCRs, they conclude that SEP-driven energetic protons are the most promising energy sources for the prebiotic production of bioorganic compounds in the atmosphere of the Hadean Earth.

### 5.3.2 Evolution

The distribution of life on Earth is captive, in part, to the distribution of the 20 or so bio-essential nutrient elements—many relatively rare—that are critical components of DNA, RNA, enzymes, and other biomolecules. Having substantially unraveled this relationship in today's oceans, biogeochemists are beginning to examine how it evolved over the ~4-billion-year history of life on Earth.

Changes in the availability of bio-essential elements must have shaped the evolution of life. For example, fossil evidence suggests that the ecological diversification of eukaryotes broadly coincided with rising redox potential of the deep oceans after ~$800 \times 10^6$ years ago and hence with increases in zinc, molybdenum, and other elements and decreases in iron, manganese, and cobalt. Bio-informatic analyses of protein-metal binding motifs encoded in genomes reveal that, relative to prokaryotes, eukaryotes require more zinc and less iron, manganese, and cobalt. Eukaryotes also require molybdenum for nitrate assimilation and can use zinc in place of cobalt in the carbonic anhydrase enzyme for carbon assimilation. Thus, it may be that eukaryotes emerged from ecological niches as bulk ocean chemistry shifted to favor their element requirements. Analogous logic may explain the rise of red eukaryotic phytoplankton after $250 \times 10^6$, years ago, after Anbar (2008).

The ineluctable requirement for the trans-iron elements molybdenum and/or tungsten in the origin of life is discussed in a paper by Schoepp-Cothenet et al. (2012). A vast number of enzymes rely on metal cofactors for catalysis and/or redox conversions. Different types of such "metallo-enzymes" appear to have evolutionary roots which reach more or less deeply into the distant past of life on Earth. The varying depths of evolutionary pedigrees likely reflect constraints imposed by paleo-geochemistry such as abundances and solubilities of metals in specific environments and eons of our planet's history. Iron-sulfur cluster-containing ferredoxins, for example, probably go back to the very origin of life. Copper enzymes, by contrast, are argued to have emerged only after the oxygenation of the environment some 2.5 billion years ago due to its insolubility under anaerobic conditions.

An evolutionary tree of key enzymes from the complex-iron-sulfur-molybdoenzyme (CISM) superfamily distinguishes "ancient" members, that is, enzymes present already in the last universal common ancestor of prokaryotes, from more recently evolved subfamilies. The majority of the presented subfamilies and, as a consequence, the molybdo-enzyme superfamily as a whole, appear to have existed in LUCA. The results are discussed with respect to the nature of bioenergetic substrates available to early life and to problems arising from the low solubility of molybdenum under conditions of the primordial Earth, after Schoepp-Cothenet et al. (2012).

Two apparently irreconcilable models dominate research into the origin of eukaryotes. In one model, amitochondrial proto-eukaryotes emerged autogenously from the last universal common ancestor of all cells. Proto-eukaryotes subsequently acquired mitochondrial progenitors by the phagocytic capture of bacteria. In the second model, two prokaryotes, probably an archaeon and a bacterial cell, engaged in prokaryotic endosymbiosis, with the species resident within the host becoming the mitochondrial progenitor. Both models have limitations. A search was undertaken for alternative routes towards the origin of eukaryotic cells in a paper by Bateman (2020). The question was addressed by considering classes of potential pathways from prokaryotic to eukaryotic cells based on considerations of cellular topology. Among the solutions identified, one, called here the "third-space model", has not been widely explored. A version is presented in which an extracellular space (the third-space), serves as a proxy cytoplasm for mixed populations of archaea and bacteria to "merge" as a transitionary complex without obligatory endosymbiosis or phagocytosis and to form a precursor cell. Incipient nuclei and mitochondria diverge by division of labor. The third-space model can accommodate the reorganization of prokaryote-like genomes to a more eukaryote-like genome structure. Nuclei with multiple chromosomes and mitosis emerge as a natural feature of the model. The model is compatible with the loss of archaeal lipid biochemistry while retaining archaeal genes and provides a route for the development of membranous organelles such as the Golgi apparatus and endoplasmic reticulum. The three models are schematically shown in Figure 5.5, after Bateman (2020).

In the autogenesis model, proto-eukaryotes, Archaea, and Bacteria all emerge from a last universal common ancestor. The proto-eukaryotes then acquire a bacterial mitochondrial precursor by phagocytosis. In prokaryotic endosymbiosis, only Archaea and Bacteria connect directly to a last universal ancestor. Eukaryotes emerge as the result of endosymbiosis of an archaeon (the host) with a eubacterium (the passenger) that subsequently evolves into the eukaryotic mitochondrion. Massive horizontal gene transfer (HGT) then transfers the majority of prokaryote-derived genes to the emerging eukaryote. In the strong third-space model, only

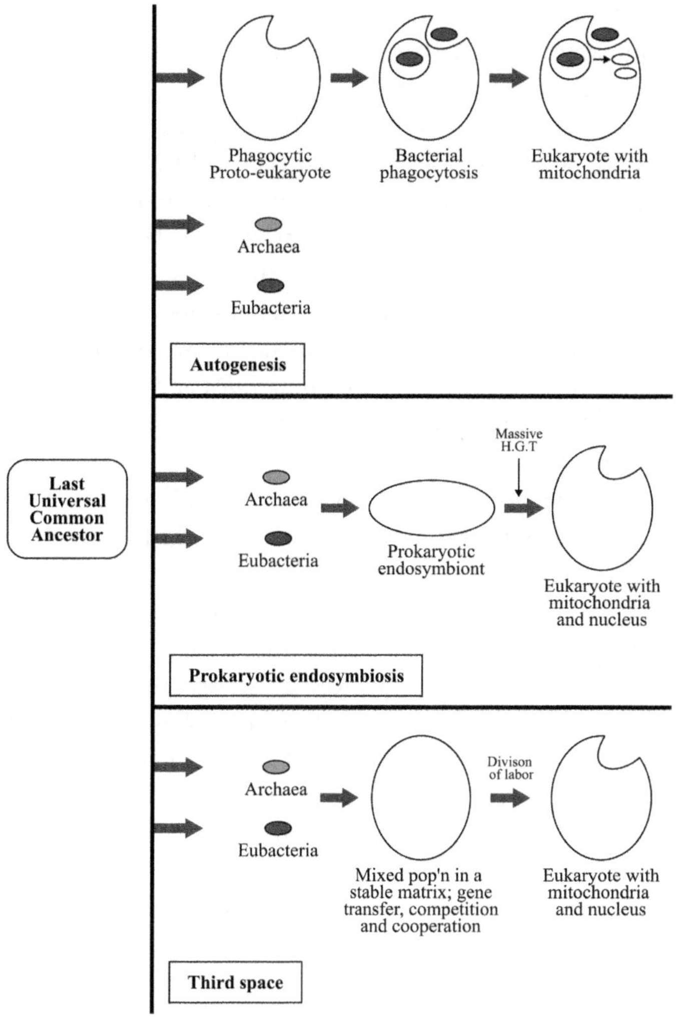


FIGURE 5.5 A comparison of autogenous and prokaryotic endosymbiotic models of eukaryogenesis with third-space model, after Bateman (2020).

Archaea and Bacteria link directly to a last universal ancestor. By co-existing in a closed and stable third-space, mitochondria and nuclei emerge by division of labor, gene transfer, and cell fusion.

Bacteria and archaea are taxonomically and metabolically the most diverse and abundant organisms on Earth. Although most model bacteria and archaea are small, some remarkably large cells, referred to as giant bacteria, are evident in at least four phyla and have cellular sizes in the range of tens or even hundreds of micrometers; see Ionescu and Bizic (2022). In a recent paper, Volland et al. (2022) studied *Thiomargarita magnifica*, a bacterium that has an average cell length greater than 9000 micrometers and is visible to the naked eye. They show that this centimeter-long bacterium represents individual cells with genetic material and ribosomes compartmentalized into a metabolically active, membrane-bound organelle.

According to Ionescu and Bizic (2022), giant bacteria are polyploid, meaning that their cells contain large numbers of genome copies ranging from tens to tens of thousands that are dispersed throughout the cell, supporting the local need for molecular machineries and overall cellular growth. With its number of genome copies being one order of magnitude above that of other giant bacteria, *Thiomargarita magnifica* accounts for the highest estimated number of genome copies within a single cell.

The discovery by Volland et al. (2022) suggests that large and more complex bacteria may be hiding in plain sight. We should keep in mind that cells of most bacteria and archaea are ~2 mm, and eukaryotic cells are usually between 10 and 20 mm, with some of the largest single-cell eukaryotes reaching 3 to 4 cm (Dumais et al. 2000).

Following Müller et al. (2022), the RNA world concept is one of the most fundamental pillars of the origin of life theory. It predicts that life evolved from increasingly complex self-replicating RNA molecules. The question of how this RNA world then advanced to the next stage, in which proteins became the catalysts of life and RNA reduced its function predominantly to information storage, is one of the most mysterious chicken-and-egg conundrums in evolution. Müller et al. (2022) showed that non-canonical RNA bases, which are found today in transfer and ribosomal RNAs and are considered relics of the RNA world, are able to establish peptide synthesis directly on RNA. The discovered chemistry creates complex peptide-decorated RNA chimeric molecules, which suggests the early existence of an RNA–peptide world from which ribosomal peptide synthesis may have emerged. The ability to grow peptides on RNA with the help of non-canonical vestige nucleosides offers the possibility of an early co-evolution of covalently connected RNAs and peptides, which then could have dissociated at a higher level of sophistication to create the dualistic nucleic acid–protein world that is the hallmark of all life on Earth.

According to Oliver et al. (2021), oxygenic photosynthesis starts with the oxidation of water to $O_2$, a light-driven reaction catalyzed by photosystem II. Cyanobacteria are the only prokaryotes capable of water oxidation, and therefore, it is assumed that the origin of oxygenic photosynthesis is a late innovation relative to the origin of life and bioenergetics. However, when exactly water oxidation originated remains an unanswered question. In this article, Oliver et al. (2021) used phylogenetic analysis to study a gene duplication event that is unique to photosystem II: the duplication that led to the evolution of the core antenna subunits CP43 and CP47. They compare the changes in the rates of evolution of this duplication with those of some of the oldest well-described events in the history of life: the duplication leading to the Alpha and Beta subunits of the catalytic head of ATP synthase and the divergence of archaeal and bacterial RNA polymerases and ribosomes. They also compare it with more recent events such as the duplication of Cyanobacteria-specific FtsH metalloprotease subunits and the radiation leading to Margulisbacteria, Sericytochromatia, Vampirovibrionia, and other clades containing anoxygenic phototrophs. They demonstrate that the ancestral core duplication of photosystem II exhibits patterns in the rates of protein evolution through geological time that are nearly identical to those of the ATP synthase, RNA polymerase, or the ribosome. Furthermore, they use ancestral sequence reconstruction in combination with comparative structural biology of photosystem subunits to provide additional evidence supporting the premise that water oxidation had originated before the ancestral core duplications. Their work suggests that photosynthetic water oxidation originated closer to the origin of life and bioenergetics than can be documented based on phylogenetic or phylogenomic species trees alone.

Life, as we know it, is (H-C-N-O) based and relies on the number of bulk (Na-Mg-P-S-Cl-K-Ca) and trace elements (Li-B-F-Si-V-Cr-Mn-Fe-Co-Ni-Cu-Zn-As-Se-Mo-I-W). The reference man is defined as being between 20 and 30 years of age, weighing 70 kg, 170 cm in height, and from a climate with an average temperature of from 10° to 20°C. He is a Caucasian and is a Western European or North American in habitat and custom. Elemental content (total body, in grams) is shown in Table 5.1, modified from reference ICRP 1975.

Global repeating events (mass extinctions of marine organisms, meteorite impacts and geomagnetic reversals) in the history of the Earth and the motion of the Sun in the galaxy are discussed in detail by Goncharov and Orlov (2003). Thirteen of 16 mass-extinction events can be described by a repetition interval of 183 ± 3 million years. This is in agreement with the anomalistic period (interval between two subsequent passages of the

TABLE 5.1 Elemental Content of Reference Man (total body/g)

| | | | | | | | | |
|---|---|---|---|---|---|---|---|---|
| H | C | N | O | | | | | |
| 7.0 × 10³ | 1.6 × 10⁴ | 1.8 × 10³ | 4.3 × 10⁴ | | | | | |
| **Na** | **Mg** | **P** | **S** | **Cl** | **K** | **Ca** | | |
| 1.0 × 10² | 1.9 × 10¹ | 7.8 × 10² | 1.4 × 10² | 9.5 × 10¹ | 1.4 × 10² | 1.0 × 10³ | | |
| **Li** | **B** | **F** | **Si** | **V** | **Cr** | **Mn** | **Fe** | |
| 6.7 × 10⁻⁴ | <2.0 × 10⁻² | 2.6 | 1.0 | <1.8 × 10⁻² | <6.6 × 10⁻³ | 1.2 × 10⁻² | 4.2 | |
| **Co** | **Ni** | **Cu** | **Zn** | **As*** | **Se** | **Mo** | **I** | **W** |
| <1.5 × 10⁻³ | 1.0 × 10⁻² | 7.2 × 10⁻² | 2.3 | 1.8 × 10⁻² | 1.3 × 10⁻² | <9.5 × 10⁻³ | 1.3 × 10⁻² | 2.0 × 10⁻⁵ |

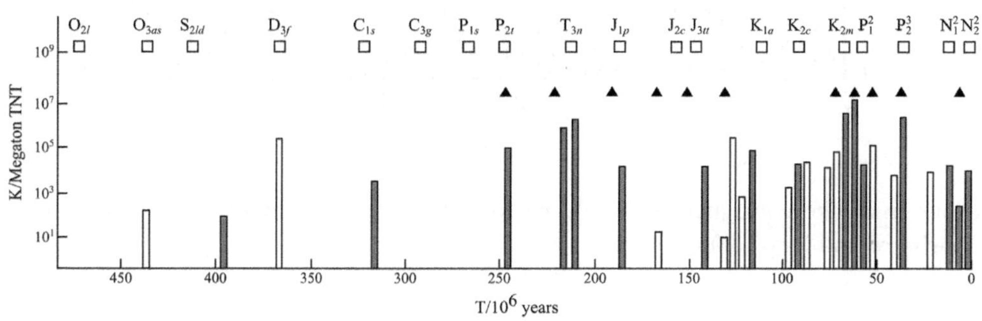


FIGURE 5.6 Dates of mass extinctions (squares), impact events (histogram), peaks in the frequency of geomagnetic reversals (triangles), and solar passages through the galactic plane in the AM model (circles). Shaded histogram columns indicate impact events that coincide with mass extinctions, after Goncharov and Orlov (2003).

Sun through the apocenter of its galactic orbit) in the model of Allen and Martos (1986). Figure 5.6 presents a histogram of the energy releases of the 72 impact events binned in intervals of $5 \times 10^6$ years. It is obvious that many of the epochs of mass extinction coincide with epochs of substantial energy release due impact of massive bodies on the Earth's surface. There some exceptions suggesting that there is not always global mass extinction after a collision of the Earth with a massive body.

The maximum in the distribution of the deviations of the dates of mass extinctions from the nearest dates of impacts of large, crater-forming bodies is close to zero, providing evidence that many such events are correlated. As a rule, extinctions follow impact events. The impacts of large bodies have occurred most often when the solar system passes through the galactic plane, while mass extinctions occur more often at some distance from the galactic plane (about 40 pc). As a rule, intervals of increases in the frequency of geomagnetic reversals coincide with dates of impacts of large bodies. At the same time, these intervals do not show a clear correlation with the dates of mass extinctions. The intensity of mass extinctions, like the energy released by impacts, is consistently higher in periods when the Sun is moving from the apocenter toward the pericenter of its orbit than when it is moving from the pericenter toward the apocenter. Thus, there is evidence for a variety of relationships between repeating global events in the Phanerozoic and the motion of the Sun in the galaxy.

To establish the epochs of maxima in the frequency of geomagnetic reversals, which may reflect the Earth's reaction to its passage through regions of various densities in the galaxy, Goncharov and Orlov (2003) used the master partial list of reversals as referenced in compilation by Schreider et al. (2011), which corresponds to the geological scale time proposed by Harland et al. (1989). Figure 5.7 shows a histogram of the geomagnetic reversal frequency binned in intervals of $5 \times 10^6$ years constructed using these data. There are two broad maxima approximately Gaussian in shape separated by $(160 \pm 30) \times 10^6$ years, which is close to the anomalistic period. In addition, some

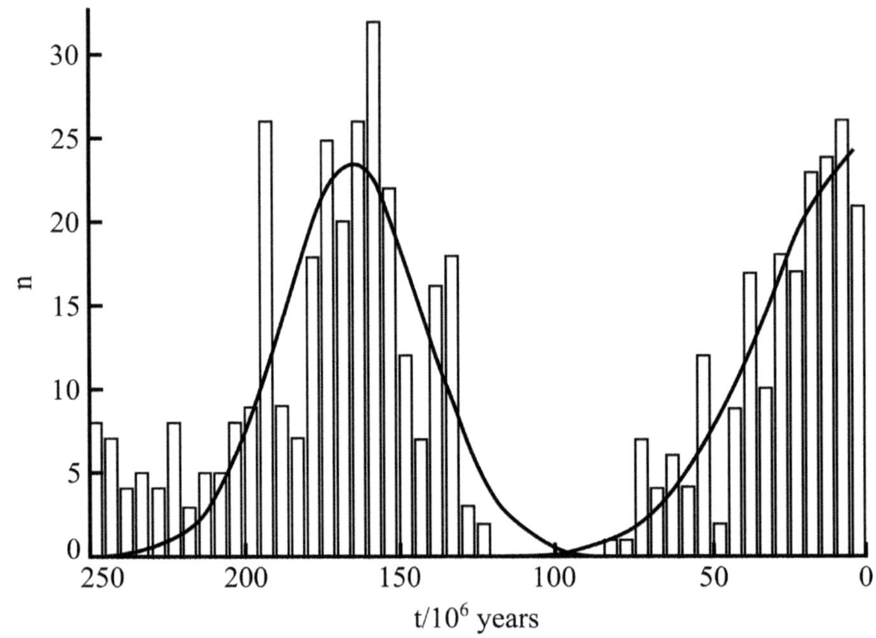


FIGURE 5.7 Distribution of 506 geomagnetic reversals in geological time. The smooth curves show an approximation using Gaussians, after Goncharov and Orlov (2003).

shorter peaks in the frequency of geomagnetic reversals are detected at the 2σ significance level with respect to their neighboring minima, as well as less significant peaks. These maxima coincide with epochs of intense energy release associated with impacts (Figure 6.7), with the most massive bodies playing a dominant role in this correlation. The only exception is the maximum of the geomagnetic-reversal frequency at $152.5 \times 10^6$ years, for which no corresponding impact events have been detected, possibly due to incompleteness in the list of such events.

The positions of the minima and maxima in Gaussian functions approximating the frequency distribution for geomagnetic reversals also agree with the times of passage of the Sun through the apocenter and pericenter, respectively, of its galactic orbit in this model. Long-period variations in the frequency of geomagnetic reversals are correlated with the orbital motion of the Sun, and increases in the frequency of geomagnetic reversals are correlated with impacts. Mass extinctions are correlated with the impacts of large bodies, whose motions may have been perturbed by clouds of interstellar material concentrated toward the galactic plane and by the shock front associated with the Perseus spiral arm through which the solar system passes; see Figure 5.8.

Dates of entries into the Perseus arm may be correlated with the most intense extinction events, such as $O_{3as}$ ($439 \times 10^6$ years ago), $P_{2t}$ ($250 \times 10^6$ years ago), and $K_{2m}$ ($65 \times 10^6$ years ago). Exits from the arm may also be correlated with the intense extinction events $D_{3s}$ ($367 \times 10^6$ years ago), $T_{3n}$ ($310 \times 10^6$ years ago), and $P^2_1$ ($56 \times 10^6$ years ago). If we associate these extinctions with epochs of entries into and exits from the Perseus spiral arm, this suggests interaction with the arm shock front as the mechanism initiating perturbations. The shortening of the interval between entries and exits during each successive orbit (72, 40, and 9 million years, respectively) may indicate that the arm is "running away" from the solar system with each new approach near the solar orbital apocenter.

Phylogenetic relationships are inferred principally from two classes of data: morphological and molecular. Currently, most phylogenies of extant taxa are inferred from molecules and when morphological and molecular trees conflict the latter are often preferred. Although supported by simulations, the superiority of molecular trees has rarely been assessed empirically. Oyston et al. (2022) tested phylogenetic accuracy using two independent data sources: biogeographic distributions and fossil first occurrences. For 48 pairs of morphological and molecular trees

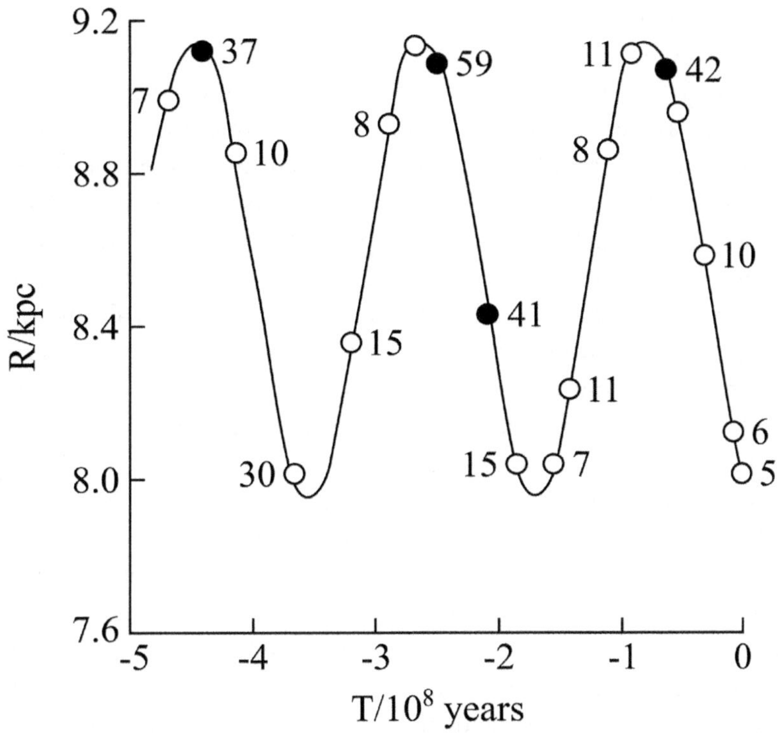


FIGURE 5.8 Time evolution of the distance between the Sun and the galactic rotational axis in the AM model and dates for mass extinctions (empty circles). The four most significant extinctions are marked by dark circles. The numbers next to the symbols indicate the percentage of species of marine organisms dying out.

they showed that, on average, molecular trees provided a better fit to biogeographic data than their morphological counterparts and that biogeographic congruence increased over research time. They found no significant differences in stratigraphic congruence between morphological and molecular trees. These results have implications for understanding the distribution of homoplasy in morphological data sets, the utility of morphology as a test of molecular hypotheses, and the implications of analyzing fossil groups for which molecular data are unavailable.

Moll et al. (2019) based on the entire elemental inventory of the otolith core region, indicating the chemical signature of the spawning area, developed a unique elemental fingerprinting index (EFI), allowing comparisons of multi-elemental chemical signatures from within and between herring juvenile areas. The elemental fingerprinting index can visualize chemical otolith signatures for each spawning area by radar plots based on the relative mean proportion of each element to the highest concentration found among the total of $10^5$ individuals included in their study. They modified a well-established ecological overlap index (Schoener index) (see Schoener 1974) to create the EFI, a novel measure of similarity between two fish individuals considering numerous chemical signatures in their otoliths.

Rambo et al. (2022) reported on genomes of six viruses that infect Asgard archaea from deep-sea sediments. Asgard archaea are globally distributed prokaryotic microorganisms related to eukaryotes; however, viruses that infect these organisms have not been described. Here, using metagenome sequences recovered from deep-sea hydrothermal sediments, they characterized six relatively large (up to 117 kb) double-stranded DNA (dsDNA) viral genomes that infected two Asgard archaeal phyla, Lokiarchaeota and Helarchaeota. These viruses encode *Caudovirales*-like structural proteins, as well as proteins distinct from those described in known archaeal viruses. Their genomes contain around 1–5% of genes associated with eukaryotic nucleocytoplasmic large DNA viruses (NCLDVs) and appear to be capable of semi-autonomous genome replication, repair, epigenetic modifications and transcriptional regulation.

Moreover, Helarchaeota viruses may hijack host ubiquitin systems similar to eukaryotic viruses. Genomic analysis of these Asgard viruses reveals that they contain features of both prokaryotic and eukaryotic viruses and provides insights into their potential infection and host interaction mechanisms.

High-throughput sequencing projects generate genome-scale sequence data for species-level phylogenies. However, state-of-the-art Bayesian methods for inferring timetrees are computationally limited to small datasets and cannot exploit the growing number of available genomes. In the case of mammals, molecular-clock analyses of limited datasets have produced conflicting estimates of clade ages with large uncertainties, and thus the timescale of placental mammal evolution remains contentious. Álvarez-Carretero et al. (2021) developed a Bayesian molecular-clock dating approach to estimate a time tree of 4705 mammal species integrating information from 72 mammal genomes. They showed that increasingly large phylogenomic datasets produce diversification time estimates with progressively small uncertainties, facilitating precise tests of macro evolutionary hypotheses. For example, they confidently rejected an explosive model of placental mammal origination in the Paleogene and showed that crown Placentalia originated in the Late Cretaceous with unambiguous ordinal diversification in the Paleocene/Eocene. Their Bayesian methodology facilitates analysis of complete genomes and thousands of species within an integrated framework, making it possible to address hitherto intractable research questions on species diversifications. This approach can be used to address other contentious cases of animal and plant diversifications that require analysis of species-level phylogenomic datasets.

Whole genome duplication (WGD) is a dramatic evolutionary event generating many new genes and which may play a role in survival through mass extinctions. Ancient WGD events have occurred across the tree of life and are especially well studied in plants, yeast, and vertebrates. These events are often hypothesized to have facilitated evolutionary success through provision of the raw genetic materials for phenotypic innovation and species diversification. A key evolutionary process after WGD is rediploidization—the transition of a polyploid, usually tetraploid, genome to a more stable diploid state. Importantly in this context, WGD events are derived from either hybridization of two different parent species (allopolyploidization) or from doubling of the same genome at the intra-species/individual level (autopolyploidization), each with distinct cytogenetic outcomes. Paddlefish and sturgeon are sister lineages that both show genomic evidence for ancient WGD. Until now this has been interpreted as two independent WGD events due to a preponderance of duplicate genes with independent histories. Redmond et al. (2023) showed that although there is indeed a plurality of apparently independent gene duplications, these derive from a shared genome duplication event occurring well over 200 million years ago, likely close to the Permian-Triassic mass extinction period. This was followed by a prolonged process of reversion to stable diploid inheritance (rediploidization), that may have promoted survival during the Triassic-Jurassic mass extinction. They showed that the sharing of this WGD is masked by the fact that paddlefish and sturgeon lineage divergence occurred before rediploidization had proceeded even halfway. Thus, for most genes, the resolution to diploidy was lineage specific. Because genes are only truly duplicated once diploid inheritance is established, the paddlefish and sturgeon genomes are thus a mosaic of shared and non-shared gene duplications resulting from a shared genome duplication event (Redmond et al. 2023).

In the ongoing debates about eukaryogenesis—the series of evolutionary events leading to the emergence of the eukaryotic cell from prokaryotic ancestors—members of the Asgard archaea play a key part as the closest archaeal relatives of eukaryotes. However, the nature and phylogenetic identity of the last common ancestor of Asgard archaea and eukaryotes remain unresolved. Eme et al. (2023) analyzed distinct phylogenetic marker datasets of an expanded genomic sampling of Asgard archaea and evaluate competing evolutionary scenarios using state-of-the-art phylogenomic approaches. They found that eukaryotes are placed, with high confidence, as a well-nested clade within Asgard archaea and as a sister lineage to Hodarchaeales, a newly proposed order within Heimdallarchaeia. Using sophisticated gene tree and species tree reconciliation approaches, the authors showed that, analogous to the evolution of eukaryotic genomes, genome evolution in Asgard archaea involved significantly more gene duplication and fewer gene loss events compared with other archaea. They also inferred that the last common ancestor of Asgard archaea was probably a thermophilic chemolithotroph and that the lineage from which eukaryotes evolved adapted to mesophilic conditions and acquired the genetic potential to support a heterotrophic lifestyle. Their work provides key insights into the prokaryote-to-eukaryote transition and a platform for better understanding the emergence of cellular complexity in eukaryotic cells.

### 5.3.3 Are We Alone?

"Are we alone in the Universe?" is the question asked through the centuries. The concept of habitable zones in the Universe is an extension of dogmas accepted by scientific community that life originated on the planet Earth and that life anywhere in Universe must be similar to it. For example, Franck et al. (2001) approach this question with the help of a segment of the Drake equation and estimated the number of "Gaias" (i.e. extra-solar terrestrial planets with a globally acting biosphere). Their answer leads to the thoroughly educated guess that there should exist $N \approx 4.8 \times 10^5$ Gaias in the Milky Way! Even the authors of this article consider this a rather large number.

The question—are we alone in the Universe?—has been discussed for at least 2500 years. Initially philosophical and religious beliefs were used as arguments, but for over 100 years, the possible transfer of bacteria and other microorganisms from planets, comets or meteoroids to other planets—panspermia—has been seriously considered. Possible mechanisms for this transfer exist, and some microorganisms are known to be extremely resistant to very low temperatures, very high vacuum, and intense high-energy radiation. Hoyle and his colleagues have published many claims that interstellar grains are partly, or even wholly, bacteria and or diatoms and that life, evolution, and many diseases would not have occurred on Earth without panspermia. The evidence and interpretations presented have not been accepted by the general body of biologists or astronomers. The present situation is that panspermia is unlikely, so far completely unsupported, but is just possible for viable, and somewhat more so for dead, microorganisms, so it should be investigated.

In practice we have produced no clear affirmation of any life beyond our own planet yet. So we can ask Fermi's famous question, "Where is everybody?" The silence of the universe beyond Earth reveals a pattern of both human limitation and steadfast curiosity. Even as ambitious programs such as search for extraterrestrial intelligence (SETI) aim to solve the technological challenges, the results have thus far turned up empty for any signs of life in the galaxy. SETI is deploying machine-learning algorithms that filter out Earthly interference and spot signals humans might miss.

Jiang et al. (2022) postulate that an existential disaster may lie in wait as our society advances exponentially towards space exploration, acting as the Great Filter: a phenomenon that wipes out civilizations before they can encounter each other, which may explain the cosmic silence. Jiang et al. (2022) proposed several possible scenarios, including anthropogenic and natural hazards, both of which can be prevented with reforms in individual, institutional, and intrinsic behaviors. They also took into account multiple calamity candidates: nuclear warfare, pathogens and pandemics, artificial intelligence, meteorite impacts, and climate change. Each of these categories has various influences but lacks critical adjustment to accommodate to the high risk. The Great Filter has the potential to eradicate life as we know it, especially as our rate of progress correlates directly to the severity of our fall.

Habitability is a widely used word in the geoscience, planetary science, and astrobiology literature, but what does it mean? In this review on habitability, we define it as the ability of an environment to support the activity of at least one known organism. We adopt a binary definition of "habitability" and a "habitable environment". An environment either can or cannot sustain a given organism. However, environments such as entire planets might be capable of supporting more or less species diversity or biomass compared with that of Earth. A clarity in understanding habitability can be obtained by defining instantaneous habitability as the conditions at any given time in a given environment required to sustain the activity of at least one known organism and continuous planetary habitability as the capacity of a planetary body to sustain habitable conditions on some areas of its surface or within its interior over geological timescales. We also distinguish between surface liquid water worlds (such as Earth) that can sustain liquid water on their surfaces and interior liquid water worlds, such as icy moons and terrestrial-type rocky planets with liquid water only in their interiors. This distinction is important since, while the former can potentially sustain habitable conditions for oxygenic photosynthesis that leads to the rise of atmospheric oxygen and potentially complex multicellularity and intelligence over geological timescales, the latter are unlikely to. Habitable environments do not need to contain life. Although the decoupling of habitability and the presence of life may be rare on Earth, it may be important for understanding the habitability of other planetary bodies, after Cockell et al. (2016).

An example of habitable zone (shading) for the solar system at three different time steps is shown in Figure 5.9. The orbits of the three terrestrial planets, Venus, Earth, and Mars, are shown. The solid lines describe the evolution of the inner and outer boundary of the HZ, after Franck et al. (2001).

We should start this discussion with the famous question Enrico Fermi asked at the Los Alamos National Laboratory in 1950: "Where are they?", following Jones (1985). A group of scientists at NASA (see Green et al. 2021) is encouraging the scientific community to establish a framework that provides a context for findings related to the search of life. The authors proposed creating a scale (see Table 5.2) for evaluating and combining different evidence that would ultimately lead to an answer to the question: Are we alone in the Universe? Fermi was referring to the enormous size and number of stars and galaxies in the sky, which has not resulted in any evidence of life. The lack of evidence of life in our galaxy resulted in what Sandberg et al. (2018) called the Fermi paradox or Fermi observation.

The main point of the Fermi paradox is that the number of sites where alien civilizations could emerge is so big that by any reasonable probability of emergence some would have been detected. This has been parametrized by the so-called Drake equation (see Drake 2015), the number of detectable civilizations being a product of seven factors

The National Academies of Sciences, Engineering, and Medicine report an Astrobiology Strategy for the Search for Life in the Universe (National Academies 2019) in which it is proposed that the utility of various life-detection measurements be evaluated primarily with respect to their potential to yield false-negative and/or false-positive interpretations.

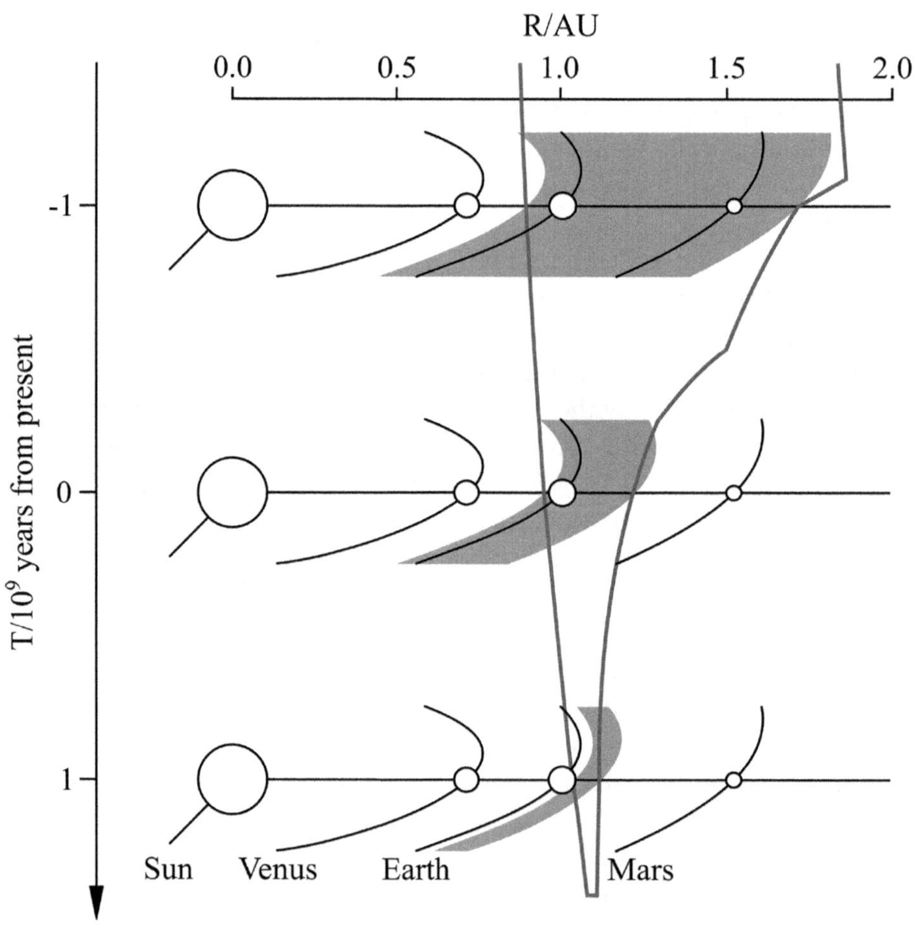


FIGURE 5.9 Habitable zone (shading) for the solar system at three different time steps, after Franck et al. (2001).

TABLE 6.2 The CoLD Scale. An Example of a Progressive Scale for Communicating the Nature of Results That May Provide Evidence for Life

| Qualitative Assessment | Measurement Indicators |
|---|---|
| Confirmation of the presence of biology | **Level 7:** Independent follow-up of predicted biological behavior in the environment |
| Alternative hypothesis eliminated | **Level 6:** Future observations that rule out alternative hypotheses proposed after original announcement |
| Independent biosignatures | **Level 5:** Additional, independent signal from biology detected |
| Abiotic false-positive discrimination | **Level 4:** All known non-biological sources of signal shown to be implausible in that environment |
| Environmental relevance | **Level 3:** Demonstration or prediction of biological production of signal in the environment of detection |
| Detection of biological signal | **Level 2:** Contamination ruled out |
| | **Level 1:** Detection of a signal known to result from a biological activity |

Here we should mention the Proceedings of the Workshop by National Academies, Rapporteur: Schmitt (2017). The workshop was organized to cover four regions of parameters space with example of the type of world in which we might find it:

- In situ detection in the solar system of life as we know it (for example, planet Mars);
- In situ detection in the solar system of life as we don't know it (for example, Titan);
- Remote detection on exoplanets of life as we know it (for example, "exo-Earths");
- Remote detection on exoplanets of life as we don't know it (location unknown).

Assuming we are not alone, advanced extraterrestrial civilizations may produce observable and distinctive techno-signatures when they migrate to escape existential threats or for other reasons. Extraterrestrial civilizations may migrate inside planetary systems, away from their post-main-sequence stars. They may also migrate in the galaxy. Extraterrestrial civilizations would most likely face serious or insurmountable technical problems when using spacecraft for interstellar migration of large populations. Therefore, they may use other methods of interstellar travel. Some extraterrestrial civilizations may build stellar engines (Forgan 2018). Some extraterrestrial civilizations may migrate to planetary systems of flyby stars; see Hansen and Zuckerman (2021).

Advanced extraterrestrial civilizations may produce observable techno-signatures when they migrate inside their planetary systems and to other planetary systems. Collectively, the search for techno-signatures produced by migrating extraterrestrial civilizations can be described as the search for migrating extraterrestrial intelligence. Romanovskaya (2022) proposed that extraterrestrial civilizations might use free-floating planets as interstellar transportation to reach, explore, and colonize planetary systems. The author also discussed possible techno-signatures and artifacts that might be produced by extraterrestrial civilizations using free-floating planets for interstellar migration and interstellar colonization, as well as strategies for the search for their techno-signatures and artifacts, together with the techno-signatures that might be produced by extraterrestrial civilizations using other methods of interstellar migration and colonization. As an example, the star GJ 433, which experienced a close flyby a few thousand years ago, was considered. Possible techno-signatures of migrating civilizations were divided into groups to highlight similarities and differences among techno-signatures produced by civilizations using different methods of migration and interstellar colonization. It follows from the comparison, for example, that interstellar migration using flybys of stars can blend with interstellar migration using free-floating planets (Romanovskaya 2022).

Tremblay et al. (2022) have done a search for techno-signatures toward the galactic center at 150 MHz; using the Murchison Wide-field Array over two nights, they integrated 7 hours of data toward the galactic center (centered on the position of Sagittarius A*) with a total field-of-view of 200 $deg^2$. In the paper, they presented a targeted search toward 144 exoplanetary systems at their best angular resolution (75 arc seconds). This represents the first techno-signature search at a frequency of 155 MHz toward the galactic center (their previous central frequencies have been lower). A blind search toward in excess of 3 million stars toward the galactic center and galactic bulge was also completed, placing an equivalent isotropic power limit $<1.1 \times 10^{19}$ W at the distance to the galactic center. No plausible techno-signatures were detected.

Discussions on what causes circular polarization (CP) observed in dusty objects can be seen quite often in scientific papers. Among the most popular explanations of CP formation are scattering of light on aligned elongated/irregular dust particles or on particles that contain homochiral molecules. If light has passed through dust in space and experienced CP formation, it could tell astronomers whether that dust contains homochiral molecules, which could be an indicator of interest to astrobiologists. Identifying homochiral molecules in space can provide important clues about the origins of life. Evidence from meteorites supports the idea that the origin of left-handed homochirality in amino acids used by biology on Earth is related to conditions in the early solar system. If the dust that formed our solar system only contained left-handed amino acids, it could explain why life on Earth developed a preference for these molecules in the first place. A survey of cosmic dust could reveal that homochirality is universal, but that doesn't mean that every system would be just like ours. It is likely that on other worlds, right-handed amino acids could dominate. It depends on the properties of the original magnetic field that aligned dust particles in star-forming regions.

Westby and Conselice (2020) presented a cosmic perspective on the search for life and examine the likely number of Communicating ExtraTerrestrial Intelligent (CETI) civilizations in our galaxy by utilizing the latest astrophysical information. Their calculation involves galactic star formation histories, metallicity distributions, and the likelihood of stars hosting Earth-like planets in their habitable zones, under specific assumptions which they describe as the astrobiological Copernican weak and strong conditions. These assumptions are based on the one situation in which intelligent, communicative life is known to exist—on our own planet. This type of life has developed in a metal-rich environment and has taken roughly $5 \times 10^9$ years to do so. They investigated the possible number of CETI civilizations based on different scenarios. At one extreme is the weak astrobiological Copernican scenario—such that a planet forms intelligent life sometime after $5 \times 10^9$ years, but not earlier. The other is the strong astrobiological Copernican scenario in which life must form between 4.5 and $5.5 \times 10^9$ years, as on Earth. In the strong scenario (under the strictest set of assumptions), they found there should be at least $36\pm^{175}{}_{32}$ civilizations within our galaxy: this is a lower limit, based on the assumption that the average lifetime, L, of a communicating civilization is 100 years (since we know that our own civilization has had radio communications for this time). If spread uniformly throughout the galaxy, this would imply that the nearest CETI is at most $17{,}000\pm^{33{,}600}{}_{10{,}000}$ lightyears away and most likely hosted by a low-mass M-dwarf star, likely far surpassing our ability to detect it for the foreseeable future and making interstellar communication impossible. Furthermore, the likelihood that the host stars for this life are solar-type stars is extremely small, and most would have to be M dwarfs, which may not be stable enough to host life over long timescales. They furthermore explore other scenarios and explain the likely number of CETI there are within the galaxy based on variations on their assumptions.

## 5.4 HABITABLE ZONES

Much of the Universe is clearly hostile to life, and only rare places offer even potential oases for its existence. Empty space, the interiors of stars, frigid gas clouds, the "surface" of gaseous planets like Jupiter—all must be lifeless. We cannot know for certain what the limits are for life's environments, but looking at what is needed to support Earth life provides a basis for estimating where in the Universe life might exist. We speculate in this manner with the understanding that we have a biased perspective—that of inhabitants of a planet that seems to provide a nearly perfect habitat. One of Earth's most basic life-supporting attributes is indeed its location, its seemingly ideal distance from the sun. In any planetary system, there are regions—distances from the central star—where a surface environment similar to the present state of Earth could occur. The favorable region or distance from the star is the basis for defining the "habitable zone" (referred to by astrobiologists as the HZ), the region in a planetary system where habitable Earth clones might exist. Since its introduction, the concept of the habitable zone has been widely adopted and has been the subject of several major scientific conferences. The defining aspect of the HZ is that it is the region where heating from the central star provides a planetary surface temperature at which a water ocean neither freezes over nor exceeds its boiling point. The actual width of the HZ depends on how Earth-like we decide a planet must be to be deemed habitable. Extreme events, such as the loss of oceans or a deep planetary freeze, may seem totally preposterous to Earthlings happily living in nearly ideal climatic conditions, but these events would surely occur if Earth were (on the one hand) slightly closer to or (on the other) slightly farther from the sun.

The "habitable zone" is defined as the distance from a star at which liquid water could exist on orbiting planets' surfaces and possibly support life. Since liquid water is essential for life on Earth, the definition of a habitable zone is based on the hypothesis that extraterrestrial life would share this requirement. Schematic presentation of the habitable zone of any star is shown in Figure 5.10.

Lammer et al. (2009) proposed four types of habitable zones:

Class I. habitable zone contains bodies on which stellar and geophysical conditions allow Earth-analog planets to evolve so that complex multi-cellular life forms may originate.

Class II. habitable zone includes bodies on which life may evolve but due to stellar and geophysical conditions that are different from Class I habitats, the planets rather evolve toward Venus or Mars type worlds where complex life forms may not develop.

Class III. Habitable zones are planetary bodies where subsurface water oceans exist which interact directly with a silicate rich core.

Class IV. Habitable zones have liquid water layers between two ice layers, or liquids above ice.

Recently, Lous et al. (2022) suggested that the concept of planetary habitability should be revisited and made more inclusive with respect to this classical definition. They argue that cold super-Earths that remain primordial, H-He-dominated atmosphere, could have surfaces that are warm enough to host liquid water. This would be due to the collision-influenced absorption of infrared light by hydrogen, which increases with pressure. It should be noted that a long-term potential for habitability of such planets has not been exploited yet.

In the search for life on extraterrestrial planets, it needs to be considered that life might manifest and thrive under conditions that would be considered extreme on Earth (Lous et al. 2022).

The provenance of oxygen on the Earth and other planets in the solar system is a fundamental issue. It has been widely accepted that the only prebiotic pathway to produce oxygen in the Earth's primitive atmosphere was via vacuum ultraviolet (VUV) photodissociation of $CO_2$ and subsequent two O atom recombination. Chang et al. (2021) provided experimental evidence of three-body dissociation (TBD) of $H_2O$ to produce O atoms in both $^1D$ and $^3P$ states upon VUV excitation using a tunable VUV free electron laser. Experimental results show that the TBD is the dominant pathway in the VUV $H_2O$ photochemistry at wavelengths between 90 and 107.4 nm. The relative abundance of water in the interstellar space with its exposure to the intense VUV radiation suggests that the TBD of $H_2O$ and subsequent O

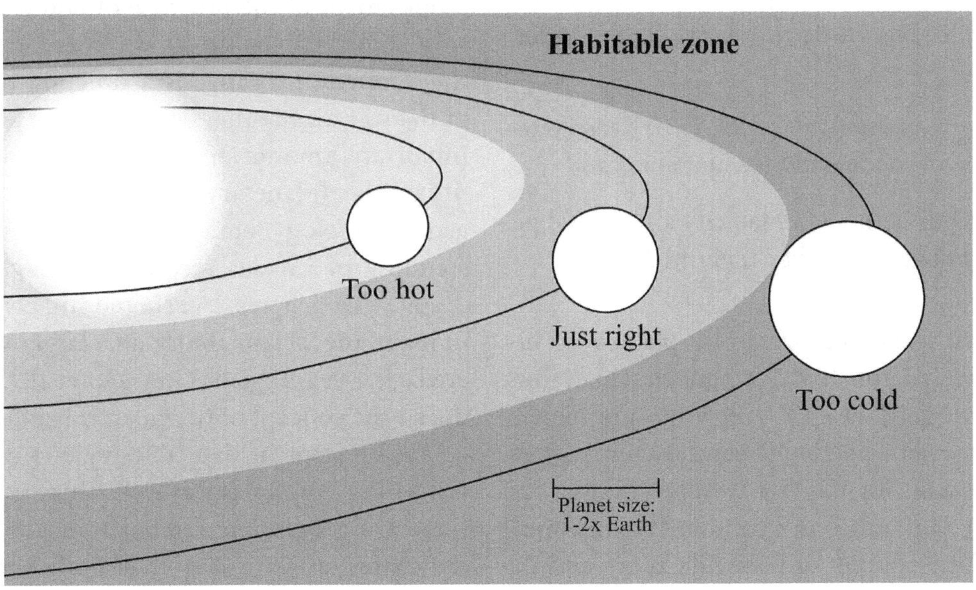


FIGURE 5.10 Schematic presentation of the habitable zone of any star.

atom recombination should be an important prebiotic $O_2$-production, which may need to be incorporated into interstellar photochemical models.

The concept of a circumstellar habitable zone (CHZ) is relatively well defined, being tightly related to the requirement of the presence of liquid water as a necessary condition for life as we know it; the corresponding temperature range is a function of the luminosity of the star and of the distance of the planet from it. On other hand, the concept of a galactic habitable zone (GHZ) considers the idea that various physical processes, which may favor the development or the destruction of complex life, may depend strongly on the temporal and spatial position in the Milky Way. For instance, the risk of a supernova explosion sufficiently close to represent a threat for life is, in general, larger in the inner galaxy than in the outer one and has been larger in the past than at present. Another example is offered by the metallicity (the amount of elements heavier than hydrogen and helium) of the interstellar medium, which varies across the Milky Way disk and which may be important for the existence of Earth-like planets. Indeed, the host stars of the ~180 extrasolar giant planets detected so far are, on average, more metal rich than stars with no planets in our cosmic neighborhood (Fischer and Valenti 2005). None of the expected fossil signatures of accretion are observed in stars with planets relative to the general sample:

1. metallicity does not appear to increase as the mass of the convective envelopes decreases;
2. subgiants with planets do not show dilution of metallicity;
3. no abundance variations for Na, Si, Ti, or Ni are found as a function of condensation temperature; and
4. no correlations between metallicity and orbital period or eccentricity could be identified.

We conclude that stars with extrasolar planets do not have an accretion signature that distinguishes them from other stars; more likely, they are simply born in higher-metallicity molecular clouds. Several other factors, potentially important for the GHZ, were discussed by Gonzalez (2005). The author also reviewed the historical development of the concept of habitable zones and the three types of habitable zones: CHZ, GHZ, and cosmic habitable age (CHA).

As broad as they are, even these three habitable zones still leave gaps. To paint a complete picture of the evolution of the habitability of the whole universe, it will be necessary to merge these presently disparate concepts. The GHZ and CHA are newer concepts and, not surprisingly, have received less attention. Both can benefit from ongoing surveys of the Milky Way and nearby and distant galaxies. Some of the most relevant results from these surveys include SN statistics, luminosity functions of galaxies and clusters of galaxies, and the evolution of the star formation rate. Continued theoretical studies of the effects of SNe and AGN outbursts on life, stellar dynamics in the Milky Way and in other galaxies, and galactic chemical evolution will also aid in refining the GHZ and CHA (Gonzales 2005).

### 5.4.1 Galactic Habitable Zones

According to Prantzos (2008), the concept of the galactic habitable zone was introduced a few years ago as an extension of the much older concept of the circumstellar habitable zone. However, the physical processes underlying the former concept are hard to identify and even harder to quantify. That difficulty does not allow us, at present, to draw any significant conclusions about the extent of the GHZ: it may well be that the entire Milky Way disk is suitable for complex life.

The modern study of the "habitability" of circumstellar environments started more than 60 years ago; see Huang (1959). The concept of a circumstellar habitable zone is relatively well defined, being tightly related to the requirement of the presence of liquid water as a necessary condition for life as we know it; the corresponding temperature range is a function of the luminosity of the star and of the distance of the planet from it. An important amount of recent work, drawing on various disciplines (planetary dynamics, atmospheric physics, geology, biology, etc.) refined considerably our understanding of various factors that may affect the CHZ; despite that progress, we should still consider the subject in its infancy. Habitability on a larger scale was considered a few years ago by Gonzalez et al. (2001), who introduced the concept of the galactic habitable zone.

The concept of habitable regions can be applied to our Milky Way galaxy as well. There is a possibility that there are geographic regions that can be plotted from the center of our galaxy that are habitable regions in a way analogous to the habitable zones around stars. Our galaxy is a spiral galaxy (the other types are elliptical and irregular galaxies). In most galaxies, the concentration of stars is highest in the center and diminishes away from the center. Spiral galaxies are dish shaped (round, but flat if viewed from the side), with branching arms when viewed from the top. However, viewed from the side, they are quite flat. Our galaxy has an estimated diameter of about 85,000 lightyears. Our sun is about 25,000 lightyears from the center, in a region between spiral arms where star density is quite low compared to the more crowded interior. In this position, we slowly orbit the central axis of the galaxy.

Like a planet revolving around a star, we maintain roughly the same distance from the galactic center, and this is fortunate. Our star—by chance—is located in the "habitable zone" of the galaxy. We suspect that the inner margins of this galactic habitable zone are defined by the high density of stars, the dangerous supernovae, and the energy sources found in the central region of our galaxy, whereas the outer regions of habitability are dictated by something quite different: not the flux of energy but the type of matter to found. At the present time, we cannot do more than crudely designate the limits of this habitable region. Its inner boundary is surely defined by celestial catastrophes occurring closer to the center, but we cannot yet estimate how close to the center of the galaxy that boundary is. Perhaps it extends 10,000 lightyears from the center, perhaps more. However, we do have at least a vague idea of the forces that impose this inner limit. Life is a very complex and delicate phenomenon that is easily destroyed by too much heat or cold and by too many gamma rays, X-rays, or other types of ionizing radiation. The center of any galaxy produces all of these. Among the lethal stellar members of any galaxy are the neutron stars called magnetars. These collapsed stars are small but astonishingly dense, and they emit X-rays, gamma rays, and other charged particles into space. Because energy dissipates as the square of distance, these objects are no threat to our planet. Closer to the center of the galaxy, however, their frequency increases. Any galactic center is a mass of stars, some lethal neutron stars, and it seems most unlikely that any form of life as we know it could exist nearby. An even greater threat comes from exploding stars known as supernovae. As stars grow old, they burn up their hydrogen and eventually collapse on themselves. Some of them then explode outward with terrific force. Any star going supernova would probably sterilize life within a radius of 1 lightyear of the explosion and affect life on planets as far as 30 lightyears away. The very number of stars in galactic centers increases the chances of a nearby supernova. Our sun and planet are protected simply by the scarcity of stars around us. The outer region of the galactic habitable zone is defined by the elemental composition of the galaxy. In the outermost reaches of the galaxy, the concentration of heavy elements is lower because the rate of star formation—and thus of element formation—is lower. Outward from the centers of galaxies, the relative abundance of elements heavier than helium declines. The abundance of heavy elements is probably too low to form terrestrial planets as large as Earth. As we shall see in the next chapter, our planet has a solid/liquid metal core that includes some radioactive material giving off heat. Both attributes seem to be necessary to the development of animal life: The metal core produces a magnetic field that protects the surface of the planet from radiation from space, and the radioactive heat from the core, mantle, and crust fuels plate tectonics, which in our view is also necessary for maintaining animal life on the planet. No planet such as Earth can exist in the outer regions of the galaxy. Not only is Earth in a rare position in its galaxy; it may also be fortunate (at least as far as having life is concerned) in being in a spiral rather than an elliptical galaxy. Elliptical galaxies are regions with little dust, which apparently exhibit little new star formation. The majority of stars in elliptical galaxies are nearly as old as the universe. The abundance of heavy elements is low, and although asteroids and comets may occur, it is doubtful that there are full-size planets.

The boundaries of the habitable zones around main-sequence stars are usually estimated using the so-called one-dimensional (1D) climate model (Kasting et al. 1993). The model is based on assumptions that we are dealing with Earth-like planets with $CO_2/H_2O/N_2$ atmospheres and that habitability requires the presence of liquid water on the planet's surface. The inner edge of the HZ is determined in this model by loss of water via photolysis and hydrogen escape. The outer edge of the HZ is determined by the formation of $CO_2$ clouds, which cool a planet's surface by increasing its albedo and by lowering the convective lapse rate. Calculations within their model results with HZ = (0.95–137) AU with climate stability ensured by feedback mechanism in which atmospheric $CO_2$ concentrations vary inversely with planetary surface temperature. The HZ evolves outward in time because the Sun increases in luminosity as it ages. A conservative estimate for the width of the 4.6 ×

$10^9$ years continuously habitable zone is CHZ = (0.95–1.15) AU.

For planets that are larger than Earth and for planets which have higher $N_2$ partial pressures, the width of HZ is greater. The Kasting et al. (1993) model predicts that the HZ around F stars is larger and occurs farther out than for our Sun; the HZ around K and M stars is smaller and occurs farther in. For a specified period of habitability, CHZs around K and M stars are wider (in log distance) than for our Sun. Kasting et al. (1993) suggested that mid-to-early K stars should be considered along with G stars as optimal candidates in the search for extraterrestrial life.

Twenty years later, Kopparapu et al. (2013) published new estimates for habitable zones around main-sequence stars. An updated 1D climate model is used to obtain new HZ widths around F, G, K, and M stars with effective temperature in the range 2600 K $\leq T_{eff} \leq$ 7200 K. They calculated HZ distances for various scenarios and presented their changes with the time. They also present CHZ boundaries within which a planet remains habitable for a specific length of time (T=5 $\times$ $10^9$ years). According to this model, the HZ for our solar system is HZ = (0.99–1.70) AU. Since the model does not include the radiative effects of clouds, the actual HZ boundaries may extend farther in both directions than the estimates given.

A number of planets have been found to reside in binary stars systems, some of which even orbit both stars; see, for example, Doyle et al. (2011), Welsh et al. (2012), and Kostov et al. (2014). Some of the challenges in assessing habitability in binary star systems arise from the fact that one has to account for two sources of radiation, possibly of different spectral type. Moreover, the distance of the planet to each star keeps changing over time in a non-trivial manner due to gravitational interactions between the planet and the two stars.Georgakarakos et al. (2021) have undertaken the challenging task of determining habitable zones in binary star systems in the presence of a giant planet. Using the concept of "dynamically informed habitable zones", one can make predictions on where to look for habitable worlds in such complex environments. Dynamically informed habitable zones have been used in the past to investigate the habitability of circumstellar planets in binary systems and Earth-like analogs in systems with giant planets. The authors extended the concept to potentially habitable worlds on circumbinary orbits and showed that habitable zone borders can be found analytically even when another giant planet is present in the system.

The James Webb Space Telescope (JWST) has had its first look at a hotly anticipated set of targets—the atmospheres of some of the seven Earth-sized planets circling the star TRAPPIST-1, just 12 parsecs (39 light years) from Earth. All seven lie in or near their star's habitable zone, where liquid water could exist, and astronomers consider them the best-known laboratory for studying what might make planets beyond the solar system suitable for life.

The TRAPPIST-1 planetary system was mapped by Gillon et al. (2017). Their observations reveal that at least seven planets with sizes and masses similar to those of Earth revolve around TRAPPIST-1. The six inner planets form a near-resonant chain, such that their orbital periods (1.51, 2.42, 4.04, 6.06, 9.1, and 12.35 days) are near-ratios of small integers. This architecture suggests that the planets formed farther from the star and migrated inwards. Moreover, the seven planets have equilibrium temperatures low enough to make possible the presence of liquid water on their surfaces.

JWST is observing all of the planets in its first year of science operations, beginning in June 2022. In ongoing research, JWST studies planetary atmospheres mainly by watching how they filter starlight as the planets pass in front of the star: particular molecules absorb the starlight in characteristic ways. Which molecules make up the atmosphere can indicate how a planet evolved and whether it might have life on its surface.

Ruderman (1974) first studied the potential threat for complex life on Earth by nearby SN explosions. He pointed out that energetic radiation from such events, in the form of hard X-rays, gamma-rays, or cosmic rays, may (partially or totally) destroy the Earth's atmospheric ozone, leaving land life exposed to lethal doses of UV fluxes from the Sun. The paper went completely unnoticed, with just one citation in about 20 years. In the last few years, however, a large number of studies were devoted to that topic (for reasons that are not quite clear to the author of this paper). Two factors are, perhaps, at the origin of that interest: the availability of complex models of Earth's atmospheric structure and chemistry and the discovery of extrasolar giant planets. The latter suggests that Earth-like planets may also be abundant in the galaxy; however, complex life on them may be a rare phenomenon, because of various cosmic threats like SN explosions. This is the basic idea underlying the concepts of GHZ (Gonzalez et al. 2001) and the Rare Earth (Ward and Brownlee 2000): complex life in the Universe may be rare not for intrinsic reasons (i.e., improbability of development of life on a planet) but for extrinsic ones related to the hostile cosmic environment.

Despite the simplicity of the idea, however, the studies devoted to the topic revealed that it is very difficult to quantify the SN threat to life (or any other cosmic threat). Studies of that kind evaluate the energetic particle flux/irradiation impinging on Earth, which could induce a significant number of gene mutations (based on our understanding of such mutations in various organisms on Earth). Knowing the intrinsic emissivity of energetic particles from a typical SN (which depends essentially on the configuration of the progenitor star and the energetics of the explosion), one may then calculate a lower limit for the distance to the SN for such fluxes/irradiations to occur. The distribution of the rates of various SN types in the Milky Way (presumably known to within a factor of a few) is then used to evaluate the frequency/probability of such events in our vicinity and elsewhere in the galaxy (Prantzos 2008).

Wagner et al. (2021) reported imaging low-mass planets within the habitable zone of α Centauri. Giant exoplanets on wide orbits have been directly imaged around young stars. If the thermal background in the mid-infrared can be mitigated, then exoplanets with lower masses can also be imaged. Authors presented a ground-based mid-infrared observing approach that enables imaging low-mass temperate exoplanets around nearby stars, in particular within the closest stellar system, α Centauri. Based on 75–80% of the best-quality images from 100 h of cumulative observations, they demonstrated sensitivity to warm sub-Neptune-sized planets throughout much of the habitable zone of α Centauri A. This is an order of magnitude more sensitive than state-of-the-art exoplanet imaging mass detection limits. They also discussed a possible exoplanet or exozodiacal disk detection around α Centauri A. However, an instrumental artifact of unknown origin cannot be ruled out. These results demonstrate the feasibility of imaging rocky habitable-zone exoplanets with current and upcoming telescopes.

Jarugula et al. (2021) reported molecular line observations in two dusty star-forming galaxies at $z = 6.9$. Dusty star-forming galaxies (DSFGs) are dust-enshrouded, intensely star-forming galaxies with tens to thousands of stars forming per year. These starburst galaxies are bright at submillimeter wavelengths, as the ultraviolet photons from the young stars are absorbed and reradiated by dust into the far-infrared (FIR) wavelengths, giving rise to total infrared luminosities (LIRs) greater than $10^{12}$ Le. Observations of these galaxies have the advantage of negative-K correction at long wavelengths, enabling galaxy selection that is independent of redshift.

SPT0311–58 is the most massive infrared luminous system discovered so far during the Epoch of Reionization (EoR). In their paper, the authors present a detailed analysis of the molecular interstellar medium at $z = 6.9$, through high resolution observations of the CO(6–5), CO(7–6), CO(10–9), [C I](2–1), and p-$H_2O(2_{1,1}-2_{0,2})$ lines and dust continuum emissions with the Atacama Large Millimeter/Submillimeter Array. The system consists of a pair of intensely star-forming, gravitationally lensed galaxies (labeled West and East). The intrinsic far-infrared luminosity is (16 ± 4) × 1012 Le in West and $(27 \pm 4) \times 10^{11}\ L_{sun}$ in East. They model the dust, CO, and [C I] using non-local thermodynamic equilibrium radiative transfer models and estimate the intrinsic gas mass to be $(5.4 \pm 3.4) \times 10^{11}\ M_{sun}$ in West and $(3.1 \pm 2.7) \times 10^{10}\ M_{sun}$ in East. Jarugula et al. (2021) found that the CO spectral line energy distribution in West and East are typical of high-redshift submillimeter galaxies (SMGs). The CO-to-$H_2$ conversion factor (αCO) and the gas depletion timescales estimated from the model are consistent with the high-redshift SMGs in the literature within the uncertainties. They found no evidence of evolution of depletion time with redshift in SMGs at $z > 3$. This is the most detailed study of molecular gas content of a galaxy in the EoR to date, with the most distant detection of $H_2O$ in a galaxy without any evidence for active galactic nuclei in the literature.

NASA estimates that there are at least $50 \times 10^9$ planets in the Milky Way and that $500 \times 10^6$ (or 1%) of these are likely in what is considered a habitable zone. The concept of habitable zones is perhaps most interesting as applied to stars other than the sun. The brightness of the star determines the location of its habitable zone, but brightness in turn depends on the star's size, type, and age. For stars more massive than our sun, the outward migration of their HZs through time is much faster and of much shorter duration. More massive stars have shorter lifetimes. The sun should be stable for nearly $10 \times 10^9$ years after its birth, but a star 50% more massive than the sun enters its red giant stage after only $2 \times 10^9$ years. When a star becomes a red giant, its brightness increases by a factor of a thousand, and the HZ retreats greatly beyond its original bounds. It has been already noted that a 1.5-solar-mass star would not be around long enough for animals to evolve at the leisurely pace enjoyed by terrestrial life. More massive stars have habitable zones farther from the star—or they may have no habitable zones at all. More stars that are massive are hotter and radiate substantially more ultraviolet light than the sun. Ultraviolet light breaks the bonds of most biological molecules, and life must be shielded from it to survive. UV also can be disastrous for the atmospheres of Earth-like planets. It is strongly absorbed at the top of such atmospheres and is a potent high-altitude heat source than can lead to escape of the atmosphere. The sun, with its effective surface temperature of 5780 Kelvin, emits less than 10% of its energy in the ultraviolet range, whereas hotter stars like Sirius radiate most of their energy in the UV. Atmospheric loss may prevent terrestrial planets with oceans and atmospheres from forming around stars that are more massive. This atmospheric problem with planets orbiting stars that are more massive is in addition to limitations imposed by their shorter stellar lifetimes.

### 5.4.2 Interstellar Panspermia

An important issue in astrobiology is whether life can be transferred from one planetary system to another. This type of transfer is generally known as panspermia and has been studied extensively by many (see, for example, Melosh 2003; Napier 2004; Wallis and Wickramasinghe 2004; Adams and Spergel 2005; Wesson 2010; Belbruno et al. 2012; Ginsburg et al. 2018). The recent discovery of the unbound interstellar objects 'Oumuamua (Meech et al. 2017) and the comet Borisov passing through our solar system emphasizes that one part of the process, astronomical objects traveling from one system to another, is not only readily realized but can also be observationally constrained. The other requirements for the transfer of life, however, are expected to be less common, and most previous estimates suggest that the likelihood of transfer is extremely low. With a focus on the astronomical aspects of the problem, a paper by Adams and Napier (2022) revisits the question of panspermia. Although a great deal of previous work has been carried out, the present treatment is motivated by recent observational input, especially the detection of unbound interstellar objects, and by recent numerical studies of the capture dynamics. They found that the odds of successful transfer of life remain low, although significant uncertainties remain.

The transfer of life between planetary systems requires a number of astronomical and biological processes to take place. First, life must arise on a suitably habitable planet in order to provide the seeds for other planetary systems. The habitable planet must then be bombarded via impacts (or experience some other cataclysmic event) in order to lift biologically active material off the planetary surface and into interstellar space. These projectiles then travel from one planetary system to another and must be captured by the recipient system(s). After capture into a bound orbit, the biologically active body must eventually find its way onto the surface of a habitable planet within its new system. Finally, the successful transfer of life requires that life actually take hold in its new environment. The first and last of these processes are biological in nature, and their odds of occurrence are largely unknown. Of course, Earth provides one example of life both arising and taking hold, but an honest assessment of the odds remains out of reach. In contrast, the dynamics of transfer can be calculated, and the corresponding odds of success can be estimated, as shown by Adams and Napier (2022).

Ginsburg et al. (2018) presented an analytic model to estimate the total number of rocky or icy objects that could be captured by planetary systems within the Milky Way resulting in panspermia in the case they harbored life. The authors estimated the capture rate of objects ejected from planetary systems over the entire phase space as well as time. Their final expression for the capture rate depends upon the velocity dispersion as well as the characteristic biological survival time and the size of the captured object. They take into account the number of stars that an interstellar object traverses, as well as the scale height and length of the Milky Way's disk. The likelihood of galactic panspermia is strongly dependent upon the survival lifetime of the putative organisms as well as the velocity of the transporter. Velocities between 10 and 100 km $s^{-1}$ result in the highest probabilities. However, given large enough survival lifetimes, even hypervelocity objects traveling at over 1000 km $s^{-1}$ have a significant chance of capture, thereby increasing the likelihood of panspermia. The authors showed that panspermia is not exclusively relegated to solar system-sized scales, and the entire Milky Way could potentially be exchanging biotic components across vast distances.

Napier (2004) considers yet another mechanism. In his scenario, meter-sized boulders ejected from the Earth by large impacts are destroyed through collisions and erosion by impacting zodiacal cloud dust particles. The time scale for such disintegration in a dense zodiacal cloud may be $\lesssim 10^4$ years. Once reduced to a critical size, the particles are rapidly ejected from the solar system by radiation pressure. The critical size for ejection is on the order of a micron, large enough to protect groups of micro-organisms within them from solar ultraviolet irradiation. Such life-bearing particles are ejected at a mean rate of $\sim 10^{20}$ per million years. During passages of the solar system through or close to dense molecular clouds, a significant proportion of the particles may be incorporated into protoplanetary systems and protected from cosmic rays within growing planetesimals. The specific number density of micro-organisms so deposited is highest in small, dense molecular clouds. On the assumption that this ejection mechanism is common in other planetary systems environmentally capable of supporting life, a "chain reaction" may seed the disk of the galaxy within a few billion years. In that case, it is unlikely that life originated on Earth.

Wesson (2010) considers yet another possibility. Astronomically, there are viable mechanisms for distributing organic material throughout the Milky Way. Biologically, the destructive effects of ultraviolet light and cosmic rays mean that the majority of organisms arrive broken and dead on a new world. The likelihood of conventional forms of panspermia must therefore be considered low. However, the information content of damaged biological molecules might serve to seed new life (necro-panspermia).

According to McKay et al. (2022), it may be that habitable planets are common but life is rare. If future advances in telescopes increasingly suggest this is so, humankind might feel motivated to seed lifeless planets with resilient terrestrial organisms or synthetic forms designed to thrive on the target planet. A useful mechanism for achieving this goal at a relatively low cost is to use interstellar comets transiting the solar system to convey microbial cargoes toward nearby planetary systems, where they could disseminate the inoculum via their dust trails. Conversely, it is conceivable that terrestrial life was deliberately seeded in this matter, a hypothesis that could be tested if we found evidence for life on other solar system bodies that displayed common basic biochemical signatures. This scenario raises a number of ethical and technological challenges that need to be addressed.

The propagation of life between planets via panspermia could occur both naturally, for example, as a consequence of impact ejection, and deliberately by an advanced civilization, a concept dubbed "directed panspermia". However, the panspermia theory does not solve the problem of life's origin (except in the steady state theory of cosmology, where life does not need an origin, as it can be co-extensive with the universe in space and time).

Historically, most discussions of panspermia assume it to be a naturally occurring random process. Several authors have suggested that interstellar comets may be naturally transporting life between stars. This proposal, coupled with the studies of the survival of spores in space, converged on an overall scenario for interstellar panspermia by comets examined by several authors. As part of this scenario, small particles or rocks containing viable spores are ejected by impacts from life-bearing planets and incorporated into passing comets.

Alongside these discussions of natural panspermia, there has been some speculation about directed panspermia by an extraterrestrial civilization. In their article, McKay et al. (2022) build on the foregoing research and propose the concept of directed panspermia using interstellar comets as opposed to dedicated spacecraft. The immediate motivation for this work is the recent observation of an interstellar comet and the associated calculations of the frequency of such transits. Their approach does not replace directed panspermia using starships, nor does it alter the arguments for naturally occurring panspermia by comets. It combines them effectively, using interstellar comets as transporters of opportunity by adding biological inoculum to the comet without attempting to change its trajectory. Their proposal has the advantage that it would reduce the cost of directed panspermia and augment the effectiveness of natural cometary panspermia.

A proposal by McKay et al. (2022) to utilize interstellar comets as vehicles to disseminate life around the cosmos raises many technical and ethical questions. Chief among them is whether humans have the right to modify the environments of other astronomical bodies, and, if they do, how much control we would have over the far downstream impact. Safeguards that fall under the general category of "planetary protection" need to be carefully assessed and protocols established for any future projects of this sort.

Whether humanity chooses to embark on directed panspermia will depend on our vision of the universe and mankind's place within it. One common perspective of the long future focuses on the development of intelligence and the rise of the intelligent machines (artificial intelligence), leading to what is termed a "post-biological" future (Dick 2003), according to which our contact with any past intelligence and the future of human actions in the universe will be via machines. Directed interstellar visitors will be machines.

Until recently, the idea that humans could literally sow the seeds of a cosmic transformation with multi-million-year downstream consequences would have been regarded as absurd. But the discovery of interstellar comets has changed all that. Although we currently lack the technology to harness these comets as biological delivery vehicles, there is no difficulty in understanding what is needed to do so and in refining the strategy to achieve the goal of seeding the galaxy with life suitably constructed to thrive in a variety of exoplanetary environments. Whether it is desirable to do so must rest with future generations.

The interstellar objects 1I/'Oumuamua and 2I/Borisov confirm the long-held expectation that bodies from one stellar system will be carried to another, allowing, in principle, interstellar panspermia. Life might be transferred between stellar systems, depending on the nature of the bodies and how they escaped their systems. 2I/Borisov appears to be a comet, with no more likelihood of carrying life than solar system comets. In contrast, the nature of 1I/'Oumuamua has been difficult to determine. Desch and Jackson (2022) reviewed various hypotheses for its origin, including ejection of $N_2$ ice from the surface of an exo-Pluto, formation in a molecular cloud by freezing of $H_2$, and a derelict solar sail of alien construction. Of these, the $N_2$ ice fragment hypothesis is uniquely falsifiable, plausible, and completely consistent with all observations. The possibility of interstellar panspermia would be made more probable if 'Oumuamua originated on a dwarf planet rather than a comet, although substantial challenges to transfer of life would remain. Of proposed mechanisms for interstellar panspermia, transfer of life via rocky meteoroids is perhaps less improbable.

In their work, Anslow et al. (2023) consider the potential of cometary impacts to deliver complex organic molecules and the prebiotic building blocks required for life to rocky exoplanets. Numerical experiments have demonstrated that for these molecules to survive, impacts at very low velocities are required. Their work shows that for comets scattered from beyond the snowline into the habitable zone, the minimum impact velocity is always lower for planets orbiting solar-type stars than M-dwarfs. Using both an analytical model and numerical N-body simulations, they show that the lowest velocity impacts occur onto planets in tightly packed planetary systems around high-mass (i.e. solar-mass) stars, enabling the intact delivery of complex organic molecules. Impacts onto planets around low-mass stars are found to be very sensitive to the planetary architecture, with the survival of complex prebiotic molecules potentially impossible in loosely packed systems. Rocky planets around M-dwarfs also suffer significantly more high-velocity impacts, potentially posing unique challenges for life on these planets. In the scenario that cometary delivery is important for the origins of life, this study predicts the presence of bio-signatures will be correlated with (1) decreasing planetary mass (i.e. escape velocity), (2) increasing stellar-mass, and (3) decreasing planetary separation (i.e. exoplanets in tightly packed systems.

### 5.4.3 Old M-Stars

Recent work has demonstrated that high levels of X-ray and UV activity on young M dwarfs may drive rapid atmospheric escape on temperate, terrestrial planets orbiting within the habitable zone. However, secondary atmospheres on planets orbiting older, less active M dwarfs may be stable and present more promising candidates for biomarker searches. In order to evaluate the potential habitability of Earth-like planets around old, inactive M dwarfs, France et al. (2020) presented new Hubble Space Telescope and Chandra X-ray Observatory observations of Barnard's Star (GJ 699), a 10-Gyr-old M3.5 dwarf, acquired as part of the Mega-MUSCLES program. Despite the old age and long rotation period of Barnard's Star, they observed two FUV ($\delta130 \approx 5000$ s; $E130 \approx 1029.5$ erg each) and one X-ray ($EX \approx 1029.2$ erg) flares and estimated a high-energy flare duty cycle (defined as the fraction of the time the star is in a flare state) of ~25%. A publicly available 5 Å to 10 μm spectral energy distribution of GJ 699 was created and used to evaluate the atmospheric stability of a hypothetical, un-magnetized terrestrial planet in the habitable zone ($r_{HZ} \sim 0.1$ au). Both thermal and non-thermal escape modeling indicate (1) the quiescent stellar XUV flux does not lead to strong atmospheric escape: atmospheric heating rates are comparable to periods of high solar activity on modern Earth, and (2) the flare environment could drive the atmosphere into a hydrodynamic loss regime at the observed flare duty cycle: sustained exposure to the flare environment of GJ 699 results in the loss of ≈87 Earth atmospheres $Gyr^{-1}$ through thermal processes and ≈3 Earth atmospheres $Gyr^{-1}$ through ion loss processes. These results suggest that if rocky planet atmospheres can survive the initial ~5 Gyr of high stellar activity, or if a second-generation atmosphere can be formed or acquired, the flare duty cycle may be the controlling stellar parameter for the stability of Earth-like atmospheres around old M stars.

France et al. (2020) analyzed new HST and Chandra X-ray Observatory observations of the ~10 Gyr mid-M dwarf Barnard's Star to place empirical constraints on the high-energy environment around a representative old M dwarf. Studies of atmospheric escape suggest that rapid atmospheric loss is likely on terrestrial planets orbiting young M dwarfs, so they were motivated to consider the stability of Earth-like atmospheres later in the main-sequence lifetimes of these stars, provided these worlds are able to outgas or capture second-generation atmospheres. The UV and X-ray data, obtained through the Mega-MUSCLES observing program, are used to estimate the currently unobservable EUV radiation that is the major driver of most atmospheric escape processes. Over approximately 25 ks of HST and 27 ks of Chandra (non simultaneous) spectroscopy, they observe three intermediate-energy flares. They used the flare and quiescent spectral energy distributions to estimate the atmospheric mass-loss rates for thermal and nonthermal processes on a hypothetical, unmagnetized terrestrial planet in the habitable zone (rHZ ~ 0.1 au) of Barnard's Star. Their modeling results indicate that the quiescent UV and X-ray flux of Barnard's Star elevate the exospheric temperature of the model planet's atmosphere relative to the modern Sun–Earth system, but it does not result in catastrophic atmospheric mass-loss rates. By contrast, the XUV flare luminosity is large enough to drive significant atmospheric expansion. Assuming steady-state illumination of the XUV flare spectrum at the measured UV/X-ray flare duty cycle (~25%), they find that the elevated XUV flux during flare periods dramatically increases the predicted escape rates for all physical processes. They find that atmospheric loss of up to ~90 Earth atmospheres per giga-year is possible for the model Earthlike planet. They add the important caveat that these results are based on a reconstruction of the high-energy spectrum and simplified escape models, and uncertainties exist in each step that could alter the overall escape rate and the quantitative conclusions of our study. While there are significant uncertainties in the absolute level of the total atmospheric escape, the broad conclusion is that if the XUV flare duty cycle is comparable to that observed in typical field age mid-M stars, then rapid atmospheric mass loss may still be likely on planets around old M dwarfs. This suggests that the XUV flare duty cycle may be one of the most critical stellar parameters when considering the habitability of M dwarf planets.

Rocky planets orbiting M-dwarf stars are among the most promising and abundant astronomical targets for detecting habitable climates. Planets in the M-dwarf habitable zone are likely synchronously rotating, such that one expects significant day–night temperature differences and potentially limited fractional habitability. Previous studies have focused on scenarios where fractional habitability is confined to the sub-stellar or "eye" region, but in their paper, Lobo et al. (2023) explore the possibility of planets with terminator habitability, defined by the existence of a habitable band at the transition between a scorching dayside and a glacial nightside. Using a global climate model, they show that for water-limited planets, it is possible to have scorching temperatures in the "eye" and freezing temperatures on the night side while maintaining a temperate climate in the terminator region, due to reduced atmospheric energy transport. On water-rich planets, however, increasing the stellar flux leads to increased atmospheric energy transport and a reduction in day–night temperature differences, such that the terminator does not remain habitable once the dayside temperatures approach runaway or moist greenhouse limits. They also show that while water abundant simulations may result in larger fractional habitability, they are vulnerable to water loss through cold trapping on the night side surface or atmospheric water vapor escape, suggesting that even if planets were formed with abundant water, their climates could become water-limited and subject to terminator habitability.

In their study, Lobo et al. (2023) explored the possibility of terminator habitability, defined as the existence of a habitable band at the transition between a scorching dayside and a glacial night side on a synchronously rotating Earth-like planet orbiting an M dwarf. In particular, they explored the viability of this configuration at the inner edge of the habitable zone to determine whether it is possible for dayside temperatures to exceed typical habitable temperatures, approaching the runaway greenhouse limit, without leading to a planet-wide runaway greenhouse state. They found that a temperate terminator is not achievable with aqua planet simulations that seek to reproduce ocean-covered planets but that it can easily occur on water-limited land planets. On aqua planets, increasing stellar flux leads to reduced day–night temperature gradients, such that the planet would tend to reach a homogeneous climate before the dayside reaches a runaway greenhouse state, never passing through a terminator habitability state. On water-limited planets, meanwhile, they find that large day–night temperature gradients are easily achievable without entering a runaway greenhouse state. They also find that the water-limited land-planet configurations may be favorable in terms of long-term climate stability, with a reduced risk of night side water cold trapping or water vapor escape. We expect that water-limited terminator habitability scenarios could be a stable configuration for ocean-covered worlds after significant water loss, especially in the case of night side cold trapping.

There are still many uncertainties regarding the water content of habitable-zone M-dwarf planets. Based on our current understanding, it is possible that water-limited planets could be abundant and possibly more common than ocean-covered worlds. Therefore, terminator habitability may represent a significant fraction of habitable M-dwarf planets. Compared to the temperate climates obtained with aqua planets, terminator habitability does offer reduced fractional habitability. Also, while achieving a temperate terminator is relatively easy on water-limited planets, constraining water availability at the terminator remains a challenge. Overall, the lack of abundant surface water in these simulations could pose a challenge for life to arise under these conditions, but mechanisms, including glacier flow, could allow sufficient surface water accumulation to sustain locally moist and temperate climates at or near the terminator. We expect that future studies, exploring a broader range of land-planet configurations, in particular those using future generations of surface and ice models, will find a wide range of habitable terminator scenarios in regimes that are intermediate to the water-limited and aquaplanet cases considered here.

An article by Farihi et al. (2022) reports quasi-continuous transiting events towards WD 1054–226 at d = 36.2 pc and V = 16.0 mag, based on simultaneous, high-cadence, multiwavelength imaging photometry using ULTRACAM over 18 nights from March 2019 to 2020. The predominant period is 25.02 h and corresponds to a circular orbit with blackbody $T_{eq}$ = 323 K, where a planetary surface can nominally support liquid water. The light curves reveal remarkable night-to-night similarity, with changes on longer time-scales, and lack any transit-free segments of unocculted starlight. The most pronounced dimming components occur every 23.1 min—exactly the 65th harmonic of the fundamental period—with depths of up to several percent and no evident color dependence. Myriad additional harmonics are present, as well as at least two transiting features with independent periods. High-resolution optical spectra are consistent with stable, photospheric absorption by multiple, refractory metal species, with no indication of circumstellar gas. Spitzer observations demonstrate a lack of detectable dust emission, suggesting that the otherwise hidden circumstellar disk orbiting WD 1054–226 may be typical of polluted white dwarfs and detected only via favorable geometry. Future observations are required to constrain the orbital eccentricity, but even if periastron is near the Roche limit, sublimation cannot drive mass loss in refractory parent bodies, and collisional disintegration is necessary for dust production.

Sagear and Ballard (2023) investigated the underlying distribution of orbital eccentricities for planets around early to mid-M dwarf host stars. They employed a sample of 163 planets around early to mid-M dwarfs across 101 systems detected by NASA's Kepler Mission. They constrained the orbital eccentricity for each planet by leveraging the Kepler light curve together with a stellar density prior, constructed using metallicity from spectroscopy, $K_s$ magnitude from 2MASS, and stellar parallax from Gaia. Within a Bayesian hierarchical framework, they extracted the underlying eccentricity distribution, assuming alternately Rayleigh, half-Gaussian, and Beta functions for both single- and multi-transit systems. They described the eccentricity distribution for apparently single-transiting planetary systems with a Rayleigh distribution with $\sigma = 0.19^{+0.04}_{-0.03}$ and for multitransit systems with $\sigma=0.03^{+0.02}_{-0.01}$. The data suggest the possibility of distinct dynamically warmer and cooler subpopulations within the single-transit distribution: The single-transit data prefer a mixture model composed of two distinct Rayleigh distributions with $\sigma_1 = 0.02^{+0.11}_{-0.00}$ and $\sigma_2 = 0.24^{+0.20}_{-0.03}$ over a single Rayleigh distribution, with 7:1 odds. They contextualized their findings within a planet formation framework by comparing them to analogous results in the literature for planets orbiting FGK stars. By combining the derived eccentricity distribution with other M dwarf demographic constraints, they estimated the underlying eccentricity distribution for the population of early to mid-M dwarf planets in the local neighborhood.

## 5.5 LIFE ON EXOPLANETS

Of the estimated 100 billion galaxies in the observable universe, only 1 in 10 can support complex life like that on Earth according to Piran and Jiminez (2014). Everywhere else, stellar explosions resulting in gamma ray bursts would regularly wipe out any life forms more elaborate than microbes. These explosions also kept the universe lifeless for billions of years after the Big Bang.

Scientists have long mused over whether a gamma ray burst could harm Earth. The bursts were discovered in 1967 by satellites designed to spot nuclear weapons tests and now turn up at a rate of about one a day. They come in two types. Short gamma ray bursts last less than a second or two; they most likely occur when two neutron stars or black holes spiral into each other. Long gamma ray bursts last for tens of seconds and occur when massive stars burn out, collapse, and explode. They are rarer than the short ones but release roughly 100 times as much energy. A long burst can outshine the rest of the universe in gamma rays. That seconds-long flash of radiation itself wouldn't blast away life on a nearby planet. Rather, if the explosion were close enough, the gamma rays would set off a chain of chemical reactions that would destroy the ozone layer in a planet's atmosphere. With that protective gas gone, deadly ultraviolet radiation from a planet's sun would rain down for months or years—long enough to cause a mass die-off.

As a copious source of gamma rays, a nearby galactic gamma ray burst (GRB) can be a threat to life. Using recent determinations of the rate of GRBs, their luminosity function, and properties of their host galaxies, Piran and Jimenez (2014) estimated the probability that a life-threatening (lethal) GRB would take place. Amongst the different kinds of GRBs, long ones are most dangerous. There is a very good chance (but no certainty) that at least one lethal GRB took place during the past 5 gigayears close enough to Earth as to significantly damage life. There is a 50% chance that such a lethal GRB took place during the last $500 \times 10^6$ years, causing one of the major mass extinction events. Assuming that a similar level of radiation would be lethal to life on other exoplanets hosting life, they explore the potential effects of GRBs to life elsewhere in the galaxy and the Universe. They find that the probability of a lethal GRB is much larger in the inner Milky Way (95% within a radius of 4 kpc from the galactic center), making it inhospitable to life. Only at the outskirts of the Milky Way, at more than 10 kpc from the galactic center, does this probability drop below 50%. When considering the Universe as a whole, the safest environments for life (similar to the one on Earth) are the lowest-density regions in the outskirts of large galaxies, and life can exist in only ≈10% of galaxies. Remarkably, a cosmological constant is essential for such systems to exist. Furthermore, because of both the higher GRB rate and galaxies being smaller, life as it exists on Earth could not take place at $z > 0.5$. Early life forms must have been much more resilient to radiation.

The modern scientific work on the search for extraterrestrial intelligence started in the late 1950s. The search for intentional radio or laser signals is still one of the most popular SETI strategies; see Huston and Wright (2022). Another possible signal could be finding a planet with an atmosphere filled with chemicals that can only be produced by technology.

Red dwarf stars are particularly interesting for finding planets. Since stellar masses are up to an order of magnitude smaller than that of solar-mass stars, the RV signals are correspondingly larger—and since stellar radii are up to a factor ten smaller, the corresponding transit signal for a given radius planet can be up to two orders of magnitude larger. Their luminosities can be a thousand times lower, meaning that temperate planets can be found in smaller orbits—making even small, rocky planets accessible through both the transit and the RV technique. This is called the red dwarf opportunity.

The recent discovery by Hu et al. (2021) and initial characterization of sub-Neptune-sized exoplanets that receive stellar irradiance of approximately Earth's raised the prospect of finding habitable planets in the coming decade, because some of these temperate planets may support liquid water oceans if they do not have massive $H_2$/He envelopes and are thus not too hot at the bottom of the envelopes. For planets larger than Earth, and especially planets in the 1.7–3.5 $R_{Earth}$ population, the mass of the $H_2$/He envelope is typically not sufficiently constrained to assess the potential habitability. Here they show that the solubility equilibria vs. thermochemistry of carbon and nitrogen gases results in observable discriminators between small $H_2$ atmospheres vs. massive ones, because the condition to form a liquid-water ocean and that to achieve the thermochemical equilibrium are mutually exclusive. The dominant carbon and nitrogen gases are typically $CH_4$ and $NH_3$ due to thermochemical recycling in a massive atmosphere of a temperate planet, and those in a small atmosphere overlying a liquid-water ocean are most likely $CO_2$ and $N_2$, followed by CO and $CH_4$ produced photo-chemically. $NH_3$ is depleted in the small atmosphere by dissolution into the liquid-water ocean. These gases lead to distinctive features in the planet's transmission spectrum, and a moderate number of repeated transit observations with the James Webb Space Telescope should tell apart a small atmosphere vs. a massive one on planets like K2–18 b. This method thus provides a way to use near-term facilities to constrain the atmospheric mass and habitability of temperate sub-Neptune exoplanets; see Hu et al. (2021).

The traditional approach of searching for extraterrestrial technology is to search in the radio, optical, or infrared bands for evidence of narrow-band signals characteristic of technological sophistication. A fundamental issue in searches of this kind is the one of target selection, as the parameter space of potential sources is enormous. Most searches are either broad in terms of source selection or focused on systems known to host exoplanets. The proposal in Hansen and Zuckerman (2021) is based on a specific hypothesis and thus defines a much smaller, focused, sample than other studies. A search for techno-signatures in these systems may also require some differences in search strategy, as traditional SETI searches assume a source located on a planet, orbiting a star. In the case of a large-scale migration, and with sufficient angular resolution, sources may also be found to be localized between the stars in question, and not solely in orbit around one star or the other.

The ease of interstellar rocket travel is an issue with implications for the long-term fate of our own and other civilizations and for the much-debated number of technological civilizations in the galaxy. We show that the physical barrier to interstellar travel can be greatly reduced if voyagers are patient and wait for the close passage of another star. For a representative time of ~1 Gyr, characteristic of the remaining time that Earth will remain habitable, one anticipates a passage of another star within ~1500 au. This lowers the travel time for interstellar migration by ~two orders of magnitude compared with calculated travel times based on distances comparable to average interstellar separations (i.e., ~1 pc) in the solar vicinity. We consider the implications for how long-lived civilizations may respond to stellar evolution, including the case of stars in wide binaries, and the difficulties of identifying systems currently undergoing a relevant close encounter. Assuming that life originates only around G-type stars but migrates primarily to lower mass hosts when the original system becomes uninhabitable, the fraction of extant technological civilizations that exist as diaspora can be comparable to the fraction that still orbit their original host stars (Hansen and Zuckerman 2021).

### 5.5.1 Detection of Extraterrestrial Life—Biosignatures

Searching for life in the Universe depends on unambiguously distinguishing biological features from background signals, which could take the form of chemical, morphological, or spectral signatures. The discovery and direct measurement of organic compounds unambiguously indicative of extraterrestrial (ET) life is a major goal of solar system exploration. Biology processes matter and energy differently from abiological systems, and materials produced by biological systems may become enriched in planetary environments where biology is operative. However, ET biology might be composed of different components than terrestrial life. As ET sample return is difficult, in situ methods for identifying biology will be useful. Mass spectrometry (MS) is a potentially versatile life detection technique, which will be used to analyze numerous solar system environments in the near future. Guttenberg et al. (2021) showed that simple algorithmic analysis of MS data from abiotic synthesis (natural and synthetic), microbial cells, and thermally processed biological materials (lab-grown organisms and petroleum) easily identifies relational organic compound distributions that distinguish pristine and aged biological and abiological materials, which likely can be attributed to the types of compounds these processes produce, as well as how they are formed and decompose. This method is independent of the detection of particular masses or molecular species samples may contain. This suggests a general method to agnostically detect evidence of biology using MS given a sufficiently strong signal in which the majority of the material in a sample has either a biological or abiological origin. Such metrics are also likely to be useful for studies of possible emergent living phenomena and palaeobiological samples (Guttenberg et al. 2021).

It is of general interest to mention the Agnostic Life Finder (ALF) instrument for large-scale screening of Martian life, as proposed by Špaček and Benner (2022). The Agnostic Life Finder is an inexpensive life-detection add-on to in situ resource utilization (ISRU). ALF exploits a well-supported view that informational genetic biopolymers in life in water must have two structural features: (1) informational biopolymers must carry a repeating charge; they must be *polyelectrolytes*, and (2) their building blocks must fit into an *aperiodic crystal structure*; the building blocks must be size-shape regular. ALF exploits the first structural feature to extract polyelectrolytes from ~10 cubic meters of mined Martian water by applying a voltage gradient perpendicularly to the water's flow. This gradient diverts polyelectrolytes from the flow toward their respective electrodes (polyanions to the anode, polycations to the cathode), where they are captured in cartridges before they encounter the electrodes. There, they can later be released to analyze their building blocks, for example, by mass spectrometry or nanopore. Upstream, Martian cells holding Martian informational polyelectrolytes are disrupted by ultrasound. To manage the (unknown) conductivity of the water due to the presence of salts, the mined water is preconditioned by electrodialysis using porous membranes.

The search for life so far is based on the assumption that life elsewhere should be similar to the life that we know on Earth, so-called terran life; see National Academy of Science (2007). The main characteristics of such life are:

- Terran life uses water as a solvent;
- It is built from cells and exploits a metabolism that focuses on the carbonyl group, C=O;
- It is thermodynamically dissipative, exploiting chemical-energy gradients;
- It exploits a two-biopolymer architecture that uses nucleic acids to perform most genetic functions and proteins to perform most catalytic functions.

Consequently, all life-seeking efforts so far are targeted to locations where the existence of liquid water is possible. These efforts are directed towards searches for structures resembling cells of terran organisms, small molecules that might be the products of carbonyl metabolism, and tests for amino acids and nucleotides similar to those found in terrestrial proteins and DNA. This is a very narrow-minded approach; therefore, the National Academy of Science 2007 Report evaluates the possibility that nonstandard biochemistry might support life in known solar system environments and extrasolar habitability zones. We find it interesting that the committee recommended further studies of chirality, particularly studies focused on the hypothesis that specific environmental conditions can favor chiral selection. We should add the environment like dust particles in a molecular cloud exposed to polarized protons from the GCR where preferential destruction of D-amino acids and L-sugars would happen.

A team of geologists (see Schreder-Gomes et al. 2022) has discovered tiny remnants of prokaryotic and algal life trapped inside crystals of halite dating back to 830 $\times 10^6$ years ago. It was suggested that the organisms trapped therein might still be alive.

In complex biological systems, there seems to exist a close metabolic interrelationship between the chemical elements K and Rb. It was shown that the excretion of each ion ought, at least in humans, to be related to the other (Meltzer et al. 1969). A similar conclusion emerges after a broad survey of the literature. It should be noted that Rb, like K, is concentrated intracellularly. There is, however, a general tendency for Rb to have higher intracellular–extracellular ratios than K, and it has been proposed that this indicates intracellular binding (Menozzi et al. 1959).

In some cases, Rb can substitute for the K requirement, where K is required for survival or growth; a complete replacement of dietary K by Rb allows normal growth for several weeks; eventually, toxic effects appear. It should be noted that the K/Rb ratio for ICRP 1975 Reference Man is $K/Rb = 2{,}03 \times 10^2$.

It seems that the K/Rb ratio holds some clue to the functioning of living matter. Even without understanding completely the mechanism of K and Rb differentiation by living cells, the value of K/Rb ratios could be used in the search for extraterrestrial life forms. Although it is difficult to present K/Rb ratios for different materials, mainly because of possible systematic analytical ratios, it seems that the K/Rb ratio is significantly higher for living matter than for inorganic material.

Gast (1965) has reported

> The ratios of potassium to rubidium determined for a series of oceanic volcanic rocks suggest that the K/Rb ratio of the upper mantle exceeds 1500. The similarity between this inferred K/Rb ratio for the mantle and that found in some achondritic meteorites suggests a similar overall composition for these materials.

Snellen et al. (2021) wrote a white paper in response to the European Space Agency (ESA) Voyage 2050 Call and recommended the ESA play a proactive role in developing a global collaborative effort to construct a large high-contrast imaging space telescope, to be launched in 2040. Such a mission will be needed to characterize a sizable sample of temperate Earth-like planets in the habitable zones of nearby sun-like stars and to search for extraterrestrial biological activity.

By now, it is known that rocky planets are very common, particularly around red dwarf stars, and planets similar to Earth are expected to orbit many of our nearest neighbors. It would be required to address several questions like whether such planets have atmospheres and whether they are wet and potentially habitable. The mission proposed by Snellen et al. (2021) should conduct a large characterization census of twin-Earths around solar-type stars in the solar neighborhood and the many other planets in these systems and assess their climate, habitability, and whether they are inhabited. This can only be done with a space telescope.

It should be kept in mind that most exoplanets have been discovered using the radial velocity technique, which measures the host star's reflex motion around the system's center of mass, or with the transit method, which targets the regular and temporal dimming of a star caused by a passing planet.

Earth, or terran, life (i.e., life as we know it) uses a variety of energy sources (e.g., solar, geothermal, or chemical) to maintain itself in a state of thermodynamic disequilibrium. Terran life is based on carbon (plus hydrogen, nitrogen, oxygen, phosphorous, and sulfur) and readily maintains covalent bonds. On Earth, water is the liquid medium of choice used to facilitate biochemical reactions and nutrient transport, hence NASA's "follow the water" strategy in the search for life and habitable environments. Finally, all known life on Earth uses DNA as the basis of a molecular system enabling Darwinian evolution. Applying life's characteristics to the search for life elsewhere, however, means identifying not simply the requirements for terran life but instead the fundamental characteristics of life. The commonly cited set of general criteria are (in decreasing order of certainty) as follows (National Research Council of National Academies 2007):

- A means to sustain thermodynamic disequilibrium;
- An environment capable of maintaining covalent bonds, especially between carbon, hydrogen, and other atoms;
- A liquid environment; and
- A self-replicating molecular system that can support Darwinian evolution.

Whether life elsewhere in the solar system or beyond might be similar to terran life remains to be determined. Speculations abound as to whether life could be based on an element other than carbon, use a solvent other than water, or employ a different type of molecular system for encoding genetic information and facilitating evolution.

Habitability refers to a set of environmental conditions capable of supporting life. On a planetary scale, this most often includes the presence of liquid water, an energy source, and the availability of the basic elements of life—commonly considered to be carbon, hydrogen, nitrogen, oxygen, phosphorous, and sulfur, or CHNOPS. At smaller spatial scales, habitability is determined by ranges in temperature (T), pressure, (P) acidity (pH), and salinity, as well as minimum energy requirements. The limits of habitability are generally defined by studying life as we know it, and our current understanding is largely explored either through experiments and models that investigate only a few parameters at a time or through empirical investigation of Earth's extreme environments. Habitability, however, also implies that environmental conditions are sustained such that life persists through time. Dynamic habitability expands this conceptual framework. First, dynamic habitability recognizes that the combined effects of multiple parameters (e.g., T, P, salinity, pH) define whether life can emerge and persist and that while one or more parameters may vary outside the canonical limit to life, it is their combined effect that causes an environment to be habitable or not. Second, dynamic habitability acknowledges that environments—and organisms—coevolve over time. At the planetary scale, characteristics of dynamic habitability include stellar evolution and its impact on the presence of liquid water over time; the evolving structure of the planet interior influencing, among other things, the magnetic field and plate tectonics; and the state of the atmosphere, which might, for example, redistribute sources of metabolic energy. On more local scales, dynamic habitability encompasses changing environmental conditions driven not only by planetary-scale dynamics but also by more local processes like fluid flow and mixing or water-rock reactions, which can alter thermal and chemical gradients and therefore control not only available energy sources but also energy gradients and fluxes. Conditions of local dynamic habitability are necessarily changed by the activity of in situ life through nutrient consumption, energy transformation, cell production, and the addition of products from metabolic activity. On Earth, conditions of global habitability have been changed by life, with the most profound example being the evolution of oxygenic photosynthesis, from *An Astrobiology Strategy for the Search for Life in the Universe* by National Academies of Sciences, Engineering and Medicine (2019).

The first result from the Webb telescope published in a peer-reviewed journal concerns the identification of carbon dioxide in an exoplanet atmosphere (JWST Transiting Exoplanet Community Early Release Science Team 2022). In their publication, they present the detection of $CO_2$ in the atmosphere of the gas giant exoplanet WASP-39b from transmission spectroscopy observations obtained with JWST as part of the Early Release Science Program (ERS). The data used in this study span 3.0–5.5 μm in wavelength and show a prominent $CO_2$ absorption feature at 4.3 μm. The overall spectrum is well matched by one-dimensional, 10× solar metallicity models that assume radiative-convective-thermochemical equilibrium and have moderate cloud opacity. These models predict that the atmosphere should have water, carbon monoxide, and hydrogen sulfide in addition to $CO_2$, but little methane.

Exoplanet WASP-39b is also often called a hot Jupiter because it has a diameter similar to Jupiter's but orbits its star much more closely than Jupiter orbits Sun, making it very hot. The planet was initially discovered in 2011 during ground-based observations; see Faedi et al. (2011). This exoplanet is not in a habitable zone since it is too close to its star. This observation satisfies half of the required life detection capability. It is generally believed that a mixture of carbon dioxide and methane could be an indicator of life (biosignature).

Ab initio calculation of equilibrium isotopic fractionations of potassium and rubidium in minerals and water, reported by Zeng et al. (2019), might be of some help in this search. Their results indicate that K and Rb form bonds of similar strengths and that the ratio between the equilibrium fractionations of K and Rb is approximately 3 to 4. In their calculations, they used first-principle approaches to calculate the equilibrium isotopic fractionation factors of K and Rb in a variety of minerals of geological relevance (orthoclase, albite, muscovite, illite, sylvite, and phlogopite). They also used molecular dynamics simulations to calculate the equilibrium isotopic fractionation factors of K in water. Under low-temperature conditions relevant to weathering of continents or alteration of seafloor basalts (~25 °C), the K isotopic fractionation between solvated K+ and illite (a proxy for K-bearing clays) is +0.24‰, exceeding the current analytical precision, so equilibrium isotopic fractionation can induce measurable isotopic fractionations for this system at low temperature. These findings, however, cannot easily explain why the $\delta^{41}K$ value of seawater is shifted by +0.6‰ relative to igneous rocks. These results indicate that part of the observed fractionation is most likely due to kinetic effects. The narrow range of mean force constants for K and Rb in silicate minerals suggests that phase equilibrium is unlikely to create large K and Rb isotopic fractionations at magmatic temperatures (at least in silicate systems). Kinetic effects associated with diffusion can, however, produce large K and Rb isotopic fractionations in igneous rocks (Zeng et al. 2019).

The search for alien life is hard because we do not know what signatures are unique to life. Marshall et al. (2021) show why complex molecules found in high abundance are universal bio-signatures and demonstrate the first intrinsic experimentally tractable measure of molecular complexity, called the molecular assembly (MA) index. To do this, the authors calculated the complexity of several million molecules and validated that their complexity could be experimentally determined by mass spectrometry. This approach allows identification of molecular bio-signatures from a set of diverse samples from around the world, outer space, and the laboratory, demonstrating it is possible to build a life detection experiment based on MA that could be deployed to extraterrestrial locations and used as a complexity scale to quantify constraints needed to direct prebiotically plausible processes in the laboratory. Such an approach is vital for finding life elsewhere in the universe or creating de-novo life in the lab (Marshall et al. 2021).

According to Summons et al. (2008), we can recognize the residual biochemicals long after life has expired and otherwise lost most life-defining features. Carbon chemistry provides a tool for identifying extant and extinct life on Earth and, potentially, throughout the Universe. In recognizing that certain distinctive compounds isolable from living systems had related fossil derivatives, organic geochemists coined the term biological marker compound or biomarker (e.g., Eglinton et al. 1964) to describe them. In this terminology, biomarkers are metabolites or biochemicals by which we can identify particular kinds of living organisms as well as the molecular fossil derivatives by which we identify defunct counterparts. The terms biomarker and molecular biosignature are synonymous.

A defining characteristic of terrestrial life is its metabolic versatility and adaptability, and it is reasonable to expect that this is universal. Different physiologies operate for carbon acquisition, the garnering of energy, and the storage and processing of information. As well as having a range of metabolisms, organisms build biomass suited to specific physical environments, habitats, and their ecological imperatives. This overall "metabolic diversity" manifests itself in an enormous variety of accompanying product molecules (i.e. natural products). The whole field of organic chemistry grew from their study and now provides tools to link metabolism (i.e. physiology) to the occurrence of biomarkers specific to, and diagnostic for, particular kinds of metabolism.

Another characteristic of living things, also likely to be pervasive, is that an enormous diversity of large molecules are built from a relatively small subset of universal precursors. These include the four bases of DNA, 20 amino acids of proteins, and two kinds of lipid building blocks. Third, life exploits the specificity inherent in the spatial, that is, the three-dimensional, qualities of organic chemicals (stereochemistry). These characteristics then lead to some readily identifiable and measurable generic attributes that would be diagnostic as biosignatures.

Measurable attributes of molecular biosignatures include:

- Enantiomeric excess
- Diastereoisomeric preference
- Structural isomer preference
- Repeating constitutional sub-units or atomic ratios
- Systematic isotopic ordering at molecular and intramolecular levels
- Uneven distribution patterns or clusters (e.g., C-number, concentration, $\delta$ 13C) of structurally related compounds.

The field of prebiotic chemistry has demonstrated that complex organic chemical systems that exhibit various life-like properties can be produced abiotically in the laboratory. Understanding these chemical systems is important for astrobiology and life detection since we do not know the extent to which prebiotic chemistry might exist or have existed on other worlds. Nor do we know what signatures are diagnostic of an extant or "failed" prebiotic system. On Earth, biology has suppressed most abiotic organic chemistry and overprints geologic records of prebiotic chemistry; therefore, it is difficult to validate whether chemical signatures from future planetary missions are remnant or extant prebiotic systems. The "biosignature threshold" between whether a chemical signature is more likely to be produced by abiotic versus biotic chemistry on a given world could vary significantly, depending on the particular environment, and could change over time, especially if life were to emerge and diversify on that world. To interpret organic signatures detected during a planetary mission, Barge et al. (2022) advocate for

1. gaining a more complete understanding of prebiotic/abiotic chemical possibilities in diverse planetary environments and
2. involving experimental prebiotic samples as analogs when generating comparison libraries for "life-detection" mission instruments.

The burden of proof for classifying a chemical signature as biological is higher for other worlds, since many of the chemical signatures that in an Earth sample would be identified as biological cannot be used unambiguously as indicators of life in a world where we do not know the extent of complex abiotic chemistry. The need for facilitating studies of prebiotic/abiotic organic standards with mission instruments means that connections between the OOL community and life detection community will be essential for planetary exploration in upcoming decades.

The major revolution in modern astronomy recognizing the universe as teeming with exoplanets, the discovery of liquid water in solar moons, and the continuing focus on Mars exploration all accelerate the re-evaluation of potential biomarkers for extraterrestrial life. Based on life on planet Earth that relies heavily on chiral molecules and especially on homochiral families, the detection of molecules with these structural properties appears in all road maps as prime indicators of extraterrestrial life. A review paper by Avnir (2021) analyzes the strengths, bounds, and potential weaknesses of relying on chirality and on homochirality as biomarkers, along with recommendations of how to practically use it. Some of the main issues presented, discussed, and answered include:

- what is the extent to which chirality can be expected to be a universal feature of life?
- is detection of chirality enough, or do we need also to detect homochirality?
- how justified is it to view life on Earth as purely homochiral?
- what are the weaknesses of the need to invent an arbitrary label of handedness (needed to define homochirality), and what are the pitfalls that emerge from these weaknesses?
- what stands behind a detected specific value of enantiomeric excess, and what affects its values as we consider old, extinct life; just emerging embryonic life; or extant but rare life?
- how can one quantify the degree of homochirality?
- what are relevant experimental approaches for detecting chirality on the ground and from a distance?

Kacar et al. (2021) discussed the evolutionary history of bio-essential elements that can guide the search for life in the Universe. Our understanding of life in the universe comes from one sample, life on Earth. Current and next-generation space missions will target exoplanets as well as planets and moons in our own solar system with the primary goal of detecting, interpreting, and characterizing indications of possible biological activity. Thus, understanding life's fundamental characteristics is increasingly critical for detecting and interpreting potential biological signatures elsewhere in the universe. Astrobiologists have outlined the essential roles of carbon and water for life, but we have yet to decipher the rules governing the evolution of how living organisms use bio-essential elements. Does the suite of life's essential chemical elements on Earth constitute only one possible evolutionary outcome? Are some elements so essential or biological functions that evolution will select for them despite low availability? How would this play out on other worlds that have different relative element abundances? When we look for life in the universe, or the conditions that could give rise to life, we must learn how to recognize it in extremely different chemical and environmental conditions from those on Earth. We argue that by exposing self-organizing biotic chemistries to different combinations of abiotic materials, and by mapping the evolutionary history of metalloenzyme bio-chemistry onto geological availabilities of metals, alternative element choices that are very different from life's present-day molecular structure might result. A greater understanding of the paleo-molecular evolutionary history of life on Earth will create a predictive capacity for detecting and assessing life's existence on worlds where alternate evolutionary paths might have been taken.

Over the course of a planet's evolution, the environmental availabilities of biologically useful elements such as transition metals can influence natural selection. For instance, the abundances of molybdenum and iron species would likely fluctuate as a function of surface oxygenation, as has occurred over Earth's history. If such metals are essential for biogeochemical processes, their availabilities would also be expected to modulate habitability at the planetary scale. The degree to which availability drives the selection of particular metals can be investigated by reconstructing ancestral metalloenzymes and investigating their paleo-molecular characteristics in a biogeochemical context. This is shown on Figure 5.11, after Kacar et al. (2021).

According to Kacar et al. (2021) metal availability is determined by planetary geophysical and geochemical processes—especially differentiation and subsequent geodynamic and tectonic evolution—and by planetary formation processes that determine bulk planetary chemical inventories. However, the distribution of metals is extremely heterogeneous on and near the Earth's

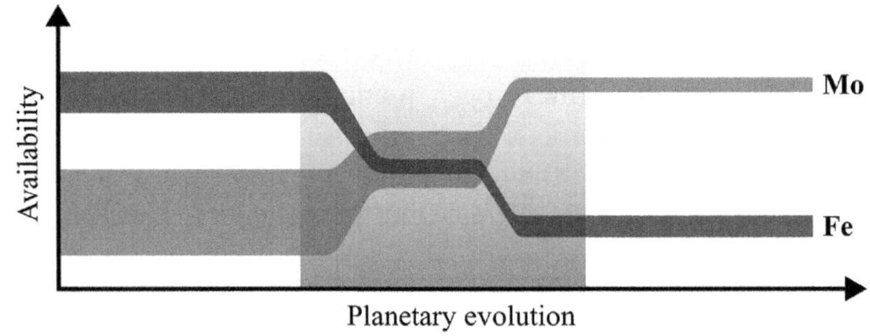


FIGURE 5.11 Over the course of a planet's evolution, the environmental availabilities of biologically useful elements such as transition metals can influence natural selection, after Kacar et al. (2021).

surface. Hence, the intersection of exoplanet geophysics and geochemistry, as well as planet formation, shapes our understanding of which planets are and are not likely to be habitable. Understanding Earth-life metal utilization dynamics can provide necessary guidance as to how we should explore the universe, allowing us to ask more sophisticated questions about the habitability of subsurface oceans on icy worlds and meso- and micro-environments, such as on Mars. Assumptions that couple global metal and geochemical abundances with the modern frequency of biological cofactor utilization may need to be reconsidered to incorporate the possibility that global metal availability and pervasive cofactor utilization can be entirely decoupled. For instance, if it does look like rare metals such as Mo are used for critical metabolic or other chemosynthetic functions, we may have to consider that use of these metals arose in fairly restricted circumstances where they were abundant or that such use evolved in unique niche environments. General assumptions about a correlation between abundance of a cofactor or substrate and its utilization and uptake in biological systems may need to be revisited through laboratory studies utilizing different metals. Studying the paleo-molecular record can help us to assess past biochemical preferences of life using data independent of the geological record. If, as hypothesized for the Mo scenario in N-fixation, laboratory-resurrected enzymes exhibit specificities toward metals that are thought to have been scarce early in Earth history, a disconnect between biological metal election and global metal abundances would be supported. This emerging field of paleo-molecular biology is necessary to understand how these two parameters, biology and geology, impacted one another in complicated and nuanced ways (Kacar et al. 2021).

According to Payne and Kaltenegger (2023), Earth provides a template of evolution for the one habitable planet we know. Earth's atmospheric composition has changed significantly throughout its history. The last $500 \times 10^6$ years, the so-called Phanerozoic Eon, which includes the origins of animals, dinosaurs, and land plants—saw oxygen rise from ≤10% to 35%. But the resulting transmission spectra are a crucial missing piece in our search for signs of life in exoplanet atmospheres. The authors simulated the atmosphere and transmission spectra of the Phanerozoic using estimates from established climate models and present the first high-resolution transmission spectra for Phanerozoic Earth. They demonstrated that the spectral biosignature pairs $O_2$ + $CH_4$ and $O_3$ + $CH_4$ in the atmosphere of a transiting Earth-like planet would indicate a biosphere, with $O_2$ and $O_3$ features potentially stronger than for modern Earth. The full model and high-resolution transmission spectra, covering 0.4–20 μm, are available online and provide a tool to plan and optimize observations, train retrieval methods, and interpret upcoming observations with ground- and space-based telescopes.

## 5.6 LIFE IN OTHER GALAXIES

Assuming Earth is a typical planet around a typical Sun in a typical galaxy, then ($50 \times 10^9$ planets) × ($80 \times 10^9$) galaxies = $4 \times 10^{21}$ planets in the observable Universe or (1%) $4 \times 10^{19}$ in the "habitable zone". This is the basis for the Drake equation, which has many arbitrary and unevaluated parameters. However, complex life may be possible in only 10% of all galaxies; radiation from gamma ray bursts may render the rest uninhabitable. Stellar explosions resulting in gamma ray bursts would regularly wipe out any life forms more elaborate than microbes. The detonations also kept the universe lifeless for billions of years after the Big Bang; see Piran and Jimenez (2014)

Astronomers have spent decades searching for galaxies sufficiently like our Milky Way that could be candidates to support advanced life. So far, they have yet to find a candidate. Astronomers' search is limited by the fact that they observe galaxies as they were in the past. The more distant the galaxy, the farther back in time astronomers see that galaxy. Also, the more distant the galaxy, the fewer details astronomers can measure in that galaxy. Nevertheless, in observing a galaxy, for example, 8 billion lightyears away, and hence seeing it as it was 8 billion years ago, astronomers can predict what that galaxy will be like 8 billion years in the future. Astronomers have yet to find a galaxy whose predicted future characteristics will sufficiently match our own galaxy where it could be a candidate to presently host advanced life. That is, the possibility for the existence of extraterrestrial physical intelligent life is remote. There is a somewhat higher possibility for finding a galaxy where microbes could conceivably exist.

The JWST has ushered in an era of unprecedented ability to characterize exoplanetary atmospheres. While there are over 5000 confirmed planets, more than 4000 TESS planet candidates are still unconfirmed, and many of the best planets for atmospheric characterization may remain to be identified. Hord et al. (2023) presented a sample of TESS planets and planet candidates that they identify as "best-in-class" for transmission and emission spectroscopy with JWST. These targets are sorted into bins across equilibrium temperature $T_{eq}$ and planetary radius $R_p$ and are ranked by transmission and emission spectroscopy metric (TSM and ESM, respectively) within each bin. In forming the target sample, they perform cuts for expected signal size and stellar brightness to remove sub-optimal targets for JWST. Of the 194 targets in the resulting sample, 103 are unconfirmed TESS planet candidates, also known as TESS Objects of Interest (TOIs). They perform vetting and statistical validation analyses on these 103 targets to determine which are likely planets and which are likely false positives, incorporating ground-based follow-up from the TESS Follow-up Observation Program (TFOP) to aid the vetting and validation process. They statistically validate 23 TOIs, marginally validate 33 TOIs to varying levels of confidence, deem 29 TOIs likely false positives, and leave the dispositions for 4 TOIs as inconclusive. Fourteen of the 103 TOIs were confirmed independently over the course of their analysis. They provide the final best-in-class sample as a community resource for future JWST proposals and observations.

NASA's all-sky survey mission, the Transiting Exoplanet Survey Satellite, is specifically designed to detect transiting exoplanets orbiting bright stars. TESS has already identified about 400 transiting exoplanets, as well as approximately 6000 candidates awaiting validation. In their work, Mistry et al. (2023) present the outcomes of the project Validation of Transiting Exoplanets using Statistical Tools (VaTEST), an ongoing endeavor dedicated to validating and characterizing new exoplanets using statistical validation tools. They have validated eight potential super-Earths using a combination of ground-based telescope data, high-resolution imaging, and the statistical validation tool known as TRICERATOPS. These validated planets bear the designations: TOI-238b ($1.61^{+0.09}_{-0.10}$ $R_{\oplus}$), TOI-771b ($1.42^{+0.11}_{-0.09}$ $R_{\oplus}$), TOI-871b ($1.66^{+0.11}_{-0.11}$ $R_{\oplus}$), TOI-1467b ($1.83^{+0.16}_{-0.15}$ $R_{\oplus}$), TOI-1739b ($1.69^{+0.10}_{-0.08}$ $R_{\oplus}$), TOI-2068b ($1.82^{+0.16}_{-0.15}$ $R_{\oplus}$), TOI-4559b ($1.42^{+0.13}_{-0.11}$ $R_{\oplus}$), and TOI5799b ($1.62^{+0.19}_{-0.13}$ $R_{\oplus}$). They also studied the synthetic transmission spectra of all eight validated planets in the HST and JWST band-pass using PLATON and PandExo. Among all these validated planets, six of them fall within the region known as "keystone planets", which makes them particularly interesting for study. Based on the location of TOI-771b and TOI-4559b below the radius valley, they characterized them as likely super-Earths, though radial velocity mass measurements for these planets will provide more details about their characterization. It is noteworthy that planets within the size range investigated herein are absent from our own solar system, making their study crucial for gaining insights into the evolutionary stages between Earth and Neptune.

## REFERENCES

Adams, F. C., and Napier, K. J. 2022. Transfer of rocks between planetary systems: Panspermia revisited. Astrobiology 22(12): 1429–1442.

Adams, F. C., and Spergel, D. N. 2005. Lithopanspermia in star-forming clusters. Astrobiology 5(4): 497–514. https://doi.org/10.1089/ast.2005.5.497.

Allen, C., and Martos, M. A. 1986. A simple, realistic model of the galactic mass distribution for orbit computations. Revista Mexicana de Astronomía y Astrofísica 13: 137–147.

Álvarez-Carretero, S., Tamuri, A. U., Battini, M., et al. 2021. A species-level timeline of mammal evolution integrating phylogenomic data. Nature, Accelerated Article Preview. https://doi.org/10.1038/s41586-021-04341-1.

Anbar, A. D. 2008. Elements and evolution. Science 322(5907): 1481–1483. https://doi.org/10.1126/science.1163100.

Anslow, R. J., Bonsor, A., and Rimmer, P. B. 2023. Can comets deliver prebiotic molecules to rocky exoplanets? Proceedings of the Royal Society A479: 20230434 (17 pages). https://doi.org/10.1098/rspa.2023.0434.

Avnir, D. 2021. Critical review of chirality indicators of extraterrestrial life. New Astronomy Reviews 92: 101596 (8 pages). https://doi.org/10.1016/j.newar.2020.101596.

Barge, L. M., Rodriguez, L. E., Weber, J. M., and Theiling, B. P. 2022. Determining the "biosignature threshold" for life detection on biotic, abiotic, or prebiotic worlds. Astrobiology 22(4): 481–493. https://doi.org/10.1089/ast.2021.0079.

Bateman, A. 2020. Division of labour in a matrix, rather than phagocytosis or endosymbiosis, as a route for the origin of eukaryotic cells. Biology Direct 15: (33 pages). htpps://doi.org/10.1186/s13062-020-00260-9.

Belbruno, E., Moro-Martın, A., Malhotra, R., and Savransky, D. 2012. Chaotic exchange of solid material between planetary systems: Implications for lithopanspermia. Astrobiology 12(8): 754–774. https://doi.org/10.1089/ast.2012.0825.

Brasier, M., Green, O., Linsay, J., and Steele, A. 2004. Earth's oldest (~3.5 Ga) fossils and the "early Eden hypothesis": Questioning the evidence. Origins of Life and Evolution of the Biosphere 34: 257–269.

Carr, B., and Rees, M. 1979. The anthropic principle and the structure of the physical world. Nature 278: 605–612. https://doi.org/10.1038/278605a0.

Carter, B. 1974. The Anthropic Principle and Large Number Coincidences. In M. Longair (ed.), Confrontation of Cosmological Theories with Observation. Dordrecht: D. Reidel.

Chang, Y., Yu, Z., An, F., et al. 2021 Three body photodissociation of the water molecule and its implication for prebiotic oxygen production. Nature Communication 12. https://doi.org/10.1038/s41467-021-22824-7.

Chyba, C. F. 1990. Impact delivery and erosion of planetary oceans in the early inner solar system. Nature 343: 129–133. https://doi.org/10.1038/343129a0.

Cleaves, II., H. J., Hystad, G., Prabhu, A., and Hazen, R. M. 2023. A robust, agnostic molecular biosignature based on machine learning. PNAS 120(41): e2307149120 https://doi.org/10.1073/pnas.2307149120.

Cockell, C. S., Bush, T., Bryce, C., et al. 2016. Habitability: A review. Astrobiology 16(1): 89-117. https://doi.org/10.1089/ast.2015.1295.

Davies, R. E. 1988. Panspermia: Unlikely, unsupported, but just possible. Acta Astronomica 17: 129–1 35.

Desch, S. J., and Jackson, A. P. 2022. Some pertinent issues for interstellar panspermia raised after the discovery of 1I/'Oumuamua. Astrobiology 22(12): 1400–1413. http://doi.org/10.1089/ast.2021.0199.

Dick, S. J. 2003. Cultural evolution, the postbiological universe and SETI. International Journal of Astrobiology 2(1): 65–74. https://doi.org/10.1017/S147355040300137X.

Dicke, R. H. 1961. Dirac's cosmology and Mach's principle. Nature 192: 440–441.

Ditzler, M. A., Popović, M., and Zajkowski, T. 2022. Chapter 5 – from Building Blocks to Cells. In R. Thombre and P. Vaishampayan (eds.), New Frontiers in Astrobiology (pp. 111–133). Elsevier Inc. https://doi.org/10.1016/B978-0-12-824162-2.00010-5.

Doyle, L. R., Carter, J. A., Fabrycky, D. C., et al. 2011. Kepler-16: A transiting circumbinary planet. Science 333(6049): 1602–1606. http://doi.org/10.1126/science.1210923.

Drake, F. 2015. The Drake Equation: Estimating the Prevalence of Extraterrestrial Life Through the Ages (Vol. 8). Cambridge: Cambridge University Press.

Dumais, J., Serikawa, K., and Mandoli, D. F. 2000. Acetabularia: A unicellular model for understanding subcellular localization and morphogenesis during development. Journal of Plant Growth Regulation 19(3): 253–264. https://doi.org/10.1007/s003440000035.

Dyson, F. 1981. Disturbing the Universe. Sloan Foundation Science Series. New York, NY: Basic Books.

Earman, J. 1987. The SAP also rises: A critical examination of the anthropic principle. American Philosophical Quarterly 24(4): 307–317.

Eglinton, G., Scott, P. M., Belsky, T., et al. 1964. Hydrocarbons of biological origin from a one-billion-year-old sediment. Science145(3629): 263–264. https://doi.org/10.1126/science.145.3629.263.

Ehrenfreund, P., and Menten, K. M. 2001. From Molecular Clouds to the Origin of Life. In G. Horneck and C. Baumstark-Kahn (eds.), Astrobiology: The Quest for the Conditions of Life (pp. 7–24). Berlin: Springer Verlag; New York: Heidelberg; London, Milan, Paris, Tokyo: Barcelonas.

Eme, L., Tamarit, D., Caceres, E. F., et al. 2023. Inference and reconstruction of the heimdallarchaeial ancestry of eukaryotes. Nature 618: 992–999. https://doi.org/10.1038/s41586-023-06186-2.

Faedi, F., Barros, S. C. C., Anderson, D. D., et al. 2011. WASP-39b: A highly inflated Saturn-mass planet orbiting a late G-type star. Astronomy & Astrophysics 531: 40 (8 pages). https://doi.org/10.1051/0004-6361/201116671.

Farihi, J., Hermes, J. J., Marsh, T. R., et al. 2022. Relentless and complex transits from a planetesimal debris disc. Monthly Notices of the Royal Astronomical Society 511: 1647–1666. https://doi.org/10.1093/mnras/stab3475.

Feng, L. 2021. Nebula-relay hypothesis: Primitive life in nebula and origin of life on earth. (11 pages). arXiv:1910.06396v4 [physics.pop-ph], 12 Jun.

Feng, L. 2022. Nebula-relay hypothesis: The chirality of biological molecules in molecular clouds. Frontiers in Astronomy and Space Sciences 9: 794067 (5 page). www.frontiersin.org/articles/10.3389/fspas.2022.794067.

Fischer, D. A., and Valenti, J. 2005. The planet-metallicity correlation. The Astrophysical Journal 622: 1102–1117.

Forgan, D. H. 2018. On the possibility of detecting class A stellar engines using exoplanet transit curves. Journal of the British Interplanetary Society 66: 144–154.

France, K., Duvvuri, G., Egan, H., et al. 2020. The high-energy radiation around a 10 Gyr M dwarf: Habitable at last? The Astronomical Journal 160: 237 (14 pages). https://doi.org/10.3847/1538-3881/abb465.

Franck, S., Block, A., von Bloh, W., et al. 2001. Planetary habitability: Is earth commonplace in the milky way? Naturwissenschaften 88: 416–426.

Friederich, S. 2019. A new fine-tuning argument for the multiverse. Foundations of Physics 49: 1011–1021. https://doi.org/10.1007/s10701-019-00246-2.

Furukawa, Y., Chikaraishi, Y., Ohkouchi, N., et al. 2019. Extraterrestrial ribose and other sugars in primitive meteorites. Proceedings of the National Academy of Sciences of the United States of America 116: 24440–24445. https://doi.org/10.1073/pnas.1907169116.

Gast, P. W. 1965. Terrestrial ratio of potassium to rubidium and the composition of earth's mantle. Science 147(3660): 858–60. https://doi.org/10.1126/science.147.3660.858.

Georgakarakos, N., Eggl, S., and Dobbs-Dixon, I. 2021. Circumbinary habitable zones in the presence of a giant planet. Frontiers in Astronomy and Space Sciences 8: 640830 (13 pages). https://doi.org/10.3339/fspas.2021.640830.

Ginsburg, I., Lingam, M., and Loeb, A. 2018. Galactic panspermia. Astrophysical Journal Letters 868: L12 (6 pages). https://doi.org/10.3847/2041–8213/aaef2d.

Gillon, M., Triaud, A., Demory, B. O., et al. 2017. Seven temperate terrestrial planets around the nearby ultra-cool dwarf star TRAPPIST-1. Nature 542: 456–460. https://doi.org/10.1038/nature21360.

Goncharov, G. N., and Orlov, V. V. 2003. Global repeating events in the history of the earth and the motion of the sun in the galaxy. Astronomy 47: 925–933.

Gonzalez, G. 2005. Habitable zones in the universe. Origins of Life and Evolution of Biospheres35: 555–606. https://doi.org/10.1007/s11084-005-5010-8.

Gonzalez, G., Brownlee, D., and Ward, P. 2001. The galactic habitable zone: Galactic chemical evolution. Icarus 152: 185–200. https://doi.org/10.1006/icar.2001.6617.

Green, J., Hoehler, T., Neveu, M., et al. 2021. Call for framework for reporting evidence for life beyond earth. Nature 598: 575–579. https://doi.org/10.1038/s41586-021-03804-9.

Greenberg, J. M. 2002. Cosmic dust and our origin. Surface Science 500: 793–822. PII: S0039-6028(01)01555–2.

Guttenberg, N., Chen, H., Mochizuki, T., and Cleaves II, H. J. 2021. Classification of the biogenecity of complex organic mixtures for the detection of extraterrestrial life. Life 11: 234 (20 pages). https://doi.org/10.3390/life11030234.

Haldane, J. B. S. 1929. The Origin of Life. In A. W. Charles (ed.), The Rationalist Annual for the Year 1929 (pp. 3–10). Reprinted in Bernal, J. D. 1967. The Origin of Life. In R. Carrington (ed.), The Weidenfeld and Nicolson Natural History (pp. 242–249). London: Readers Union.

Hansen, M. S., and Zuckerman, B. 2021. Minimal conditions for survival of technological civilizations in the face of stellar evolution. The Astronomical Journal 161(3): 145 (6 pages). https://doi.org/10.3847/1538-3881/abd547.

Harland, W. B., Armstrong, R. L., Cox, A. V., et al. 1990 A Geologic Time Scale 1989. Cambridge: Cambridge University. Press.

Hoefs, J. 2009. Geochemical fingerprints: A critical appraisal. European Journal of Mineralogy 22: 3–15. Published online December 2009.

Hord, B. J., Kempton, E. M.-R., Thomas Mikal-Evans, T., et al. 2023. Identification of the top TESS objects of interest for atmospheric characterization of transiting exoplanets with JWST. (24 pages). arXiv:2308.09617v1 [astro-ph.EP], 18 Aug.

Hu, R., Damiano, M., Scheucher, M., et al. 2021. Unveiling shrouded oceans on temperate sub-Neptunes via transit signatures of solubility equilibria vs. gas thermochemistry. arXiv:2108.04745v1 [astro-ph.EP]. https://doi.org/10.48550/arXiv.2108.04745.

Huang, S.-S. 1959. The problem of life in the universe and the mode of star formation. Publications of the Astronomical Society of the Pacific 71(422): 421–424.

Huston, M., and Wright, J. 2022. Signatures of alien technology could be how humanity finds extraterrestrial life. The Conversation, 21 Oct.

ICRP, International Commission on Radiological Protection. 1975. Report of the Task Group on Reference Man. In ICRP Publication No. 23. Oxford: Pergamon Press.

Ionescu, D., and Bizic, M. 2022. Giant Bacteria. In eLS, John Wiley&Sons,Ltd.https://doi.org/10.1002/9780470015902.a0020371.pub2.

Jelen, B. I., Giovannelli, D., and Falkowski, P. G. 2016. The role of microbial electron transfer in the coevolution of the biosphere and geosphere. Annual Review of Microbiology 70: 45–62. https://doi.org/10.1146/annurev-micro-102215-095521.

Jiang, J. H., Rosen, P. E., Lu, K., et al. 2022. Avoiding the "great filter": Extraterrestrial life and humanity's future in the universe. http://arxiv.org/abs/2210.10582.

Jones, E. M. 1985. Where is everybody? An account of Fermi's question. Technical Report LA-10311-MS, Los Alamos National Laboratory. www.osti.gov/accomplishments/documents/fullText/ACC0055.pdf.

Jarugula, S., Vieira, J. D., Weiss, A., et al. 2021. Molecular line observations in two dusty star-forming galaxies at z = 6.9. The Astrophysical Journal 921: 97 (26 pages). https://doi.org/10.3847/1538-4357/ac21db.

JWST Transiting Exoplanet Community Early Release Science Team. 2022. Identification of carbon dioxide in an exoplanet atmosphere. Nature—Accelerated Article Preview (24 pages). https://doi.org/10.1038/s41586-022-05269-w.

Kacar, B., Garcia, A. K., and Anbar, A. D. 2021. Evolutionary history of bioessential elements can guide the search for life in the universe. ChemBioChem 22: 114–119. https://doi.org/10.1002/cbic.202000500.

Kasting, J. F., Whitmire, D. P., and Reynolds, R. T. 1993. Habitable zones around main sequence stars. Icarus 101: 108–128. https://doi.org/10.1006/icar.1993.1010.

Kloprogge, J. T., and Hartman, H. 2022. Clays and the origin of life: The experiments. Life 12: 259 (54 pages). https://doi.org/10.3390/life12020259.

Kobayashi, C., Karakas, A. I., and Lugaro, M. 2020. The origin of elements from carbon to uranium. The Astrophysical Journal 900: 179 (33 pages). https://doi.org/10.3847/1538-4357/abae65.

Kobayashi, K., Ise, J.-I., Aoki, R., et al. 2023. Formation of amino acids and carboxylic acids in weakly reducing planetary atmospheres by solar energetic particles from the young sun. Life 13: 1103 (17 pages). https://doi.org/10.3390/life13051103.

Komiya, T., Yamamoto, S., Aoki, S., et al. 2015. Geology of the Eoarchean, > 3.95 Ga, nulliak supracrustal rocks in the Saglek block, northern Labrador, Canada: The oldest geological evidence for plate tectonics. Tectonophysics 662: 40–66.

Kopparapu, R. K., Ramirez, R., Kasting, J. F., et al. 2013. Habitable zones around main sequence stars: New estimate. The Astrophysical Journal 765: 131 (16 pages). https://doi.org/10.1088/0004-637X/65/2/131.

Kostov, V. B., McCullough, P. R., Carter, et al. 2014. Kepler-413b: A slightly misaligned, Neptune-size transiting circumbinary planet. The Astrophysical Journal 784: 14 (18 pages). https://doi.org/10.1088/0004-637X/784/1/14.

Krasnokutski, S. A., Chuang, K.-J., Jäger, C., et al. 2022. A pathway to peptides in space through the condensation of atomic carbon. Nature Astronomy 6: 381–386. https://doi.org/10.1038/s41550-021-01577-9.

Lammer, H., Bredehöft, J. H., Coustenis, A., et al. 2009. What makes a planet habitable? The Astronomy and Astrophysics Review 17: 181–249.

Lobo, A. H., Shields, A. L., Palubski, I. Z., et al. 2023. Terminator habitability: The case for limited water availability on M-dwarf planets. The Astrophysical Journal 945: 161 (13 pages). https://doi.org/10.3847/1538-4357/aca970.

Lous, M. M., Helled, R., and Mordasini, C. 2022. Potential long-term habitable conditions on planets with primordial H-He atmospheres. Nature Astronomy 6: 819–827. https://doi.org/10.1038/s41550-022-01699-8.

Marshall, S. M., Mathis, C., Carrick, E., et al. 2021. Identifying molecules as biosignatures with assembly theory and mass spectrometry. Nature Communications 12: 3033 (9 pages). https://doi.org/10.1038/s41467-021-23258-x.

McKay, C. P., Davies, P. C. W., and Worden, S. P. 2022 Directed panspermia using interstellar comets. Astrobiology 12: 1443–1451. https://doi.org/10.1089/ast.2021.0188.

Meech, K. J., Weryk, R., Micheli, M., et al. 2017. A brief visit from a red and extremely elongated interstellar asteroid. Nature 552: 378–381. https://doi.org/10.1038/nature25020.

Melosh, H. J. 2003. Meteorites (and Life?) between stellar systems. Astrobiology 3(1): 207–215. https://doi.org/10.1089/153110703321632525.

Meltzer, H. L., Taylor, R. M., Platman, S. R., and Fieve, R. R. 1969. Rubidium: A potential modifier of affect and behavior. Nature 223: 321–322.

Menozzi, P., Norman, D., Polleri, A., et al. 1959. Specific intracellular binding of rubidium by rat diaphragm muscle. The Proceedings of the National Academy of Sciences 45: 80–88.

Michalski, J. R., Goudge, T. A., Crowe, S. A., et al. 2022. Geological diversity and microbiological potential of lakes on Mars. Nature Astronomy 6: 1133–1141. https://doi.org/10.1038/s41550-022-01743-7.

Michalski, J. R., Onstott, T. C., Mojzsis, S. J., et al. 2018. The Martian subsurface as a potential window into the origin of life. Nature Geoscience 11: 21–26.

Mistry, P., Prasad, A., Maity, M., et al. 2023. VaTEST III: Validation of 8 Potential Super-Earths from TESS Data. In Cambridge Large (pp. 1–25). arXiv:2311.00688v1 [astro-ph.EP].

Mojzsis, S. J., Arrhenius, G., McKeegan, K. D., et al. 1996. Evidence for life on earth before 3,800 million years ago. Nature 384: 55–59.

Moll, D., Kotterba, P., Jochum, K. P., et al. 2019. Elemental inventory in fish otoliths reflects natal origin of Atlantic herring (Clupea harengus) from Baltic Sea juvenile areas. Frontiers in Marine Science 6: Article 191 (11 pages). https://doi.org/10.3389/fmars.2019.00191.

Müller, F., Escobar, L., Xu, F., et al. 2022. A prebiotically plausible scenario of an RNA-peptide world. Nature 605: 279–284.

Napier, W. M. 2004. A mechanism for interstellar panspermia. Monthly Notices Royal Astronomical Society 348(1): 46–51. https://doi.org/10.1111/j.1365-2966.2004.07287.x.

National Academies of Sciences, Engineering, and Medicine. 2019. An Astrobiology Strategy for the Search for Life in the Universe. Washington, DC: The National Academies Press. https://doi.org/10.17226/25252.

National Research Council of National Academies. 2007. The Limits of Organic Life in Planetary Systems. Washington, DC: The National Academies Press.

Nutman, A. P., McGregor, V. R., Friend, et al. 1996. The itsaq gneiss complex of southern west Greenland; The world's most extensive record of early crustal evolution (3900–3600 Ma). Precambrian Research 78(1–3): 1–39.

Oba, Y., Takano, Y., Furukawa, Y., et al. 2022. Identifying the wide diversity of extraterrestrial purine and pyrimidine nucleobases in carbonaceous meteorites. Nature Communication 13: 2008 (10 pages). https://doi.org/10.1038/s41467-022-29612-x.

Oliver, T., Sánchez-Baracaldo, P., Larkum, A. W., et al. 2021. Time-resolved comparative molecular evolution of oxygenic photosynthesis. BBA—Bioenergetics 1862: 148400 (20 pages). https://doi.org/10.1016/j.babio.2021.184400.

Oparin A. 1924. Proiskhozhedenie Zhizni. Moscow: Moscovksii Rabotchii, Moscow. The origin of Life, first version in English published in 1967.

Oyston, J. W., Wilkinson, M., Ruta, M., et al. 2022. Molecular phylogenies map to biogeography better than morphological ones. Communications Biology 5: 521 (12 pages). https://doi.org/10.1038/s42003–022–03482-x.

Payne, R. C., and Kaltenegger, L. 2023. Oxygen bounty for earth-like exoplanets: Spectra of earth through the phanerozoic. Monthly Notices of the Royal Astronomical Society 527(1): 1151–1155. https://doi.org/10.1093/mnrasl/slad147.

Piran, T., and Jimenez, R. 2014. Possible role of gamma ray bursts on life extinction in the Universe. Physical Review Letters 11302. https://doi.org/10.1103/PhysRevLett.113.231102.

Piran, T., and Jiminez, R. 2014. On the role of GRBs on life extinction in the universe. arXiv:1409.2506 [astro-ph.HE].

Potapov, A., Bouwman, J., Jäger, C., and Henning, T. 2022. Dust/ice mixing in cold regions and solid-state water in the diffuse interstellar medium. Nature Astronomy 5: 78–85. https://doi.org/10.1038/s41550-020-01214-x.

Prantzos, N. 2008. On the "galactic habitable zone". Space Science Reviews 135: 313–322.

Preiner, M., Xavier, J. C., Vieira, A. do N., et al. 2019. Catalysts, autocatalysis and the origin of metabolism. Interface Focus 9: 20190072 (8 pages). http://doi.org/10.1098/rsfs.2019.0072.

Rambo, I. M., Langwig, M. V., Leão, P., et al. 2022. Genomes of six viruses that infect asgard archaea from deep-sea sediments. Nature Microbiology 7: 953–961. https://doi.org/10.1038/s41564-022-01150-8.

Redmond, A. K., Casey, D., Gundappa, M. K., et al. 2023. Independent rediploidization masks shared whole genome duplication in the sturgeon-paddlefish ancestor. Nature Communication 14: 2879 (15 pages). https://doi.org/10.1038/s41467-023-38714-z.

Rivilla, V. M., Jiménez-Serra, I., Martın-Pintado, J., et al. 2021. Discovery in space of ethanolamine, the simplest phospholipid head group. PNAS 118(22): e2101314118 (pages 8). https://doi.org/10.1073/pnas.2101314118.

Rivilla, V. M., Jiménez-Serra, I., Martin-Pintado, J., et al. 2022. Molecular precursors of the RNA-world in space: New nitriles in the G+0.693–0.027 molecular cloud. Frontiers in Astronomy and Space Science 9: 876870 (18 pages).

Romanovskaya, I. K. 2022. Migrating extraterrestrial civilizations and interstellar colonization: Implications for SETI and SETA. International Journal of Astrobiology (25 pages). https://doi.org/10.1017/S1473550422000143.

Ruderman, M. 1974. Possible consequences of nearby supernova explosions for atmospheric ozone and terrestrial life. Science 184: 1079–1081. https://doi.org/10.1126/science.184.4141.1079.

Sagear, S., and Ballard, S. 2023. The orbital eccentricity distribution of planets orbiting M dwarfs. PNAS 120(23): e2217398120 (23 pages). https://doi.org/10.1073/pnas.2217398120.

Sandberg, A., Drexler, E., and Ord, T. 2018. Dissolving the Fermi paradox. arXiv:1806.02404v1 [physics.pop-ph], 06 Jun.

Sauterey, B., Charnay, B., Affholder, A., et al. 2022. Early Mars habitability and global cooling by $H_2$-based methanogens. Nature Astronomy. https://doi.org/10.1038/s41550-022-01786-w.

Schmitt, J. R., Rapporteur. 2017. Searching for Life Across Space and Times. Proceeding of a Workshop by National Academies Science, Engineering, Medicine. Washington, DC: The National Academies Press. https://doi.org/10.17226/24860.

Schoener, T. W. 1974. Resource partitioning in ecological communities. Science 185: 27. https://doi.org/10.1126/science.185.4145.27.

Schoepp-Cothenet, B., von Lis, R., Phillippot, P., et al. 2012. The ineluctable requirement for the trans-iron elements molybdenum and/or tungsten in the origin of life. Scientific Reports 2: 263 (5 pages). https://doi.org/10.1038/srep00263.

Schopf, J. W. 1993. Microfossils of the early archean apex chert: New evidence of the antiquity of life. Science 260: 640–646.

Schopf, J. W., Kitajima, K., Spicuzza, M. J., Kudryaytsev, A. B., and Valley, J. W. 2017. SIMS analyses of the oldest known assemblage of microfossils document their taxon-correlated carbon isotope compositions. PNAS 115(1): 53–58. https://doi.org/10.1073/pnas.1718063115.

Schreder-Gomes, S. I., Benison, K. C., and Bernau, J. A. 2022. 830-million-year-old microorganisms in primary fluid inclusions in halite. Geology. https://doi.org/10.1130/G49957.1.

Schreider, A. A., Schreider, Al. A., Varga, P., and Denis, C. 2011. Variations of the Earth's magnetic field in the Phanerozoic. Oceanology 51(3): 506–510. Original Russian Text published in Okeanologiya 51(3): 537–541.

Sephus, C. D., Fer, E., Garcia, A. K., et al. 2022. Earliest photic zone niches probed by ancestral microbial rhodopsins. Molecular Biology and Evolution 39(5): msac100. https://doi.org/10.1093/molbev/msac100.

Snellen, I. A. G., Snik, F., Kenworthy, M., et al. 2021. Detecting life outside our solar system with a large high-contrast-imaging mission. Experimental Astronomy. https://doi.org/10.1007/s10686-021-09803-y.

Summons, R. E., Albrecht, P., McDonald, G., and Moldowan, J. M. 2008. Molecular biosignatures. Space Science Reviews 135: 133–159.

Susskind, L. 2005. The Cosmic Landscape: String Theory and the Illusion of Intelligent Design. New York: Back Bay Books.

Špaček, J., and Benner, S. A. 2022. Agnostic life finder (ALF) for large-scale screening of Martian life during in situ refueling. Astrobiology 22(10): 1255–1263. http://doi.org/10.1089/ast.2021.0070.

Tashiro, T., Ishida, A., Hori, M., et al. 2017. Early trace of life from 3.95 ga sedimentary rocks in labrador, Canada. Nature 549: 516–518.

Trail, D., Elsila, J., Müller, U. F., et al. 2022. Rethinking the search for the origins of life. Eos 103: (8 pages). https://doi.org/10.1029/2022EO220065.

Tremblay, C. D., Price, D. C., and Tingay, S. J. 2022. A Search for technosignatures toward the galactic centre at 150 MHz. arXiv:2202.03324v1 [astro-ph.GA], 7 Feb.

Volland, J.-M., Gonzalez-Rizzo, S., Gros, O., et al. 2022. A centimeter-long bacterium with DNA contained in metabolically active, membrane-bound organelles. Science 376: 1453–1458.

Wagner, K., Boehle, A., Pathak, P., et al. 2021. Imaging low-mass planets within the habitable zone of α Centauri. Nature Communication 12: 922 (7 pages). https://doi.org/10.1038/s41467-021–21176–6.

Wallis, M. K., and Wickramasinghe, N. C. 2004. Interstellar transfer of planetary microbiota, monthly. Notices Royal Astronomical Society 348(1): 52–61. https://doi.org/10.1111/j.1365-2966.2004.07355.x.

Ward, P. D. and Brownlee, D. 2000. Rare Earth: Why Complex Life is Uncommon in the Universe. Copernicus Books; New York: Springer-Verlag, Inc.

Welsh, W. F., Orosz, J. A., Carter, J. A., et al. 2012. Transiting circumbinary planets Kepler-34 b and Kepler-35 b. Nature 481: 475–479. https://doi.org/10.1038/nature10768.

Wesson, P. S. 2010. Panspermia, past and present: Astrophysical and biophysical conditions for the dissemination of life in space. Space Science Review 156: 239–252. 2 https://doi.org/10.1007/s11214–010–9671-x.

Westby, T., and Conselice, C. J. 2020. The astrobiological Copernican weak and strong limits for intelligent life. The Astrophysical Journal 896: 58 (18 pages). https://doi.org/10.3847/1538-4357/ab8225.

Yang, C., Omont, A., Martín, S., et al. 2023. Sunrise: The rich molecular inventory of high-redshift dusty galaxies revealed by broadband spectral line surveys. Astronomy & Astrophysics 680: A95 (39 pages). https://doi.org/10.1051/0004-6361/202347610.

Zeng, H., Rozsa, V. F., Nie, N. X., et al. 2019. Ab initio calculation of equilibrium isotopic fractionations of potassium and rubidium in minerals and water. ACS Publications. Collection. https://doi.org/10.1021/acsearthspacechem.9b00180.

## ADDITIONAL READINGS

Bejan, A. 2016. Life and evolution as physics. Communicative & Integrative Biology 9 (3): e1172159 (13 pages). https://doi.org/10. 1080/19420889.2016.1172159.

Cabrera, T., and Rodriguez, C. L. 2023. Runaway and hypervelocity stars from compact object encounters in globular clusters. arXiv:2302.03048v1 [astro-ph.GA], 6 Feb.

Cleland, C. E., and Chyba, C. F. 2002. Defining "life". Origins of Life and Evolution of the Biosphere 32(4): 387–393.

Delhaize, J., Heywood, I., Prescott, M., et al. 2021. Mightee: Are giant radio galaxies more common than we thought? Monthly Notices of the Royal Astronomical Society 501: 3833–845.

Deng, H., Mayer, L., and Helled, R. 2021. Formation of intermediate-mass planets via magnetically controlled disk fragmentation. Nature Astronomy Letters. https://doi.org/10.1038/s41550-020-01297-6.

Di Stefano, R., and Ray, A. 2016. Globular clusters as cradles of life and advanced civilizations. The Astrophysical Journal 827: 54 (12 pages). http://doi.org/10.3847/0004-637X/827/1/54.

Forbes, D. A., Bastian, N., Gieles, M., et al. 2018. Globular cluster formation and evolution in the context of cosmological galaxy assembly: Open questions. Proceedings Royal Society A474: 20170616 (27 pages). http://dx.doi.org/10.1098/rspa.2017.0616.

Giullani, A., Drysdale, R. N., Woodhead, J. D., et al. 2022. Perturbation of the deep-Earth carbon cycle in response to the Cambrian Explosion. Science Advances 8: eabj1325 (11 pages).

Gómez-Márquez, J. 2021. What is life? Molecular Biology R. https://doi.org/10.1007/s11033-021-06594-5.eports.

Grimaldi, C., Lingam, M., and Balbi, A. 2021. Feasibility of detecting interstellar panspermia in astrophysical environments. arXiv:2105.03295v1 [astro-ph.EP], 07 May.

Gronstal, A. L. 2014. Light scattering on dust holds clues to habitability. Astrobiology Magazine, www.astrobio.net/alien-life/light-scattering-dust-holds-clues-habitability/.

Gurnett, D. A., Kurth, W. S., Stone, E. C., et al. 2021. A foreshock model for interstellar shocks of solar origin: Voyager 1 and 2 observations. The Astronomical Journal 161: 11 (11 pages).

Hammouda, T., Boyer, M., Frossard, P., and Cartier, C. X. 2022. The message of oldhamites from enstatite chondrites. Progress in Earth and Planetary Science 9: 13 https://doi.org/10.1186/s40645-022-00471-w.

Hinkley, S., Lacour, S., Marleau, G.-D., et al. 2022. Direct discovery of the inner exoplanet in the HD206893 system (13 pages). arXiv:2208.04867v4 [astro-ph.EP]. https://doi.org/10.48550/arXiv.2208.04867.

Jedamzik, K., and Pospelov, M. 2009. Big bang nucleosynthesis and particle dark matter. New Journal of Physics 11: 105028 (20 pages). https://doi.org/10.1088/1367-2630/11/10/105028.

Kobayashi, C. Tsujimoto, T., and Nomoto, K. 2000. The history of the cosmic supernova rate derived from the evolution of the host Galaxies. arXiv:astro-ph/9908005v2, 17 Mar.

Kudella, P. W., Tkachenko, A. V., Salditt, A., et al. 2021. Structured sequences emerge from random pool when replicated by templated ligation. PNAS 118(8): e2018830118 (8 pages). https://doi.org/10.1073/pnas.2018830118.

Larriestra, F., Covo, M. E., Cuneo, R., et al. 2015. The K/Rb ratio and its geological and petrophysical implications: A test case of the San Jorge Gulf Basin, Argentina. Conference paper, Adapted from Extended Abstract Prepared in Conjunction with Poster Presentation at AAPG Annual Convention & Exhibition 2015. Search and Discovery Article # 41644.

Leseigneur, G., Filipi, J.-J., Baldovini, N., and Meierhenrich, U. 2022. Absolute configuration of aliphatic hydrocarbon enantiomers identified by gas chromatography: Theorized application for isoprenoid alkanes and the search of molecular biosignatures on Mars. Symmetry 14: 326 (12 pages). https://doi.org/10.3390/sym14020326.

Mariscal, C. 2021. Life. In E. Zalta (ed.), Stanford Encyclopedia of Philosophy (winter 2021 Edition). https://plato.stanford.edu/archives/win2021/entries/life/.

Martin, J. S., Smith, N. A., and Francis, C. D. 2013. Removing the entropy from the definition of entropy: Clarifying the relationship between evolution, entropy, and the second law of thermodynamics. Evolution: Education and Outreach 6: 30 (9 pages). https://doi.org/10.1186/1936-6434-6-30.

Morowitz, H. J., Srinivasan, V., Smith, E. 2010. Ligand field theory and the origin of life as an emergent feature of the periodic table of elements. The Biological Bulletin 219(1): 1–6. https://doi.org/10.1086/BBLv219n1p1.

Nakano, H., Hirakawa, N., Matsubara, Y., et al. 2020. Precometry organic matter: A hidden reservoir of water inside the snow line. Scientific Reports 10: 7755. https://doi.org/10.1038/s41598-020-64815-6.

National Academies of Sciences, Engineering, and Medicine. 2023. Origins, Worlds, and Life: A Decadal Strategy for Planetary Science and Astrobiology 2023–2032. Washington, DC: The National Academies Press. https://doi.org/10.17226726522.

Oba, Y., Takano, Y., Furukawa, Y., et al. 2022. Identifying the wide diversity of extraterrestrial purine and pyrimidine nucleobases in carbonaceous meteorite. Nature Communications 13: 2008 (10 pages). https://doi.org/10.1038/s41467-022-29612-x.

Ollila, A. M., Newsom, H. E., Clark III, B., et al. 2014. Trace element geochemistry (Li, Ba, Sr, and Rb) using curiosity's chemcam: Early results for Gale Crater from Bradbury landing site to Rocknest. Journal of Geophysical Research Planets 119: 255–285.

Papineau, D., She, Z., Dodd, M. S., et al. 2022. Metabolically diverse primordial communities in Earth's oldest seafloor-hydrothermal jasper. Science Advances 8: eabm2296 (15 pages).

Qureshi-Hurst, E. 2021. Is simplicity that simple? An assessment of Richard Swinburne's argument from cosmic fine-tuning. Theology and Science 19(4): 379–389.

Schopf, J. W. 1999. Cradle of Life (p. 367). Princeton, NJ: Princeton University Press.

Spaargaren, D. H. 1991. The biological use of chemical elements: Selection on environmental availability and electron configuration. Oceanologica Acta 14: 569–574.

Timmermann, A., Yun, K.-S., Raia, P., et al. 2022. Climate effects on archaic human habitats and species successions. Nature 604(7906): 495–501; https://doi.org/10.1038/s41586-022-04600–9.

Valković, V. 1990. Origin of Trace Element Requirements by Living Matter. In B. Gruber and J. H. Yopp (eds.), Symmetries in Science (pp. 213–242). New York, NY: Plenum Press.

Yang, X., Kimmig, J., Zhai, D., et al. 2021. A juvenile-rich palaeocommunity of the lower Cambrian Chengjiang biota sheds light on paleo-boom or paleo-bust environments. Nature Ecology & Evolution 5(8). https://doi.org/10.1038/s41559-021-01490-4.

CHAPTER 6

# Physical, Philosophical, and Religious Considerations

## *The End of Life and Death of the Universe*

### 6.1 INTRODUCTION

As we come to understand the Universe with better precision and more comprehensively, many questions which were previously pondered by philosophical and religious thought leaders grow to have definitive answers. However, the information we possess within our observable Universe is now and will always be finite and limited, implying that there is a fundamental limit to what is knowable. As long as we remain curious about the unknown, there will always be a place for philosophy and religion both, independently of whatever becomes scientifically known.

Many of today's most exciting mysteries, from the origin of life to extraterrestrial intelligence to quantum gravity to the puzzles of dark matter and dark energy, currently lie beyond the realm of what is scientifically well understood.

There are religious and ethical conceptions that we have about the Universe, which is typically what we understand as the realm of theology. Theology in general deals with questions such as purpose, right and wrong, and an authoritative source that sets forth some tenets that must be accepted as incontrovertibly true.

Science attempts to answer questions that start with "how", venturing to explain and predict what the outcome (or sets of possible outcomes) of a physical system, initially set up with certain conditions, will be. On the other hand, theology attempts to answer questions that ask "why", pondering questions that surpass definitive knowledge and offering confident—albeit controversial to many—answers to those inquiries.

DOI: 10.1201/9781003366607-6

Many questions that were once considered to fall into the realm of theology, where we lacked definitive knowledge, have now become scientific questions that have definitive answers. Scientifically, we now know:

- how planet Earth emerged during the formation of our solar system some 4.5 billion years ago;
- how life evolved and various plants and animals emerged throughout the ages on planet Earth;
- how recent and ancient events shaped our planet's geological, atmospheric, and hydrological history; and
- how the stars, galaxies, and larger structures in our Universe formed and grew up from a more uniform, smaller, denser, hotter past.

Yet at the interface of these two fields, science and theology, beyond our definite knowledge but without an appeal to an authoritative source, lies philosophy.

Philosophy is, in some sense, the ultimate warzone. Dealing with the interface, and the limits, of both science and religion, philosophy seeks to probe questions that science cannot (yet) answer. However, unlike religion, philosophy approaches these questions with appeals to reason and logic and attempts to use these tools to explore questions whose answers are not yet known but may someday be.

Where our scientific knowledge is insufficient and where theological answers fail to compel and convince us, philosophy remains a useful endeavor. Questions concerning consciousness, the purpose of the Universe, whether reality is objective or is observer dependent; of whether the laws of nature and the physical constants of the Universe are unchanging with time or whether they are mutable, are all realms where philosophy may be of use to the intellectually curious.

In 1981, many of the world's leading cosmologists gathered at the Pontifical Academy of Sciences, a vestige of the coupled lineages of science and theology located in an elegant villa in the gardens of the Vatican. Stephen Hawking chose the august setting to present what he would later regard as his most important idea: a proposal about how the universe could have arisen from nothing.

Before Hawking's talk, all cosmological origin stories, scientific or theological, had invited the rejoinder, "What happened before that?" The Big Bang theory, pioneered some 50 years before Hawking's lecture by the Belgian physicist and Catholic priest Georges Lemaître, who later served as president of the Vatican's academy of sciences—rewinds the expansion of the universe back to a hot, dense bundle of energy. However, the question of where the initial energy came from remains.

The Big Bang theory had other problems. Physicists understood that an expanding bundle of energy would grow into a crumpled mess rather than the huge, smooth cosmos that modern astronomers observe. In 1980, the year before Hawking's talk, the cosmologist Alan Guth realized that the Big Bang's problems could be fixed with an add-on: an initial, exponential growth spurt known as cosmic inflation, which would have rendered the universe huge, smooth, and flat before gravity had a chance to wreck it. Inflation quickly became the leading theory of our cosmic origins. Yet the issue of initial conditions remained: What was the source of the minuscule patch that allegedly ballooned into our cosmos, and of the potential energy that inflated it?

Hawking, in his brilliance, saw a way to end the interminable groping backward in time: He proposed that there is no end, or beginning, at all. According to the record of the Vatican conference, the Cambridge physicist, then 39 and still able to speak with his own voice, told the crowd, "There ought to be something very special about the boundary conditions of the universe, and what can be more special than the condition that there is no boundary?"

The "no-boundary proposal" (see Hartle and Hawking 1983) envisions the cosmos having the shape of a shuttlecock with a diameter of zero at its bottommost point and gradually widening on the way up; the universe, according to the no-boundary proposal, smoothly expanded from a point of zero size. Hartle and Hawking (1983) derived a formula describing the whole shuttlecock—the so-called "wave function of the universe" that encompasses the entire past, present, and future at once—making moot all contemplation of seeds of creation, a creator, or any transition from a time before.

"Asking what came before the Big Bang is meaningless, according to the no-boundary proposal, because there is no notion of time available to refer to", Hawking said in another lecture at the Pontifical Academy in 2016, a year and a half before his death. "It would be like asking what lies south of the South Pole".

Hartle and Hawking (1983) published the wave function of the universe: The quantum state of a spatially closed universe can be described by a wave function, which is a functional on the geometries of compact three-manifolds and on the values of the matter fields on these manifolds. The wave function obeys the Wheeler–DeWitt second-order functional differential equation; see DeWitt (1967). They put forward a proposal for the wave function of the "ground state" or state of minimum excitation: the ground-state amplitude for a three-geometry is given by a path integral over all compact positive-definite four-geometries which have the three-geometry as a boundary. The requirement that the Hamiltonian be Hermitian then defines the boundary conditions for the Wheeler–DeWitt equation and the spectrum of possible excited states. To illustrate this, they calculate the ground and excited states in a simple mini super-space model in which the scale factor is the only gravitational degree of freedom; a conformal invariant scalar field is the only matter degree of freedom and $\Lambda > 0$. The ground state corresponds to de Sitter (1917) space in the classical limit. There are excited states, which represent universes which expand from zero volume, reach a maximum size, and then collapse but which have a finite, very small probability of tunneling through a potential barrier to a de Sitter–type state of continual expansion. The path-integral approach allows us to handle situations in which the topology of the three-manifold changes. They estimate the probability that the ground state in their mini-super-space model contains more than one connected component of the space-like surface.

Considering all of this, it is obvious that it took quite a bit more than seven days to create the universe, as we know it today! Now when we have intelligent life forms, us, we have to attempt to understand how information content can be included in an accounting of the energy flux of the biosphere have led to the conclusion that, in information transmission, one component, the semantic content, or "the meaning of the message", adds no thermodynamic burden over and above costs arising from coding, transmission and translation; see Crofts (2007). For all life forms, the message of the genotype encoded in DNA specifies the phenotype and hence the organism that is tested against the real world through the mechanisms of Darwinian evolution. For human beings, communication through language and similar abstractions provides an additional supra-phenotypic vehicle for semantic inheritance, which supports the cultural heritages around which civilizations revolve.

Next, we present the thinking of Crofts (2007). The following three postulates provide the basis for discussion of a number of themes that demonstrate some important consequences:

1. Information transmission through either pathway has thermodynamic components associated with data storage and transmission.
2. The semantic content adds no additional thermodynamic cost.
3. For all semantic exchange, meaning is accessible only through translation and interpretation, and has a value only in context.

The following formulations of Crofts (2007) are still acceptable today and should be considered when the evolution of life on planet Earth is discussed.

- For both pathways of semantic inheritance, translational and copying machineries are imperfect. Consequently, both pathways are subject to mutation and to evolutionary pressure by selection.
- The emergent properties of life are dependent on a processing of semantic content. The translational steps allow amplification in complexity through combinatorial possibilities in space and time. The initial interpretational steps include protein synthesis, molecular recognition, and catalytic potential that facilitate structural and functional roles.
- All living things show a behavior that indicates awareness of time, or chronognosis. The ~4 billion years of biological evolution have given rise to forms with increasing sophistication in sensory adaptation. This has been linked to the development of an increasing chronognostic range and an associated increase in combinatorial complexity.
- Development of a modern human phenotype and the ability to communicate through language led to the development of archival storage and invention of the basic skills, institutions, and mechanisms that allowed the evolution of modern civilizations. The translational machineries of the human mind, the "mutation" of ideas therein, and the "conversations" of our social intercourse have allowed a limited set of symbolic descriptors to evolve into an exponentially expanding semantic heritage.
- The previous three postulates open interesting epistemological questions. An understanding of topics such as dualism, the élan vital, the status of hypothesis in science, memetics, the nature of consciousness, the role of semantic processing in the survival of societies, and Popper's three worlds require recognition of an insubstantial component. By recognizing a necessary linkage between semantic content and a physical machinery, we can bring these perennial problems into the framework of a realistic philosophy. It is suggested, following Popper, that the ~4 billion years of evolution of the biosphere represent an exploration of the nature of reality at the physicochemical level, which, together with the conscious extension of this exploration through science and culture, provides a firm epistemological underpinning for such a philosophy (from Crofts 2007).

## 6.2 PHYSICAL CONSIDERATIONS

Many scientists consider and argue that life emerged on planet Earth. We shall mention only some. With the statement that "life arose on Earth sometimes in the first few hundred million years after the young planet has cooled to the point that it could support water based organisms on its surface", the authors Spiegel and Turner

(2011) had to come to the conclusion that life originated independently on other places in Universe. They actually tried to infer something about the probability per unit time of abiogenesis (meaning "the original evolution of life or living organism from inorganic or inanimate substances") on an Earth-like planet elsewhere in the Universe. They introduce a term, symbol λ, the probability of abiogenesis.

According to their model, Earth became a hospitable location for life very fast after its formation, and in time it transitioned from being merely habitable to being inhabited. They performed a Bayesian analysis of the astro-biological implications of life's early emergence on Earth. Their conclusion was that if we should find evidence of life that arose independently of us, we would have considerably stronger grounds to conclude that life is probably common in our galaxy (Spiegel and Turner 2011).

According to Kipping (2020), life emerged on Earth within the first quintile of its habitable window, but technological civilization did not blossom until its last. In his work, the author interpreted the chronology of these two events in Bayesian framework by considering that the evolutionary timescale is itself an unknown that needs to be inferred rather than fiducially set. According to the author, the early emergence of life on Earth is naively interpreted as meaning that if we reran the events, life would generally reappear quickly. However, since the timescale for intelligence evolution is long, a quick start of life is simply a necessary byproduct of our existence and not evidence for general rapid abiogenesis rate.

Based on these two works, one can see that it is not possible to reject the hypothesis that Earth is the only location of intelligent life in the Universe (Whitmire 2022). In addition, the author concluded that we necessarily had to find ourselves on a planet where abiogenesis occurred, which is an arbitrary statement, as shown by the rest of my text.

The usual theory of inflation breaks down in eternal inflation. We derive a dual description of eternal inflation in terms of a deformed Euclidean Compact Field Tester (CFT) located at the threshold of eternal inflation. The partition function gives the amplitude of different geometries of the threshold surface in the no-boundary state. Its local and global behavior in dual toy models shows that the amplitude is low for surfaces which are not nearly conformal to the round three-sphere and essentially zero for surfaces with negative curvature. Based on this, we conjecture that the exit from eternal inflation does not produce an infinite fractal-like multiverse but is finite and reasonably smooth (Hawking and Hertog 2018).

Modern theories of the Big Bang predict that our local universe came into existence with a brief burst of inflation; in other words, a tiny fraction of a second after the Big Bang itself, the universe expanded at an exponential rate. It is widely believed, however, that once inflation starts, there are regions where it never stops. It is thought that quantum effects can keep inflation going forever in some regions of the universe so that globally, inflation is eternal. The observable part of our universe would then be just a hospitable pocket universe, a region in which inflation has ended and stars and galaxies formed.

"The usual theory of eternal inflation predicts that globally our universe is like an infinite fractal, with a mosaic of different pocket universes, separated by an inflating ocean", said Hawking in an interview in 2017.

> The local laws of physics and chemistry can differ from one pocket universe to another, which together would form a multiverse. But I have never been a fan of the multiverse. If the scale of different universes in the multiverse is large or infinite the theory can't be tested.

In their paper, Hawking and Hertog (2018) claim that eternal inflation as a theory of the big bang is wrong. "The problem with the usual account of eternal inflation is that it assumes an existing background universe that evolves according to Einstein's theory of general relativity and treats the quantum effects as small fluctuations around this", said Hertog. "However, the dynamics of eternal inflation wipes out the separation between classical and quantum physics. As a consequence, Einstein's theory breaks down in eternal inflation".

"We predict that our universe, on the largest scales, is reasonably smooth and globally finite. So it is not a fractal structure", said Hawking.

The theory of eternal inflation that Hawking and Hertog (2018) put forward is based on string theory: a branch of theoretical physics that attempts to reconcile gravity and general relativity with quantum physics, in part by describing the fundamental constituents of the universe as tiny vibrating strings. Their approach uses the string theory concept of holography, which postulates that the universe is a large and complex hologram: physical reality in certain 3D spaces can be mathematically reduced to 2D projections on a surface. They developed a variation of this concept of holography to project out the time dimension in eternal inflation. This enabled them to describe eternal inflation without having to rely on Einstein's theory. In the new theory, eternal inflation is reduced to a timeless state defined on a spatial surface at the beginning of time.

As the chemistry expands beyond molecules to systems, new synthetic targets appear. Among these, life represents the ultimate synthetic challenge. Otto (2022) proposed a new approach by assuming that the synthesis of life requires the functional integration of different subsystems that harbor the different characteristics that are essential to life. The most important of these are self-replication, metabolism, and compartmentalization. Integrating these features into a single system, maintaining this system out of equilibrium, and allowing it to undergo Darwinian evolution should ideally result in the emergence of life.

Belluati et al. (2023) have reported artificial cell synthesis using biocatalytic polymerization-induced self-assembly. Artificial cells are biomimetic microstructures that mimic functions of natural cells, can be applied as building blocks for molecular systems engineering, and host synthetic biology pathways. They reported enzymatically synthesized polymer-based artificial cells with the ability to express proteins. Artificial cells were synthesized using biocatalytic atom transfer radical polymerization-induced self-assembly, in which myoglobin synthesizes amphiphilic block co-polymers that self-assemble into structures such as micelles, worm-like micelles, polymersomes, and giant unilamellar vesicles (GUVs). The GUVs encapsulate cargo during the polymerization, including enzymes, nanoparticles, micro-particles, plasmids, and cell lysate. The resulting artificial cells act as microreactors for enzymatic reactions and for osteoblast-inspired bio-mineralization. Moreover, they can express proteins such as a fluorescent protein and actin when fed with amino acids. Actin polymerizes in the vesicles and alters the artificial cells' internal structure by creating internal compartments. Thus, biocatalytic atom transfer radical polymerization-induced self-assembly-derived GUVs can mimic bacteria as they are composed of a microscopic reaction compartment that contains genetic information for protein expression upon induction.

Belluati et al. (2023) demonstrated how an enzymatic reaction could induce the formation of polymeric vesicles with increasing levels of functional complexity, from simple microcapsules to microreactors and complex cell mimics that are capable of producing their own proteins, thanks to co-encapsulated plasmids as genetic information. An enzyme could self-encapsulate within a compartment formed by a non-natural membrane that the enzyme synthesized itself, entrapping all the complex components deriving from cell lysates, which in turn could express specific proteins and modify the structure of the vesicles, producing internal compartments or transforming them into polymer-mineral hybrids. Cell mimics were produced in the most biosimilar process possible, showcasing the versatility and robustness of bioCRP. Moreover, bioPISA proved to be a very efficient and biocompatible encapsulation method for a range of cargo. Although bioPISA-derived GUVs are still simplified mimics of actual cells, bioPISA will pave the way for hybrid systems combining synthetic polymers with the diversity of natural molecules, offering a plethora of applications in synthetic biology, sensing, catalysis, and biomaterials.

### 6.2.1 Homochirality

The selection of a single molecular handedness, or homochirality, across all living matter is a mystery in the origin of life. The seminal model by Frank (1953) showed how chiral symmetry breaking can occur in nonequilibrium chemical networks. However, an important shortcoming in this classic model is that it considers a small number of species, while there is no reason for the prebiotic system, in which homochirality first appeared, to have had such a simple composition. Furthermore, this model does not provide information on what could have been the size of the molecules involved in this homochiral prebiotic system. Here, we show that large molecular systems are likely to undergo a phase transition toward a homochiral state, because they contain a large number of chiral species. Using chemo-informatics tools, Laurent et al. (2021) quantify how abundant chiral species are in the chemical universe of all possible molecules of a given length. They proposed that model described in Frank (1953) should be extended to include a large number of species in order to possess the transition toward homochirality, as confirmed by numerical simulations. Finally, using random matrix theory, they proved that large nonequilibrium reaction networks possess a generic and robust phase transition toward a homochiral state.

Laurent et al. (2021) presented a scenario that explains why a large complex molecular system tends to become chiral. Based on fundamental properties of phase transitions, confirmed by numerical simulations of a non-equilibrium reaction network, their scenario is both general and robust. Details of the reaction network should not matter, nor the precise way in which the system is driven out of equilibrium, provided the system is large enough. In addition, the homochiral state of their model does not need to be all D or L across all species, in agreement with the observation that, for instance, amino acids are L-chiral, but sugars are D-chiral. The reaction scheme need not satisfy mass-action law, and there is also no requirement that the system be fully well mixed; it could be compartmentalized. As an illustration of this idea, they study two diffusively coupled chemical reactors, identical to the one considered so far. For low coupling, the two compartments undergo separately the homochiral transition, while at high coupling, they reach the same homochiral state. Compartmentalized systems enrich the scenarios for the transition to homochirality, because, on the one hand, compartmentalization significantly broadens the diversity of available autocatalytic networks, which can be built with a limited number of compounds, and, on the other hand, in such systems, the number of species is effectively increased, which favors their mechanism.

The scenario proposed by Laurent et al. (2021) offers a universal pathway toward homochirality, potentially unifying many previous approaches on this issue. In the context of the origin of life, they also find that there is no need for very long and complex molecules for this homochiral state to emerge. The transition can already occur in a prebiotic world containing molecules with about ten heavy atoms.

## 6.3 PHILOSOPHICAL CONSEQUENCES OF EXPERIMENTAL CONFIRMATION OF PROPOSED HYPOTHESES

Scientific research progresses by the dialectic dialogue between hypothesis building and the experimental testing of these hypotheses. Although scientists can rely on an increasing set of sophisticated experimental methods for hypothesis testing, the importance of hypothesis building in science should not be neglected. Brüssow (2022) discussed the role of hypotheses in science in great detail. His work presents a primer into philosophical thoughts on hypothesis building and illustrates it with two hypotheses that played a major role in the history of science (the parallel axiom and the fifth element hypothesis). It continues with philosophical concepts on hypotheses as a calculus that fits observations, the need for plausibility and for explicatory power imposing a strong selection on theories. Galilei (1564–1642) introduced and James (1842–1907) and Poincaré (1854–1912) later justified the reductionist principle in hypothesis building. Waddington (1905–1975) stressed the feed-forward aspect of fruitful hypothesis building, while Poincaré called for a dialogue between experiment and hypothesis and distinguished false, true, fruitful, and dangerous hypotheses. Theoretical biology plays a much smaller role than theoretical physics because physical thinking strives for a unification principle across the universe, while biology is confronted with a breathtaking diversity of life forms and its historical development on a single planet.

An article by Resnick (1959) consists of a critical examination of an argument which purports to prove that many scientific hypotheses held to be probable are actually certain. The argument rests on the assumption that since the non-philosopher would say of many scientific hypotheses that they are certain and would deny that the best-established hypotheses are merely probable, philosophers who say that no scientific hypotheses are certain must be mistaken. Examination reveals that the argument fails to take account of the technical nature of the claim that even the best-established hypotheses are probable.

Classic analyses of hypothesis testing assume that people should pick the test with the largest probability of falsifying their current hypothesis, while experiments have shown that people tend to select tests consistent with that hypothesis. Using two different normative standards, Austerweil and Griffiths (2008) prove that seeking evidence predicted by current hypothesis is optimal when the hypotheses in question are deterministic and other reasonable assumptions hold. They tested this account with two experiments using a sequential prediction task, in which people guess the next number in a sequence. Experiment 1 shows that people's predictions can be captured by a simple Bayesian model. Experiment 2 manipulates people's beliefs about the probabilities of different hypotheses and shows that they confirm whichever hypothesis they are led to believe is most likely.

Scientists often use hypotheses to frame experiments. A hypothesis is simply a statement of the expected outcome of an experimental test. Moreover, based on the results of a carefully designed experiment, we accept or reject our hypotheses, generating knowledge as a result. The hypothesis is constructed in advance of the experiment; it is therefore unproven in its original form. The very idea of "proof" of a hypothesis is problematic on philosophical grounds because the hypothesis is established to be falsified, not verified. The second framework for experimental design involves building a model as an explanation for a data set. A model is distinct from a hypothesis in that it is constructed after data are derived. In contrast to the hypothesis, the model must be held up for verification—its success is determined by its ability to predict a particular outcome. Furthermore, an unsuccessful, or not fully successful, model need not be scrapped in the way that the alternative framework urges the rejection of a falsified hypothesis, but it may instead serve as the starting point for a suitably refined successor. The concept of a model's "verification" requires an acceptance of "inductive reasoning"—a form of logic that allows the scientist to both generalize a particular result and say that the same result will occur in the future—which itself has been criticized. It should be noted that many scientists use the term "hypothesis" when they mean "model", Glass and Hall (2008).

### 6.3.1 Universality of Life

Universality means the quality or state of being universal (existing everywhere or involving everyone). There is no evidence or reason to assume that life elsewhere in the universe must have the same RNA/DNA biochemistry as ours, and we are in the process of creating artificial, life-like systems in the form of artificially intelligent machines.

Furthermore, much larger systems such as supraorganisms—insect colonies, forests, bird flocks, our societies, our human species, our planetary biosphere, our solar system, galaxies, and even the universe as a whole—have many of the same properties as biological organisms. They consist of smaller, interacting components. What makes a system "alive"? The answer has to do with how the parts of the system interact—how the actions of one component cause or affect actions of other components of the system. A system can be called "alive" if its individual components are interacting in coordinated ways that result in effects that can be attributed to the system as a whole, as more than the sum of its parts.

This is exactly the N = 1 problem. Having only one life sample (hence the N = 1), how can we make plausible inferences about what life on other planets might be like if we have only one sample of life and do not know which characteristics are universal and which are contingent? To make a comparison, it would be like doing a cognitive psychology experiment on a single subject and expecting the results to be projectable to all human beings as well. Without terms of comparison, it is not easy to tell which characteristics of the study subject are common to all humans and which are idiosyncratic, according to Di Lauro and Bedetti (2023). Their paper explores the universality of life characteristics as we know them on Earth and addresses the "N = 1 problem"—the challenge of making inferences about extraterrestrial life based on a single sample of life. They discuss proposal of a "Shadow Biosphere" by Davies et al. (2009), a hypothetical second biosphere with different biochemical and/or molecular features with respect to our known biosphere. This work critically examines this proposal, considering criticisms and possible cosmic scenarios, such as the impact-bottleneck hypothesis, suggesting that catastrophic events, like the late heavy bombardment, may have influenced the evolution of life on Earth and the potential existence of a shadow biosphere. They also explore the habitability of celestial bodies in our solar system, such as Europa, Enceladus, and Titan, as potential targets for testing the universality of life characteristics. The conclusion highlights the importance of ongoing research on these environments to provide insights into the existence of a shadow biosphere and the universality of life beyond Earth.

According to Davies et al. (2009), extraterrestrial life might differ from known life, and considerable thought has been given to possible signatures associated with weird forms of life on other planets. So far, however, very little attention has been paid to the possibility that our own planet might also host communities of weird life. If life arises readily in Earth-like conditions, as many astrobiologists contend, then it may well have formed many times on Earth itself, which raises the question whether one or more shadow biospheres have existed in the past or still exist today. In this paper, the authors discuss possible signatures of weird life and outline some simple strategies for seeking evidence of a shadow biosphere

Cleland (2012) discusses a challenging idea: life without definitions. The most popular approach among both philosophers and scientists for answering the question "What is life?" is to provide a "definition" of life. In his article, the author explores a variety of different definitional approaches, both traditional and non-traditional, that have been used to "define" life. He argues that all of them are deeply flawed. It is his contention that a scientifically compelling understanding of the nature of life presupposes an empirically adequate scientific theory (vs. definition) of life; as he argues, scientific theories are not the sort of thing that can be encapsulated in definitions. Unfortunately, scientists are currently in no position to formulate even a tentative version of such a theory. Recent discoveries in biology and biochemistry have revealed that familiar Earth life represents a single example that may not be representative of life. If this is the case, life on Earth today provides an empirically inadequate foundation for theorizing about life considered generally. The author sketched a strategy for procuring the needed additional examples of life without the guidance of a definition or theory of life and closed with an application to NASA's fledgling search for extraterrestrial life.

## 6.4 RELIGIOUS CONSIDERATIONS

The teleological argument has received a resurgence in recent years. This is thanks to cosmological data, which appear to offer new evidence to indicate that the existence of life is the result of contrivance in the face of staggering improbability. Swinburne (1997) argues that cosmic fine-tuning gives us good evidence to believe in the creator God of classical theism, who has good reasons to create human agents. Qureshi-Hurst (2021) assessed Swinburne's use of the "principle of simplicity" as a criterion for selecting the best explanation in this particular case, arguing that it is not as compelling as Swinburne claims.

*The Existence of God*, the book in which Swinburne constructs probabilistic arguments for the existence of God, is arguably his most successful work. Its chapter on teleological arguments will form the basis of the subsequent discussion. He argues that scientific insights regarding the origin of the universe and its fine structure constitute good evidence for theism—in effect, that physics leads to God.

Teleological arguments from cosmic fine-tuning begin by observing the fundamental features of the universe that make possible the evolution and existence of complex life and conducting thought experiments in which these features are different; see Qureshi-Hurst (2021). The universe is perfectly suited to the flourishing of complex life, but "not all initial conditions or laws of nature would lead to, or even permit, the existence of human bodies", argues Swinburne (1997). "If only a very narrow range of laws and initial conditions allow such an evolution, then we may say that the universe is fine-tuned for this evolution". Out of the set of all possible universes, a very small proportion is life conducive. If scientific theories provide the ultimate explanation of our universe, such that God is not responsible for the processes the theory describes, Swinburne argues that such fine-tuning is a priori very improbable.

Fine-tuning discourse begins just after the so-called Planck era, approximately $10^{-43}$ seconds after the Big Bang, when the universe was in an extremely hot dense state. Physicists are currently unable to model the universe in this epoch, as general relativity and quantum mechanics break down before this time. Quantum gravitational laws governed, and these are as of yet unknown to the scientific community. Scientific understanding begins in the inflationary period, during which the universe grew by at least a factor of $10^{25}$ in 10–30 seconds or less, from smaller than a sub-atomic particle to roughly the size of a grapefruit. This was an extremely critical period at which point the conditions we observe in the present-day were set. The four fundamental forces separated: gravitation, electromagnetic forces between charged particles, the weak nuclear force giving rise to nuclear processes such as decay of neutrons into protons, and the strong nuclear force binding the nuclei of atoms together. Sub-atomic particles and anti-particles were formed, creating the building blocks for the chemical elements crucial for the evolution of life.

A critical factor in the formation of stars and galaxies, necessary places for life to evolve, are the physical constants (so named because they have the same value every-where and at every time). Of all the possible values of physical constants, only a very narrow range could give rise to a universe, which could support life.

An example, the cosmological constant $\Lambda$, first arose when Einstein's 1915 field equations predicted that the universe was expanding. As this predated Hubble's discovery of an expanding universe, Einstein inserted $\Lambda$ to stabilize his equations and eliminate expansion. It is the value of the energy density of the vacuum of space, and acts as a counter balance to gravity. It determines both the rate of the expansion of the universe and gravitational condensations such as stars and galaxies. As the universe was in fact expanding, scientists now assume that Λ has a value of zero. Its minimum possible value is 120 orders of magnitude smaller that its limit (6.128)—a truly staggering range. That it is likely zero is extremely important for the evolution of life. Whilst a change in the value of most constants would have prevented life as we know it emerging, if Λ differed only slightly from its actual value, then this would have prevented the emergence of life itself. The universe would either have expanded too rapidly for the formation of stars and galaxies, or collapsed back in on itself soon after the Big Bang. Due to this highly unlikely state of affairs, many have been motivated to try to establish some deep underlying principle that requires Λ to be zero.

A further feature, which Swinburne (1997) notes as being fine-tuned, relates to carbon. As far as we know, carbon is necessary for all forms of complex intelligent life. An abundance of carbon requires a delicate balance of the four fundamental forces. For example, an increase or decrease of more than 0.5% in the strength of the strong force or more than 4% in the strength of the electromagnetic force would lead to negligible amounts of carbon and oxygen being formed, thus making life very unlikely. These variations represent an "infinitesimal range within the range of logically possible values of the forces".

The data reveal that, "there are several small islands within the space of possible values of constants and variables within which human life could evolve". Based on all the data, Swinburne claims that it is extremely unlikely that life-conducive conditions have arisen accidentally. His argument is probabilistic. Each set of possible constants and variables, he argues, is roughly equally probably a priori. Therefore, the probability of any one set of conditions occurring before referring to observations is roughly constant for all possible values. Only a tiny minority of possible values facilitates the production of life, so any randomly occurring universe is far more likely to be sterile than fertile. Therefore, he argues, "given standard theory and no more fundamental explanation thereof (physical or theistic), tuning is a priori immensely improbable", after Qureshi-Hurst (2021).

Here is an interesting question we can ask ourselves. Has science discovered God? A quote from (https://jesusonline.com/discover-evidence/god/science-discovered-god/): "Einstein didn't believe it was possible. Stephen Hawking said it might be the greatest scientific discovery of all time".

Because of the sophisticated coding within DNA many scientists are speculating that the universe appears to be part of a grand design. One cosmologist put it this way: "Many scientists, when they admit their views, incline toward the teleological or design argument" (Harrison 1985). Surprisingly, many scientists who are talking about God have no religious belief whatsoever. (An atheist believes God does not exist. An agnostic believes we cannot know.)

So, what are some stunning discoveries that have scientists suddenly speaking of God? Three revolutionary discoveries from the fields of astronomy and molecular biology stand out:

1. The universe had a beginning.
2. The universe is just right for life.
3. DNA coding reveals intelligence.

Perhaps the most vocal adversary of a beginning to the universe was British astronomer Sir Fred Hoyle, who sarcastically nicknamed the creation event a "Big Bang". He stubbornly held to his steady state theory that the universe has always existed. So did Einstein and other scientists until the evidence for a beginning became overwhelming. The "elephant in the room" implication of a beginning is that something or Someone beyond scientific investigation must have started it all.

Finally, in 1992, COBE satellite experiments proved that the universe really did have a one-time beginning in an incredible flash of light and energy. Although some scientists called it the moment of creation, most preferred referring to it as the Big Bang.

Many scientists were struck by the parallel of a one-time creation event from nothing with the biblical creation account in Genesis 1:1 (In the beginning God created the heaven and the Earth). Prior to this discovery, many scientists regarded the biblical account of creation from nothing as unscientific.

Another agnostic, George Smoot, the Nobel Prize–winning scientist in charge of the COBE experiment, also admits to the parallel. "There is no doubt that a parallel exists between the Big Bang as an event and the Christian notion of creation from nothing"; see Smoot and Davidson (1993)

Cosmologists, who specialize in the study of the universe and its origins, soon realized that a chance cosmic explosion could never bring about life any more than a nuclear bomb would—unless it was precisely engineered to do so. Moreover, that meant a designer must have planned it. They began using words like, "Superintellect", "Creator", and even "Supreme Being" to describe this designer.

Physicists calculated that for life to exist, gravity and the other forces of nature needed to be just right, or our universe could not exist. Had the expansion rate been slightly weaker, gravity would have pulled all matter back into a "big crunch". We are not talking about merely a one or 2% reduction in the universe's expansion rate. Hawking (1996) points out, "If the rate of expansion one second after the big bang had been smaller by even one part in $10^{16}$, the universe would have re-collapsed before it ever reached its present size".

In addition, for life to exist, the conditions in our solar system and planet need to be just right. For example, we all realize that without an atmosphere of oxygen, none of us would be able to breathe. Moreover, without oxygen, water could not exist. Without water, there would be no rainfall for our crops. Other elements such as hydrogen, nitrogen, sodium, carbon, calcium, and phosphorus are also essential for life.

However, that alone is not all that is needed for life to exist. The size, temperature, relative proximity, and chemical makeup of our planet, sun, and moon also need to be just right. In addition, there are dozens of other conditions that needed to be exquisitely fine-tuned or we would not be here to think about it. Hawking (1999), an agnostic, writes, "The remarkable fact is that the values of these numbers seem to have been very finely adjusted to make possible the development of life".

Although Einstein was not religious and did not believe in a personal God, he called the genius behind the universe "an intelligence of such superiority that, compared with it, all the systematic thinking and acting of human beings is an utterly insignificant reflection"; see Einstein (1931).

Davies (1988) states, "There is for me powerful evidence that there is something going on behind it all. It seems as though somebody has fine-tuned nature's numbers to make the Universe. . . . The impression of design is overwhelming".

DNA's intricate complexity caused its co-discoverer, Crick (1981), to believe that it could never have originated on earth naturally. Crick, an evolutionist who believed life is too complex to have originated on earth and must have come from outer space, wrote,

> An honest man, armed with all the knowledge available to us now, could only state that in some sense, the origin of life appears at the moment to almost be a miracle, so many are the conditions which would have had to have been satisfied to get it going.

The coding behind DNA reveals such intelligence that it staggers the imagination. A mere pinhead of DNA contains information equivalent to a stack of paperback books that would encircle the earth 5000 times. Moreover, DNA operates like a language with its own extremely complex software code. Microsoft founder Bill Gates says that the software of DNA is "far, far more complex than any software we have ever developed".

We shall finish this section with a view from the National Academy of Science underlining that science is not the only way of acquiring knowledge about ourselves and the world around us. Humans gain understanding in many other ways, such as through literature, the arts, philosophical reflection, and religious experience. Scientific knowledge may enrich aesthetic and moral perceptions, but these subjects extend beyond science's realm, which is to obtain a better understanding of the natural world. The claim that equity demands balanced treatment of evolutionary theory and special creation in science classrooms reflects a misunderstanding of what science is and how it is conducted. Scientific investigators seek to understand natural phenomena by observation and experimentation. Scientific interpretations of facts and the explanations that account for them therefore must be testable by observation and experimentation.

Creationism, intelligent design, and other claims of supernatural intervention in the origin of life or of species are not science because they are not testable by the methods of science. These claims subordinate observed data to statements based on authority, revelation, or religious belief. Documentation offered in support of these claims is typically limited to the special publications of their advocates. These publications do not offer hypotheses subject to change in light of new data, new interpretations, or demonstration of error. This contrasts with science, where any hypothesis or theory always remains subject to the possibility of rejection or modification in the light of new knowledge. No body of beliefs that has its origin in doctrinal material rather than scientific observation, interpretation, and experimentation should be admissible as science in any science course. Incorporating the teaching of such doctrines into a science curriculum compromises the objectives of public education. Science has been greatly successful at explaining natural processes, and this has led not only to increased understanding of the universe but also to major improvements in technology and public health and welfare. The growing role that science plays in modern life requires that science, and not religion, be taught in science classes, concluded the National Academy of Sciences (1999).

### 6.4.1 Astrotheology

Astrotheology is that branch of theology that provides a critical analysis of the contemporary space sciences combined with an explication of classic doctrines such as creation and Christology for the purpose of constructing a comprehensive and meaningful understanding of our human situation within an astonishingly immense cosmos, according to Peters (2019). Within the growing field of religion and science, astrotheology and its sister subject, astroethics, represent critical responses to the excitement permeating the space sciences, especially astrobiology. The excitement arises because of new searches for microbial life within the solar system and the discovery of exoplanets elsewhere in the Milky Way, some of which may be homes for extraterrestrial intelligent civilizations.

Astrochristology, as a subfield within the more comprehensive astrotheology, speculates on the implications of what astrobiology and related space sciences learn about our future space neighbors. Confirmation of the existence of extraterrestrial intelligent civilizations living on exoplanets will force Christian theologians to decide on two issues. The first issue deals with the question: should Christians expect many incarnations, one for each inhabited exoplanet, or will the single incarnation in terrestrial history suffice? The second issue deals with the question: why is there an incarnation in the first place? Does the divine presence in the historical Jesus mark a divine attempt to fix a broken creation, or does it mark a divine self-communication that would occur with or without creation's fall into sin and death? Sorting these issues out is one task for astrochristology. The position of Peters (2016) is to affirm both a single incarnation on Earth valid for cosmic redemption from the brokenness of creation in its present state.

Astrotheology interprets astrobiology, of course; yet it covers much more. Astrotheology is that branch of theology which provides a critical analysis of the contemporary space sciences combined with an explication of classic doctrines such as creation and christology for the purpose of constructing a comprehensive and meaningful understanding of our human situation within an astonishingly immense cosmos, Peters (2011). In this case, it is christology that we could label it astrochristology.

What about sin and estrangement? Would an extraterrestrial civilization be fallen and in need of redemption? This is where the Christian vocabulary regarding sin enters with words such as sin and evil. "Sin is the estrangement between God and humans instigated by human defiance or abnegation. Evil is the disorder within humans individually and among them collectively", writes Heim (2001). Might alien civilization exist within a condition similar to ours: fallen, estranged, subject to sin? Does the extraterrestrial situation need redemption? Redemption is more than mere revelation. Redemption constitutes the soteriological or atoning work of incarnation, the work of reconciliation, not merely the revelatory presence of God. Can a theologian affirm that Jesus' redemptive work within Earth's history is efficacious for the entire cosmos without belting out a prideful Earth anthem? This is the question of planetary chauvinism we now turn.

Pannenberg (1991) expressed confidence in the universal efficacy of Jesus' atoning work because his incarnation is the incarnation of the universal Logos, the principle by which all of creation is generated and held together. Like thoughts originating in our mind, the Logos originates in God. And as our thoughts come to expression in speech, when God speaks the Son becomes differentiated from the Father, and the world with all of its particularity comes into existence. This is the divine nature of the Son, the universal ground of all finite reality. With a world-picture that includes the entire universe we perceive the grandeur, and with faith we feel the gratitude. As audacious as it may sound, the Christian world-picture includes a vision of a coming new creation for the whole of the cosmos. The Christian claim is that what happened to Jesus of Nazareth on the first Easter models what will happen to the entirety of God's creation in the future.

This is what the concept of the incarnation is intended to convey. It would be misleading if we think of incarnation in terms of a heavenly being taking a vacation on planet Earth or a spiritual being dressing temporarily in human skin. Rather, incarnation is an abbreviated cipher for the entire life, death, and resurrection of Jesus, and it is nested inextricably within God's promise of renewal of all that exists in the creation. Astrochristology delivers a promise that extends well beyond Earth; it includes all the stars and all our space neighbors.

### 6.4.2 Astrobiology as a Contemporary Theology

Religion is one of those aspects to which the human experience that the astrobiology field can extend by exploring questions like: How did life appear in our universe? Was it "created"? How did life develop? Is it possible that life elsewhere could have also been "created"?

Of the $8 \times 10^9$ people on Earth, an overwhelming majority of humans claim to adhere to some type of religion. By focusing on the religion of Catholicism in light of astrobiology's discoveries regarding creation of the universe, creation of life, and even creation of intelligent life, Borland (2017) tried to show that religious beliefs and science do not have to contradict one another, and they may even fill in gaps each seems to have when standing alone. He explored the implications that possible future astrobiological discoveries may have on Catholicism.

Following this reasoning, one has to start from the Big Bang, because it is a widely accepted scientific theory stating that the universe originated from a miniscule "ball" of electrons, protons, and photons that expanded, cooled, and eventually joined to form our universe and all of the galaxies, planets, stars, and so on within it. From this discovery, the assumption that there was an ultimate consolidation at some point prior to this expansion logically followed. Scientists have concluded that the expansion and cooling process took under three minutes, after which the joining of protons and neutrons began. The universe continued to expand as a hot cloud of gas for hundreds of thousands of years. After a lengthy time of expansion, galaxies formed and stars formed from gas clouds collapsing within those galaxies. The Big Bang is hypothesized to have occurred $13.8 \times 10^9$ years ago, but our particular galaxy is aged at $13.2 \times 10^9$ years old, and our sun is much younger at about $4.6 \times 10^9$ years old. According to the nebular hypothesis, Earth, like the other planets in our solar system, was created from a cloud whose matter was drawn inward on itself, eventually fusing molecules like hydrogen and helium together and forming the Sun at the hot center. The materials in the once disk-like cloud not consumed by this collapse correspond to the planets of the solar system. That is not all if we are to consider the maturation of these planets to what we "know" of them today. About $(4–3.8) \times 10^9$ years ago, the Late Heavy Bombardment took place; the alignment of Jupiter and Saturn's orbits caused a shift in the planetary orbits of Neptune and Uranus that eventually sent small icy bodies and asteroids hurtling into the inner solar system towards planet like our Earth. These bombardments would eventually create the planet in its current form, and some scientists propose that the collisions with certain bodies like asteroids are responsible for the delivery of organic chemicals and water supply to Earth.

Interestingly, once the true meaning of each is acknowledged, it can be shown that the claims being made do not actually contradict one another. It is quite possible that God created the universe out of love, seeing everything he created as good and beautiful, as Catholicism suggests, while implementing the details science points to. God could have created that "miniscule ball of electrons, protons, and photons that eventually expanded and cooled to form our universe and all of the galaxies, planets, stars, etc. within it". Because the 7 days of creation were not meant to be taken as a literal representation of history but rather a form of organized telling of a meaningful story, it is still possible that the scientific timeline of creation stands alongside Genesis' 7-day story. Both offer different but not necessarily contradicting truths about the same reality of creation, following Borland (2017).

All of this is well and good, but why should we care about this intersection of theology and science when it comes to the creation of life? The simple answer is that neither can fully explain what life is, and both bring something unique to the complex question. Science seeks to find what distinguishes life from non-life and what material pre-cursors must be present for life to take place. Countless definitions have been proposed, but none seems to be all encompassing enough. Some describe life in a sense of boundaries: life vs. non-life. Others use a central principle that is most essential or most interesting when it comes to life, and yet others operationally define life as if it is something that can be tested for. Theology and philosophy, on the other hand, seek answers describing the meaning of life. Why was life created? What is the purpose or ultimate end of life? Could life have some immaterial aspect to it? An immaterial end? What does it mean to be fully alive? By bridging these two aspects of life, science and religion enhance how we can comprehend life in its entirety. Science allows us to see how things are manifested in our world while theology provokes thoughts on why things are manifested in that way.

The presence of intelligent life and the religious implications of finding intelligent life elsewhere are some of the aspects of astrobiology that are particularly controversial as astrobiology research progresses. The current hypothesis that technological civilizations exist currently drives the search for extraterrestrial intelligence—SETI—and their techno-signatures.

For Catholicism, a characteristic specific to humanity when comparing it to other species is its uniqueness. Genesis describes how God created humans "in his image and likeness", giving them rational intellect and free will. In order to fully unpack this, the first thing we must do is define intelligence. Intelligence is a higher, spiritual cognitive power whose functions are attention, conception, judgement, reasoning, reflection, and self-consciousness. Modern-day psychology may argue that intelligence can be simplified to sense cognitions, but that is not what we mean here. The difference becomes apparent when we evaluate animal "intelligence" and human intelligence. An animal such as a dolphin or chimpanzee have been said to be very intelligent, but they do not have abstract thought. They do not imagine the future. They do not try out new ideas or ponder why they exist. They may know that something is right or wrong by association or conditioning with negative or positive reinforcement, but they will not reflect on why something is right or wrong, if a hypothetical situation would be right or wrong. This rationality regardless of association/conditioning is something that distinguishes human intelligence, and this is the gift God gave humans by making them "in his image and likeness", as seen in Genesis.

Moving forward, then, would the discovery of another intelligent species similar to humanity in intelligence and free will destroy humanity's uniqueness or its purpose? Catholicism already holds the belief that there are intelligent beings similar in intelligence to humans: angels. The Catechism of the Catholic Church describes angels as "spirits" who are "messengers of God", meaning they do not have bodies as humans do. "As purely spiritual creatures, angels have intelligence and will: they are personal and immortal creatures, surpassing in perfection all visible creatures, as the splendor of their glory bears witness". However, this does not destroy the uniqueness of humanity, which was created body and soul, material and immaterial, in the "image and likeness of God" himself. It also follows that whether another intelligent species is found or not found on Earth, the presence of human life can still be significant and special. Life in itself is an awesome concept that even today we struggle to fully comprehend.

> the history of sin and salvation recorded in the two testaments of the Bible is not a history of the universe; it is a particular religious history on one planet. Earlier centuries saw wider intimations of Christianity, because the universe as they knew it centered on earth. Christians believed that Jesus was the Word of God come to this earth as a man. But the central importance of Jesus for us does not necessarily imply anything about other races on other planets. Jesus' teaching and life brings eschatology but not astronomy.

After analyzing aspects of crossover between science and religion through the lenses of astrobiology and Catholicism, Borland (2017) has shown that the two do not have to stand in opposition to one another. First, he explored creation as explained by astrobiology and as explained by Catholicism to find that they not only can stand side by side but also fill in the gaps the other cannot fill, such as how historically things were created and why things were created in such a way. Next, he was able to explore the scientific search for extraterrestrial life stemming from the field of astrobiology and the implications that discoveries may have on Catholic teaching in areas including human uniqueness, the incarnation, and the sacraments. From what he has discussed in light of the little outside discussion that has been done on these possibilities, he found that it is still possible for astrobiology and Catholicism to agree even if intelligent extraterrestrial life is found elsewhere. In order to show this relationship even further, he encouraged theologians and scientists alike to further dive into this conversation on what a discovery of intelligent life elsewhere could mean for both fields. As a Catholic who acknowledges how the church is in favor of ethical scientific research as a way to discover more about the Creator, His creation, and His constant hand in the continuing of creation, Borland (2017) argues that no matter what is discovered by true science will be welcomed by the Church and further enlightened through the insight the Church can offer beyond the natural sciences.

### 6.4.3 Response to Possible Extraterrestrial Life

The detection of life elsewhere, intelligent or not, would go down as one of the most profound discoveries in human history. Harris-Wyrick (2022) discuss how humans might respond to extraterrestrial life. In addition, NASA has recently co-sponsored a study at Princeton University's Center for Theological Inquiry with two dozen theologians from many religions who discussed how humans might respond to the discovery of extraterrestrial life.

For centuries, the Catholic Church had a position that extraterrestrial life, at least intelligent life, could not exist because the Son of God only came to Earth to save God's creatures. In their minds, that meant only Earth possessed life. In the eyes of the early Church, no planets existed beyond our solar system. Those individuals who disputed the Church's stance, like Giordano Bruno, faced strong retaliation. Among other "heretical" statements, Bruno said, "In space, there are countless constellations, suns and planets; we see only the suns because they give light; the planets remain invisible, for they are small and dark. There are also numberless earths circling around their suns". For this, among other heresies, Bruno was condemned to death in 1600.

We now know of many thousands of planets beyond our solar system, and astronomers believe that number likely exceeds one hundred billion. So how might religions react to the discovery of life out there? Study participant the Rev. Dr. Andrew Davison, of the University of Cambridge, wrote in his forthcoming book on the study, *Astrobiology and Christian Doctrine*, "The headline findings are that adherents of a range of religious traditions report that they can take the idea in their stride".

How humanity will react to the discovery of extraterrestrial life is discussed by Kwon et al. (2017). Speculation on this topic abounds, but empirical research is practically non-existent. The authors report the results of three empirical studies assessing psychological reactions to the discovery of extraterrestrial life using the Linguistic Inquiry and Word Count (LIWC) text analysis software. They examined language use in media coverage of past discovery announcements of this nature, with a focus on extraterrestrial microbial life. A large online sample ($N = 501$) was asked to write about their own and humanity's reaction to a hypothetical announcement of such a discovery (Study 1). An independent, large online sample ($N = 256$) was asked to read and respond to a newspaper story about the claim that fossilized extraterrestrial microbial life had been found in a meteorite of Martian origin (Study 2). Across these studies, authors found that reactions were significantly more positive than negative and more reward vs. risk oriented. A mini-meta-analysis revealed large overall effect sizes (positive vs. negative affect language: $g = 0.98$; reward vs. risk language: $g = 0.81$). They also found that people's forecasts of their own reactions showed a greater positivity bias than their forecasts of humanity's reactions (Study 1) and that response to reading an actual announcement of the discovery of extraterrestrial microbial life showed a greater positivity bias than responses to reading an actual announcement of the creation of man-made synthetic life (Study 2). Taken together, this work suggests that our reactions to a future confirmed discovery of microbial extraterrestrial life are likely to be positive.

Here is something we might think about: Neal (2014) introduces the issue of extraterrestrials to the study of disasters, their prevention, and their management. With increasing scientific support for the existence of planets able to support life and development of astrobiology, the author argues that limiting the scope of disaster research to terrestrial matters is increasingly intellectually untenable. In order to rectify this situation, and begin a debate within the field, the author develops five theoretical scenarios concerning the likelihood of, and the threats posed by, extraterrestrial life forms. Doing so highlights possible new themes in disaster and risk management research, hoping that this may close the gap between scholarly work on risk management and the scientific findings of astrobiology.

While humanity has not yet observed any extraterrestrial intelligence (ETI), contact with ETI remains a possibility. Contact could occur through a broad range of scenarios that have varying consequences for humanity. In their paper, Baum et al. (2011) analyze a broad range of contact scenarios in terms of whether contact with ETI would benefit or harm humanity. This type of broad analysis could help preparation for actual contact with ETI even if the details of contact do not fully resemble any specific scenario. Their analysis suggests some immediate practical recommendations for humanity. One recommendation is that messages to extraterrestrials should be written cautiously. Another recommendation is that humanity should avoid giving off the appearance of being a rapidly expansive civilization. A final recommendation is that preparations for ETI encounter, whether through METI, SETI, human explorations of space, or any other form, should consider the full breadth of possible encounter scenarios. Indeed, perhaps the central conclusion of the analysis presented here is that ETI contact could proceed in a wide range of ways. It is inappropriate and inadequate to blindly assume that any one specific scenario would result from contact. Until such contact occurs, we simply do not know what would happen. Given the uncertainty, the broad scenario analysis presented here is an important step towards helping us think through and prepare for possible contact.

Wisian and Traphagan (2020) published a paper, "The Search for Extraterrestrial Intelligence: A Realpolitik Consideration". Here are some of their thoughts. In the vigorous academic debate over the risks of the search for extra terrestrial intelligence and active messaging extra terrestrial intelligence, a significant factor has been largely overlooked. Specifically, the risk of merely detecting an alien signal from passive SETI activity is usually considered negligible. The history of international relations viewed through the lens of the realpolitik tradition of realist political thought suggests, however, that there is a measurable risk of conflict over the perceived benefit of monopoly access to ETI communication channels. This possibility needs to be considered when analyzing the potential risks and benefits of contact with ETI. This view is criticized by Wright et al. (2023). They argue that sufficient reason is not given to justify treating this potential scenario as action guiding over other candidate geopolitical responses. Furthermore, even if one assumes that a realpolitik response is the most relevant geopolitical response, they show that it is highly unlikely that a nation could successfully monopolize communication with ETI. Instead, the real threat that the authors identify is based on the perception by state actors that an information monopoly is likely. However, as we show, this perception is based on an overly narrow contact scenario. Overall, they comment on Wisian and Traphagan's (2020) argument and resulting recommendations on technical, political, and ethical grounds. Ultimately, they find that not only are these recommendations unlikely to work, they may also precipitate the very ills that they foresee. As an alternative to these recommendations, they recommend transparency and data sharing (which are consistent with currently accepted best practices), further development of post-detection protocols, and better education of policymakers in this space.

## 6.5 END OF LIFE ON EARTH?

Life on Earth has proven to be remarkably resilient, surviving and recolonizing the planet after snowball glaciations, abrupt warming episodes, and catastrophic meteor impacts. However, in the distant future, a hot climate, which life may not survive, probably awaits the planet. At present the Sun is increasing in brightness by 1% every ~110 million years; see Gough (1981). Increased solar radiation should at some point trigger a moist and then runaway greenhouse climate, ending any hope for continued surface habitability. Surface habitability of a planet is traditionally defined as the ability to support liquid water

One-dimensional climate models predict that a catastrophic thermal runaway will be triggered by a 6% increase in the solar constant above its present level (Wolf and Toon 2014). Their simulations using a three-dimensional climate model with fixed carbon dioxide and methane indicate that surface habitability may be maintained at significantly larger solar constants. A 15.5% increase in the solar constant yields global mean surface temperatures of 312.9 K (39.75° C), well short of moist and runaway greenhouse states. Their results imply that Earth's climate may remain safe against both water loss and thermal runaway limits for at least another $1.5 \times 10^9$ years and probably for much longer (Wolf and Toon 2014).

The results of Kopparapu et al. (2013), however, place the present day Earth very close to the inner edge of habitability. That is, according to their model, the water-loss (inner border of habitable zone) and maximum greenhouse (outer border of habitable zone) limits for our solar system are at 0.99 and 1.70 AU, respectively.

Earth's climate has remained continuously habitable throughout 3 or 4 billion years. This presents a puzzle (the "habitability problem") because loss of habitability appears to have been more likely. Solar luminosity has increased by 30% over this time, which would, if not counteracted, have caused sterility. Furthermore, Earth's climate is precariously balanced, potentially able to deteriorate to deep-frozen conditions within as little as 1 million years. Tyrrell (2020) presented results from a novel simulation in which thousands of planets were assigned randomly generated climate feedbacks. Each planetary setup was tested to see if it remained habitable over a period of 3 billion years. The conventional view attributes Earth's extended habitability solely to stabilizing mechanisms. The simulation results shown revealed instead that chance also plays a role in habitability outcomes. Earth's long-lasting habitability was therefore most likely a contingent rather than an inevitable outcome.

In the following, we list just a few possible doomsday events that could permanently extinguish all life on Earth—and the last one is likely unavoidable. Sloan et al. (2017) explored the statistics of events that completely sterilize an Earth-like planet with planet radii in the range $(0.5–1.5) \times R_{\oplus}$ and temperatures of ~300 K, eradicating all forms of life. They considered the relative likelihood of complete global sterilization events from three astrophysical sources—supernovae, gamma-ray bursts, large asteroid impacts, and passing stars. To assess such probabilities, they considered what cataclysmic event could lead to the annihilation of not just human life but also extremophiles through the boiling of all water in Earth's oceans. Surprisingly they found that although human life is somewhat fragile to nearby events, the resilience of Ecdysozoa such as *Milnesium tardigradum* renders global sterilization an unlikely event.

Considering asteroid impact apocalypse, one has to recall the situation when a a large asteroid struck the Gulf of Mexico some $66 \times 10^6$ years ago. This event influenced the existence of dinosaurs, as well as most other species on Earth at the time. The impact of this event was perhaps the single most important event in human history. Without that asteroid strike, dinosaurs might have continued to rule the Earth, leaving us mammals still hiding in the shadows.

It is difficult to hope that humans will always be on the winning side of such random events. A future asteroid could just as easily take out every person on Earth. Fortunately, that is unlikely to happen anytime soon. Based on the geological record of cosmic impacts, according to NASA (2023), Earth is hit by a large asteroid roughly every $(50–100) \times 10^6$ years. However, smaller asteroid impacts do happen all the time. There are now over 100 ring-like structures on Earth recognized as definite impact craters. Most of them are not obviously craters, their identity masked by heavy erosion over the centuries, but the minerals and shocked rocks present make it clear that impact was their cause. There are now more than 150 asteroids known that come nearer to the Sun than the outermost point of Earth's orbit. These range in diameter from a few meters up to about 8 kilometers. A working group chaired at NASA Ames Research Center estimates that there are some 2100 such asteroids larger than 1 kilometer and perhaps 320,000 larger than 100 meters, the size that caused the Tunguska event and the Arizona Meteor Crater. An impact by one of these larger meteors in the wrong place would be a catastrophe, but it would not threaten civilization. However, the working group concluded that an impact by an asteroid larger than 1–2 kilometers could degrade the global climate, leading to widespread crop failure and loss of life. Such global environmental catastrophes, which place the entire population of the Earth at risk, are estimated to take place several times per million years on average. A still larger impact by an object larger than about 5 kilometers is damaging enough to cause mass extinctions.

However, what are the chances that our planet will ever be struck by an asteroid massive enough to wipe out all life on Earth? Simulations published by Sloan et al. (2017) suggest it would take a truly gigantic space rock to accomplish such a job. Killing all life on Earth would require an impact that literally boils away the oceans; only asteroids like Pallas and Vestathe, the solar system's largest, are big enough to do that. There is a new impact scenario proposal that at the end of its formation, Earth collided with another planet-sized body. This produced a great deal of debris in Earth's orbit, which in turn coalesced into the Moon. The impacting planet would later be named "Theia", after the Greek goddess who was the mother of the Moon. However, nowadays, collisions of such large objects are extremely unlikely.

For a more likely glimpse of an Earth-altering cataclysm, that of death by deoxygenation, we need to look to the distant past. Nearly $2.5 \times 10^9$ years ago, a period called the Great Oxidation Event resulted in the breathable atmosphere we now depend on. An eruption of cyanobacteria, sometimes called blue-green algae, filled our atmosphere with oxygen, creating a world where multicellular life forms could take hold, and where creatures like humans could ultimately breathe. However, one of Earth's great die-offs, an event 450 million years ago called the Late Ordovician mass extinction, likely happened because the inverse took place. The planet saw a sudden drop in oxygen levels that lasted for several million years. What could have caused such an extreme event? During the Ordovician period, the continents were one jumbled mass called Gondwana. Most life on Earth still lived in the oceans, but plants were beginning to emerge on land. Then, near the end of the Ordovician, a sweeping climate shift left the supercontinent covered with glaciers. That global cooling alone was enough to start killing off species. Then a second pulse of the extinction ramped up as oxygen levels plummeted. Scientists see evidence of this shift in seafloor samples collected from around the world. Some researchers think that the glaciers were responsible for fundamentally changing the layers of the oceans, which have unique temperatures and specific concentrations of elements like oxygen. Yet the exact cause of the oxygen drop is still up for debate. Whatever the cause, the end result is that more than 80% of life on Earth died during the Late Ordovician mass extinction.

Even if a sudden spate of global cooling sparked the Late Ordovician mass extinction, what set that in motion in the first place? Over the years, numerous astronomers have suggested that to be a gamma ray burst (GRB). GRBs are mysterious events that seem to be the most violent and energetic explosions in the universe and are probably tied to extreme supernovae. However, we have not seen a burst close enough to us to fully understand what is going on. So far, GRBs have only been spotted in other galaxies. However, if one did happen in the Milky Way, as has likely happened in the past, it could cause a mass extinction on Earth. A GRB pointed in our direction might last just 10 seconds or so, but it could still destroy at least half Earth's ozone in that short period. A GRB would wipe out the lifeforms that live in the upper levels of the ocean, which currently contribute significant amounts of oxygen to our atmosphere. In addition, gamma rays also break apart atmospheric oxygen and nitrogen. These gases are converted into nitrogen dioxide, which is more commonly known as the smog that blocks out the Sun above heavily polluted cities. Having this smog blanketing the entire Earth would block out sunshine and start a global ice age.

Yet another issue which needs to be considered is the ozone layer: the unique role of ozone is in absorbing certain wavelengths of incoming solar ultraviolet light. This was recognized in the latter part of the 19th century. Interest in ozone stems from the fact that such absorption of solar radiation is important in determining not only the thermal structure of the stratosphere but also the ecological framework for life on the Earth's surface. Decreased ozone results in increased ultraviolet transmission, which can affect the health of humans, animals, and plants. For a review of concepts and history of stratospheric ozone depletion, see Solomon (1999). Stratospheric ozone depletion through catalytic chemistry involving human-made chlorofluorocarbons is an area of focus in the study of geophysics and one of the global environmental issues of the 20th century, and it remains so; see Gonçalves Jr et al. (2023), who studied the stratospheric ozone depletion in the Antarctic region that triggered intense changes in sea salt aerosol geochemistry.

There are number of possible events in the Milky Way galaxy, which can damage or even rip off the protective ozone layer from Earth and lead to the mass extension of all life on the planet. One such event is a nearby supernova explosion, as already discussed by Ruderman (1974). Hard X-ray pulses or increased cosmic radiation originating in nearby supernova explosions may be capable of temporarily removing most of the earth's atmospheric ozone cover even when direct radiation effects at the earth's surface are negligible. Consequently, terrestrial life may be subject to relatively huge solar ultraviolet fluxes every few hundred million years.

Collision between two neutron stars could be deadly for life on Earth, although a collision between two neutron stars, tightly bound on a decaying orbit, appears to be a relatively rare event, $1{:}10^{10}$. Such a violent star collision, called a kilonova, could devastate our planet due to a lethal spew of radiation (X-rays, gamma rays, cosmic rays) that are emitted from this event. The powerful collision of ultra-dense neutron stars creates a particle blast that would decimate our planet's ozone layer and make it vulnerable to ultraviolet radiation for the next 1000 years: an extinction-level event.

The mergers of binary compact objects such as neutron stars and black holes are of central interest to several areas of astrophysics, including as the progenitors of gamma-ray bursts, sources of high-frequency gravitational waves (GWs) and likely production sites for heavy element nucleosynthesis via rapid neutron capture (the r-process). Levan et al. (2023) presented observations of the exceptionally bright gamma-ray burst GRB 230307A. They showed that GRB 230307A belongs to the class of long-duration gamma-ray bursts associated with compact object mergers and contains a kilonova similar to AT2017gfo, associated with the gravitational-wave merger GW170817. They obtained James Webb Space Telescope mid-infrared (mid-IR) imaging and spectroscopy 29 and 61 days after the burst. The spectroscopy shows an emission line at 2.15 microns, which they interpret as tellurium (atomic mass A = 130), and a very red source, emitting most of its light in the mid-IR due to the production of lanthanides. These observations demonstrate that nucleosynthesis in GRBs can create r-process elements across a broad atomic mass range and play a central role in heavy element nucleosynthesis across the Universe.

According to Ozaki and Reinhard (2021), Earth's modern atmosphere is highly oxygenated and is a remotely detectable signal of its surface biosphere. However, the lifespan of oxygen-based bio signatures in Earth's atmosphere remains uncertain, particularly for the distant future. The authors used a combined biogeochemistry and climate model to examine the likely timescale of oxygen-rich atmospheric conditions on Earth. Using a stochastic approach, they obtained the mean future lifespan of Earth's atmosphere, with oxygen levels more than 1% of the present atmospheric level, to be $(1.08 \pm 0.14) \times 10^9$ years ($1\sigma$). The model projects that a deoxygenation of the atmosphere, with atmospheric $O_2$ dropping sharply to levels reminiscent of the Archaean Earth, will most probably be triggered before the inception of moist greenhouse conditions in Earth's climate system and before the extensive loss of surface water from the atmosphere. The authors found that future deoxygenation is an inevitable consequence of increasing solar fluxes, whereas its precise timing is modulated by the exchange flux of reducing power between the mantle and the ocean–atmosphere–crust system. Their results suggest that the planetary carbonate–silicate cycle will tend to lead to terminally $CO_2$-limited biospheres and rapid atmospheric deoxygenation, emphasizing the need for robust atmospheric bio-signatures applicable to weakly oxygenated and anoxic exoplanet atmospheres and highlighting the potential importance of atmospheric organic haze during the terminal stages of planetary habitability.

It seems to be an accepted assumption that human migration into space is inevitable. However, almost 60 years of scientific studies of the effects of space on Earth life suggest this is not a given. Life on Earth evolved in the context of conditions that are unique to Earth and are not duplicated anywhere else in our solar system. The science indicates that life-sustaining conditions on Earth could be the very things that inhibit our ability to live off Earth. A paper by Irons and Irons (2023) combines 100 years of scientific development of a theory of ecological thermodynamics with classical mechanics theory and analytical models of self-restoring heat engines to explain how the Sun and Earth have evolved into islands of order in the entropy of space. An explanation is provided regarding how naturally occurring conservative force fields engage a diversity of natural resources in semi-reversible cycles that build a high-exergy ecosphere. The science infers that the ability to establish a human settlement in space without Earth-like self-restoring order, capacity, and organization will result in settlement sustainment challenges. Historical evidence of Earth settlements with disrupted ecosystems point to the following possibilities. Supply chains would disappear, market resources would be depleted, advancement in human pursuits would be disrupted, social and governance systems would falter or collapse, human population numbers would decline, genetic diversity in the human genome would be lost, average human individual biomass would decrease, and human knowledge and understanding would be forgotten. What does it mean to have a location in space outside of Earth be "like Earth"? The results of research are presented as a pancosmorio theory of human sustainability that is developed using the scientific philosophy methodology of adductive reasoning. Four analytical models of space ecosphere sustainability and five hypotheses with proposed tests for falsifiability are provided, including a theorem that suggests a limit to human expansion into space. A new quantitative method of human sustainability is developed from theories of network ecology, providing orthogonal properties of an ecosystem network stability function based upon an ecosystem network production function. Conclusions are made regarding the potential for sustainable development in space using balanced sustainability. Insights are provided regarding human endeavors on the Moon and Mars, as well as the Fermi paradox.

There is a maximum number of astronomical units from the Sun out to which a terraformed system is basally sustainable (assuming dissipative structures and area conditionals are met) based upon meeting the basal power conditional with solar power panel area in use and the availability of non-solar power. The pancosmorio theorem is used to determine the combination of power from solar panels and power from nonsolar-power sources that is needed to meet the basal power conditional to enable achieving and maintaining the organization necessary for level 1 sustainability at the distance it is planned to be from the Sun.

Hayes and Gallagher (2022) reported the detection of a significant sudden ionospheric disturbance in the D-region of Earth's ionosphere (~60–100 km), which was associated with the massive γ-ray burst GRB 221009A that occurred on October 9, 2022. They identified the disturbance over northern Europe—a result of the increased ionization by X- and γ-ray emission from the GRB-using very low frequency radio waves as a probe of the D-region. These observations demonstrate that an extra-galactic GRB ($z \sim 0.151$) can have a significant impact on the terrestrial atmosphere and illustrates that the Earth's ionosphere can be used as a giant X- and γ-ray detector. Indeed, these observations may provide an insight into the impacts of GRBs on the ionospheres of planets in our solar system and beyond.

### 6.5.1 Earth Cannot Support Life Forever

Jiang et al. (2021) developed a foundational model based on trends built from empirical data of space exploration and computing power through the first six-plus decades of the Space Age, which projects earliest possible launch dates for human-crewed missions from cis-lunar space to selected solar system and interstellar destinations. The model uses computational power, expressed as transistors per microprocessor, as a key broadly limiting factor for deep space missions' reach and complexity. The goal of this analysis is to provide a projected timeframe for humanity to become a multi-world species through off-world colonization, and in so doing all but guarantees the long-term survival of the human race from natural and human-caused calamities that could befall life on Earth. Beginning with the development and deployment of the first nuclear weapons near the end of World War II, humanity entered a "Window of Peril", which will not be safely closed until robust off-world colonies become a reality. Findings by Jiang et al. (2021) suggest the first human-crewed missions to land on Mars, selected Asteroid Belt objects, and selected moons of Jupiter and Saturn can occur before the end of the 21st century. Launches of human-crewed interstellar missions to exoplanet destinations within roughly 40 lightyears of the solar system are seen as possible during the 23rd century and launch of intra-galactic missions by the end of the 24th century. An aggressive and sustained space exploration program, which includes colonization, is thus seen as critical to the long-term survival of the human race.

It is becoming increasingly clear that planets are sometimes torn from the grasp of their central stars and hurtled into the darkness of space. The most common sources of such planetary ejection are interactions between giant planets. Although the orbits of the planets in our solar system have not changed appreciably for billions of years, they do interact with each other, and the shapes of their orbits do vary. Planetary systems in general are not necessarily gravitationally stable for time scales of billions of years. If Saturn were closer to Jupiter or if it were more massive, the long-term game of gravitational cat and mouse that planets play could lead to ejection of one of these planets, and it would escape into the galaxy. If Saturn were lost, then Jupiter would stay trapped in solar orbit, but its orbit would be oddly elliptical. Some of the giant planets recently discovered orbiting other stars have highly elliptical orbits, and the past ejection of a long-lost partner may have been the cause. Planets can also be ejected from binary star systems where two stars (and their planets) orbit each other. Although ejected planets might not be hospitable to life, the outlook is much more favorable for large moons orbiting ejected planets. If somehow a Jupiter with its four large moons could be ejected into interstellar space, it might provide a very interesting habitat not only for the continuation of microbial life but for its possible evolution as well; see Ward and Brownlee (2003).

For nearly all of Earth history, life was limited to creatures so small that they are invisible to the naked eye. Casual inspection during all that time would have suggested that it was a failed planet. In other planetary systems, primitive life might flourish but never advance to the point where forests and flying animals even get a serious chance to evolve. Stars with short lifetimes, unstable planetary atmospheres, changes in orbital or spin axis, massive extinctions, impacts, crustal catastrophes, the cessation of plate tectonics, or any of a whole raft of other problems could prevent the evolution of advanced life or its prolonged survival. Moreover, on Earth itself, complex life has thrived only for the last

10% of the planet's existence. Perhaps the most predictable aspect of advanced life (if it exists) on other planets around other stars is that its tenure is limited and that eventually any such life—and even some of the planets—will perish. Like individual organisms, planets and their grand environments have life spans. All planets with life eventually become extinct. This final outcome may be brought about by external sources such as impacts or a nearby supernova, by internal effects such as atmospheric or biological catastrophe, or (if all else fails) by increase in the brightness of the central star. This will be the ultimate fate of Earth: Life on our planet will eventually be roasted out of existence. The sun is slowly getting brighter. It is now 30% brighter than it was in the early history of the planet. Over the next $4 \times 10^9$ years, it will double in brightness. Even if life survives this travail, it will soon be stilled. About $4 \times 10^9$ years from now, the sun will begin to expand rapidly in size, and its brightness will dramatically increase. The sun will become a red giant, as did the stars Antares in the constellation Scorpio and Betelgeuse in the constellation Orion. In a $10^9$-year time span, its brightness will increase over 5000 times. At the very beginning of this process, Earth's oceans will vaporize, driving our precious water supply into space. In the final stages of its transformation into a red giant, the sun will expand to the point where it will nearly reach the orbit of Earth. The Universe will be one living planet poorer.

## 6.6 HOW WILL THE SUN AGE?

Our Sun, like almost all of the stars that host a planet are main-sequence stars that sustain their energy emission through the nuclear reactions ongoing in their cores, which can last for ($100 \times 10^6$ – many $\times 10^9$) years, depending on their mass. When such a star exhausts the hydrogen fuel in its core, it expands enormously, shatters any close-enough planet, and becomes a white dwarf. Thereafter, what remains of the planetary system may move close enough to the star to become subject to collisions and to strong tidal forces, grinding the remaining planetary cores; see Farihi et al. (2018).

Theoretical models of the orbital evolution of planetary systems indicate that possibly large (a few to hundreds of kilometers in diameter) rocky bodies might survive the last stages of stellar evolution toward the white dwarf phase. However, because of the chaotic motion present in the disk surrounding some 20 to 25% of white dwarfs (known as "polluted" white dwarfs), there is a continuous fall of rocky, planetary material on the stellar surface, which could be detected by the presence of metal absorption lines in the stellar spectrum; see Fossati (2019).

One such polluted white draft, G29–38, has been studied by Cunningham et al. (2022). They estimated an instantaneous accretion rate of $M = 1.63^{+1.29-}{}_{0.40} \times 10^9$ g s$^{-1}$ independent of the stellar atmosphere model.

Vissapragada et al. (2022) presented evidence of tidally driven inspiral in the Kepler-1658 (KOI-4) system, which consists of a giant planet (1.1 $R_J$, 5.9 $M_J$) orbiting an evolved host star (2.9$R_{Sun}$, 1.5$M_{Sun}$). Using transit timing measurements from Kepler, Palomar/WIRC, and TESS, they showed that the orbital period of Kepler-1658b appears to be decreasing at a rate of $131_{-22}^{+20}$ ms yr$^{-1}$, corresponding to an infall timescale P/(dP/dt)~$2.5 \times 10^6$ years. The authors also considered other explanations for the data, including line-of-sight acceleration and orbital precession, but found them to be implausible. The observed period derivative implies a tidal quality factor was found to be in good agreement with theoretical predictions for inertial wave dissipation in subgiant stars. Additionally, while it probably cannot explain the entire inspiral rate, a small amount of planetary dissipation could naturally explain the deep optical eclipse observed for the planet via enhanced thermal emission. As the first evolved system with detected inspiral, Kepler-1658 is a new benchmark for understanding tidal physics at the end of the planetary life cycle. This is the first time that a direct evidence for a planet spiraling towards its evolved star was observed. Since the Sun will evolve into a red giant star eventually, the tidally driven processes seen by Vissapragada et al. (2022) could result in the decay of the Earth's orbit towards the Sun.

Relevant to this subject could be also a paper by De et al. (2023) describing an infrared transient from a star engulfing a planet. Planets with short orbital periods (roughly under 10 days) are common around stars like the Sun. Stars expand as they evolve, and thus we expect their close planetary companions to be engulfed, possibly powering luminous mass ejections from the host star. However, this phase has never been directly observed. The authors reported observations of ZTF SLRN-2020, a short-lived optical outburst in the galactic disk accompanied by bright and long-lived infrared emission. The resulting light curve and spectra share striking similarities with those of red novae, a class of eruptions now confirmed to arise from mergers of binary stars. Its exceptionally low optical luminosity (approximately $10^{35}$ erg $s^{-1}$) and radiated energy (approximately $6.5 \times 10^{41}$ erg) point to the engulfment of a planet of fewer than roughly ten Jupiter masses by its Sun-like host star. It is estimated that the galactic rate of such sub-luminous red novae to be roughly between 0.1 and several per year. According to the authors, future galactic plane surveys should routinely identify these, showing the demographics of planetary engulfment and the ultimate fate of planets in the inner solar system.

## 6.7 DEATH OF THE UNIVERSE: WHAT COMES AFTER IT?

From cosmic inflation to a primordial particle soup to the expanding, cooling aftermath, the Universe went through many important stages in its history. About 6 $\times 10^9$ years ago, however, a new form of energy began to dominate the Universe's expansion: dark energy, which now determines our cosmic fate. The era we are living in, where dark energy dominates the Universe's expansion, is the final one our Universe will ever experience. Here is why we are already living through the beginning of the ultimate end.

A Universe that has been around longer will have expanded more. It will be cooler in the future and was hotter in the past; it was gravitationally uniform in the past and is clumpier now; it was smaller in the past and will be much, much larger in the future. By applying the laws of physics to the Universe and comparing the possible solutions with the observations and measurements we have obtained, we can determine both where we came from and where we are headed. We can extrapolate our history all the way back to the beginning of the hot Big Bang. We can extrapolate our current Universe into the far distant future as well and foresee the ultimate fate that awaits everything that exists.

Six different eras of the Universe, as described by Siegel (2022), are:

1. Inflationary era: which preceded and set up the hot Big Bang. Prior to the hot Big Bang, the Universe was filled only with a form of energy inherent to space itself: a form of energy that caused the Universe to expand both extremely rapidly and relentlessly, in an exponential fashion. It abruptly stopped some 13.8 billion years ago, and inflation ended. All of that energy, once inherent to space itself, got converted into particles, antiparticles, and radiation. With this transition, the inflationary era ended, and the hot Big Bang began.

2. Primordial Soup era: from the start of the hot Big Bang until the final four types of interactions occurrence in the early Universe. Once the expanding Universe is filled with matter, antimatter, and radiation, it's going to cool. Whenever particles collide, they will produce whatever particle–antiparticle pairs are allowed by the laws of physics. The primary restriction comes only from the energies of the collisions involved, as the production is governed by $E = mc^2$. As the Universe cools, the energy drops, and it becomes harder and harder to create more massive particle–antiparticle pairs, but annihilations and other particle reactions continue unabated. A few seconds after the Big Bang, the antimatter is all gone, leaving only matter behind. A few minutes after the Big Bang, stable deuterium is formed, and nucleosynthesis of the light elements occurs.

3. Plasma era: from the end of non-scattering nuclear and particle interactions until the Universe cools enough to stably form neutral matter. Once those light nuclei form, they are the only positively charged objects in the Universe, and they are everywhere. They are balanced by an equal amount of negative charge in the form of electrons. These two species of particle would find one another immediately, forming atoms and paving the way for stars. Every time an electron and a nucleus bind together, a high-enough energy photon comes along and blasts them apart. It is not until the Universe cools dramatically, from billions of degrees to just thousands of degrees, that neutral atoms can finally form.

4. Dark Ages era: from the formation of neutral matter until the first stars and galaxies reionize the intergalactic medium of the Universe completely. Filled with neutral atoms, at last, gravitation can begin the process of forming structure in the Universe. In order to end these dark ages, the intergalactic medium needs to be reionized. That requires enormous amounts of star formation and tremendous numbers of ultraviolet photons and that requires time, gravitation, and the start of the cosmic web. The first major regions of reionization take

place (200–250) × 106 years after the Big Bang, but reionization does not complete, on average, until the Universe is 550 × 106 years old. At this point, the star-formation rate is still increasing, and the first massive galaxy clusters are just beginning to form.

5. Stellar era: from the end of reionization until the gravity-driven formation and growth of large-scale structure ceases, when dark energy density dominates matter density. Once the dark ages are over, the Universe becomes transparent to starlight. The great recesses of the cosmos are now accessible, with stars, star clusters, galaxies, galaxy clusters, and the great, growing cosmic web. The Universe is dominated, energy-wise, by dark matter and normal matter, and the gravitationally bound structures continue to grow larger and larger.

   The star-formation rate rises, peaking about 3 billion years after the Big Bang. At this point, new galaxies continue to form, existing galaxies continue to grow and merge, and galaxy clusters attract more and more matter into them. But the amount of free gas within galaxies begins to drop, as the enormous amounts of star formation have used up a large amount of it. Slowly but steadily, the star-formation rate drops.

6. Dark Energy era: the final stage of our Universe, where the expansion accelerates and disconnected objects speed irrevocably and irreversibly away from one another. The large-scale structure in the Universe ceases to grow. The objects that were gravitationally bound to one another before dark energy's takeover will remain bound, but those that were not yet bound by the onset of the dark energy age will never become bound. Instead, they will simply accelerate away from one another, leading lonely existences in the great expanse of nothingness. The individual bound structures, like galaxies and groups/clusters of galaxies, will eventually merge to form one giant elliptical galaxy. The existing stars will die; new star formation will slow down to a trickle and then stop; gravitational interactions will eject most of the stars into the intergalactic abyss. Planets will spiral into their parent stars or stellar remnants, owing to decay by gravitational radiation. Even black holes, with extraordinarily long lifetimes, will eventually decay from Hawking radiation.

We already entered this final era billions of years ago. Most of the important events that will define our Universe's history have already occurred. The far distant fates of the Universe offer a number of possibilities, but if dark energy is truly a constant, as the data indicates, it will continue to follow the curve (thin solid line), leading to the long-term scenario frequently described in *Starts with a Bang*: of the eventual heat death of the Universe; see Figure 6.1. If dark energy evolves with time, a Big Rip or a Big Crunch are still possible, but we don't have any evidence indicating that this evolution is anything more than idle speculation. The steady state model, like the perfect cosmological principle, is ruled out.

The Big Rip scenario will occur if we find that dark energy increases in strength while remaining negative in direction over time. In order, galaxy groups and clusters will dissociate; galaxies themselves will be ripped apart; the solar system will eject its planets from the outside in; and then individual planets, moons, molecules, atoms, and even subatomic particles will be destroyed, all in the final instants before space and time are ripped apart as well; see Teaford (2015). While this is a theoretical possibility if the Universe collapses into a "Big Crunch", observations suggest the current cosmic expansion will continue indefinitely.

The universe might meet its end in a big freeze, a big crunch, or a big rip. However, whether time ends with the demise of the cosmos depends on whether it is even real after all. The second law of thermodynamics may be formulated in many different ways, but it predicts a dull state of equilibrium, the inevitable "heat death" of the universe. Indeed, Clausius' (1867) statement, "the entropy of the universe tends to a maximum", may be regarded as one form of the second law of thermodynamics.

### 6.7.1 Cold Death of the Universe

Our cosmos' final fate is a long and frigid affair that astronomers call the Big Freeze, or Big Chill. It is a fitting description for the day when all heat and energy are evenly spread over incomprehensibly vast distances. At this point, the universe's final temperature will hover just above absolute zero, following Betz (2020). Eventually, black holes will be the last remaining matter in the

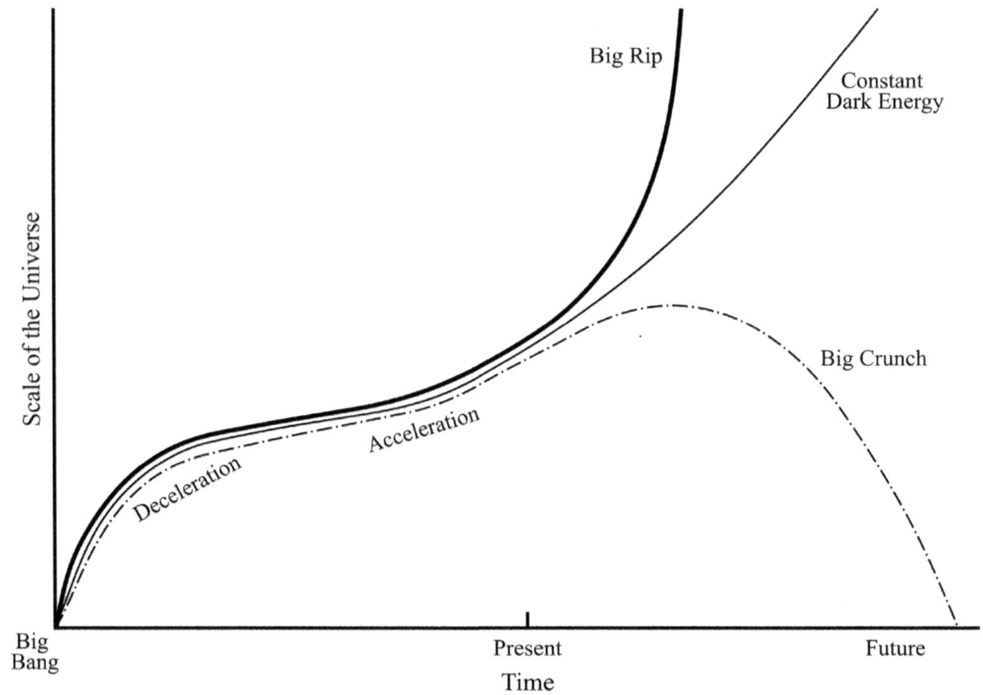


FIGURE 6.1 The far distant fates of the Universe offer a number of possibilities, after NASA/GSFC (2022).

universe. However, what happens after that? White dwarfs, brown dwarfs, and neutron stars are expected to eventually die through a process known as proton decay, when the subatomic particles they are made of literally fall apart. The process of proton decay will take countless years, but matter itself will eventually rot away; see Ohlsson (2023).

Cosmologists predict this will occur late in the Degenerate Era, as the half-life (the time it takes for half of a substance to decay) of a proton is thought to be about $10^{34}$ years. Moreover, when the last remnants of stars rot away at the particle scale, only black holes will remain, dominating what is left of the universe. Still, even these monsters will not last forever. Although the popular conception is that "nothing can escape a black hole, not even light", scientists are not entirely sure that is true. Astronomers believe black holes do emit radiation—in particular, Hawking radiation, named after Stephen Hawking, who first proposed the idea. Although Hawking radiation has yet to be detected, if black holes do leak the radiation, it would provide a mechanism by which they could die off—literally evaporating into the cosmos.

However, even for tiny black holes, this process would still take an absurd amount of time. For a stellar-mass black hole, it could take up to $10^{64}$ years, and for the largest supermassive black holes, it might take as long as a couple googol (1 googol = $10^{100}$) years, or possibly even longer. Astronomers simply do not have the observational evidence to know for sure.

After the last black hole has faded away, it is hard to even comprehend what the universe will be like. The concepts of space and time barely have any real meaning once the last structures have disappeared. The period following the demise of black holes is known as the Dark Era, which is expected to begin sometime around $10^{101}$ years after the Big Bang, though its start depends on how long black holes last. So when and if this era ends is anybody's guess. During the Dark Era, the universe will consist of only a few subatomic particles and potentially dark matter, which is a little-understood substance that does not absorb, emit, or reflect electromagnetic radiation and may not decay at all. Whatever remains, however, will be very spread out. That is because as the universe cools, it will likely continue to expand. Scientists still debate how much the universe can balloon up, but by the time of the Dark Era, even a volume of space larger than our current observable universe might only contain a single solitary subatomic particle.

However, despite interactions between subatomic particles being incredibly rare, the occasional collision should still occur. In the absence of protons and neutrons, an electron will sometimes slam into a positron—the positively charged, antimatter counterpart of an electron. This may briefly form an atom of the bizarre element positronium, which is unstable and will quickly destroy itself when the matter and antimatter annihilate each other.

Many cosmologists think the universe will continue to cool, eventually playing out the so-called Big Freeze, when there is no heat remaining anywhere. The cosmos will eventually reach a point of total disorder, or maximum entropy. The second law of thermodynamics, which states that the entropy of a closed system (like the entire universe) can only increase, will have finally reached its logical conclusion.

### 6.7.2 Heat Death of the Universe

The heat-death of the universe is when the universe has reached a state of maximum entropy. This happens when all available energy (such as from a hot source) has moved to places of less energy (such as a colder source). Once this has happened, no more work can be extracted from the universe. Since heat ceases to flow, no more work can be acquired from heat transfer. This same kind of equilibrium state will also happen with all other forms of energy (mechanical, electrical, etc.). Since no more work can be extracted from the universe at that point, it is effectively dead, especially for the purposes of humankind.

This concept is quite different from what is commonly referred to as "cold death". Cold death is when the universe continues to expand forever. Because of this expansion, the universe continues to cool down. Eventually, the universe will be too cold to support any life; it will end in a whimper. The opposite of cold death, as you can see, is not heat death, but actually the big crunch. The big crunch occurs when the universe has enough matter density to contract back on itself, eventually shrinking to a point. This shrinking will cause the temperature to rise, resulting in a very hot end of the universe.

A couple trillion ($10^{12}$) years from now, the universe will have expanded so much that no distant galaxies will be visible from our own Milky Way, which will have long since merged with its neighbors. Eventually, $100 \times 10^{12}$ years from now, all star formation will cease, ending the Stelliferous Era (see Adams and Laughlin 1999) that has been running since not long after our universe first formed.

Much later, in the so-called Degenerate Era, galaxies will be gone, too. Stellar remnants will fall apart. In addition, all remaining matter will be locked up inside black holes. In fact, black holes will be the last surviving sentinels of the universe as we know it. In the Black Hole Era, they will be the only "normal" matter left. However, eventually, even these titans will disappear, too.

Stephen Hawking predicted that black holes slowly evaporate by releasing their particles into the universe. First, the smaller, solar-mass black holes will vanish. Moreover, by a googol years into the future, Hawking radiation will have killed off even the supermassive black holes.

No normal matter will remain in this final "Dark Era" of the universe, which will last far longer than everything that came before it. Moreover, the second law of thermodynamics tells us that in this period, all energy will ultimately be evenly distributed. The cosmos will settle at its final resting temperature, just above absolute zero, the coldest temperature possible.

## 6.8 THE END OF TIME

First, let us ask ourselves: What is time, exactly? Nothing is more mysterious and elusive than time. It looks like the most powerful force in the universe, carrying everybody from birth to death. However, what exactly is it? Saint Augustine (died AD 430) summed up the problem this way: "If nobody ask me, I know what the time is, but if I am asked then I am at a loss what to say".

Physicists have a tendency not to ask this question at all. This is a consequence of positions of Newton and Einstein on the subject. They defined the way physicists think about space, time, and motion. Each of them created a representation of the world with unsurpassed clarity. However, having seen their way to a structure of things, they did not bother about its foundation. Their theories contain wonderful truths, but they both take time as given!

Richard Feynman once quipped that "Time is what happens when nothing else does". But Barbour (1999) disagrees: if nothing happened, if nothing changed, then time would stop. For time is nothing but change. It is change that we perceive occurring all around us, not time. Put simply, time does not exist. In this highly provocative volume, Barbour (1999) presents the basic evidence for a timeless universe and shows why we still experience the world as intensely temporal. It is a book that strikes at the heart of modern physics. It casts doubt on Einstein's greatest contribution, the spacetime continuum, but also points to the solution of one of the great paradoxes of modern science, the chasm between classical and quantum physics. He argues that the holy grail of physicists—the unification of Einstein's general relativity with quantum mechanics—may well spell the end of time.

Barbour (1999) writes with remarkable clarity as he ranges from the ancient philosophers Heraclitus and Parmenides, through the giants of science Galileo, Newton, and Einstein, to the work of the contemporary physicists John Wheeler, Roger Penrose, and Steven Hawking. Along the way he treats us to enticing glimpses of some of the mysteries of the universe and presents intriguing ideas about multiple worlds, time travel, immortality, and, above all, the illusion of motion. *The End of Time* is a vibrantly written and revolutionary book. It turns our understanding of reality inside out.

Paul Dirac attempted to unify Einstein's general theory of relativity with quantum theory. His result led him to doubt how fundamental the four-dimensional requirement in physics is. In other words, Dirac was doubting the idea of fusion of space and time into spacetime. Our view of the world includes two very different theories, classical physics describes large things, and quantum physics describes small things. However there problems with this. First, Einstein's theory of gravity is incompatible with the principles of quantum mechanics, while Newton's dynamics and theory of electromagnetism are not. It is proved that the last two can be transformed into quantum theories. So far, all attempts to apply the process of quantization to general relativity and create quantum gravity have failed.

Newton's time is absolute. It flows with perfect uniformity forever, and nothing affects its flow. Space is also absolute. Both space and time are more fundamental than things. Newton could imagine an empty world but not a world without space and time. All the laws cannot stand alone without initial conditions; this is because the past determines the future, not vice versa; causality does not work from the future to the past. However, both Newton's and Eistein's laws work equally well in both directions. Time reversal is almost always preserved!

Only two possibilities exist to explain time's arrow. Either the universe was created in a special state, and its initial order has been degrading ever since, or it has existed forever, and at some time in the recent past it entered a state of very low entropy from which it is now emerging. The second possibility is entirely compatible with the laws of physics. That was the source of the so-called anthropic principle that if the world were not in a highly structured but extremely unlikely state, we should not exist and be here to observe it.

According to Einstein's general relativity, which is our best current description of space and time, the only place where time, and also space, ends is in a so-called singularity. This involves gravitational forces becoming so intense that space and time lose all meaning. However, do singularities exist? According to Kerr (2023), there is no proof that black holes contain singularities when they are generated by real physical bodies. Penrose (1965) claimed 60 years ago that trapped surfaces inevitably lead to light rays of finite affine length (FALL). Penrose (1965) and Hawking (1972) then asserted that these must end in actual singularities. When they could not prove this, they decreed it to be self evident. It is shown that there are counterexamples through every point in the Kerr metric. These are asymptotic to at least one event horizon and do not end in singularities.

## REFERENCES

Adams, F. C., and Laughlin, G. 1999. The Five Ages of the Universe. New York, NY: The New York Times Company.

Austerweil, J., and Griffiths, T. L. 2008. A rational analysis of confirmation with deterministic hypotheses. Proceedings of the Annual Meeting of the Cognitive Science Society 30. https://escholarship.org/uc/item/6n60k9zn.

Barbour, J. 1999. The End of Time: The Next Revolution in Physics. Reprint Edition, Oxford, New York: Oxford University Press.

Baum, S. D., Haqq-Misra, J. D., and Domagal-Goldman, S. D. 2011. Would contact with extraterrestrials benefit or harm humanity? A scenario analysis. Acta Astronautica 68(11–12): 2114–2129.

Belluati, A., Jimaja, S., Chadwick, R. J. et al. 2023. Artificial cell synthesis using biocatalytic polymerization-induced self-assembly. Nature Chemistry. https://doi.org/10.1038/s41557-023-01391-y.

Betz, E. 2020. The big freeze: How the universe will die. Astronomy. www.astronomy.com/science/the-big-freeze-how-the-universe-will-die/.

Borland, A. 2017. Connections between Catholicism and science: Astrobiology edition. University of St. Thomas, Minnesota, Biology Undergraduate Projects 2. (21 pages). https://core.ac.uk/download/pdf/217157878.pdf.

Brüssow, H. 2022. On the role of hypotheses in science. Microbial Biotechnology 15(11): 2687–2698. https://doi.org/10.1111/1751-7915.14141.

Clausius, R. 1867. The Mechanical Theory of Heat—with Its Applications to the Steam Engine and to Physical Properties of Bodies (J. van Voorst, ed.). London:. English Translations of Nine Papers.

Cleland, C. E. 2012. Life without definitions. Synthese 185: 125–144. https://doi.org/10.1007/s11229-011-9879-7.

Crick, F. H. C. 1981. Life Itself. Toronto: Simon & Schuster. ISBN 0-671-25562-2.

Crofts, A. R. 2007. Life, information, entropy, and time: Vehicles for semantic inheritance. Complexity 13(1): 14–50. https://doi.org/10.1002/cplx.20180.

Cunningham, T., Wheatley, P. J., Trembley, P.-E., et al. 2022. A white dwarf accreting planetary material determined from X-ray observations. Nature 602: 219–222.

Davies, P. 1988. The Cosmic Blueprint. New York: Simon/Schuster.

Davies, P. C. W., Benner, S. A., Cleland, C. E., et al. 2009. Signatures of a shadow biosphere. Astrobiology 9(2): 241–249. https://doi.org/10.1089/ast.2008.0251.

De, K., MacLeod, M., Karambelkar, V. et al. 2023. An infrared transient from a star engulfing a planet. Nature 617: 55–60. https://doi.org/10.1038/s41586-023-05842-x.

De Sitter, W. 1917. On the Relativity of Inertia. Remarks Concerning Einstein's Latest hypothesis. In KNAW Proceedings, 19 II (pp. 1217–1225). Amsterdam: Huygens Institute-Royal Netherlands Academy of Arts and Sciences.

DeWitt, B. S. 1967. Quantum theory of gravity. I. The canonical theory. Physical Review 160(5): 1113–1148. https://doi.org/10.1103/PhysRev.160.1113.

Di Lauro, V., and Bedetti, M. 2023. Are the characteristics of life as we know it universal? Scienza e Filosovia 30: 69–85.

Einstein, A. 1931. The World As I See It (vol. 84). Originally published in Forum and Century, in 1931 and now found in the Collection Ideas and Opinions.

Farihi, F. J., van Lieshout, R., Cauleyossati, P. W., et al. 2018. Dust production and depletion in evolved planetary systems. Monthly Notices of the Royal Astronomical Society 481(2): 2601–2611.

Fossati, L. 2019. A glance into the end of a planetary system. Science 364(6435): 25–26.

Frank, F. C. 1953. On spontaneous asymmetric synthesis. Biochimica et Biophysica Acta 11: 459–463. https://doi.org/10.1016/0006-3002(53)90082-1.

Galilei, G. (1564–1642). The two sciences. Great Books of the Western World, GBWW 26: 200.

Glass, D. J., and Hall, N. 2008. A brief history of the hypothesis. Cell 134(3): 378–381. https://doi.org/10.1016/j.cell.2008.07.033.

Gonçalves, Jr., S. J., Evangelista, H., Weis, J. et al. 2023. Stratospheric ozone depletion in the Antarctic region triggers intense changes in sea salt aerosol geochemistry. Commun Earth Environ 4: 77 (11 pages). https://doi.org/10.1038/s43247-023-00739-z.

Gough, D. O. 1981. Solar interior structure and luminosity variations. Solar Physics 74: 21–34.

Harrison, E. 1985. Masks of the Universe (pp. 252–263). New York: Collier Books, Macmillan.

Harris-Wyrick, W. 2022. Theologians discuss how humans might respond to extraterrestrial life. The Oklahoman, February 12.

Hartle, J. B., and Hawking, S. W. 1983. Wave function of the Universe. Physical Review D 28(12): 2960–2975. https://doi.org/10.1103/PhysRevD.28.2960.

Hawking, S. 1996. The Illustrated a Brief History of Time (p. 248). New York: Bantam.

Hawking, S. 1999. A Brief History of Time. Norwalk CO: The Easton Press.

Hawking, S. W. 1972. Black holes in general relativity. Communications in Mathematical Physics 25: 152–166. https://doi.org/10.1007/BF01877517.

Hawking, S. W., and Hertog, T. 2018. A smooth exit from eternal inflation?. Journal of High Energy Physics 147: (14 pages). https://doi.org/10.1007/JHEP04(2018)147.

Hayes, L. A., and Gallagher, P. T. 2022. A significant sudden ionospheric disturbance associated with gamma-ray burst GRB 221009A. Research Notes of the American Astronomical Society 6(10): 222. https://doi.org/10.3847/2515–5172/ac9d2f.

Heim, M. S. 2001. The Depth of the Riches: A Trinitarian Theology of Religious Ends (Sacra Doctrina: Christian Theology for a Postmodern Age (Sacra)). Grand Rapids, MI: William B. Eerdmans—Lightning Source.

Irons, L. G., and Irons, M. A. 2023. Pancosmorio (world limit) theory of the sustainability of human migration and settlement in space. Frontiers in Astronomy and Space Sciences 10: 1081340 (24 pages). https://doi.org/10.3389/fspas.2023.1081340.

James, W. (1842–1907). Principles of psychology. Great Books of the Western World, GBWW: 862–866.

Kerr, R. 2023. Do Black Holes have Singularities? (820 pages). https://doi.org/10.13140/RG.2.2.34286.38723.

Kipping, D. 2020. An objective Bayesian analysis of life's early start and our late arrival. PNAS-USA 117: 11995–12003.

Kopparapu, R. K., Ramirez, R., Kasting, J. F., et al. 2013. Habitable zones around main sequence stars: New estimate. The Astrophysical Journal 765: 131 (16 pages). https://doi.org/10.1088/0004-637X/65/2/131.

Kwon, J. Y., Bercovici, H. L., Cunningham, K., and Varnum, M. E. W. 2017. How will we react to the discovery of extraterrestrial life? Frontiers in Psychology 8: 2308. https://doi.org/10.3389/fpsyg.2017.02308.

Laurent, G., Lacoste, D., and Gaspard, P. 2021. Emergence of homochirality in large molecular systems. PNAS 118(3): e2012741118 (6 pages). https://doi.org/10.1073/pnas.2012741118.

Levan, A., Gompertz, B. P., and Salafia, O. S. 2023. Heavy element production in a compact Object merger observed by JWST. Nature (34 pages). https://doi.org/10.1038/s41586–023-06759–1.

NASA. 2023. The probability of collisions with earth. Downloaded, December 18. https://www2.jpl.nasa.gov/sl9/back2.html.

NASA. 2022. NASA/GSFC, Goddard Space Flight Center. Downloaded, December 20.

National Academy of Sciences. 1999. Science and Creationism: A View from the National Academy of Sciences, Second Edition. Washington, DC: The National Academies Press. https://doi.org/10.17226/6024.

Neal, M. 2014. Preparing for extraterrestrial contact. Risk Management 16(2): 63–87. https://doi.org/10.1057/rm.2014.4.

Ohlsson, T. 2023. Proton decay. Nuclear Physics B 993: 116268 (14 pages). https://doi.org/10.1016/j.nuclphysb.2023.116268.

Otto, S. 2022. An Approach to the de novo synthesis of life. Accounts of Chemical Research 55(2): 145–155. https://doi.org/10.1021/acs.accounts.1c00534.

Ozaki, K., Reinhard, C. T. 2021. The future lifespan of earth's oxygenated atmosphere. Nature Geoscience 14: 138–142. https://doi.org/10.1038/s41561-021-00693-5.

Pannenberg, W. 1991. Systematic Theology. Volume I (G. W. Bromiley, Trans., pp. xiii + 473). Grand Rapids, MI: Wm. B. Eerdmans Publishing Co.

Penrose, R. 1965. Gravitational collapse and space-time singularities. Physical Review Letters 14(3): 57–59. https://doi.org/10.1103/PhysRevLett.14.57.

Peters, T. 2011. The Implications of the discovery of extra-terrestrial life for religion. The Royal Society, Philosophical Transactions A 369(1936): 644–55. http://rsta.royalsocietypublishing.org/content/369/1936.toc.

Peters, T. 2016. Astrobiology and astrochristology. Zygon, Journal of Religion & Science 51(2): 480–496. https://doi.org/10.1111/zygo.12249.

Peters, T. 2019. Astrotheology. Published online: 23 May. https://doi.org/10.1093/acrefore/9780199340378.013.10.

Poincaré, H. (1854.-1912b). Science and hypothesis. Great Books of the Western World, GBWW 56: 40–52.

Qureshi-Hurst, E. 2021. Is simplicity that simple? An assessment of Richard Swinburne's argument from cosmic fine-tuning. Theology and Science 19(4): 379–389. https://doi.org/10.1080/14746700.2021.1982250.

Resnick, L. 1959. Confirmation and hypothesis. Philosophy of Science 26(1): 25–30. www.jstor.org/stable/185572.

Ruderman, M. A. 1974. Possible Consequences of nearby supernova explosions for atmospheric ozone and terrestrial life. Science 184(4141): 1079–1081. https://doi.org/10.1126/science.184.4141.1079.

Siegel, E. 2022. The Universe is already in its sixth and final era. Starts with a Bang, 5 Jan. https://bigthink.com/starts-with-a-bang/universe-final-era/Downloaded 19.12.2023.

Sloan, D., Batista, R. A., and Loeb, A. 2017. The resilience of life to astrophysical events. Nature-Scientific Reports s 7: 5419 (5 pages). https://doi.org/10.1038/s41598–017–05796-x.

Smoot, G., and Davidson, K. 1993. Wrinkles in Time. London: Little Brown Company.

Solomon, S. 1999. Stratospheric ozone depletion: A review of concepts and history. Reviews of Geophysics 37(3): 275–316. https://doi.org/10.1029/1999RG900008.

Spiegel, D. S., and Turner, E. L. 2011. Bayesian analysis of the astrobiological implications of life's early emergence on earth. PNAS 109(2): 395–400. https://doi.org/10.1073/pnas.1111694108.

Swinburne, R. 1997. Simplicity as Evidence for Truth. Milwaukee: Marquette University Press.

Swinburne, R. 2009. The Universe Makes It Probable That There Is A God. In M. Y. Stewart (ed.), Science and Religion in Dialogue (vol. 1, pp. 203–234). Chichester: Wiley-Blackwell.

Teaford, J. 2015. New model of cosmic stickiness favors "big rip" demise of universe. Vanderbilt University Research News, Jun. 30 https://news.vanderbilt.edu/2015/06/30/new-model-of-cosmic-stickiness-favors-%E2%80%9Cbig-rip%E2%80%9D-demise-of-universe/.

Tyrrell, T. 2020. Chance played a role in determining whether Earth stayed habitable. Communications Earth & Environment 1: 61 (10 pages). https://doi.org/10.1038/s43247-020-00057-8.

Vissapragada, S., Chontos, A., Greklek-McKeon, M., et al. 2022. The possible tidal demise of Kepler's first planetary system. The Astrophysical Journal Letters 941: L31 (8 pages). https://doi.org/10.3847/2041-8213/aca47e.

Waddington, C. H. (1905–1975). The nature of life. Great Books of the Western World, GBWW 56: 697–699.

Ward, P. D., and Brownlee, D. 2003. Rare Earth: Why Complex Life is Uncommon in the Universe. New York: Copernicus Books, Springer-Verlag, Inc.

Whitmire, D. P. 2022. Abiogenesis: The Carter argument reconsidered. International Journal of Astrobiology 1–6. https;//doi.org/10.1017/S1473550422000350.

Wisian, K. W., and Traphagan, J. W. 2020. The search for extraterrestrial intelligence: A realpolitik consideration. Space Policy 52: 101377. https://doi.org/10.1016/j.spacepol.2020.101377.

Wolf, E. T., and Toon, O. B. 2014. Delayed onset of runaway and moist greenhouse climates for Earth. Geophysical Research Letters 41: 167–172. https;//doi.org/10.1002/2013GL058376.

Wright, J. T., Haramia, C., and Swiney, G. 2023. Geopolitical implications of a successful SETI program. Space Policy 63: 101517. https://doi.org/10.1016/j.spacepol.2022.101517.


## ADDITIONAL READING


Adams, F. C., and Laughlin, G. 1997. A dying universe: The long term fate and evolution of astrophysical objects. Reviews of Modern Physics 69(2): 337–372. https://link.aps.org/doi/10.1103/RevModPhys.69.337.

Alharthy, A., and Kassandrov, V. V. 2020. On a crucial role of gravity in the formation of elementary particles. Universe 6: 193 (14 pages). https://doi.org/10.3390/universe6110193.

Canup, R. M. 2019. Giant impact hypothesis: An evolving legacy of Apollo. Astronomy. www.astronomy.com/.

Jedamzik, K., and Pospelov, M. 2009. Big Bang nucleosynthesis and particle dark matter. New Journal of Physics 11: 105028 (20 pages). https://doi.org/10.1088/1367-2630/11/10/105028.

Jiang, J. H., Rosen, P. E., and Fahy, K. A. 2021. Avoiding the "great filter": A projected timeframe for human expansion off-world. https://doi.org/10.48550/arXiv.2108.01730v1.

Longino, H. E. 1979. Evidence and hypothesis: An analysis of evidential relations. Philosophy of Science 46(1): 35–56. https://doi.org/10.1086/288849.

National Research Council of National Academies. 2007. The Limits of Organic Life in Planetary Systems. Washington, DC: The National Academies Press.

Parhizkar, A., and Galitski, V. 2022. Strained bilayer graphene, emergent energy scales, and moiré gravity. Physical Review research 4: L022027 (5 pages). https://doi.org/10.1103/PhysRevResearch,2.L022027.

Ricciardi, A., Cassey, P., Leuko, S., and Woolnough, A. R. 2021. Planetary biosecurity: Applying invasion science to prevent biological contamination from space travel. BioScience 72(3): 247–253.

Seckbach, J., and Gordon, R. 2020. Theology and Science: From Genesis to Astrobiology. Singapore: World Scientific Publishing Co Pte Ltd.

Swinburne, R. 2004. The Existence of God (Second Edition). Oxford: Clarendon Press.

Totani, T. 2020. Emergence of life in an inflationary universe. Scientific Reports 10: 1671. Nature Research. https://doi.org/10.1038/s41598-020-58060-0.

Vopson, M. M. 2021. Estimation of the information contained in the visible matter of the universe. AIP Advances 11: 105317 (5 pages). https://doi.org/10.1063/5.0064475.

Westby, T., and Conselice, C. J. 2020. The Astrobiological Copernican weak and strong limits for intelligent life. The Astrophysical Journal 896: 58 (18 pages). https;//doi.org/10.3847/1538–4357/ab8225.c.

Wolchover, N. 2019. Physicists debate Hawking's idea that the universe had no beginning. Quanta Magazine, June 6. www.quantamagazine.org/physicists-debate-hawkings-idea-that-the-universe-had-no-beginning-20190606/.

# Index

L



M



N



O



P



Q



R

For Product Safety Concerns and Information please contact our EU representative GPSR@taylorandfrancis.com Taylor & Francis Verlag GmbH, Kaufingerstraße 24, 80331 München, Germany

**Batch number: 10731594**

Printed by Printforce, the Netherlands